Springer Natural Hazards

The Springer Natural Hazards series seeks to publish a broad portfolio of scientific books, aiming at researchers, students, and everyone interested in Natural Hazard research. The series includes peer-reviewed monographs, edited volumes, textbooks, and conference proceedings. It covers all categories of hazards such as atmospheric/climatological/oceanographic hazards, storms, tsunamis, floods, avalanches, landslides, erosion, earthquakes, volcanoes, and welcomes book proposals on topics like risk assessment, risk management, and mitigation of hazards, and related subjects.

More information about this series at http://www.springer.com/series/10179

Sebastiano D'Amico
Editor

Moment Tensor Solutions

A Useful Tool for Seismotectonics

Springer

Editor
Sebastiano D'Amico
Department of Geosciences
University of Malta, Msida Campus
Msida
Malta

ISSN 2365-0656 ISSN 2365-0664 (electronic)
Springer Natural Hazards
ISBN 978-3-030-08437-0 ISBN 978-3-319-77359-9 (eBook)
https://doi.org/10.1007/978-3-319-77359-9

Printed on acid-free paper

This Springer imprint is published by the registered company Springer International Publishing AG part of Springer Nature
The registered company address is: Gewerbestrasse 11, 6330 Cham, Switzerland

Preface

In the last decades, the use of waveforms recorded at local-to-regional distances has increased considerably and waveform modeling is largely used to estimate faulting parameters, evaluate stress and strain patterns as well as contribute to the understanding of regional and local tectonic feature and seismotectonic settings. It also has implications for seismic verification efforts. Moment tensors provide a general theoretical framework to describe seismic sources and this is not restricted to earthquake sources, but covers also other types of seismic sources such as explosions, implosions, rock falls, landslides, as well as ruptures driven by fluid and gas injections. Moment tensors have the potential to complement more traditional source parameter estimations (e.g., magnitude or focal solutions from first motion polarities). Modeling regional seismograms for constraining moment tensors is widely accepted and largely documented by extensive literature. Modeling regional waveforms also provide a good constraint in accurately determining source mechanism, magnitude, and depth. Thus, the understanding of reliable determination seismic source and of its uncertainty can play a key role in contributing to geodynamic investigation, seismic hazard assessment, and earthquake studies. Waveform modeling has been used also to estimate faulting parameters of small-to-moderate-sized earthquakes. The present book focuses both on the explanation of the basic theory and methods about moment tensor solutions and their role in the modern seismology. Some chapters are devoted to investigate uncertainties in the final solution, while others are written with the main goal of collecting state-of-the-art papers related to different seismotectonic settings in several areas of the planet.

The ultimate goal of the book is to encourage discussions and future research in the seismotectonic field. In addition, it will serve both as an instructional book and platform to disseminate state-of-the-art research and results in the field.

This volume cannot be considered as a textbook, but surely it can be used at undergraduate and graduate level.

Msida, Malta Sebastiano D'Amico

Acknowledgements

I am grateful to all the authors for their close cooperation while preparing their contributions. I also gladly acknowledge all the referees, belonging to a number of research institutions located worldwide. Their careful reading and constructive suggestions contribute to the standard of the final versions of each manuscript collected in this book. A special thank goes to Johanna Schwarz, Claudia Mannsperger, and all the editorial staff for their professional assistance and technical support during the entire publishing process.

Msida, Malta Sebastiano D'Amico

Contents

ISOLA Code for Multiple-Point Source Modeling—Review

Jiří Zahradník and Efthimios Sokos

1 Introduction

ISOLA software package has been developed to invert local or regional full-wave seismograms for single- and multiple-point source models. The code was introduced in 2003; since then it has been continually upgraded, and, presently, it can be considered a well-established tool, used worldwide. Originally, the code name came from 'isolated asperities', to be resolved at fault planes of large earthquakes. However, with time, the code has been adapted for very diverse applications, ranging from Mw 0.3 to Mw 9. Many research papers based on usage of ISOLA have been published (see References). Almost every new application is challenging—hence the code is continually updated. The objective of this work is to explain the basic principles of the method, review code status, demonstrate a few examples to attract new users, and shortly touch also future development. The code is free, and can be downloaded together with manual and test examples from http://seismo.geology.upatras.gr/isola/ (last accessed March 2018).

2 Basics

ISOLA software serves for modeling an extended seismic source as a point source, or a series of point sources. The point source contributions to an earthquake source model are called *subevents*. Their moment tensors (MT) are calculated by *least-squares*, using a full, deviatoric (i.e. zero-trace), or DC-constrained mode (i.e. requesting the double couple part to be close to 100%). Alternatively, 100% DC

J. Zahradník
Faculty of Mathematics and Physics, Charles University, Prague, Czech Republic
e-mail: jiri.zahradnik@mff.cuni.cz

E. Sokos (✉)
Seismological Laboratory, Department of Geology, University of Patras, Patras, Greece
e-mail: esokos@upatras.gr

S. D'Amico (ed.), *Moment Tensor Solutions*, Springer Natural Hazards,
https://doi.org/10.1007/978-3-319-77359-9_1

focal mechanisms of the subevents can be kept fixed (prescribed), and inversion is performed only for position, time and moment of the subevents.

The position and time of subevents are calculated by a *space-time grid search*. The spatial grids are either linear or planar, e.g. along horizontal planes, or assumed fault planes. The subevent moment-rate time function, also called elementary time function is supposed to be known (delta function, or a single triangle of a prescribed duration). Alternatively, the time function can be calculated from waveform data, assuming a known focal mechanism. If a source appears (at low frequencies) as a single dominant subevent, we speak about centroid position and centroid moment tensor determination (CMT). The name centroid reminds that it is 'center of gravity' of slip on the fault. For small earthquakes, centroid and hypocenter are (within errors) identical.

Green's functions, including near-, intermediate- and far-field terms, are calculated by the discrete wavenumber method (broadly used code AXITRA of Bouchon 1981 and Coutant 1989), using a 1D velocity model (parallel layers with constant parameters). In this way, all body waves and surface waves existing in the velocity model are automatically taken into account. Alternatively, several source-station, path-dependent, 1D models can be used in ISOLA, if such models are available.

The waveform agreement between observed and synthetic seismograms is quantified by their *correlation* (corr) or by *variance reduction* (VR): $VR = 1 - (res^2/data^2)$, where res^2 denotes misfit (i.e. the sum of squared residuals between observed and synthetic data), while $data^2$ is the sum of squared observed data. The two measures are related: $corr^2 = VR$ (Eq. 13 of Křížová et al. 2013). Resolution of the moment tensor is expressed by condition number (CN); it will be discussed in detail. Uncertainty of MT (including, for example, scatter plot of nodal lines, or DC% histograms) is computed from the covariance matrix of the source parameters; see Appendix of Zahradník and Custódio (2012). Stability of the inversion with respect to space position and time of subevents is tested by repeatedly removing stations, or individual components (jackknifing). The resulting focal mechanisms are checked for their agreement with polarities. Alternatively, waveform inversion can be pre-constrained by polarities (see CSPS method below).

Subevents are calculated by two methods: (i) *Iterative deconvolution* is a standard method (Kikuchi and Kanamori 1991; Sokos and Zahradník 2008). Initially, a first subevent fitting data as well as possible is found, the corresponding synthetics are subtracted from real data, then a second subevent is found, etc. Temporal variation of each subevent is that of the elementary time function. The application yields a single (best-fitting) set of subevents. (ii) *Joint inversion of source pairs* is a newer method (Zahradník and Sokos 2014), suitable if the studied earthquake seems to be basically composed of two dominant subevents. We systematically inspect all possible combinations of two trial source positions on a spatial grid, and for each member of the source pair we calculate the moment-rate function. The time function is modeled as a series of equally shifted elementary time functions whose relative weights are calculated by non-negative least squares (NNLS) (Lawson and Hanson 1974). In this way, time variation of each subevent may be considerably more complex than if it is expressed by a single elementary time function. This method yields a suite of subevent pairs, not only the best-fitting solution. Both methods have been validated in several applications, e.g. Sokos et al. (2012), Quintero et al. (2014), Hicks and Rietbrock (2015), Sokos et al. (2015, 2016) and references therein. Each of the two methods has its pros and cons as shown by comparisons in several of the cited papers. For example, as thoroughly discussed in Zahradník and Sokos (2014), the NNLS approach may perform better for extended sources observed with imperfect station coverage. The standard iterative deconvolution seems useful for identification of variable focal mechanisms participating in complex fault ruptures. More comparisons of both methods in future would be useful.

A very difficult problem is the MT inversion of earthquakes for which only few stations can be modeled, either because the other stations are too distant, or too noisy. It is often the case of *small earthquakes*. In situations like that a special method can be used, i.e. *Cyclic Scanning of the Polarity Solutions*, shortly CSPS, combining a few waveforms at near stations with polarities at many stations (Fojtíková and Zahradník 2014; Zahradník et al. 2015). The idea is to pre-constrain the solution by means of the polarities. The polarities themselves often provide a strongly non-unique focal mechanism. Then, systematically scanning the polarity-satisfying solutions, we select among them those focal mechanisms which provide good waveform fit at the available near stations. Note that CSPS assumes 100% DC sources, because non-double-couple components of small events are rarely reliable.

ISOLA is an "all-in-one" package, containing many mutually coupled Fortran codes for computer speed and hundreds of Matlab codes, forming a compact Graphic User Interface (GUI) for user's comfort. Its great advantage over other existing MT codes is that it combines all necessary tools for processing input data and output results, such as, for example (Fig. 1):

- Converting records from several standard seismic formats
- Defining parameters for instrumental correction
- Removing instrument response (instrumental correction)
- Calculating signal-to-noise ratio
- Trying several filters to find the usable frequency range
- Visually checking the records to avoid instrumental disturbances
- Rotating horizontal components (if not recording N, E)
- Preparing velocity models in ISOLA format and plotting them
- Choosing various geometries of trial source positions
- Calculating (full-wave) Green's functions
- Inverting for MT in several modes (full, deviatoric, DC-constrained)
- Choosing station-dependent frequency ranges
- Selecting/de-selecting stations or their components for inversion
- Simulating waveform data (forward problem), e.g. for synthetic tests
- Visualizing space and time variation of the correlation between real and synthetic seismograms
- Plotting waveform fit, beachballs and polarities
- Plotting space-time distribution of subevents

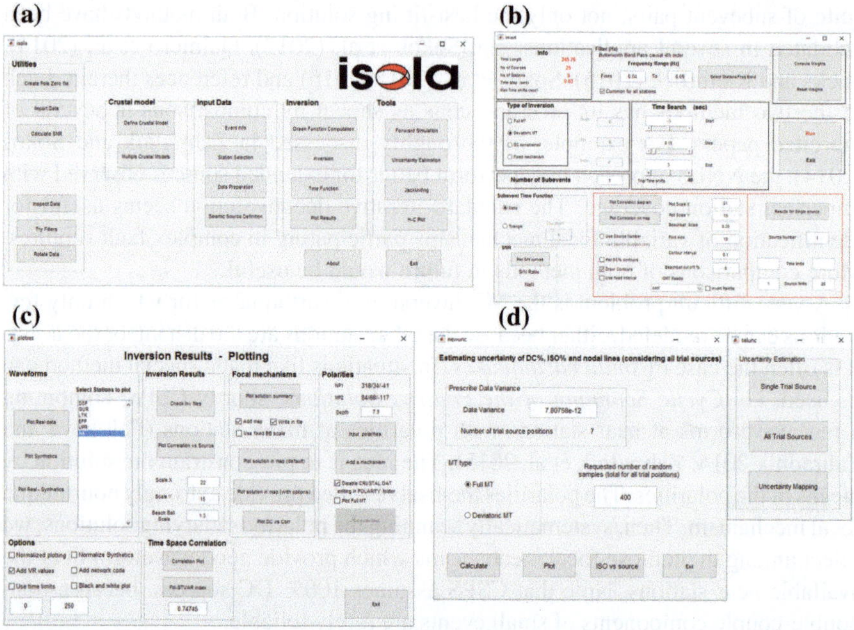

Fig. 1 An example of a few screenshots of the ISOLA GUI. **a** The front window—basic panel, **b** inversion, **c** plotting, **d** uncertainty analysis. Many other capabilities are available

- Plotting time functions
- Plotting uncertainty measures (e.g. suites of nodal lines, histograms of DC% and ISO%)

ISOLA will be preferred by users who like easy and intuitive GUI. The GUI environment is not just substitute for editing files; it is a complex system controlling all processing steps in a sophisticated way which reduces risk of many errors. Indeed, it provides an internal check (invisible to user) of consistency of the files; output files from some operations are easily used as input for the others. If batch files are needed, they don't need to be written by user, but they are automatically created by the GUI. Many warning messages and 'hints' are available. All (intermediate and final) results come in form of plots, in an almost publication-ready quality.

ISOLA is particularly useful for getting a deep insight into all processing steps and for recognizing possible variability (uncertainty) of the results. As such this software is best suited for a detailed investigation of selected earthquake sequences; for example—calculating a multiple-point source model for a mainshock, and single-point source MT solutions for largest aftershocks.

Although routine "manual" processing of many events, in a single-point source mode, by ISOLA takes place at several national seismological centers (e.g. in Greece, Turkey, Iran, Colombia, Romania), it is likely that routine processing will be better solved in future by fully automated codes. The existing automated codes, closely related to ISOLA, are two: SCISOLA, which is a SeisComP implementation by Triantafyllis et al. (2016), and Bayesian ISOLA of Vackář et al. (2017). The latter makes use of the data covariance matrix derived from real pre-event noise, which automatically identifies and suppresses noisy components and noisy frequency ranges.

ISOLA users might be surprised why so much attention is paid to various uncertainty measures. It is because checking the resulting MT's and multiple-point source solutions, as much as possible, is of fundamental importance. A great *danger* is that ISOLA will (always) provide "some" solution, but sometimes the solution may be *physically meaningless*, and it is upon the user to detect low reliability of the results by means of all available tools of this package. This task is far from being trivial. This topic will be thoroughly discussed throughout the whole text. Some practical hints can be found at the end, in the section on 'Frequently asked questions'.

Technical aspects, such as installation of the code, as well as the role and usage of the individual 'windows' of the Matlab GUI, are explained in ISOLA manual (Zahradník and Sokos 2016). The manual also contains fully documented test examples (input and output data) helping user to understand all practical use of the code. As with any code, certain training is of course invaluable. That is why very popular ISOLA training courses have been organized by various institutions, e.g. in Costa Rica (2011 and 2016), Brazil (2013), Colombia (2013), Turkey (2014). Due to these courses, many interesting international collaborations have been established.

3 Short Outline of Theory

This paragraph mainly serves as definition of the terms and quantities used in ISOLA. For complete theory, see Chap. 4 of Aki and Richards (2002). Let us start with assuming a point source (a vectorial position $\mathbf{x}$), a set of stations (position $\mathbf{y}$) and a 1D velocity model. Further we select 6 *elementary moment tensors,* i = 1, 2, …, 6. In literature, there are various sets of the 6 tensors, cf. Kikuchi and Kanamori (1991) or Bouchon (1981); we use the latter (Eq. 2 of Křížová et al. 2013). The first five tensors represent double-couple mechanisms, while the 6th one represents a pure volume change. See Table 1.

Temporal variation of the moment rate is assumed to be known, e.g. delta function. For each source-station combination ($\mathbf{x}, \mathbf{y}$), and a certain frequency band, we calculate 6 three-component *elementary seismograms* $G_i(\mathbf{x}, \mathbf{y}, t)$, i = 1, 2, …, 6; each one is a convolution of Green's tensor and elementary moment tensor. For simplicity, notation of their 3 components (N, E, Z) and discretized time samples is omitted.

Seismogram due to *arbitrary moment tensor* is a linear combination of the elementary seismograms, i.e. $d(\mathbf{x}, \mathbf{y}, t) = \sum_1^6 a_i G_i(\mathbf{x}, \mathbf{y}, t)$, where the a_i coefficients have unique relation to the moment tensor and scalar moment M_0 (Eqs. 3 and 4 of Křížová et al. 2013). In matrix notation, $\mathbf{d} = \mathbf{G}\,\mathbf{a}$, where $\mathbf{d}$ is a column vector of N waveform values (all stations, each one in three components, all time samples), $\mathbf{G}$ is a N × 6

Table 1 Six elementary moment tensors used in ISOLA

No.	strike (°)	dip (°)	rake (°)	
1	0	90	0	
2	270	90	-90	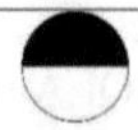
3	0	90	90	
4	90	45	90	
5	0	45	90	
6	-	-	-	

matrix whose columns are $G_1, \ldots, G_6$ (each one ordered in same way as **d**), and **a** is a column vector of the above introduced 6 a-coefficients.

The moment-tensor (MT) determination is a *linear* inverse problem in which **d** represents a given (measured) data vector, **G** is a known (numerically calculated) matrix, and **a** is a parameter vector of 6 components to be found. A special case arises if assuming $a_6 = 0$, i.e. vanishing volume change; then we consider only 5 parameters a_i, $i = 1, 2, \ldots, 5$, and 5 elementary seismograms $G_1, \ldots, G_5$; this is the so-called deviatoric MT inversion. In most of practical applications $N \gg 6$, so $\mathbf{d} = \mathbf{G}\,\mathbf{a}$ is an over-determined linear system of algebraic equations. Optionally, a non-linear constraint is applied in ISOLA to make determinant of MT close to zero, i.e. enforcing MT to represent a pure double couple (DC). This is called DC-constrained MT inversion, particularly useful for modeling complex tectonic events as possibly composed from pure-shear subevents.

The over-determined system $\mathbf{d} = \mathbf{G}\,\mathbf{a}$, resulting from the waveform inversion, is inconsistent, i.e. it has no exact solution. An approximate solution can be found by the least-squares method, i.e. $a = (\mathbf{G}^T\mathbf{G})^{-1}\,\mathbf{G}^T\,\mathbf{d}$, provided the inverse matrix $(\mathbf{G}^T\mathbf{G})^{-1}$ exists; here T denotes transposition and -1 stands for matrix inversion. Equivalent condition is that the $\mathbf{G}^T\mathbf{G}$ (square) matrix is regular, i.e. it has a non-zero determinant. The determinant is zero valued if one of the singular values of matrix **G** is zero. In practice, this is almost never the case, but some singular value may be 'small' (in some relative sense).

Ratio of the largest and smallest singular values w of matrix **G** is called condition number, $CN = w_{max}/w_{min}$. The CN can be easily calculated by means of eigenvalues e_{max}, e_{min} of matrix $\mathbf{G}^T\mathbf{G}$: $CN = \sqrt{e_{max}/e_{min}}$. Small e_{min}, i.e. large CN value, signalize that the problem is ill-conditioned; thus the MT cannot be reliably resolved. A situation like that often appears if only few stations are available (e.g. 1 station) and/or sources are very shallow. At shallow depths some MT components are poorly resolved because the related Green's function components are vanishing at the free surface (Henry and Das 2002). Typically, mainly non-DC components of moment tensor suffer from the limited resolvability, i.e. those components of the earthquake moment tensor are most uncertain. A closer insight into resolvability of the linear MT inversion problem is provided by covariance matrix of model parameters **cov** = $\sigma^2 (\mathbf{G}^T \mathbf{G})^{-1}$, where σ^2 is the data variance. Having some moment tensor retrieved in the inversion process, that MT is understood as a statistical 'mean'. The covariance matrix allows estimation of a possible variability around the mean. Therefore, we generate random ensemble of moment tensors corresponding to the multivariate normal distribution described by the mean and covariance matrix. The obtained ensemble then, serves for constructing histograms of the parameters of interest, e.g. strike, dip and rake angles (hereafter s/d/r), DC%, ISO%; also, scatter-plots of possible nodal lines can be generated easily in the uncertainty tool of the GUI. In ISOLA we must assume some value of σ^2, same for all stations. In more sophisticated methods, σ^2 is derived, either from noise (Vackář et al. 2017), or from assumed imprecision of velocity models (Halló and Gallovič 2016). Therefore, our uncertainty analysis is meaningful only in *relative* sense, like this: Assuming a fixed σ^2 value and comparing **cov**, or CN, for several source-station configurations, several velocity models, or several frequency ranges we compare their MT resolvability. No observed waveform data are needed for this analysis, just the **G** matrix. Note that an ill-posed waveform inversion problem (typically CN > 10) may have a very good match between data and synthetics, e.g. VR = 0.9, but such a good match in no sense guarantees that the focal mechanism is correct.

Having obtained a MT, ISOLA makes its traditional decomposition and calculates the so-called percentages of Double Couple (DC), Compensated Linear Vector Dipole (CLVD) and Isotropic component (ISO) (Eq. 8 of Vavryčuk 2001). As mentioned above, any departure from high DC (close to 100%) is to be considered with great care, because it may be a pure artifact. On the other hand, from practical point of view, it is important that if same data are processed in the full-, deviatoric-, or DC-constrained mode, featuring very different CLVD and ISO, their strike/dip/rake angles are often very similar. This indicates that the DC-part of the solution is fortunately robust.

Inaccurate knowledge of the source position is solved by repeating the linear MT inversion in a grid of predefined (trial) source positions, searching for the best-fitting centroid position and time. The MT uncertainty at the best-fitting source position is evaluated using **cov** matrix (explained above). Alternatively, to include also effects of the uncertain source position, we make random sampling of the total MT probability density function by properly combining samples from all trial positions (Eq. 7 of Vackář et al. 2017).

Using ISOLA for retrieving more than a single subevent, and/or a more complex temporal variation of the source, needs a more detailed explanation. We briefly explain two methods currently used in the GUI, i.e. iterative deconvolution and NNLS method.

Iterative Deconvolution. The observed waveforms d (i.e. seismograms d(t) in a set of stations) are represented by a sum of K point source synthetics, $d = s_1 + s_2 + \cdots + s_K$. Each synthetic is due to a different source position, time, and moment tensor which have to be found. However, all moment tensors have the same temporal variation (given by the assumed elementary time function E(t), e.g. delta function). The K subevents are searched *successively*. In the first step, the entire wavefield d is optimally approximated exclusively by the first point source. The position and time of the first subevent is grid searched, while moment tensor is calculated by the least squares. As a rule, the first calculated subevent has the largest moment. The synthetic seismograms s_1 are subtracted from the data, i.e. 'new data' is obtained $d_1 = d - s_1$. Next inverted waveform is d_1, thus second subevent is obtained, its contribution s_2 being again subtracted, getting $d_2 = d_1 - s_2$, etc. Several simple hints are useful to quickly guess when to stop the iterations: (i) Subevent (K + 1) should not have its moment much smaller (e.g. 5× smaller) than subevent (K). (ii) The cumulative moment of the subevents should increase. If a well guaranteed value of the total moment is available (e.g. that of GCMT project, http://www.globalcmt.org/, last accessed in June 2017), the cumulative moment should not exceed that value. (iii) If focal mechanisms are free, and they strongly vary among subevents, these mechanisms have probably no physical meaning. Then, searching subevents with a constant (prescribed) focal mechanism may be more relevant; the mechanism obtained from a previous low-frequency single-point source model is a good choice for this. (iv) The cumulative variance reduction of subevent(K + 1) should be reasonably greater than that of subevent(K). Statistical significance of the latter can be justified by F-test, discussed later in Box 2.

NNLS inversion. The observed waveforms, d, are again supposed to be a superposition of K point-source subevents, $d = s_1 + s_2 + \cdots + s_K$. Focal mechanisms (100% DC) of the subevents are assumed to be known. Their position is either known or searched (see below). Time variation T(t) of moment rate of each subevent is generally not the same; it is represented by a series of J elementary time functions E(t), with constant mutual time shifts τ, i.e. $T(t) = \sum_1^J n_i E(t - \tau_i)$, the goal is to calculate the $n_1, n_2, \ldots, n_J$ (non-negative) coefficients for all K sources, i.e. J × K numbers, *simultaneously*, so that we fit real data by the sum $s_1 + s_2 + \cdots + s_K$. This inverse problem is solved by the non-negative least squares method of Lawson and Hanson (1974), shortly NNLS. As a result, time function of each subevent is obtained. For extended sources, particularly interesting is the time function at centroid position, which often describes temporal complexity of the whole source process (Zahradník and Sokos 2014; Sokos et al. 2016). In ISOLA GUI we also allow for a case that position of the subevents is not known. This application, useful for large events (e.g. M > 6), is restricted in the current GUI to two subevents only. A grid of trial source position over fault plane is designed, and the NNLS method is repeated systematically for all possible source pairs. The output is a suite of source pairs (and their time functions) which fit observed data within a selected threshold.

All calculations in ISOLA operate with observed and synthetic seismograms band-pass filtered in same frequency range. A causal, i.e. 'one-way', 4th order Butterworth filter is used (code XAPiir by Harris 1990). The causal filter has been selected recently (after years of using non-causal filtration) because it avoids appearance of the filtered signal before arrival time. Thus, the causal filter simplifies processing of near stations, whose signal arrives close to origin time, t = 0. By observed seismograms we mean seismograms corrected for instrument, i.e. with instrument response removed, also done in ISOLA.

4 Overview of Selected Applications

To demonstrate broad applicability of ISOLA, we review here several published studies, in which the code was used. Possible interpretation of observations featuring large *non-DC components* by a source model composed from a few purely 100% DC sources were proposed (Zahradník et al. 2008b; Adamová et al. 2009). One of very few ISOLA applications proving possible reliability of a significant ISO component was that by Křížová et al. (2013). Studies of MT uncertainties started in Zahradník and Custódio (2012). The latter paper, followed by Michele et al. (2014), represented pioneering steps, demonstrating possible *mapping of the MT uncertainty* over a geographic territory of interest, without need of any real event, similarly to commonly applied analyses of the location capabilities of seismic networks. Later, the 'uncertainty mapping' tool became a part of ISOLA GUI and was applied by Fojtíková et al. (2016) in their network upgrade planning.

The ability of ISOLA to successfully retrieve *multiple-point source models* from near-regional records has been tested by comparisons with finite fault slip inversions, both on synthetic and real data (Zahradník and Gallovič 2010; Gallovič and Zahradník 2011, 2012). ISOLA correctly retrieved main subevents at same places where the slip inversion identified major slip patches. It was also the case of an "extreme" application of ISOLA to the M9 Tohoku 2011 earthquake (Zahradník et al. 2011). Advantages of the joint NNLS inversion of the position and size of two sources, available in ISOLA as possible replacement of iterative deconvolution, were demonstrated for a Mw 7.1 event in Turkey by Zahradník and Sokos (2014), and then the joint inversion of the source pairs was successfully used by Quintero et al. (2014), Hicks and Rietbrock (2015). The latter is an example where the authors comprehensively used many ISOLA tools.

To list at least a few, there were several applications retrieving focal mechanisms of many events in a certain seismic region for purposes of *seismo-tectonic interpretations*: Chile, Agurto et al. (2012); Martinique region, Gonzalez et al. (2017); Turkey, Cambaz and Mutlu (2016); Karakoram-Himalaya, Hazarika et al. (2017); Greece, Serpetsidaki et al. (2010) to name a few. Numerous MT's from some of these studies led to inversion of focal mechanisms into *stress field*. A representative example of the latter is Carvalho et al. (2016), based on relatively rare waveform inversion of eleven *microearthquakes* (Mw 0.8–1.4) in Central Brazil, where data are sparse but extremely important for understanding tectonic stresses. Application of ISOLA to such small events was made possible thanks to availability of local stations (<8 km), where frequencies as high as 2 Hz could be successfully modeled. Even smaller earthquakes (Mw 0.3) were inverted in ISOLA by Benetatos et al. (2013).

Quite a lot of efforts in development of ISOLA GUI were devoted to simple visual evaluation of waveforms, important for identifying suitable frequency range. Low frequencies are needed for MT inversions because they are less sensitive to inaccuracies of available velocity models, but the low frequency range often suffers from the low signal-to-noise ratio. For example, almost no strong motion acceleration record is usable below 0.03 Hz (except stations close to faults of very large events). Broad-band records, although featuring much less instrumental noise, may contain special *low-frequency disturbances*. The latter were intensively studied by Zahradník and Plešinger (2005, 2010); later, such disturbances (called 'mice') were encountered worldwide by many ISOLA users. Use of disturbed records may even damage the whole MT inversion, because these records may have very large (false) amplitudes, not obvious after band-pass filtering the records. The disturbances cannot be simply filtered out. A lot of efforts were spent on 'cleaning' (or correcting) records, but finally we gave up inclusion of such (unstable) tools into ISOLA. Instead, we emphasized the need to identify the disturbances, and remove the damaged records from the processing. Nowadays, this can be done either manually, or—for certain type of disturbances—also automatically Vackář et al. (2015).

ISOLA enabled us to recognize several multiple events, e.g. *earthquake doublets*. In fact, the first earthquake processed by ISOLA at all was a M6.2 Lefkada 2003 earthquake (Ionian Islands, western Greece), shown to be composed from two earthquakes, Mw 5.9 and 5.8, separated of each other in space and time by 40 km and 14 s, respectively (Zahradník et al. 2005). Interestingly, this 40-km long segment of Cephalonia-Lefkada Transform Fault, unbroken in 2003, has been silent up to 2015, when it produced a M6.4 earthquake (Sokos et al. 2016). This finding is an example of how detailed MT inversions may help in elucidating existing *segmentation of major fault zones*, and thus contribute to understanding of their future potential and seismic hazard.

For an application of ISOLA to full moment tensor study of the North Korea 2017 *nuclear test* and the associated uncertainties, the reader is referred to Liu et al. (2018).

In the following two paragraphs, some of the applications are somewhat expanded. We selected same data as in ISOLA manual (Zahradník and Sokos 2016), keeping even same names: 'Example Greece' and 'Example Brazil'. While in the manual we focused on technical issues (how to run the calculations, which 'buttons' of the GUI should be used, what are the input/output files), here we rather demonstrate a broad range of the code capabilities, mostly those which have not yet been often published, and we emphasize physical meaning of the results.

5 Example Greece

In this paragraph we demonstrate ISOLA capabilities on an example of Lefkada Mw 6.2, 2015 earthquake in Greece, published by Sokos et al. (2016). We start from low-frequency single-point analysis. Trial sources are designed in a grid of 7 × 7

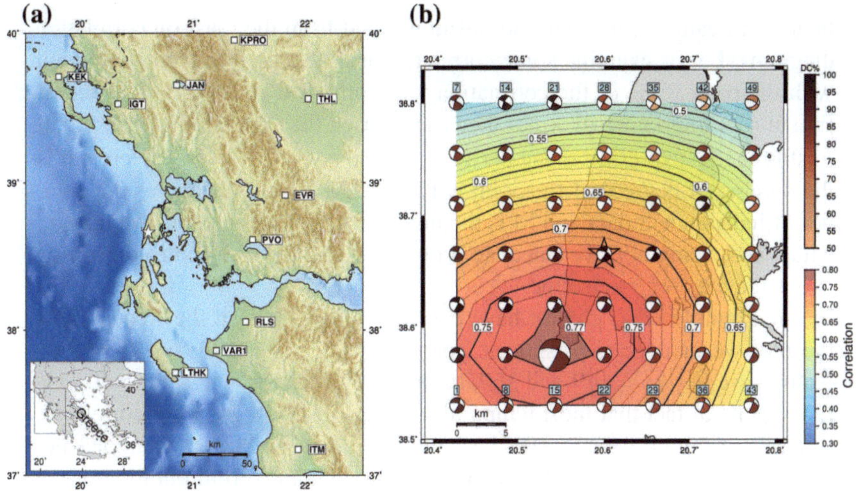

Fig. 2 Lefkada Mw 6.4 earthquake, 2015; single-point source solution. **a** Locations of the used stations (white squares). Star depicts the mainshock epicenter. Inset shows the study area with respect to Greece. **b** Grid search of centroid position, showing the Latitude-Longitude variation of the correlation between observed and synthetic waveforms for 0.01–0.05 Hz (the correlation plot). The grid is situated at a depth of 5 km. The calculated moment tensors are shown by beachballs, color-coded according their double-couple percentage (DC%). The centroid is represented by the largest beachball

points in a horizontal plane at a depth of 5 km; this depth was previously indicated by grid search below epicenter as the preferred source depth. Green functions are calculated up to fmax = 0.2 Hz, and the elementary time function is delta function.

Low frequencies, single-point source model (CMT). Initially, waveform inversion is performed in the frequency range 0.01–0.05 Hz (which we denote as "low-frequency range", or LF). Figure 2 shows the used stations and the obtained correlation between observed and synthetic seismograms as a function of the trial source position, serving to identify the centroid position and its mechanism.

The best-fitting trial source (no. 16)—the centroid—is considerably shifted from the epicenter, 10 km towards south and 5 km towards west. This position, together with the corresponding strike, dip and rake angles (24°, 81°, −148°) constitute our CMT solution. The significant difference between hypocenter and centroid positions is due to the finite-source extent, where hypocenter (nucleation of the rupture) and centroid (center of gravity of slip on fault) are generally not the same. The centroid time is 7.5 s after hypocenter time.

Box 1: Chi-square (χ^2) test

Here we explain the Chi-square test considered for possible future inclusion in the GUI. This test (Eq. 5.31 of Shearer 2009) is useful for checking statistical significance of the centroid position found from the Lat-Lon correlation diagrams. Let us examine some correlation isoline, corr_iso, and compare it with the peak value of the correlation, corr_opt. The χ^2 statistic, defined as [misfit/(optimum misfit)] × ndf then can be used for evaluations. The ratio of misfits can be expressed using correlations, $\chi^2 = \frac{1-corr\_iso^2}{1-corr\_opt^2} ndf$. Here ndf is the number of degrees of freedom; ndf = N − M, where N denotes the number of independent data and M is the number of free parameters. We take χ^2 value, and, using ndf, we calculate probability that the true centroid position is outside the corr_iso isoline.

In our example (Fig. 2b) we have corr_opt = 0.78 and, in the 'red' part of the plot we can see two isolines: corr_iso = 0.77 and 0.75. For these two isolines we get χ^2 = 149 and 160, respectively. Now how to setup ndf: Due to the general fact that most temporal samples of seismogram are correlated, as a rule of thumb we consider the 'number of data' to be just 5 samples per component, hence 15 per station. The number 5 comes from duration of the dominant wave group, divided by predominant period (Adamová et al. 2009). Then total N can be estimated as 15 × number of stations = 150, while M = 7 (i.e., 5 MT components + 1 depth + 1 time); ndf = 143. Using χ^2 and ndf, we calculate the probability that true position of centroid is out of the 0.77 and 0.75 isoline; we obtain probability of 35% and 16%, respectively. In other words, if accepting a 35% risk of failure, we can say that the centroid is inside the 0.77 isoline. If we are more conservative, and accept only the 16% risk, then we must accept greater uncertainty of the source position, i.e. admit its position somewhere inside a broader area, especially inside the 0.75 isoline. If user is surprised by saying that one component is equivalent to just 5 data points, and believes that this number should be increased e.g. 10-times, then ndf = 1493 and χ^2 values increase to 1552 and 1667. Then the same isolines of corr = 0.77 and 0.75 correspond to probability 14% and 0.09%, and even the conservative user will claim that centroid is within the 0.77 isoline. The example demonstrated difficulties due to missing objective definition of ndf. Nevertheless, in both cases (N = 5 or 50) we see that epicenter is outside the region encompassing the likely centroid!

Although for this event, located on a well-known fault (Cephalonia Transform Fault) we know the likely fault plane (i.e. the nodal plane striking at ~20°), it is instructive to use the H-C tool of ISOLA (Zahradník et al. 2008a). Considering two planes passing through centroid, and having strike and dip of nodal planes, we find that hypocenter is situated 0.5 and 6.5 km from these two planes. Proximity (0.5 km) to the plane striking at 24° is the confirmation that this nodal plane is the fault plane.

Higher frequencies, multiple-point source model. To possibly resolve more source details, we increase maximum frequency of the inversion. Now we invert waveforms in the range 0.03–0.08 Hz. Knowing that the fault was very likely related to the (almost vertical) fault striking at 24°, we design a new model, composed of trial sources distributed along a horizontal line (Fig. 3) at a depth of 5 km.

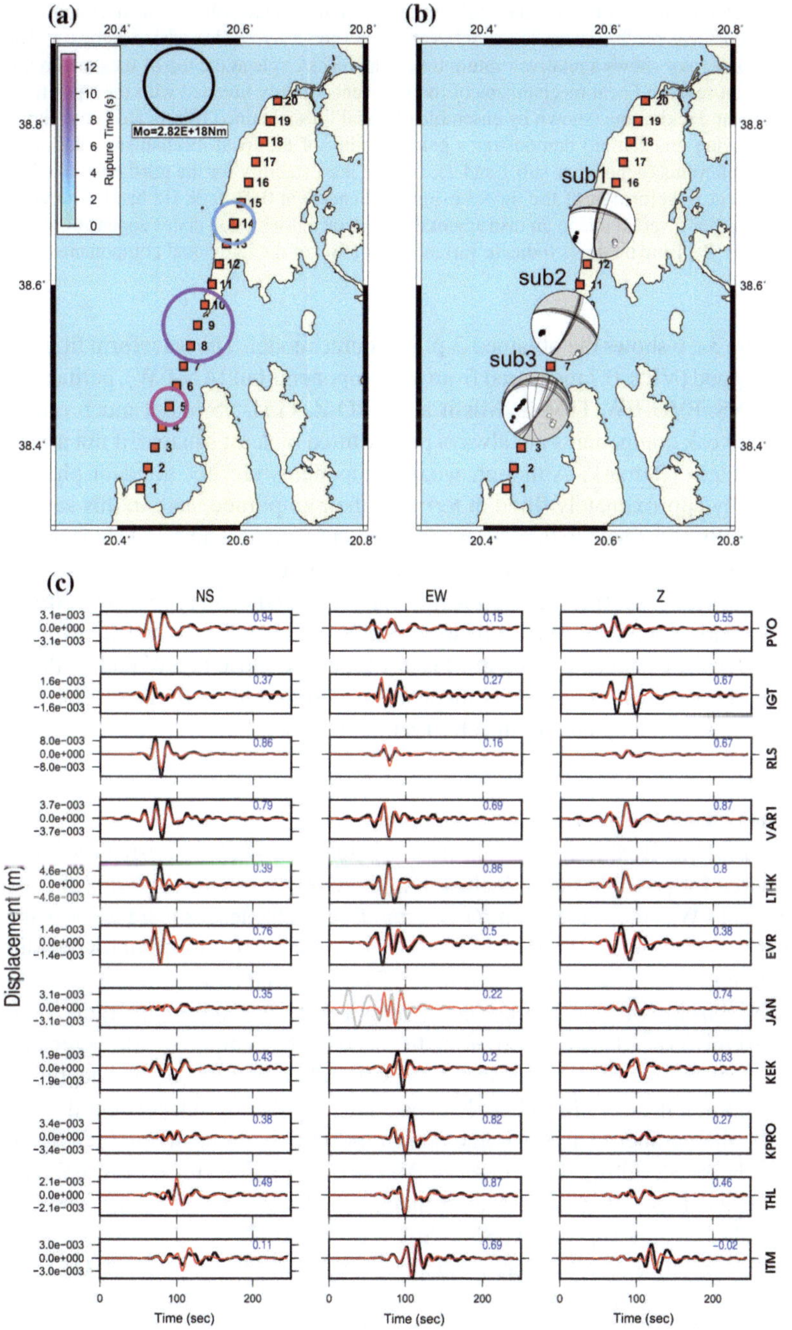
(a)
(b)
(c)
20.4°
20.6°
20.8°
38.8°
38.6°
38.4°
Rupture Time (s)
0
2
4
6
8
10
12
Mo=2.82E+18Nm
sub1
sub2
sub3
NS
EW
Z
PVO
IGT
RLS
VAR1
LTHK
EVR
JAN
KEK
KPRO
THL
ITM
Displacement (m)
Time (sec)
0
100
200
0.94
0.15
0.55
0.37
0.27
0.67
0.86
0.16
0.67
0.79
0.69
0.87
0.39
0.86
0.8
0.76
0.5
0.38
0.35
0.22
0.74
0.43
0.2
0.63
0.38
0.82
0.27
0.49
0.87
0.46
0.11
0.69
-0.02

◂**Fig. 3** Lefkada Mw 6.4 earthquake, 2015; multiple-point source solution. **a** Inversion with DC-constrained focal mechanisms. Three subevents are shown by circles whose radius scales with moment and color shows a relative rupture time. The black circle at the top of the panel represents the moment scale. **b** Focal mechanisms of the subevents, jointly inverted with their position, time and moment. Jackknifing (shown by ensemble of nodal lines obtained by repeated inversions, each time removing one station) demonstrates great stability of the focal mechanism solution for the first two subevents (denoted as sub 1 and 2), and the least stability for the third one (sub 3). **c** The displacement waveform fit of the model using frequencies in 0.03–0.08 Hz band. The waveform JAN-EW is not inverted due to an instrumental disturbance; for station codes appearing at the right refer to Fig. 2a. Blue numbers indicate variance reduction of the individual components

Figure 3a, b shows the obtained 3-point source model. The waveform fit in Fig. 3c is fairly good (VR = 0.7 measured from all components but JAN-EW), perhaps except LTHK-NS, PVO-EW, ITM-Z. Misfit at KPRO-Z, ITM-NS is not much important, because weak components are always poorly fitted by least squares, if not artificially up-weighted. [Remark: Although weak components are not fitted in phase, they are usually approximately fitted in terms of their amplitude, and, in this sense, they significantly contribute to the retrieval of correct focal mechanism. That is why weak components should not be generally excluded from inversion.] The largest subevent has its position (no. 9) close to centroid. The second (smaller) subevent corresponds to a 3-s earlier moment release near epicenter (no. 14). These two subevents have similar focal mechanisms. The third (late) subevent, close to SW end of the tested line, is the smallest one, and its mechanism is different, most likely not reliable. Below we shall complement this observation by F-test.

Box 2: F-test

Similarly to Box 1, we explain here a potentially useful statistical test considered for possible future inclusion in the GUI. In relation to Fig. 3 the user can ask: Was the source composed from three episodes, or just one or two of them are significant? Statistical F-test (e.g., Menke 2012) is useful for deciding where to stop iterative deconvolution. The test needs a ratio of the L2-norm misfits from two calculations, containing n and n + 1 subevents: ratio = misfit(n)/misfit(n + 1). The ratio of misfits needs to be compared with percentiles (critical points) of the F distribution with ndf(n) and ndf(n + 1) degrees of freedom, ndf = N − M, where N denotes the number of independent data and M is the number of free parameters. In most applications it can be assumed that $N \gg M$, ndf(n) = ndf(n + 1) = N. Here, as in χ^2 test, the N is estimated as 15×number of stations (i.e., 5 independent samples per component); for details, see also Suppl. Text S5 in Sokos et al. (2016).

In the present example, there is misfit(1) = 0.0028766, misfit(2) = 0.0020073, misfit(3) = 0.0017929. We find misfit(1)/misfit(2) = 1.43 and misfit(2)/misfit(3) = 1.12. With 10 stations, we have N = 150. The 0.95 critical point for 150 degrees of freedom is 1.31, while the 0.70 critical point is 1.09. Therefore

subevent 2 represents a statistically significant improvement against subevent 1 (since 1.43 > 1.31) at the 95% confidence level, while subevent 3 is an improvement against Sub 2 at the 70% confidence level only. The latter result indicates that subevent 3 may be physically meaningless.
Similarly to Chi-square test in Box 1, we have a problem with objectively defining N. If increasing the number of independent points per component from 5 to 50 (which is still considerably less than the total number of samples per trace, 8192), N increases to 1500 and the 0.95 critical point drops from 1.31 to 1.09. This time, even subevent 3 is a statistically significant improvement compared to two subevents.

Joint inversion of waveforms for source pairs. The iterative deconvolution has revealed two subevents, the large and late at point no. 9, and the early and small at point no. 14. We want to analyze reliability of this result, to see its possible variability. In the GUI tool called Time Function (Free Source Pairs) we prescribe the fixed focal mechanism of two sources (taking s/d/r angles of our CMT solution). Further, we constrain the total moment of any source pair to be the same as our previously obtained CMT moment value. The time function at every source is assumed to be composed from 12 triangles, duration 12 s each, shifted by 1 s. The result (ISOLA graphic output) is shown in Fig. 4a demonstrating various source pairs, all fitting waveforms with high VR, i.e. between 0.95 VRopt and VRopt, where VRopt denotes the largest VR value. The plot clearly shows that it is unlikely to have the two main source contributions at positions 1–7 or 18–20. On the other hand, several combinations may exist between positions 8–17. To inspect one of the pairs, e.g. points 9 and 17, the GUI provides their time function in Fig. 4b.

6 Example Brazil

This example is related to Mw 3 earthquake which occurred in Central Brazil. In this example, we analyze a small earthquake recorded at just three near stations MR07 (37 km, 233°), MR08 (30 km, 233°), and CAN3 (84 km, 50°), where the epicentral distances and azimuths are in the brackets. Later, in the inversion, we shall use only two of them, MR07 and CAN3. It is because using both stations MR07 and MR08, situated close to each other, would be equivalent to using just one and give it weight 2 compared to the third station CAN3. There are 9 trial sources below epicenter, starting at depth 2 km, step 1 km. Green's functions (and elementary seismograms) are calculated up to 1.5 Hz. The inversion is in the range of 0.3–0.8 Hz for MR07 and 0.5–0.8 Hz for CAN3; the low-frequency limit is higher for CAN3 because frequencies <0.5 Hz were noisy.

Plot of correlation versus depth and time (Fig. 5a) indicates two strips, i.e. the bands of pink color, featuring comparable correlation values, but with flipped P-T axes. Focal mechanism corresponding roughly to origin time (which is formally

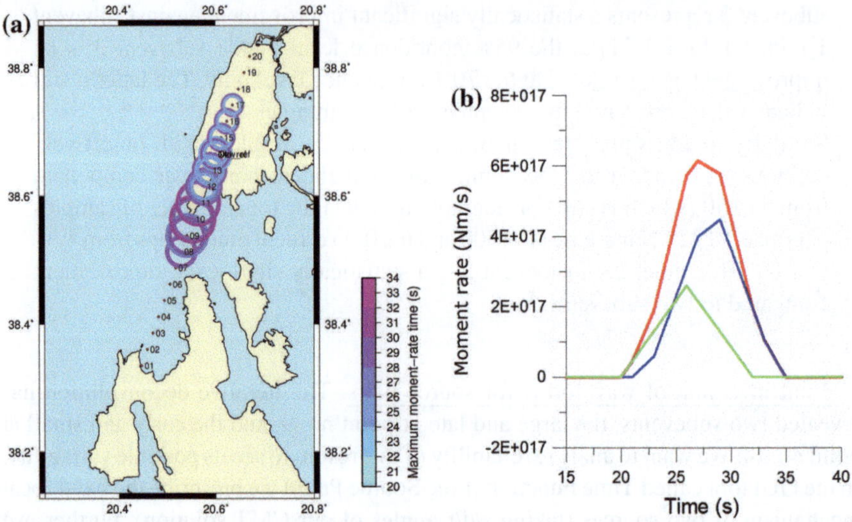

Fig. 4 Lefkada Mw 6.4 earthquake, 2015; NNLS line-source modeling. **a** An ensemble of various source pairs fitting waveforms almost equally well. The circles are sized proportional to moment and color-coded according their rupture time (formally increased by 20 s; origin time of the earthquake is plotted as 20 s). Trial source positions 1–20 are shown. **b** The moment-rate time function for a selected source pair, that of trial positions 9 and 17, plotted in blue and green, respectively. Source rupture starts with weak early episode (green = position 17) and continues later with major moment release (blue = position 9). The total moment-rate function is shown in red

denoted as 20 s) is compared with the first motion polarities (Fig. 5b). The only strongly unfitted polarity is the compression at PMNB. Station BDFB, situated close to PMNB on the focal sphere, has an opposite polarity (discussed later). Waveform match (VR = 0.55) is not very good, considering that we are inverting only two stations. Also, DC = 53% is low, likely indicating a problem with velocity model. The most alarming parameter is CN = 13; its large value indicates poor reliability of the solution. This can be further proven by means of the Uncertainty tool: We prescribe Data Variance σ^2 equal to an estimate of posterior data variance from the previous inversion. Choosing the optimum source position 3 (corresponding to depth 4 km), the covariance matrix (Fig. 6a) shows a lot of trade-offs between the individual source parameters (a-coefficients). That is why small changes of strike/dip/rake angles, and depth, may largely tradeoff with non-DC components, producing same data fit with very different values of DC%. Histogram of DC% (Fig. 6b) and the scatter nodal-line plot (Fig. 6c) confirm a poor MT resolvability.

CSPS method. To increase reliability of the focal mechanism, at least its DC-part, we apply the CSPS method. The method makes use of previous calculations in FOCMEC code of Snoke (2003) (outside of ISOLA), based on polarities at five stations MR07, MR08, CAN3, PMNB and BDFB. All of them were read from original unfiltered, uncorrected records without any doubt (clear onsets); moreover, the 3-component polarity reading confirmed their consistency with station azimuths.

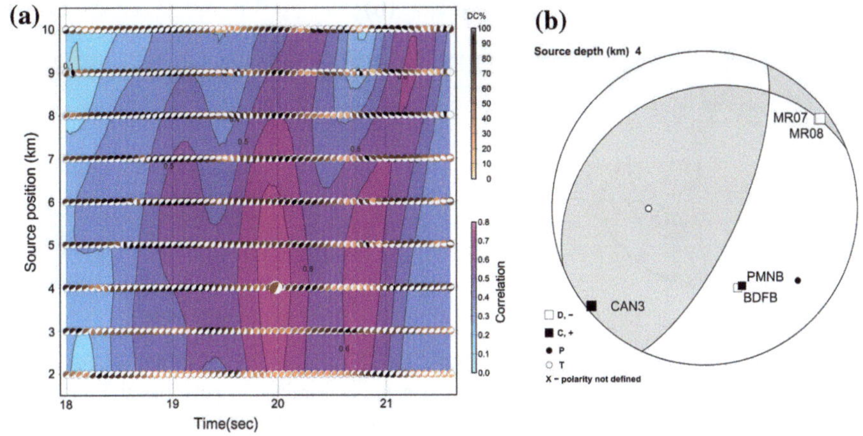

Fig. 5 Small earthquake in central Brazil. **a** Plot of correlation between observed and synthetic seismograms as a function of source depth and time. Origin time is formally placed at t = 20 s. The best-fitting solution is shown by the largest beachball. **b** Posterior polarity check. Two more stations are added: BDFB (distance 247 km, azimuth 158°), PMNB (602 km, 155°). The stations MR07 and MR08 overlap each other on the focal sphere, and the same takes place with BDFB and PMNB

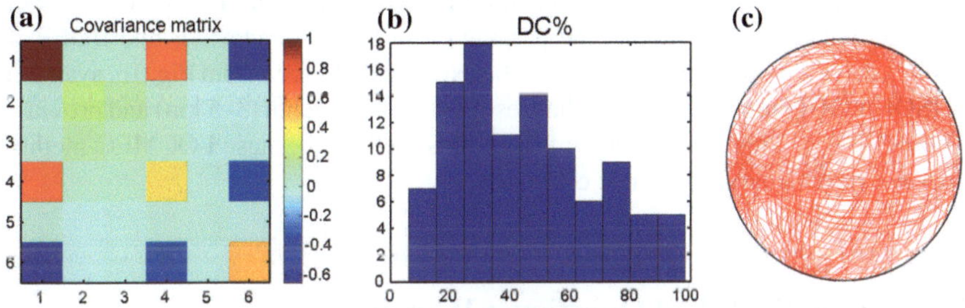

Fig. 6 Small earthquake in central Brazil—uncertainty analysis. **a** Covariance matrix for full-MT solution, numbers 1–6 refer to the inverted parameters a1–a6. Large values on the diagonal indicate the most uncertain parameters. If a column (or row) contain several large values, they indicate strong trade-off between the parameters. **b** Histogram of DC%. **c** Scatter-plot of nodal lines

Nevertheless, we ran FOCMEC with one polarity misfit allowed. It is because of the opposite polarity at stations BDFB and PMNB, which are situated close to each other on the focal sphere. If we do not allow any polarity error, a nodal line will pass between the two stations, and the FOCMEC solution would be 'over-constrained'. Initially FOCMEC provides a set of s/d/r angles and then ISOLA systematically uses all of them in waveform inversion at MR07 and CAN3 stations in a fixed-mechanism mode (100% DC); it means that ISOLA takes every strike, dip, and rake triplet, and inverts waveforms only for the source depth, time and moment. Finally, user can see not only the best-fitting solution, but also some other well-fitting (or acceptable)

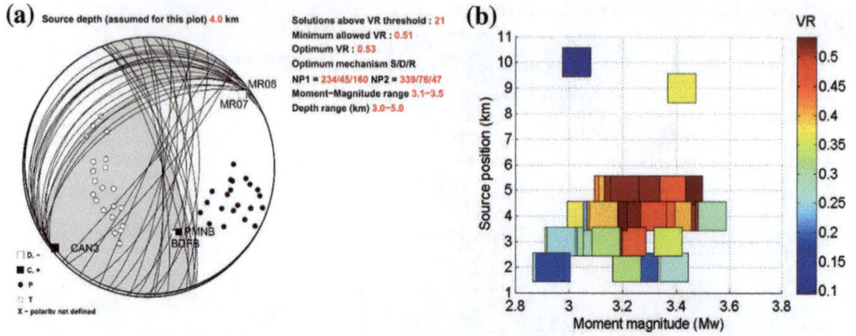

Fig. 7 Small earthquake in central Brazil—CSPS method. **a** Ensemble of well-fitting focal mechanisms obtained by waveform inversion at 2 stations, pre-constrained by polarities at 4 stations. **b** Variance reduction as a function of the source position and moment magnitude

solutions. To this goal, we define a threshold limit, to provide solutions between 0.95 VRopt and VRopt (Fig. 7a).

As seen in Fig. 7a, the combination of 6 polarities and 2 waveforms (MR07 and CAN) quite efficiently constrained the focal mechanisms. The best-fitting solution is shown by shaded sectors. It is very important to notice that this solution (as well as some others) has one nodal line passing close to stations PMNB and BDFB. It implies that their opposite polarities can be explained. As seen from Fig. 7b, available as ISOLA GUI output, the method has resolved well the depth (3–5 km) and provided Mw = 3.1–3.5. This is advantage of the CSPS method, because FOCMEC method does not provide any magnitude estimate.

7 Frequently Asked Questions; Warnings

There are certain critical issues which might severely degrade ISOLA results. Some of them are discussed here.

Why sometimes strange (spurious) signals appear during processing? Many users are trying to use records whose *start time* is greater than origin time. As an example, we can imagine old triggered strong-motion acceleration records. We strongly discourage such an attempt. ISOLA automatically processes data starting at origin time (t = 0) and if there is a gap between t = 0 and (later) start time, this gap is filled with zeros; thus, spurious signals can be generated at t = 0 and/or at the start time. ISOLA issues a warning for such zero padding.

Sometimes, spurious signals are generated at t = 0 even if the start time of the record is smaller than origin time. This is often the case when we want to use records of small earthquakes and the low-frequency *noise is amplified* by instrumental correction. The spurious signal at t = 0 is particularly critical in case of earthquakes recorded at small epicentral distances, where the seismic signal is close to t = 0. A helpful trick is to artificially decrease origin time in the very beginning of ISOLA processing. For example, a 20-s decrease of true origin time is equivalent to shifting the earthquake signal 20 s to the right. (User might have observed such a technical shift in some figures, e.g. in Fig. 5a.) This time, if a spurious signal appears at t = 0, it is far from the useful signal, hence the signal is not damaged. That is why we recommend that records always start well before origin time, e.g. 30 s. Another, even much simpler possible trick is to apply a pre-filter to the instrumentally corrected signals, if they suffer from large low-frequency noise. The latter operation is non-standard and only approximate; users may request some help from the authors.

Some spurious signals in the initial part of records may appear also due to improper (too short) *time window* selected for processing in ISOLA. The time window (same for all stations) must be as large as to encompass complete waveform at the most distant station. If not, later parts of the signal appear, due to temporal aliasing effect, in the initial part, i.e. between origin time and the first arrival.

Spurious signals might appear also at the *end of record*. That is why we recommend that records always end after the time window used in ISOLA.

Low-frequency disturbances ('mice', or 'flings'), already mentioned above, can be regarded as spurious signals, too (Zahradník and Plešinger 2010; Vackář et al. 2015). Use of disturbed records may damage the whole MT inversion, because these records may have very large (false) amplitudes, not obvious after band-pass filtering the records. That is why quality control of waveforms, before entering ISOLA, is extremely important. The disturbances may often be easily recognized if full-band records (without any filtration), without instrumental correction, are baseline corrected and integrated.

All components with spurious signals must be removed from inversion. Contrarily, the sole fact that some components or whole stations cannot be matched in the inversion is not indicative enough for their removal (de-selection) from the inversion!

Why observed waveforms cannot be matched? Spurious signals (discussed above) are the most frequent cause. Another typical problem is that some users do not know very well the *instrumental parameters* (poles, zeros, normalization constant, conversion from counts to m/s), or make mistakes in their use. ISOLA needs the constants in rad/s, not in Hz as provided by some instrument manufacturers. The response provided to ISOLA must be always response of the instrument for input *velocity*. The latter applies to processing of any records, i.e. those from short-period, broad-band or strong-motion instruments. For technical details of this type, the reader is referred to Appendix of the ISOLA manual (Zahradník and Sokos 2016).

In regions were seismicity is infrequent, and/or networks are sparse, the analyzed events may have severely *biased epicenter*. At the same time, having only few waveforms, the ISOLA tool for horizontal grid-searching of the source (like in Fig. 2b) may be inefficient tool for correcting the inaccurate epicenter. This drawback can be often recognized if waveforms cannot be fitted in any frequency range and with any velocity model.

Similar situation—impossibility to match real data—may occur if the stations are few, and some of them have a *gross error in instrumental parameters*. The latter can be identified by a trick: Inverting each station separately, which of course is not reliable for MT, can provide the Mw value much greater for some stations than for the others (e.g. M6 instead M4); thus, we may get indication of a problematic instrumental gain.

Velocity models used in ISOLA may be critical. If synthetic waveforms are considerably 'shorter' than real, i.e. the synthetics miss many later arriving phases present in real record, it is an indication that velocity model is lacking relatively shallow subsurface layer(-s) of low velocity. This is often the case if velocity models used in ISOLA were derived from travel times, they provide location with relatively low travel-time residuals, but may fail in explaining waveforms, especially their surface waves. Models previously tested for agreement with surface wave dispersion are usually better. In any case, repeated ISOLA runs with *several velocity models* are always useful for checking stability of the results.

Users often worry about *Q-factors* (attenuation). ISOLA is basically 'low-frequency method'. In most applications, we work at frequencies below ~0.1 Hz. In this range, waveforms are less sensitive to Q; as such, Qp = 300, Qs = 150 in the whole crust can be usually used.

A typical mistake is that user overestimates the accuracy of his/her hypocenter depth and searches centroid in a *narrow range*, but with an unnecessarily fine grid. For example, having location depth 10 km, some users seek source in ISOLA using trial depths 9.5, 9.6, …, 10.5 km. Instead, we must inspect a broader range of trial depths, possibly with larger steps, e.g., 5, 7.5, …, 15 km. The same applies for temporal grid search.

MT inversion in ISOLA is solved by the least squares. This method, implicitly, fits larger amplitudes better. That is why some stations (or certain station components), having small amplitudes, are not fitted very well. This is normal situation, and there is *no reason to remove weak-amplitude components* from inversion. Even if fitted only roughly, in terms of amplitude, these components significantly contribute to the retrieval of correct focal mechanism. The problem of poorly fitted small amplitudes at some stations might be solved, to some extent, by implementing certain weighting (either using Compute Weights in the Inversion window, or by manually editing allstat.dat file in the invert folder); however, we cannot generally recommend it.

A special case arises when a single station is very near, e.g. at the epicentral distance of 10 km, while the others are at 100–300 km. In case like that, it is always recommended to make inversion both with and without the near station; it is in fact equivalent to a user-defined weighting (i.e. down-weighting the near station due to its very large amplitudes). Nevertheless, it is useful to keep such station in the analyzed station set, which means that synthetics are predicted for this station, although the station is not inverted. The *near station* may bias the inversion not only because of its large amplitude, but also because, in case of larger event, it may be affected by source finiteness (if the epicentral distance is smaller than the fault size). Another potential problem with the nearest station is that it is most vulnerable to the epicenter error. On the other limit, as regards the most *distant stations*, they are usually matched the worst; it is because the effects of the inaccurate velocity model increase with distance. As a rule, poorly fitted distant stations can be removed from inversion almost without any effect upon the solution.

A near station may help to reveal a problem with inaccurate epicenter position. To this goal, it is recommended to read the first-motion *polarities on all three components* and to check if they are consistent with azimuth of the station. Alternatively, if epicenter is correct, this check may reveal a problem with *non-precise orientation of horizontal sensors*.

Naturally, user cannot expect that all first-motion polarities will be always well predicted by the retrieved MT. Some polarities are simply not matched because they were read from noisy records, i.e. those polarities themselves were not guaranteed. Only very clear, possibly sharp impulsive onsets should be used. Unfortunately, this is rarely the case of the Moho-guided (head) waves. Some polarities are well guaranteed, but their projection on focal sphere is problematic due to inaccurate takeoff angles in imprecise velocity model. Note that takeoff angle is a high-frequency, ray-theory concept; as such, it is vulnerable to departures of real Earth from simple 1D velocity models; more vulnerable than the low-frequency waveform inversion. For example, highly unlikely are takeoff angles of very shallow sources calculated at distant stations as head waves due to shallow discontinuities (e.g. at 2 km), simply because shallow discontinuities are strongly laterally variable. Many polarity mismatches can be explained if the takeoff angles are understood as uncertain (Zahradník et al. 2015). Another possibility is to treat the polarity agreement/disagreement in statistical sense (Dias et al. 2016).

How to choose the inverted frequency band? Choosing proper frequency range for the inversion is an important issue. Theoretically, low frequencies are more welcome because synthetics are less suffering from *inaccurate velocity model*. High frequencies are welcome because they help to get more source details (e.g. to construct a multiple-point model). These two requirements must be properly balanced. In practice, the low-frequency limit is given by *natural or instrumental noise*. For example, for magnitude ~6 at regional stations, the signal amplitude is usually well above the noise at frequencies >0.01 Hz, but for magnitude ~2 at local stations we need to start inversion perhaps at 0.5 Hz. The high-frequency limit is given by the epicentral distance. Indeed, as a rule, due to inherent inaccuracy of common velocity models we cannot model waveforms at epicentral distances greater than *10 MSW*, where MSW is the minimum shear wavelength. Considering standard shear wave speed in the crust of 3 km/s, then at 0.1 and 1.0 Hz we have MSW = 30 and 3 km, respectively. Therefore, for magnitude ~6 we can make inversion up to 0.1 Hz at stations with epicentral distances less than 300 km. However, using frequencies up to 1 Hz we must restrict the inversion to the stations at distances less than 30 km. These numbers should be considered as a rough estimate only. Each case needs a careful consideration, or repeated tests in various frequency ranges and several velocity models.

Applying the above discussed rules, it may happen that user disposes just with a very *narrow interval* of usable frequencies. As an undesired consequence, the correlation diagrams (e.g. in Fig. 5a) have several parallel bands, or '*strips*', characterized by almost same correlation values, but the strips have *flipped P-T axes*. Which mechanism is the correct one? To answer the question, the MT solution must be complemented by a polarity check, at least in few stations, optimally those which are situated close to centers of the positive/negative sectors on the focal sphere. If the solution disagrees with polarities, user has to select the other strip, even if it has a (slightly) lower correlation. To this goal, he has to repeat the inversion, narrowing the temporal grid search so that the code identifies the 'correct' strip. However, if the 'correct' strip is far from origin time, e.g. >10 s, we have an indication that velocity model is inaccurate, and the procedure may be inapplicable.

How many stations are needed to get a reliable MT? There is *no simple answer* to this simple question. Sometimes, correct strike, dip, and rake angles are obtained even with very few stations, e.g. two, or just one. Cases like that should however be considered as 'especially favorable'. In no way user can believe that the single station solution is correct just because it matched observed records with large variance reduction, e.g. 0.9! Getting large VR for few records should not be a problem, but it does not guarantee correct mechanism. The question why a solution obtained from few stations is good or not is partially answered by condition number. A rule of thumb is that if CN is larger than 5 or 10, the problem is ill-posed, hence the solution may be wrong. Fortunately, a wrong solution is often indicated by the overall instability of the solution, e.g. in the jackknifing tests, or in correlation graphs. That's why the latter have been supplemented by various stability indices (FMVAR, STVAR, introduced in Sokos and Zahradník 2013). The indices should be also used with caution, because they are somewhat dependent on the subjectively chosen extent of the space-time grid. The indices have always a good sense when comparing two inversions in the same space-time region, particularly in such a case where two or more strips of high correlation are avoided (see paragraph above).

Although we said that few stations might be dangerous, ISOLA *does not need a very good azimuthal coverage*, especially if velocity models are highly reliable. The latter might be the case of the velocity models fitting well various data sets in the studied region, including travel times, dispersion curves, etc. Dense azimuthal coverage is needed in methods using less information from the records, e.g. peak P and S amplitudes, or ratios, like in FOCMEC; use of waveforms may be less demanding. Anyway, as in any physical problem, more data are always welcome (although bringing often more complications). As in many previous places we again emphasize that especially for non-DC components, extremely good data are needed.

How relevant is the moment magnitude Mw of ISOLA? The moment magnitude Mw is calculated from scalar moment M_0 by standard formula (Hanks and Kanamori 1979). Hence the quality of Mw depends on M_0. So how good is our M_0? As explained in the theoretical section, ISOLA calculates M_0 from moment tensor (as its norm). The moment tensor is calculated by the least-squares method; therefore it can be shown that M_0 is equal to scalar product between observed data o and synthetic data s, where s are determined for unit moment; symbolically, $M_0 = (s.o)/(s.s)$, see Eq. (9) of Zahradník and Gallovič (2010). Therefore, M_0 is proportional to s.o, which is the zero-lag correlation. If o and s are matched perfectly (except a multiplicative constant), o = const.s, then obviously, we correctly retrieve M_0 = const. On the other hand, if s and o are not well correlated (not matching their waveforms), then s.o is low, hence also the moment M_0 is low.

Naturally, wrong M_0 may also appear either due to wrong velocity model and/or wrong depth. Indeed, considering for simplicity only S waves, M_0 is proportional to the product $\rho\ \beta^3$ at the assumed source depth, where ρ is density and β is the shear velocity. As a rule, seismic data are insensitive to density, but, in this sense, M_0 is an exception. If user does not know anything about density, ISOLA GUI is assisting during preparation of the 'velocity model', by providing an empirical formula (e.g. Talwani et al. 1959) to convert P-velocity α (in km/s) into density ρ (in g/cm^3): $\rho = 1.7 + 0.2\alpha$. The only goal is to avoid some totally inappropriate values of ρ.

Is Green's function calculation free of any problem? Fortunately, just quite rarely user might encounter numerical problems in calculating Green's functions. These may appear as various spurious impulses, sometimes related to t = 0 or t = TL, where TL is the considered time window length. The problems may appear if sources are very shallow, e.g. <1 km, mainly if at the same time the velocity model features very low velocities below the Earth surface, e.g. $\beta < 2$ km/s. Some of the problems can be solved by changing the internally defined parameters of the discrete wavenumber method, which is a non-standard operation. User should ask the authors for special help. Greens' function in ISOLA cannot be calculated for zero source depth.

Why ISOLA includes the Forward Simulation tool? From time to time, users need *synthetic tests*. They want to generate synthetic data (by prescribing a given moment tensor), perhaps add some noise, and invert these data with the intention to understand the resolution of their problems. Users often forget that synthetic data without noise must be inverted with extremely good match, e.g. VR = 0.99, or 1.0. If obtaining VR = 0.9, which is a high value when processing real data, such VR is too low in synthetic noise-free inversion, and it signalizes a problem. Very often the problem is due to inappropriate preparation of the synthetic data, in particular due to certain filtering. Optimally, synthetic data must be prepared without any filtration, i.e. in a range starting at zero frequency, and ending at Nyquist frequency (or at least at a frequency much greater than fmax frequency of the Green function).

User may want to compare the inversion result with a solution of some agency. Although he/she can use the fixed-mechanism inversion mode, that mode does not enable use of a given (prescribed) scalar moment M_0 (hence also Mw). Prescription of complete focal mechanism, including M_0, is possible in the forward simulation tool of ISOLA. A great advantage is that the tool makes it possible to calculate synthetics for an *arbitrary full moment tensor*, chosen by the user, e.g. a pure volume change. Moreover, in contrast to inversion (which is always made in displacement), we may choose to forward-simulate velocity.

Another reason for using forward simulation is to get a feeling how much seismograms can vary due to change of the *velocity model*.

Still another reason is curiosity to see low-frequency waveform features, missing or obscured in real data due to limited instrumental frequency band, or due to noise. In particular, it is interesting to learn about near-field effects present in low-pass synthetics, such as 'ramps' between P and S onsets, and/or static displacements. If the latter is significant (of the order of centimeters, or more) it signalizes usefulness to complement the seismic analyses by GPS or InSAR geodetic data.

Miscellaneous: Some ISOLA tools, very useful in certain applications, might be totally *misleading* in others. A typical example is the so-called H-C tool; Zahradník et al. (2008a). It serves for identifying fault plane among the two nodal planes. It is intended for events of M > 5–6, where centroid and hypocenter position might be displaced of each other at a distance considerably greater than their own errors. Then the tool may correctly identify that plane, passing through centroid, which encompasses hypocenter, so that plane is the fault. However, when using the method for weak events in situations where there is no physical reason for having centroid significantly displaced from hypocenter, and/or their errors are large, the method is naturally misleading.

ISOLA is not intended for routine use of stations at distances >1000 km, where *sphericity of the Earth* may be an issue for deep earthquakes. However the problem can be overcome (Zahradník et al. 2017), please ask authors for special help.

At the end, we would like to acknowledge the fact that many beginners are able to start with ISOLA quite intuitively, without reading any instruction. However, this is *not* the best practice, and users like that often need to solve unnecessary complications. Most complications can be avoided if putting more attention to a preliminary study. We strongly emphasize the need to read ISOLA manual and ISOLA related papers, e.g. those referenced at the end.

8 Outlook

Closely related to the problem how to find a suitable frequency range are recent efforts of Fabio Dias, Brazil. To circumvent 'manual' search of the range, he proposed the so-called frequency range tests FRT, where ISOLA is repeatedly used in many different ranges and results are jointly processed (Dias et al. 2016). The FRT method is a candidate for possible future inclusion it the GUI.

Statistical tools (F-test, Chi-square test), discussed in Boxes 1 and 2, but currently not included in the GUI represent other candidates.

Recent studies of small earthquakes in sparse networks that we performed in co-operation with colleagues in Argentina (Celeste Bollini) and Brazil (Juraci Carvalho) have indicated a need to solve the problem with limited applicability of a single velocity model at various stations at large epicentral distances. The solution consists in releasing strict requirement to fit waveforms (simply because we are unable to fit them). Instead, we might want to invert amplitude spectra, which are independent of unknown time shifts caused by inaccurate velocity model, or invert envelopes of the full waveforms. This study and implementation of these methods in the GUI is in progress (Zahradník and Sokos 2018).

New codes have been written also for inverting static GPS displacements into a suite of points or finite-extent sources. If several GPS observations are available for large earthquake (M ~ 7) near the fault, these codes might help in designing the trial point-source grids for inversion of seismic data in ISOLA.

Our plans include extension to 3D velocity models. The idea is that forward problem in 3D (Green's functions and elementary seismograms) are calculated in some finite-difference code outside ISOLA. Then it is relatively straightforward to use the 3D elementary seismograms even already in the existing ISOLA codes. Please ask for help, if you have 3D elementary seismograms and want to use them for the MT inversion of single- or multiple-point source models.

Finite-extent slip inversion is not planned for inclusion in ISOLA, because excellent codes for this purpose (Gallovič et al. 2015) are available elsewhere (Github: http://fgallovic.github.io/LinSlipInv/, last accessed June 2017). Note that the slip inversion is well applicable only to larger events (M ~ 6) for which we dispose with many (dozens) strong-motion records from near-fault stations.

Also not planned is automation, because this task is fully covered by SCISOLA (Triantafyllis et al. 2016), and Bayesian ISOLA (Vackář et al. 2017), already mentioned above.

Acknowledgements The authors sincerely thank to Ronnie Quintero, Lucas Barros, Patricia Pedraza and Didem Cambaz who organized ISOLA training courses. The courses provided important feedback to the authors. We also thank to users, worldwide, for huge number of e-mail questions that helped us to make the code more user friendly. ISOLA includes a modified version of the discrete-wave number code AXITRA of O. Coutant, filter XAPiir encoded by D. Harris, the NNLS inversion code of C.L. Lawson and R.J. Hanson, the MT-decomposition code by J. Šílený, several codes from Numerical recipes (Press 1992) and Matlab user file repository. Plots are created using GMT (Wessel and Smith 1998). For acknowledgment of data used as Example Greece, see Sokos et al. (2016). The Example Brazil is based on unpublished data provided by Lucas Barros and Juraci Carvalho, University of Brasilia. E.S. acknowledges financial support by HELPOS project, "Hellenic Plate Observing System" (MIS 5002697). J.Z. was supported by CzechGeo/EPOS (LM2015079) and the Czech Science Foundation (GACR-18-06716J).

References

Adamová P, Sokos E, Zahradník J (2009) Problematic non-double-couple mechanism of the 2002 Amfilochia Mw 5 earthquake, Western Greece. J Seismol 13:1–12. https://doi.org/10.1007/s10950-008-9112-4

Agurto H, Rietbrock A, Ryder I, Miller M (2012) Seismic-afterslip characterization of the 2010 Mw 8.8 Maule, Chile, earthquake based on moment tensor inversion. Geophys Res Lett. https://doi.org/10.1029/2012gl053434

Aki K, Richards PG (2002) Quantitative seismology. University Science Books

Benetatos C, Málek J, Verga F (2013) Moment tensor inversion for two micro-earthquakes occurring inside the Háje gas storage facilities, Czech Republic. J Seismol 17:557–577. https://doi.org/10.1007/s10950-012-9337-0

Bouchon M (1981) A simple method to calculate Green's functions for elastic layered media. Bull Seismol Soc Am 71:959–971

Cambaz MD, Mutlu AK (2016) Regional moment tensor inversion for earthquakes in Turkey and its surroundings: 2008–2015. Seismol Res Lett 87:1082–1090. https://doi.org/10.1785/0220150276

Carvalho J, Barros LV, Zahradník J (2016) Focal mechanisms and moment magnitudes of micro-earthquakes in central Brazil by waveform inversion with quality assessment and inference of the local stress field. J South Am Earth Sci 71:333–343. https://doi.org/10.1016/j.jsames.2015.07.020

Coutant O (1989) Program of numerical simulation AXITRA. University of Joseph Fourie, Grenoble, France

Dias F, Zahradník J, Assumpção M (2016) Path-specific, dispersion-based velocity models and moment tensors of moderate events recorded at few distant stations: examples from Brazil and Greece. J South Am Earth Sci 71:344–358. https://doi.org/10.1016/j.jsames.2016.07.004

Fojtíková L, Zahradník J (2014) A new strategy for weak events in sparse networks: the first-motion polarity solutions constrained by single-station waveform inversion. Seismol Res Lett 85:1265–1274. https://doi.org/10.1785/0220140072

Fojtíková L, Kristeková M, Málek J et al (2016) Quantifying capability of a local seismic network in terms of locations and focal mechanism solutions of weak earthquakes. J Seismol. https://doi.org/10.1007/s10950-015-9512-1

Gallovič F, Zahradník J (2011) Toward understanding slip inversion uncertainty and artifacts: 2. Singular value analysis. J Geophys Res Solid Earth 116:B02309. https://doi.org/10.1029/2010JB007814

Gallovič F, Zahradník J (2012) Complexity of the Mw 6.3 2009 L'Aquila (central Italy) earthquake: 1. Multiple finite-extent source inversion. J Geophys Res Solid Earth. https://doi.org/10.1029/2011jb008709

Gallovič F, Imperatori W, Mai PM (2015) Effects of three-dimensional crustal structure and smoothing constraint on earthquake slip inversions: case study of the *Mw* 6.3 2009 L'Aquila earthquake. J Geophys Res Solid Earth 120:428–449. https://doi.org/10.1002/2014JB011650

Gonzalez O, Clouard V, Bouin M-P, Zahradník J (2017) Centroid moment tensor solutions along the central Lesser Antilles for 2013–2015. http://dx.doi.org/10.1016/j.tecto.2017.06.024

Hallo M, Gallovič F (2016) Fast and cheap approximation of Green function uncertainty for waveform-based earthquake source inversions. Geophys J Int 207:1012–1029. https://doi.org/10.1093/gji/ggw320

Hanks TC, Kanamori H (1979) A moment magnitude scale. J Geophys Res 84:2348. https://doi.org/10.1029/JB084iB05p02348

Harris D (1990) XAPiir: a recursive digital filtering package. Lawrence Livermore National Laboratory. United States

Hazarika D, Paul A, Wadhawan M et al (2017) Seismotectonics of the Trans-Himalaya, Eastern Ladakh, India: constraints from moment tensor solutions of local earthquake data. Tectonophysics 698:38–46. https://doi.org/10.1016/j.tecto.2017.01.001

Henry C, Das S (2002) The Mw 8.2, 17 February 1996 Biak, Indonesia, earthquake: rupture history, aftershocks, and fault plane properties. J Geophys Res Solid Earth 107:ESE 11-1–ESE 11-16. https://doi.org/10.1029/2001jb000796

Hicks SP, Rietbrock A (2015) Seismic slip on an upper-plate normal fault during a large subduction megathrust rupture. Nat Geosci 8:955–960. https://doi.org/10.1038/NGEO2585

Kikuchi M, Kanamori H (1991) Inversion of complex body waves—III. Bull Seismol Soc Am 81:2335–2350

Křížová D, Zahradník J, Kiratzi A (2013) Resolvability of isotropic component in regional seismic moment tensor inversion. Bull Seismol Soc Am 103:2460–2473. https://doi.org/10.1785/0120120097

Lawson CL, Hanson RJ (1974) Solving least square problems. New Jersey, 340 pp

Liu J, Li L, Zahradník J, Sokos E, Liu C, Tian X (2018) North Korea's 2017 test and its nontectonic aftershock. Geophys Res Lett 45. https://doi.org/10.1002/2018GL077095

Menke W (2012) Geophysical data analysis: discrete inverse theory. Elsevier/Academic Press

Michele M, Custódio S, Emolo A (2014) Moment tensor resolution: case study of the Irpinia Seismic Network, Southern Italy. Bull Seismol Soc Am 104:1348–1357. https://doi.org/10.1785/0120130177

Press WH (1992) Numerical recipes in FORTRAN: the art of scientific computing. Cambridge University Press

Quintero R, Zahradník J, Sokos E (2014) Near-regional CMT and multiple-point source solution of the September 5, 2012, Nicoya, Costa Rica Mw 7.6 (GCMT) earthquake. J South Am Earth Sci 55:155–165. https://doi.org/10.1016/j.jsames.2014.07.009

Serpetsidaki A, Sokos E, Tselentis G-A, Zahradník J (2010) Seismic sequence near Zakynthos Island, Greece, April 2006: identification of the activated fault plane. Tectonophysics 480:23–32. https://doi.org/10.1016/j.tecto.2009.09.024

Shearer PM (2009) Introduction to seismology. Cambridge University Press

Snoke JA (2003) FOCMEC: FOCal MEChanism determinations. In: Lee WHK, Kanamori H, Jennings PC, Kisslinger C (eds) International handbook of earthquake and engineering seismology. Academic Press, San Diego, pp 1629–1630

Sokos EN, Zahradník J (2008) ISOLA a Fortran code and a Matlab GUI to perform multiple-point source inversion of seismic data. Comput Geosci 34:967–977. https://doi.org/10.1016/j.cageo.2007.07.005

Sokos E, Zahradník J, Kiratzi A et al (2012) The January 2010 Efpalio earthquake sequence in the western Corinth Gulf (Greece). Tectonophysics 530–531:299–309. https://doi.org/10.1016/j.tecto.2012.01.005

Sokos E, Zahradník J (2013) Evaluating centroid-moment-tensor uncertainty in the new version of ISOLA software. Seismol Res Lett. https://doi.org/10.1785/0220130002

Sokos E, Kiratzi A, Gallovič F et al (2015) Rupture process of the 2014 Cephalonia, Greece, earthquake doublet (Mw 6) as inferred from regional and local seismic data. Tectonophysics 656:131–141. https://doi.org/10.1016/j.tecto.2015.06.013

Sokos E, Zahradník J, Gallovič F et al (2016) Asperity break after 12 years: the Mw 6.4 2015 Lefkada (Greece) earthquake. Geophys Res Lett 43:6137–6145. https://doi.org/10.1002/2016GL069427

Talwani M, Sutton GH, Worzel JL (1959) A crustal section across the Puerto Rico trench. J Geophys Res 64:1545–1555. https://doi.org/10.1029/JZ064i010p01545

Triantafyllis N, Sokos E, Ilias A, Zahradník J (2016) Scisola: automatic moment tensor solution for SeisComP3. Seismol Res Lett 87:157–163. https://doi.org/10.1785/0220150065

Vackář J, Burjánek J, Zahradník J (2015) Automated detection of long-period disturbances in seismic records; MouseTrap code. Seismol Res Lett 86:442–450. https://doi.org/10.1785/0220140168

Vackář J, Burjánek J, Gallovič F et al (2017) Bayesian ISOLA: new tool for automated centroid moment tensor inversion. Geophys J Int 210:693–705. https://doi.org/10.1093/gji/ggx158

Vavryčuk V (2001) Inversion for parameters of tensile earthquakes. J Geophys Res 106:16339–16355. https://doi.org/10.1029/2001JB000372

Wessel P, Smith WHF (1998) New, improved version of generic mapping tools released. EOS Trans Am Geophys Union 79:579. https://doi.org/10.1029/98EO00426

Zahradník J, Gallovič F (2010) Toward understanding slip inversion uncertainty and artifacts. J Geophys Res. https://doi.org/10.1029/2010JB007414

Zahradník J, Plešinger A (2005) Long-period pulses in broadband records of near earthquakes. Bull Seismol Soc Am 95:1928–1939. https://doi.org/10.1785/0120040210

Zahradník J, Plešinger A (2010) Toward understanding subtle instrumentation effects associated with weak seismic events in the near field. Bull Seismol Soc Am 100:59–73. https://doi.org/10.1785/0120090087

Zahradník J, Serpetsidaki A, Sokos E, Tselentis G-A (2005) Iterative deconvolution of regional waveforms and a double-event interpretation of the 2003 Lefkada Earthquake, Greece. Bull Seismol Soc Am 95:159–172. https://doi.org/10.1785/0120040035

Zahradník J, Gallovič F, Sokos E et al (2008a) Quick fault-plane identification by a geometrical method: application to the Mw 6.2 Leonidio Earthquake, 6 January 2008. Greece. Seismol Res Lett 79:653–662. https://doi.org/10.1785/gssrl.79.5.653

Zahradník J, Sokos E, Tselentis G-A, Martakis N (2008b) Non-double-couple mechanism of moderate earthquakes near Zakynthos, Greece, April 2006; explanation in terms of complexity. Geophys Prospect 56:341–356. https://doi.org/10.1111/j.1365-2478.2007.00671.x

Zahradník J, Gallovič F, Sokos E, Tselentis G-A (2011) Preliminary slip model of M9 Tohoku earthquake from strong-motion stations in Japan—an extreme application of ISOLA code. http://www.emsc-csem.org/Files/event/211414/ISOLA_Report_tohoku.pdf. Accessed 21 June 2017
Zahradník J, Custodió S (2012) Moment tensor resolvability: application to Southwest Iberia. Bull Seismol Soc Am 102:1235–1254. https://doi.org/10.1785/0120110216
Zahradník J, Sokos E (2014) The Mw 7.1 Van, Eastern Turkey, earthquake 2011: two-point source modelling by iterative deconvolution and non-negative least squares. Geophys J Int 196:522–538. https://doi.org/10.1093/gji/ggt386
Zahradník J, Fojtíková L, Carvalho J et al (2015) Compromising polarity and waveform constraints in focal-mechanism solutions; the Mara Rosa 2010 Mw 4 central Brazil earthquake revisited. J South Am Earth Sci 63:323–333. https://doi.org/10.1016/j.jsames.2015.08.011
Zahradník J, Sokos E (2016) ISOLA User's Guide, Training course in Jaco, Costa Rica, 2016, 30 pp. http://geo.mff.cuni.cz/~jz/for_ISOLA/
Zahradník J, Sokos E (2018) Fitting waveform envelopes to derive focal mechanisms of moderate earthquakes. Seismol Res Lett, https://doi.org/10.1785/0220170161
Zahradník J, Čížková H, Bina CR et al (2017) A recent deep earthquake doublet in light of long-term evolution of Nazca subduction. Sci Rep 7:45153. https://doi.org/10.1038/srep45153

Seismic Moment Tensors in Anisotropic Media: A Review

Václav Vavryčuk

1 Introduction

Seismic anisotropy is a common property of rocks and geological structures in the Earth crust and in the upper mantle (Babuška and Cara 1991; Rabbel and Mooney 1996; Silver 1996; Savage 1999). It may be caused by sediment layering, by stress-aligned systems of microcracks, cracks or fractures, or by the textural ordering of rock-forming minerals. Anisotropy affects propagation of seismic waves as well as radiation of waves by seismic sources. So far seismologists have focused mainly on studying wave propagation in anisotropic media (e.g., Musgrave 1970; Helbig 1994; Červený 2001; Vavryčuk 2001b, 2003), and on observing the effects of anisotropy on seismic waves (Babuška and Cara 1991) as the directional variation of seismic velocities or shear-wave splitting, detected and measured in situ (Kaneshima et al. 1988; Crampin 1993; Savage 1999) or in the laboratory on rock samples (Kern and Schmidt 1990; Pros et al. 1998; Mainprice et al. 2000; Svitek et al. 2014).

However, equally important is the way in which anisotropy affects the radiation of waves from seismic sources. This comprises calculating the Green's functions (Burridge 1967; Ben-Menahem and Sena 1990a, b; Gajewski 1993; Vavryčuk 1997; Pšenčík 1998; Červený 2001; Vavryčuk 2007), seismic moment tensors and focal mechanisms in anisotropic media (Šílený and Vavryčuk 2000, 2002; Vavryčuk 2005). For example, Kawasaki and Tanimoto (1981) and Vavryčuk (2005) pointed out that shear faulting in anisotropic media can produce mechanisms with non-double couple (non-DC) components. Since the non-DC mechanisms of earthquakes are frequently observed (Miller et al. 1998), the problem, whether anisotropy contributes to them or not, is not only of theoretical interest but also of practical relevance. Anisotropy as a possible origin of non-DC mechanisms has been mentioned and discussed also by other authors (Kawakatsu 1991; Frohlich 1994; Julian et al. 1998; Vavryčuk 2002, 2004, 2006; Rössler et al. 2003; Vavryčuk et al. 2008).

V. Vavryčuk (✉)
Institute of Geophysics, Czech Academy of Sciences, Boční II/1401, 141 31 Praha 4, Czech Republic
e-mail: vv@ig.cas.cz

S. D'Amico (ed.), *Moment Tensor Solutions*, Springer Natural Hazards,
https://doi.org/10.1007/978-3-319-77359-9_2

Since the effects of anisotropy on focal mechanisms and moment tensors are still not well known and often neglected, I present a review of basic properties of moment tensors in anisotropic media. The review covers theory, numerical modelling and several applications to earthquakes on various scales including acoustic emissions measured in the lab, microearthquakes in the upper crust and large deep-focus earthquakes in a subducting slab. The goal is to point out errors introduced when seismic anisotropy is neglected in moment tensor inversions, and a possibility to study seismic anisotropy using accurately determined moment tensors.

2 Theory

2.1 Moment Tensor in Anisotropic Media

Moment tensor **M** of a seismic source in an anisotropic medium is expressed as (Aki and Richards 2002, Eq. 3.19)

$$M_{ij} = uSc_{ijkl}\nu_k n_l, \tag{1}$$

where u is the slip, S is the fault area, c_{ijkl} are the elastic stiffness parameters of the medium surrounding the fault, $\boldsymbol{\nu}$ is the slip direction, and **n** is the fault normal. Introducing source tensor **D** as

$$D_{kl} = \frac{uS}{2}\left(\nu_k n_l + \nu_l n_k\right), \tag{2}$$

and taking into account the symmetry of c_{ijkl}, we can express Eq. (1) in the following form

$$M_{ij} = c_{ijkl} D_{kl}, \tag{3}$$

which resembles the Hooke's law

$$\tau_{ij} = c_{ijkl} e_{kl}, \tag{4}$$

expressing the relation between stress and strain tensors τ_{ij} and e_{ij} in elastic media. Therefore, Eq. (3) can be called as "the generalized Hooke's law at the source". Source tensor **D** defines geometry of dislocation at the source being an analogue to strain tensor e_{ij}. Moment tensor **M** defines the equivalent body forces acting at the source being an analogue to stress tensor τ_{ij}.

Equation (3) can equivalently be expressed in matrix form as

$$\mathbf{m} = \mathbf{C}\mathbf{d}, \tag{5}$$

where **C** is the 6×6 matrix of the elastic parameters in the two-index Voigt notation, where the pairs of the subscripts in the four-index tensor c_{ijkl} are substituted in the following way: $11 \to 1, 22 \to 2, 33 \to 3, 23 \to 4, 13 \to 5$ and $12 \to 6$ (see Musgrave 1970, Eq. 3.13.4). Quantities **m** and **d** are the 6-vectors defined as

$$\mathbf{m} = (M_{11}, M_{22}, M_{33}, M_{23}, M_{13}, M_{12})^T, \tag{6}$$

$$\mathbf{d} = uS\,(n_1\nu_1, n_2\nu_2, n_3\nu_3, n_2\nu_3 + n_3\nu_2, n_1\nu_3 + n_3\nu_1, n_1\nu_2 + n_2\nu_1)^T. \tag{7}$$

Source tensor **D** is expressed by the components of vector **d** as follows:

$$\mathbf{D} = \frac{1}{2}\begin{bmatrix} 2d_1 & d_6 & d_5 \\ d_6 & 2d_2 & d_4 \\ d_5 & d_4 & 2d_3 \end{bmatrix}. \tag{8}$$

The moment tensor **M** of a seismic source in an isotropic medium reads (Aki and Richards 2002, Eq. 3.21)

$$M_{ij} = \lambda D_{kk}\delta_{ij} + 2\mu D_{ij}, \tag{9}$$

with the trace

$$M_{kk} = (3\lambda + 2\mu)\, D_{kk}, \tag{10}$$

where λ and μ are the Lamé constants describing the isotropic medium surrounding the fault, and δ_{ij} is the Kronecker delta.

2.2 *Eigenvalues and Eigenvectors of Tensors M and D*

Tensor **M** has a diagonal form expressed by three eigenvalues M_1, M_2 and M_3

$$\mathbf{M}^{diag} = \begin{bmatrix} M_1 & 0 & 0 \\ 0 & M_2 & 0 \\ 0 & 0 & M_3 \end{bmatrix}, \text{ where } M_1 \geq M_2 \geq M_3. \tag{11}$$

In isotropic media, Eq. (11) reads

$$\mathbf{M}^{diag} = uS\begin{bmatrix} (\lambda + \mu)\,\mathbf{n}\cdot\boldsymbol{\nu} + \mu & 0 & 0 \\ 0 & \lambda\mathbf{n}\cdot\boldsymbol{\nu} & 0 \\ & & (\lambda + \mu)\,\mathbf{n}\cdot\boldsymbol{\nu} - \mu \end{bmatrix}. \tag{12}$$

where $\mathbf{n}\cdot\boldsymbol{\nu}$ is the scalar product of two unit vectors **n** and $\boldsymbol{\nu}$,

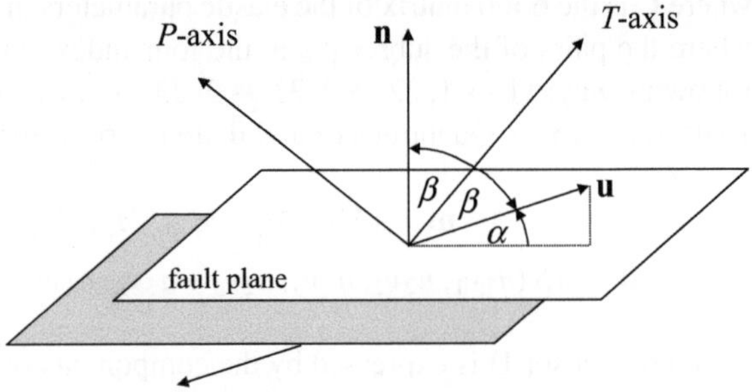

Fig. 1 Model of a shear-tensile earthquake. Vector **u** is the slip vector, vector **n** is the fault normal, and α is the slope angle. Angle β is defined as $\beta = (90° - \alpha)/2$

$$\mathbf{n} \cdot \boldsymbol{\nu} = n_1 \nu_1 + n_2 \nu_2 + n_3 \nu_3. \tag{13}$$

A diagonal form of tensor **D** is independent of elastic properties of the medium being expressed as

$$\mathbf{D}^{diag} = \begin{bmatrix} D_1 & 0 & 0 \\ 0 & D_2 & 0 \\ 0 & 0 & D_3 \end{bmatrix} = \frac{uS}{2} \begin{bmatrix} \mathbf{n} \cdot \boldsymbol{\nu} + 1 & 0 & 0 \\ 0 & 0 & 0 \\ 0 & 0 & \mathbf{n} \cdot \boldsymbol{\nu} - 1 \end{bmatrix}. \tag{14}$$

The determinant of **D** is zero and the trace of **D** is

$$D_{kk} = uS\,(\mathbf{n} \cdot \boldsymbol{\nu}) = uS \sin\alpha, \tag{15}$$

where α is the slope angle defined as the deviation of the slip vector **n** from the fault plane (Vavryčuk 2001a, b, 2011).

The eigenvectors of tensors **M** and **D** specify the coordinate systems in which these tensors are diagonalized. The eigenvectors of **M** are denoted as the P, T and B axes and correspond to minimum eigenvalue M_3, maximum eigenvalue M_1, and intermediate eigenvalue M_2, respectively. Physically, the P, T and B axes are directions of the maximum compressional, maximum tensional and intermediate stresses generated at the source.

The eigenvectors of moment tensor **M** are generally different from those of source tensor **D**. However, the both coordinate systems coincide in isotropic media. In this case, the B axis is perpendicular to fault normal **n** and slip direction $\boldsymbol{\nu}$. The P and T axes lie in the plane defined by vectors **n** and $\boldsymbol{\nu}$ (Fig. 1) being expressed as

$$\mathbf{t} = \frac{\mathbf{n} + \boldsymbol{\nu}}{|\mathbf{n} + \boldsymbol{\nu}|}, \mathbf{b} = \mathbf{n} \otimes \boldsymbol{\nu}, \mathbf{p} = \frac{\mathbf{n} - \boldsymbol{\nu}}{|\mathbf{n} - \boldsymbol{\nu}|}, \tag{16}$$

where **p**, **t**, and **b** are the directional vectors of the P, T and B axes, respectively, and symbol $\otimes$ denotes the vector product.

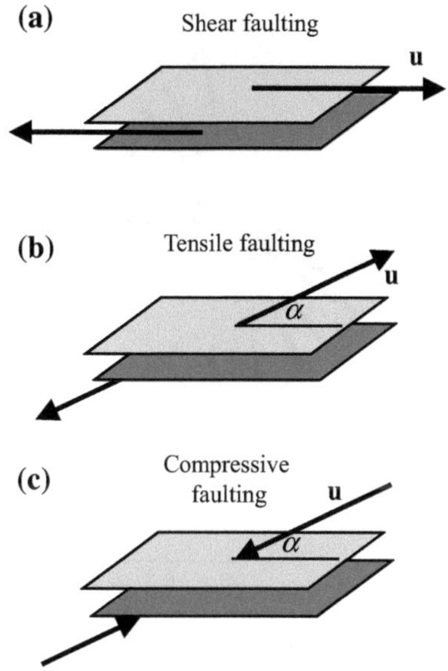

Fig. 2 Model of shear (**a**), tensile (**b**) and compressive (**c**) faulting. Vector **u** is the slip vector, α is the slope angle

2.3 *Tensile Faulting in Isotropic Media Versus Shear Faulting in Anisotropic Media*

For shear faulting, slope angle α is zero (Fig. 2a) and Eqs. (12) and (14) for **M** and **D** further simplify in isotropic media to

$$\mathbf{M}^{diag} = \mu u S \begin{bmatrix} +1 & 0 & 0 \\ 0 & 0 & 0 \\ 0 & 0 & -1 \end{bmatrix}, \mathbf{D}^{diag} = \frac{uS}{2} \begin{bmatrix} +1 & 0 & 0 \\ 0 & 0 & 0 \\ 0 & 0 & -1 \end{bmatrix} \tag{17}$$

being known as the double-couple (DC) tensors.

For tensile/compressive faulting (Fig. 2b, c), slope α is non-zero and tensors **M** and **D** of a source in isotropic media contain also non-double-couple (non-DC) components. The non-DC components are usually decomposed into the isotropic (ISO) and compensated linear vector dipole (CLVD) parts (see Vavryčuk 2015). For tensor **D**, the relative amounts of ISO and CLVD depend on the slope angle; for tensor **M**, the ISO and CLVD depend on the slope and on the ratio of the *P*- and *S*-wave velocities *vp*/*vs*.

For moment tensors in anisotropic media, the problem is more involved. In anisotropic media, even shear faulting on a planar fault produces a generally non-DC moment tensor. This is caused by the elasticity tensor in Eq. (3) which is formed by 21 elastic parameters. However, the moment tensor is simplified for anisotropy of higher symmetry. For example, under orthorhombic symmetry, **M** is also a pure DC provided that fault normal **n** coincides with the symmetry axis and slip **u** lies in the

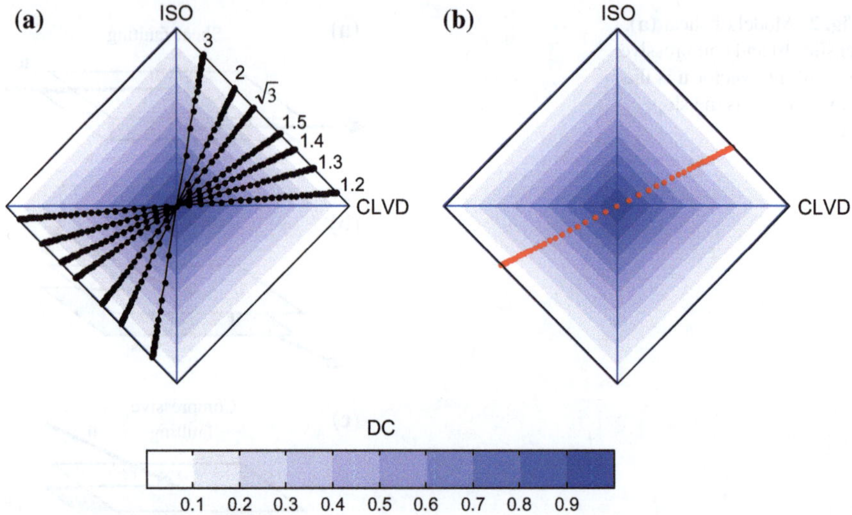

Fig. 3 The source-type plots for the moment tensors (**a**) and source tensors (**b**) for shear-tensile-compressive faulting in isotropic media. Red dots—the source tensors, black dots—the moment tensors. Isotropic media in (**a**) are characterized by various values of the v_P/v_S ratios (the values are indicated in the plot). The dots in (**a**) and (**b**) correspond to the sources with a specific value of the slope angle (i.e. the deviation of the slip vector from the fault). The slope angle ranges from −90° (pure compressive crack) to 90° (pure tensile crack) in steps of 3°. For a detailed explanation of properties of the source-type plot, see Vavryčuk (2015). After Vavryčuk (2015)

symmetry plane (but not along the symmetry axis). However, the orientation of the DC can deviate from the plane defined by the fault normal and slip direction (see Vavryčuk 2005).

The properties of the moment and source tensor decompositions for shear and tensile sources in isotropic and anisotropic media are illustrated in Figs. 3 and 4. Figure 3 shows the source-type plots for shear-tensile-compressive sources with a variable slope angle α (i.e. the deviation of the slip vector from the fault, see Vavryčuk 2011) situated in an isotropic medium. The plot shows that the scale factors of the ISO and CLVD components are linearly dependent for both moment and source tensors. For the moment tensors, the line direction depends on the *vp*/*vs* ratio (Fig. 3a); for the source tensors, the line is independent of the properties of the elastic medium and the ISO/CLVD ratio is always 1/2 (Fig. 3b). This property of the source tensors is preserved even for anisotropic media.

On the contrary, the moment tensors can behave in a more complicated way in anisotropic media. Figure 4 shows the ISO and CLVD components of the moment tensors of shear (Fig. 4a) or shear-tensile (Fig. 4b) faulting in the Bazhenov shale (Vernik and Liu 1997). This complicated behaviour prevents a straightforward interpretation of moment tensors in terms of physical faulting parameters. Therefore, first the source tensors must be calculated from moment tensors and then interpreted.

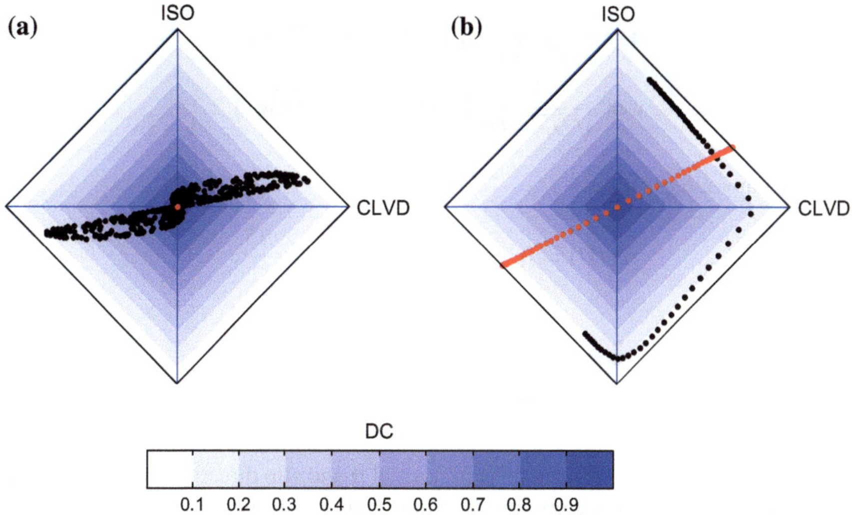

Fig. 4 The source-type plots for the moment and source tensors for shear (**a**) and shear-tensile-compressive (**b**) faulting in anisotropic media. Red dots—the source tensors, black dots—the moment tensors. The black dots in (**a**) correspond to 500 moment tensors of shear sources with randomly oriented faults and slips. The black dots in (**b**) correspond to the moment tensors of shear-tensile sources with strike = 0°, dip = 20° and rake = −90° (normal faulting). The slope angle ranges from −90° (pure compressive crack) to 90° (pure tensile crack) in steps of 3°. The medium in (**a**) and (**b**) is transversely isotropic with the following elastic parameters (in 10^9 kg $m^{-1}s^{-2}$): $c_{11} = 58.81$, $c_{33} = 27.23$, $c_{44} = 13.23$, $c_{66} = 23.54$ and $c_{13} = 23.64$. The medium density is 2500 kg/m^3. The parameters are taken from Vernik and Liu (1997) and describe the Bazhenov shale (depth of 12,507 ft.). After Vavryčuk (2015)

2.4 Inversion for Geometry of Faulting

In order to determine fault normal **n** and slip direction **ν** from moment tensor **M** and from matrix of elastic parameters **C**, we have to calculate vector **d** from Eq. (5),

$$\mathbf{d} = \mathbf{C}^{-1}\mathbf{m}, \tag{18}$$

and subsequently we construct source tensor **D** using Eq. (8). Diagonalizing **D**, we obtain eigenvalues D_1 and D_3, and eigenvectors $\mathbf{e}_1$, $\mathbf{e}_2$, and $\mathbf{e}_3$. The slope angle α between the fault plane and slip direction is determined from the trace of **D**, see Eq. (15)

$$\sin\alpha = \frac{1}{uS}\mathrm{Tr}\,(\mathbf{D}) = \frac{D_1 + D_3}{D_1 - D_3}. \tag{19}$$

Vectors **n** and **ν** are determined from the eigenvectors and eigenvalues of **D**

$$\mathbf{n} = \frac{1}{\sqrt{D_1 - D_3}} \left(\sqrt{|D_1|}\mathbf{e}_1 + \sqrt{|D_3|}\mathbf{e}_3 \right), \tag{20}$$

$$\boldsymbol{\nu} = \frac{1}{\sqrt{D_1 - D_3}} \left(\sqrt{|D_1|}\mathbf{e}_1 - \sqrt{|D_3|}\mathbf{e}_3 \right). \tag{21}$$

For shear sources ($\mathbf{n} \perp \boldsymbol{\nu}$),

$$\mathrm{Tr}\,(\mathbf{D}) = 0,\ D_1 = \frac{uS}{2} \text{and}\ D_3 = -\frac{uS}{2}, \tag{22}$$

hence

$$\mathbf{n} = \frac{1}{\sqrt{2}} \left(\mathbf{e}_1 + \mathbf{e}_3 \right), \boldsymbol{\nu} = \frac{1}{\sqrt{2}} \left(\mathbf{e}_1 - \mathbf{e}_3 \right). \tag{23}$$

It follows from the symmetry of fault normal **n** and slip direction $\boldsymbol{\nu}$ in Eq. (2) that the solution for **n** and $\boldsymbol{\nu}$ is ambiguous and the plus and minus signs in (20–21) and (23) can be interchanged.

2.5 Inversion for Anisotropy

If elastic parameters of the medium are not known, we can invert moment tensors **M** jointly for the source tensors **D** of individual earthquakes and for elastic parameters **C** in the focal zone using the following equation

$$\mathbf{Cd} = \mathbf{m}. \tag{24}$$

The right-hand side of Eq. (24) represents observations (i.e. a set of moment tensors), and the left-hand side of Eq. (24) is unknown being a product of unknown elastic parameters **C** and unknown geometry of faulting **d** for a set of earthquakes. The elastic parameters **C** are common for all earthquakes, while vector **d** is specific for each earthquake.

Since source tensor **D** is formed by a dyad of vectors $\boldsymbol{\nu}$ and **n**, it should always have one zero eigenvalue, and subsequently its determinant must be zero:

$$\mathrm{Det}\,(\mathbf{D}) = 0. \tag{25}$$

If faulting is shear, tensor **D** is constrained to have also zero trace

$$\mathrm{Trace}\,(\mathbf{D}) = uS\ (\mathbf{n} \cdot \boldsymbol{\nu}) = 0. \tag{26}$$

Equations (25–26) can be used for defining the misfit function in the inversion for anisotropy. If we know moment tensors of many earthquakes that occurred at the

Table 1 Elastic parameters of the dry and water-filled crack models

Model	C_{11}	C_{22}	C_{33}	C_{44}	C_{55}	C_{66}	C_{12}	C_{13}	C_{23}
Dry cracks	53.51	53.51	33.35	14.28	14.28	17.86	17.79	12.32	12.32
Water-filled cracks	56.62	56.62	56.11	14.28	14.28	17.86	20.90	20.75	20.75

Elastic parameters C_{ij} are in 10^9 kg m^{-1} s^{-2}

same source area, we can invert for elastic parameters c_{ijkl} minimizing the sum of absolute values of Det(**D**) for all earthquakes. This can be applied to shear as well as non-shear earthquakes. If we are confident that the studied earthquakes are shear, we can minimize the sum of absolute values of Det(**D**) and Trace(**D**) for all earthquakes. The method can be modified to be applicable also to the inversion of moment tensors constrained to have the zero trace (see Vavryčuk 2004).

The extent and quality of a set of moment tensors limits the number of anisotropic parameters, which can be inverted for. A general triclinic anisotropy is described by 21 elastic parameters c_{ijkl}. However, two of them must always be fixed to overcome the problem of coupling between elastic parameters c_{ijkl}, slip u and fault area S in Eq. (26) under shear faulting. Hence, for isotropy, which is described by two parameters, no information on the medium can be gained from moment tensors of shear earthquakes, but the vp/vs ratio can be determined from moment tensors of tensile/compressive earthquakes (Fig. 3a).

3 Numerical Modelling

A sensitivity of moment tensors to seismic anisotropy is exemplified on two models displaying effective transverse isotropy (TI) produced by presence of preferentially aligned cracks. Figure 5 shows phase velocities for the dry and water-filled crack models (see Hudson 1981; Shearer and Chapman 1989) and Table 1 summarizes their elastic parameters. Anisotropy strength of the P, SV and SH waves is 23.5, 1.3, and 11.2% for dry cracks, and 3.5, 11.0 and 11.2% for the water-filled cracks.

3.1 *Non-DC Components Produced by Shear Faulting*

As mentioned above, shear faulting in anisotropic media can generate non-DC components in moment tensors. The non-DC components depend on type and strength of anisotropy and on the orientation of faulting. Figure 6 shows the ISO and CLVD percentages for a fixed geometry of faulting: the fault normal is along the z-axis and the slip along the x-axis. Such faulting generates no ISO and CLVD for TI with

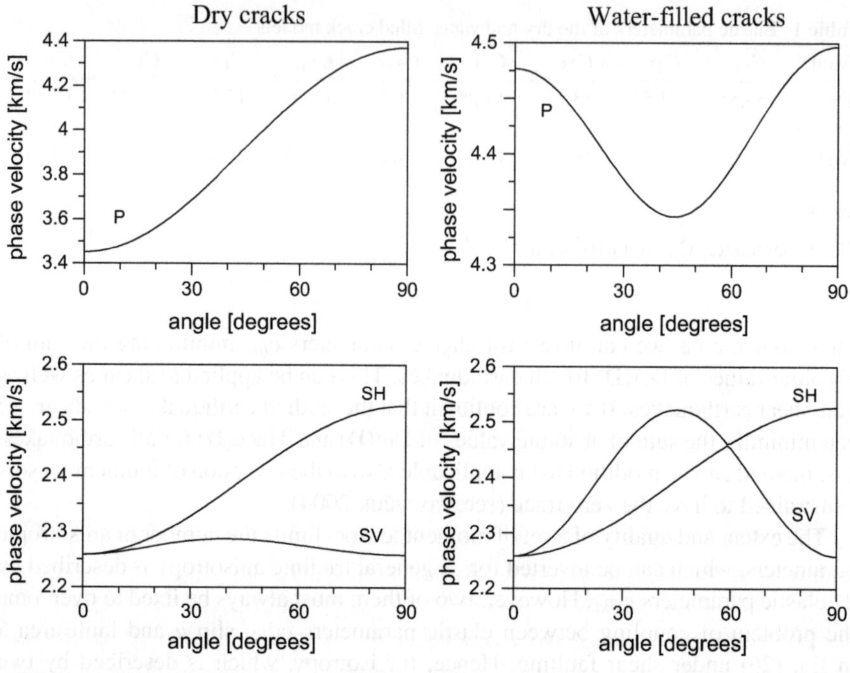

Fig. 5 Phase velocities of *P* (upper plots) and *S* (lower plots) waves as a function of the deviation of the wave normal from the symmetry axis for the model of dry cracks (left-hand plots) and water-filled cracks (right-hand plots)

a vertical symmetry axis. However, if the symmetry axis is inclined, the ISO and CLVD become non-zero. The values of the ISO and CLVD are shown as a function of direction of the symmetry axis for the dry crack model (Fig. 6, left-hand plots) and the water-filled crack model (Fig. 6, right-hand plots).

For the model of dry cracks, the percentages of the ISO and CLVD are in the intervals: (–20.7, 20.7) and (–16.1, 16.1), respectively (see Table 2). For the model of water-filled cracks, the percentages of the ISO and CLVD are in the intervals: (–0.6, 0.6) and (–19.9, 19.9), respectively. Hence, the shear source produces remarkable non-DC components in both anisotropy models. For dry cracks, both ISO and CLVD are high. On the contrary, water-filled cracks produce a high CLVD, but almost zero ISO. Despite the different percentages of the ISO in both models, their directional variation is similar. Interestingly, the directional variations of the CLVD are quite different for both models, the variation for water-filled cracks being more complicated than for dry cracks.

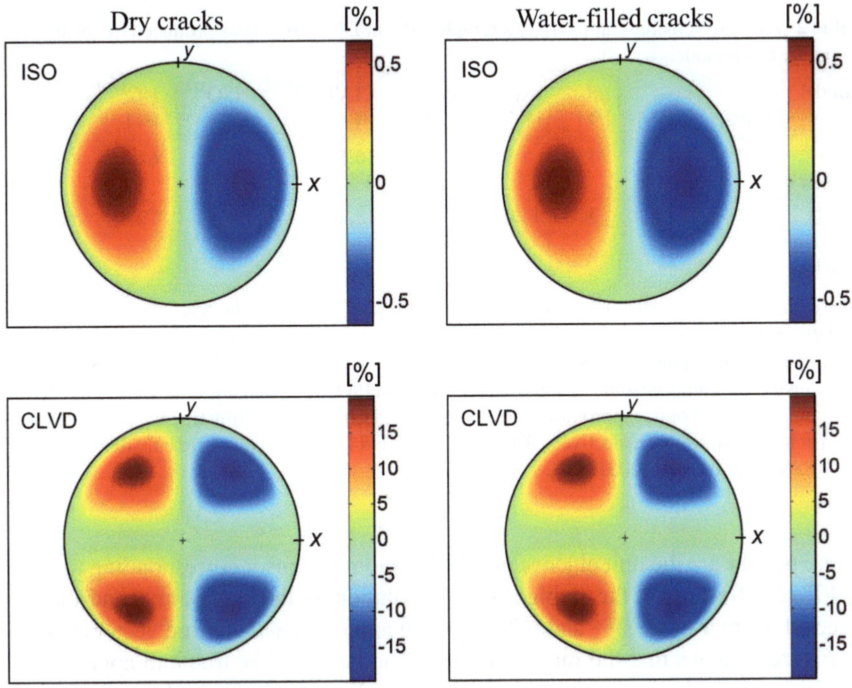

Fig. 6 The percentages of the non-DC components generated by shear faulting in the dry crack model (left-hand plots) and water-filled crack model (right-hand plots) with an inclined symmetry axis. Geometry of faulting is fixed: $\mathbf{n} = (0, 0, 1)^T$, $\boldsymbol{\nu} = (1, 0, 0)^T$. Points inside the circle correspond to TI with a varied orientation of the symmetry axis. The plus sign marks the TI with the vertical symmetry axis, the points along the circle correspond to the TI with the horizontal symmetry axes. The colour indicates the value of the non-DC component. Equal-area projection is used. The CLVD and ISO percentages are calculated by using Eq. (8) of Vavryčuk (2001a)

3.2 *Spurious Rotation of a Fault Normal and Slip Direction*

Fault plane solutions are usually calculated under the assumption of an isotropic focal area. If the focal area is anisotropic, the procedure yields distorted results. The errors owing to neglecting of anisotropy are illustrated for the dry crack model (Fig. 7, left-hand plots) and for the water-filled crack model (Fig. 7, right-hand plots). The upper/lower plots in the figures show the deviation between the true and approximate fault normals/slip directions. The approximate values were calculated from the eigenvectors of the moment tensor using the following standard formulae:

$$\mathbf{n}^{approx} = \frac{1}{\sqrt{2}}(\mathbf{p} + \mathbf{t}), \tag{27}$$

$$\boldsymbol{\nu}^{approx} = \frac{1}{\sqrt{2}}(\mathbf{p} - \mathbf{t}), \tag{28}$$

Table 2 Average velocities, anisotropy strength and non-DC components produced by shear faulting in the crack models

Model	v_P [km/s]	v_S [km/s]	a_P [%]	a_{SV} [%]	a_{SH} [%]	$CLVD^{MAX}$ [%]	ISO^{MAX} [%]	DC^{MIN} [%]	δ^{MAX} [°]
Dry cracks	3.92	2.33	23.5	1.3	11.2	16.1	20.7	64.3	6.4
Water filled cracks	4.42	2.39	3.5	11.0	11.2	19.9	0.6	79.8	6.4

a_P, a_{SV}, a_{SH} denote the anisotropy strength for the *P*, *SV* and *SH* waves The percentage of anisotropy strength is defined as $a = 200 \left(v^{MAX} - v^{MIN}\right)/\left(v^{MAX} + v^{MIN}\right)$, where v^{MAX} and v^{MIN} are the maximum and minimum phase velocities of the respective wave. $CLVD^{MAX}$, ISO^{MAX}, DC^{MIN} and δ^{MAX} are the maximum absolute values of the CLVD and ISO, the minimum value of the DC and the maximum deviation of the approximate fault normal and slip direction from true values observed in the specified anisotropy, respectively. The DC, CLVD and ISO percentages are calculated by using Eqs. (8a–c) of Vavryčuk (2001a)

where **p** and **t** are the unit eigenvectors of the moment tensor corresponding to the *P* and *T* axes. Because of the ambiguity of the solution for the fault normal and slip, we selected the pair of $\mathbf{n}^{approx}$ and $\mathbf{v}^{approx}$ which approximated the true vectors better.

Figure 7 shows that the maximum deviation between the true and approximate fault normals and slips attains coincidently a value of 6.4° for both crack models. This indicates that the deviation is not sensitive to strength of the *P*-wave anisotropy, which is different in the models, but rather to strength of the *S*-wave anisotropy. The maximum deviation of 6.4°, corresponding to the *S*-wave anisotropy of 11.2%, is not very high, but it can still introduce a non-negligible bias in carefully determined focal mechanisms. Similarly as for the CLVD, the directional variation of the deviations is more complicated for water-filled cracks than for dry cracks.

4 Applications

Let us illustrate the sensitivity of moment tensors to seismic anisotropy on three real datasets covering a broad range of scales: (1) acoustic emissions generated in the lab, (2) microearthquakes produced by fluid injection in the Earth's crust, and (3) large deep-focus earthquakes in the Tonga subduction zone.

4.1 Acoustic Emissions in Anisotropic Rocks

Laboratory experiments are advantageous for the analysis of sensitivity of moment tensors to seismic anisotropy because: (1) they are carried out under controlled conditions; (2) the physical parameters of the rock sample can be accurately monitored;

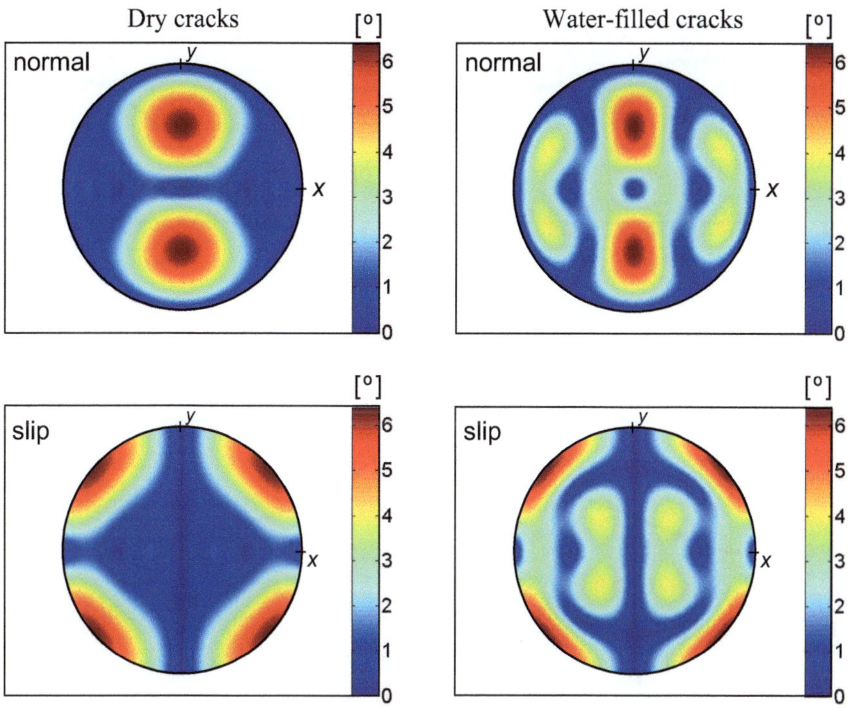

Fig. 7 The deviation between the true and approximate fault normals (upper plots), and between the true and approximate slip directions (lower plots) for the dry crack model (left-hand plots) and water-filled crack model (right-hand plots) with an inclined symmetry axis. The centre of the circle corresponds to the TI with the vertical symmetry axis, the points along the circle correspond to the TI with the horizontal symmetry axes. The colour indicates the angular deviation. Equal-area projection is used. The deviation is in degrees

(3) the rock samples can display various levels of anisotropy, and (4) the loading of the rocks samples produces large sets of acoustic emissions (AEs) suitable for a robust statistical analysis.

Stierle et al. (2016) analysed acoustic emission data measured by Stanchits et al. (2006) during triaxial compression experiments on a granite sample. The cylindrical sample (diameter 50 mm, length 100 mm) was subjected to a differential stress cycle at horizontal confining pressure of 40 MPa. During the cycle, the sample was loaded by axial vertical compression up to a maximum differential stress of 500 MPa. The AE activity and velocity changes were monitored by twelve *P*-wave and eight *S*-wave piezoelectric sensors. The ultrasonic measurements of the *P*-wave velocity revealed that the originally isotropic sample became anisotropic under axial loading. The *P*-wave velocity decreased in the horizontal direction and slightly increased in the vertical direction during loading (Fig. 8a). Concurrently, the *P*-wave attenuation increased in the horizontal direction and slightly decreased in the vertical direction (Fig. 8d). After unloading, the sample became again isotropic. The strength of the *P*-wave anisotropy developed during the loading cycle increased up to 24%.

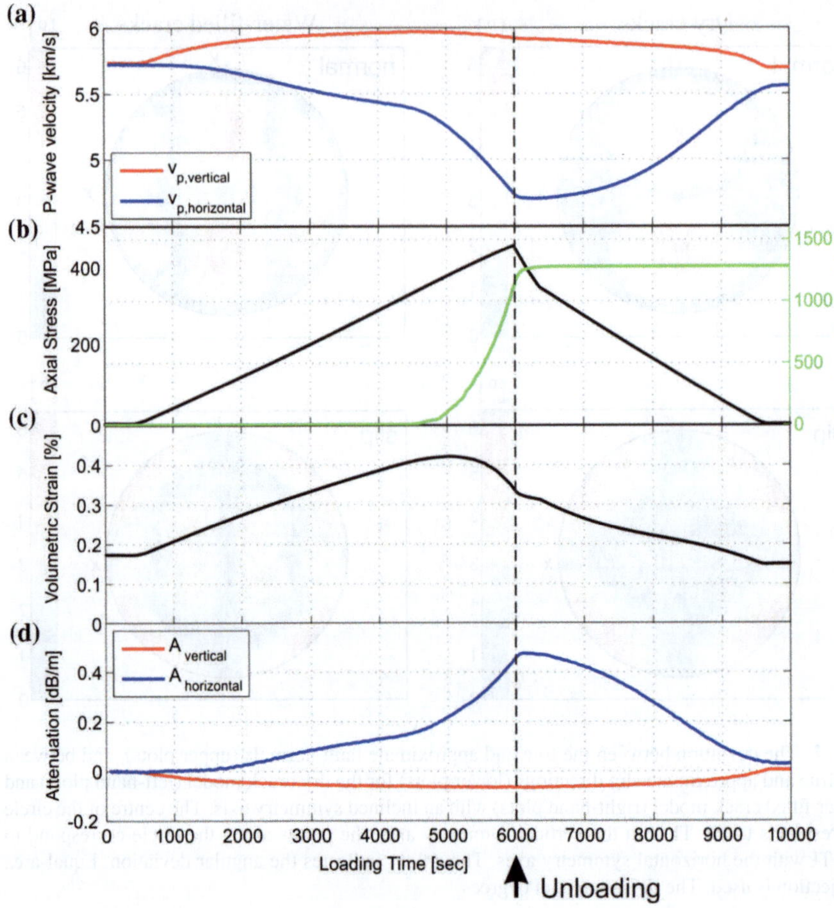

Fig. 8 **a** Vertical and horizontal *P*-wave velocities, **b** applied axial stress and acoustic emission rate, and **c** volumetric strain measured by Stanchits et al. (2006). Plot **d** shows the relative horizontal attenuation (blue line) and the relative vertical attenuation (red line) derived from ultrasonic transmission data after Stanchits et al. (2003, their Eq. 3). The horizontal confining pressure was 40 MPa during the loading cycle. After Stiele et al. (2016)

The moment and source tensors of observed AEs (Fig. 9) were calculated using three velocity models: anisotropic attenuating model, isotropic attenuating model and anisotropic elastic model. The parameters of velocity anisotropy of the sample were taken from ultrasonic measurements (Fig. 8a). The results revealed that neglecting the velocity anisotropy had a significant impact on the retrieved moment tensors. The P/T axes of the source tensors (Fig. 10b, upper plot) were highly scattered. By contrast, if anisotropy was considered and the source tensors correctly calculated (Fig. 10a, c, upper plots), the P/T axes were physically reasonable and in correspondence with the applied stress regime.

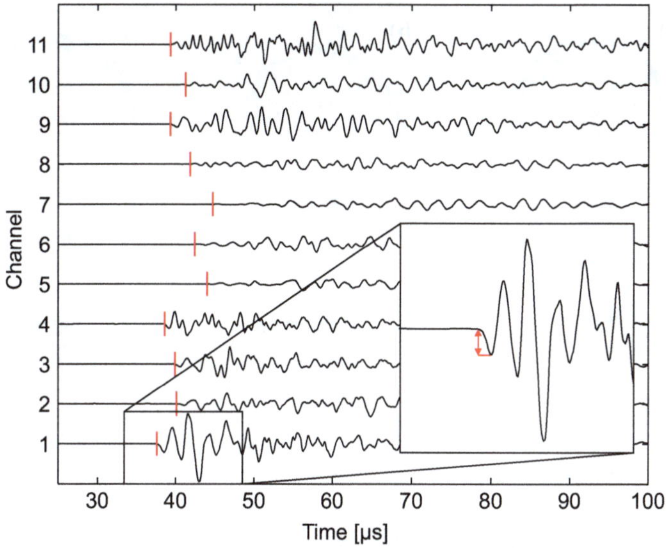

Fig. 9 Waveforms of an AE. Red vertical lines mark *P*-wave arrivals. The inset shows how *P*-wave amplitudes, used in the moment tensor inversion, were measured. After Stiele et al. (2016)

The non-DC components of the moment and source tensors were stable in all three inversions (Fig. 10, source-type plots). The ISO and CLVD components were predominantly positive indicating that mainly tensile (opening) cracks were activated. The scatter in the CLVD component was higher than that in the ISO component indicating that the errors in the velocity model mainly disturbed the CLVD component. Stierle et al. (2016) also proved that the moment tensor inversion applied to a large dataset of AEs can be utilized to determine anisotropic attenuation parameters of the rock sample.

4.2 *Microearthquakes Induced During the 2000 Fluid-Injection Experiment at the KTB Site, Germany*

The site of the KTB superdeep deep drilling borehole in Germany is characterized by a complex and heterogeneous crystalline crust (Emmerman and Lauterjung 1997) consisting of inclined alternating felsic and mafic layers, with mainly biotite gneiss and amphibolite (Rabbel et al. 2004). Field mapping, regional geophysics, and borehole results indicate that the region can be viewed as a block of steeply dipping foliated rocks with a uniform N330°E strike (Berkhemer et al. 1997; Okaya et al. 2004). In such rocks, preferred orientations of minerals prevail, and the crust may display a significant anisotropy which might be as high as 10–15% for *P* waves, and similar or even higher for *S* waves (Babuška and Cara 1991).

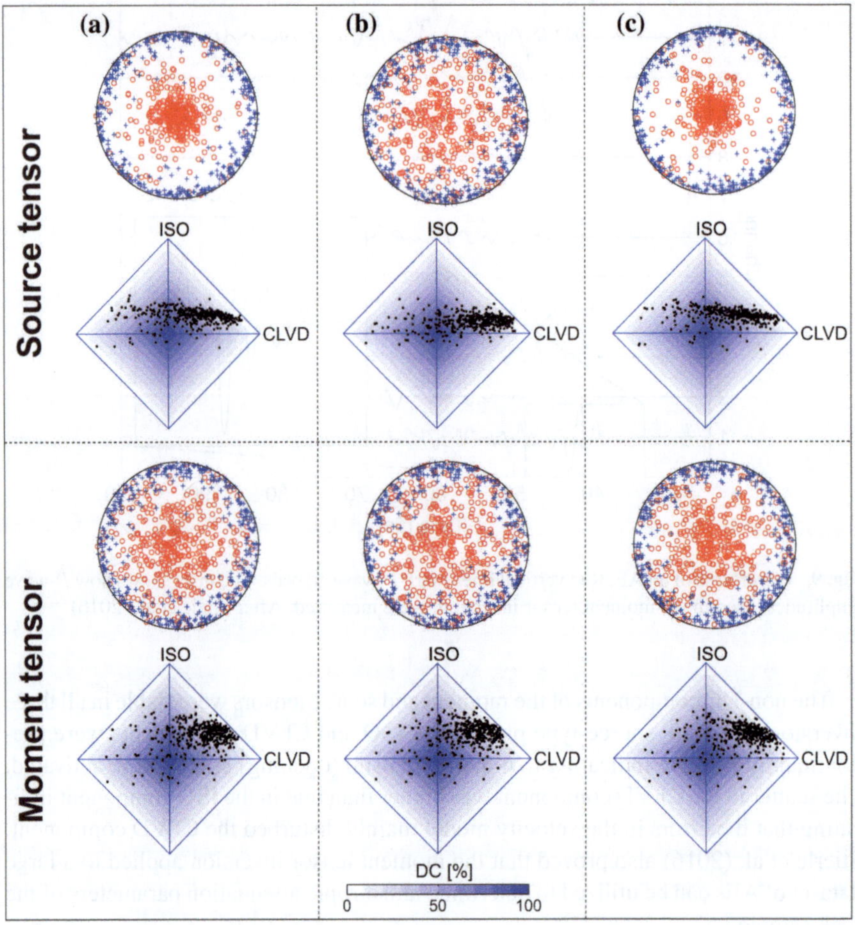

Fig. 10 Moment and source tensor inversions of observed AEs. The *P* axes (red circles in focal spheres) and *T* axes (blue plus signs in focal spheres) and the non-DC components (black dots in diamond plots) are calculated for the source tensors (upper plots) and moment tensors (lower plots) of AEs in: **a** anisotropic attenuating model, **b** isotropic attenuating model, and **c** anisotropic elastic model. The CLVD and ISO percentages are calculated by using Eq. (8) of Vavryčuk (2001a). After Stierle et al. (2016)

In 2000, a 60-day long-term fluid injection experiment was performed at the KTB site (Baisch et al. 2002). About 4000 m^3 of water were injected into the well head to induce microseismicity. The entire borehole was pressurized and the well head pressure gradually increased during the experiment from 20 to 30 MPa. The seismicity was monitored by a surface network of 40 three-component seismic stations and by one downhole three-component sensor at a depth of 3.8 km, situated at the nearby pilot hole (Fig. 11). A total of 2799 induced microearthquakes were detected at the downhole sensor, and 237 of them were located using records at the surface

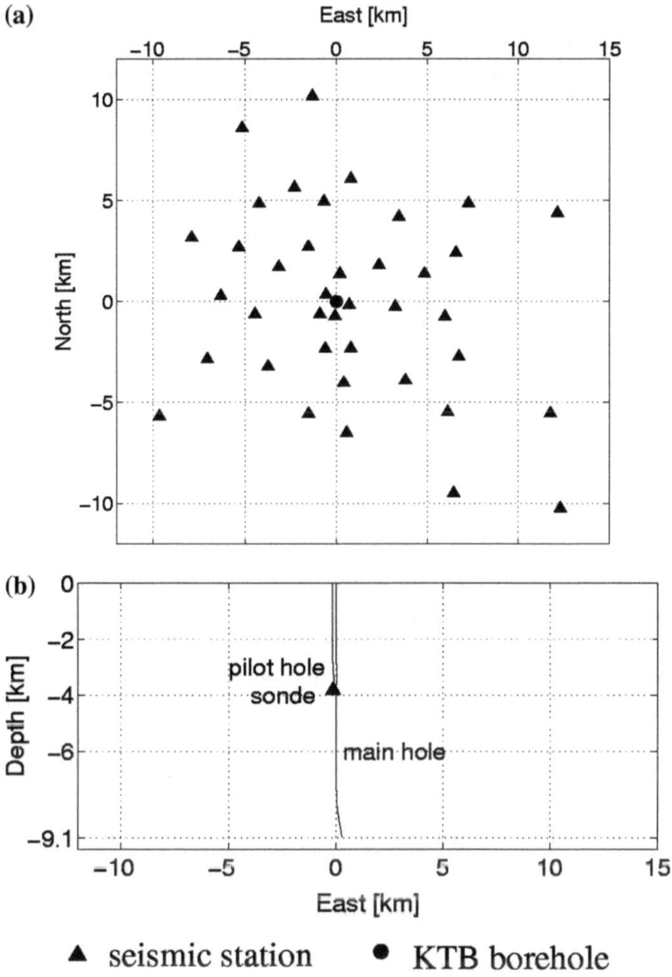

Fig. 11 The temporary seismic network operating during the 2000 injection experiment. **a** Map view of the network. The position of the KTB main hole is indicated by the dot. **b** Cross view of the main and pilot holes (view from the south). After Bohnhoff et al. (2004)

stations (Baisch et al. 2002). Fault plane solutions have been calculated for 125 events by Bohnhoff et al. (2004) and full moment tensors for 37 selected events by Vavryčuk et al. (2008). The retrieved moment tensors displayed significant non-DC components.

Vavryčuk et al. (2008) adopted four alternative anisotropy models, published by Jahns et al. (1996) and Rabbel et al. (2004) obtained from VSP data, sonic logs and from laboratory measurements, and inverted the non-DC components of the moment tensors for the optimum orientation of anisotropy. The anisotropy orientation was sought over a sphere in a 5° grid of spherical angles. The misfit function was

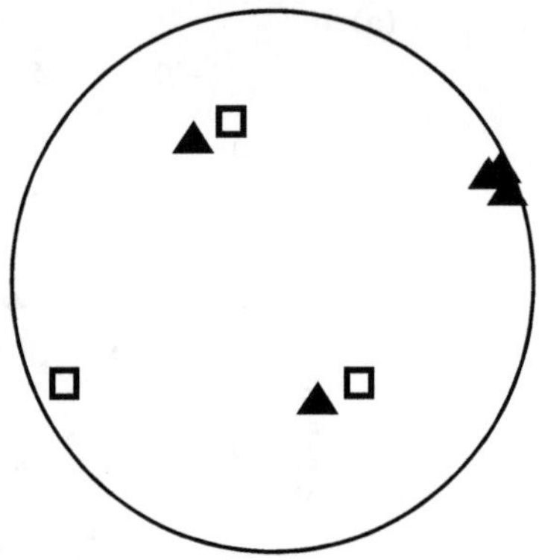

Fig. 12 A comparison of retrieved orientations of anisotropy axes from moment tensors for several alternative models of anisotropy with the orientations of anisotropy published by Rabbel et al. (2004, a model for the depth range of 0–8 km). After Vavryčuk et al. (2008)

calculated by using Eq. (25) as the sum of determinants of source tensors of all earthquakes under study. Hence, no a priori assumption about any specific type of faulting was made, and the inversion was applicable not only to shear but also to non-shear earthquakes.

The moment tensors retrieved by Vavryčuk et al. (2008) contained about 60% of the DC component and 40% of the non-DC components. The ISO and CLVD components were mutually uncorrelated. The errors of the non-DC components produced by noise and by limitations of input data were suppressed by analysing only the most reliable moment tensors determined for excellent ray coverage of the focal sphere. The optimum orientation of the symmetry plane of transverse isotropy inferred from the non-DC components was nearly vertical (see a nearly horizontal symmetry axis in Fig. 12) with a strike typical for many major lithological units in the area (Rabbel et al. 2004; Okaya et al. 2004). After removing the anisotropy effects from the non-DC components, the distribution of the ISO of the source tensors significantly narrowed (Fig. 13). The ISO and CLVD of the source tensors became correlated with the correlation coefficient of 0.6 for the most confident source tensors (Vavryčuk et al. 2008). This indicated that the non-DC components originated jointly in seismic anisotropy and in tensile faulting due to the fluid injection.

4.3 *Deep-Focus Earthquakes in the Tonga Subduction Zone*

Deep-focus earthquakes are particularly suitable for studying the sensitivity of moment tensors to seismic anisotropy because: (1) they occur in subducting slabs, which are assumed to be anisotropic (Fukao 1984; Kendall and Thomson 1993; Hiramatsu et al. 1997; McNamara et al. 2002); (2) the surrounding mantle is nearly isotropic, (3) the moment tensors of deep earthquakes are determined with a high accuracy; and (4) they often display non-DC components (Dziewonski et al. 2001, 2003; Sipkin 1986). It is speculated that anisotropy in the slab is caused by an alignment of metastable olivine and its polymorphs wadsleyte and ringwoodite, or the

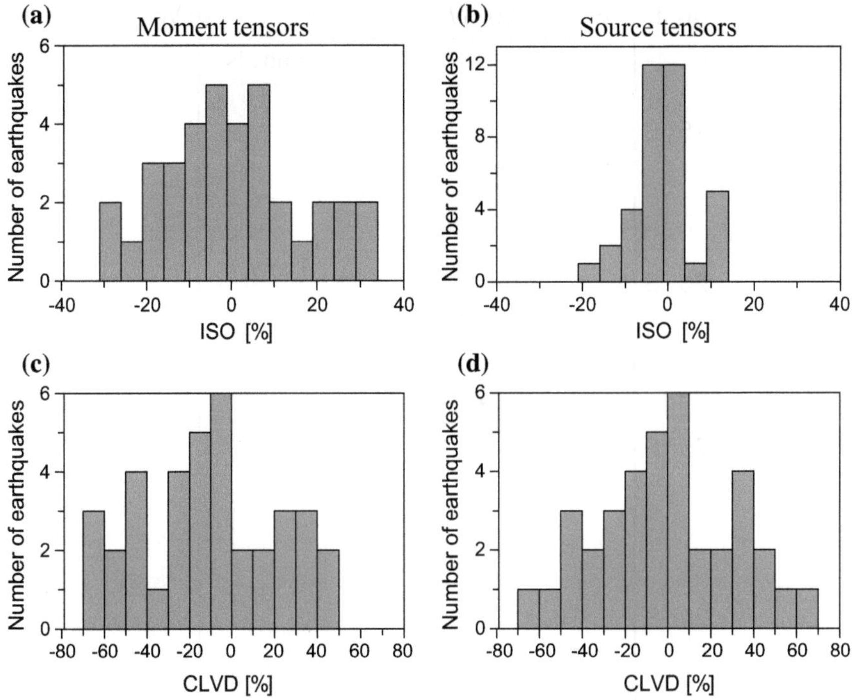

Fig. 13 Histograms of the percentages of the ISO (**a**, **b**) and CLVD (**c**, **d**) components of the moment tenors (left-hand plots) and source tensors (right-hand plots). The CLVD and ISO percentages are calculated by using Eq. (8) of Vavryčuk (2001a). For details, see Vavryčuk et al. (2008)

ilmenite form of pyroxene (Anderson 1987; Mainprice et al. 2000). The intra-slab anisotropy can also be induced by strain due to large stresses generated when a rigid slab encounters the 670 km discontinuity (Wookey et al. 2002).

Vavryčuk (2004, 2006) analysed the moment tensors of deep earthquakes in the Tonga-Kermadec subduction zone, which is the most active zone in the world and offers a largest dataset of moment tensors of deep earthquakes. The deep part of the Tonga subduction zone consists of two differently oriented slab segments (Fig. 14a): the northern segment within latitudes 17–19° S, and the southern segment within latitudes 19.5–27° S. Both segments are seismically active at depths from 500 to 700 km. The mechanisms of deep-focus earthquakes reported in the Harvard moment tensor catalogue (Dziewonski et al. 2001, 2003) contain CLVD components that behave differently in both segments. The mean absolute value of the CLVD is 12% for the northern segment and 16% for the southern segment (Fig. 14b). The complex behaviour of the CLVD is explained by spatially dependent seismic anisotropy in the slab.

The inversion for anisotropy from the non-DC components of moment tensors performed by Vavryčuk (2004, 2006) pointed to orthorhombic anisotropy in the both

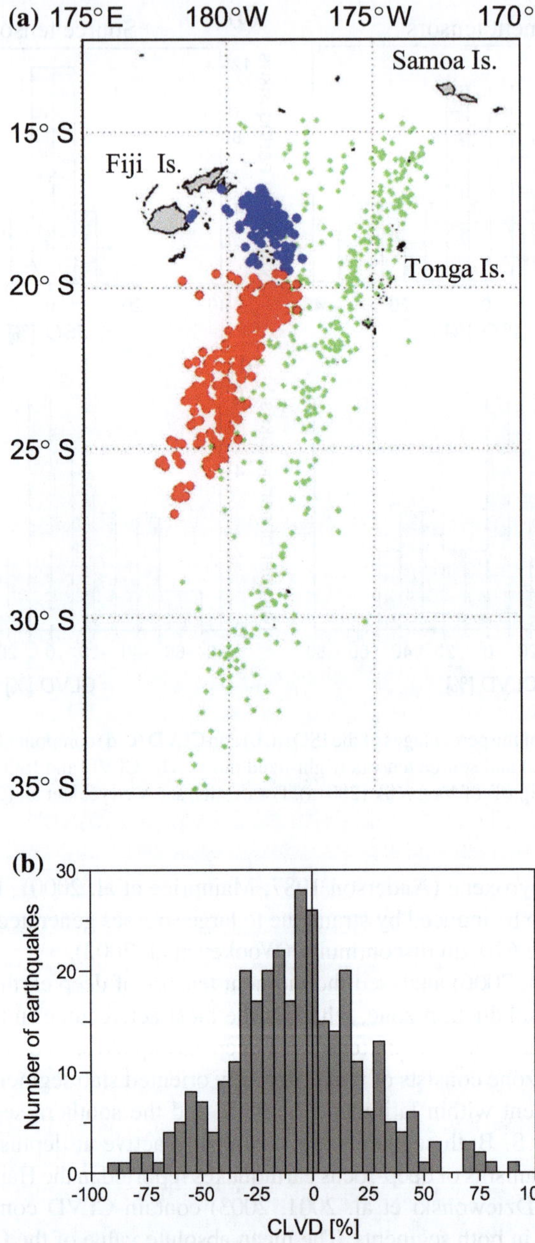

Fig. 14 **a** Epicentres of earthquakes in the Tonga subduction zone. Blue and red dots mark the deep-focus earthquakes with depth > 500 km in the northern and southern segment, respectively. Green dots mark the other earthquakes in the region (depth > 100 km). **b** Histogram of the CLVD of moment tensors for the deep earthquakes in the southern cluster. The CLVD percentages are calculated by using Eqs. (8a–c) of Vavryčuk (2001a). After Vavryčuk (2004)

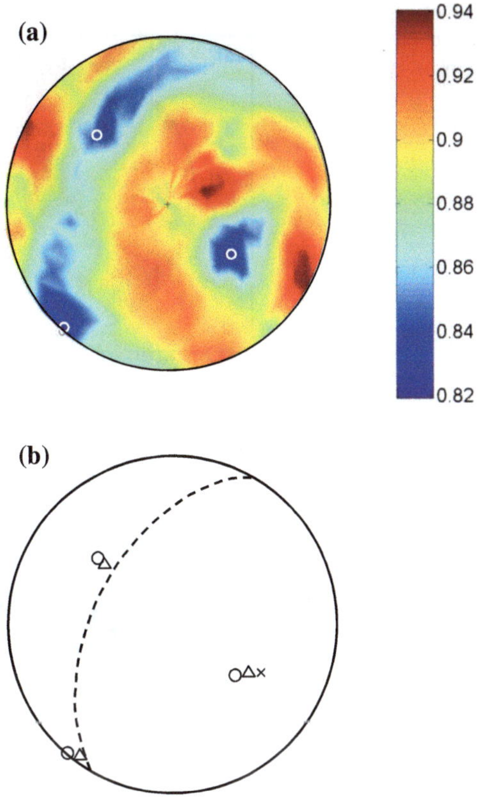

Fig. 15 Inversion for orthorhombic anisotropy using moment tensors of deep-focus earthquakes in the southern segment. **a** Misfit function for the symmetry axes of orthorhombic anisotropy normalized to the misfit for an isotropic medium and displayed in the lower hemisphere equal-area projection. The optimum directions of the symmetry axes of anisotropy are marked by circles. **b** A comparison of slab, stress and anisotropy orientations. The directions of the slab normal (x), of the principal stress axes (triangles) calculated from focal mechanisms, and of the symmetry axes of anisotropy (circles) calculated from moment tensors. The lower-hemisphere equal-area projection is used. The dashed line shows the intersection of the slab with the hemisphere Modified after Vavryčuk (2004)

slab segments. The anisotropy was of a uniform strength of 5–7% for the *P* waves and of 9–12% for the *S* waves, and was oriented according to the orientation of each segment and the stress acting in it (Fig. 15 for the southern segment). The spatial variation of velocities was roughly similar in both segments (Fig. 16 for the southern segment). The retrieved anisotropy might have several possible origins. It can be: (1) intrinsic, caused by preferentially aligned anisotropic minerals such as wadsleyite, ringwoodite, ilmenite or others, (2) effective, caused, for example, by intra-slab layering, or (3) partly apparent, produced by systematic errors in the moment tensors due to neglecting 3D slab geometry and the slab/mantle velocity contrast when calculating the Green functions in the moment tensor inversion.

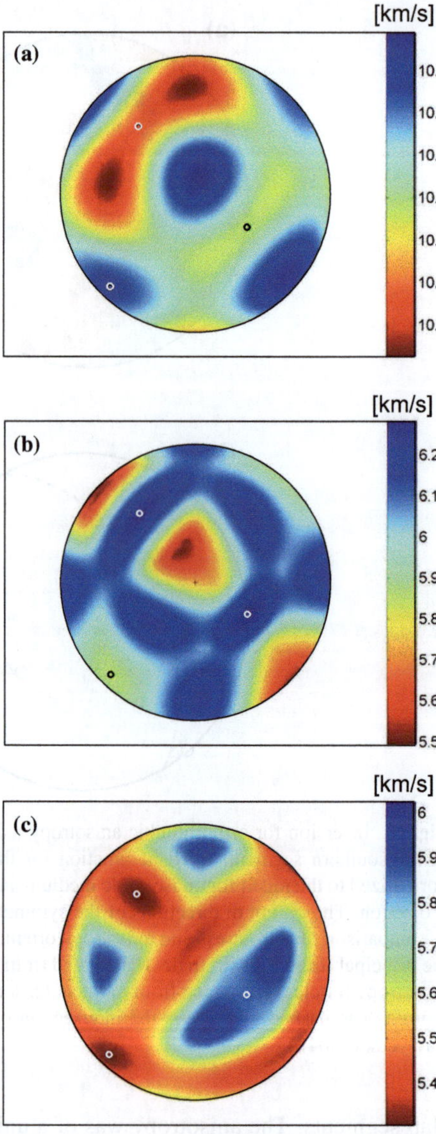

Fig. 16 Spatial variation of the velocity in the southern slab segment predicted by the optimum anisotropy model for the P (**a**), $S1$ (**b**) and $S2$ (**c**) waves. Lower hemisphere equal-area projection is used. Directions of the symmetry axes are marked by circles. After Vavryčuk (2004)

5 Discussion and Conclusions

Moment tensors are quite sensitive to seismic anisotropy of rocks. Neglect of anisotropy in the moment tensor inversion reduces the accuracy of the retrieved DC and non-DC components of the moment and source tensors. The errors in the non-DC components are mainly projected into the CLVD component, which usually displays about three times higher scatter than the ISO component.

Shear faulting on planar faults in anisotropic media can produce non-DC mechanisms. The amount of the CLVD and ISO depends on strength and symmetry of anisotropy and on the orientation of faulting. The fault plane solutions in anisotropy are characterized by the same ambiguity in identifying a fault normal and slip direction as in isotropy. The moment tensor corresponding to shear faulting in anisotropy has eigenvectors (P, T and B axes) that can deviate from those in isotropy. If anisotropy is neglected at a focal area, the fault plane and slip calculated under the isotropic assumption can deviate from true ones.

Shear faulting in anisotropic rocks present in the Earth's crust, Earth's mantle or in the subduction zones can produce mechanisms with significant non-DC components. The CLVD can typically attain values up to 30%, and the ISO up to 15%. The fault plane solutions calculated under the assumption of isotropy typically deviate from the true solutions by the angle of 5°–10°. However, for strongly anisotropic rocks such as some shales or shists, the CLVD and ISO can be much higher, and the isotropic procedure for determining the fault plane solution can completely fail (Vavryčuk 2005).

If the focal zone is anisotropic, the moment tensors can be utilized for determining parameters of anisotropy. The inversion of moment tensors for anisotropy is advantageous because it yields a local value of anisotropy just in the focal area. Hence it should, in principle, be capable of retrieving the anisotropy of a focal area that is surrounded by a differently anisotropic or by isotropic medium. The case when the surrounding medium is differently anisotropic is particularly complicated for standard methods, because the effects of focal anisotropy can easily be masked by those of the surrounding medium. In addition, the inversion of moment tensors for anisotropy is very robust. If a large set of high-quality moment tensors is available, the inversion can retrieve the orientation and strength of anisotropy even for low anisotropy symmetries as for orthorhombic symmetry. For example, the method was capable to retrieve the orientation and strength of the P, $S1$ and $S2$ anisotropy in the Tonga subduction zone (Vavryčuk 2004, 2006).

Acknowledgements The study was supported by the Grant Agency of the Czech Republic, Grant No. 16-19751J.

References

Anderson DL (1987) Thermally induced phase changes, lateral heterogeneity of the mantle, continental roots and deep slab anomalies. J Geophys Res 92:13968–13980

Aki K, Richards PG (2002) Quantitative seismology. University Science Books, Sausalito

Babuška V, Cara M (1991) Seismic anisotropy in the Earth. Kluwer Academic Publisher, London

Baisch S, Bohnhoff M, Ceranna L, Tu Y, Harjes H-P (2002) Probing the crust to 9 km depth: fluid injection experiments and induced seismicity at the KTB superdeep drilling hole, Germany. Bull Seism Soc Am 92:2369–2380

Ben-Menahem A, Sena AG (1990a) Seismic source theory in stratified anisotropic media. J Geophys Res 95:15395–15427

Ben-Menahem A, Sena AG (1990b) The elastodynamic Green's tensor in an anisotropic half-space. Geophys J Int 102:421–443

Berckhemer H, Rauen A, Winter H, Kern H, Kontny A, Lienert M, Nover G, Pohl J, Popp T, Schult A, Zinke J, Soffel HC (1997) Petrophysical properties of the 9-km-deep crustal section at KTB. J Geophys Res 102:18337–18361

Bohnhoff M, Baisch S, Harjes H-P (2004) Fault mechanisms of induced seismicity at the superdeep German Continental Deep Drilling Program (KTB) borehole and their relation to fault structure and stress field. J Geophys Res 109:B02309. https://doi.org/10.1029/2003jb002528

Burridge R (1967) The singularity on the plane lids of the wave surface of elastic media with cubic symmetry. Q J Mech Appl Math 20:40–56

Crampin S (1993) A review of the effects of crack geometry on wave propagation through aligned cracks. Can J Expl Geophysics 29:3–17

Červený V (2001) Seismic ray theory. Cambridge University Press, Cambridge

Dziewonski AM, Ekström G, Maternovskaya NN (2001) Centroid moment tensor solutions for July–September 2000. Phys Earth Planet Inter 124:9–23

Dziewonski AM, Ekström G, Maternovskaya NN (2003) Centroid moment tensor solutions for October–December 2000. Phys Earth Planet Inter 136:145–164

Emmerman R, Lauterjung J (1997) The German Continental Deep Drilling Program KTB: overview and major results. J Geophys Res 102:18179–18201

Frohlich C (1994) Earthquakes with non-double-couple mechanisms. Science 264:804–809

Fukao Y (1984) Evidence from core-reflected shear waves for anisotropy in the Earth's mantle. Nature 309:695–698

Gajewski D (1993) Radiation from point sources in general anisotropic media. Geophys J Int 113:299–317

Helbig K (1994) Foundations of anisotropy for exploration seismics. Pergamon, New York

Hiramatsu Y, Ando M, Ishikawa Y (1997) ScS wave splitting of deep earthquakes around Japan. Geophys J Int 128:409–424

Hudson JA (1981) Wave speeds and attenuation of elastic waves in material containing cracks. Geophys J R astr Soc 64:133–150

Jahns E, Rabbel W, Siegesmund S (1996) Quantified seismic anisotropy at different scales: a case study from the KTB crustal segment. J Geol Wiss 24:729–740

Julian BR, Miller AD, Foulger GR (1998) Non-double-couple earthquakes, 1 Theory. Rev Geophys 36:525–549

Kaneshima S, Ando M, Kimura S (1988) Evidence from shear-wave splitting for the restriction of seismic anisotropy to the upper crust. Nature 335:627–629

Kawakatsu H (1991) Enigma of earthquakes at ridge-transform-fault plate boundaries—distribution of non-double-couple parameter of Harvard CMT solutions. Geophys Res Lett 18:1103–1106

Kawasaki I, Tanimoto T (1981) Radiation patterns of body waves due to the seismic dislocation occurring in an anisotropic source medium. Bull Seismol Soc Am 71:37–50

Kendall J-M, Thomson C (1993) Seismic modelling of subduction zones with inhomogeneity and anisotropy—I Teleseismic P-wavefront tracking. Geophys J Int 112:39–66

Kern H, Schmidt R (1990) Physical properties of KTB core samples at simulated in situ conditions. Sci Drill 1:217–223

Mainprice D, Barruol G, Ben Ismail W (2000) The seismic anisotropy of the Earth's mantle: from single crystal to polycrystal. In: Forte A, Liebermann RC, Masters G, Stixrude L, Karato S (eds) Earth's deep interior, mineral physics and tomography from the atomic to the global scale. Geophysical monograph series. American Geophysical Union, Washington, pp 237–264

McNamara AK, van Keken PE, Karato S (2002) Development of anisotropic structure in the Earth's lower mantle by solid-state convection. Nature 416:310–314

Miller AD, Foulger GR, Julian BR (1998) Non-double-couple earthquakes, 2 observations. Rev Geophys 36:551–568

Musgrave MJP (1970) Crystal acoustics. Holden Day, San Francisco

Okaya D, Rabbel W, Beilecke T, Hasenclever J (2004) P wave material anisotropy of tectono-metaorphic terrane: an active source seismic experiment at the KTB super-deep drill hole, south-east Germany. Geophys Res Lett 31:L24620. https://doi.org/10.1029/2004gl020855

Pros Z, Lokajíček T, Klíma K (1998) Laboratory approach to the study of elastic anisotropy on rock samples. Pure Appl Geophys 151:619–629
Pšenčík I (1998) Green functions for inhomogeneous weakly anisotropic media. Geophys J Int 135:279–288
Rabbel W, Beilecke T, Bohlen T, Fischer D, Frank A, Hasenclever J, Borm G, Kück J, Bram K, Druivenga G, Lüschen E, Gebrande H, Pujol J, Smithson S (2004) Superdeep vertical seismic profiling at the KTB deep drill hole (Germany): seismic close-up view of a major thrust zone down to 85 km depth. J Geophys Res 109:B09309. https://doi.org/10.1029/2004jb002975
Rabbel W, Mooney WD (1996) Seismic anisotropy of the crystalline crust: what does it tell us? Terra Nova 8:16–21
Rössler D, Rümpker G, Krüger F (2003) Ambiguous moment tensors and radiation patterns in anisotropic media with applications to the modeling of earthquake mechanisms in W-Bohemia. Stud Geophys Geod 48:233–250
Savage MK (1999) Seismic anisotropy and mantle deformation: what have we learned from shear wave splitting? Rev Geophys 37:65–106
Shearer PM, Chapman CH (1989) Ray tracing in azimuthally anisotropic media—I. Results for models of aligned cracks in the upper crust. Geophys J Int 96:51–64
Šílený J, Vavryčuk V (2000) Approximate retrieval of the point source in anisotropic media: numerical modelling by indirect parametrization of the source. Geophys J Int 143:700–708. https://doi.org/10.1046/j1365-246x200000256x
Šílený J, Vavryčuk V (2002) Can unbiased source be retrieved from anisotropic waveforms by using an isotropic model of the medium? Tectonophysics 356:125–138. https://doi.org/10.1016/s0040-1951(02)00380-3
Silver PG (1996) Seismic anisotropy beneath the continents: probing the depths of geology. Ann Rev Earth Planet Sci 24:385–432
Sipkin SA (1986) Interpretation of non-double-couple earthquake mechanisms derived from moment tensor inversion. J Geophys Res 91:531–547
Stanchits SA, Lockner DA, Ponomarev AV (2003) Anisotropic changes in P-wave velocity and attenuation during deformation and fluid infiltration of granite. Bull Seism Soc Am 93(4):1803–1822
Stanchits S, Vinciguerra S, Dresen G (2006) Ultrasonic velocities, acoustic emission characteristics and crack damage of basalt and granite. Pure App Geophys 163(5–6):975–994
Stierle E, Vavryčuk V, Kwiatek G, Charalampidou E-M, Bohnhoff M (2016) Seismic moment tensors of acoustic emissions recorded during laboratory rock deformation experiments: sensitivity to attenuation and anisotropy. Geophys J Int 205(1):38–50. https://doi.org/10.1093/gji/ggw009
Svitek T, Vavryčuk V, Lokajíček T, Petružálek M (2014) Determination of elastic anisotropy of rocks from P- and S-wave velocities: numerical modelling and lab measurements. Geophys J Int 199(3):1682–1697. https://doi.org/10.1093/gji/ggu332
Vavryčuk V (1997) Elastodynamic and elastostatic Green tensors for homogeneous weak transversely isotropic media. Geophys J Int 130:786–800. https://doi.org/10.1111/j.1365-246x.1997.tb01873.x
Vavryčuk V (2001a) Inversion for parameters of tensile earthquakes. J Geophys Res 106:16339–16355. https://doi.org/10.1029/2001jb000372
Vavryčuk V (2001b) Ray tracing in anisotropic media with singularities. Geophys J Int 145:265–276. https://doi.org/10.1046/j0956-540x200101387x
Vavryčuk V (2002) Non-double-couple earthquakes of 1997 January in West Bohemia, Czech Republic: evidence of tensile faulting. Geophys J Int 149:364–373. https://doi.org/10.1046/j.1365-246x.2002.01654.x
Vavryčuk V (2003) Parabolic lines and caustics in homogeneous weakly anisotropic solids. Geophys J Int 152:318–334. https://doi.org/10.1046/j1365-246x200301845x
Vavryčuk V (2004) Inversion for anisotropy from non-double-couple components of moment tensors. J Geophys Res 109:B07306. https://doi.org/10.1029/2003jb002926
Vavryčuk V (2005) Focal mechanisms in anisotropic media. Geophys J Int 161:334–346. https://doi.org/10.1111/j1365-246x200502585x

Vavryčuk V (2006) Spatially dependent seismic anisotropy in the Tonga subduction zone: a possible contributor to the complexity of deep earthquakes. Phys Earth Planet Inter 155:63–72. https://doi.org/10.1016/jpepi200510005

Vavryčuk V (2007) Asymptotic Green's function in homogeneous anisotropic viscoelastic media. Proc R Soc A 463:2689–2707. https://doi.org/10.1098/rspa20071862

Vavryčuk V (2011) Tensile earthquakes: theory, modeling, and inversion. J Geophys Res 116(B12):B12320. https://doi.org/10.1029/2011jb008770

Vavryčuk V (2015) Moment tensor decompositions revisited. J Seismol 19(1):231–252. https://doi.org/10.1007/s10950-014-9463-y

Vavryčuk V, Bohnhoff M, Jechumtálová Z, Kolář P, Šílený J (2008) Non-double-couple mechanisms of micro-earthquakes induced during the 2000 injection experiment at the KTB site, Germany: a result of tensile faulting or anisotropy of a rock? Tectonophysics 456:74–93. https://doi.org/10.1016/jtecto200708019

Vernik L, Liu X (1997) Velocity anisotropy in shales: a petrological study. Geophysics 62:521–532

Wookey J, Kendall J-M, Barruol G (2002) Mid-mantle deformation inferred from seismic anisotropy. Nature 415:777–780

The Frequency-Domain Moment-Tensor Inversion: Retrieving the Complete Source Moment-Tensor Spectra and Time Histories

Xiaoning Yang, Brian W. Stump and Mason D. Macphail

1 Introduction

When the wavelength of a seismic signal of interest is much longer than the dimension of the internal seismic source that generates the signal, whether it is an earthquake, an underground explosion or an underground mine collapse, the seismic source may be represented by a symmetric second-order moment tensor. The observed seismic signal is then the convolution of the source moment tensor with the earth's impulse response, or the Green's function between the source and the receiver, which records the signal. This relationship is universally used to retrieve the source moment tensor from observed seismic data for seismic-source characterizations. Assuming a step source time function, research organizations use seismic body waves, surface waves and the earth's free oscillation to routinely calculate moment tensors of global earthquakes (e.g., Ekström et al. 2012; Pondrelli et al. 2011; Tsuruoka et al. 2009). For detailed source analysis, seismologists also attempt to recover source time functions that are not a step function using either time-domain methods (e.g., Sipkin 1982; Šílený et al. 1992) or a frequency-domain approach (e.g., Stump and Johnson 1977).

Here we describe the frequency domain moment tensor inversion method of Stump and Johnson (1977) including its mathematical formulation, the inversion method and error assessments. We also provide some examples to illustrate the application of the method.

Los Alamos National Laboratory unlimited release number: LA-UR-17-22976.

X. Yang (✉)
Los Alamos National Laboratory, Los Alamos, NM 87545, USA
e-mail: xyang@lanl.gov

B. W. Stump · M. D. Macphail
Southern Methodist University, Dallas, TX 75275, USA

S. D'Amico (ed.), *Moment Tensor Solutions*, Springer Natural Hazards,
https://doi.org/10.1007/978-3-319-77359-9_3

2 Theory

The basic relationship between the ground displacement and the seismic source excitation is the representation theorem. The theorem states that ground displacement u_n ($n=1, 2, 3$) at location $\mathbf{x}$ and time t due to the equivalent body-force density f_i, which includes contributions from the body force, the initial condition and the boundary condition, is

$$u_n(\mathbf{x},t) = \int_{-\infty}^{\infty}\left[\iiint_V G_{ni}(\mathbf{x},t-\tau;\boldsymbol{\xi},0)f_i(\boldsymbol{\xi},\tau)dV(\boldsymbol{\xi})\right]d\tau, \qquad (1)$$

where G_{ni} is the Green's function and the volume V contains all nonzero f_i (Aki and Richards 2002). Einstein summation is assumed for all equations in this article. By expanding the Green's function in a Taylor series about the coordinate origin, we have

$$G_{ni}(\mathbf{x},t-\tau;\boldsymbol{\xi},\mathbf{0}) = \sum_{m=0}^{\infty}\frac{1}{m!}\xi_{j_1}\ldots\xi_{j_m}G_{ni,j_1\ldots j_m}(\mathbf{x},t-\tau;\mathbf{0},0), \qquad (2)$$

where $G_{ni,j_1\cdots j_m}$ is the mth spatial derivative of the Green's function. If we define the force moment tensor of order $m+1$ as

$$M_{ij_1\ldots j_m}(\mathbf{0},\tau) = \iiint_V \xi_{j_1}\ldots\xi_{j_m}f_i(\boldsymbol{\xi},\tau)dV(\boldsymbol{\xi}), \qquad (3)$$

Equation (1) becomes

$$\begin{aligned} u_n(\mathbf{x},t) &= \int_{-\infty}^{\infty}\left[\sum_{m=0}^{\infty}\frac{1}{m!}G_{ni,j_1\ldots j_m}(\mathbf{x},t-\tau;\mathbf{0},0)M_{ij_1\ldots j_m}(\mathbf{0},\tau)\right]d\tau \\ &= \sum_{m=0}^{\infty}\frac{1}{m!}\int_{-\infty}^{\infty}G_{ni,j_1\ldots j_m}(\mathbf{x},t-\tau;\mathbf{0},0)M_{ij_1\ldots j_m}(\mathbf{0},\tau), \qquad (4) \\ &= \sum_{m=0}^{\infty}\frac{1}{m!}\int_{-\infty}^{\infty}G_{ni,j_1\ldots j_m}(\mathbf{x},t;\mathbf{0},0) * M_{ij_1\ldots j_m}(\mathbf{0},t) \end{aligned}$$

where * denotes temporal convolution (Stump and Johnson 1977; Julian et al. 1998).

If we assume that the linear momentum of the body-force system is conserved during the source process, which is usually true for sources internal to the earth such as earthquakes and underground explosions, then from Eq. (3),

$$M_i(\mathbf{0}, \tau) = \iiint\limits_V f_i(\boldsymbol{\xi}, \tau) dV(\boldsymbol{\xi}) = 0.$$

In addition, if the volume of the body-force system is much smaller than the wavelength of the seismic wave that the source generates, terms with orders higher than $m = 1$ in the Green's function expansion (Eq. 2) can be neglected (Stump and Johnson 1977). As a result, Eq. (4) is reduced to

$$u_n(\mathbf{x}, t) = G_{ni,j}(\mathbf{x}, t; \mathbf{0}, 0) * M_{ij}(\mathbf{0}, t). \quad (5)$$

For internal seismic sources, the angular momentum of the source system is usually conserved. It can be shown that for such sources, the source moment tensor M_{ij} is symmetric with six independent components (Aki and Richards 2002). If we Fourier transform Eq. (5) into the frequency domain, convolution becomes multiplication and we obtain

$$u_n(\mathbf{x}, f) = G_{ni,j}(\mathbf{x}, f; \mathbf{0}, 0) M_{ij}(\mathbf{0}, f). \quad (6)$$

Equation (6) is the basic equation that we use to describe the frequency-domain moment-tensor-inversion method.

3 Methodology

To invert Eq. (6) for the source moment tensor M_{ij}, we use ground-motion data recorded by multiple receivers at different azimuths and distances from the source. In order to facilitate the inversion, we first partition ground-motion data into radial, transverse and vertical components with the projection of the direction from the source to the receiver on a horizontal plane as the radial direction. We then represent the ground motion as a function of the source moment tensor defined in the source coordinate system. We describe the procedure in detail below.

Without loss of generality, we abbreviate Eq. (6) to

$$u_n = G_{ni,j} M_{ij}. \quad (n = 1, 2, 3) \quad (7)$$

We assume that Eq. (7) is defined in a coordinate system with its origin at the source, and with x_1-axis positive to the north, x_2-axis positive to the east and x_3-axis positive down. We call this system the source coordinate system. Considering a receiver that is located at θ-degree azimuth from the source, we establish another coordinate system that has the same origin as the source coordinate system, but with axes in radial ($x_{1'}$), transverse ($x_{2'}$) and vertical ($x_{3'}$) directions. For this coordinate system, the horizontal direction θ degrees from the north is radial positive, the horizontal direction 90° clockwise from radial positive is transverse positive and

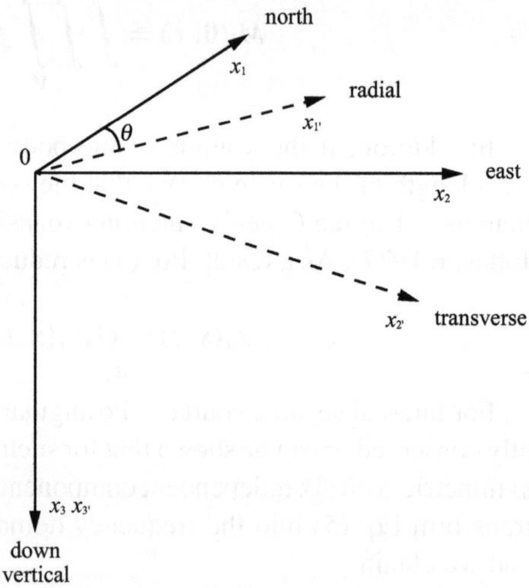

Fig. 1 Definition of and relationship between source and receiver coordinate systems

down is vertical positive (Fig. 1). We refer to this system as the receiver coordinate system.

In the receiver coordinate system, Eq. (7) becomes

$$u_{n'} = G_{n'i',j'} M_{i'j'}. \tag{8}$$

According to the transformation law for a second-order tensor,

$$M_{i'j'} = \alpha_{ii'} \alpha_{jj'} M_{ij}, \tag{9}$$

where α_{ij} are direction cosines from the transformation matrix between the source and the receiver coordinate systems:

$$\boldsymbol{\alpha} = \begin{bmatrix} \alpha_{11'} & \alpha_{12'} & \alpha_{13'} \\ \alpha_{21'} & \alpha_{22'} & \alpha_{23'} \\ \alpha_{31'} & \alpha_{32'} & \alpha_{33'} \end{bmatrix} = \begin{bmatrix} \cos\theta & -\sin\theta & 0 \\ \sin\theta & \cos\theta & 0 \\ 0 & 0 & 1 \end{bmatrix}. \tag{10}$$

Using Eq. (10) to expand Eq. (9), we have

$$\begin{aligned}
M_{1'1'} &= M_{11}\cos^2\theta + 2M_{12}\cos\theta\,\sin\theta + M_{22}\sin^2\theta \\
M_{1'2'} &= (M_{22} - M_{11})\sin\theta\,\cos\theta - M_{12}\left(\sin^2\theta - \cos^2\theta\right) \\
M_{1'3'} &= M_{13}\cos\theta + M_{23}\sin\theta \\
M_{2'2'} &= M_{11}\sin^2\theta - 2M_{12}\cos\theta\,\sin\theta + M_{22}\cos^2\theta \\
M_{2'3'} &= M_{23}\cos\theta - M_{13}\sin\theta \\
M_{3'3'} &= M_{33}
\end{aligned} \tag{11}$$

Substituting Eq. (11) into the expansion of Eq. (8) yields

$$
\begin{aligned}
u_{n'} = {} & M_{11}[\cos^2\theta G_{n'1',1'} + \sin^2\theta G_{n'2',2'} - \cos\theta\ \sin\theta(G_{n'1',2'} + G_{n'2',1'})] \\
& + M_{12}[2\cos\theta\ \sin\theta(G_{n'1',1'} - G_{n'2',2'}) + (\cos^2\theta - \sin^2\theta)(G_{n'1',2'} + G_{n'2',1'})] \\
& + M_{13}[\cos\theta(G_{n'1',3'} + G_{n'3',1'}) - \sin\theta(G_{n'2',3'} + G_{n'3',2'})] \\
& + M_{22}[\sin^2\theta G_{n'1',1'} + \cos^2\theta G_{n'2',2'} + \sin\theta\ \cos\theta(G_{n'1',2'} + G_{n'2',1'})] \\
& + M_{23}[\sin\theta(G_{n'1',3'} + G_{n'3',1'}) + \cos\theta(G_{n'2',3'} + G_{n'3',2'})] \\
& + M_{33}G_{n'3',3'}
\end{aligned}
, \tag{12}
$$

which expresses the ground displacement in the receiver coordinate system in terms of the source moment tensor in the source coordinate system. If we denote $x_{1'}$ as r, $x_{2'}$ as t and $x_{3'}$ as z, we arrive at the equation that we use for the moment-tensor inversion:

$$
\begin{aligned}
u_r = {} & M_{11}[\cos^2\theta G_{rr,r} + \sin^2\theta G_{rt,t} - \cos\theta\ \sin\theta(G_{rr,t} + G_{rt,r})] \\
& + M_{12}[2\cos\theta\ \sin\theta(G_{rr,r} - G_{rt,t}) + (\cos^2\theta - \sin^2\theta)(G_{rr,t} + G_{rt,r})] \\
& + M_{13}[\cos\theta(G_{rr,z} + G_{rz,r}) - \sin\theta(G_{rt,z} + G_{rz,t})] \\
& + M_{22}[\sin^2\theta G_{rr,r} + \cos^2\theta G_{rt,t} + \sin\theta\ \cos\theta(G_{rr,t} + G_{rt,r})] \\
& + M_{23}[\sin\theta(G_{rr,z} + G_{rz,r}) + \cos\theta(G_{rt,z} + G_{rz,t})] \\
& + M_{33}G_{rz,z} \\
u_t = {} & M_{11}[\cos^2\theta G_{tr,r} + \sin^2\theta G_{tt,t} - \cos\theta\ \sin\theta(G_{tr,t} + G_{tt,r})] \\
& + M_{12}[2\cos\theta\ \sin\theta(G_{tr,r} - G_{tt,t}) + (\cos^2\theta - \sin^2\theta)(G_{tr,t} + G_{tt,r})] \\
& + M_{13}[\cos\theta(G_{tr,z} + G_{tz,r}) - \sin\theta(G_{tt,z} + G_{tz,t})] \\
& + M_{22}[\sin^2\theta G_{tr,r} + \cos^2\theta G_{tt,t} + \sin\theta\ \cos\theta(G_{tr,t} + G_{tt,r})] \\
& + M_{23}[\sin\theta(G_{tr,z} + G_{tz,r}) + \cos\theta(G_{tt,z} + G_{tz,t})] \\
& + M_{33}G_{tz,z} \\
u_z = {} & M_{11}[\cos^2\theta G_{zr,r} + \sin^2\theta G_{zt,t} - \cos\theta\ \sin\theta(G_{zr,t} + G_{zt,r})] \\
& + M_{12}[2\cos\theta\ \sin\theta(G_{zr,r} - G_{zt,t}) + (\cos^2\theta - \sin^2\theta)(G_{zr,t} + G_{zt,r})] \\
& + M_{13}[\cos\theta(G_{zr,z} + G_{zz,r}) - \sin\theta(G_{zt,z} + G_{zz,t})] \\
& + M_{22}[\sin^2\theta G_{zr,r} + \cos^2\theta G_{zt,t} + \sin\theta\ \cos\theta(G_{zr,t} + G_{zt,r})] \\
& + M_{23}[\sin\theta(G_{zr,z} + G_{zz,r}) + \cos\theta(G_{zt,z} + G_{zz,t})] \\
& + M_{33}G_{zz,z}
\end{aligned}
. \tag{13}
$$

To use Eq. (13) in the moment-tensor inversion, we need to calculate 3-component Green's functions for the 6 moment-tensor components, which amounts to a total of 18 Green's functions. If the earth medium is horizontally homogeneous and isotropic, which we often assume, Eq. (13) can be further simplified. This is because for such an earth model, we have

$$
\begin{aligned}
G_{tr,r} &= G_{tt,t} = G_{tz,z} = G_{tr,z} + G_{tz,r} = G_{rr,t} + G_{rt,r} \\
&= G_{rt,z} + G_{rz,t} = G_{zr,t} + G_{zt,r} = G_{zt,z} + G_{zz,t} = 0.
\end{aligned}
$$

This conclusion can be proven by expanding Eq. (4.29) of Aki and Richards (2002) and realizing that when the source moment tensor and the ground motion are expressed in the same receiver coordinate system, for a horizontally homogeneous and isotropic earth model, the ray path between the source and the receiver is always in the r-z plane and the direction cosine γ_t is zero. With this observation, Eq. (13) becomes

$$\begin{aligned}
u_r = {} & M_{11}[\cos^2\theta G_{rr,r} + \sin^2\theta G_{rt,t}] \\
& + M_{12}[2\cos\theta\ \sin\theta(G_{rr,r} - G_{rt,t})] \\
& + M_{13}[\cos\theta(G_{rr,z} + G_{rz,r})] \\
& + M_{22}[\sin^2\theta G_{rr,r} + \cos^2\theta G_{rt,t}] \\
& + M_{23}[\sin\theta(G_{rr,z} + G_{rz,r})] \\
& + M_{33}G_{rz,z} \\
u_t = {} & -M_{11}[\cos\theta\ \sin\theta(G_{tr,t} + G_{tt,r})] \\
& + M_{12}[(\cos^2\theta - \sin^2\theta)(G_{tr,t} + G_{tt,r})] \\
& - M_{13}[\sin\theta(G_{tt,z} + G_{tz,t})] \\
& + M_{22}[\sin\theta\ \cos\theta(G_{tr,t} + G_{tt,r})] \\
& + M_{23}[\cos\theta(G_{tt,z} + G_{tz,t})] \\
u_z = {} & M_{11}[\cos^2\theta G_{zr,r} + \sin^2\theta G_{zt,t}] \\
& + M_{12}[2\cos\theta\ \sin\theta(G_{zr,r} - G_{zt,t})] \\
& + M_{13}[\cos\theta(G_{zr,z} + G_{zz,r})] \\
& + M_{22}[\sin^2\theta G_{zr,r} + \cos^2\theta G_{zt,t}] \\
& + M_{23}[\sin\theta(G_{zr,z} + G_{zz,r})] \\
& + M_{33}G_{zz,z}
\end{aligned} \quad . \tag{14}$$

Next, we express Eq. (14) in terms of the 10 specific, sometimes called canonical, Green's functions. First, we define 4 types of sources in the source coordinate system. We require that the time dependence of all 4 sources is an impulse that starts at $t = 0$. Its Fourier transform is then 1. The first source is a left-lateral vertical strike-slip source with its fault in the $x_1 - x_3$ plane. Its moment tensor representation is

$$\mathbf{M}_{SS}(f) = \begin{bmatrix} 0 & 1 & 0 \\ 1 & 0 & 0 \\ 0 & 0 & 0 \end{bmatrix}. \tag{15a}$$

The second source is a vertical dip-slip source with the fault in the $x_1 - x_3$ plane and the slip direction of the footwall, the wall on the positive x_2 side of the fault, is vertically downwards. Its moment tensor is

$$\mathbf{M}_{DS}(f) = \begin{bmatrix} 0 & 0 & 0 \\ 0 & 0 & -1 \\ 0 & -1 & 0 \end{bmatrix}. \tag{15b}$$

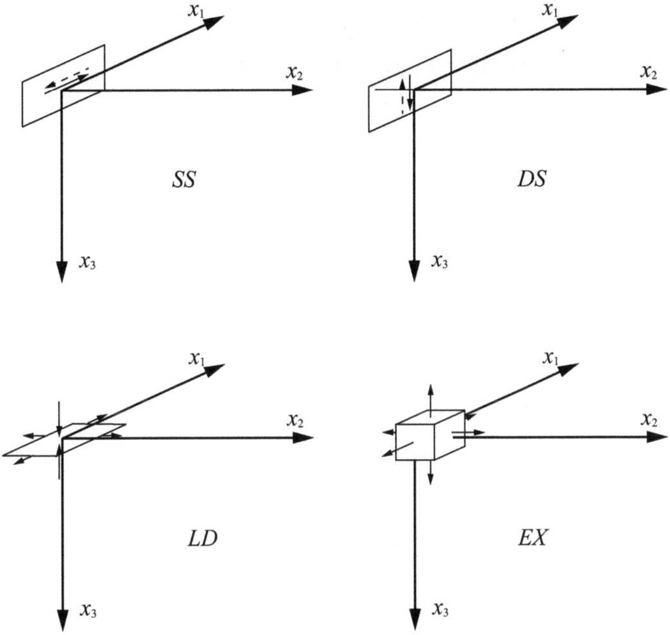

Fig. 2 Depiction of the four sources: strike slip (*SS*), dip slip (*DS*), CLVD (*LD*) and explosion (*EX*)

The third source is a compensated linear vector dipole (CLVD) with its vertical dipole twice as large as its horizontal dipoles. The moment tensor of this source is

$$\mathbf{M}_{LD}(f) = \begin{bmatrix} 0.5 & 0 & 0 \\ 0 & 0.5 & 0 \\ 0 & 0 & -1 \end{bmatrix}. \tag{15c}$$

The fourth source is an isotropic explosion with the moment tensor

$$\mathbf{M}_{EX}(f) = \begin{bmatrix} 1 & 0 & 0 \\ 0 & 1 & 0 \\ 0 & 0 & 1 \end{bmatrix}. \tag{15d}$$

The four sources are pictured in Fig. 2.

When we record seismic signals or compute synthetic seismograms from these four sources at specific azimuths, we obtain the 10 so-called canonical Green's functions. When the receiver is at an azimuth of 45° ($\pi/4$) from the strike-slip source, from Eqs. (14) and (15a), the radial- and vertical-component ground displacements are

$$
\begin{aligned}
SS_r = u_r &= M_{12}[2\cos\left(\frac{\pi}{4}\right)\sin\left(\frac{\pi}{4}\right)(G_{rr,r} - G_{rt,t})] = G_{rr,r} - G_{rt,t} \\
SS_z = u_z &= M_{12}[2\cos\left(\frac{\pi}{4}\right)\sin\left(\frac{\pi}{4}\right)(G_{zr,r} - G_{zt,t})] = G_{zr,r} - G_{zt,t}
\end{aligned}. \quad (16a)
$$

When the receiver is at an azimuth of 0° from the strike-slip source, the transverse component is

$$
SS_t = u_t = M_{12}[(\cos^2(0) - \sin^2(0))(G_{tr,t} + G_{tt,r})] = G_{tr,t} + G_{tt,r}. \quad (16b)
$$

For the dip-slip source, the radial and vertical components at an azimuth of 90° ($\pi/2$) are

$$
\begin{aligned}
DS_r = u_r &= M_{23}[\sin\left(\frac{\pi}{2}\right)(G_{rr,z} + G_{rz,r})] = -(G_{rr,z} + G_{rz,r}) \\
DS_z = u_z &= M_{23}[\sin\left(\frac{\pi}{2}\right)(G_{zr,z} + G_{zz,r})] = -(G_{zr,z} + G_{zz,r})
\end{aligned}. \quad (16c)
$$

The transverse component from the dip-slip source at the azimuth of 0° is

$$
DS_t = u_t = M_{23}[\cos(0)(G_{tt,z} + G_{tz,t})] = -(G_{tt,z} + G_{tz,t}). \quad (16d)
$$

For CLVD and explosion sources, there is no transverse ground motion and the azimuth of the receiver can be arbitrary. For the CLVD source, we have

$$
\begin{aligned}
LD_r = u_r &= M_{11}[\cos^2\theta G_{rr,r} + \sin^2\theta G_{rt,t}] + M_{22}[\sin^2\theta G_{rr,r} + \cos^2\theta G_{rt,t}] + M_{33}G_{rz,z} \\
&= \frac{1}{2}(G_{rr,r} + G_{rt,t}) - G_{rz,z} \\
LD_z = u_z &= M_{11}[\cos^2\theta G_{zr,r} + \sin^2\theta G_{zt,t}] + M_{22}[\sin^2\theta G_{zr,r} + \cos^2\theta G_{zt,t}] + M_{33}G_{zz,z} \\
&= \frac{1}{2}(G_{zr,r} + G_{zt,t}) - G_{zz,z}
\end{aligned}. \quad (16e)
$$

For the explosion source, we have

$$
\begin{aligned}
EX_r = u_r &= M_{11}[\cos^2\theta G_{rr,r} + \sin^2\theta G_{rt,t}] + M_{22}[\sin^2\theta G_{rr,r} + \cos^2\theta G_{rt,t}] + M_{33}G_{rz,z} \\
&= G_{rr,r} + G_{rt,t} + G_{rz,z} \\
EX_z = u_z &= M_{11}[\cos^2\theta G_{zr,r} + \sin^2\theta G_{zt,t}] + M_{22}[\sin^2\theta G_{zr,r} + \cos^2\theta G_{zt,t}] + M_{33}G_{zz,z} \\
&= G_{zr,r} + G_{zt,t} + G_{zz,z}
\end{aligned}. \quad (16f)
$$

When we substitute Eq. (16a–16f) into Eq. (14), we obtain

$$
\begin{aligned}
u_r = {} & M_{11}\left[\frac{SS_r}{2}\cos 2\theta + \frac{EX_r + LD_r}{3}\right] \\
& + M_{12}[SS_r \sin 2\theta] \\
& - M_{13}[DS_r \cos\theta] \\
& - M_{22}\left[\frac{SS_r}{2}\cos 2\theta - \frac{EX_r + LD_r}{3}\right] \\
& - M_{23}[DS_r \sin\theta] \\
& - M_{33}\left[\frac{2}{3}\left(LD_r - \frac{EX_r}{2}\right)\right] \\
u_t = {} & - M_{11}\left[\frac{SS_t}{2}\sin 2\theta\right] \\
& + M_{12}[SS_t \cos 2\theta] \\
& + M_{13}[DS_t \sin\theta] \\
& + M_{22}\left[SS_t \frac{\sin 2\theta}{2}\right] \\
& - M_{23}[DS_t \cos\theta] \\
u_z = {} & M_{11}\left[\frac{SS_z}{2}\cos 2\theta + \frac{EX_z + LD_z}{3}\right] \\
& + M_{12}[SS_z \sin 2\theta] \\
& - M_{13}[DS_z \cos\theta] \\
& - M_{22}\left[\frac{SS_z}{2}\cos 2\theta - \frac{EX_z + LD_z}{3}\right] \\
& - M_{23}[DS_z \sin\theta] \\
& - M_{33}\left[\frac{2}{3}\left(LD_z - \frac{EX_z}{2}\right)\right]
\end{aligned} \tag{17}
$$

Equations (13) and (17) are the basic equations that are used to invert for the frequency-domain source moment tensor, or the moment-tensor spectra. With observations at multiple receivers, these equations can be expressed in the matrix form as

$$\mathbf{d} = \mathbf{G}\mathbf{m}, \tag{18}$$

where **d** contains Fourier-transformed ground motion, **G** contains frequency-domain Green's functions and **m** are frequency-domain moment-tensor components. With digital data, all spectra in Eq. (18) are for discrete frequencies.

Equation (18) can be solved for **m** at each frequency with standard least-squares inversion methods such as the singular-value-decomposition method (SVD) (e.g., Menke 1989), in which **G** is decomposed as

$$\mathbf{G} = \mathbf{U}\boldsymbol{\Lambda}\mathbf{V}^{\mathrm{H}},$$

where **U** is composed of eigenvectors of $\mathbf{G}\mathbf{G}^{\mathrm{H}}$ and **V** is composed of eigenvectors of $\mathbf{G}^{\mathrm{H}}\mathbf{G}$. Superscript ^{H} denotes conjugate transpose. Diagonal matrix $\boldsymbol{\Lambda}$ contains nonzero square roots of corresponding eigenvalues, called singular values. With the SVD method, the source moment tensor is estimated as

$$\mathbf{m}^{\text{est}} = \mathbf{V}\boldsymbol{\Lambda}^{-1}\mathbf{U}^{\text{H}}\mathbf{d}. \tag{19}$$

The complete source moment-tensor spectra are obtained by combining inversion results at individual frequencies. Time histories of the moment tensor are obtained by the inverse Fourier transform of the moment-tensor spectra.

For statistically independent ground-motion data with a uniform variance σ_d^2, the covariance of the estimated moment tensor is (Menke 1989)

$$\text{cov}(\mathbf{m}^{\text{est}}) = \sigma_d^2\mathbf{V}\boldsymbol{\Lambda}^{-2}\mathbf{V}^{\text{H}}. \tag{20}$$

Equation (20) shows that the covariance and the variance of the estimated moment tensor depend on the singular-value matrix. Small singular values result in large moment-tensor variances.

The time history of a seismic source, whether it is an earthquake or an explosion, often has a static offset as time approaches infinity. The Fourier transform of this component approaches infinity as the frequency goes to zero. This low-frequency behavior makes the frequency-domain inversion results using ground displacements unstable. To stabilize the inversion at low frequencies, we usually use ground velocity, instead of ground displacement, in the moment-tensor inversion. As a result, the time derivative of the moment-tensor spectra, or the moment-rate-tensor spectra, is retrieved.

4 Examples

Because the purpose of using the frequency-domain moment-tensor-inversion technique is to retrieve the complete source moment-tensor spectra including its high-frequency component, this method is often employed to characterize small sources where the point-source approximation is more appropriate for high-frequency data. These sources include small earthquakes, e.g., induced seismicity, and man-made sources, such as underground explosions or mine collapses. In this section, we first give two examples demonstrating the technique in characterizing an underground chemical explosion and a unique underground-mine collapse. The final example illustrates the importance of accurate velocity models to inversion results.

4.1 An Underground Chemical Explosion

The first example is the moment-tensor inversion using near-source seismograms from an underground chemical explosion (Yang and Bonner 2009). The explosion

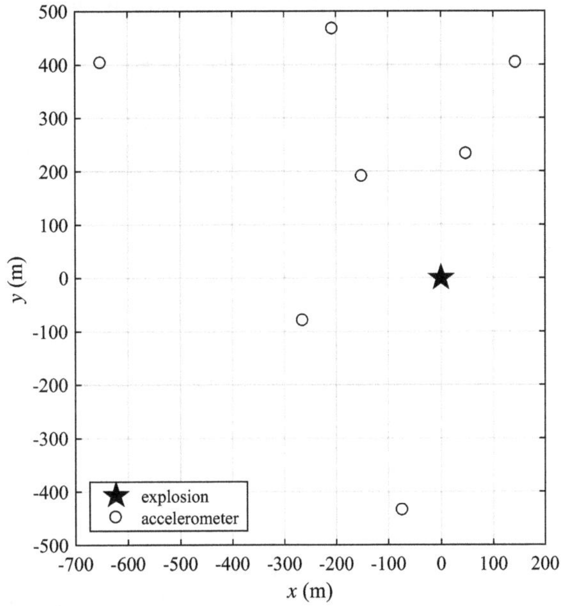

Fig. 3 Locations of the explosion source and accelerometers

source was constructed by drilling 5 holes in a cross pattern to a depth of 40 m, filling the holes with 10 m of explosives, and stemming the holes to the surface. The total weight of the explosive is 5484 kg. The arm length of the cross drill pattern is about 3.5 m. Explosives in the 5 holes were detonated simultaneously resulting in an effective single, contained explosion source. Seismic signals from this explosion were recorded by an accelerometer array around the source. Figure 3 plots the locations of the explosion and accelerometers. The source-receiver distances are from 201 to 668 m. The accelerometers cover about half of the complete 360° azimuths. An optimal azimuthal coverage would span 360° for a complete sampling of the source radiation pattern. Considering the fact that the radiation pattern from an explosion source should be relatively uniform and the fact that three-component waveforms are utilized, this azimuthal coverage should be sufficient. Deployment of receivers at different distances is also desired for improved sampling of the source focal sphere.

Figure 4 displays the ground-velocity seismograms determined by integrating ground accelerations recorded at the receivers shown in Fig. 3. These are typical of seismic signals we see from explosive sources at close distances. The raw signals are generally short (<1 s). The main signal is the *P*-wave. If we low-pass filter the signals, short-distance, fundamental-mode Rayleigh wave (R_g) becomes significant. Even though explosive sources are supposed to generate mainly compressional waves, we usually see appreciable shear waves on the transverse component due to the potential asymmetry of the sources and/or media heterogeneity.

To invert these ground motions for the source moment tensor, we need to calculate Green's functions between the source and receivers. This in turn requires a proper

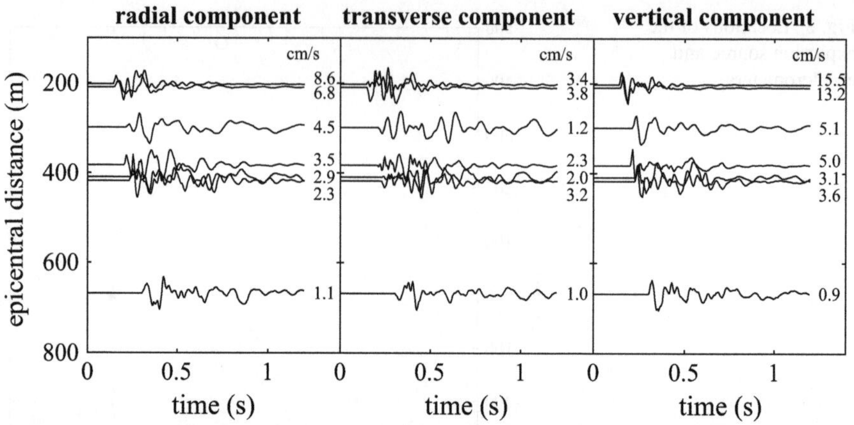

Fig. 4 Ground velocities from the explosion. Amplitudes are normalized. Maximum absolute amplitudes are marked to the right of the traces

earth model with P-wave velocity, S-wave velocity, density, P-wave quality factor Q_p and shear-wave quality factor Q_s. In many cases, because source signals are recorded at close distances (e.g., <2 km), the medium heterogeneity is not severe and a laterally homogeneous, one-dimensional (1D) earth model can be used. This assumption is good for this experiment where the explosion was detonated in sedimentary rocks with velocity changes mainly in the vertical direction. Generally, different parameters of the earth model are derived from different data types. In this example, P-wave velocity structure of the earth model was developed using P-wave arrival times from a refraction survey and from seismometers installed around the test site (Leidig et al. 2005). S-wave velocities were derived from surface-wave analysis, and density data were from laboratory sample tests. Shear-wave quality factor was from analysis of surface-wave amplitudes, whereas that of P-wave was based on a theoretical relationship between Q_s and Q_p (Müller 1985).

For 1D earth models, a variety of computation methods can be used to calculate synthetic seismograms, or Green's functions. In this example, we use a method called the reflectivity method (Müller 1985) for the calculation. The method is a slowness-integration method that generates complete seismograms including both near-field and far-field terms. Other than an accurate earth model, the key aspect of getting proper Green's functions using the method is to select appropriate input parameters, such as slowness-integration interval, slowness sampling and length of the time trace, for the problem at hand so that important seismic phases are captured in the Green's function and numerical noise is minimized.

After Green's functions for all receivers are obtained, they are substituted into Eq. (18) and a moment-tensor inversion can then be performed. Because the 1D model is an approximation of the real earth structure, errors such as differences in phase arrival times between the Green's functions and the observed seismograms exist. One way to reduce the effect of model error is to manually align P-wave arrival times

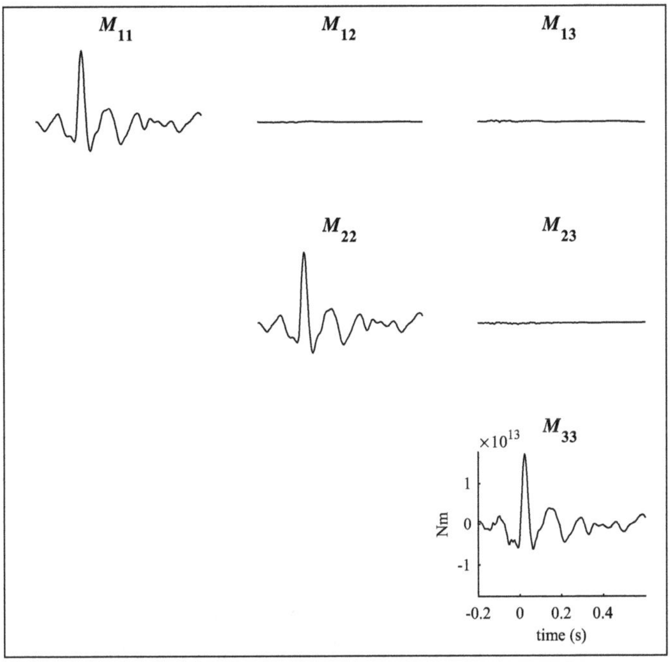

Fig. 5 Moment-tensor time histories of the explosion source from the inversion

between the Green's function and the observed signal before the Fourier transform. This procedure has been used in regional moment-tensor inversion studies (e.g., Zhao and Helmberger 1994), and it is a common practice in frequency-domain moment-tensor inversions. Even though the method does not correct for arrival-time error of later phases or amplitude errors, it still improves the inversion result. An additional measure that can be taken is to multiply the seismograms and the Green's functions by corresponding epicentral distances to some power. This is to increase amplitudes of seismograms at longer distances so that they provide similar contributions to the inversion result as close-in seismograms do. Multiplication by distance itself is a usual choice.

Figure 5 plots the moment tensor time histories of the explosion source from the inversion of seismograms shown in Fig. 4. Both manual alignment of *P*-wave arrival times and multiplication of seismograms by distance were applied before the inversion. The time histories provide detailed information about the source process including its mechanism, strength and evolution over time. It shows that the explosion-source moment tensor is dominated by its diagonal components. The source signal consists mainly of a single pulse that starts at time zero. All three diagonal components have similar wave shapes and amplitudes.

The isotropic nature of the source can be seen more clearly in its moment-rate spectra shown in Fig. 6. The amplitudes of the diagonal moment-rate spectra are

Fig. 6 The moment-rate spectra of the explosion source

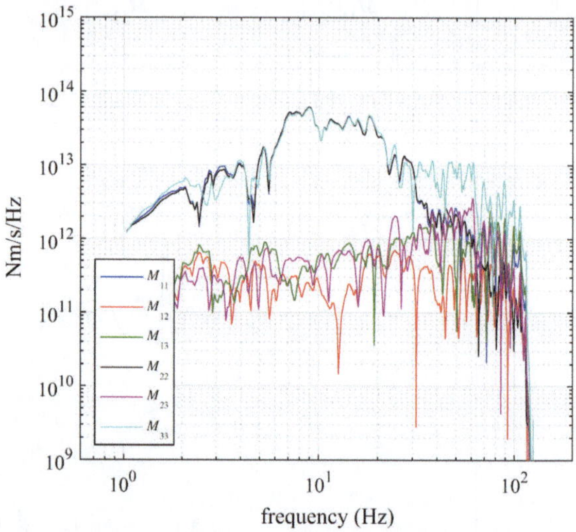

almost identical between 5 and 25 Hz. Below 5 Hz, the diagonal-component spectra display an amplitude plateau possibly containing the static component, or the long-period level, of the source signal, which is used to estimate the source moment. When we integrate ground accelerations to ground velocities, we enhance the long-period noise in the data. As a result, we usually need to high-pass filter the seismograms. Because of the filtering, we see the roll-off of spectral amplitudes toward low frequencies. Below 20–30 Hz, off-diagonal-component spectral amplitudes are more than an order of magnitude lower than those of diagonal components. This is consistent with time-domain observations (Fig. 5).

Figure 7 plots the relative variance of the source moment-tensor spectra calculated using Eq. (20), assuming a unit data variance. The variance can be used to evaluate the reliability of the source spectra from the inversion. The variances of the diagonal moment-tensor components track each other closely for frequencies below 30 Hz. Above 30 Hz, the variance of M_{33} begins to increase significantly from those of M_{11} and M_{22}. We see corresponding deviation of M_{33} spectral amplitude in Fig. 6 for the same high frequencies. The correlation implies that the deviation could be caused by errors in modeling M_{33} at high frequencies and a physical interpretation of the deviation should be done with caution. Variances of off-diagonal components also increase with increasing frequency, meaning their spectra at high frequencies are not as well constrained either.

To further assess the quality of the inversion result, we usually compare observed seismograms used in the inversion with those predicted by the inversion result. Figure 8 shows such a comparison. Based on the correlation coefficient values, we conclude that the moment tensor from the inversion accurately reproduces observed signals particularly for vertical-component data and at closer distances.

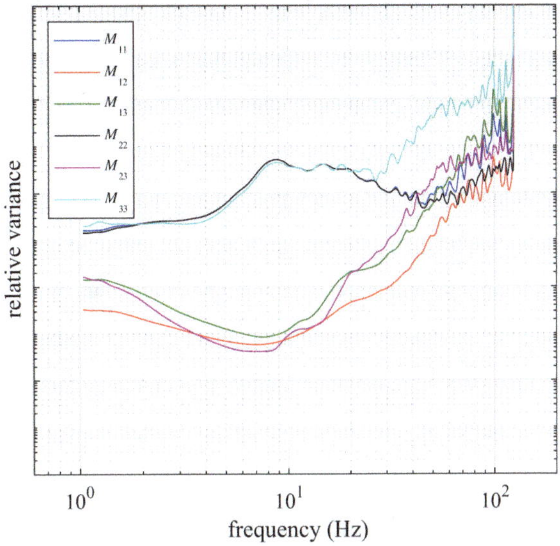

Fig. 7 Relative variance of moment-tensor spectra shown in Fig. 6

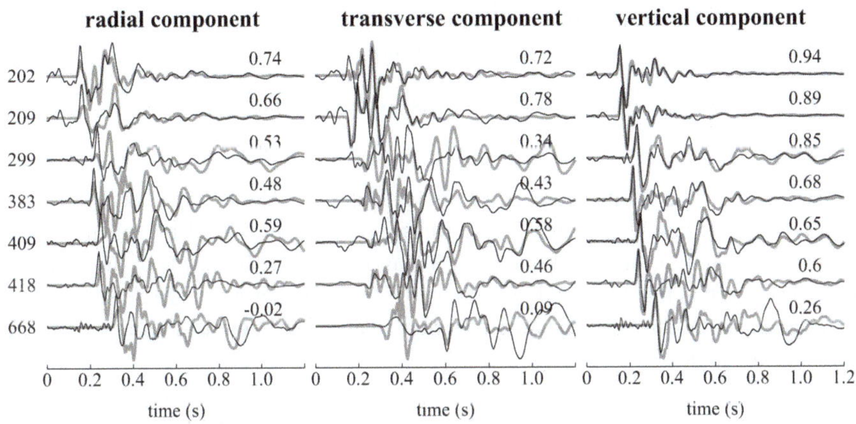

Fig. 8 A comparison between observed seismograms and synthetic seismograms predicted by the source moment tensor. Thick gray lines are observed seismograms. Trace amplitudes are normalized. Correlation coefficients between observed and predicted seismograms are marked above corresponding traces. Epicentral distances in meters are given on the far left of the figure

4.2 An Explosion-Induced Mine Collapse

In this example, we present the moment-tensor-inversion result of a mine collapse (Yang et al. 1998). The mine is an underground copper mine where the room-and-pillar method had been used to extract ore. The collapse was the intentional goal of an experiment where multiple pillars supporting a mined-out opening were explosively destroyed simultaneously. After the explosion-induced removal of pillars, the ceiling

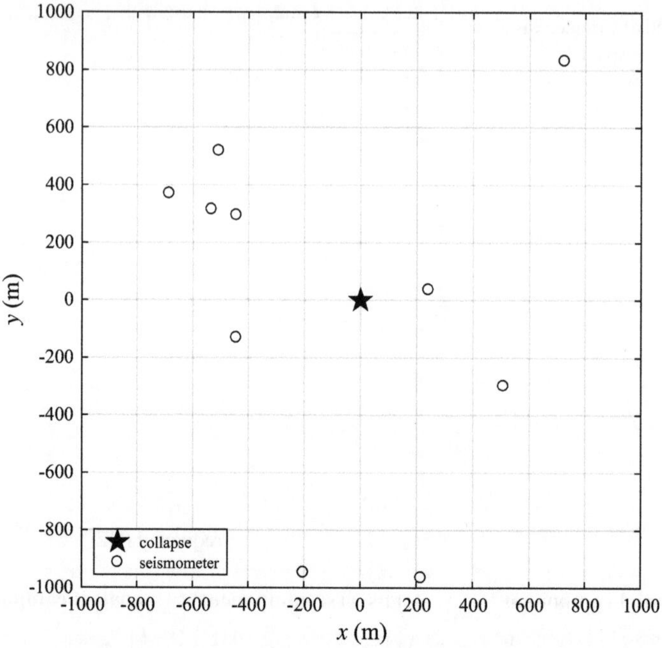

Fig. 9 Locations of the collapse and seismometers that recorded the collapse

collapsed upon the sudden loading of its own weight. The collapse occurred at a depth of 320 m below the earth's free surface and was recorded by seismometers deployed on the free surface. Figure 9 is a plan view of the locations of the collapse and seismometers that recorded the collapse. For this experiment, the seismometers have better azimuthal and distance coverage. The experiment design resulted in a better-constrained source moment tensor.

Figure 10 shows the mine-collapse moment-tensor time histories from the inversion. Again, the moment tensor is dominated by its diagonal components. This time, however, the three diagonal moment-tensor components do not have the same amplitudes. Among them, the vertical dipole M_{33} has the largest amplitude. This is consistent with the moment-tensor representation of a horizontal tension-crack model (Aki and Richards 2002).

Because we were able to retrieve the complete time history of the collapse source including its initiation and evolution, we gain better insight into the source process that would otherwise not be obtainable. One observation about this collapse from its moment-tensor time history is that the source started as an expansion source with positive onsets for all diagonal components. This is in contrast to accidental mine collapses that occur naturally as the result of the slow strain accumulation and final rock failure. Those collapses all start as implosional sources with dilatational seismic-wave first motions and negative onsets for their diagonal moment-tensor components (e.g., Taylor 1994; Pechmann et al. 1995). This unique collapse charac-

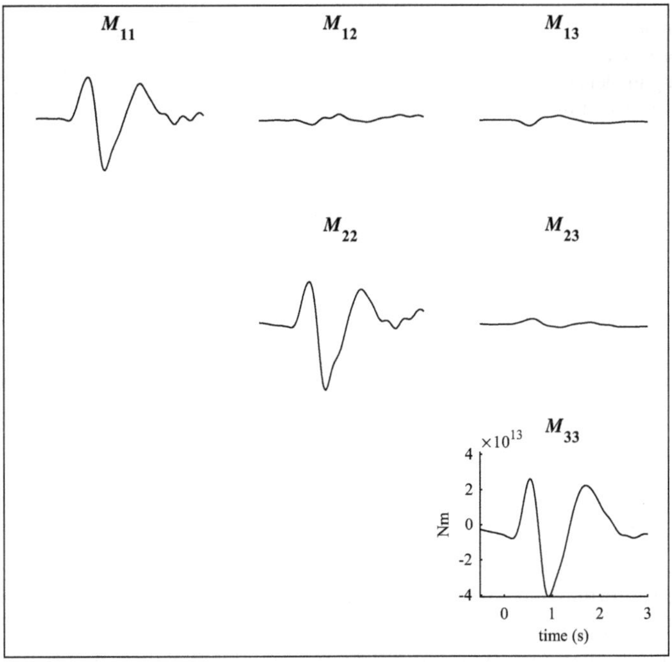

Fig. 10 Moment-tensor time histories of the explosion-induced mine collapse from the inversion

teristic motivated Yang et al. (1998) to propose a new model for the source process of this kind of explosion-induced mine collapses. This example illustrates the special power of the frequency-domain moment-tensor inversion technique.

4.3 Effects of Green's Functions on the Moment-Tensor Inversion

Because the process of moment-tensor inversion is essentially removing the propagation effects from observed seismograms to retrieve the source signal, it is apparent that accurate Green's functions that adequately represent the true path effects are critical to the success of the inversion. In this final example, we show how uncertainty in the velocity model used to calculate Green's functions can affect the resulting moment tensor.

This example is from another field experiment involving a small, contained underground chemical explosion. The explosion was recorded by seismometers within 700 m of the source. Refraction surveys were conducted around the test site during the experiment. As a result, the *P*-wave velocity structure was well constrained. The *S*-wave velocity, on the other hand, could not be determined reliably from available

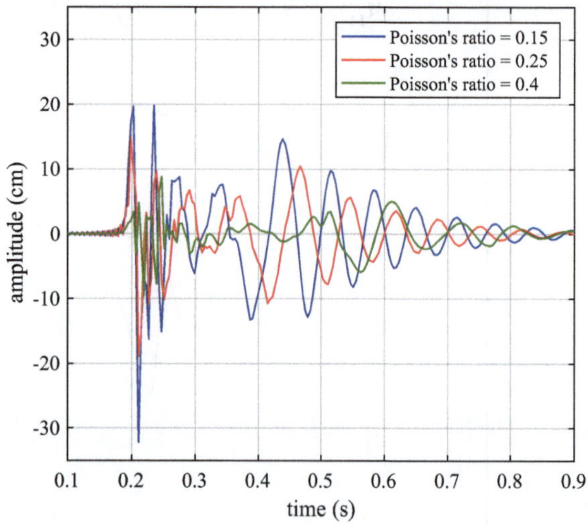

Fig. 11 Synthetic seismograms calculated using velocity models with different Poisson's ratios

data. As an alternative, *S*-wave velocities of the structure model were calculated based on *P*-wave velocities and different Poisson's ratios.

Figure 11 compares synthetic seismograms calculated using velocity models with different *S*-wave velocities corresponding to Poisson's ratios of 0.15, 0.25 and 0.4. Several observations can be made from the figure. As the Poisson's ratio increases, *S*-wave velocity decreases. This delays the arrival time of the surface wave (the long-period waves after about 0.35 s) significantly. In addition, the dominant frequency of the surface wave shifts to lower frequencies. Even though the quality factor Q of the velocity model remains the same, the amplitudes of the seismograms are reduced due to the decrease of the *S*-wave velocity. This is true not only for the surface-wave amplitude, but for the *P*-wave amplitude as well.

Figure 12 shows the effect of changing the *S*-wave velocity on the resulting moment tensor. We plot only results for Poisson's ratios of 0.15 and 0.4 as two end-member examples. For the Poisson's ratio of 0.15, the moment tensor has strong M_{13} and M_{23} signals. Among the diagonal components, M_{33} is much larger. As the Poisson's ratio increases to 0.4, the moment tensor looks much more like an isotropic-source moment tensor with dominant diagonal components that have similar amplitudes. The difference between the two moment tensors in terms of their source mechanisms is apparent, which poses a challenge to the reliable and quantitative interpretation of the source process.

Lessons learned from this example are that it is critically important to have an accurate velocity model for a moment-tensor inversion. Efforts should be made to constrain not only the *P*-wave velocity structure, but also *S*-wave-velocity and attenuation structures. In certain situations, a 3D model may be the only viable choice and Eq. (13), instead of Eq. (17), should be used.

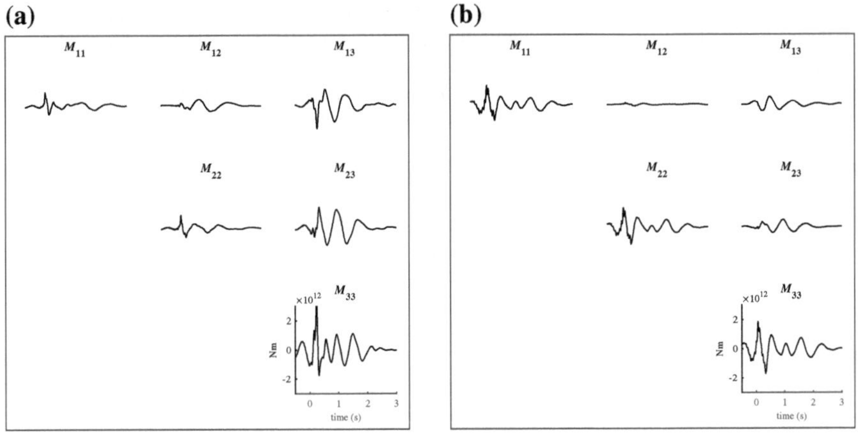

Fig. 12 Moment tensors from two inversions using Green's functions from two velocity models with Poisson's ratios of 0.15 (**a**) and 0.4 (**b**)

5 Conclusions

In this article, we described a frequency-domain moment-tensor inversion technique. Often used with near-source (e.g., <2 km) observations from small seismic sources, the method retrieves the complete source moment-tensor spectra and time histories, revealing detailed source information.

We provided examples illustrating the application and the advantage of the method. The last example highlights one of the issues that require attention in order to conduct successful frequency-domain moment-tensor inversions.

References

Aki K, Richards PG (2002) Quantitative Seismology, 2nd edn. University Science Books, Sausalito

Ekström G, Nettles M, Dziewoński AM (2012) The global CMT project 2004–2010: centroid-moment tensors for 13,017 earthquakes. Phys Earth Planet Int 200–201:1–9. https://doi.org/10.1016/j.pepi.2012.04.002

Julian BR, Miller AD, Foulger GR (1998) Non-double-couple earthquakes. 1. Theory. Rev Geophys 36(4):525

Leidig MR, Bonner JL, Reiter DT (2005) Development of a velocity model for Black Mesa, Arizona, and the southern Colorado Plateau from multiple data sets. Bull Seismol Soc Am 95(6):2136–2151

Menke W (1989) Geophysical data analysis: discrete inverse theory (rev edn). Academic Press, San Diego

Müller G (1985) The reflectivity method: a tutorial. J Geophys 58:153–174

Pechmann JC, Walter WR, Nava SJ et al (1995) The February 3, 1995, ML 5.1 seismic event in the trona mining district of southwestern Wyoming. Seism Res Lett 66(3):25–34

Pondrelli S, Salimbeni S, Morelli A et al (2011) European-Mediterranean regional centroid moment tensor catalog: solutions for 2005–2008. Phys Earth Planet Int 185(3–4):74–81. https://doi.org/10.1016/j.pepi.2011.01.007

Šílený J, Panza GF, Campus P (1992) Waveform inversion for point source moment tensor retrieval with variable hypocentral depth and structural model. Geophys J Int 109(2):259–274. https://doi.org/10.1111/j.1365-246X.1992.tb00097.x

Sipkin SA (1982) Estimation of earthquake source parameters by the inversion of waveform data: synthetic waveforms. Phys Earth Planet Int 30(2):242–259. https://doi.org/10.1016/0031-9201(82)90111-X

Stump BW, Johnson LR (1977) The determination of source properties by the linear inversion of seismograms. Bull Seismol Soc Am 67(6):1489–1502

Taylor SR (1994) False alarms and mine seismicity: an example from the Gentry mountain mining region, Utah. Bull Seismol Soc Am 84(2):350

Tsuruoka H, Kawakatsu H, Urabe T (2009) GRiD MT (grid-based real-time determination of moment tensors) monitoring the long-period seismic wavefield. Phys Earth Planet Int 175(1–2):8–16. https://doi.org/10.1016/j.pepi.2008.02.014

Yang X, Bonner JL (2009) Characteristics of chemical explosive sources from time-dependent moment tensors. Bull Seismol Soc Am 99(1):36–51. https://doi.org/10.1785/0120080243

Yang X, Stump BW, Phillips WS (1998) Source mechanism of an explosively induced mine collapse. Bull Seismol Soc Am 88(3):843–854

Zhao LS, Helmberger DV (1994) Source estimation from broad-band regional seismograms. Bull Seismol Soc Am 84(1):91–104

Berkeley Seismic Moment Tensor Method, Uncertainty Analysis, and Study of Non-double-couple Seismic Events

Douglas S. Dreger

1 Introduction

For the past two decades, the Berkeley Seismological Laboratory has implemented and utilized as part of its routine operations (100 to ~1200 km) a regional distance, complete waveform seismic moment tensor inversion method (Romanowicz et al. 1993; Pasyanos et al. 1996). This method is being used by several regional distance seismic monitoring organizations around the world as well as independent researchers, and is available from the Northern California Earthquake Data Center (NCEDC) FTP site (ftp://ncedc.org/outgoing/dreger).

Although the method was initially developed for use at regional distances to study earthquakes ranging in magnitude from Mw 3.5 to 8+ (e.g. Romanowicz et al. 1993; Pasyanos et al. 1996; Fukuyama and Dreger 2000; Guilhem et al. 2011), it has also been used in studies of smaller earthquakes at closer distances. The code itself is simply a waveform fitting algorithm and thusly its application is not bounded, and really limited only by the appropriateness of the velocity model and the method used to compute the Green's functions used in the inversion. This complete waveform method has been used to study micro (Mw < -2.0) alpine glacier ice quakes (Walter et al. 2009, 2010), small earthquakes (typically M < 2) in geothermal energy (Boyd et al. 2015, 2017) and hydrocarbon storage (Nayak and Dreger 2014) environments, earthquakes related to volcanic processes (Dreger et al. 2000; Minson et al. 2007; Tkalcic et al. 2009), events related to mining activities (Ford et al. 2008) and for the discrimination of nuclear explosions from natural seismicity (Ford et al. 2009; Chiang et al. 2014).

D. S. Dreger (✉)
Department of Earth and Planetary Science, University of California, Berkeley, CA, USA
e-mail: dreger@seismo.berkeley.edu

S. D'Amico (ed.), *Moment Tensor Solutions*, Springer Natural Hazards,
https://doi.org/10.1007/978-3-319-77359-9_4

2 Method Overview

The Berkeley Time Domain Moment Tensor INVerse code (TDMT_INV) matches complete three component waveforms to estimate the seismic moment tensor, solving $u_i(\vec{x}, t) = M_{pq} \frac{\partial}{\partial x_q} G_{ip}(\vec{x}, t)$, where i, p and q are indices referring to the component of motion, the direction of force, and the direction of a spatial derivative applied to the Green's function solution to the elastodynamic equation of motion for a given force direction. The applied derivatives therefore relate the observed ground motion to force-couples and force-dipoles. The moment tensor, M, represents the scalar weight of these force-couple and force-dipole components of the Green's function. It is general in that it encapsulates the set of double-couple, compensated linear-vector-dipole (CLVD), dipole, cracks, and volumetric sources, and all combinations of them, although it does not account for applied body forces, which would require an additional term. As written, this equation is for a synchronous point-source implying that all moment tensor elements have the same impulsive time history. Because, the inversion is linear, and commonly over-determined for waveform inversion applications, a unique estimate of the seismic moment tensor can be found. However, it is important to understand that while the moment tensor estimate is itself unique that its decomposition into various combinations of fundamental mechanism types such as double-couples, tensile or closing cracks, CLVDs, volumetric sources, and combinations of these is not unique. Therefore, care is needed in deciding on how to decompose the moment tensor for interpretation, as well as for understanding the resolution and uncertainties of the various components of the moment tensor. In addition, some types of mechanism combinations such as shear-tensile cracks cannot easily be decomposed from the moment tensor and non-linear inversion or grid searches for such specific models may be needed (e.g. Minson et al. 2007).

At local to regional distances the data is typically bandpass filtered between 0.02 and 0.10 Hz, which allows the use of relatively simple 1D velocity models for describing the source to receiver wave propagation. For example, at the minimum period of 10 s the surface waves due to the central valley of California, or the deep tertiary basins in Santa Clara and San Pablo Bay are effectively removed from the data allowing the modeling of the records with 1D velocity models quite well. For studies of smaller earthquakes recorded at close distances higher frequencies may be employed. Essentially the method assumes that the 1D seismic velocity models used to compute Green's functions adequately describe the wave propagation between source to receiver, which is typically satisfied at long period corresponding to wave lengths comparable to the path length. Long period regional distance seismic phases that are important in the inversion include the P_{nl}, S_n, their related depth phases and mode conversions, and Love and Rayleigh waves. In the long-period passband near- and intermediate-terms of the solution to the elastodynamic equation of motion are important and must be included in the calculation of the Green's functions. Currently we utilize the frequency wavenumber integration program provided by Robert Herrmann in his Computer Programs for Seismology 3.3 package (Herrmann 2013) to compute the needed Green's functions. Figure 1 shows an example of the waveform

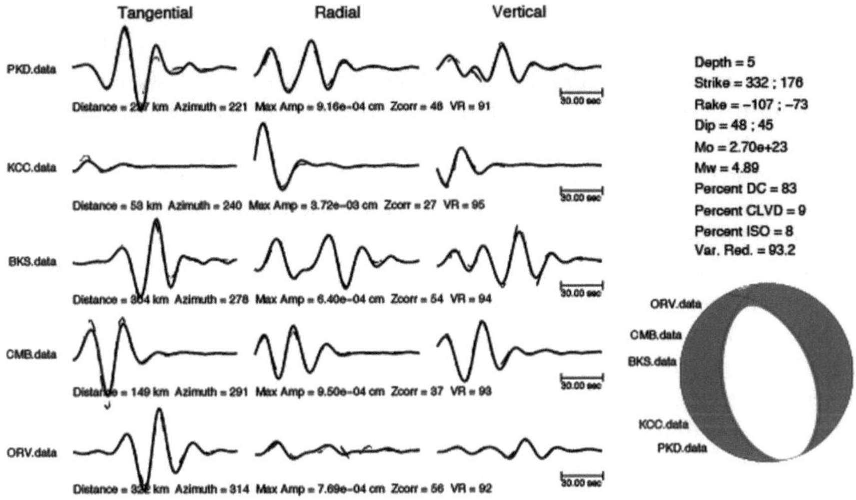

Fig. 1 Example of seismic moment tensor inversion for a Mw 4.9 earthquake located in the Long Valley Caldera of eastern California. The data (solid) and synthetic (dashed) have been band pass filtered between 0.02 and 0.05 Hz using an acausal Butterworth filter. Var. Red. is the variance reduction which is a L2-norm based goodness of fit measure that is sensitive to both phase and amplitude, where 100% is perfect fit

Table 1 Event information

Event ID	Date time (UTC)	Latitude	Longitude	Depth (km)	Mw
EVT1[a]	07/15/1998 04:53:19.3	37.564	−118.806	5	4.92
EVT2[a]	11/30/1997 21:17:05.4	37.634	−118.946	5	5.02
EVT3[b]	09/09/2016 00:30:01.4	41.287	129.078	1	4.93

[a]Origin time and location from Dreger et al. (2000)
[b]Origin time and location from USGS ANSS catalog (accessed July 26, 2017)
Depth and Mw are from the moment tensor inversion. Mw is computed using the Bowers and Hudson (1999) total moment

fit that can be achieved at regional distances in the 0.02–0.05 Hz passband for a California earthquake (EVT1, Table 1).

With TDMT_INV it is assumed that the models are well calibrated such that the overall wave packets, relative timing between P, S and surface waves is approximately correct and that the overall time alignment can be accomplished with the single shift parameter for each station's time series. In some cases, separate velocity models may be used to account for more severe cases of lateral heterogeneity in velocity structure (Tkalcic et al. 2009). Our simplified method for optimizing time alignment of data and Green's functions was done to better enable automated applications carried out in near-realtime (e.g. Pasyanos et al. 1996). For more complex regions or where the velocity structure may not be as well calibrated the 'Cut and Paste' (CAP) method of Zhu and Helmberger (1996) is a good alternative since it allows for different time shifts between the P, S and surface wave packets thereby accounting for higher degrees of lateral variability in structure.

Minson and Dreger (2008) describes the theoretical setup of the general moment tensor inverse problem for TDMT_INV. Jost and Herrmann (1989), Miller et al. (1998) and Julian et al. (1998) provide excellent general overviews of seismic moment tensor inversion, the decomposition of the seismic moment tensor and moment tensors for different types of source mechanisms.

Much of the recent development has focused on improving capabilities for assessing solution uncertainties for applications where significant non-double-couple components to the seismic moment tensor are possible. In Ford et al. (2009), the source-type representation of the seismic moment tensor (Hudson et al. 1989) is utilized to illustrate the solution in terms of the ratios of the eigenvalues of the solution. In Hudson et al. (1989), two parameters are defined, $T = -2\frac{m_3}{|m_1|}$ and $k = \frac{M^{iso}}{|m_1|+|M^{iso}|}$, where for a general moment tensor $M = \begin{pmatrix} m_{xx} & m_{xy} & m_{xz} \\ m_{xx} & m_{yy} & m_{yz} \\ m_{xz} & m_{yz} & m_{zz} \end{pmatrix}$ $M^{iso} = 1/3\left(m_{xx} + m_{yy} + m_{zz}\right)$ and $|m_1| \geq |m_2| \geq |m_3|$, are the ranked deviatoric eigenvalues. The deviatoric parameter T ranges between -1 and 1 where the extreme values are $\pm$CLVD and a zero value is a double-couple. The parameter k also ranges between -1 and 1 where -1 represents a pure spherically symmetric implosion (volume decrease) and 1 represents an explosion (volume increase). The locations of tensile cracks and linear vector dipoles are typically identified on such plots (e.g. Hudson et al. 1989). Mechanisms such as shear tensile cracks are located over a range of T and k depending upon the geometry of the tensile opening and the shear dislocation, and the relative moment of each component (e.g. Julian et al. 1998). The two primary advantages of source-type representations are (1) that the projection enables a straightforward representation to allow documentation of uncertainties in the solution, and (2) when examining different classes of events the respective populations separate in the source-type space. The source-type projections however do not convey the orientation of a given mechanism. For example there are an infinite number of double-couple orientations located at the double-couple point (DC) in the source-type projection. Nevertheless, the projection is useful for identifying different classes or types of earthquakes, where for example, in Fig. 2, the populations of western United States earthquakes, underground cavity collapses and nuclear explosions are found to separate well in the Hudson et al. (1989) source-type projection. This separation is found to be significant when considering the uncertainties obtained by bootstrapping residuals.

The Network Sensitivity Solution (NSS) (Ford et al. 2012) was developed to assess the sensitivity of moment tensor solutions to seismic recording topology, the resolution of the source-type, uniqueness, and whether there are tradeoffs in source type. The NSS method considers a large number of uniformly distributed (order of 100 million) eigenvalue ratios (values of T, and k) with random eigenvector

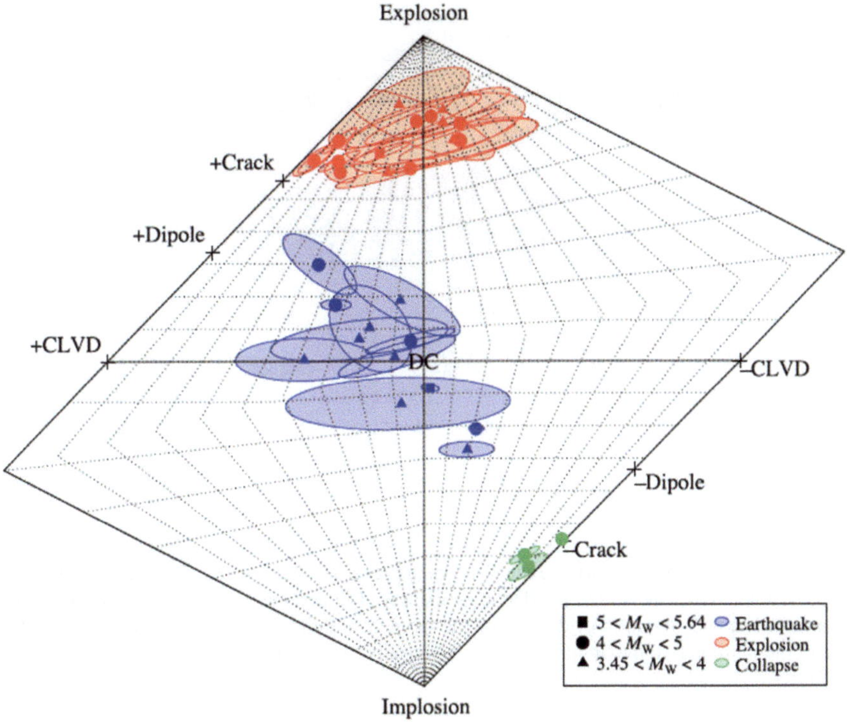

Fig. 2 Populations of western United States earthquakes (blue), underground cavity collapses (green), and nuclear explosions (red) are plotted with uncertainty ellipses obtained from residual bootstrap. The locations of idealized source types (explosions, Crack, Dipole, CLVD and double-couples (DC)) are noted. From Ford et al. (2009)

orientations, which are then tested against the data to map the maximum fit surface in the source-type space. Different types of norms can be considered when preparing the plot. Beyond the uncertainty obtained from random fluctuations in data that can be assessed by a residual bootstrap (e.g. Ford et al. 2009; Fig. 2) the map of the maximum-fit surface illustrates the sensitivity of the data to estimate the moment tensor parameters and the surface may or may not have a Gaussian shape depending upon many factors including the network topology and the appropriateness of the velocity model used to compute the Green's functions. In order to develop a true maximum-fit surface the Source-Type Inversion was developed (Nayak and Dreger 2015) in which for regularly specified source-type (values of T and k) non-linear inversions are performed to find the optimally oriented eigenvectors for each source-type. The NSS or Source-Type-Inversion can be carried out in other source-type projections (e.g. Vavryčuk 2011; Tape and Tape 2012; Aso et al. 2016). Vavryčuk (2015) points out that uniform distributions in one source-type projection may not be uniform in another, and therefore for a sampling approach like the NSS method this needs to be taken into consideration. The Source-Type-Inversion (Nayak and

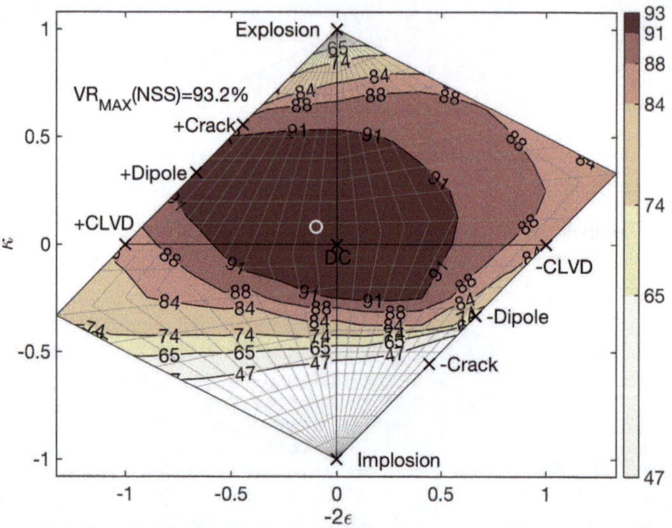

Fig. 3 Source-type inversion for the event shown in Fig. 1. The white-circle shows the best fit solution. The color shows the variance reduction goodness of fit parameter representing the maximum fit surface. Kappa is the ratio of the volumetric moment and the total moment, and epsilon is the ratio of the minimum to maximum deviatoric eigenvalues in the absolute sense (e.g. Hudson et al. 1989)

Dreger 2015) can be applied in any source-type projection as it does not rely on a distribution of values in the space, but rather is constructed from a regular grid which can be optimized for the different types of source-type projections.

The maximum-fit surface for EVT1 is shown in Fig. 3. The best solution (white circle) shows a slight volume increase component, but the difference in the level of fit between it and deviatoric solutions is not large enough to be statistically significant. An F-test assuming the correlation length of the applied bandpass filter (e.g. Dreger et al. 2000) finds that the additional degree of freedom of the full moment tensor compared to the deviatoric moment tensor is only statistically significant at the 49% level. The full moment tensor and deviatoric moment tensor solutions are essentially indistinguishable from one another in terms of the goodness of fit.

One method to compute uncertainties is to bootstrap data-synthetic residuals. First, residuals in the waveform fit are estimated from the best-fitting model. The residuals are then randomized, filtered in the same passband as the data, scaled to the maximum raw residual value, and then added to the data in a sequence of realizations. Typically, we use approximately 2000 realizations of the randomized residuals. The perturbed data in each realization is then inverted. For EVT1 the solutions from a 2000 realization bootstrap are shown in Fig. 4 as gray dots. The 1 and 2 standard deviation ellipses were estimated by fitting a multivariate normal distribution to the bootstrap data. The results show that a double-couple (DC) mechanism lies within the uncertainty.

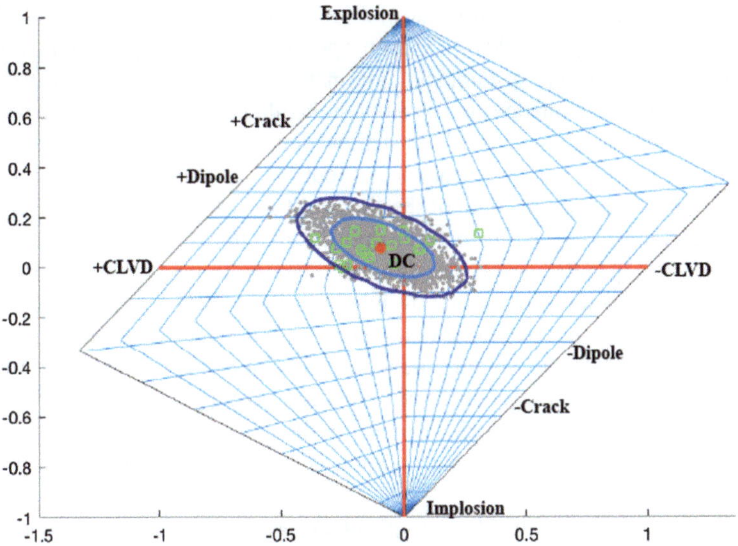

Fig. 4 Source-type plot of solution uncertainty for EVT1 calculated in different ways. Bootstrap of 2000 realizations of residuals from the best fit model (gray circles). The green squares show the 25 permutations of 4-, 3- and 2-station subsets of stations from the original 5-station set of three-component waveform data. The best fitting 5-station solution is shown as the red circle. 1 and 2 standard deviation fits of a normal distribution to the bootstrap results are shown as the two ellipses

Another method to compute uncertainties is apply to Jackknife resampling of the data used in the inversion. This is done by considering all permutations of 4-, 3- and 2-station inversions sampled from the original 5-station set of observations. The Jackknife results show a similar spread in solutions, but it is interesting that the trend is different than for the bootstrap.

For EVT1 based, on the F-test result, the bootstrap and Jackknife test distributions and the plot of the best-fit surface this event is identified as being principally double-couple. It would be possible to estimate the uncertainties in other parameters such as the scalar seismic moment, strike, rake and dip of the double-couple portion of the moment tensor if such information would be useful for a given study.

3 Examples of Non-double-couple Events

3.1 Non-double-couple Earthquake at Long Valley Caldera

In June of 1997 a period of unrest began at the Long Valley Caldera in eastern California, which continued into early 1998 (e.g. Dreger et al. 2000). The unrest included an increase in the inflation rate of the resurgent dome, the numbers of M > 1.2 earthquakes increasing 20-fold to 100 events per week, a doubling of the flux of CO_2 emitted from the volcanic system, and culminating in November 1997 with a large, two-week dilational strain detected by a borehole dilatometer interpreted to be due to the emplacement of a magmatic dike at shallow depth (Dreger et al. 2000). The strain transient was accompanied by a sequence of four M > 4.5 events. EVT1 was a part of this unrest, but was located south of the 'South Moat' of the Caldera and is likely a tectonic earthquake responding to the change in stress due to the resurgent activity in the Caldera. In Fig. 5 the waveform fit and moment tensor solution are compared for one of the events that occurred on November 30, 1997 (EVT2, Table 1). This event appears to have a large volumetric component with about 40% of the total seismic moment corresponding to volume increase. The Source-Type-Inversion (Fig. 5b) shows a distribution of the goodness of fit that is markedly different than for EVT1, and in fact the F-test applied to EVT2 indicates that the additional degree of freedom afforded by the full moment tensor solution is statistically significant at the 99% level.

Figure 6 shows the residual bootstrap and the Jackknife test results in the source-type space for EVT2. The gray circles show the elliptical distribution of solutions from 2000 residual bootstraps. The distribution indicates that the volumetric component is well recovered, but that there is considerable uncertainty in the ratio of double-couple and CLVD which ranges from 25–90% double-couple.

The Jackknife test was carried out by performing inversions for all permutations of 4-, 3-, and 2-station sets from the total 5 stations used to estimate the moment tensor. The Jackknife permutations, the green squares in Fig. 6, are consistent with the bootstrap results in that the volumetric component is well resolved, but the proportion of double-couple to CLVD is poorly resolved and is observed to range from 50–100% double-couple. In Fig. 7 the deviatoric and full moment tensor solutions for one of the 3-station sets shows that the additional degree of freedom in the full moment

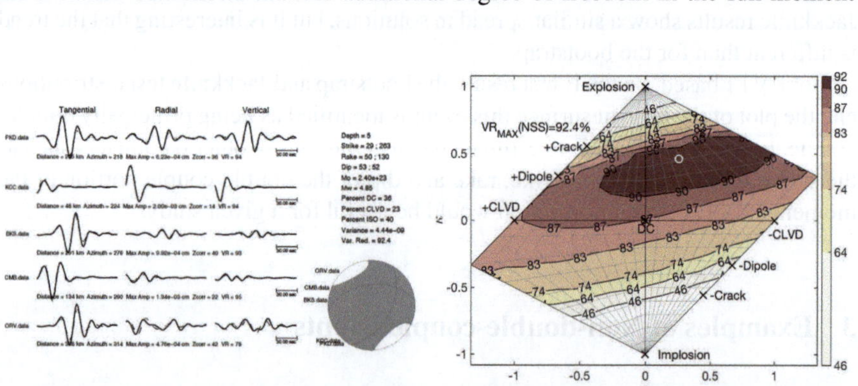

Fig. 5 (left) Observed (solid) and synthetic (dashed) displacement waveforms bandpass filtered between 0.02 and 0.05 Hz using an acausal Butterworth filter. (right) Source-type inversion result. The white circle shows the best fit solution and the color shows the variance reduction goodness of fit parameter

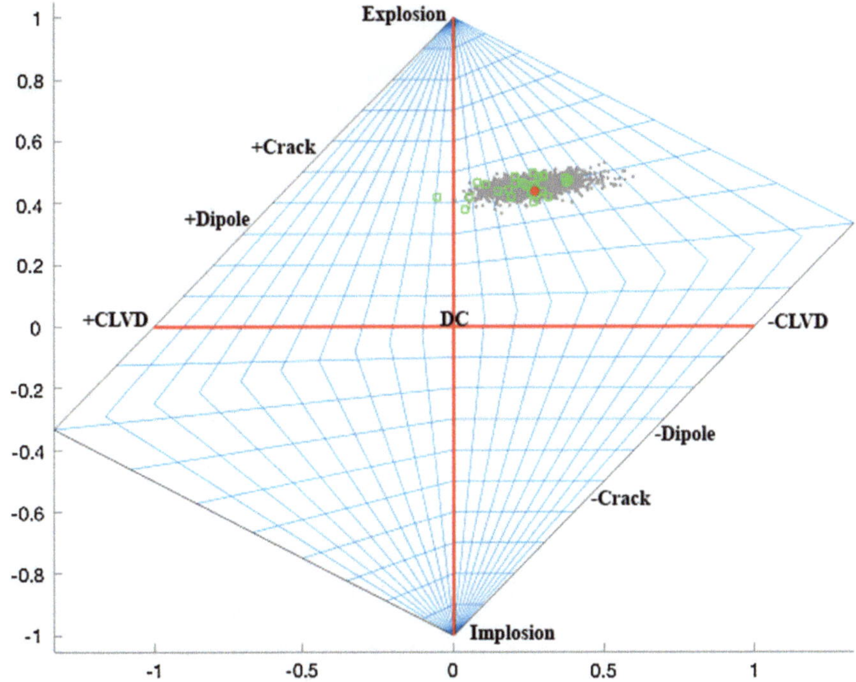

Fig. 6 Bootstrap of 2000 realizations of residuals from the best fit model (gray circles). The green squares show the 25 permutations of 4-, 3- and 2-station subsets of stations from the original 5-station set of three-component waveform data. The best fitting 5-station solution is shown as the red circle

tensor solution not surprisingly improves the fit for the 3-stations (KCC, BKS and CMB) used in the inversion, but also importantly the full moment tensor solution significantly improves the fit at the two unused (PKD and ORV) stations. The same behavior is observed for other permutations of stations. This type of cross-validation of results provides a heightened level of credibility of the recovered volumetric component of the seismic moment tensor.

It is notable that the obtained solution, including uncertainty, is not located near the location of a shear-tensile crack model, in which the intermediate axis of the shear failure lies in the tensile plane and tensile opening is coplanar. Such solutions would lie near the location of the positive crack in the source-type space, and extend along the line to the double-couple for all combinations of geometry and weighting of shear to tensile failure (e.g. Julian et al. 1998). It is possible that the solution is representative of a more complex shear-tensile failure in which the tensile fault is canted by approximately 40–60° from the intermediate axis of the shear plane (e.g. Julian et al. 1998), however it is not possible to resolve a precise mechanism from the moment tensor solution since by nature the decomposition of the moment tensor is non-unique.

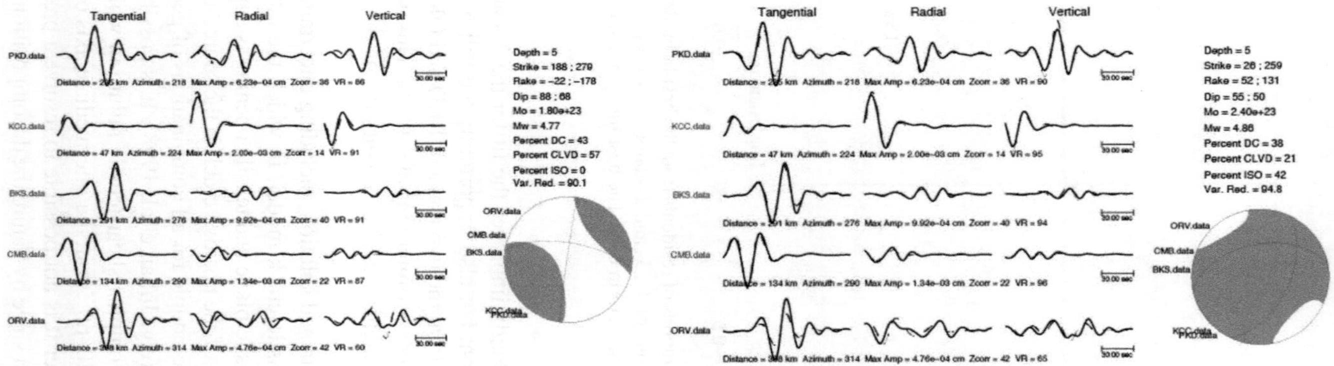

Fig. 7 (left) Deviatoric inversion of the three stations identified in red (KCC, BKS and CMB). The other two stations (PKD and ORV) show forward predictions of waveforms for the 3-station inversion result. (right) The full moment tensor inversion shows a significant improvement in fit to the three stations used in the inversion, as well as for PKD and ORV, which were not inverted and are an independent forward test of the 3-station moment tensor solution

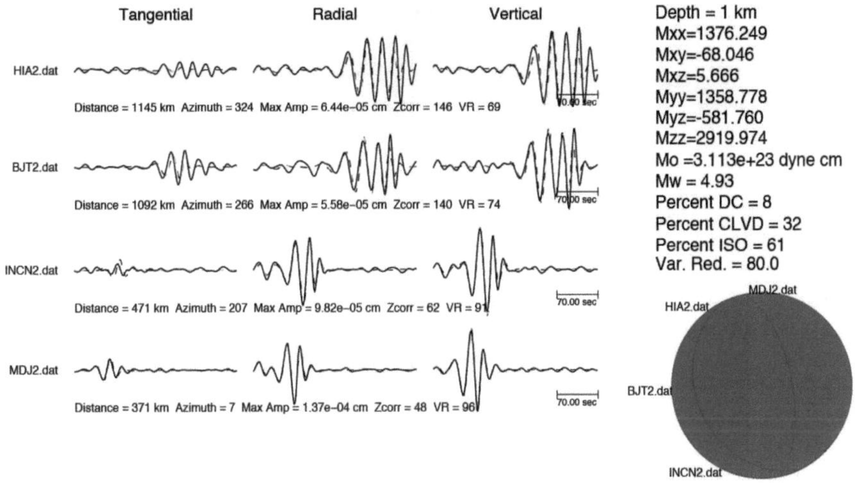

Fig. 8 Regional distance long-period displacement (0.02–0.05 Hz) moment tensor inversion for the September 8, 2016 DPRK nuclear test. The best fitting moment tensor solution is dominantly isotropic with 61% of the total moment explosive. The Bowers and Hudson (1999) total moment was used to compute Mo and Mw. The distance, azimuth, and maximum amplitude of the traces is given

3.2 *September 9, 2016 DPRK Nuclear Test*

Regional distance seismic moment tensor analysis has been shown to effectively discriminate nuclear explosions from earthquakes (Ford et al. 2009 and Fig. 2). Ford et al. (2012) showed that positive identification of the explosive source signature can be obtained by simultaneous inversion of regional waveforms and teleseismic P-waveforms. Chiang et al. (2014) showed that incorporation of regional distance first motions with regional long-period complete waveforms can effectively identify explosions using the source-type inversion method.

On September 9, 2016, the Democratic People's Republic of Korea (DPRK) carried out its 5th nuclear test, EVT3 (Table 1). Applying the complete three-component waveform inversion assuming a focal depth of 1 km results in the solution shown in Fig. 8. The primary wave type contributing to the solution are Rayleigh waves, particularly at the more distant stations (HIA 1145 km, BJT 1092 km). The closer stations INCN and MDJ also show long period Pnl energy which helps to constrain the mechanism. All four stations show moderate levels of Love wave radiation indicating a component of tectonic release in the event, or conversion of P to S energy through wave propagation.

The residual bootstrap covers a broader region of the source-type space (Fig. 9) than shown for the two previous earthquakes. This is due to the relatively lower signal to noise, and the corresponding larger residuals. Most of the 2000 realizations are located close to the best solution, which is squarely in the region found for other

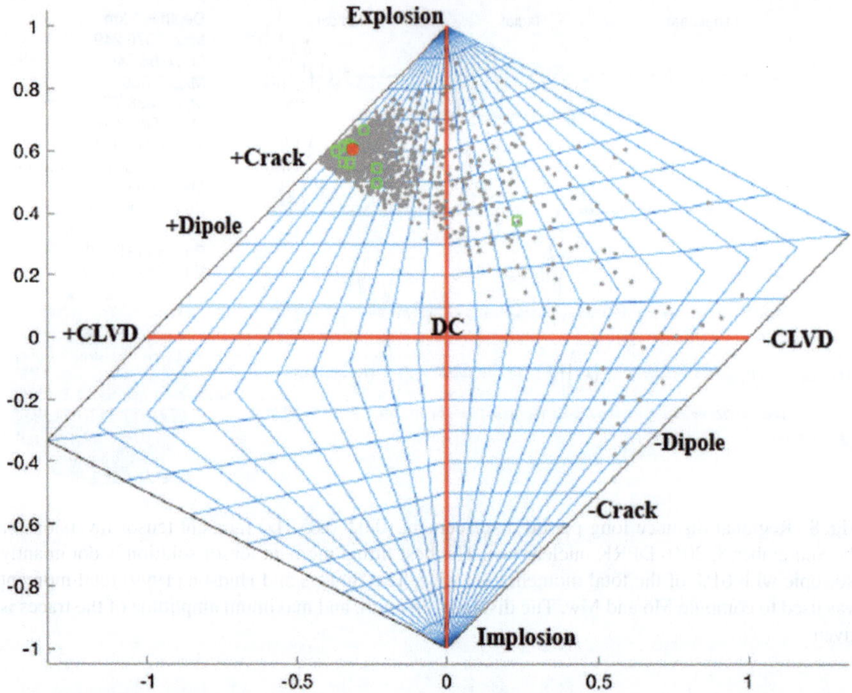

Fig. 9 Bootstrap of 2000 realizations of residuals from the best fit model (gray circles) for the September 8, 2016 DPRK nuclear test. The green squares show the 25 permutations of 3-, and 2-station subsets of stations from the original 4-station set of three-component waveform data. The best fitting 4-station solution is shown as the red circle

nuclear tests (e.g. Fig. 2; Ford et al. 2009), however there are a number of solutions which trend toward the deviatoric line and a—CLVD. The distribution is clearly non-Gaussian in contrast to what was found for EVT1 and EVT2. This illustrates a danger of using strictly Gaussian statistics to evaluate the uncertainty of moment tensor inversion results, which in this case would substantially underestimate the uncertainty in source-type.

In fact, the source-type-inversions (Nayak and Dreger 2015) show (Fig. 10) that for the waveform observations only there is the well-known theoretical tradeoff between pure explosive mechanisms and a vertically oriented, compressive CLVD (e.g. Ford et al. 2012). This tradeoff results from the radiation patterns of surface waves being the same for explosions and a—CLVD with a vertically oriented major vector-dipole in compression. For both types of sources, and combinations of them the Rayleigh wave has an isotropic phase with azimuth, and there is no excitation of Love waves. Given the large propagation distances, the shallow depth of the source, and the dominance of surface wave energy in the waveforms the focal sphere is essentially sampled at close to horizontal takeoff angles. Thus, there is a broad region of the source type space that can fit the surface waves equally well, which

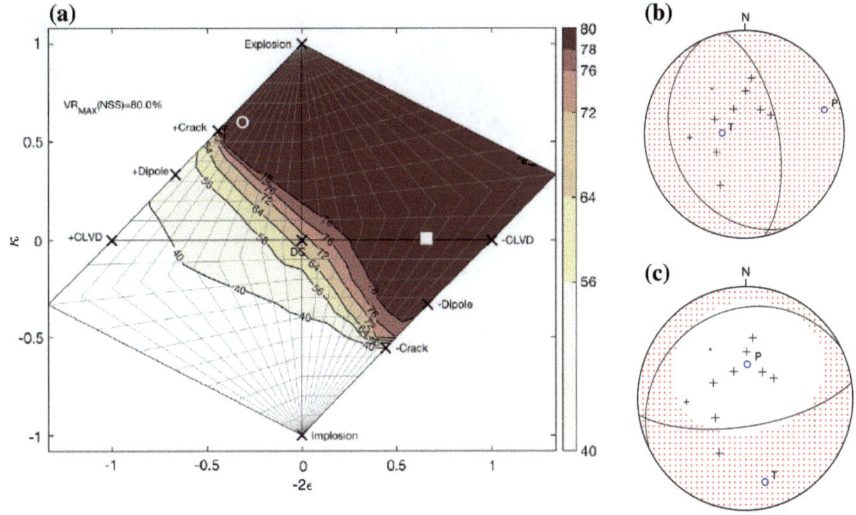

Fig. 10 **a** Source-type inversion of long-period waveforms only. Color shows the goodness of fit parameter (100 being perfect fit). The white circle shows the best solution and the square shows the best deviatoric solution. **b** Comparison of first-motion polarities (black plusses indicating upward first-motion) with the best moment tensor solution (white circle). **c** Comparison of first-motion polarities with the best deviatoric solution (white square). Note that based on the goodness of fit parameter alone the waveform data cannot distinguish between the white circle and square solutions. This is due to the vertical CLVD in compression having an identical surface wave radiation pattern as the pure explosion model

extends to the deviatoric line in the projection. The first-motion polarities, which are all positive are not consistent with the best deviatoric solution (Fig. 10c). Joint Source-Type-Inversion (Nayak and Dreger 2015) of high-frequency P-wave polarity observations from the four regional stations as well as other more distant stations with adequate signal to noise from the GSN network together with the regional long-period waveforms further constrains the source-type space such that the deviatoric line is ruled out and the event is discriminated as being primarily explosive in nature (Fig. 11).

4 Discussion and Conclusions

Local and regional distance seismic moment tensor analysis is a powerful tool for the estimation of source parameters for natural earthquake and man-made seismic signals. Seismic moment tensors can provide insight into the underlying mechanisms producing the seismic radiation, and can be used to distinguish between different types of seismic events (e.g. double-couple earthquakes, cavity collapse, shear-tensile failure, and explosions).

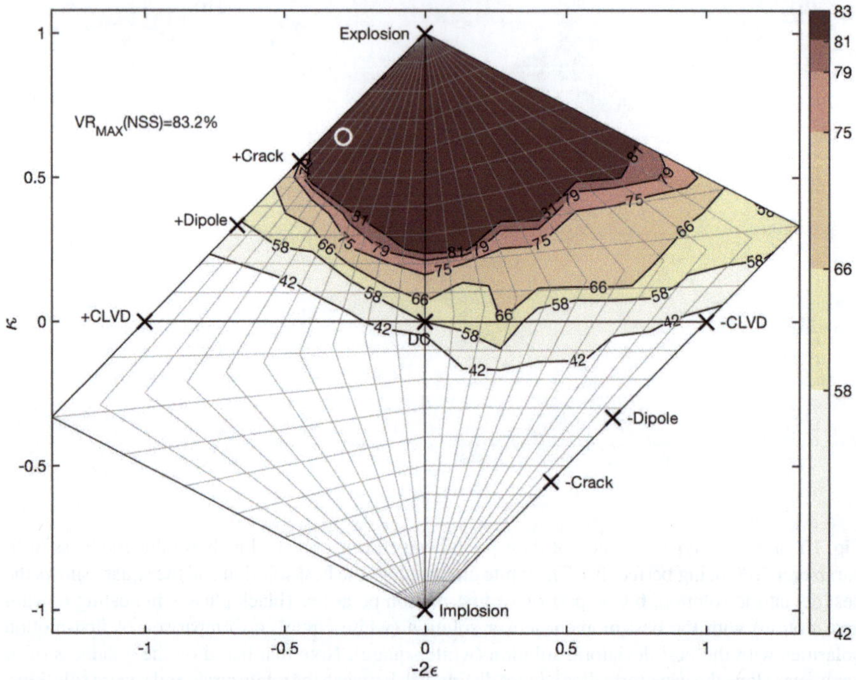

Fig. 11 Source-type inversion results obtained from joint inversion of long-period waveforms and high-frequency first-motion polarities for the September 9, 2016 DPRK nuclear test

At the Berkeley Seismological Laboratory, we have been applying regional distance seismic moment tensor analysis in near-realtime using data from the Berkeley Digital Seismic Network, and over the years have developed tools for evaluating solution uncertainty in cases deviating from the typical tectonic earthquake case. We have found that a battery of methods is needed to assess the resolution and uncertainty of moment tensor solutions, particularly for those cases in which there are large non-double-couple components. For example, the F-test is useful for evaluating if higher degree of freedom inversions, those allowing for either a deviatoric mechanism (double-couple plus CLVD, 5 moment tensor parameters), or a deviatoric mechanism with an isotropic component (6 moment tensor parameters) are statistically significant given the random fluctuations or noise in the data. This test can serve as a first-pass test, and can give robust results where large non-double-couple components are present and where there is a substantial change in the fit to the data when they are included. It can also be applied to events with no volumetric component to assess whether the deviatoric solution (5 moment tensor parameters) is statistically significant compared to a pure double-couple (only 4 source parameters) as shown in Templeton and Dreger (2006). For the examples given in this paper the F-test indicates that the relatively small volume increase of EVT1 is not statistically significant, but that for EVT2 it is highly so. EVT2 is a case where the volume increase component is large (greater than 40% of the total seismic moment) and therefore it is contributing substantially to the synthetic waveform, thereby providing the ability to use the F-test to assess statistical significance. It must be pointed out however that in cases where the volumetric component might not be large (10–20%) but real, the F-test fails as a reliable measure because at such small contributions to the synthetic waveform there will not be a large enough difference in fit between a deviatoric and full moment tensor inversion (e.g. Boyd et al. 2015). In Boyd et al. (2015) it was found that for some Geysers, California earthquakes with waveforms only it was not possible to distinguish between a deviatoric mechanism (CLVD dominated) and the general or full moment tensor inversion yielding a 10–20% isotropic component (volume increase). Boyd et al. (2015) found that when P-wave first-motions were combined with the waveforms the joint dataset required the small volume increase evident in the full moment tensor inversion results, despite the low F-test results for waveforms only. This is similarly the case for EVT3 where the source-type inversion of waveform only data reveals that a solution with a large explosive component is indistinguishable in terms of fit to the surface wave data with a deviatoric solution with no volume increase. In this case the F-test assigns a significance of only 57.9%. On the other hand, both the full and deviatoric solutions when compared to the pure double-couple are statistically significant at 99.9%. Thus it is clear that the event is not a double-couple, however there is considerable range on the deviatoric line where natural events are often located that fits the long-period waveform data well. When both waveforms and first-motions were considered together (Fig. 11) the full moment tensor solution compared to the best deviatoric solution is statistically significant at the 99.9% level.

Thus, it is good practice to evaluate solution uncertainty. We use two methods for doing this. One is to randomly bootstrap the residuals between the data and synthetics for the best fitting solution. Typically, 2000 realizations of such random residuals applied to the data for reinversion are sufficient to map the uncertainty in the moment tensor solution in terms of the scalar seismic moment estimate and derived parameters such as the strike, rake and dip of the major double-couple. The use of the Hudson et al. (1989) source-type diagram was used to illustrate the uncertainty in the solution in terms of the deviatoric parameter T, and the volumetric parameter k. Other projections could have been used just as well. The three examples provided show that for EVT1 the solution is indistinguishable from a pure double-couple mechanism whereas for both EVT2 and EVT3 they are, where a pure double-couple does not lie within the distribution mapped by the bootstrap inversions. The case for EVT3 illustrated that the uncertainty may not behave as Gaussian and therefore it is probably better to use techniques that allow for non-Gaussian distributions. The second approach is to use a Jackknife of subsets of the data used in the inversion. For a waveform inversion, this method is useful for potentially revealing station specific bias, or possibly path bias for 1 or more source-station paths. It can be a useful tool for identifying where there might be problems in the assumption of the velocity model used to compute the Green's functions for the inversion. In the three cases presented in this paper the Jackknife tests were generally consistent with the uncertainty mapped from the residual bootstrap, and no station or path biases were found.

While the estimation of uncertainties using the two standard approaches is useful, there can still be problems if there may be a significant bias, non-uniqueness or tradeoff in the results. To identify such potential problems we find that maps of the goodness of fit parameter in the source-type space can be especially useful (e.g. Ford et al. 2010; Nayak and Dreger 2015). For the three events presented in this paper the mapped goodness of fit surface in the source-type space are all different, indicating that there are fundamental differences in the source-types of the events. EVT1 is consistent with a double-couple mechanism, or a more typical deviatoric moment tensor consisting of a double-couple and minor CLVD as observed for many earthquakes. EVT2 and EVT3 are clearly different and the double-couple and deviatoric moment tensor solutions do not adequately fit the data. However, all three best-fit maps show that the respective regions of acceptable solutions is typically larger than the regions delineated with either bootstrapped residuals or Jackknife variability. The best-fit map for EVT3 clearly shows a broad region of acceptable solutions in terms of waveform fit that are combinations of either a dominant volume increase or—CLVD, where the major vector dipole of the CLVD is in compression and vertically oriented. Both of these mechanisms generate no Love waves (which are very weak for the DPRK test), and Rayleigh waves with azimuthally independent phase. The Jackknife test fails, and the bootstrap only very weakly captures evidence of this tradeoff with 40 or so points out of the 2000 tests, only about 2% of the cases (Fig. 9). This poses a significant problem for the source-type identification or discrimination of nuclear explosions. However, the nature of the tradeoff, notably the vertical orientation of the—CLVD, fortunately offers the possibility of using teleseismic P-waveforms (Ford et al. 2010), or regional and teleseismic first-motion observations (Chiang et al. 2014) to sample the focal sphere at takeoff angles that eliminates the tradeoff allowing for the identification of an event as primarily explosive in nature.

Acknowledgements This paper is an overview of two decades of work by my students. In particular I acknowledge the efforts Sarah Minson, Hrvoje Tkalcic, Dennise Templeton, Sean Ford, Aurelie Trillia, Sierra Boyd, and Avinash Nayak in the development of the various methods that are now in routine usage. Figures were made with GMT Wessel and Smith (1991).

References

Aso N, Ohta K, Ide S (2016) Mathematical review on source-type diagrams. Earth Planets Space 68:52. https://doi.org/10.1186/s40623-016-0421-5

Bowers D, Hudson JA (1999) Defining the scalar moment of a seismic source with a general moment tensor. Bull Seism Soc Am 89(5):1390–1394

Boyd OS, Dreger DS, Lai VH, Gritto R (2015) A systematic analysis of seismic moment tensor at the geysers geothermal field, California. Bull Seism Soc Am 105(6). https://doi.org/10.1785/0120140285

Boyd OS, Dreger DS, Gritto R, Garcia J, Hartline C (2017) Analysis of seismic moment tensors and in-situ stress during enhanced geothermal system development at The Geysers geothermal field, California. (submitted to) Geophys J Int
Chiang A, Dreger DS, Ford SR, Walter WR (2014) Source characterization of underground explosions from combined moment tensor and first motion analysis. Bull Seism Soc Am 104(4):1587–1600
Dreger DS, Tkalcic H, Johnston M (2000) Dilational processes accompanying earthquakes in the Long Valley Caldera. Science 288:122–125
Ford S, Dreger D, Walter W (2008) Source characterization of the August 6, 2007 Crandall Canyon mine seismic event in central Utah. Seism Res Lett 79:637–644
Ford SR, Dreger DS, Walter WR (2009) Identifying isotropic events using a regional moment tensor inversion. J Geophys Res 114:B01306. https://doi.org/10.1029/2008JB005743
Ford SR, Dreger DS, Walter WR (2010) Network sensitivity solutions for regional moment tensor inversions. Bull Seism Soc Am 100:1962–1970
Ford SR, Walter WR, Dreger DS (2012) Event discrimination using regional moment tensors with teleseismic-P constraints. Bull Seism Soc Am 102:867–872. https://doi.org/10.1785/0120110227
Fukuyama E, Dreger D (2000) Performance test of an automated moment tensor determination system for the future "Tokai" earthquake. Earth Planets Space 52:383–392
Guilhem A, Dreger DS (2011) Rapid detection and characterization of large earthquakes using quasi-finite-source Green's functions in continuous moment tensor inversion. Geophys Res Lett 38(L13318). https://doi.org/10.1029/2011gl047550
Hudson JA, Pearce RG, Rogers RM (1989) Source type plot for inversion of the moment tensor. J Geophys Res 9(B1):765–774
Herrmann RB (2013) Computer programs in seismology: an evolving tool for instruction and research. Seism Res Lett 84:1081–1088. https://doi.org/10.1785/022011096
Jost ML, Herrmann RB (1989) A student's guide to and review of moment tensors. Seismol Res Lett 60(2):37–57
Julian BR, Miller AD, Foulger GR (1998) Non-double-couple earthquakes (1) theory. Rev Geophys 36(4):525–549
Miller AD, Foulger GR, Julian BR (1998) Non-double-couple earthquakes (2) observations. Rev Geophys 36(4):551–568
Minson SD, Dreger RB, Kanamori H, Larsen K (2007) Seismically and geodetically determined non-double-couple source mechanisms from the 2000 Miyakejima volcanic earthquake swarm. J Geophys Res 112:B10308. https://doi.org/10.1029/2006JB004847
Minson S, Dreger D (2008) Stable inversions for complete moment tensors. Geophys J Int 174:585–592
Nayak A, Dreger DS (2014) Moment tensor inversion of seismic events associated with the sinkhole at Napoleonville Salt Dome, Louisiana. Bull Seism Soc Am 104:1763–1776
Nayak A, Dreger DS (2015) Source-type specific inversion of moment tensors. Bull Seism Soc Am 105:2987–3000. https://doi.org/10.1785/0120140334
Pasyanos ME, Dreger DS, Romanowicz B (1996) Towards real-time determination of regional moment tensors. Bull Seism Soc Am 86:1255–1269
Romanowicz BD, Dreger MP, Urhammer R (1993) Monitoring of strain release in central and northern California using broadband data. Geophys Res Lett 20:1643–1646
Tape W, Tape C (2012) A geometric setting for moment tensors. Geophys J Int 190:476–498
Templeton DC, Dreger DS (2006) Non-double-couple earthquake in the Long valley volcanic region. Bull Seism Soc Am 96:69–79
Tkalcic H, Dreger DS, Foulger GR, Julian BR (2009) The puzzle of the 1996 Bardarbunga, Iceland, earthquake: no volumetric component in the source mechanism. Bull Seism Soc Am 99:3077–3085. https://doi.org/10.1785/0120080361
Walter F, Dreger DS, Clinton JF, Deichmann N, Funk M (2010) Evidence for near-horizontal tensile faulting at the base of Gornergletscher, a Swiss alpine glacier. Bull Seism Soc Am 100(2):458–472

Walter F, Clinton JF, Deichmann N, Dreger DS, Minson SE, Funk M (2009) Moment tensor inversions of icequakes on Gornergletscher, Switzerland. Bull Seism Soc Am 99-2A. https://doi.org/10.1785/0120080110

Wessel P, Smith WHF (1991) Free software helps map and display data. EOS Trans AGU 72:441

Vavryčuk V (2011) Tensile earthquakes: theory, modeling, and inversion. J Geophys Res 116(B12320). https://doi.org/10.1029/2011jb008770

Vavryčuk V (2015) Moment tensor decompositions revisited. J Seism 19(1):231–252. https://doi.org/10.1007/s10950-014-9463-y

Zhu L, Helmberger DV (1996) Advancement in source estimation techniques using broadband regional seismograms. Bull Seism Soc Am 86(5):1634

Estimating Stability and Resolution of Waveform Inversion Focal Mechanisms

S. Scolaro, C. Totaro, D. Presti, Sebastiano D'Amico, G. Neri and B. Orecchio

1 Introduction

The main aim of this study is to describe several tools for testing the stability and resolution of waveform inversion focal mechanisms already successfully adopted for crustal earthquakes occurred in the Calabrian Arc region, southern Italy (Fig. 1). It is well known that focal mechanism quality can decrease for low magnitude earthquakes and that, on overall, several factors can influence the results of seismic waveform inversion, for example seismic network coverage, earth model uncertainties and inaccurate earthquake location (Valentine and Trampert 2012; Chen et al. 2013; Brandmayr et al. 2013; Silwal and Tape 2016). Earthquake focal mechanisms can be considered as a primary tool for studying the interactions between earthquakes, seismic faults, and active tectonics (Anderson et al. 1993). Information coming from focal mechanisms is unfortunately confined to the time period of instrumental record-

S. Scolaro · C. Totaro · D. Presti · G. Neri · B. Orecchio
Department of Mathematics, Computer Sciences, Physics, and Earth Sciences, University of Messina, Viale F. Stagno D'Alcontres, 31, 98166 Messina, Italy
e-mail: silscolaro@unime.it

C. Totaro
e-mail: ctotaro@unime.it

D. Presti
e-mail: dpresti@unime.it

G. Neri
e-mail: geoforum@unime.it

B. Orecchio
e-mail: orecchio@unime.it

S. D'Amico (✉)
Department of Geosciences, University of Malta, Msida MSD 2080, Malta
e-mail: sebastiano.damico@um.edu.mt

S. D'Amico (ed.), *Moment Tensor Solutions*, Springer Natural Hazards,
https://doi.org/10.1007/978-3-319-77359-9_5

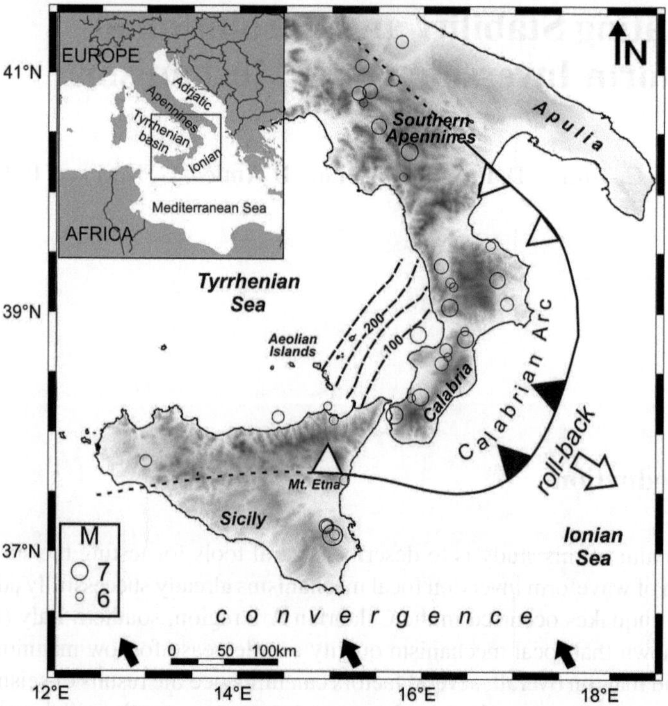

Fig. 1 Map view of southern Italy. The solid curve with the sawtooth pattern indicates the present-day location of the Ionian subducting system. According to recent literature, black sawteeth indicate the continuous subducting slab while white sawteeth the plate boundary segments where slab has already undergone detachment (see, among others, Neri et al. 2009, 2012; Orecchio et al. 2014). The white arrow shows the sense of the subducting slab rollback. The black arrows indicate the present motion of Africa relative to Europe (Nocquet 2012 and references therein). Circles show the locations of the earthquakes of magnitude 6.0 and larger that have occurred after 1000 A.D. according to the CPTI15 catalog (Rovida et al. 2016; http://emidius.mi.ingv.it/CPTI15). Thick dashed lines are depth contour lines of the Wadati-Benioff zone (Faccenna et al. 2011). In the upper right inset we report the study area (black box) in the wider regional framework

ings and processing of the historical seismograms (i.e. early XX century seismograph recordings) is often difficult (Batlló et al. 2008; Palombo and Pino 2013). Reliable source information is generally available for earthquakes of M $\geq$ 5.5 that occurred after the inception of the World-Wide Standardized Seismic Network (WWSSN) in the early 1960s.

The methods traditionally used to compute focal mechanism solutions are based on the polarity of P-wave first motion. First-motion focal solutions reflect only the initial stages of faulting and strongly suffer from both uncertainty on velocity models used to reconstruct the wave path and inadequate azimuthal coverage of seismic networks (Lay and Wallace 1995; Pondrelli et al. 2006; Scognamiglio et al. 2009; D'Amico et al. 2010; Presti et al. 2013). Moreover, errors in first-motion observations may occur because of station polarity reversals or incorrect direct P-arrival picks due to low signal-to-noise ratio. Much more powerful methods capable to furnish more stable and reliable focal mechanisms with respect to the traditional techniques are those based on waveform inversion (e.g. CMT, RCMT, TDMT). The Harvard Centroid Moment Tensor database (CMT; http://www.globalcmt.org) provides robust and reliable seismic source mechanisms through the inversion of long period ($T > 45$ s) body-waves and very-long period ($T > 132$ s) surface waves recorded at the global scale for earthquakes occurred since 1976 with $M_w > 4.5$ (Ekström et al. 2012). The European-Mediterranean Centroid Moment Tensor (RCMT; http://www.bo.ingv.it/RCMT/) procedure is based on the inversion of intermediate and long period surface waves recorded at regional and teleseismic distances (Pondrelli et al. 2002, 2004, 2006, 2007, 2011). The Time Domain Moment Tensor (TDMT; http://earthquake.rm.ingv.it/tdmt.php) algorithm performs long-period full waveform inversion for local and regional events with magnitude $M_w \geq 3.5$ (Dreger 2003; Dreger and Helmberger 1993; Scognamiglio et al., 2009).

During the last years, our research team made continuous processing and improvements on Calabrian Arc earthquake focal mechanisms in order to increase their reliability and to expand the temporal and magnitude range of focal mechanism databases (Neri et al. 2003, 2004, 2005; D'Amico et al. 2010, 2011, 2013; Presti et al. 2013; Totaro et al. 2013, 2015, 2016). We provided in Totaro et al. (2016) the most updated database including 438 crustal earthquake focal mechanisms for southern Italy (Fig. 2). Most of these focal mechanisms (344) have been computed by using the waveform inversion method Cut and Paste (CAP, Zhao and Helmberger 1994; Zhu and Helmberger 1996). This method have shown to furnish reliable and high-quality focal mechanism solutions also for relatively low-magnitude earthquakes (down to a minimum of ca. 2.6) not reported in the national catalogues and often not well resolved by using P-wave first motions (D'Amico et al. 2010, 2011). Because of their frequent occurrence, these small earthquakes are particularly important for characterizing local tectonics and constraining stress orientations.

The Calabrian Arc (Fig. 1) is the result of the convergence between Africa and Europe in the central Mediterranean (Billi et al. 2011; Faccenna et al. 2004; Rosenbaum and Lister 2004). This area is characterized by very heterogeneous seismotectonic regimes along its length (Cristofolini et al. 1985; Montone et al. 2004; Totaro et al. 2016) and has been the site of destructive earthquakes ($M > 6$) that occurred both in recent and historical times (Galli et al. 2008; Neri et al. 2006). The tectonic framework of the Calabrian Arc is complicated by the presence of a narrow subducting slab beneath Calabria (Neri et al. 2009; Orecchio et al. 2014 Selvaggi and Chiarabba 1995) and of two active volcanic districts: the Mt Etna in eastern Sicily and Aeolian Islands in southeastern Tyrrhenian (Carminati et al. 2010; Peccerillo 2003). In the study area, different lithospheric units with changing thickness, composition and velocity have been detected, even if the exact location of their boundaries, together with the effective role of slow-rate Africa-Eurasia convergence and residual Ionian slab rollback on regional geodynamics are still matter of debate (Carafa et al. 2015;

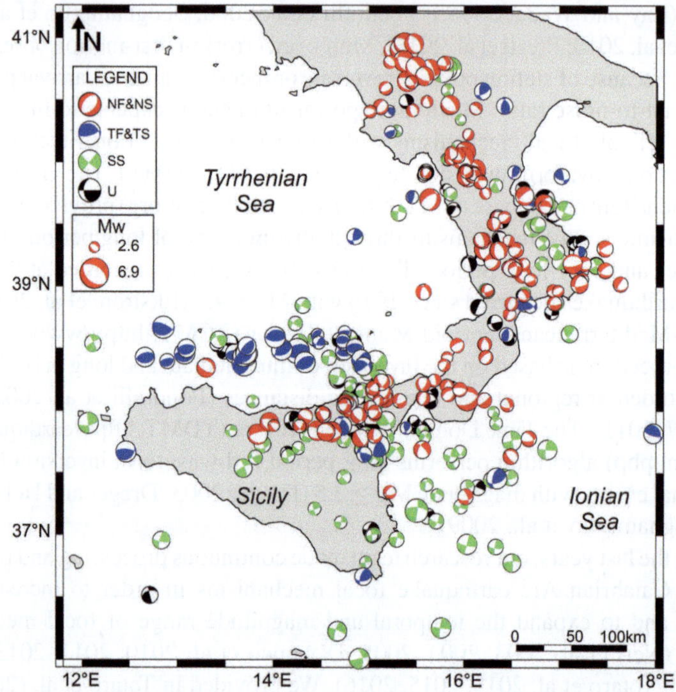

Fig. 2 Crustal earthquake focal mechanism catalogue for the study area. Different colors identify different types of mechanisms following Zoback's (1992) classification based on values of plunges of P and T axes: red = normal faulting (NF) or normal faulting with a minor strike-slip component (NS); green = strike-slip faulting (SS); blue = thrust faulting (TF) or thrust faulting with a minor strike-slip component (TS); black = unknown stress regime (U). "U" includes all focal mechanisms which do not fall in the other five categories (Zoback 1992). The beach ball size is proportional to the earthquake magnitude (see legend)

Faccenna et al. 2014; Gallais et al. 2013; Peròuse et al. 2012; Devoti et al. 2008). Well constrained focal mechanisms obtained in the last years provide a key element to study the regional tectonic processes in the Calabrian Arc region characterized by high heterogeneity in terms of seismotectonics and kinematics.

2 Data

Since it is widely accepted that waveform inversion focal solutions in the study area are much better constrained than P onset polarity ones (see, among others, Presti et al. 2013; Scognamiglio et al. 2009; Pondrelli et al. 2006), all the focal mechanisms selected in the present work are waveform inversion solutions (i) computed by the CAP method or (ii) coming from Italian centroid moment tensor (ItCMT, i.e. Centroid Moment Tensors computed from the Italian region http://rcmt2.bo.ingv.it/Italydataset.html), and (iii) time domain moment tensor (TDMT) catalogs (http://cnt.rm.ingv.it/tdmt). In particular, our database (Fig. 2) consists of 438 waveform inversion focal mechanisms coming from catalogues (104 solutions) and computed by the CAP method (334 solutions). Concerning the data coming from catalogues, most of them are from the Italian CMT catalog (time interval 1976-present, $M_w \geq 4$; Pondrelli et al. 2006) obtained by merging the existing global CMTs and European-Mediterranean RCMTs data for the Italian region. For the period 2006–2015 the database also includes focal solutions computed by using the Time-Domain Moment Tensor ($M_w \geq 3.5$). The CAP focal mechanisms have been estimated for earthquakes of magnitude $M_w \geq 2.6$ that originated at depths shallower than 40 km in the study region between January 2006 and October 2015. The CAP method allows to compute reliable and high-quality focal mechanism solutions also for relatively low-magnitude earthquakes ($2.6 \leq M_w \leq 3.5$) usually not reported in the national catalogues and often not well resolved by using P-wave first motions (D'Amico et al. 2010, 2011; Orecchio et al. 2015; Totaro et al. 2016). Thus, it permitted to strongly increase the amount of data available for the southern Italy region. Since the 90% of our dataset is composed by focal mechanisms computed by applying the CAP method, we carried out several analyses aiming to evaluate stability and resolution of the algorithm. In addition, we also performed different tests in order to estimate error on focal mechanism parameters.

3 The CAP Inversion Method

In the CAP method (Zhao and Helmberger 1994, 1996), each waveform is broken up into P_{nl} (P_n followed by train of crust-trapped reflected/converted P-SV) and surface wave segments, which are weighted differently during the inversion procedure. The use of different portion of the waveform increases the stability of the final solution since different phases are sensitive to different parts of crustal structure and have different amplitude decay with distances. The surface waves, although large in amplitudes, are easily influenced by shallow crustal heterogeneities whereas P_{nl} waves are controlled by the averaged crustal velocity structure and are therefore more stable.

In order to invert the data, waveforms are converted in ground velocity and preferred to ground displacement mainly because the majority of the events have magnitude smaller than four and we needed to avoid the influence of long-period noise embedded in ground displacements. Furthermore, working with ground velocity rather than ground displacement reduces the influence of a low frequency site or instrument noise on the deconvolution. The same frequency bands have been used to filter synthetic and observed ground velocities, in detail 0.02–0.1 Hz for surface waves and 0.05–0.3 Hz for P_{nl} waves. All these features make the CAP method effective for earthquakes over a wide range of magnitudes (down to a minimum of 2.6; D'Amico et al. 2010, 2011; Zhu et al. 2006) as also proven by several tests and comparisons (D'Amico et al. 2010, 2011; Tan et al. 2006; Zhao and Helmberger 1994; Totaro et al. 2016).

4 Stability Tests and Resolution Estimates

The use of CAP in the Calabrian Arc region has allowed to estimate focal mechanism solutions also for low magnitude events (down to a minimum of 2.6) and therefore to significantly increase the number of focal mechanisms based on waveform inversion method (i.e., 90% of the dataset comes from CAP inversions). Such a relevant increase has important implications for better constrain local stress conditions and geodynamic interpretations in the study area (Totaro et al. 2016 and references therein). Hence it is necessary to carefully check quality and stability of newly-added waveform inversion solutions estimated by CAP method. Starting from these considerations and by also taking into account concerned literature information (see e.g., D'Amico et al. 2010, 2011; Presti et al. 2013; Orecchio et al. 2014) we present in this study several resolution and stability tests aimed to properly verify the robustness of CAP results.

We report the results of several tests performed on a subset of 5 earthquakes chosen as representative of different network condition, magnitude value, location area and focal depth.

For each earthquake we observed how the moment tensor varies as function of focal depth in order to evaluate its stability around the global misfit minimum. The depth increment in the grid search is 5 km and for each depth we report the best-fit solution obtained by searching over the full space of orientations and magnitudes and the relative misfit value.

Then, we repeated the inversion procedure by considering different seismic network distribution. Seismic network geometry is fundamental in the earthquake analysis. In particular, in our study region the network configuration is often limited by the presence of wide off-shore sectors and the substantial lack of OBS (Ocean Bottom Seismometer) data. This factor can reduce the quality of solutions and therefore we verify CAP results even with not-optimal azimuthal coverage and few available records by using earthquake located both on-shore and off-shore in the Calabrian Arc area.

Also, we investigated the influence of epicentral errors on the waveform inversion. Non-linear earthquake locations performed in the study area using the method by Presti et al. (2004, 2008) and consequent hypocentral error evaluation indicate that the mean epicentral uncertainty is in the range of 4–6 km (Orecchio et al. 2014). Then we forced the epicenter to lie 5 km away from the true location in order to test the solution even taking into account the mean uncertainty on earthquake location.

An important role in waveform inversion procedure is also played by the seismic velocity model that is used for the calculation of Green's Functions basically for

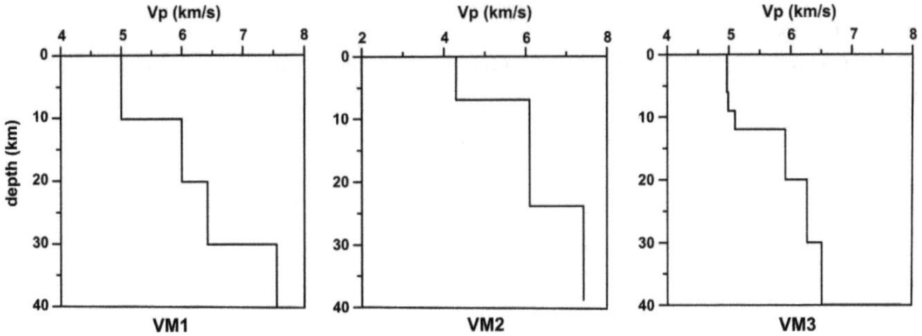

Fig. 3 Different velocity models used to compute the Green's functions for tests on the influence of possible crustal structure heterogeneities in the study area (see also plot c in Fig. 4)

the double-couple mechanism estimates. Each focal mechanism of the most updated CAP catalog (Totaro et al. 2016) has been computed by using a specific 1D velocity model for each target area. To take into account the lithospheric heterogeneities of the Calabrian Arc region we used the most detailed 3D velocity models available from the literature (Barberi et al. 2004; Orecchio et al. 2011; Totaro et al. 2014) to compute theoretical travel times for properly defined target area and to build from these specific 1D velocity models (D'Amico et al. 2011). Even if the time-shift allowed in the CAP algorithm can partly reduce the influence introduced by the velocity model uncertainties, we further verified the stability of the solution with respect to velocity structure by using different velocity models representative of structure heterogeneities of the study region (Fig. 3).

Generally, waveform inversion methods give a standard error for each focal mechanism parameter (e.g., strike, dip and rake) derived from linearized techniques. It provides an important and useful measure of quality of focal mechanism solution but, as shown by several authors, linearized inversion methods tend to underestimate formal errors on focal mechanism parameter (Tan et al. 2006; Bevington and Robinson 2003). We present a procedure aimed to assess more reliable confidence limits of estimated strike, dip and rake. Following the approach described by Stich et al. (2003), we used a grid search for error analysis in the full range of focal parameter space. For each earthquake a set of "artificial" focal mechanisms has been obtained by moving around the best-fit solution in all directions of the focal parameter space with a sampling step of 10°. Then we estimated the misfit for all artificial focal mechanisms and compared these values with the global minimum misfit of the best solution obtained by CAP. In this way it is possible to observe how the misfit value changes with respect to strike, dip and rake, respectively. This comparison can assess the confidence limits and the range of potential alternative solutions over fault plane parameters, allowing us to define the accuracy of the focal mechanism solution. According to Stich et al. (2003), we assumed that the uncertainty region of the solution includes all the artificial focal mechanisms having misfit <10% above the global minimum. By application of this procedure to CAP moment tensor solutions

Table 1 Events used in th study

ID event	DATE yy/mm/dd	TIME hh:mm:ss	Lat (°)	Lon (°)	Depth (km)	Strike	Dip	Rake	M_w
1	20090701	17:58:54	38.34	15.01	2	40	90	19	3.1
2	20111119	10:19:16	16.00	37.81	14	121	70	−25	3.4
3	20140323	18:31:52	37.47	16.48	38	177	61	21	3.6
4	20140708	05:02:43	39.90	16.12	2	347	51	−83	2.9
5	20150329	10:48:46	38.09	16.21	12	52	76	−83	3.5

we are able to estimate that our focal mechanism solutions are characterized by fault parameter errors of the order of 8°–10° (Totaro et al. 2016).

5 Results and Discussion

In Fig. 4 we report the results of the above described tests performed for the 5 events listed in Table 1 and chosen to fairly represent dataset heterogeneities. For each event the epicenter location (black star) and the recording seismic network (triangles) are shown on plots (a). The earthquakes are located both on-shore and off-shore (ID 2, 4 and 1, 3, 5 in Fig. 4, respectively) in the Calabrian Arc area and they are also characterized by different network coverage. We display on plots (b) the best focal mechanism solution in the waveform misfit versus depth curve obtained by the grid search procedure. From plot (b) we can observe that, in general, the focal solution does not change significantly near the minimum misfit value. Only for the earthquake ID 3, located in the Ionian Sea, the curve of waveform misfit is almost flat around the minimum indicating that the network coverage does not provide a tight depth constraint.

As already mentioned, we assess inversion results also by using different station configurations as indicated in the left column of plots (c) of Fig. 4 where we also report the respective best focal mechanism solution. In all cases the focal mechanisms are very similar to that obtained from the inversion with the real station network. Even in this case some differences between the focal mechanisms calculated by test and the true one are visible for earthquake ID 3 and may be probably related to the position of the epicenter with respect to the recording stations. By also taking into account these small differences this test shows the good stability of CAP results even in case of poor station distribution or quite low magnitude. By way of example see, events ID 3, ID 4, and ID 5: for these earthquakes, in fact, the seismic network configuration has an azimuthal gap as large as 180° and also the simulation in extremely bad conditions characterized by only 2 recording stations show very stable solutions. Plots (c) of Fig. 4 also report (right side) the focal mechanisms obtained by varying of 5 km the epicenter location (grey star in plots a). It clearly appears that a mislocation compatible with the hypocenter location uncertainties estimated does not produce

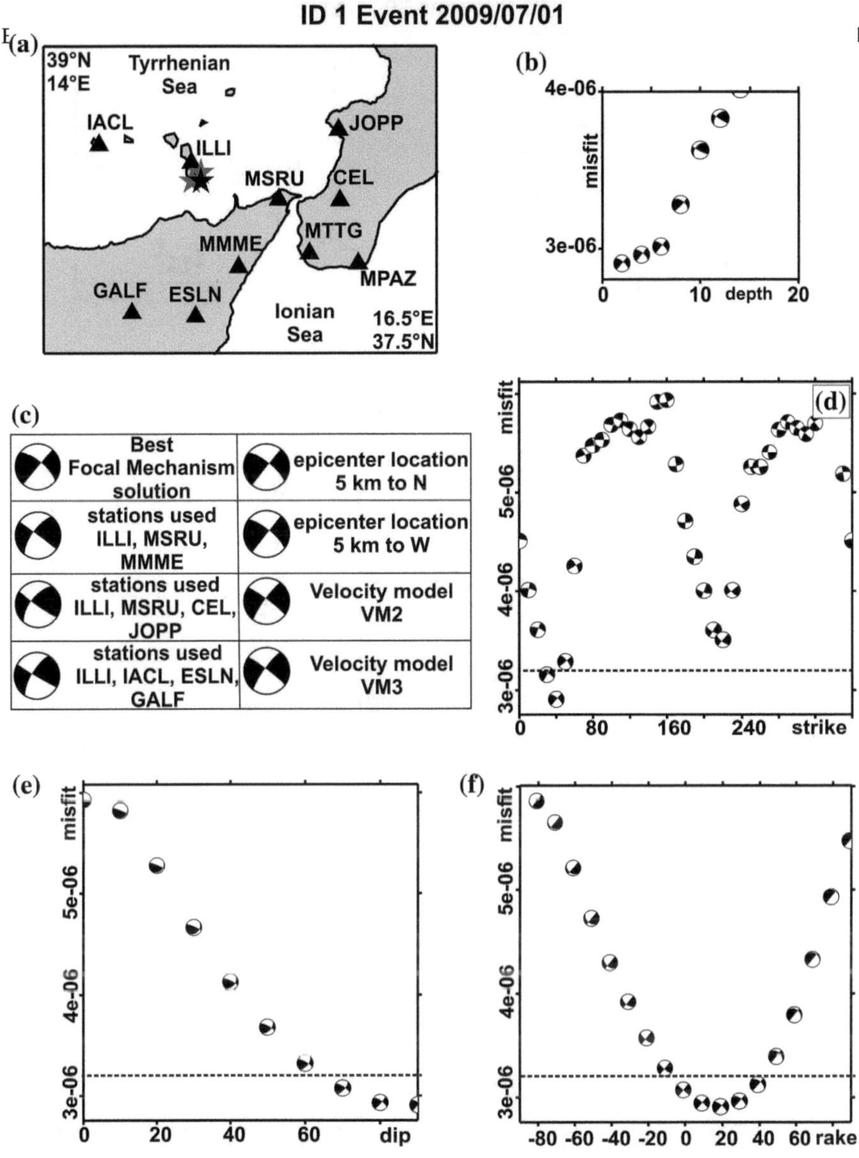

Fig. 4 The figure shows results obtained after tests performed to verify the stability and the error on CAP focal mechanism solutions. Plot **a** map reporting recording stations (black triangles) and epicentral location (black star) used in the waveform inversion procedure together with two biased epicentral locations used for synthetic tests (grey stars) for each event. Plot **b** misfit error as function of depth. Plot **c** reports the best focal mechanisms solution for each earthquake and the results of different tests performed by changing the recording networks (also using very unfavourable conditions), the velocity models for the study area (see Fig. 3) and by forcing the epicenters to lie 5 km away from the "true" locations. Plots **d**, **e**, **f** reports the graph of RMS versus strike, dip and rake, respectively. The dashed line marks the 10% threshold of RMS. The minimum shown in each diagram is the best solution of the event

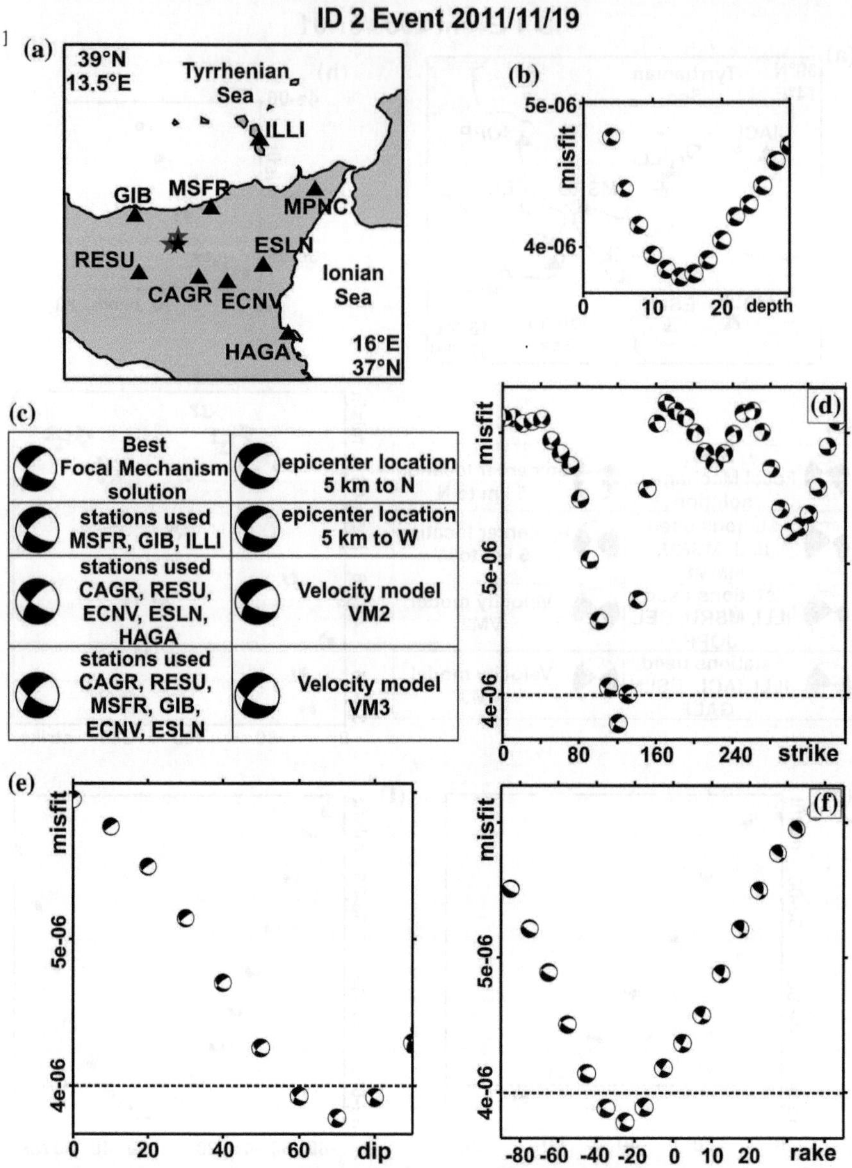

Fig. 4 (continued)

significant differences between true and simulated solution, further supporting the stability of our results. In the same plots we compare the solutions estimated using different 1D velocity models, VM1, VM2 and VM3 (Fig. 3), used for computation of the focal mechanisms. Even in these examples, the high stability of the mechanisms is evident. More pronounced differences affect the results of event ID 3, probably due to inaccuracy of the model VM2 and VM3 for the Ionian Sea area. Plots (d), (e) and (f) of Fig. 4 report the misfit values versus strike, dip and rake respectively. These

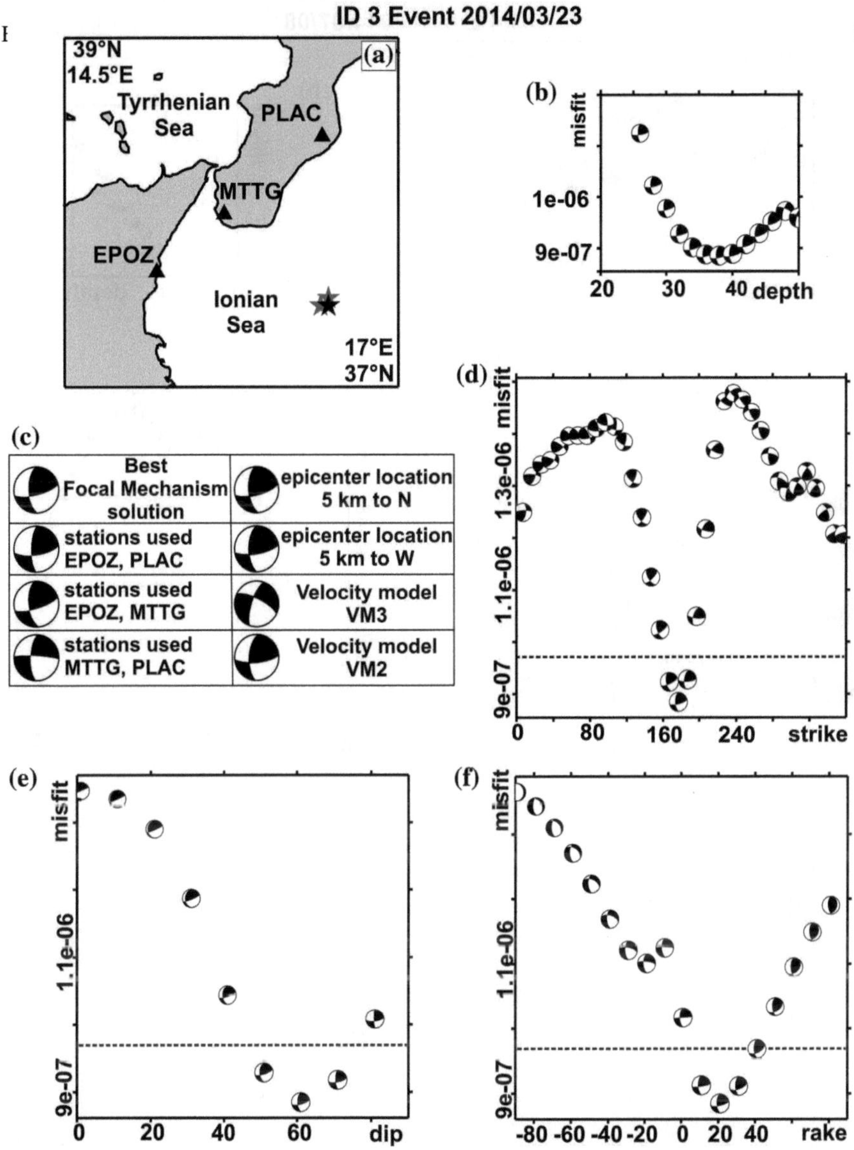

Fig. 4 (continued)

diagrams useful to study the uncertainties of the focal mechanism solution have been built by estimating the misfit values throughout the space of moment tensor orientations. As the uncertainly volume is defined by misfit lower than 10° respect to the best moment tensor solution all the focal mechanisms under the dashed line are acceptable solutions. Usually, relying on focal mechanisms that follow the 10% threshold we can estimate that our focal parameter errors are on average of 8°–10°, like in the examples here reported.

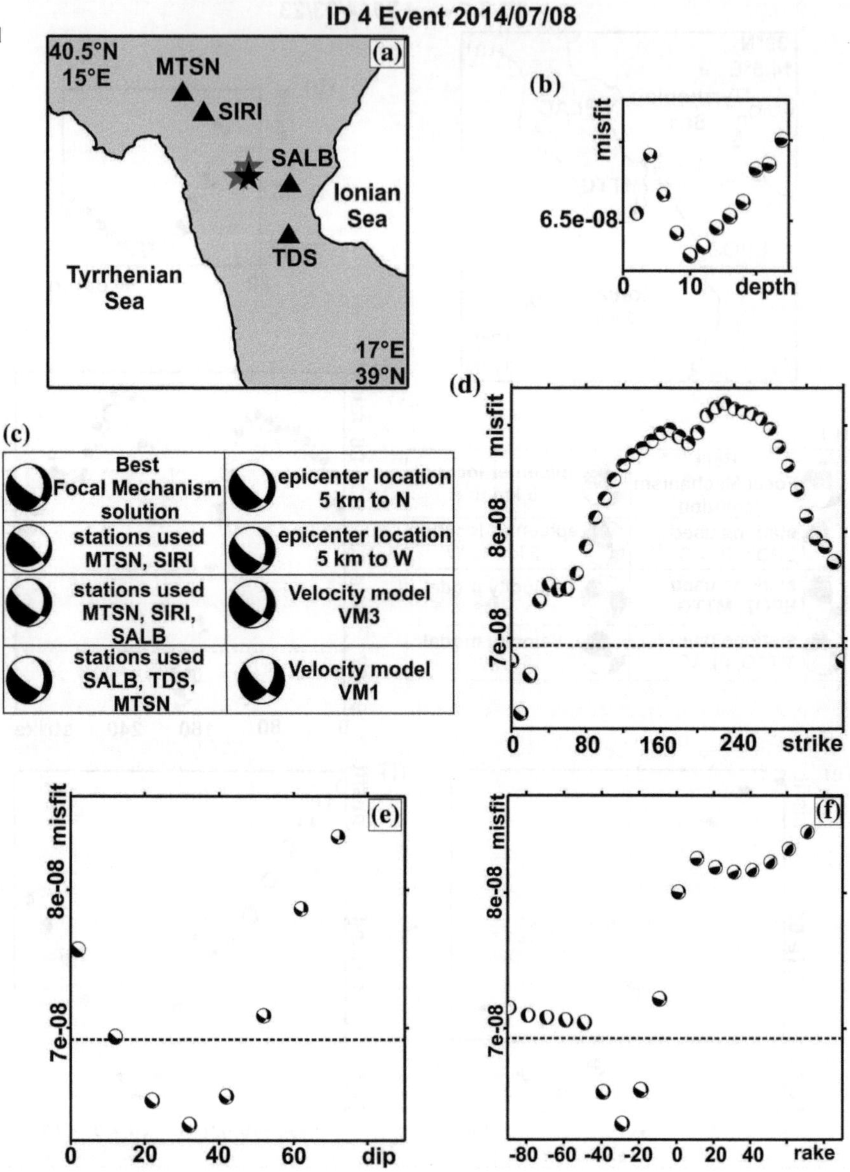

Fig. 4 (continued)

6 Concluding Remarks

In order to check the method robustness, the quality and stability of CAP focal mechanism solutions have been proven by mean of several tests also for low magnitude earthquakes. Following previous investigations carried out in the study region, we performed tests taking into account (1) recording station geometry, (2) different

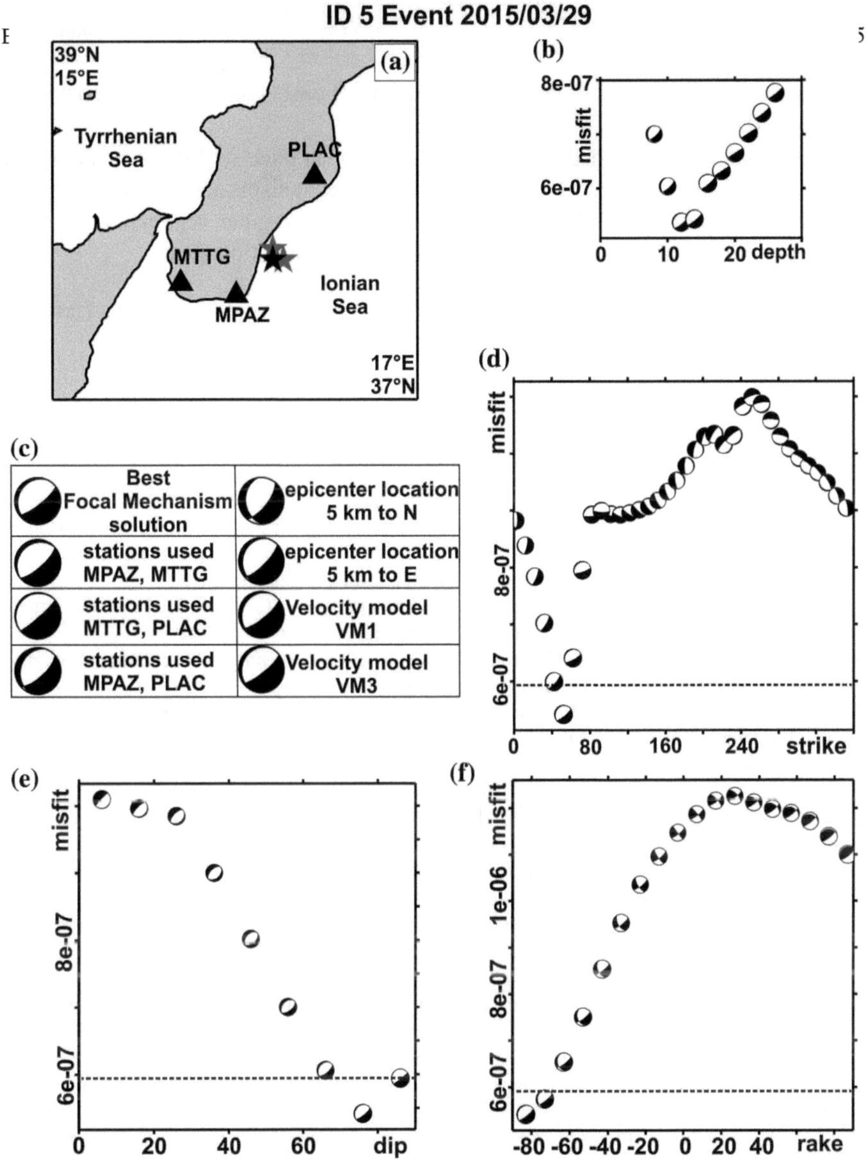

Fig. 4 (continued)

velocity models, (3) the misfit error as a function of depth, (4) epicenter uncertainty, and (5) strike, dip and rake variations as function of waveform misfit. Through these tests we verified that CAP solutions are robustly determined and just a few stations provide enough information to properly constrain the earthquake focal mechanism. Furthermore, the application of CAP method can provide good-quality solutions in a magnitude range (i.e. $2.6 \leq M_w \leq 3.5$) not properly represented in the Italian national catalogues and where the solutions estimated from P-onset polarities are often poorly constrained.

The procedures described in this paper could be applied to different datasets in order to verify the robustness of estimated focal mechanisms and, consequently, to properly improve the knowledge of the seismotectonic regime, regional stress field features as well as the seismic hazard of different investigation areas.

Acknowledgements Some Figures were created using the Generic Mapping Tools (GMT) by Wessel and Smith (1991).

References

Anderson H, Webb T, Jackson J (1993) Focal mechanisms of large earthquakes in the south-island of New Zealand—implications for the accommodation of Pacific-Australia plate motion. Geophys J Int 115:1032–1054. https://doi.org/10.1111/j.1365-246X.1993.tb01508.x

Barberi G, Cosentino MT, Gervasi A, Guerra I, Neri G, Orecchio B (2004) Crustal seismic tomography in the Calabrian Arc region, south Italy. Phys Earth Planet Inter 147:297–314

Batlló J, Stich D, Macià R (2008) Quantitative analysis of early seismograph recordings. In: Historical seismology. Springer, Dordrecht, pp 385–402

Bevington PR, Robinson DK (2003) Data Reduction and error analysis. McGraw-Hill, New York

Billi A, Faccenna C, Bellier O, Minelli L, Neri G, Piromallo C, Presti D, Scrocca D, Serpelloni E (2011) Recent tectonic reorganization of the Nubia-Eurasia convergent boundary heading for the closure of the western Mediterranean. Bull de la Société Géologique de Fr 182(4):279–303

Brandmayr E, Romanelli F, Panza GF (2013) Stability of fault plane solutions for the major N Italy seismic events in 2012. Tectonophysics 608:525–529

Carafa M, Barba S, Bird P (2015) Neotectonics and long-term seismicity in Europe and the Mediterranean region. J Geophys Res Solid Earth 120(7):5311–5342

Carminati E, Lustrino M, Cuffaro M, Doglioni C (2010) Tectonics, magmatism and geodynamics of Italy: what we know and what we imagine. J Virtual Explorer 36. http://dx.doi.org/10.3809/jvirtex.2010.00226

Chen W, Wang D, Wei S (2013) A study on the uncertainties of the centroid depth of the 2013 Lushan earthquake from teleseimic body wave data. Earthq Sci 26(3–4):161–168

Cristofolini R, Ghisetti F, Scarpa R, Vezzani L (1985) Character of the stress field in the Calabrian arc and southern Apennines (Italy) as deduced by geological, seismological and volcanological information. Tectonophysics 117(1):39–58

D'Amico S, Orecchio B, Presti D, Zhu L, Herrmann RB, Neri G (2010) Broadband waveform inversion of moderate earthquakes in the Messina Straits, southern Italy. Phys Earth Planet Inter 179(3–4):97–106

D'Amico S, Orecchio B, Presti D, Gervasi A, Zhu L, Guerra I, Neri G, Herrmann RB (2011) Testing the stability of moment tensor solutions for small earthquakes in the Calabro-Peloritan Arc region (southern Italy). Bollettino di Geofisica Teorica ed Applicata 52(2):283–298

D'Amico S, Lombardo G, Panzera F (2013) Seismicity of the Mediterranean region and mitigation of earthquake losses. Phys Chem Earth 63:1–2. https://doi.org/10.1016/j.pce.2013.07.001

Devoti R, Riguzzi F, Cuffaro M, Doglioni C (2008) New GPS constraints on the kinematics of the Apennines subduction, Earth Planet. Sci Lett 273:163–174

Dreger DS, Helmberger DV (1993) Determination of source parameters at regional distances with three-component sparse network data. J Geophys Res Solid Earth 98(B5):8107–8125

Dreger D (2003) TDMT_INV: time domain seismic moment tensor INVersion. In: International handbook of earthquake and engineering seismology, vol 81B, p 1627

Ekström G, Nettles M, Dziewonski AM (2012) The global CMT project 2004–2010: Centroid-moment tensors for 13,017 earthquakes. Phys Earth Planet Inter 200–201. https://doi.org/10.1016/j.pepi.2012.04.002

Faccenna C, Piromallo C, Crespo-Blanc A, Jolivet L, Rossetti F (2004) Lateral slab deformation and the origin of the western Mediterranean arcs. Tectonics 23(1)

Faccenna C, Molin P, Orecchio B, Olivetti V, Bellier O, Funiciello F, Minelli L, Piromallo C, Billi A (2011) Topography of the Calabria subduction zone (southern Italy): clues for the origin of Mt Etna. Tectonics 30, TC1003. http://dx.doi.org/10.1029/2010TC002694

Faccenna C, Becker TW, Auer L, Billi A, Boschi L, Brun JP, Capitanio FA, Funiciello F, Horvàth F, Jolivet L, Piromallo C, Royden L, Rossetti F, Serpelloni E (2014) Mantle dynamics in the Mediterranean. Rev Geophys 52(3):283–332

Gallais F, Graindorge D, Gutscher MA, Klaeschen D (2013) Propagation of a lithospheric tear fault (STEP) through the western boundary of the Calabrian accretionary wedge offshore eastern Sicily (Southern Italy). Tectonophysics 602:141–152

Galli P, Galadini F, Pantosti D (2008) Twenty years of paleoseismology in Italy. Earth Sci. Rev. 88:89–117

Lay T, Wallace TC (1995) Modern global seismology, vol 58. Academic press

Montone P, Mariucci MT, Pondrelli S, Amato A (2004) An improved stress map for Italy and surrounding regions (central Mediterranean). J Geophys Res Solid Earth 109(B10)

Neri G, Barberi G, Orecchio B, Mostaccio A (2003) Seismic strain and seismogenic stress regimes in the crust of the southern Tyrrhenian region. Earth and Planet Sci Lett 213(1):97–112

Neri G, Barberi G, Oliva G, Orecchio B (2004) Tectonic stress and seismogenic faulting in the area of the 1908 Messina earthquake, south Italy. Geophys Res lett 31(10)

Neri G, Barberi G, Oliva G, Orecchio B (2005) Spatial variations of seismogenic stress orientations in Sicily, south Italy. Phys Earth Planet Inter 148(2):175–191

Neri G, Oliva G, Orecchio B, Presti D (2006) A possible seismic gap within a highly seismogenic belt crossing Calabria and eastern Sicily, Italy. Bull Seismol Soc Am 96(4A):1321–1331

Neri G, Orecchio B, Totaro C, Falcone G, Presti D (2009) Subduction beneath southern Italy close the ending: Results from seismic tomography. Seismol Res Lett 80(1):63–70

Neri G, Marotta AM, Orecchio B, Presti D, Totaro C, Barzaghi R, Borghi A (2012) How lithospheric subduction changes along the Calabrian arc in southern Italy: Geophysical evidences. Int J Earth Sci 101(7):1949–1969. https://doi.org/10.1007/s00531-012-0762-7

Nocquet JM (2012) Present-day kinematics of the Mediterranean: a comprehensive overview of GPS results. Tectonophysics 579:220–242. https://doi.org/10.1016/j.tecto.2012.03.037

Orecchio B, Presti D, Totaro C, Guerra I, Neri G (2011) Imaging the velocity structure of the Calabrian Arc region (South Italy) through the integration of different seismological data. Bollettino di Geofisica Teorica ed Applicata 52:625–638

Orecchio B, Presti D, Totaro C, Neri G (2014) What earthquakes say concerning residual subduction and STEP dynamics in the Calabrian Arc region, south Italy. Geophys J Int 199(3):1929–1942

Orecchio B, Presti D, Totaro C, D'Amico S, Neri G (2015) Investigating slab edge kinematics through seismological data: The northern boundary of the Ionian subduction system (south Italy). J Geodyn 88:23–35

Palombo B, Pino NA (2013) On the recovery and analysis of historical seismograms. Ann Geophys 56(3):0326

Peccerillo A (2003) Plio-quaternary magmatism in Italy. Episodes 26:222–226

Perouse E, Chamot-Rooke N, Rabaute A, Briole P, Jouanne F, Georgiev I, Dimitrov D (2012) Bridging onshore and offshore present-day kinematics of central and eastern Mediterranean: implications for crustal dynamics and mantle flow. Geochem Geophys Geosyst 13(9)

Pondrelli S, Morelli A, Ekström G, Mazza S, Boschi E, Dziewonski AM (2002) European-Mediterranean regional centroid-moment tensors: 1997–2000. Phys Earth Planet Inter 130(1):71–101

Pondrelli S, Morelli A, Ekstrom G (2004) European-Mediterranean regional centroid-moment tensor catalog: solutions for years 2001 and 2002. Phys Earth Planet Inter 145(1–4):127–147

Pondrelli S, Salimbeni S, Ekström G, Morelli A, Gasperini P, Vannucci G (2006) The Italian CMT dataset from 1977 to the present. Phys Earth Planet Inter 159(3):286–303

Pondrelli S, Salimbeni S, Morelli A, Ekström G, Boschi E (2007) European-Mediterranean regional centroid moment tensor catalog: solutions for years 2003 and 2004. Phys Earth Planet Inter 164(1):90–112

Pondrelli S, Salimbeni S, Morelli A, Ekström G, Postpischl L, Vannucci G, Boschi E (2011) European-Mediterranean regional centroid moment tensor catalog: solutions for 2005–2008. Phys Earth Planet Inter 185(3):74–81

Presti D, Troise C, De Natale G (2004) Probabilistic location of seismic sequences in heterogeneous media. Bull Seismol Soc Am 94(6):2239–2253

Presti D, Orecchio B, Falcone G, Neri G (2008) Linear versus non-linear earthquake location and seismogenic fault detection in the southern Tyrrhenian Sea, Italy. Geophys J Int 172:607–618

Presti D, Billi A, Orecchio B, Totaro C, Faccenna C, Neri G (2013) Earthquake focal mechanisms, seismogenic stress, and seismotectonics of the Calabrian Arc, Italy. Tectonophysics 602:153–175

Rosenbaum G, Lister GS (2004) Neogene and Quaternary rollback evolution of the Tyrrhenian Sea, the Apennines, and the Sicilian Maghrebides. Tectonics 23(1)

Rovida A, Locati M, Camassi R, Lolli B, Gasperini P (2016) CPTI15, the 2015 version of the Parametric Catalogue of Italian Earthquakes, Istituto Nazionale di Geofisica e Vulcanologia. http://emidius.mi.ingv.it/CPTI/, https://doi.org/10.6092/ingv.it-cpti15

Selvaggi G, Chiarabba C (1995) Seismicity and P-wave velocity image of the southern Tyrrhenian subduction zone. Geophys J Int 121(3):818–826

Scognamiglio L, Tinti E, Michelini A (2009) Real-time determination of seismic moment tensor for the Italian region. Bull Seismol Soc Am 99(4):2223–2242

Silwal V, Tape C (2016) Seismic moment tensors and estimated uncertainties in southern Alaska. J Geophys Res Solid Earth 121(4):2772–2797

Stich D, Ammon CJ, Morales J (2003) Moment tensor solutions for small and moderate earthquakes in the Ibero—Maghreb region. J Geophys Res Solid Earth 108(B3)

Tan Y, Zhu L, Helmberger D, Saikia C (2006) Locating and modeling regional earthquakes with two stations. J Geophys Res 111:B01306

Totaro C, Presti D, Billi A, Gervasi A, Orecchio B, Guerra I, Neri G (2013) The ongoing seismic sequence at the Pollino Mountains, Italy. Seismol Res Lett 84(6):955–962

Totaro C, Koulakov I, Orecchio B, Presti D (2014) Detailed crustal structure in the area of the southern Apennines-Calabrian Arc border from local earthquake tomography. J Geodyn 82:87–97

Totaro C, Seeber L, Waldhauser F, Steckler M, Gervasi A, Guerra I, Orecchio B, Presti D (2015) An intense earthquake swarm in the southernmost Apennines: fault architecture from high-resolution hypocenters and focal mechanisms. Bull Seismol Soc Am

Totaro C, Orecchio B, Presti D, Scolaro S, Neri G (2016) Seismogenic stress field estimation in the Calabrian Arc region (south Italy) from a Bayesian approach. Geophys Res Lett 43(17):8960–8969

Valentine AP, Trampert J (2012) Assessing the uncertainties on seismic source parameters: towards realistic error estimates for centroid-moment-tensor determinations. Phys Earth Planet Inter 210:36–49

Zhao LS, Helmberger DV (1994) Source estimation from broadband regional seismograms. Bull Seismol Soc Am 84(1):91–104

Zhu L, Helmberger DV (1996) Advancement in source estimation techniques using broadband regional seismograms. Bull Seismol Soc Am 86(5):1634–1641

Zhu L, Akyol N, Mitchell BJ, Sozbilir H (2006) Seismotectonics of western Turkey from high resolution earthquake relocations and moment tensor determinations. Geophys Res Lett 33(7)

Zoback ML (1992) First- and second-order patterns of stress in the lithosphere: the world stress map project. J Geophys Res 97:11703–11728

Data and Sharing Resources

http://www.globalcmt.org/CMTsearch.html
http://rcmt2.bo.ingv.it/
http://cnt.rm.ingv.it/tdmt

The Method of Cataclastic Analysis of Discontinuous Displacements

Yu. L. Rebetsky and A. Yu. Polets

1 Introduction

Studying regularities of the spatial distribution and temporal variations of tectonic stress is one of the most important issues in a number of disciplines of the Earth sciences. In geodynamics, the problem of understanding of the stress state in the Earth's crust and in the lithosphere is associated with the need to explain the mechanism of formation of the tectonic structures of various scale levels. In seismology, this is a problem of studying the formation mechanism of the earthquake source at the afterschock stage and the development of the post seismic relaxation at the aftershock stage. In geology, a stress state helps to establish interrelations of the formation conditions of the complex structures, discontinuous structures (slip faults) and other deformation structures with mineral deposits. In mining and oil production, stress data provide a safe and efficient exploration of natural resources.

The methods of studying the distribution patterns of tectonic stresses in the earth's crust can be divided into methods of mathematical modeling of stress fields and deformations (physical and mathematical modeling) and experimental methods for studying the tectonic stresses in conditions of naturally occurring rocks. The object of our study will be the methods of the study of natural tectonic stresses on the basis of data directly or indirectly characterizing the deformation process of the rock mass.

It should be noted that the main feature of the problem, which solution will be considered below, consists in the determination of scale levels of the studied stresses. It will be about the tectonic stresses with their different averaging scales of the first meters to tens and even hundreds of kilometers. Such range of the scales is defined by the initial data which gives data on the stress state of a local area of a geological massif,

Y. L. Rebetsky (✉)
Tectonophysics, The Schmidt Institute of Physics of the Earth of the Russian Academy of Sciences (IPE RAS), Moscow, Russia
e-mail: reb@ifz.ru

A. Y. Polets (✉)
Seismology, Institute of Marine Geology and Geophysics Far Eastern Branch of the Russian Academy of Sciences (IMGG FEB RAS), Yuzhno-Sakhalinsk, Russia
e-mail: polec84@mail.ru

S. D'Amico (ed.), *Moment Tensor Solutions*, Springer Natural Hazards,
https://doi.org/10.1007/978-3-319-77359-9_6

sedimentary cover rocks, crust and lithosphere. The uniqueness of the considerable problem is precisely that in such a wide range of scales, the analysis of the state of stress of the earth's interior is carried out according to a unified methodology.

Up to the present time, there were two data sources of such multiscale stresses. The first source is geological data on discontinuous shear displacements of the type—the *slickenside of fault* (the development of the methods began with this data). The second source is seismological data on the of earthquake focal mechanisms (earthquake source model–slip fault). After the development of 3D-seismic methods of reservoir geophysics, it allow us to see the massif structure in detail in depth, it is possible to say that the methods of studying tectonic stresses penetrate into the geophysical data sources.

The problem of studying tectonic stresses should be attributed to the junction of two scientific disciplines such as mechanics and geology. On the one hand, this problem requires knowledge in obtaining geological and geophysical data and knowledge of geological and geophysical property data. On the other hand, this problem requires ability to work in terms of mechanics and tensor analysis. It turned out that, the works in the field of study the tectonic stresses made it possible to create specific approaches in the use of stress tensor analysis, which had not previously been applied in classical mechanics and geomechanics.

It is especially noteworthy that such a well-known scientific discipline as geomechanics has never studied the tectonic stresses in such a wide range of scales. Its knowledge area was the mechanical properties at the scale of the samples according to data from laboratory experiments, as well as local data on the stresses in the rock massif obtained through in situ mining (Hydraulic fracturing, strain gauging, well logging, etc.).

The problem of determining the stresses in a natural massif from data on discontinuous displacements is the main inverse problem in tectonophysics. In the methods are being developed, active faults and discontinuities should be considered as a kind of dynamometers, stress tensometers and deformographs, which are generally used in laboratory rock deformation experiments. Thus, the methods for inversion of the tectonic stresses of a megascopic scale are equivalent, in a certain way, to those instruments for studying stresses deformations, which are available to experimenter in the laboratory modeling, but for objects of incommensurably smaller scales.

This chapter presents a Method of Cataclastic Analysis (MCA) of discontinuous displacements for calculation of tectonic stresses and corresponding semi-brittle deformations of rock massif (Rebetsky et al. 2012, 2016; Rebetsky and Tatevossian 2013). This method includes determination of all components of stress tensors and semi-brittle (faulting) strain increments, together with the estimation of strength parameters of brittle rock massif at the scale level, corresponding to the most representative dimension of faults used for the stress inversion. The main accent of this method is made on the difference in the behavior of continuous solid samples and natural massifs, possessing numerous defects, such as, surfaces of decreased strength, in conditions of their deformation.

This method should be considered as the development of the methodology of studying natural tectonic stress fields, which is based on methods of inversion of principal stress axes orientations and seismotectonic deformations, created in the works by E. Anderson, M. V. Gzovsky, Yu. V. Riznichenko, J. Brune, E. Carey, J. Angelier, O. I. Goustchenko, S. L. Yunga, P. N. Nikolaev, V. V. Stepanov, J. Gephart, V. D. Parfenov, L. M. Rastsvetaev, L. A. Sim, S. I. Sherman, and others; methods of estimation of strength (stress values) of parts of the Earth's crust, developed in the papers by R. Sibson, G. Renelli, D. Murphy, and others; and the results of laboratory testing of rock samples, described in the papers by K. Mogi, J. Byerlee, J. Handin, R. Stesky, F. Rummel, A. N. Stavrogin, and others.

It is proposed to construct an algorithm for calculating stresses and deformations of the megascopic scale, based on the initial data characterizing the fact of displacement along the planes of faults and cracks. At the same time, if the construction of the algorithm for calculating residual deformations requires only formalization of the transition from the dislocation type of deformation to the continuum dislocation, then to calculate the stresses we need to take some physical concepts about the deformation process and the possible character of the relationship between deformations and stresses.

It should be taken into account that the calculation of the stress tensors parameters, in the form stated above, predetermines averaging over a large megascopic volume (hundreds of meters, the first tens of kilometers) and time (hours and days for aftershock sequences, months and years for earthquakes, hundreds or thousands of years using geological data on cracks).

Therefore, at the considered averaging scale level, the determined stresses and deformations are in fact the generalized forces and the generalized residual deformations—integrants of actually existing tectonic stresses and deformations in the rock massif.

2 The Main Principles of the Method of Cataclastic Analysis of Discontinuous Displacements

The method of cataclastic analysis of discontinuous displacements is applicable for the calculation of modern stress and paleostress. In the first case, seismological data on the earthquake focal mechanisms are used, and in the second case—geological data on the spatial orientation of the fault planes and direction of slickenline. These data characterize the discontinuous structure and the structural-kinematic state of the rock massif. Since the main distinguishing feature of the MCA is the universality of the algorithm, going forward, we will use a term "structural kinematic data" (SKD) on faults for all types of data.

Further in the text, we will use references to seismological data and the relationship between the earthquakes parameters and the stress-strain state of the Earth's interior. However, if instead of parameters that determine the orientation of the poles of the earthquake nodal planes we substitute parameters, characterizing normal to the fault

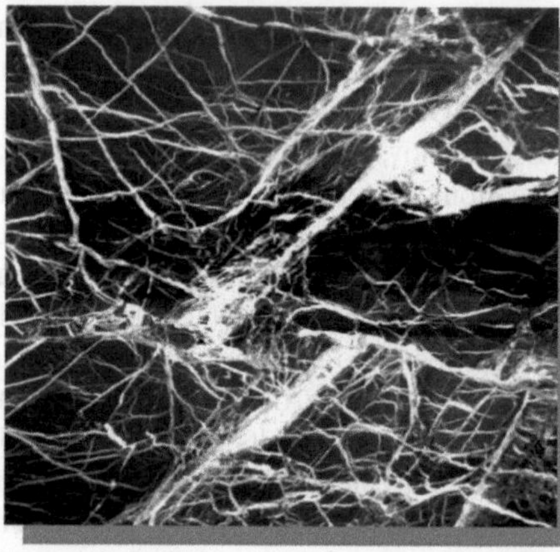

Fig. 1 A natural object showing a large number of closely located existing defects (mineralized cracks)

plane and slickenline vector (rock outcrop), in proper equations, then all the main algorithms of the MCA are stored.

2.1 *Model of the Medium and Process of Semi-brittle Deformation*

2.1.1 An Elastic Body with a Lot of Slip Faults—The Averaging Scale Is Lower Than the Megascopic

In constructing theory of calculating the stress tensor parameters on the SKD set, we will adopt a model of elastic body, containing a set of intercrossing defects in the form of planes of reduced strength. It is *preexisting faults and cracks* (Fig. 1). This body capable of brittle failure if reached a certain limiting conditions and it is a medium model of the average scale of lower than the megascopic. Stresses of this averaging determine the elastic deformations of the body and formation or activation of faults and cracks.

2.1.2 Dualism of Analysis of the Slip Faults Set

Geo-environment has many different (in scale and in genesis) defects in the form of the differently oriented and intercrossed densely packed in the space of reduced strength surfaces, which are formed primarily by *preexisting faults and cracks*, cleavage, stratification and foliation. During the rock deformation process, the transformation of some elastic deformations into irreversible (residual), causing the mechanical energy dissipation in the corresponding volume of the geo-environment, is carried out mostly due to shifts along the already existing surfaces of reduced strength (pre-existing faults).

Beginning with a certain averaging scale of time (determined by the number of ruptures) and space (determined by the most representative linear size of the ruptures), the formation of a new or activation of the previously formed slip fault can be regarded as a *faulting deformation* microact (in time). When many microacts united together, they form the process of *faulting flow* of a rock massif. This deforming process is similar to *plasticity flow*. Thus, depending on the scale of averaging, the fault or crack can be considered both as an act of brittle fracture or activation, and as a faulting deformation microact.

2.1.3 The Semi-brittle Body—The Megascopic Averaging Scale

Consideration of a single slip fault as a micro-temporal act of *semi-brittle deformation* for perfectly elastic models is well founded. For such a model, due to the ideal elasticity, every single displacement act along the fault plane is without formation of residual deformations, and it only leads to a redistribution of the elastic deformations and to release of energy. However, displacements along the set of intersecting slip faults lead to fixation of a certain part of the deformations. In this case, residual deformations and internal self-stresses are present, even after all external loading forces have been removed.

Thus, both elastic and residual deformations will present at the appropriate average scale for the chosen model of the geo-environment. Accumulation of these deformation occurs with increasing of intensity of the external loading and involving the number of defects and newly formed cracks in brittle failure. Such geo-environment can be considered as an *semi-brittle body* on a megascopic scale level (Batdorf and Budyansky 1949), and megascopic stresses determine the character of the development of semi-brittle deformations. On a megascopic scale level, such a medium can be considered as an semi-brittle body, and megascopic stresses determine the way of the semi-brittle deformations development.

2.1.4 Newly Formed Slip Faults

It is suggested, that new faults and cracks should cracks in areas for which *brittle strength limit* is exceeded (Fig. 2). According to the principles of fracture mechanics, the planes of newly formed cracks are being formed subparallel to the axis of the intermediate principal stresses. Their normals deviate from the position of the normals of the pair of planes of maximum shear stresses toward the algebraically larger principal stress axis. Respectively, the plane of the slip faults deviates towards the axis of the deviatoric stresses of the maximum compression.

It is obvious that, for natural rock massif, laying of new surfaces of fault take place with involving of already existing areas of reduced strength. Because of this, the strength of the rock massif should be less than the strength of rocks obtained in

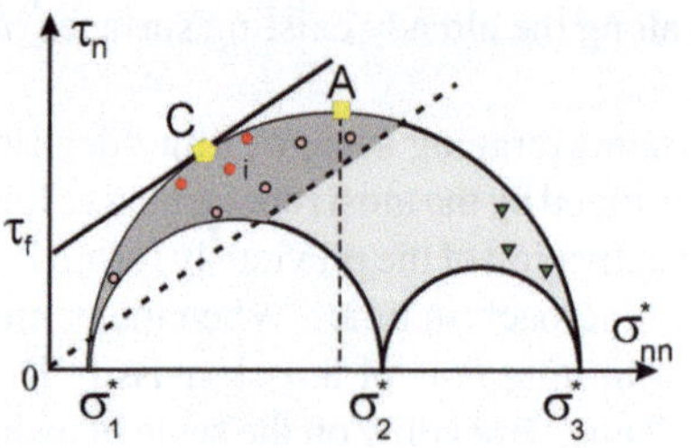

Fig. 2 Mohr diagram and points of stress state for active slip faults (red circles) and active slip faults (green triangles). Yellow pentagon (C) is stress state of new faulting and yellow rectangle (A) is stress state of maximum shear stress plane. Strong line is the brittle strength, broken line is the minimum surface resistance of Coulomb friction (tension stress is positive)

experiments on defect-free samples (Paul 1968; Nikolaevsky 1996). Moreover, this is achieved more by decreasing of the value of internal cohesion of a certain average scale.

2.1.5 Activation of the Preexisting Slip Faults

Since the resistance of inner static friction of already existing, but partially healed faults is less than limiting condition of unbroken areas of the rocks. Therefore, it follows from an analysis of the Mohr diagram that limiting condition of resistance of surface static friction can be overcome for some range of planes orientation of *already existing faults* (Fig. 2). At the same time the planes of active slip faults may deviate in different directions from the plane position defined by the strength limit of internal friction, forming a range of admissible orientation (Rebetskii 2003).

With reasonable certainty it can be assumed that for a certain range of normal stresses acting in the upper part of the tectonosphere, the coefficient of Coulomb friction is close to a constant value (Byerlee 1967). In this case, the limiting condition of Coulomb friction in newly formed faults depends on the normal stress and the ultimate cohesion (limit of strength at zero normal stresses) which can vary from zero (for newly formed faults) to the values of strength limit of the brittle failure for completely healed faults (Byerlee 1978).

2.1.6 Elastic Unloading Area

An *elastic unloading areas* are formed around of each slip faults as a result of the relative movement of its surfaces (planes). The stress state changes significantly within the elastic unloading area. Cumulative inner elastic energy reduced, as a result intensity of the tensor of average deformation and stresses also reduce in the elastic unloading area.

It can be shown that residual deformation which is being accumulated in the geo-environment due to displacement at rupturing (discontinuous residual deformation) corresponds to the value of average elastic deformation relieved in the elastic unloading area (Kostrov and Das 1988). Elastic problem solving for slip faults (Osokina 1988) or anomaly inclusions (Dobrovolsky 2009) shows that the reduction of stresses disturbed by the crack in 10–30 times, at distances of 5–10 lengths of the crack. However, the size of the influence area of the anomaly stress state rises sharply, reaching 50–100 characteristic dimensions of the anomaly, in the case of semi-brittle behavior of the geo-environment (Rebetsky and Lermontova 2016).

2.1.7 Interaction of Closely Located Cracks

At high fracture density, closely located cracks begin to influence on each other after crossing the elastic unloading area of the previously activated crack (Fig. 1). Thus, it is impossible to reject the possibility of mutual influence of the cracks, even in the case of noncrossing cracks. It should be noticed that in a number of previously developed the stress inversion methods was a statement about the possibility of activating a large number of already existing defects of strength in rock massif.

This only explains the observated fact (Angelier 1975; Goustchenko 1975) that the slip faults are formed along planes that deviate from the planes of the maximum shear stresses (or planes with the Coulomb internal friction angle). In methods (J. Angelier and O. Goustchenko) there is an implicit condition: faults do not influence at each other. This condition follows from the principles of the slip theory of plasticity by Batdorf-Budiansky (Batdorf and Budiansky 1949). So, the MCA algorithm assumes *interaction of faults and cracks* in the process of semi-brittle flow.

2.1.8 Initial Sampling of SKD

The condition about the mutual influence of the cracks in the process of semi-brittle flow was instrumental in the MCA algorithm development. This condition actually determined the principle of forming of the calculation of the residual deformations in the overlapping zone of elastic unloading areas of the slip faults—in interaction zone of the cracks. Cracks contribution to the total tensor of residual deformations must be summed up at this particular *area of the cumulative influence of the cracks.* And only this way calculated tensor can be defined as the tensor of generalized residual or semi-brittle deformations.

2.1.9 Steady-State Deformation Process

It is believed that for the steady-state stage of the deformation process, the tensor of increments of semi-brittle deformations tends to become similar to a tensor of elastic deformations at every instant (average scale exceeds the size of the defects—discontinuities, cracks). Such a process of deformation appears for a constant way of applying external and internal loads and uniform rheological properties of a geo-environment, with corresponding average scale (Fig. 3). The tensor of elastic deformations, as well as the stress tensor, is determined by the boundary conditions and the geo-environment properties.

The corresponding tensors of the steady-state stage will be defined as the *short-* and *long*-period components of the stress and deformation tensors (depending on the size of the time averaging). This component of the identified tensors is the main characteristic. Stress tensors and deformation tensors of different averaging scale, with increasing time window and linear averaging scale will converge to this component, with further increase of the loading.

2.1.10 Homogeneous Sample of SKD

The assumption of the regime of steady-state deformation is an essential condition, allowing, in the framework of the MCA, to elaborate the algorithm of finding the parameters of the macrostress tensors and tensors of semi-brittle strains (increments of seismotectonic deformations). It replaces the postulate introduced in Goustchenko's kinematic analysis method—(Goustchenko 1996) and the right dihedral method (Angelier 1984) of the coincidence of the fault plane a slip direction with the shear stress direction, the right dihedral method. This postulate allows for a set of activated in the investigated volume slip faults to select a set of stress states that satisfy the principle of energy dissipation for each faults on the required macrostress tensor.

Thereafter, we will define such states as possible states of stress, and the corresponding sets of SKD as homogeneous samples of the SKD. Thus, a homogeneous sample of SKD determines a set of data on the orientation of the rapture plane and direction of the relative displacement of its sides or on the earthquake focal mech-

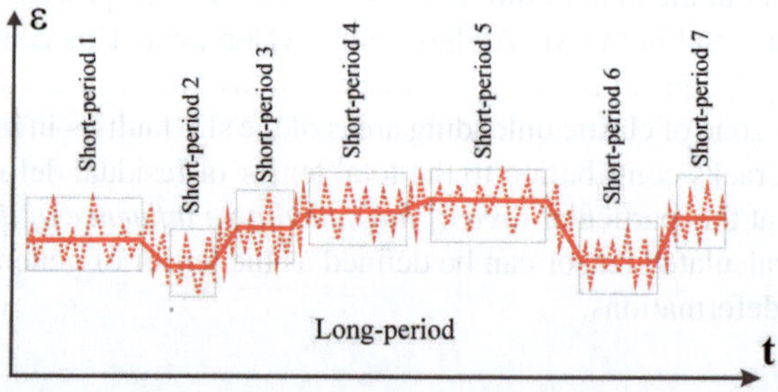

Fig. 3 The difference between the steady state and non-steady state phases of long-period deformation. A straight section of the smooth line shows that the shape of the ellipsoid of full deformations (elastic and irreversible) and its orientation are constant (a *short-period component*). A thin broken line shows the deviations of the ellipsoid of irreversible deformations, which are formed as a result of each act of brittle fracture from the ellipsoid of full deformations (*ultra-short* period component)

anism data that correspond to the conditions of quasi-homogeneity of deformation of the volume where they were fixed for a certain time period. In the MCA, there are criteria that allow one to form a homogeneous sample of SKD by filtering the fractures in the initial sample set.

2.1.11 Tensor of Semi-brittle Deformations for the Faulting Rock Massif

Residual deformation in the rocks can be formed due to cracks and ruptures according to the available data on the level of macro- and mega-scopic deviatoric stresses in the Earth's crust. Actual plastic deformations, occurring as a result of dislocations movement in the crystals and grains, are in the rocks at great depth below Moho (Nikolaevskiy 1996).

Therefore, under the tensor of semi-brittle deformations, we mean the total contribution of discontinuous displacements (averaged over a volume). A dissipation of elastic energy is being occurred within it. In the MCA, the parameters of such a tensor are calculated from data on fractures forming a homogeneous sample of SKD.

2.1.12 Time Variations of the Macrostresses

Essential procedure for calculation of residual deformations is not only spatial averaging, but also time averaging, because the accumulation of residual deformations due to discontinuous displacements occurs over a certain time interval. Thus, within the framework of our study, each point of the geo-environment is a center of four-dimensional volume. In the surrounding activating slip faults areas, the average stress state for unloading volume deviates from the *long-period* component. Thus there is a local change of the stress field in space and time—the *ultrashort disturbance*.

We will define the stress tensor, corresponding to this disturbance, as an *ultrashort* component of the stress tensor. This stress field component of the megascopic averaging scale is the most unstable component in time. Stresses corresponding to this component experience fast (immediately after the rupture displacement), frequent (due to the proximity of the different scale faults) and large changes (because of the difference in the shape of the tensor of external stresses and the tensor of relieved stresses, in consequence of rupture displacement. The deviations of the stress tensor from the long period condition can cover all large areas and act for a relatively long period of time (the first months and years), forming a *short-period* disturbance of the stress field with increasing density of activated and already existed fractures. The stress tensor corresponding to this disturbance condition we will define as *short-period* component of the stress tensor. Activated faults in such time period do not correspond to a long period stress field, and their location and displacement character of their sides are determined by the short period component of the stress tensor.

2.1.13 Influence of the Linear Dimensions of the Slip Faults

The other side of the considerable problem is the linear dimensions of the faults and cracks (the energy level of earthquakes). Stresses and seismotectonic deformations are calculated on the basis of this data. Osokina's studies (1988) show the dependence of the mechanism of small secondary fractures from their linear size, which is due to the hierarchical level of the stress field, responsible for fractures activation of the corresponding length. Therefore, in order to create SKD sample we need criteria that make it possible to determine the quasi-homogeneity of the stress state of different hierarchical levels.

At the same time, the data separation criterion for the scale levels should be not only the linear dimensions of the faults (it is possible such a state for which different linear size of faults will give equivalent results if we use them for the stress inversion), but also the correspondence of their structural-kinematic data to the reconstructed stress tensor. Such criteria exist in the MCA. These criteria are based on the energy limitations of the method.

2.2 *Energy Principles of the MCA*

2.2.1 Energy Dissipation of an Elastic Deformation

Destruction can occur only when the process of rupture is accompanied by the release of energy. It is follows from the perspective of the semi-brittle deformation of the geo-environment as a result of displacements along a set of slip faults. Upon that, we will assume from the main principles of the plasticity theory that:

- the direction of the average displacement along the single fault coincides with the direction of the vector projection of elastic macrodeformations onto this plane (averaged over the space volume, with a characteristic dimension corresponding to the dimensions of the newly formed fault line) for the tensor that operated here at the time of its appearance- an ultrashort period component;
- displacements along planes of the slip fault set (with regard to the active tensor of elastic macrodeformations and macrostress tensor at this time) may occur arbitrarily for a set of simultaneously activated fractures, but so that a release of mechanical energy occur as a result of such a multiple semi-brittle deformation microact (in time), i.e. the internal elastic energy of the geo environment should decreased. At this time, the total tensor of residual deformations tends to maximally approximate with the tensor of long-period macrodeformations in form and orientation of the principal stress axes. It will provide the maximum efficiency of removing of internal energy.

In considering a series of slip faults activated for a certain time period, it can be said that there is an average tensor of generalized stresses (for this time period). Each act of activation of the fault for this tensor leads to a decrease of internal energy accumulated in geo environment. This proposition was confirmed experimentally (Bridgman 1949), and it is the main energy principle for development of the MCA algorithm for the tectonic stress inversion.

Further it will be shown that the rule of dissipation of internal elastic energy determines as possible stress states those for which there is an acute angle between the direction of the observed displacement and the vector of shear stress on the fault plane. This relationship can be formalized in the form of inequalities systems.

2.2.2 Note on the Wallace-Bott Hypothesis

The main principles allowing to realize the mathematical formulation of the inverse problem of tectonophysics, within the framework of the dislocations theory, were represented in the studies (Wallace 1951; Bott 1959). It was proposed that average shear stresses τ_n along the plane of fault or crack, acting prior to its activation or formation, coincide in orientation with the direction of the averaging relative displacement $[u_s]$, which is realized in the shift process of the fault side. From this follows the minimal spatial average scale of the study stresses, corresponding to the most representative length of the used faults for reconstruction. It should be noted that this proposition indirectly determines the properties of the geo environment (it corresponds to the isotropic body). The propositions of the R. Wallace and M. Bott can be written in the following form:

$$\tau_n = \sigma_{ns};\ \sigma_{nm} = \tau\left[(1-\mu_\sigma)\,\ell_{n1}\ell_{m1} - (1+\mu_\sigma)\,\ell_{n3}\ell_{m3}\right] = 0. \tag{1}$$

where the ℓ_{ni} and ℓ_{mi} (i = 1, 2, 3) are the direction cosines of the normal vector to the rupture ***n***, to the shear fracture plane and to the vector ***m*** in the coordinate system *associated with the* principal stresses axes of the source field.

Expression (1) determines that the component of the averaged shear stresses, acting before the slip fault activation, is zero in the shear fracture plane, in the direction of the vector ***m***. This direction is normal to the future direction of the relative displacement of the sides ***s***. Invariants of the stress tensor: the τ (modulus of the maximum shear stresses) and the *Lode-Nadai* coefficient μ_σ or *stress* ratio ($R = (1-\mu_\sigma)/2$), which *characterizes* the *shape* of the *stress ellipsoid*) are related to the principal stresses (uniaxial compression—pure shear—uniaxial tension):

$$\tau = \frac{1}{2}(\sigma_1 - \sigma_3);\quad \mu_\sigma = 2\frac{\sigma_2-\sigma_3}{\sigma_1-\sigma_3} - 1,\ \sigma_1 \ge \sigma_2 \ge \sigma_3, \tag{2}$$

where the σ_i $(i = 1, 2, 3)$ are the principal stresses. Hereinafter, we will use the concepts of classical mechanics. In the frame of the classical mechanics the compressive stress is negative, and the tensile stress is positive. This is different from the concepts used in mining, where the compressive stress is positive.

In contrast to the assumption made in (Bott 1959; Wallace 1951), it is assumed in the MCA that the direction of the average displacement along the cleavage plane (vector s) may not coincide with the direction of the tangential stress (vector t) of the sought stress tensor (Fig. 4).

In contrast to the proposition of the R. Wallace and M. Bott, in the MCA, it is considered that the direction of the average displacement along the fault plane (vector s) may not coincide with the direction of the shear stresses (vector t) of the unknown stress tensor (Fig. 4).

The direction of the average displacement **s** must coincide with the direction of the macrodeformation of the elastic shear along this plane for the tensor which existed in the corresponding volume just prior to the slip fault activation—the ultra-short period component of the tensor according to the principles of the MCA. At the same time, the short-period component of the stress tensor and tensor of deformation is unknown.

2.2.3 The Principle of Maximum Dissipation of Internal Elastic Energy for the Unknown Stress Tensor

Formulated above energy dissipation principles for each slip fault, from a homogeneous sample set of SKD, allow one to select noncontroversial possible stress states. In the MCA, the calculation of the stress tensor parameters and increments of seismotectonic deformations is carried out independently on the same set of data. Meanwhile, there are equations for calculation the parameters of the tensor increments for seismotectonic deformations. The equations calculate these parameters on the strength of SKD. Also there are inequations for calculation the stress tensor parameters. These inequalities limit the stress tensor arbitrary. Basically, arbitrary orientations of the principal stress axes can be reduced to a certain predetermined minimum for a large number of data.

MCA uses principle which is in some ways, similar to the well-known in the theory of plasticity *von Mises's principle of maximum plastic dissipation.* In MCA, it is applied to *agreement the* determined parameters of the tensors increments for seismotectonic deformations, stress tensors and to find the only one of all possible states

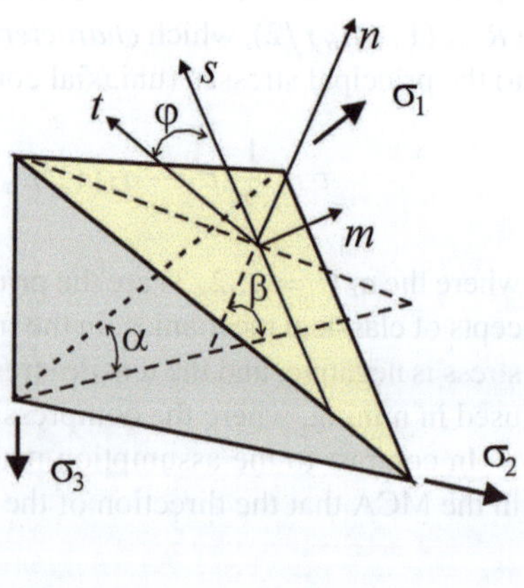

Fig. 4 The mismatch of the displacement vector (***s***) along the fault (***n***—pole of the fault) and direction of the shear stresses (***t***) on the slip fault plane ($\mathbf{m} = \mathbf{n} \times \mathbf{t}$). α, β, φ—azimuth, inclination of the pole of the fault and the displacement angle from strike of the fault

of stress. It is expected that maximum energy dissipation rate of elastic macrodeformations for a homogeneous sample set of SKD is achieved on the tensor of the unknown macrostresses. In accordance with the principle of dissipation positivity, from all possible state of stress permitted for the homogeneous sample set of SKD, in the capacity of unknown state of stress is determined such a state for which the SKD sample set delivers the maximum release of internal elastic energy.

3 The Homogeneous SKD Sample and the Tensor Increments for Seismotectonic Deformation

In the MCA, the problem of calculating the components of the increment tensors of seismotectonic deformations and macro-stresses, as well as the formation of a homogeneous sample of SKD are interrelated tasks that are solved in parallel during the reconstruction. The specific calculating features of the components of the increment tensor for seismotectonic deformations are given on the assumption that for each fault are realized the requirements of energy principles. The formalization principles will be given in the next paragraph. Thus, in fact, it is considered that a homogeneous sample of SKD already exists.

3.1 The Criteria for Creating Homogeneous SKD Sample

Using a model of an initial fracture ideally elastic body as a model of geo environment at the micro-scale level leads to the fact that the model of geo environment with accumulation of residual deformations is obtained (i.e. semi-brittle or cataclastic body) at the macro-scale level of averaging. That is why we will use the fundamental principles of the theory of plasticity during development of the principles of forming *homogeneous samples* set of SKD. Within the framework of the theory of plasticity, there are principles for interconnection between the components of the stress tensors, tensors of deformations and their increments. Calculation of the stress tensor components and increments of seismotectonic deformations should be carried out on the basis of homogeneous sample (set of slip faults or earthquake focal mechanisms). Its formation in the MCA is preceded by the creation of an *initial sample* of SKD. For geological data, this problem is solved by the geologist, at the initial stage of the research, during forming a sample of SKD on the basis of the morphological view of the structures in the rock outcrop. In the case of reconstruction of the modern tectonic stresses on the basis of seismological catalogs of earthquake focal mechanisms, this problem should be solved in the process of calculation on the basis of analysis of geological and tectonic maps or with the involvement of physical laws that allow one to estimate the possible compatibility of the activity of mechanisms included in the SKD sample set.

However, the initial sample of SKD, which is formed on the basis of spatial location of the events and their energy capacity (magnitude) can not yet be considered as a homogeneous sample of SKD, which characterizing the homogeneous deformation phase of the investigated volume. Formation of the homogeneous sample of SKD is the priority issue in the MCA. It should be solved before the reconstruction of stresses and deformations. Criteria that allow one to select data from the initial sample set into the homogeneous sample of SKD are constructed on the basis of energy principles of the theory of plasticity.

3.1.1 Drucker's Postulate

Postulate of positiv work of the actual stress on the increments of plastic deformations is realized for an semi-brittle body with an associated flow law:

$$\sigma_{ij} d\varepsilon_{ij}^{p} \geq 0. \tag{3}$$

An acute angle between the stress vector and the vector of the plastic deformation increments corresponds to this postulate in the six-dimensional stress space. Expression (3) is also one of the consequences of the Drucker's postulate (Drucker 1959), which requires that additional stresses produce non-negative work during a loading process (material stability), for a complete cycle of additional loading and unloading, additional stresses do positive work if plastic strains occur, and zero when the strains are purely elastic.

It should be noted that here and below the positive sign of the normal (principal) stresses corresponds to the extension. Similarly, the positive sign of longitudinal deformations or their increments corresponds to the deformation of elongation. This positive-negative sign convention is accepted in classical mechanics. This is the opposite of the criterion used in mining.

Extend this postulate of positive work (3) of the unknown stresses for each microact (in time) of the semi-brittle deformation, i.e. for each event with the α number from the initial SKD sample set, we can write:

$$\sigma_{ij} d\varepsilon_{ij}^{\alpha} \geq 0. \tag{4}$$

3.1.2 Corollary of the Elastic Energy Dissipation Requirement for Each Slip Fault

Inequality (4) can be rewritten in the coordinate axes associated with the principal axes of the stress tensor ($\sigma_i, i = 1, 2, 3$). Using representations of the principal stresses through invariants of the stress tensor

$$\sigma_1 = -p + \tau\,(1 - \mu_\sigma/3)\,,\ \sigma_2 = -p + 2\tau\mu_\sigma/3,$$
$$\sigma_3 = -p - \tau\,(1 + \mu_\sigma/3)\ \text{for}\ p = -\,(\sigma_1 + \sigma_2 + \sigma_3)\,/3 \tag{5}$$

and the condition ($d\varepsilon_{ij}^{\alpha} = 0$) of incompressibility at the stage of plastic flow, instead of (4), we will get:

$$\left[(1-\mu_\sigma)\,d\varepsilon_{11}^{\alpha} - (1+\mu_\sigma)\,d\varepsilon_{33}^{\alpha}\right]\tau \geq 0. \tag{6}$$

Here $d\varepsilon_{11}^{\alpha}$ and $d\varepsilon_{33}^{\alpha}$ are the increments of longitudinal deformations caused by discontinuous displacement, in the direction of the principal stresses σ_1 and σ_3, respectively. In formula (6), *p*—is an isotropic pressure.

Since, $\tau > 0$ and the value of the Lode-Nadai coefficient is $-1 \leq \mu_\sigma \leq 1$, then (6) is satisfied for any previously unknown values μ_σ when

$$d\varepsilon_{11}^{\alpha} \geq 0, \quad d\varepsilon_{33}^{\alpha} \leq 0. \tag{7}$$

For a wide class of materials, inequalities (7) determine the range of the possible values of the components of the increment tensor of plastic deformations which formed in the direction of the principal stress axes and satisfying the principle of the internal dissipation elastic energy.

The contribution from each slip fault for the increment tensor of seismotectonic deformation should be an elongation deformation, in the direction of action of the axis of maximum deviatoric tension, and deformation of shortening in the direction of action of the axis of maximum deviatoric compression, according to (7).

We pass from (7) to the inequalities, using earthquake focal mechanism data or data on the planes of the faults and the directions of the slip

$$\ell_{\mathrm{n}1}^{\alpha}\ell_{\mathrm{s}1}^{\alpha} \geq 0, \quad \ell_{\mathrm{n}3}^{\alpha}\ell_{\mathrm{s}3}^{\alpha} \leq 0. \tag{8}$$

Here the direction cosines of the normal vectors to the fault ℓ_{ni}^{α} and slip along it ℓ_{si}^{α} (the cosines of the corresponding angles which determine the position of the poles *n* and s of the nodal planes of the earthquake focal mechanisms) are also recorded in the coordinate system associated with the orientation of the principal axes σ_1 and σ_3 of the unknown macro-stress tensor.

The inequality (8) determines the geometric locus of points on opposite quadrants, formed by the nodal planes on the lower or upper hemispheres of the earthquake focal mechanism, as an permissible orientation of the axes of the algebraically maximum σ_1 and minimum σ_3 principal stresses. The σ_3 axis should be in the quadrant of *P*-axis of focal mechanism, and the σ_1 axis should be in the quadrant *T*-axis of the focal mechanism (Fig. 5).

In the case, when the centroid-moment-tensor (CMT) data are used for reconstruction of the principal stresses the restriction on the possible orientation of unknown stress axes should be determined from the inequalities:

$$dm_{11}^{\alpha} \geq 0, \quad dm_{33}^{\alpha} \leq 0, \tag{9}$$

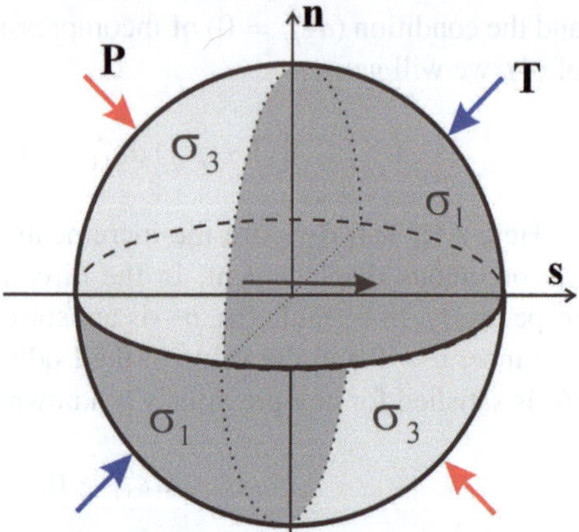

Fig. 5 Areas of possibility orientation principal axes σ_1 and σ_3 for one earthquake focal mechanism. *P* and *T*—axis of focal mechanism of earthquake, *n, s*—poles of nodal planes

where dm_{ii}^{α} ($i = 1, 3$) are the centroid moment tensor components of the earthquake with the index α in the coordinate system associated with the direction of the principal axes of the unknown macrostresses.

The inequalities (7) reflect the relation between the displacements along the fracture surface and change in the energy of the investigated volume. Therefore, we will define inequalities (8) and (9) as the criterion of dissipation elastic energy on each fault in relation to the tensor of unknown macro-stresses.

Inequalities (8) coincide with the well known limitations of the kinematic method (Goustchenko 1996) and the right dihedral method (Angelier 1984). These limitations are obtained as a corollary of the postulate of the coincidence of the shear stress and slip directions. The postulate determines the investigate volume as isotropic, in addition to the requirement that is no mutual influence of the faults of the same scale. Despite the obvious inconsistencies between the postulates of these two methods and the actual state of the geo environment, and the character of semi-brittle deformation process, the results of the stress inversion (on the basis of earthquake focal mechanisms) for the seismically active regions corresponded to the well-known tectonic conceptualization about the current stage of the deformation process in these regions. This analysis showed that the inequalities (8), which are used in the kinematic method and the right dihedral method, have the fundamental roots.

3.1.3 Corollary of the Ordering of Plastic Flow Principle

It will be required the realization of the ordering principle of elastic-plastic deformation (Chernykh 1998) on the unknown macro-stresses tensor, for the contribution to the increment tensor of the seismotectonic deformations from each slip fault:

$$d\varepsilon_{11}^{\alpha} \geq d\varepsilon_{22}^{\alpha} \geq d\varepsilon_{33}^{\alpha}. \tag{10}$$

The inequalities (10) apply more hard restrictions on the deviation of the individual stress contributions to the increment tensor of semi-brittle deformations $d\varepsilon_{ij}^{\alpha}$ from the principal axes orientation of the unknown macro-stresses σ_i $(i = 1, 2, 3)$.

It should be noted that for inequalities (7) and (10), the contribution from each fault to the increment tensor of seismotectonic deformations should provide an elongation deformation in the direction of the axis of the deviatoric stresses of the maximum extension, and a shortening deformation in the direction of maximum deviatoric compression.

From inequality (10) follows an additional limitation: from each slip fault may occur deformation elongation and shortening in the direction of the intermediate principal stress σ_2. In this case, the deformation amplitudes of the elongation should be less than the deformation in the direction of the axis σ_1, and the deformation amplitude of the shortening should be less than the deformation in the direction of the axis σ_3.We will rewrite inequality (10) for the purposes of using data on the earthquakes focal mechanisms or slip faults or centroid-moment-tensor solutions:

$$\ell_{\mathrm{n}1}^{\alpha}\ell_{\mathrm{s}1}^{\alpha} \geq \ell_{\mathrm{n}2}^{\alpha}\ell_{\mathrm{s}2}^{\alpha} \geq \ell_{\mathrm{n}3}^{\alpha}\ell_{\mathrm{s}3}^{\alpha}, \tag{11}$$

$$dm_{11}^{\alpha} \geq dm_{22}^{\alpha} \geq dm_{33}^{\alpha} \text{ for } dm_{11}^{\alpha} + dm_{22}^{\alpha} + dm_{33}^{\alpha} = 0. \tag{12}$$

It should be noted that when (11) and (12) are satisfied the inequalities (8) and (9) are automatically satisfied due to the orthogonality of the vectors $\boldsymbol{s}^{\alpha}$ and $\boldsymbol{n}^{\alpha}$.

Thus, the inequalities (11), (12), that follow from the monotonicity condition of the tensor components of the stresses and deformations apply more hard restrictions on the character of displacements along the fault, relative to the unknown macro-stresses tensor, than the inequalities that follow from the energy dissipation condition. We will define the inequalities (11) and (12) as the criterion of ordering of semi-brittle strain (Rebetsky 1999).

3.2 *Cumulative Area of the Elastic Unloading for a Set of Faults*

3.2.1 Interaction of the Elastic Unloading Areas of Slip Fault Set and a Criterion of Spatial Proximity of the Faults

The accumulated residual deformations cannot be distributed to arbitrary volumes according to the distribution analysis of the relieved elastic deformations within the old activated or newly created fault. Residual deformations formed as a result of each slip along the fault must be distributed only within the area of elastic unloading of the certain shear.

Thus, the summation of residual deformations should be made only within the mutual intercrossing area of the elastic unloading volumes of many earthquake focal mechanisms, i.e. within the cumulative area. Figure 6 shows the area of elastic

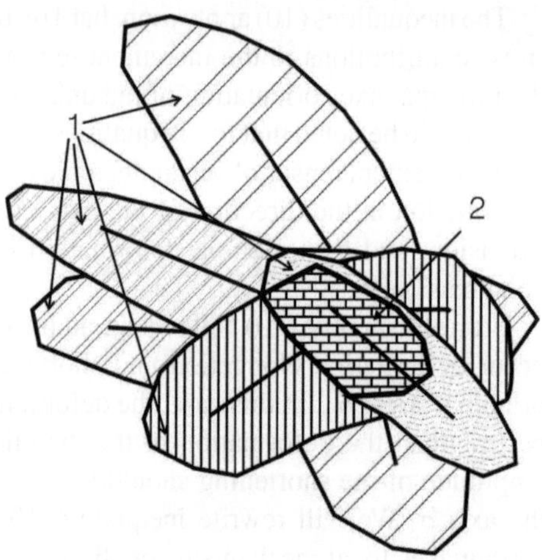

Fig. 6 Calculation of the increment tensor for seismotectonic deformations in the cumulative area (2) of mutual intersection of the elastic unloading (1) areas, which is formed within the earthquakes source areas (faults—black line)

unloading which was formed as a result of several earthquakes in the two-dimensional space. The summation of the deformations, that relieved in the process of earthquakes, is within this area. The above statement is the first criterion of the MCA. It form the basis of the initial sample and then homogeneous sample of SKD. We will define this criterion as criterion of spatial proximity.

It should be noted that, for geological data on discontinuous displacements, the issue of forming the initial sample set is solved by a geologist in the field during collecting data within the a limited geological outcrop. When the geological outcrop is chosen a geologist should analyzed it and indicate the areas of quasi-homogeneous deformation (for example, different limbs of the fold or the fault blocks).

Choosing a geological discovery, the geologist must make his analysis, highlighting the areas of quasi-homogeneous deformation. For example, different fold wings or fracture blocks. In the future, similar preliminary practice of forming the initial sample of SKD will allow one to solve the problem of separating the shifts into homogeneous phases in time. Such a problem is particularly acute for the geological indicators of deformations.

3.2.2 Calculation of the Macro-deformations and Macro-rotation Tensors in the Cumulative Area

It can be written the following formula for the increment tensor of seismotectonic deformations S_{ij} and the rotation tensor π_k for the sample of SKD that is formed in accordance with the criterion of spatial proximity:

$$S_{ij} = \sum_{\alpha=1}^{A} \frac{U^\alpha \Omega^\alpha}{2\Delta V_{Ue}^\alpha} \left(n_i^\alpha s_j^\alpha + n_j^\alpha s_i^\alpha\right), \quad \pi_k = \sum_{\alpha=1}^{A} \frac{U^\alpha \Omega^\alpha}{2\Delta V_{Ue}^\alpha} \left(n_i^\alpha s_j^\alpha - n_j^\alpha s_i^\alpha\right), \; k \neq i, j, \tag{13}$$

where U^α—the amplitude of the shear displacement of the rupture sides, Ω^α—the square of the rupture, ΔV^α_{Ue}—elastic influence zone of the rupture, $n^\alpha_i s^\alpha_j$ and $n^\alpha_i s^\alpha_j$—the direction cosines of the poles of two nodal planes of an earthquake source mechanism or the poles to the rupture and the displacement vector along it ($\alpha = 1 \ldots A$—the number of rupture in the homogeneous sample of SKD).

In the MCA it is shown that (13) can be rewritten as:

$$S_{ij} = 0.5\sum_{\alpha=1}^{A} \gamma^\alpha_{ns}\left(n^\alpha_i s^\alpha_j + n^\alpha_j s^\alpha_i\right),\ \pi_k = 0.5\sum_{\alpha=1}^{A} \gamma^\alpha_{ns}\left(n^\alpha_i s^\alpha_j - n^\alpha_j s^\alpha_i\right), \quad (14)$$

where γ^α_{ns}—the average for the elastic unloading area relieved shear deformation. It is, determined with allowance for correlation of its dimensions and the characteristic dimensions of the earthquake source L^α as follows:

$$\gamma^\alpha_{ns} = \frac{U^\alpha \Omega^\alpha}{\Delta V^\alpha_{Ue}} \approx 0.01 - 0.1\frac{U^\alpha}{L^\alpha}. \quad (15)$$

Exponential law was determined between the observable values U^α and L^α in work (Shteynberg 1983). It was done on the basis of processing of a large number of parameters of the various type of earthquakes that occurred in different parts of the Earth. It was found out that the exponent of power is changed in the range from 0.44 to 0.79. A similar analysis was carried out later in the work (Wells and Coppersmith 1994), it allowed to redefine the value of the ratio U^α/L^α. The value of the ratio U^α/L^α depends significantly on the type of mechanism realized in the earthquake source. However, in general, relationship between U^α and L^α is close to linear relationship (the coefficient of the logarithm of the rupture length is about 1) in a double logarithmic coordinate system. It reflects in rather well localized and stretched diagonally cloud of points. At a first approximation, it can be calculated the average value γ^α_{ns} (it is approximately about $1\text{–}3 \times 10^{-5}$), using the ratio between the maximum sip value and its average value, obtained in works (Starr 1928; Knopoff 1958; Osokina 1988), expression (15) and data from the work (Wells and Coppersmith 1994). This value is much less than the value of the elastic deformations of the strongest rocks.

If we take elastic modulus equal to 5×10^5 bar (50 GPa) than the average relieved stresses can be estimated in the first bars—tens of bars within the elastic unloading area. This value $\gamma^\alpha_{ns} = \gamma_{ns} = \tilde{\gamma}$ may be assumed constant and be taken off the sum sign in expressions (14) for the faults from a homogeneous sample set of SKD, characterizing the quasi-homogeneous state of the investigated volume under specific loading conditions. It should be emphasized that the value $\tilde{\gamma}(p, \tau, \mu_\sigma, G)$ may differ for neighboring volumes which have same structure and same rheological properties, but different quasi-homogeneous loading conditions. The value $\tilde{\gamma}$, we will define as the geodynamic parameter of the seismotectonic flow.

3.3 The Tensor Increments for Seismotectonic Strains

3.3.1 Normalization of the Tensor

The following formula can be written for *tensor increments of seismotectonic strains* on the basis of (14) and the made above conclusion:

$$\bar{S}_{ij} = I_S \sum_{\alpha=1}^{A} \left(n_i^\alpha s_j^\alpha + n_j^\alpha s_i^\alpha\right). \tag{16}$$

where I_s is the normalizing factor determined on the basis of the condition $\sqrt{0.5\bar{S}_{ij}\bar{S}_{ij}} = 1$. Thus, $\bar{S}_{ij}$ is the normalized tensors of increments for seismotectonic strains with the value of the shear intensity equal to 1. For simplicity we will define S_{ij} *as* the tensor of increments for seismotectonic strains, keeping in mind the above mentioned.

3.3.2 Using CMT Solutions

In the case of calculating the tensor increments for seismotectonic strains on the basis of the centroid moment tensor solutions

$$\bar{S}_{ij} = I_S \sum_{\alpha=1}^{A} \mathrm{m}_{ij}^\alpha, \tag{17}$$

where m_{ij}^a is the centroid moment tensor of the earthquake with the index α. Only four of the six tensor components of the residual macro-deformation can be determined from the results of the reconstruction according to (16) and (17).

This means that three Euler angles that determine the orientation of the principal axes of the tensor and the Lode-Nadai coefficient characterizing the parameters of the deformation ellipsoid will be known after summing the SKD from a homogeneous sample. The value of maximum shift of semi-brittle deformations is characterized by the parameter γ and it remains unknown. This value should be determined on the basis of other data or using additional assumptions.

4 The Calculation of the Stress Tensor (the Principal Stress Axes and the Lode—Nadai Coefficient). The First Stage of the MCA

The possibility of forming a homogeneous sample of SKD, on the basis of the principles following from the theory of elastic-plastic deformation, which adapted to the analysis of the deformation process of a fractured structure is shown in the previous paragraph. The stress field at the level preceding the considered one is essentially inhomogeneous. That is why in the MCA, the energy principles similar to those that were formulated to forming the homogeneous samples of SKD are used. The concept of generalized stresses for a fractured structure is formulated on the basis of these principles.

4.1 *Requirement of Elastic Energy Dissipation—Graphical Algorithm of Realization*

It was shown above that inequalities (8) and (9) make it possible to separate the events from the initial sample of SKD, leaving only the events that satisfy to these criteria. The inequalities (8) and (9) was proposed to use directly for determining the position of the principal stress axes in work (Carey and Bruneier 1974; Angelier and Mechler 1977). This approach is based on the limitation for possible orientation of the principal stress axes responsible for the displacement in the earthquake source. This imitation was shown in the work (McKenzie 1969).

The inequality (8) determines the geometric locus of points on opposite quadrants, formed by the nodal planes on the lower or upper hemispheres of the earthquake focal mechanism, as an permissible orientation of the axes of the algebraically maximum σ_1 and minimum σ_3 principal stresses. The σ_3 axis should be in the quadrant of P-axis of focal mechanism, and the σ_1 axis should be in the quadrant T-axis of the focal mechanism. These areas are represented by a corresponding type of hatching (Fig. 7a, b) on the unit lower hemisphere for two different earthquakes.

Thus, the use of only one event from a homogeneous sample makes it possible to halve the range of enumeration for the possible location of the principal stress axes. If two earthquakes are included in homogeneous sample set it is necessary to sum the corresponding quadrants of these earthquake to determine the possible position of the principal stress axes of the tensor (Fig. 7c). Those points on the lower hemisphere, which get into intersection of spatial quadrants of the focal mechanisms are forming a new geometric set. It determines possible location area of exit of the corresponding principal axes (areas: 1 and 2, Fig. 7c). The set of points that get into heteronymous areas are forming a geometric locus where the axes of the principal deviatoric stresses of the maximum compression and tension cannot appear on the hemisphere (area 3 in Fig. 7c).

It should be pointed out that the result of the analysis that was carried out in work (McKenzie 1969) is correct for isotropic medium, i.e. in the case when the principal axes of the tensors of elastic deformations ε_{ij}^{e} and the stress tensor σ_{ij} are coaxial. Dan McKenzie and E. Carey did not to take into account this moment. In fact, in the general case of anisotropic media, the corresponding the principal axes of elastic deformations should be in the quadrants of compression and tension. In the MCA, in the case of steady deformation (tensors ε_{ij}^{e} and $d\varepsilon_{ij}^{p}$ are similar), inequalities (8)

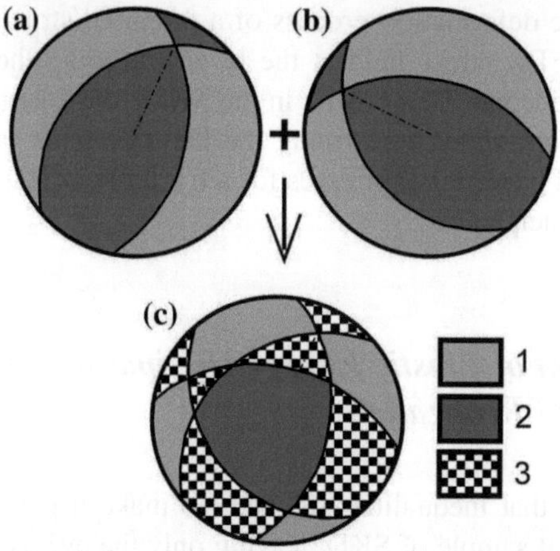

Fig. 7 Summation of the possible location areas of exit of the axes of algebraically maximum and minimum principal stresses on the lower hemisphere according to the inequalities (8): **a, b** the quadrants of compression (1) and tension (2) for two different earthquake focal mechanisms, **c** the result of summing the quadrants of compression and tension for two mechanisms. 1—the area of summation only the quadrants of compression, 2—the area of summation only the quadrants of tension, 3—the area of summation the quadrants of compression and tension

and (9) determine distribution of the principal axes of increments for seismotectonic strains S_1 and S_3

The possible exit area of the principal axes of the macrostress tensor to the lower hemisphere can be narrowed by summing the homonym quadrants to relatively small size (the solid angle is less than 15–20°), depending on the number of events in the homogeneous sample of SKD and the variation of their mechanisms (the spread in the orientation of the normal vectors and movements for the fracture share).

Figure 8 shows the final result of the summation of the quadrants of compression and tension for eight events from the initial sample of SKD. One hemisphere determines the possible area of the axes of the maximum deviatoric extension, and the other one the possible area of the axes of maximum deviatoric compression.

Due to the fact that a homogeneous sample set of SKD allow one to calculate the parameters of the increment tensor of seismotectonic strains on the basis of formula (16), the penetration of the principal axes of the tensor S_i ($i = 1, 3$) into the admissible areas on the hemispheres is a validity test of SKD homogeneous sample formation. We will define the areas of mutual intersection of the homonym quadrants as the areas of admissible exit of the principal axes of the increment tensor for seismotectonic strains to the hemispheres.

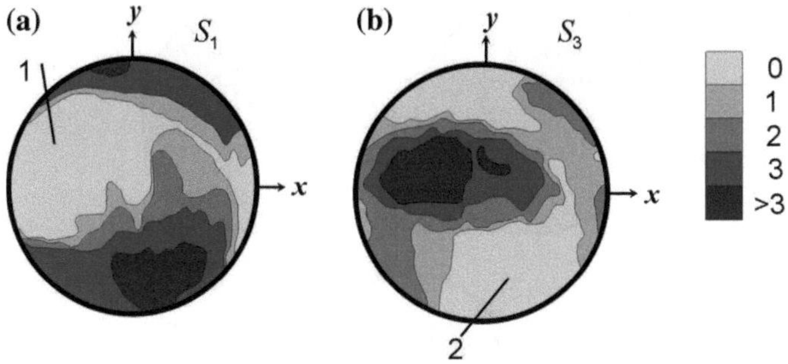

Fig. 8 The result of summation on the lower hemispheres of the quadrants of compression (**a**) and tension (**b**) (see Fig. 7) for finding possibility areas of the principal increments of seismotectonic strains for eight events according to inequalities (8). 1—the possible area of the axis of maximum elongation S_1; 2—the possible area of the axis of the maximum shortening S_3. Shades of gray characterize the numbers of earthquake inconsistent by inequalities (8)

4.2 The Graphic Way of Realizing the Principle of Ordering of Semi-brittle Deformation

The inequalities (11) and (12) make it possible to localize the admissible exit areas of the axes of the tectonic macro-stress tensor in hemispheres in a manner analogous to that considered above. It is important to note the following fact.

The admissible exit areas of algebraically maximum and minimum the principal stresses coincide with the quadrant distribution (as in the case of using inequalities (8), see Fig. 9a, b) for each individual event from a homogeneous sample.

The areas of admissible position of the principal stress axes are marked on the hemispheres according to inequalities (11), will always be inside areas determined by inequalities (8), see Fig. 9, when summing a set of focal mechanisms from a homogeneous sample.

We will define the areas on hemispheres that satisfied inequalities (11) as the areas of admissible exit of the principal stress axes of the macrostress tensor σ_1 and σ_3 to the hemispheres. The localization of the areas of admissible exit of the principal stress axes of the macrostresses to the hemispheres must be carried out on the basis of inequalities (12) if the centroid moment tensor solutions are used for the reconstruction of the state of stress.

4.3 The Determination of the True Orientation of the Principal Axes of the Unknown Stress Tensor

On the one hand, the above mentioned restrictions, on the axes orientation of the unknown tensors of macro-stress and the increments of seismotectonic deformations

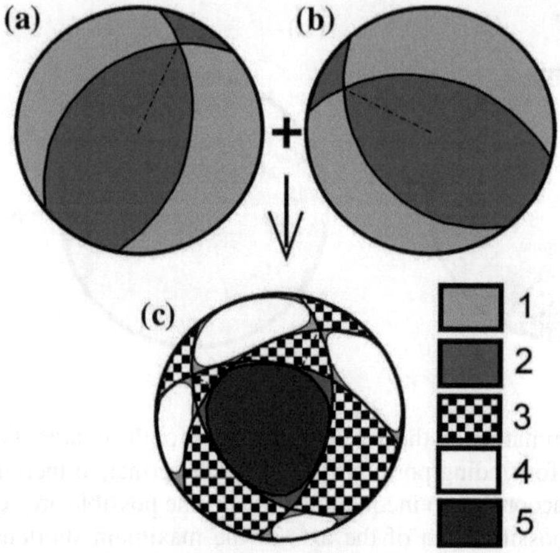

Fig. 9 The summation of the possible exit areas of the axes of algebraically maximum and minimum principal stresses to the lower hemisphere according to the inequalities (11): **a, b** the quadrants of compression (1) and tension (2) for two different earthquake focal mechanisms; **c** the result of summation of quadrants for two mechanisms. 1, 2—the areas of summation only the quadrants of compression and extension according to inequalities (8), 3—the areas of summation quadrants of compression and extension according to inequalities (8), 4, 5—the areas inside the quadrants of compression (4) and tension (5) which are satisfied inequalities (11)

with reference to the set of SKD, make it possible to select the range of the possible stress-strain states that best satisfy them. On the other hand, a homogeneous sample of SKD, will characterize the phase of homogeneous deformation of the investigated volume.

Data on the tensor increments for seismotectonic deformations and principle of the maximum dissipation of internal elastic energy, (which is similar to the Von Mises principle in the theory of plasticity), should be used for searching the only possible (correlated) values of the stress tensor.

The Von Mises (Mises 1928) principle determines that, from all possible states, the desired state is that one for which achieved the maximum internal energy dissipation (accumulated in the elastic deformations) on the known tensor of the increments of plastic deformations.

$$\left(\sigma_{ij} - \tilde{\sigma}_{ij}\right) d\varepsilon_{ij}^{p} \geq 0. \quad (18)$$

Where σ_{ij} and $\tilde{\sigma}_{ij}$ are the required (correlated) and possible stress tensors, respectively. By $\tilde{\sigma}_{ij}$ we shall mean the tensor whose axes may have the orientation satisfied by the inequalities (10) or (12) for a homogeneous sample of SKD. We also will use the made above assumption about the similarity of the tensors $d\varepsilon_{ij}^{p}$ and S_{ij}. In this case, the inequality

$$\left(\sigma_{ij} - \tilde{\sigma}_{ij}\right) S_{ij} \geq 0 \quad (19)$$

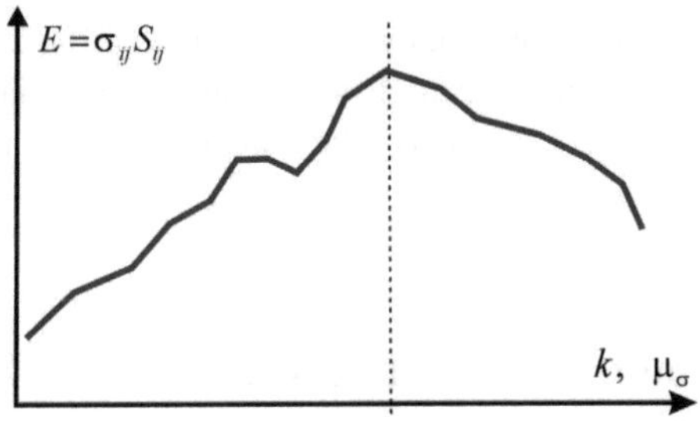

Fig. 10 The scheme is illustrated algorithm of finding a principal stress axis orientation and Lode-Nadai coefficient μ_σ on the base of hypothesis of maximum energy dissipation. k—number of the points on unit half-sphere into areas σ_1 and σ_3 after processing all earthquakes from homogeneous sample

allows us to select the orientation of the principal axes of the macro-stress tensor in the admissible exit areas of the principal stress axes of the maximum deviatoric extension and compression on the hemisphere, and also to select the value of the Lode-Nadai coefficient (it's range of values variations is given by inequality $|\mu_\sigma| \leq 1$).

The inequality (19) determines the maximum for the (in applications to the calculated characteristics of the stress-strain state)

$$E = \sigma_{ij} S_{ij} \tag{20}$$

where

$$\sigma_{ij} = -\left(p - \frac{2}{3}\tau\mu_\sigma\right)\delta_{ij} + \tau\left[(1-\mu_\sigma)\,\ell_{1i}\ell_{1j} - (1+\mu_\sigma)\,\ell_{3i}\ell_{3j}\right], \quad S_{ii} = 0$$

by a discrete search with some step for the Euler angles, (determining the values of the direction cosines ℓ_{ki} ($k = 1, 3$) of the desired principal stress axes in the Cartesian coordinate system), and varying with a small step of values μ_σ from +1 to −1. The search for orientation of the principal axes is carried out within the allocated areas of their possible exit to the corresponding hemispheres (Fig. 10). It is impossible to determine the values of confining pressure and the modulus of the maximum shear stresses according to (19) and (20).

The harmonization of the stress tensors parameters and increments of seismotectonic strains is the main procedure of the MCA. It is carried out after all the events from the initial sample set have been tested for homogeneity or after forming the admissible exit areas of the principal axes of the unknown tensors on the unit hemispheres.

The harmonization is carried out on the basis of the principle of the maximum dissipation of the internal elastic energy, which makes it possible to choose the only one stress state from all possible stress states (the admissible exit areas on the hemispheres determine only possible stress states). Various situations are possible as a result of using this principle, including those where the calculated tensors of stresses and increments of seismotectonic strains are not coaxial (anisotropy of the medium properties, and the coefficients of the state of the medium are equal (isotropy of the medium properties)).

5 The Calculation of the Relative Values of the Stresses on the Basis of Principles of the Fracture Mechanics. The Second Stage of the MCA

Algorithm for calculating four of the six stress tensor components (orientations of the principal axes of the tensor, determined by the three Euler angles, and the coefficient of the form of the stress tensor or the Lode-Nadai coefficient) was formulated in the previous paragraph.

Two parameters of the stress tensor such as confining pressure and the modulus of the shear stresses remain unknown after the calculations. These parameters are especially important from the point of view of studying geodynamic processes and seismic zoning are unknown. In accordance with the above approach, the obtained results will be interpreted as the results of the first stage of stress analysis.

Nevertheless, in many cases, the field of trajectories of the principal axes of the tensor and the field of the values of the Lode-Nadai coefficient allow one to suggest, with some degree of certainty, the way of external loading of the investigated area of the earth's crust.

These data are extremely important for mathematical or physical modeling, since they make it possible to formulate the boundary conditions of loading. At the same time, the stress tensor parameters, which remain unknown, are the key parameters when estimating the intensity of local loading, the forecast of earthquake preparation areas and areas of increased fracturing. In fact, the values of stresses remain unknown. Algorithm for calculating the relative stress values is presented below. It is based both on the results of laboratory experiments on the destruction of rock samples, and on the fundamental principles of continuum mechanics. Since the calculation of these components is carried out after determining the orientation of the principle axes, the values of the Lode-Nadai coefficient, and also after forming a homogeneous sample set of SKD, the presented calculation algorithm is the second stage of the stress inversion.

In some ways, the proposed approach is based on the intuitive idea that the activation character of the existed fractures should depend on the value of isotropic pressure. At low intensity of all-round compression the range of orientation of newly activated fractures should be greater than for the same set of fractures, but with a higher value of all-round compression.

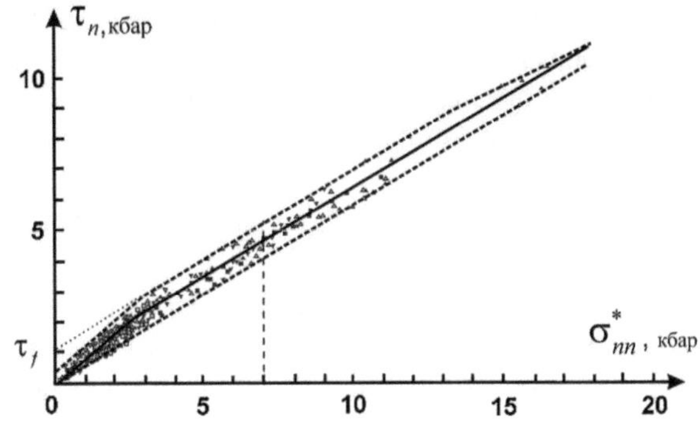

Fig. 11 The results of laboratory experiments on the role of friction in the destruction of rock samples at high pressure (according to Byerlee 1978 with addition dotted lines restricting the cloud of points (experiments): the line of internal strength and line of minimum surface resistance of Coulomb friction)

5.1 *Breaking Strength of the Rock and Minimum Frictional Resistance to Ruptures*

5.1.1 Experiments on the Destruction of Fractured Rock Samples

The principles for determining the relationship between the spherical and deviator components of the stress tensor are based on the results of brittle failure experiments of rock samples of different types, both initially solid and containing faults or prepared surfaces of reduced strength, performed within the engineering studies (industrial and civil), as well as special geophysical studies (Mogi 1964, 1966; Byerlee 1967, 1968, 1978; Handin 1969; Stesky 1978; Rummel et al. 1978). These experiments made it possible to identify two characteristic lines on the Mohr diagram that determine the breaking strength of internal friction of non-destructive rock samples and the minimum resistance to shear due to friction at the boundaries of already existing faults-(further we will call it the minimum strength friction (Fig. 11).

The indicated lines limit the top and bottom of the main cloud of points (obtained in the experiments) on the parametric area of the Mohr diagram (in the coordinate axes: the normal and shear stresses on the rupture plane) (see Fig. 11). The upper convex line is the limiting envelope. The points obtained in experiments with initially solid samples tend to this line. The lower line determines the states on the rupture sides which correspond to the minimum and zero values of surface cohesion. In a number of experiments it was shown that the line of minimum strength of Coulomb friction comes to the origin of coordinates (Handin 1969; Stesky 1978; Rummel et al. 1978).

It should be noted that Beyrlee's experiments were carried out for the samples under dry conditions, i.e. without the fluid addition. In reality, the fluid pressure simplify the formation of slip fault in the fractured-pore space of the rocks i.e. the pressure on the shear bands is reduced.

The equations for these lines can be represented in the following form (taking into account the weakening effect of the fluid pressure (Terzaghi 1943):

$$\tau_{Cz} = \tau_n + k_f \sigma^*_{nn} = \tau_f, \ \tau_n + k_s \sigma^*_{nn} = 0 \text{ при } \sigma^*_{nn} = \sigma_{nn} + p_{fl}. \tag{21}$$

Here τ_{Cu}—Coulomb stress determined by the level of the shear stresses (τ_n) and effective (taking into account the fluid pressure) normal stresses (σ^*_{nn}) acting on the plane of the formed or activated fracture. Strength parameters k_f and k_s are the coefficients of internal and surface friction, respectively; and τ_f is the value of internal cohesion or the breaking strength at shear of the intact rock samples. k_f и τ_f are the functions of the σ^*_{nn} in accordance with the results of experiments. This dependence is especially evident at low (less than 3 kbar) and high (more than 15 kbar) values of pressure on the fault plane. The upper limit line gradually flattens out at pressures of more than 20 kbar.

It is important to note that there is an extended area for which the upper bounding line is approximately parallel to the lower line. It can be assumed that $k_f \cong k_s$, $\tau_f \cong$ const for this area. Such an interpretation is possible for a section of the horizontal axis, where the normal pressures on the fracture planes are from 3 to 15 kbar according to the results of experiments.

5.1.2 Minimum Frictional Strength of Existing Faults

This interpretation of the results of the experiments is accepted as a basis of the method for calculating the values of the macro-stress tensor. We will assume that slip along flanks of partially resistive faults is realized with dry friction according to the Coulomb law with a constant value of the surface friction coefficient (k_s = const) for the critical states, (the critical states correspond to the activation process on partially resistive fractures). In addition, the surface cohesion for newly activated faults can vary from the minimum (zero) value to the maximum value that equal to the internal cohesion value of undestroyed massif ($\tau_s \leq \tau_f$). Thus the difference of the limit states lies in the different cohesion values $0 \leq \tau_s^i \leq \tau_f$ for faults with the same value σ^*_{nn} on their flanks.

All possible limit states on the bands of newly formed or activated ruptures can be represented in the following form:

$$\tau_{Cu} = \tau_n + k_s \sigma^*_{nn} = \tau_s^i \text{ when } 0 \leq \tau_s^i \leq \tau_f \text{ и } \sigma^*_{nn} < 0. \tag{22}$$

If the Coulomb stress τ_{Cu} is less than the cohesion τ_s^i for the partially resistive faults, the fault will not activated, although it may lay within the brittle fracture band. But if, the cohesive strength decreases and becomes equal to τ_{Cu} the fault will active.

5.1.3 The Relationship Between the Proposed Approach and the Theory of J. Byerlee

In constructing a theory of failure—determining the stresses responsible for brittle fracture of the rocks, J. Byerlee drew a median line through a cloud of experimental points, averaging the results of observations for initially solid samples, as well as for samples with prepared surfaces of reduced strength and fractures. This approach is applicable for forward modeling to determine the areas of intensive destruction in the Earth's crust. But it is unproductive for the inverse problem, when it is necessary to reconstruct the stress values responsible for the deformation process on the basis of activated fractures data. In this case, the all set of experimental data should be used. It follows that, destruction processes should be characterized by two laws: the limit of internal friction for solid areas of rocks—the Mohr's envelope, and the law of dry friction for already existing but partially resistive faults and fractures—the line of minimum resistance due to friction starts at the origin.

Thus, the medium area of the Mohr diagram (Fig. 9) is taken as the starting point in developing the calculation method. This is the area where both limiting lines are parallel. The strength value τ_s^i may take different values along differently oriented planes of fractures (without exceeding the limit of cohesion value τ_f) according to (22). Probably, the value τ_f may be different for different types of the earth's crust and it should depend on the history of tectonic development of the investigated area and its current structural-tectonic state. In any case, the value τ_f for the rock massif should be less than the value of maximum cohesion obtained from experiment on the destruction of rock samples approximately (1 kbar).

It should be noted that the value of the coefficient of surface friction, acting on the fracture planes in the rock massif, may also differ from the values obtained in the experiments on the samples ($k_s = 0.85$; $k_s = 0.6$).

5.2 *Mohr Circle Diagram and Areas of Admissible Position of Brittle Fractures*

5.2.1 Representation of Stresses on Ruptures on the Mohr Diagram

The possibility of solving the problem, which was formulated at the beginning of this paragraph, will be estimated using the limiting condition (22). To do this, we depict on the Mohr diagram the areas for which slip faults activation is possible with the cohesion variation on their wings $0 \leq \tau_s^i \leq \tau_f$ (*i* the number of fractures in the homogeneous sample). The size and type of the area of admissible solutions (the permissible orientation of the activated fractures) depend not only on the slope of the limiting lines, (that determine the range of possible conditions on the rupture), but also on the type of the state of stress (the values of the Lode-Nadai coefficient) (Fig. 12).

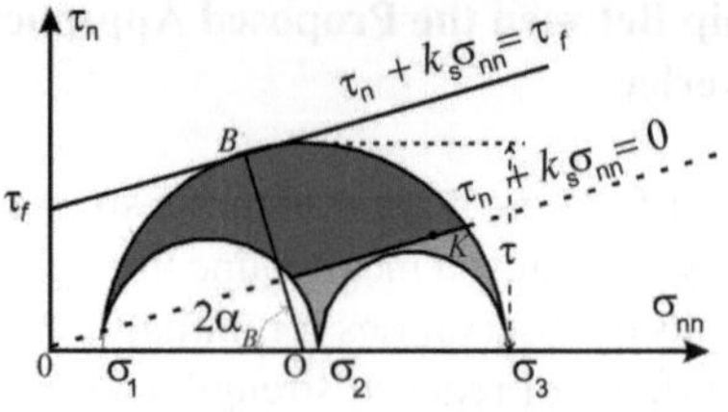

Fig. 12 The Mohr circle diagram and the line of minimum resistance due to friction on existing faults (dotted line) and the line of brittle strength (solid line) for the general case of stress state. Dark grey area is the area of the permissible position of the fractures planes (with cohesion $0 \leq \tau_s^i \leq \tau_f$). Light gray areas—other areas of possible states at arbitrary planes

It should be pointed out that the center of the Mohr circle (point O, Fig. 12) will correspond to the value of the compressive effective normal stress acting on the areas of maximum shear stresses, which according to the expressions (5) can be represented as:

$$\sigma_O^* = -\left(p^* + \frac{\mu_\sigma}{3}\tau\right) \text{ при } p^* = p - p_{fl}, \tag{23}$$

where p^* is the effective pressure (in this case, the effective pressure is the tectonic pressure minus the pore fluid pressure).

5.2.2 The Uniaxial State of Deviatoric Stresses

We will analyze the Mohr circle diagram (Fig. 12) to estimate the influence of the stress values on the distribution of the fracture planes. This is easier to do for the case when the type of the stress tensor is uniaxial compression or uniaxial tension. In this case, the points characterizing the stresses on inclined surfaces lie on a large Mohr circle. The shear stresses ($\tau_n = \sigma_{nt}$) and normal (σ_{nn}^*) stresses acting on inclined planes (***t*** is the vector determining the direction of action of shear stresses on the plane with the normal in the form of the vector n) are determined according to the expressions:

$$\tau_n = \tau \sin 2\alpha; \; \sigma_{nn}^* = -p^* + \left(\cos 2\alpha - \frac{\mu_\sigma}{3}\right)\tau, \tag{24}$$

where α is the angle between the axis σ_1 and the normal n to the plane of the fracture.

In the case of seismological data, it is believed that the orientation of the rupture plane (we shall return to the issue, how to be determine this plane) and slip direction in the earthquake source area are known. It is in the same way as in the case of reconstruction on the basis of geological shear fracture data. It is also believed that the stresses reached their limiting values (when new formation of ruptures becomes

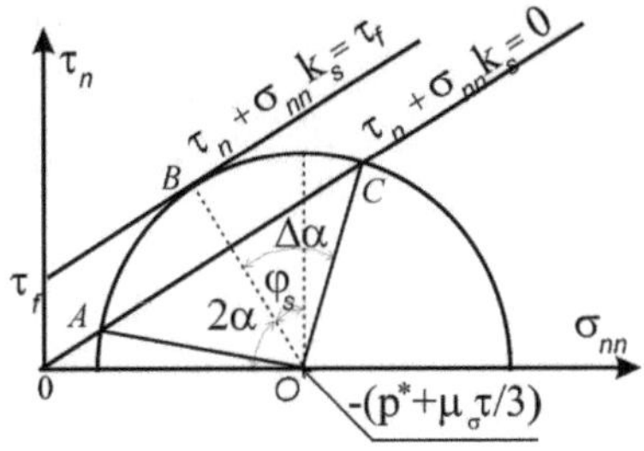

Fig. 13 The Mohr diagram for the case of a uniaxial state of stress. φ_s—the angle of internal and surface friction, $\alpha = \alpha_B$—the angle between the axis of the algebraically maximum principal stress and the pole to the plane of internal friction

possible in the undistracted parts of the rocks), as a result of loading the earth's crust. In this case, the activation of all range of orientations of the fractures, that get into the area between (Fig. 13) the below line of minimum resistance due to friction and the above line of brittle strength (arc *AC*), becomes possible. Suppose that we have succeeded in fixing the position of the planes of slip faults, which poles of the faults has a minimum and maximum deviation from the direction of the axis of deviatoric compression (points *A* and *C*, Fig. 13).

We assume that the actual geo-mechanical characteristics of the rock massif are unknown (i.e. characteristics that satisfy the averaging scale which correspond to the length of the activated ruptures. We will also assume that the orientation of the principal stress axes and the value of the Lode-Nadai coefficient were determined as a result of applying the MCA algorithm to the chosen set of fractures (the first stage of the MCA). The problem consists in calculating the values τ, p^* and in determining the coefficient of surface friction of the geo environment k_s from data on the slope angles of the sip fault planes, in relation to the principal axes of the stress tensor that were reconstructed at the first stage.

According to the above assumption, the line passing through the extreme points *A* and *C* on the Mohr diagram (see Fig. 13) should be considered as the line of minimum resistance to dry friction of existing fractures. Using (24), the limiting condition for the frictional resistance for ruptures (22) that correspond to these points can be written in the following form:

$$\tau \sin 2\alpha_i + k_s\left[-p^* + \tau\left(\cos 2\alpha_i - \frac{\mu_\sigma}{3}\right)\right] = 0 \text{ при } i = A, C. \tag{25}$$

Whence it follows that

$$\frac{p^*}{\tau} = \frac{\sin 2(\alpha_C - \alpha_A)}{\sin 2\alpha_C - \sin 2\alpha_A} - \frac{\mu_\sigma}{3} \tag{26}$$

by virtue of the fact that

$$2\alpha_A = 2\alpha_B - \Delta\alpha,\ 2\alpha_C = 2\alpha_B + \Delta\alpha \text{ and } \alpha_B = (\alpha_A + \alpha_C)/2. \tag{27}$$

Then the second expression in (33) can be rewritten in a more compact form:

$$\frac{p^*}{\tau}=\frac{\sqrt{1+k_s^2}}{k_s}\left(\cos\Delta\alpha-\frac{\mu_\sigma}{3}\right) \text{ при } k_s=\text{tg } \varphi_s=\text{ctg}2\alpha_B, \tag{28}$$

where $\varphi_s = 90 - 2\alpha_B$—angle of internal friction.

The limit state for point B (the tangency point of the brittle fracture line with the Mohr circle) can be written in the following form:

$$\tau \sin 2\alpha_B + k_s\left[-p^* + \tau\left(\cos 2\alpha_B - \frac{\mu_\sigma}{3}\right)\right] = \tau_f. \tag{29}$$

Point B on the Mohr diagram, (in accordance with the assumption of equality of the coefficients of surface and internal friction), corresponds both to the limit of brittle strength for newly created fractures and to the limiting resistance due to friction on already existing faults, the cohesion of which has been restored to the value of internal cohesion τ_f of undistracted areas.

Expression (29), together with (28), makes it possible to determine the value of the invariants of the stress tensor

$$\tau = \frac{\sin 2\alpha_B}{1-\cos\Delta\alpha}\tau_f,\ p^* = \frac{\cos\Delta\alpha - \frac{\mu_\sigma}{3}\cos 2\alpha_B}{(1-\cos\Delta\alpha)\,k_s}\tau_f. \tag{30}$$

Thus, in this case of a uniaxial stress state, the data on the orientation of the two planes of faults (which deviate most strongly from each other) are enough both for determining the coefficients of surface and internal friction of rocks, and for estimating the relative values of maximum shear stresses (up to an unknown value of internal cohesion) and isotropic effective pressure.

5.3 *Determination of Relative Values* p* *and* τ

5.3.1 Reduced Stresses

The calculation algorithm will be constructed for the general case of the state of stress, based on the assumption about possibility of forming faults of arbitrary orientation that formed in accordance with criterion (22). It is considered that the sample set of earthquake focal mechanisms data (faults), (calculation of the principal stress orientation and the Lode-Nadai coefficient are performed on the basis of this sample), is enough representative and it contain the definitions of the faults planes for the all possible range of their orientations within the investigated volume (it also includes fractures with the cohesion value $\tau_s = 0$). The case of reconstruction from seismological data on the earthquake focal mechanisms, when we do not know the actual fault plane will be considered below.

The problem will be formulated in the same way as it was done for the uniaxial state of stress under the assumption that the value of the coefficient of surface friction is known. In this case, a point on the Mohr diagram should be chosen in order to draw the line of minimum resistance. The point (K, Fig. 12) characterizes the state of stress on the surface of one of the faults. The length of the perpendicular that drawn from the center of the circle will be minimal for the point. The values of the stresses on the faults surface can be represented as

$$\sigma^i_{nn} = -(p + \tau\mu_\sigma/3) + \tau\tilde{\sigma}^i_{nn};\ \tau^i_n = \tau\tilde{\tau}^i_n, \tag{31}$$

where $\tilde{\tau}^i_n = \tilde{\sigma}^i_{nt}$ и $\tilde{\sigma}^i_{nn}$—the reduced stresses and

$$\tilde{\sigma}^i_{nj} = (1-\mu_\sigma)\,\ell^i_{1n}\ell^i_{1j} - (1+\mu_\sigma)\,\ell^i_{3n}\ell^i_{3j} + \delta_{nj}\mu_\sigma,\ j = n, t, \tag{32}$$

where ℓ^i_{kn} and ℓ^i_{kt} are the direction cosines of the normal vector $\boldsymbol{n}$ to the fault plane (with the i number from the homogeneous sample), and the vector of the shear stresses $\boldsymbol{t}$ on this plane, in the coordinate system associated with the principal stress axes (k = 1, 2, 3). Henceforward, the superscript at stresses identifies the point on the Mohr diagram.

The introduced definition of the reduced stresses is very important, since it allows one to construct Mohr diagrams for the homogeneous samples set of SKD without data on the values of pressure and the maximum shear stress. The diagrams characterize the limiting states on the faults surfaces.

5.3.2 Calculation of the Relative Values of the Effective Pressure and the Maximum Shear Stresses

The ratio of the value of the confining pressure to the modulus of the maximum shear stresses can be determined using expressions (22) and (31). For this purpose, we will assign to a slip fault the minimum possible value of surface cohesion ($\tau_s^K = 0$), which has a minimum perpendicular length that drawn from the center of the circle (point K, Fig. 12),

$$\frac{p^*}{\tau} = \frac{1}{k_s}\left(\tilde{\tau}^K_n + k_s\tilde{\sigma}^K_{nn}\right) - \mu_\sigma/3. \tag{33}$$

The limiting expression (22) is valid for the point B, in which the line of maximum strength of rocks (the limit of internal friction) touches the Mohr circle. The expression (22) is written for the maximum value of cohesion $\tau_s^B = \tau_f$. Using it on the basis of this expression and expressions (31) and (33), we will obtain:

$$\left\langle\frac{\tau}{\tau_f}\right\rangle = \frac{1}{\mathrm{cosec}2\alpha_s - \left(\tilde{\tau}^K_n + k_s\tilde{\sigma}^K_{nn}\right)};\ \left\langle\frac{p^*}{\tau_f}\right\rangle = \frac{\left(\tilde{\tau}^K_n + k_s\tilde{\sigma}^K_{nn}\right) - k_s\mu_\sigma/3}{k_s\left[\mathrm{cosec}2\alpha_s - \left(\tilde{\tau}^K_n + k_s\tilde{\sigma}^K_{nn}\right)\right]}, \tag{34}$$

where $\alpha_s = \frac{1}{2} \arctan \frac{1}{k_s}$.

Expressions (34) allow one to calculate the relative values of the effective pressure $\langle p^*/\tau_f \rangle$ and the modulus of the maximum shear stresses $\langle \tau/\tau_f \rangle$ on the basis of the results of the first stage of MCA reconstruction and data on the values of the coefficient of surface friction k_s. We will use the angle brackets for the normalized values of the effective pressure and maximum shear stresses.

It should be remembered, that we don't know which of the two nodal planes is the actual fault plane. Within the framework of MCA, there is a criterion for determining the actual fault plane, which directly follows from the Coulomb condition (22). It corresponds to the energy approach that developed in the MCA.

It should be noted that, in the construction of the calculation algorithm, the assumption about the core role of the total value of the shear stress $\tau_n = \sigma_{nt}$ acting on the plane of the activated fault was brought in the calculation algorithm, instead of the shear stress acting in the direction of the slip vector σ_{ns}. It is the total shear stresses acting on the fault plane are responsible for overcoming friction (destruction of the roughness of the bands of the fault, that impede movement).

The deviation of the slip direction along the rupture plane from the shear stress direction may be due to anisotropy in brittle failure zone (the existence of the large roughness besides small one, oriented roughness—fault plane corrugation) or it may be due to the kinematic limitation, caused by rupture structure of the geo environment (cutting by other faults).

5.4 *Criteria for Selecting One of the Nodal Planes as the Actual Fault Plane*

The knowledge of the orientation of the actual fault plane is a necessary condition for using the expressions (34). Two nodal planes (for double-couple earthquake focal mechanism) represent two variants of position of such a plane in the stress reconstruction from seismological data. There are data on the position of the actual fault planes only for earthquakes, which faults break the surface or for sufficiently large earthquakes. The MCA includes new criteria for selecting one of the nodal planes as the actual fault plane.

5.4.1 Criteria for Selecting the Actual Fault Plane

We will assume that the slip fault corresponds to the nodal plane with the normal $\boldsymbol{n}$ that satisfies the condition:

$$\left(\tilde{\tau}_n^i + k_s \tilde{\sigma}_{nn}^i\right) - \left(\tilde{\tau}_s^i + k_s \tilde{\sigma}_{ss}^i\right) > 0. \tag{35}$$

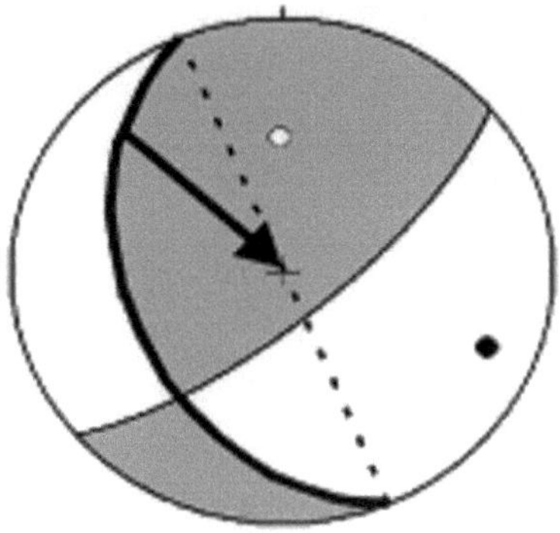

Fig. 14 The double-couple earthquake source mechanism of the 1994 Shikotan earthquake with the selected actual fault plane (in accordance with the criterion (35)) and slip direction (lower hemisphere)

In other words, from two nodal planes of the earthquake focal mechanism, the actual fault plane is that plane for which the Coulomb stresses are greater than for other.

The nodal planes delivering points to the left upper part of the summary Mohr diagram are more preferable as the actual fault plane according to maximumal. Criterion (35) works formally, in the case, when the nodal planes are close to the shear stresses planes, because the points of these nodal planes on the Mohr diagram are located near the vertical axis. In fact, the criterion (35) predetermines that the actual fault plane is that one which has the greater value of cohesion τ_s^i. As will be shown later, this means that the value of the stress drop for the chosen plane will also be greater.

5.4.2 Examples of Using Criteria for Selecting the Actual Fault Plane

For example, the criterion (35) was applied for choosing the actual fault plane for the Shikotan earthquake of 4th October, 1994 (M_b =7.4, the coordinates of the earthquake were—43.71 N, 147.33 E, depth 33 km). The performed calculations, based on catalog data of Global CMT Project (http://www.globalcmt.org), made it possible to monitor the source area before and after the earthquake. These results are presented in Table 1. The best approximation of the CMT solution for the analyzed earthquake determines the position of the poles of two nodal planes with azimuths and dip angles for **n** 68° and 49° and ***s*** 320° and 16°, respectively. The double-couple earthquake source mechanism is shown in Fig. 14.

As can be seen from the data (Table 1), the orientation of the principle axes of the macro stress tensor has stabilized a few years before the catastrophic earthquake (since 1989). The Lode-Nadai coefficient varied from 0.07 to 0.18 for the entire monitoring period and just before the earthquake the Lode-Nadai coefficient was 0.08. The calculations allow us to obtain the values of the reduced stresses for both nodal planes according to expressions (32): $\tilde{\sigma}_{nn} = 0.585, \tilde{\sigma}_{ss} = -0.645, \tilde{\tau}_n = 0.731, \tilde{\tau}_s = 0.764$.

Table 1 The results of stress monitoring for a volume with a center of 43.5° E and 147.5° N

Date	Azimuth σ_1	Dip σ_1	Azimuth σ_3	Dip σ_3
1983.04	316.4	64.8	129.4	22.1
1983.08	345.9	65.2	121.7	17.9
1985.03	316.4	64.8	129.4	22.1
1985.03	315.6	73.1	127.8	16.7
1985.03	304.5	68.5	129.3	21.5
1987.06	345.9	65.2	136.1	21.9
1988.03	345.9	65.2	136.1	21.9
1988.12	345.9	65.2	136.1	21.9
1989.08	14.1	65.2	119.7	7.1
1989.12	14.1	65.2	119.7	7.1
1990.07	17.8	71.1	131.5	7.9
1991.07	17.8	71.1	131.5	7.9
1992.07	17.8	71.1	131.5	7.9
1993.11	**17.8**	**71.1**	**131.5**	**7.9**
1994.10	316.4	64.8	129.2	25.1
1994.11	360	60	129.4	22.1
1994.12	360	60	129.4	22.1

The obtained result coincides with the determination of the fault plane, performed in work (Arefiev and Delui 1998) on the basis of the analysis of aftershock sequences in the area near the source zone of this earthquake. We think that there is a curvature of the contact zone between the oceanic and continental plates at this area and it caused this type of destruction. As a consequence of the above analysis, the choice of the actual fault plane is associated with a greater efficiency of discharging the accumulated internal energy.

Another example of using the criterion (35) is the analysis of sub sources of the Spitak earthquake of 7th December, 1988 (M = 6.9—according to the Richter scale). Table 2 shows the parameters of sub-sources and the main shock (F). The mechanism at the focus of the earthquake was different from the double couple and it had a coefficient of the form of the centroid moment tensor of 0.33. Figure 15 shows: a—the best approximation of the CMT solution; b—the mechanisms of the sub-sources (Arefiev 2003).

The results of the reconstruction allowed us to determine the orientation of the principle stress axes of the stress tensor for the study region: the azimuth and the dip angle of the principal stress—σ_1: 252° and 6°, убрать—σ_3: 157° и 37°, and the Lode-Nadai coefficient $\mu_\sigma \approx 0.3$. The mechanisms of sub-sources were analyzed for the purpose of selecting the actual fault in accordance with the presented approach here. The results of the analysis with the value of the coefficient of the surface friction are shown in table.

Table 2 Parameters of the sub-sources of the Spitak earthquake (Arefiev 2003)

#	Strike	Dip	Slip	$M_{o,}$ 10^{18} Nm
F	290	53	118	0.3
1	295	53	118	4.9
2	322	85	150	3.0
3	300	53	115	1.0
4	305	90	175	1.6
5	305	68	115	2.2

Table 3 The reduced stresses and the criterion (35) for the sub-sources

i	σ_{nn}^{i}	σ_{ss}^{i}	$\tilde{\tau}_{n}^{i}$	$\tilde{\tau}_{s}^{i}$	Criterion (35)
F	−0.260	0.270	0.948	0.180	0.46
1	−0.261	0.277	0.962	0.182	0.39
2	0.828	0.182	0.522	0.520	0.15
3	−0.263	0.281	0.960	0.183	0.45
4	0.503	0.059	0.760	0.870	0.15
5	0.217	0.296	0.971	0.111	0.79

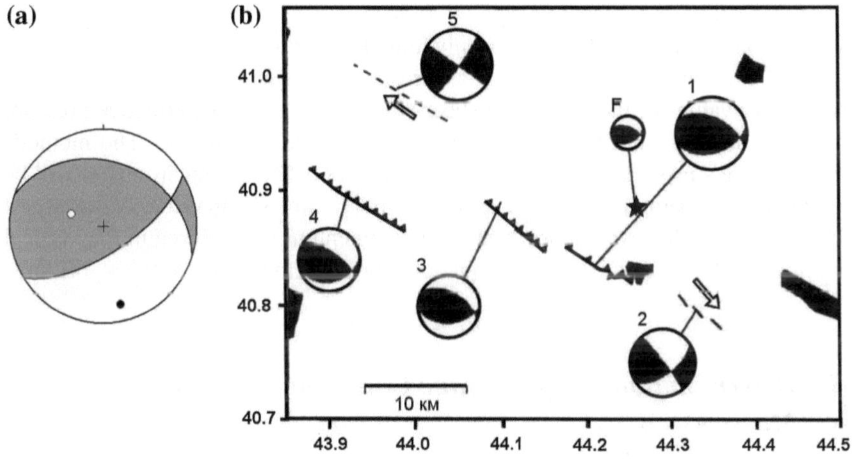

Fig. 15 The earthquake focal mechanism of the Spitak's earthquake (**a**) and a fault map with corresponding earthquake focal mechanisms of the Spitak sub-sources (**b**) (Arefiev, 2003). The asterisk is the epicenter of the main shock

Figure 16 shows the results of the analysis of the actual fault planes of the sub-sources on Mohr diagram in the form of reduced stresses. All the planes that were identified as actual fault planes (on the basis of the analysis of aftershock sequences (Arefiev 2003) satisfy the criterion (35). It can be seen (Fig. 16) that the vectors of reduced stresses determine the position of the points near the largest Mohr circle, and all auxiliary planes (exception is # 4) are located near the central part of the Mohr

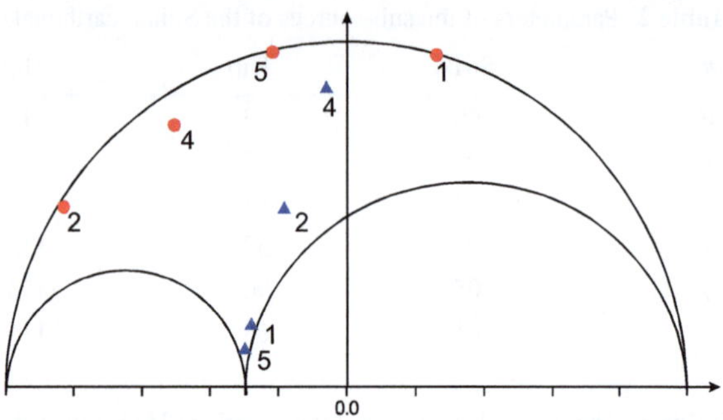

Fig. 16 Analysis of the actual fault planes of the sub-sources of the Spitak's earthquake on the Mohr diagram. Circles—points for nodal planes that satisfy the criterion (35); triangles—points for nodal planes that do not satisfy criterion (35). Point F and point 3 (Table 3) coincide with point 1

diagram, thus that determines the impossibility of realization of these planes for the analyzed stress field.

6 Stress Calculation and Estimation of the Rock Massive Strength Parameters. Third and Fourth Stages of MCA

As noted above, it is possible to obtain only four of the six independent parameters that determine the symmetric stress tensor as a result of the first stage of reconstruction. The relative values of the maximum shear stresses τ/τ_f and the effective pressure p^*/τ_f are determined after the second stage of the reconstruction. The methods for estimating stress values and *strength parameters* of the geo environment will be presented in this paragraph (the third stage of MCA). Stress drop values of the largest earthquakes and the law of conservation of momentum of the force in the vertical direction will be used as additional equations for implementation of the algorithms.

6.1 Determination of the Average Cohesion of the Rock Massive

6.1.1 Algorithm for Estimating the Stress Values on the Basis of Stress Drop Values of the Largest Earthquakes

There are earthquakes with wide range of magnitudes in the region where tectonic stresses are reconstructed. The most effective range of magnitudes for stress reconstruction is the range in which earthquakes with the largest and the low magnitudes are excluded according to the practice of applying the MCA algorithm and the STRESSseism program. The best range is that, for which the difference between the magnitudes is 2–3 units. In the case, when there are many earthquakes with low

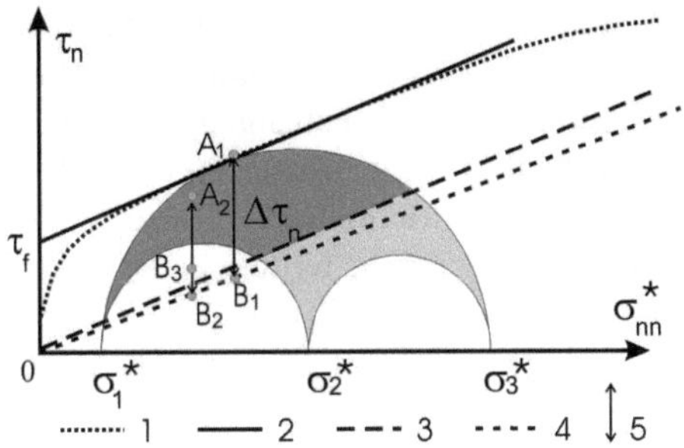

Fig. 17 Geometric analysis of the stress drop on the Mohr's diagram 1—a possible view of the maximum curvilinear envelope, which characterize the limit of the effective internal strength for a rock massif of an average scale that correspond to the scale of the stress parameters reconstructed with the help of MCA; 2—linear approximation of the limiting curvilinear envelope, used in the MCA; 3—line of static friction of rocks; 4—line of kinematic friction of rocks; Area of dark gray color—the geometrical place of points for stresses on the plane of cracks, which can be activated for a given stress tensor (light gray fill—other stress states on arbitrarily oriented planes); 5—the value of the drop of shear stresses points A_1 and A_2

magnitudes the additional stress inversion can be done for these earthquakes, but the difference between the low magnitudes also should be 2–3 units in the range.

Thus, there are some earthquakes which were not used in the reconstruction. The source size of these earthquakes is much more than the source size of those which were used in the reconstruction. Within the framework of the MCA, it is proposed to use such earthquakes to estimate the absolute values of the stresses. The availability of data on the values of shear stresses relieved in the earthquake source for such earthquakes is a necessary condition for the application of the algorithm. These data are the result of the analysis of seismic records based on a complex of seismological methods. Therefore in the MCA, they can be considered as additional. These data form the basis of the third stage of the reconstruction of the stress tensor parameters. It should be noted that, the data on shear stresses relieved in the source area of strong California earthquake were used to estimate the tectonic stress values in the work (Hardebeck and Hauksson 2001).

Such approach became possible, because the MCA allows us to estimate the relative values of the stress drop of the largest earthquakes after the first two stages of reconstruction of the stress parameters. The characteristic linear source dimension of these earthquakes is not less than the characteristic linear dimension of the averaging window for the reconstructed stress tensor parameters. The algorithm for this estimation is illustrated in Fig. 17. This figure shows the vertical segment connecting the point of contact of the limiting envelope with the largest Mohr circle of the state of stress in the area which includes the earthquake source and a line of kinematic friction. The length of this segment (A_1B_1) is equal to the value of the difference of shear stress before and after the earthquake, i.e. ($\Delta\tau_n = \left|\tau_n^0 - \tau_n^1\right|$), for the case when the slip direction coincides with the shear stress direction, which was here before the earthquake.

The possibility of such a geometric interpretation is based on the fact that follows from the theory of destruction (Rice 1980; Osokina 1988): the normal stresses on a fault plane (shear type fracture) remain unchanged before and after fault activation ($\sigma_{nn}^0 = \sigma_{nn}^1 = \sigma_{nn} \leq 0$), as well as shear stresses on the fault (after fault activation) are equal to the value of kinematic friction ($\tau_n^1 = -k_k \sigma_{nn}^*$). The expression for the stress drop value normalized by τ_f can be written in the following form:

$$\begin{aligned} \langle \Delta\tau_n/\tau_f \rangle &= \bar{\tau}_n \langle \tau/\tau_f \rangle + k_k \left(\bar{\sigma}_n \langle \tau/\tau_f \rangle - \langle p^*/\tau_f \rangle \right) \text{ where} \\ \bar{\tau}_n &= (1-\mu_\sigma)\,\ell_{1n}\ell_{1s} - (1+\mu_\sigma)\,\ell_{3n}\ell_{3s} > 0, \\ \bar{\sigma}_n &= (1-\mu_\sigma)\,\ell_{1n}^2 - (1+\mu_\sigma)\,\ell_{3n}^2 + 2\mu_\sigma/3 < 0. \end{aligned} \tag{35}$$

Here $\bar{\tau}_n$, $\bar{\sigma}_n$—the reduced deviatoric stresses on the fault plane, ℓ_{in}—the direction cosines of the normal to the fault plane of the analyzed earthquake with the principal stress axes σ_i ($i = 1, 3$), and k_k—the coefficient of kinematic friction.

The stresses normalized by τ_f are marked with angle brackets in expression (35), as it was done in the previous paragraph. Thus, the right-hand side of the expression (35) is completely defined after the first ($\bar{\tau}_n$, $\bar{\sigma}_n$) and second ($\langle \tau/\tau_f \rangle$,$\langle p^*/\tau_f \rangle$) stages of the reconstruction.

In the MCA algorithm, implemented in calculations below, the value of static friction k_s and effective internal friction k_f were assumed at 0.6, and the value of kinematical friction k_k was 0.5. The possibility of using expression (35) implies the constancy of fluid pressure in each of the calculated domains at the moment of rupture activation ($p_{fl}^0 = p_{fl}^1$). The first summand in (35) characterizes the values of shear stresses, acting along the fault plane before the earthquake (point A_I in Fig. 17), and the second one does the shear stresses τ_n^1, which will act here after the earthquake (point B_I in Fig. 17).

After the calculation of the relative value $\langle \Delta\tau_n/\tau_f \rangle$ by the data of the first two stages of MCA, we might estimate τ_f. As additional data, we use the value of stresses $\Delta\tau_n$ discharged during the large earthquake:

$$\tau_f = \frac{\Delta\tau_n}{\langle \Delta\tau_n/\tau_f \rangle}. \tag{36}$$

The value $\Delta\tau_n$ on the right of (36) is defined with the use of data on the scalar value of seismic moment M_o and from the earthquake source geometry (Kasahara 1981):

$$\Delta\tau_n = \chi M_o/(WL^2), \tag{37}$$

where L and W are the length (the greatest of the subhorizontal dimensions) and the width of the earthquake source, and χ is a parameter dependent on the type of displacement in the earthquake source (normal fault, revers, or strike-slip fault) and on its geometrical form ($0.65 \leq \chi \leq 1.85$).

It is shown in (Kostrov 1975; Kostrov and Das 1988) that if the energy associated with change in the area of the growing rupture surface is negligible and we consider that shear stresses after the beginning of displacement along the rupture have not changed, but remained equal to the kinematic friction, then shear stresses discharged in the earthquake source could be defined using the data on released seismic energy E_s:

$$\Delta\tau_n = 2\mu E_s / M_o, \tag{38}$$

where μ is the shear elastic modulus (for the crust $\mu \approx$ 3–5 10^4 MPa), E_s and M_o are presented in joules.

It should be noted there that there is a proportion in seismology that allows us to estimate the socalled apparent stresses $\bar{\tau}_n = \left|\tau_n^0 + \tau_n^1\right|/2$, which is the average stress between the two states at the rupture, on the basis of data on seismic energy:

$$\eta\bar{\tau}_n = \mu E_s / M_o, \tag{39}$$

where $\eta = \Delta\tau_n / 2\bar{\tau}_n$ (Kostrov 1975; Kostrov and Das 1988) is the so called seismic efficiency. Expression (39) is often used for definition of $\bar{\tau}_n$. But, due to the great uncertainty of the η value, such estimations cannot consider reliable.

The value E_s in expression (39) could be obtained using correlation dependences with the magnitude of earthquake M_e, which characterizes energy radiated in seismic waves:

$$\log\ E_s = 3M_e/2 + 4.35. \tag{40}$$

If the destruction is implemented along the previously existing rupture, which has not restore its cohesive strength to the undistracted state ($\tau_s < \tau_f$), then on the Mohr diagram the vertical segment, which characterizes stress drops, begins from the point lying below the limit envelope, but higher than the line of minimum resistance due to dry static friction (the segment A_2B_2 in Fig. 17). In the case when the slip direction along the rupture does not coincide with the shear stress direction acting on the plane before activation, the vertical segment does not reach the line of resistance due to kinematic friction (the segment A_2B_3 in Fig. 17).

The stress parameters substituted into expression (35) should correspond the averaging period, which precedes the strong earthquake, and data on its stress drop will be used for determining τ_f. For example, using the catalogue of earthquake focal mechanisms with the range of magnitudes of 4.5–7 allows us to tell that the reconstructed parameters of the stress tensor may correspond to the averaging scale of 20–100 km. The exact value depends on the density of the distribution of earthquake foci, and the range of magnitudes of 2.5–6 may correspond to the averaging scale of 5–30 km. In this case strong earthquakes, whose stress drop can be used for estimation of internal cohesion, should not have magnitudes less than 6.5 and 5.5 correspondingly, i.e., the focal area should not be less than the stress averaging window. The algorithm of the third stage of MCA allows us to estimate the value of effective internal cohesion of rock τ_f, and then, using the data on relative values of maximum shearing stress and effective pressure (the second stage of MCA), transit to their absolute values.

$$\tau = \left\langle \frac{\tau}{\tau_f} \right\rangle \tau_f, \quad p^* = \left\langle \frac{p^*}{\tau_f} \right\rangle \tau_f. \tag{41}$$

It should be noted that the deviatoric component of the stress tensor is defined after the third stage of the MCA, but the spherical component of the stress tensor is not defined. The effective pressure p^* includes two unknown values - isotropic tectonic pressure and fluid pressure in the fractured-pore space.

6.1.2 Examples of Estimation of the Value of Effective Internal Cohesion of Rock Massif

For calculation of τ_f, we will use the data on values of the stress drops of the Simushir earthquake November 15, 2006 ($M_w = 8.3$). The stresses acting in the crust of the northwestern flank of the Pacific seismoactive area before the Simushir earthquake was implemented on a 3D grid for the layer with the median surface at a depth of 20 km. The grid step was 0.2° on the basis of earthquake focal mechanisms data of Global CMT Project. The most representative range of earthquake magnitudes from this catalogue for the studied part of the crust is $5.5 > M_w > 6.5$, and the distribution density of their foci allowed us to tell about the averaging scale of stress parameters of 50–70 km.

To estimate τ_f, the seismological data on energy parameters of the Simushir earthquake event were used. The value of energy released in seismic waves E_s and the seismic moment M_o of the Simushir earthquake were correspondingly 7.4×10^{16} and 3.4×10^{21} J, according to the data of Global CMT Project. The fault parameters of the Simushir earthquake, according to the CMT Project, were the following: Strike1 = 215°, Dip1 = 15°; Strike2 = 33°, Dip2 = 75° (thrust faulting).

The results of seismic inversion (http://earthquake.usgs.gov/eqcenter/eqachives/poster/2006/20061115_image.php) show that the actual fault plane was the first nodal plane, This plane corresponded to the gentle dipping (15°) of the subducted oceanic plate under the subcontinental plate and had a length along the Trench (W) of about 400 km, while in the cross direction (L) of about 130 km (solid line contour in Fig. 18). The analysis of distribution of the aftershock sequences, made in (Tikhonov et al. 2007), gives correspondingly 300 and 60 km (thin line contour in Fig. 18).

Using the data on energy released during the Simushir earthquake, through expression (38) we find $\Delta\tau_n \approx 1.7$ MPa (17 bar). If we use expression (37) and data on the earthquake source geometry (with $\chi = 1$, because the earthquake source area is not isometric), then we will obtain $\Delta\tau_n \approx 0.5$ MPa (by the Global CMT Project data), $\Delta\tau_n \approx 1.4$ MPa (by works Tikhonov et al. 2007). The presented results show that calculation of $\Delta\tau_n$ by the CMT Project data on the earthquake source dementions understates the value, while the estimation of source geometry in the work (Tikhonov et al. 2007) gives a better correspondence to the energergy parameters of the earth-

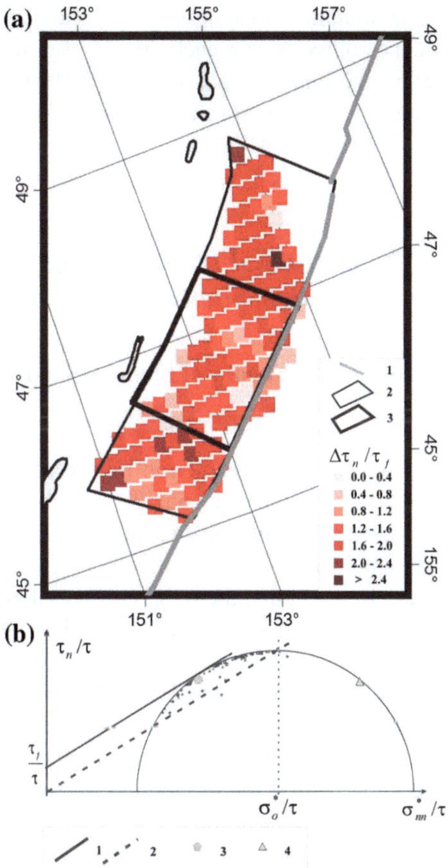

Fig. 18 **a** Distribution of the values of shear stresses normalized by τ_f for Simushir earthquake calculated within the framework of the third stage algorithm of the MCA. 1—oceanic Trench, 2—geometry of the used in our calculation (Tikhonov et al. 2007), 3—geometry of the Simushir earthquake source by the Global CMT Project data. **b** For the generalizing Mohr diagram, there is the position of points with values of the normalized normal and shear stresses along the source areas of these earthquakes. Normalization is made by the value of maximum shear stresses corresponding to the domain with parameters of the stress tensor, which is the closest to the beginning of earthquake rupture. Here the solid line is the limit of effective strength (1); the dashed line is the line of minimum dry friction (2); the pentagon (3) and triangle (4) with light gray fill correspond to the stresses along the fault and auxiliary nodal planes of focal mechanisms Simushir earthquake

quake. In the following calculations, we will use the source geometry given in these works. The value of stress drop in the source area of Simushir earthquake can be obtained, with a precision to valuation at the unknown value of effective adhesion of rock massif τ_f, using the results of the first two stages of MCA reconstruction and expression (35). For the source area of Simushir earthquake, 90 domains exist for which the data on stress tensor parameters were obtained (Fig. 18).

Using the data of these domains and summing the contribution in stress drop for the parts of the earthquake source within each domain, on the basis of expression (35) and data on the earthquake source dementions, we find the average value of ratio $\langle \Delta\tau_n/\tau_f \rangle = 1.41$. Next, using the data on values of the stress drop, we find the value of effective cohesion $\tau_f = 1.0 - 1.2$ MPa.

6.2 Estimation of the Isotropic Pressure Using the Equations of Equilibrium of Shallow Shells

6.2.1 Defining Equations of the Theory of Shallow Shells and Thick Plates

The results of the three stages of reconstruction by the MCA algorithm make it possible to determine all the stress tensor components (the principal stresses are reduced ($\sigma_i^*, i = 1, 2, 3$) by the level of the fluid pressure). Thus, the tectonic pressure p and the fluid pressure p^* remain unknown, and effective pressure p^* is the difference between these two values. To determine them, it is proposed to use the fact that the stress data obtained for discrete grid nodes of the geo-medium must satisfy the notion of the stress field.

The notion of stress field in mechanics is understood as the continuity of the stress components, which requires them to satisfy the differential equations of conservation of the momentum of the force. Violation of this requirement is possible at the contact boundaries of essentially different properties of the mediums and at the fault boundaries in the case of a displacement jump.

According to the results of the natural stress inversion, as a rule, data on stresses are obtained in averaging scale comparable to the thickness of the crust or its main layers. Therefore, the equation of the law of momentum conservation of forces, or the equation of equilibrium, can be written in a simplified form, used in the theory of shallow shells and thick plates (Timoshenko and Goodier 1970):

$$\frac{\partial \widehat{\sigma}_{xx}}{\partial x} + \frac{\partial \widehat{\sigma}_{xy}}{\partial y} + \frac{T_{xz}^{\xi} - T_{xz}^{\varsigma}}{h} = 0,$$

$$\frac{\partial \widehat{\sigma}_{xy}}{\partial x} + \frac{\partial \widehat{\sigma}_{yy}}{\partial y} + \frac{T_{yz}^{\xi} - T_{yz}^{\varsigma}}{h} = 0,$$

$$\frac{\partial \widehat{\sigma}_{xz}}{\partial x} + \frac{\partial \widehat{\sigma}_{yz}}{\partial y} + (\kappa_{xx}\widehat{\sigma}_{xx} + \kappa_{yy}\widehat{\sigma}_{yy}) + \frac{N_{zz}^{\xi} - N_{zz}^{\varsigma}}{h} + \widehat{\gamma} = 0, \quad (42)$$

where z—a vertical axis directed toward the center of the earth, $h(x, y) = \xi(x, y) - \zeta(x, y)$—thickness of the layer, changing laterally and estimated taking into account the irregularities of the surface of its roof ζ and bottom ξ, κ_{xx} and κ_{yy}—the curvature of the shell in the proper directions, $\widehat{\gamma}(x, y)$—the average specific weight of the rocks of the upper crust, is equal to

$$\widehat{\gamma} = \frac{1}{h} \int\limits_{\zeta(x,y)}^{\xi(x,y)} \gamma(x, y, z) dz, \tag{43}$$

$\widehat{\sigma}_{ij}$—the average stresses are defined as:

$$\widehat{\sigma}_{ij} = \frac{1}{h} \int\limits_{\zeta(x,y)}^{\xi(x,y)} \sigma_{ij} dz, i, j = x, y, z, \tag{44}$$

and $T^{\xi}_{xz} = \sigma_{xz}|^{\xi}$, $T^{\xi}_{yz} = \sigma_{yz}|^{\xi}$, $N^{\xi}_{zz} = \sigma_{zz}|^{\xi}$ и $T^{\varsigma}_{xz} = \sigma_{xz}|^{\varsigma}$, $T^{\varsigma}_{yz} = \sigma_{yz}|^{\varsigma}$, $N^{\varsigma}_{zz} = \sigma_{zz}|^{\varsigma}$—the stresses applied to the bottom and roof of the layer under consideration. It should be noted that when equations were written in the form (42), the condition of absence of the loading on the daylight surface of the earth's crust is used. An approximate representation of the equilibrium equations in the form (42) is valid for a slow change along the lateral thickness of the layer.

We shall accept the linear distribution law along the z axis for the normal and shear stresses acting on the horizontal planes. In this case, the average stress in the layer can be related to the stresses acting along its roof and the bottom by simple expressions:

$$\widehat{\sigma}_{xz} = 0.5\left(T^{\zeta}_{xz} + T^{\xi}_{xz}\right), \ \widehat{\sigma}_{yz} = 0.5\left(T^{\zeta}_{yz} + T^{\xi}_{yz}\right), \ \widehat{\sigma}_{zz} = 0.5\left(N^{\zeta}_{zz} + N^{\xi}_{zz}\right). \tag{45}$$

Further, we will put a tilde over the stress tensor parameters, realizing that these stress tensor parameters determine the average value of the stresses. These average stresses (in the accepted linear approximation of the peculiarities of distribution of the stresses in depth) can be regarded as stresses which related to the middle surface of the layer (shell). We shall also neglect the sphericity of the Earth's surface, assuming in the last expression of the set of equations (42), that $\kappa_{xx} = 0$, $\kappa_{yy} = 0$, writing the equilibrium equations, going from the theory of shallow shells to the theory of thick plates.

6.2.2 Calculation of the Value of the Confining Pressure *p*

Calculation of stresses in a layered shell should be carried out sequentially from the upper layer to the lower layers. This is due to the fact that on the roof of the upper layer of the shell (see Fig. 19) the normal and shear stresses are zero and, consequently, here

$$\widehat{\sigma}_{xz} = 0.5T^{\xi}_{xz}, \ \widehat{\sigma}_{yz} = 0.5T^{\xi}_{yz}, \ \widehat{\sigma}_{zz} = 0.5N^{\xi}_{zz}. \tag{46}$$

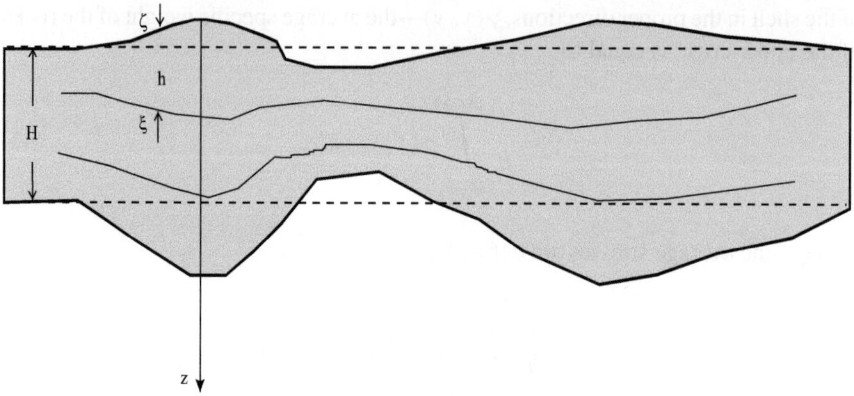

Fig. 19 Approximation scheme of layers of the earth's crust by shallow shells, H—the average thickness of the earth's crust

Using the expressions (31), (32), we will replace the stresses σ_{ij} in the expressions (42), through the reduced stresses $\tilde{\sigma}_{ij}$ and values p, τ,

$$-\frac{\partial p}{\partial x} + \widehat{\sigma}_{xx}\frac{\partial \tau}{\partial x} + \widehat{\sigma}_{xy}\frac{\partial \tau}{\partial y} + \left(\frac{\partial \widehat{\sigma}_{xx}}{\partial x} + \frac{\partial \widehat{\sigma}_{xy}}{\partial y} + \frac{2\widehat{\sigma}_{xz}}{h}\right)\tau = 0,$$

$$-\frac{\partial p}{\partial y} + \widehat{\sigma}_{xy}\frac{\partial \tau}{\partial x} + \widehat{\sigma}_{yy}\frac{\partial \tau}{\partial y} + \left(\frac{\partial \widehat{\sigma}_{xy}}{\partial x} + \frac{\partial \widehat{\sigma}_{yy}}{\partial y} + \frac{2\widehat{\sigma}_{yz}}{h}\right)\tau = 0,$$

$$-\frac{2p}{h} + \widehat{\sigma}_{xz}\frac{\partial \tau}{\partial x} + \widehat{\sigma}_{yz}\frac{\partial \tau}{\partial y} + \left(\frac{\partial \widehat{\sigma}_{xz}}{\partial x} + \frac{\partial \widehat{\sigma}_{yz}}{\partial y} + \frac{2\widehat{\sigma}_{zz}}{h}\right)\tau + \widehat{\gamma} = 0. \quad (47)$$

It should be reminded that the reduced values of the stresses are determined by the values of the direction cosines of the angles of the principle stresses with the coordinate axes $j = x, y, z$ (ℓ_{ij}) and the Lode-Nadai coefficient (μ_σ) and they are calculated from the results of the first stage of the reconstruction. The parameters of the unknown stress tensor refer to the middle of the shell layer.

The set of equation (47) can be written in an orthogonal curvilinear coordinate system: one coordinates is vertical, and the other two are in the horizontal plane. We will consider that one of the coordinate directions coinciding with the shear stress acting on the horizontal planes (Fig. 20). We will define these stresses as underthrusting shear stresses, i.e. they are responsible for under-thrust faults. The last equation in the (47) can be written in the form, using such a representation of the law of momentum conservation in the vertical direction:

$$p = \widehat{\gamma}\frac{h}{2} + \tau\widehat{\sigma}_{zz} + \frac{h}{2}\frac{\partial}{\partial s}\left(\widehat{\sigma}_{sz}\tau\right). \quad (48)$$

where s—coordinate along the shear stresses acting on horizontal planes (see Fig. 20).

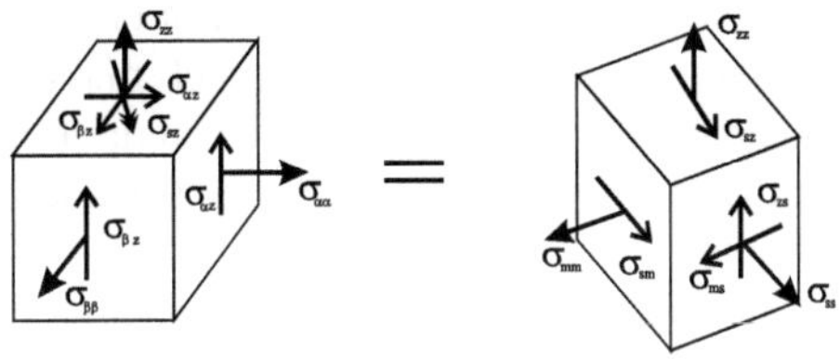

Fig. 20 The loading scheme of a parallelepiped, connected with the direction of the shear stress on the middle surface of the shell

It should be pointed out that the expression (48) is more accurate than the calculation of tectonic pressure on the basis of R. Sibson hypothesis (Sibson 1974) in which the vertical stresses acting on horizontal planes are equated with the pressure exerted by the weight of the overlying rocks. It should be noted that the Sibson's condition can be obtained from (48) by putting $\widehat{\sigma}_{sz} = 0$.

The value of the isotropic pressure of the investigated area of the earth's crust can be estimated using the data on the relative value of the maximum shear stress obtained after the second stage of the MCA, data on the s direction for underthrusting shear stresses on the horizontal plane (the first stage of the MCA), and also assuming the values of the maximum shear stresses (the third stage of the MCA). After that, the value of the fluid pressure (it is average for the scale of the obtained stress data) can be determined:

$$p_{fl} = p - p^{*}. \tag{49}$$

Note that one of the other two equations of conservation of momentum in the horizontal direction can be used to verify the accuracy of estimation of stresses during the first three stages of reconstruction, rewriting it in the s direction:

$$\left(-p + \widehat{\sigma}_{xx}\tau\right)_{s_1} = \left(-p + \widehat{\sigma}_{xx}\tau\right)_{s_0} - \frac{2}{h}\left(\tau\widehat{\sigma}_{xz}\right)_{s_0}. \tag{50}$$

Here the subscripts are behind the brackets: s_0 and s_1 are two consecutive coordinates of the points along the s trajectory for which verification is performed.

7 Conclusion

Some theoretical principles of geomechanics and tectonophysics were presented in this work. These principles are subject to the problem of creating a methodology for studying the stress-strain state and the mechanical properties of rock massif. It is important to note that the developing approach is based on the difference in solid samples and natural massifs, (which have a multitude of defects in the form of low strength surfaces), deformation behavior. The developed methodological principles for reconstructing the parameters of tectonic stresses and increments for seismotectonic deformations can be considered as a tool, which uses for this purpose active tectonic faults and fractures. It can be said that the creation of methods for studying the natural stresses in the rock massif is equivalent to the laboratory tools: (a dynamometer and a ruler, which are used by experimenter in laboratory modeling). At the same time, the scale of the solid samples and natural massifs are incommensurable.

The basis of the algorithm of calculation of the stress-strain state parameters is composed of approaches, following from a number of fundamental principles of the continuum mechanics and mechanics of rupture destruction. These approaches allowed the formulation of energetic criteria, defining the concepts of generalized stresses responsible for the activation of old faults and creation of new ones, and released generalized deformations, formed in geo-environment results of faulting displacements. The totality of these criteria characterizes the stage of the steady-state quasi-plastic fl owing, during which loading conditions and stresses, acting in natural massifs, remain stable at rather long time intervals in the process of deformation. This stage is characterized by the associated flow law. The energetic criteria determine the requirements for the dissipation of internal energy on the discontinuous displacements and the monotonic elasto-plastic deformation, which is a consequence from the Von Mises's maximum principle. Energetic criteria composed the basis of the method of cataclastic analysis and allowed formalization of the principles of creation of homogeneous samplings of structural-kinematic data on slip faults (earthquake source mechanisms).

In the framework of the MCA, it was possible to show that a number of well-known methods of stress inversion (defining inequalities of the kinematic method by Gushchenko (Goustchenko 1996) and the right dihedral method by J. Angelier (Angelier and Mechler 1977)) directly follow from the condition of a decreasing internal energy after discontinuous displacements. This fact made it possible to revise one of the main principles of the Batdorf-Budyansky's dislocation theory of plasticity (Batdorf and Budiansky 1949), about the coincidence of the slip direction and the shear stresses (acting before the fault), which was used in a number of methods for reconstruction of the tectonic stresses. It is established that the faults of identical hierarchical level influence on each other. Displacements for simultaneously activated and closely located faults, are determined not only by the initial stress field, but also by the fault configuration. Analysis of displacements in the source areas of the largest earthquakes showed that they form the tensor of the released deformations, which is maximally approximated (in the form and orientation of the principle axes) to the stress tensor that was before the earthquake. It is in this case, the most effective internal energy dissipation in the vicinity of the earthquake source area is achieved.

Theoretical analysis of the properties and peculiarities of quasi-plastic flow, developing under the action of loading in fractured media, have shown, that its character changes depending on the intensity of deviatoric components of the stress tensor and on the value of the effective pressure (accounting for the fluid pressure). The earlier stage of fracturing deformation, during which new ruptures or faults are not formed, is resulted from the activation of the existing fault structure and is characterized by the non-associated flowing law, whereas, at the stage of the maximum effectiveness of quasi-plastic deformation, the flowing law tends to become the associated one. These peculiarities lead to the fact that, at the initial stage of quasi-plastic flowing, only a part of energetic criteria of the method of cataclastic analysis hold (criterion of internal energy dissipation on rupture displacements). The fulfillment of all these criteria of the method of cataclastic analysis takes place at the limiting stage, for which in natural massifs, along with activation of pre-existing faults, new ruptures are also formed.

A certain dual nature of the discontinuous deformations, (which on the one hand, represent an act of destruction, and on the other hand, characterize the deformation properties of the continuum), predetermines sequence of the reconstruction stages of the stress-strain state parameters. In the MCA, during the first stage: the energy principles of the continuum mechanics are used, the formation of homogeneous SKD samples is performed, the parameters characterizing the ellipsoids of the stress tensor and quasiplastic deformations are calculated: the orientation of the main semi-axes (principle stresses and increments for seismotectonic deformations) and their ratio (Lode-Nadai coefficient). During the second stage: these stress parameters and the formed homogeneous samples are the basis for determining the relative stress values. Here, the principles of the mechanics of destruction that define the natural massifs as an initially fractured Coulomb medium are used.

It is important to note that formulation of the conditions of fracturing of natural massif (creation of new faults and activation of previously existed ones) is accepted in the Method of Cataclastic Analysis in its general form, satisfying the results of experiments on rock sample fracturing. Defining parameters of Coulomb's condition were considered as unknown. The determination of these strength parameters and the stress values is performed during the third stage of the MCA. The principles of the continuum mechanics are again used, in particular, the law of momentum conservation in the vertical direction, written in the approximation of shallow shells.

The homogeneous samples set of SKD are the most important element of the investigations. They are formed as the results of the first stage of reconstruction based on the energetic criteria. In fact, it is the homogeneous samples set of SKD made it possible to make a qualitative jump from the determination of the orientation of the principal axes of the stress tensors and increments for seismotectonic deformations to the calculation of the value of the spherical and the intensity of the deviatoric components of these tensors. It can also be drawn an analogy with laboratory experiments in which in order to estimate the strength parameters of brittle rock samples it is necessary to have the measures of the morphological parameters of the forming discontinuous structures, with respect to the loading stresses.

At the second stage of calculations, the key issue for the estimation of modem stresses from data on earthquake source mechanisms is the method of determining the fault plane, suggested in this work. This method is based on the application of the Coulomb's criterion within the frameworks of the limiting state theory for selecting one of the two nodal planes, which is realized as the fault plane in the earthquake source. According to the suggested method, that of the nodal planes is selected, for which the maximum release of the shear stresses is achieved. The effectiveness of such method of selecting the realized plane of source is verified on the strongest earthquake. The use of the Coulomb failure criterion together with the analysis of earthquakes dynamic data on the Mohr's circle made it possible to develop principles for estimating the rock mass strength parameters, such as the coefficients of static and kinetic friction.

The determination of the stress values is possible with the use of additional equations including seismological data on the dynamic source parameters of the largest earthquakes (shear stress drop) and the law of momentum conservation in the vertical direction. Previously, this type of data has already been used in a number of works on the study of natural stresses (R. Sibson, J Hardebeck).

Currently, the computer programs that allow working with both seismological (STRESSseism) and geological (STRESSgeol) data on discontinuous displacements have been created on the basis of the MCA algorithm. The performed reconstruction of the modern state of stress for the intracontinental orogens regions (Rebetsky et al. 2012) and active continental margins (Rebetsky and Tattevossian 2013) allow us to identify a number of important regularities.

References

Angelier J (1975) Sur l'analyse de measures recueilles dans des sites failles: l'utilite d'une confrontation entre les methods dynamiques et cinematiquues. Bull Soc Geol France 281:1805–1808

Angelier J (1984) Tectonic analysis of fault slip data sets. Geophys Res 89(B7):5835–5848

Angelier J, Mechler P (1977) Sur une methode graphique de recherche des contraintes principales egalement utilisable en tectonique et en seismologie: la methode des diedres droits. Bull Soc Geol Fr V XIX 6:1309–1318

Arefiev SS, Delui B (1998) The focal zone of the Shikotan earthquake of 1994: the choice of the active plane. Phys Earth 6:64–74

Arefiev SS (2003) Epicentral seismological studies. Moscow, Akademkniga. p 374 (in Russian)

Batdorf SB, Budiansky B (1949) A mathematical theory of plasticity based on the concept of slip. NASA Technical Note 1871

Bott MHP (1959) The mechanics of oblique slip faulting. Geol Mag 96:109–117

Bridgman W (1949) Studies in large plastic flow and fracture, with special emphasis on the effects of hydrostatic pressure. McGraw-Hill Book Co., Inc., New York, p 362

Byerlee JD (1967) Frictional characteristics of granite under high confining pressure. J Geophys Res 72(14):3639–3648

Byerlee JD (1978) Friction of rocks. Pure Appl Geophys 116:615–626

Byerlee JD (1968) Brittle-ductile transition in rocks. J Geophys Res 73(14):4741–4750

Carey E, Bruneier B (1974) Analyse theorique et numerique d'un modele mecanique elementaire applique a l'etude d'une populaton de failles. C R Acad Sci Paris D 279:891–894

Chernykh KF (1998) An introduction to modern anisotropic elasticity. Begell Publishing House, New York, p 282

Dobrovol'sky IP (2009) Mathematical theory of preparation and prediction of earthquakes. FIZMATLIT. Moscow, p 240 (in Russian)

Drucker DC (1959) A definition of stable inelastic material. Trans ASME, J Appl Mech 101–106

Goustchenko OI (1975) Kinematic principle of reconstruction of directions of main stresses (on geological and seismological data) DAN USSR Ser Geophys 225 (3):557–560 (in Russian)

Goustchenko OI (1996) Seismotectonic stress-monitoring of the lithosphere (structural-kinematic principle and basic elements of the algorithm) DAN USSR Ser Geophys 346 (3):399–402 (in Russian)

Handin J (1969) On the Colombo—Morh failure criterion. J Geophys Res 74(22):5343–5348
Hardebeck JL, Hauksson E (2001) Crustal stress field in southern California and its implications for fault mechanics. J Geophys Res 106(B10):21859–21882
Kasahara K (1981) Earthguake mechanies. Cambriage University Press, p 284
Knopoff L (1958) Energy release in earthquakes. Geophys J MNRAS 1: 44–52
Kostrov BV (1975) Mechanics of tectonic earthquake source. Nauka Наука, Moscow, p 176 (Russia)
Kostrov BV, Das S (1988) Principles of earthquake source mechanics. In: Cambridge monographs on mechanics and applied mathematics. Cambridge University, Cambridge, New York, Port Chester, Melbourne, Sydney, p 286
McKenzie Dan P (1969) The relation between fault plane solutions for earthquakes and directions of the principal stresses. Bull Seism Soc Am 59(2):591–601
Mises RV (1928) Angew Math Mech 8:161–185
Mogi K (1964) Deformation and fracture of rocks under confining pressure (2) compression test on dry rock sample Bul Earthq Res Inst Univ Tokyo 42(Part 3):491–514
Mogi K (1966) Pressure dependence of rock strength and transition from brittle fracture to ductile flow. Bull Earthq Res Inst Univ Tokyo 44:215–232
Nikolaevskiy VN (1996) Geomechanics and fluidodynamics. Kluwer, Dordrecht, p 349
Osokina DN (1988) Hierarchical properties of a stress field and its relation to fault displacements. J Geodyn 10:331–344
Paul B (1968) Macroscopic criteria for plastic flow and brittle fracture. In: Liebowitz H (ed) Fracture. An advanced treatise, vol 2. Academic Press, New York, pp 313–496
Rebetskii YuL (2003) Development of the method of cataclastic analysis of shear fractures for tectonic stress estimation. Dokl Earth Sci 388(1):72–76
Rebetsky YuL (1999) Methods for reconstructing tectonic stresses and seismotectonic deformations based on the modern theory of plasticity. Dokl Earth Sci 365A(3):370–373
Rebetsky YuL, Lermontova AS (2016) Accounting for the supercritical state of the geoenvironment and the problem of the long-range effect of earthquake foci. Vestnik Kraunts Nauka Zemle 32:115–123 (In Russia)
Rebetsky YuL, Polets AYu, Zlobin TK (2016) The state of stress in the earth's crust along the northwestern flank of the Pacific seismic focal zone before the Tohoku earthquake of 11 March 2011. Tectonophysics 685:60–76. https://doi.org/10.1016/j.tecto.2016.07.016
Rebetsky YuL, Sycheva NA, Kuchay OA, Tatevossian RE (2012) Development of inversion methods on fault slip data: stress state in orogenes of the central Asia. Tectonophysics 581:114–131. https://doi.org/10.1016/j.tecto.2012.09.027
Rebetsky YuL, Tatevossian RE (2013) Rupture propagation in strong earthquake sources and tectonic stress field. Bull Soc Geol Fr 184(4–5):335–346
Rice J (1980) The mechanics of earthquake rupture. In: Dziewonski AM, Boschi E (eds) Physics of the earth's interior. North-Holland, Amsterdam, pp 555–649
Rummel F, Alheid HJ, Frohn C (1978) Dilatancy and fracture induced velocity changes in rock and their relation to friction sliding pure and app. Geophys 116:743–764
Sibson RH (1974) Frictional constraints on thrust, wrench and normal faults. Nature 249:542–544
Starr AT (1928) Slip in crystal and rupture in solid due to shear. Proc Camb Phil Soc 24:489–500
Shteynberg VV (1983) On focal parameters and seismic effect of earthquakes. Izv Academy Sci USSR Phys Earth (7): 536–548
Stesky RM (1978) Rock friction-effect of confining pressure, temperature, and pore pressure Pure App. Geophys 116:691–704
Terzaghi K (1943) Theoretical soil mechanics. Wiley, New York, p 506
Tikhonov IN, Vasilenko NF, Prytkov AS, Spirin AI, Frolov DI (2007) Catastrophic simushir earthquakes November 15, 2006 and January 13, 2007. Probl Seismic sound Far East and East Sib: Intern Sci Simp Yu-Sakhalinsk: Izd IMGG FEB RAS. From 27–28 (Russian)

Timoshenko SP, Goodier JN (1970) Theory of elasticity. McGraw-Hill Book Company, New Yurk, p 576
Wallace RE (1951) Geometry of shearing stress and relation to faulting. J Geol 59:18–130
Wells DL, Coppersmith KJ (1994) New empirical relationship among magnitude, rupture length, rupture width, rupture area, and surface displacement. Bull Seism Soc Am 84(4):974–1002

Challenges in Regional Moment Tensor Resolution and Interpretation

Simone Cesca and Sebastian Heimann

1 Introduction

The seismic moment tensor (Gilbert 1970) is a representation of a seismic source, described mathematically by a symmetric tensor of second order. Its entries are generalized couples, with trace element corresponding to dipoles of different orientations and off-trace components to single couples. The moment tensor is typically interpreted upon its decomposition into subterms, representing different physical processes. Jost and Hermann (1989), further extended by Minson and Dreger (2008), reviewed proposed moment tensor decompositions. Typically, the moment tensor is decomposed first into an isotropic term, describing the amount of volumetric change at the seismic source, and a deviatoric term. This decomposition is unique. The choice of a further decomposition of the deviatoric term is instead subjective. The most used decomposition is into a double couple (DC), a model able to describe shear faulting, and a compensated linear vector dipole (CLVD), a net-volumetric source deformation model. The DC model identifies two candidate planes for the fault orientation, referred as fault and auxiliary planes; the true rupture plane cannot be discriminated under a point source approximation. The decomposition as the sum of DC and CLVD has been often adopted for tectonic earthquakes and whenever the earthquake is expected to be modelled by a DC source. The appearance of a remaining CLVD term has been attributed to poor monitoring conditions or data quality, as well as to real source processes, e.g. multiple sources, tensile faulting, collapses, rockbursts near-source heterogeneities or unmodelled wave propagation effects.

To be submitted to "*Moment Tensor Solutions—A Useful Tool for Seismotectonics*", Springer.

S. Cesca (✉) · S. Heimann
GFZ German Research Centre for Geosciences Potsdam, Helmholtzstrasse 7, 14467 Potsdam, Germany
e-mail: cesca@gfz-potsdam.de

S. D'Amico (ed.), *Moment Tensor Solutions*, Springer Natural Hazards,
https://doi.org/10.1007/978-3-319-77359-9_7

The seismic moment tensor provides important information on the seismic source parameters and the geometry of the rupture process. We can express it as the product of a scalar, the scalar moment, and a normalized matrix. The first term provides information on the released moment and moment magnitude, and could be used to assess the average slip or the rupture area, while the second term contains information on the rupture type, the rupture geometry and the role of shear, tensile and isotropic components. The moment tensor decomposition allows to discuss each moment tensor contribution. For example, the analysis of the double couple term can be used to investigate the orientation of the fault and auxiliary planes.

The moment tensor inversion relies on the classical formulation for the corresponding forward problem, which is the computation of synthetic seismograms for a given source model (moment tensor) and a given propagation problem between source and receivers (described by Green's functions). In the inverse problem, synthetic seismograms for a fixed propagation medium are compared to observations, namely recordings of the target earthquake on the Earth surface, and the procedure iterated for different moment tensor configurations. The moment tensor solution which optimizes the fit of observations is assumed as the earthquake source model.

Several methods have been implemented to perform the moment tensor inversion, either fitting first motion polarities, body wave amplitudes, amplitude ratios, full waveforms or amplitude spectra. Automated regional moment tensor inversion, here defined as the source inversion of moderate earthquakes (Mw 3.0–6.0) from regional distances (0–1000 km) using broadband data, mostly relies on the fit of full waveforms, either in the time or frequency domain. The literature focusing on the implementation and automation of regional moment tensor inversion routines is broad; relevant references include Romanowicz et al. (1993), Dreger and Helmberger (1993), Kawakatsu (1995), Pasyanos et al. (1996), Braunmiller et al. (2002), Pondrelli et al. (2002), Kubo et al. (2002), Ichinoise et al. (2003), Stich et al. (2003), Bernardi et al. (2004), Rueda and Mezcua (2005), Cesca et al. (2006), Konstantinou et al. (2010) and Cambaz and Mutlu (2016).

The resolution and stability of moment inversions are affected by a poor knowledge of the velocity model (Romanowicz et al. 1993; Š´ılený 2004; Cesca et al. 2006), noisy seismic data (Š´ılený et al. 1996; Sokos and Zahradnik 2013), limited number of seismic stations or unfavourable azimuthal coverage (Cesca et al. 2006; Ford et al. 2010; Vera Rodriguez et al. 2011; Domingues et al. 2013). As a result of these effects, the instability of moment tensor solutions may be combined with the determination of spurious non-DC terms. Thus, the interpretation of non-DC terms as real source features is often difficult; the relevance of the problem was illustrated by Cesca et al. (2013), showing a significant amount of events in the Global CMT catalogue with large CLVD components.

The challenges in the resolution of full moment tensor inversion for regional applications was discussed by Dufumier and Rivera (1997) and Krizova et al. (2013). For regional applications, the resolution of moment tensor components may become critical for shallow sources. Bukchin et al. (2010) found that the moment tensor cannot be uniquely determined from surface wave records, dominating the full waveform recordings, when wavelengths are much larger than the source depth. In such conditions, only four independent element of the moment tensor are reliably determined, while M_{xz} and M_{yz} are poorly constrained (Bukchin et al. 2010; Valentine and Trampert 2012). The problem does not only affect the isotropic term, but also the deviatoric and DC terms which are most used for interpretation. Bukchin et al. (2010) further derived a formal description of families of double couples which are poorly resolvable. The ambiguous determination of a double-couple source may also affect its polarity, as illustrated by Selby et al. (2005) and Cesca et al. (2011) for a shallow induced earthquake in the North Sea. In a similar way, Cesca et al. (2017), discussed the poor resolution of isotropic and vertical CLVD terms. Valentine and Trampert (2012) suggested to consider source parameters uncertainties, by considering as plausible models all those which fit the observations sufficiently well. Such approach should help to overcome intrinsic limitations of the standard approach only considering the best fitting solution: on one side, the specific best solution may strongly depend on chosen data and inversion methodology, while on the other side, former studies illustrate that specific perturbations of source parameters do not alter significantly the predicted synthetic seismograms and/or estimated misfits.

Source parameters uncertainties provide important constrains to moment tensor interpretation. However, they should not be considered independently. In the following of this work, we will discuss important trade-offs among the resolution of different parameters, due to the fact that similar synthetic seismograms and data fits can be obtained for different source parameters configurations. Exploring the range of well fitting solutions of an inverse problem may show that their diversity arise by multiple source parameters trade-offs. This manuscript explores this problem, combining new synthetic tests and significant examples from recent real data applications. A second topic we wish to discuss here, regards the role of the moment tensor decomposition towards the interpretation of moment tensor results. Finally, the selected real data examples are illustrative of the problem of moment tensor interpretation in presence of complex faulting. Considered real data cases have been recently published and concern moment tensor sources of nuclear explosions (Cesca et al. 2017), mining induced seismicity (Rudziński et al. 2015), and volcanic caldera collapse (Gudmundsson et al. 2016).

2 Moment Tensor Resolution, a Synthetic Test

We make use here of a moment tensor optimization algorithm recently developed at GFZ Potsdam, implemented in the *grond* package (http://gitext.gfz-potsdam.de/heimann/grond/, last visited 1.3.2017). The algorithm can be used to derive point source parameters based on the fit of waveforms in the time domain, amplitude spectra in the frequency domain, or a combination of both. To solve the moment tensor inverse problem we employ a stochastic global optimization procedure, scanning the space of source parameters within selected intervals. In the present work, we consider the following source parameters: latitude, longitude, depth, origin time and six moment tensor components (assuming a full moment tensor representation). We will consider the source duration to be fixed and model waveforms below the corner frequency, with a point source approximation. Based on the source parametrization, we will also discuss derived source parameters, such as the scalar moment, the moment magnitude, the strike, dip and rake of the planes associated to the double-couple term, and the isotropic, CLVD and DC percentage after the standard moment tensor decomposition. The algorithm is here applied in two steps: in the first training period (here 10000 iterations) source models are chosen randomly, in the second phase (here up to 600000 iterations) the search for new models becomes increasingly sharper close to the best fitting solutions, while decreasing the number or random tested models. The adopted moment tensor inversion is time consuming, when compared to standard inversion routines, which rely e.g. on a linear inversion in the time domain, grid search and/or gradient descent in the frequency domain, but it is extremely useful to map source parameter uncertainties and the range of acceptable solutions, as it does not provide only a best solution but an ensemble of source models with acceptable data fit.

One of the first application of the optimization method has been dedicated to the 2016 nuclear explosions in North Korea (Cesca et al. 2017). We adopt here the source and receiver geometry from this case study, to generate a number of synthetic waveform datasets, which are helpful to discuss the problem of moment tensor resolution for shallow seismic sources. We will assume here that shallow (1 km depth) seismic sources are located at the January 2016 North Korea explosion site, and seismic signals recorded at regional distances, using true locations of broadband seismic stations. The monitoring setup is illustrated in Fig. 1. In the following, synthetic displacement seismograms and their amplitude spectra will be generated using a local velocity model (model MDJ2, Ford et al. 2009), already used for moment tensor inversion in this region. Clean synthetic seismograms are superposed to white noise; the noise contamination is weak here, since we wish to discuss the intrinsic moment tensor ambiguities rather than the limitations due to the seismic noise.

For each dataset we will perform two moment tensor optimizations. The first one is performed in the time domain, fitting low frequency (0.02–0.04 Hz) vertical, radial and transversal displacement full waveforms for stations up to 1200 km epicentral distances. The second one uses an amplitude spectra approach (see e.g. Cesca et al. 2010, 2013) applied on the same data, which are combined with the fit of few displacement traces recorded at the closest stations; these traces only contribute with a small weight to the misfit function, and are needed to constrain the moment tensor polarity, which is not solved in a pure amplitude spectra inversion.

We first consider the problem of a moderate tectonic event, choosing two pure DC focal mechanisms, one corresponding to 45° dipping normal faulting (from now on referred as NF, with strike 0°, dip 45°, rake −90° and moment magnitude Mw 4.5) and one to a pure strike slip (labelled SS, with strike 0°, dip 90°, rake 0°, Mw 4.5). Results for the time domain inversion are shown in Fig. 2. Depth, magnitude and source types (vertical pressure axis for the strike-slip source, and horizontal pressure axis for the normal fault) are in both cases accurately resolved. The source uncertainties maps into the source type diagrams (upper plots), with some uncertainty in the CLVD and isotropic component. Spurious non-DC components can be retrieved in both cases,

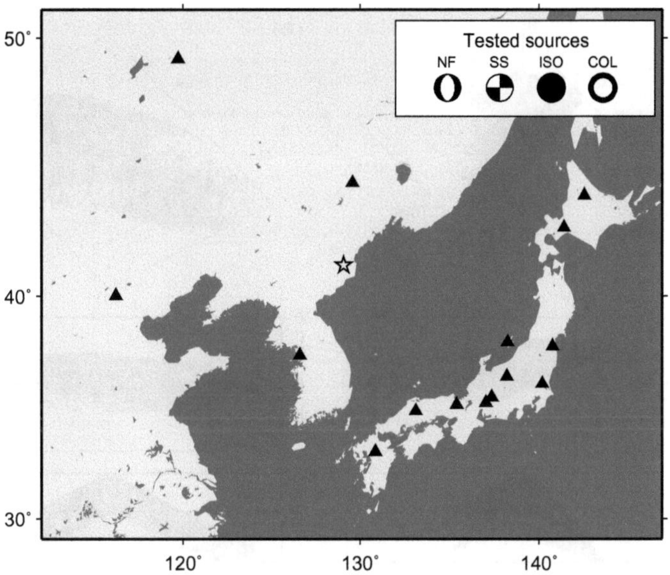

Fig. 1 The location and monitoring setup for the synthetic tests resembles the real case study of the January 2016 North Korea explosion (Cesca et al. 2017). A star denotes the location of the four synthetic sources (top right legend and mechanisms). Triangles denote locations of seismic stations

a well known possible effect due to seismic noise or a poor velocity model. We can further observe additional interesting features. First, the non-DC terms are differently characterized for the two sources: spurious CLVD and isotropic terms with consistent signs are mostly found for the NF source, but they often have opposite signs for the SS source. Second, the correct solution may not correspond to the mean solution in the distribution of acceptable ones: e.g. for the NF case, solutions with positive isotropic term are more abundant than those with negative non-DC terms.

We consider two cases of non-DC sources (Fig. 3). The first mechanism corresponds to a pure explosion (ISO; $M_{xx} = M_{yy} = M_{zz} = 1.0$; Mw 4.5), the second one to a vertical collapse (COL; $M_{xx} = M_{yy} = -0.3$; $M_{zz} = -1.0$; Mw 4.5). As discussed in Cesca et al. (2017) the poor resolution of the M_{zz} component for shallow sources is accompanied by a trade off among horizontal dipoles and the vertical one. As a result, solutions with very different decompositions can satisfactorily fit the data: these solutions sample the Hudson plot (Hudson et al. 1989) in the region comprises between a pure isotropic source and a pure negative vertical CLVD. The broad variety of solutions presents significant differences in terms of magnitude and depth, which are very poorly resolved. In particular, the magnitude uncertainty (up to 0.3) is significant. The orientation of minor spurious DC components (note that the small DC contribution can be fully removed by considering noise free synthetic data) is also

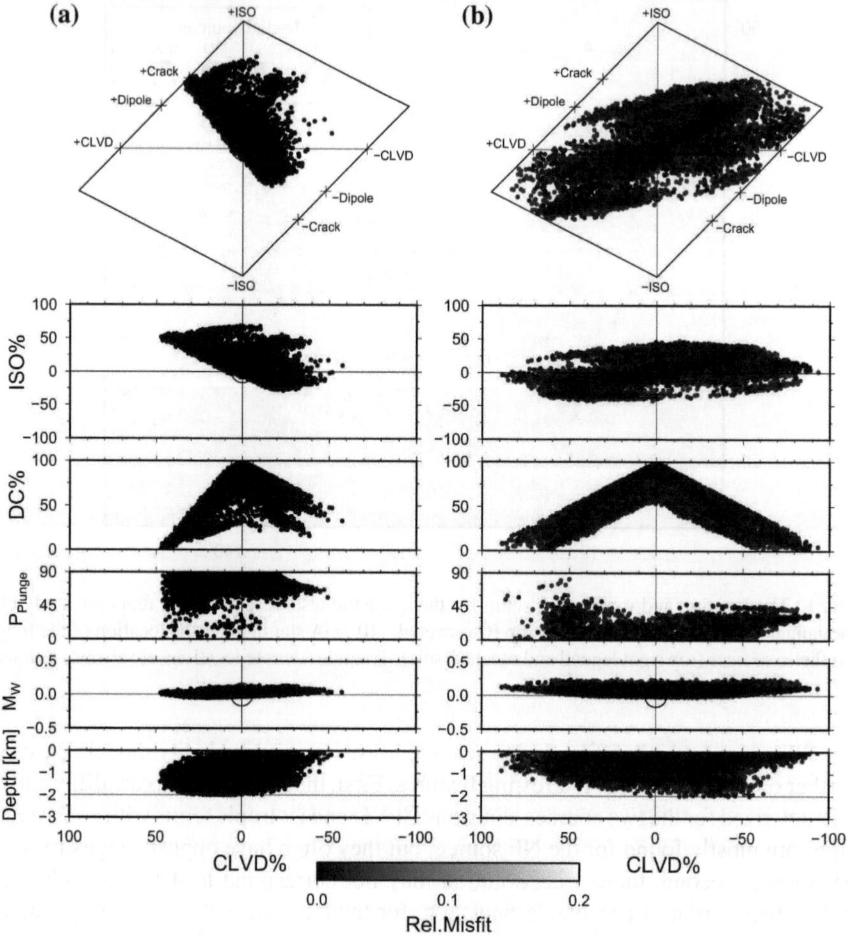

Fig. 2 Summary of synthetic inversion results for pure DC sources: normal faulting, NF (**a**), and strike slip, SS (**b**). From top to bottom graphics illustrate the Hudson diagram, the isotropic percentage, DC percentage, plunge of the pressure axis, moment magnitude and centroid depth as a function of the CLVD percentage. Each plot shows the best 10000 solutions with a relative misfit (difference between absolute misfit and misfit value for the best solution, small solid circles), which value is used to define the color scale (color bar). A large empty circle denotes the parameter of the reference synthetic source

affected by the solution heterogeneities: in particular, ~45° dipping normal faulting appear in correspondence to solutions dominated by negative vertical CLVD, constrained by the vertical orientation of the pressure axis.

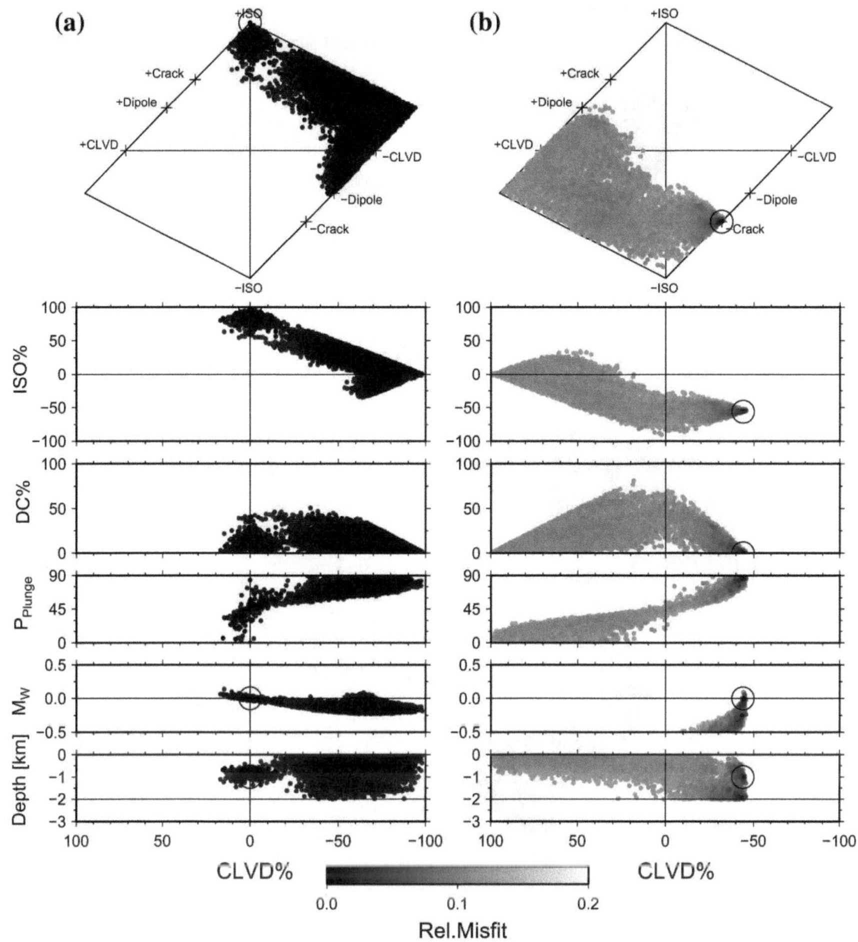

Fig. 3 Summary of synthetic inversion results for non-DC sources: positive isotropic source, ISO (**a**), and vertical collapse, COL (**b**). From top to bottom graphics illustrate the Hudson diagram, the isotropic percentage, DC percentage, plunge of the pressure axis, moment magnitude and centroid depth as a function of the CLVD percentage. Each plot shows the best 10000 solutions with a relative misfit (difference between absolute misfit and misfit value for the best solution, small solid circles), which value is used to define the color scale (color bar). A large empty circle denotes the parameter of the reference synthetic source

Figure 4 compares time domain and amplitude spectra inversions for the 4 study cases, limited to source type diagrams. Both inversion produce similar results. The ambiguity among potential MT solutions originates in the fact that all these sources produce similar synthetic seismograms. The figure illustrates that the "source similarity" persists in the spectral domain, and that the ambiguity is affecting different type of full waveform based moment tensor inversion approaches.

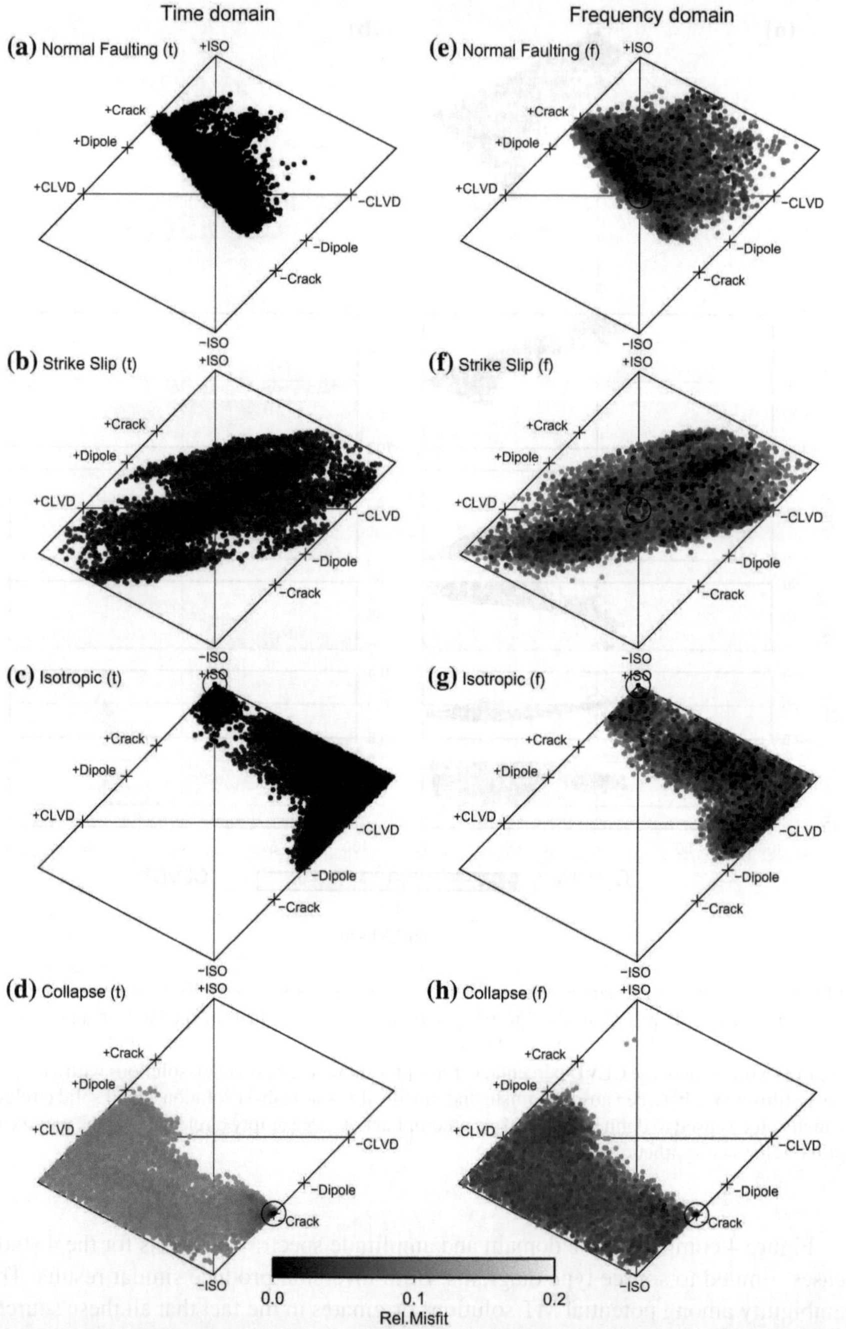

Fig. 4 The comparison of synthetic inversion results in the time (fitting full waveform displacements, left) and frequency (full waveform amplitude spectra, right) are shown from top to bottom for the four considered seismic sources: normal fault (NF), strike slip (SS), isotropic (ISO) and collapse (COL). Symbols are defined according to Figs. 2 and 3

3 Trade-Offs, the 2016 North Korea Explosion Case

As a first, real data application, we consider the January 2016 nuclear explosion in North Korea, and mostly refer to the results presented by Cesca et al. (2017), where more details are given and both the January and September 2016 explosions discussed. The North Korea explosion case is also interesting because it took place in the vicinity of former explosion tests, performed in 2006, 2009 and 2013 (see Ford et al. 2009; Murphy et al. 2013; Barth 2014; Vavryčuk and Kim 2014 for more details on earlier explosions). The seismic source model expected for an explosion is a pure positive isotropic source. Waveforms of different explosions recorded by the same receivers show high waveform similarities, except for the amplitude, which testify a repeating source mechanism and an increasing moment release from 2006, until the 2016 explosions. One would then expect that published solutions would agree on a similar moment tensor solution, except for the scalar moment estimate. Cesca et al. (2017) revealed this was not the case. The source type diagram with solutions proposed for the 2009 (Ford et al. 2009; Barth 2014), 2013 (Barth 2014; Vavryčuk and Kim 2014) and 2016 (Cesca et al. 2017) explosions are illustrated in Fig. 5. They show a large scatter and appear to be inconsistent, considering the similarity of waveforms used for the inversion and the similar, known source type. However, if we consider these solutions in the light of the recognised ambiguities for an isotropic source (Fig. 3), they can be better understood. The moment tensor solutions are, in fact, very different, and they could describe different physical processes, from an explosion to the closing of a horizontal tensile crack. The synthetic seismograms they generate, in change, are similar, and it is thus very challenging to discriminate the true source, dominated by an isotropic component. Even more, in the Hudson plot the true source is located at the edge of the cloud of possible solutions. The ambiguity finally originates from the large uncertainty of M_{zz} and its trade off with horizontal dipoles. The set of acceptable diagonal MT configurations shows important differences with respect to the scalar moment, which further explains the inconsistency of resolved magnitudes e.g. for the 2009 and 2013 explosions, which were studied by different authors with inconsistent results (Ford et al. 2009; Barth 2014; Vavryčuk and Kim 2014).

4 Temporal Complexity of the Seismic Source, the Rudna Mine Case

Sometimes the retrieval of inconsistent moment tensor solutions for the same event originates in the complex nature of the seismic source, in combination with the way we observe and analyse the seismologically measurable effects of the rupture process, which is the seismic data. As an illustrative example, we consider here a mining induced event, which took place on March 19th, 2013, in the Rudna copper mine, within the Lubin–Głogów Copper District. The event was studied and its

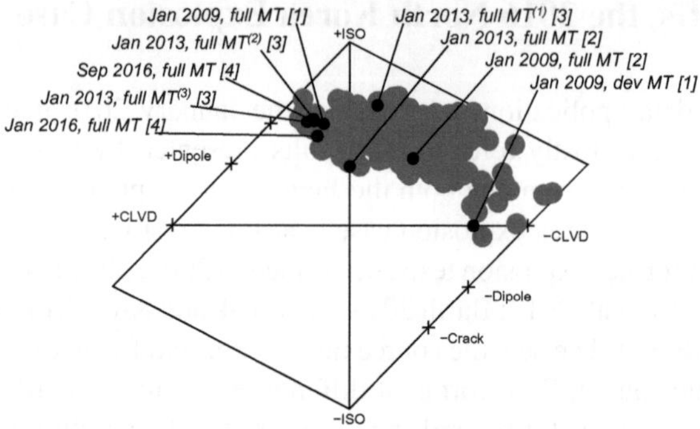

Fig. 5 Example of moment tensor ambiguity due to source parameters trade-offs, obtained for a real data case: a nuclear explosion test conducted in January 2016 in North Korea (figure elaborated after Cesca et al. 2017). A broad region of the Hudson diagram is covered by well fitting moment tensor solutions for the January 2016 explosion (gray circles). This broad region explains the retrieval of discordant moment tensor solutions by Ford et al. (2009) [1], Barth (2014) [2], Vavryčuk and Kim (2014) [3] and Cesca et al. (2017) [4] for a number of explosions in the region, either using a full (full MT) or deviatoric moment tensor (dev MT). Vavryčuk and Kim (2014) proposed three different solutions: (1) full MT with 1 km source depth, (2) full MT with 2 km source depth and (3) improved full MT (upon additional fit of the transversal/radial surface waves amplitude ratio) with 2 km source depth

source modelled in detail by Lizurek et al. (2014) and Rudziński et al. (2015), who discussed how the rupture nucleates as a small thrust fault earthquake, but continues in a large tabular collapse. Evidence of both processes was revealed by a visual inspection within the mine, shortly after the event took place. Independent moment tensor inversions were performed using different networks, data and methodologies at local (both in-mine and surface installations) and regional distances.

The moment tensor derived based on the in-mine installation could only rely on first motion polarities and amplitudes, due to saturation problems in the near field. This solution is characterized by a dominant thrust faulting, accompanied by smaller but non negligible vertical positive CLVD and isotropic components. Using a surface installation of the LUMINEOS network at local distances, Rudziński et al. (2015) derived a moment tensor characterized by a similar DC component, plus a stronger implosive component and negligible CLVD. Finally, using a full waveform-based approach (combination of amplitude spectra and time domain fit) on broadband seismometers at regional distances, the moment tensor showed a combination of a dominant negative isotropic term plus a negative subvertical CLVD. It is noteworthy that the regional solution remains very stable, assuming different velocity models (Rudziński et al. 2015). The comparison of the three moment tensor solutions is illustrated in Fig. 6. Their large discrepancy, accompanied in the original manuscript by a discussion on differences among spectral parameters derived by local and regional data, has been interpreted by means of a complex rupture model. The rupture process includes at least two main phases. The first one, which we refer to here as the nucleation phase, is less energetic and corresponds to the activation of a thrust fault.

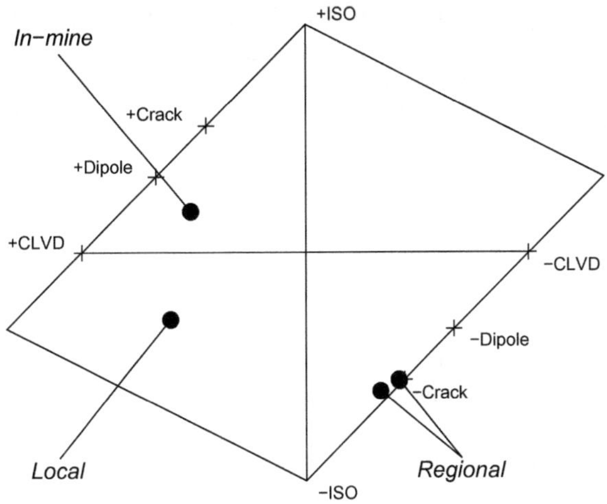

Fig. 6 Summary of moment tensor inversion results proposed for the Rudna mining induced event of March 19th, 2013 (Rudziński et al. 2015). Plotted solutions (black circles) were obtained based on the in-mine network, using first motion polarity and amplitude only, the local surface LUMINEOS network and the broadband network at regional distances (for the regional inversion, three similar solutions are plotted, corresponding to the adoption of three velocity models to compute synthetic seismograms)

The presence of a local fault was reported, but it was classified as normal, implying a significant rotation of the local stress field. The second phase corresponded to a tabular collapse, what was also confirmed during the postseismic inspection by the observation of rockfall and tunnel floor uplift. The local in-mine moment tensor analysis, based on first motion polarity only, can only depict the nucleation phase, indeed dominated by a thrust mechanism. The regional moment tensor inversion, on the opposite, can only reveal the most energetic source term, in this case the later tabular collapse, and is therefore characterized by a major sub-vertical negative CLVD. Using observations at intermediate distances, from the LUMINEOS surface network, the MT solution appears as a combination of the two processes. In conclusion, a complex rupture model can explain all apparently inconsistent features of the local and regional MT solutions.

5 Spatial Complexity of the Seismic Source, the Bardarbunga Collapse Case

The complexity of the seismic source process may also concern the superposition of multiple moment tensors. This case is well known for large earthquakes, where the rupture may occur on multiple planes of different orientation, along a curved surface or along a planar surface with variable slip direction (see e.g. Ide et al. 2005 for a review). However, rupture complexity and slip heterogeneity are often investigated for large earthquakes only and resolved through finite source inversion approaches, considering that the size of the rupture area limits the adoption of a point source approximation. We discuss here the case of a caldera collapse, referring to the recent outstanding case of the Bardarbunga volcano, Iceland (Gudmundsson et al. 2016). During caldera collapse, the fracture process can be located at shallow sources in the vicinity of the caldera rim. During the 2014–2015 collapse of the Bardarbunga caldera, for example, two seismicity clusters were located at depths of 0–9 km, close beneath the northern and southern rims of the caldera. Hypocenter distributions show steep to sub-vertical faulting at the rim, a spatial pattern which is consistent with the fracturing processes taking place above a deflating magma chamber. The regional moment tensor solutions were interpreted as the superposition of sub-vertical faulting at the rim, plus a deeper source contribution (Gudmundsson et al. 2016). The resulting moment tensor from a ring fault geometry has been discussed by Shuler et al. (2013a) and proposed as a possible source model for different volcanic environments (Shuler et al. 2013b).

We simulate here the rupture process considering a perfect ring fault of circular shape. Its failure is modeled by the superposition of DC sources along the rim. The circular surface is discretized into 36 segments. In each simulation, the discretized segments pertaining to a ring segment are simultaneously activated with a homogenous moment release. We consider an inward dipping rupture surface, so that DC are modelled by pure normal faulting. Different dip angles and lateral length of the activated ring-fault segment (i.e. simulating partial to full ring sources) are tested. The resulting moment tensor is obtained by superposing the moment tensors of contributing sources, and then decomposed with a standard decomposition into DC and CLVD terms. No isotropic term appears, since all contributing sources are purely deviatoric.

Results are illustrated in Fig. 7. A DC term is often resolved, except for the case of a full ring fault. Its orientation, however, deviates from the mechanism at the center of the ring segment (we refer as the central focal mechanism), as soon as the segment length grows. In some cases, the orientation of the DC term is even perpendicular to the central focal mechanism. A steep vertical CLVD appears as soon as the ring segment grows. It is worth noting that the final scalar moment is smaller or equal to the sum of the contributing scalar moments, due to the destructive interference of waveforms generated by different (in some case opposite) focal mechanisms along the ring fault. In the most extreme case, a vertical full ring fault model, opposite moment tensor contributions from the opposite side of the ring-fault cancel out completely.

6 Source Decomposition Challenges

A final problem we wish to discuss concerns the correct interpretation of moment tensor inversion results, as those obtained from of any of the previously discussed cases. In most of these cases, MT solutions include large relevant non-DC terms,

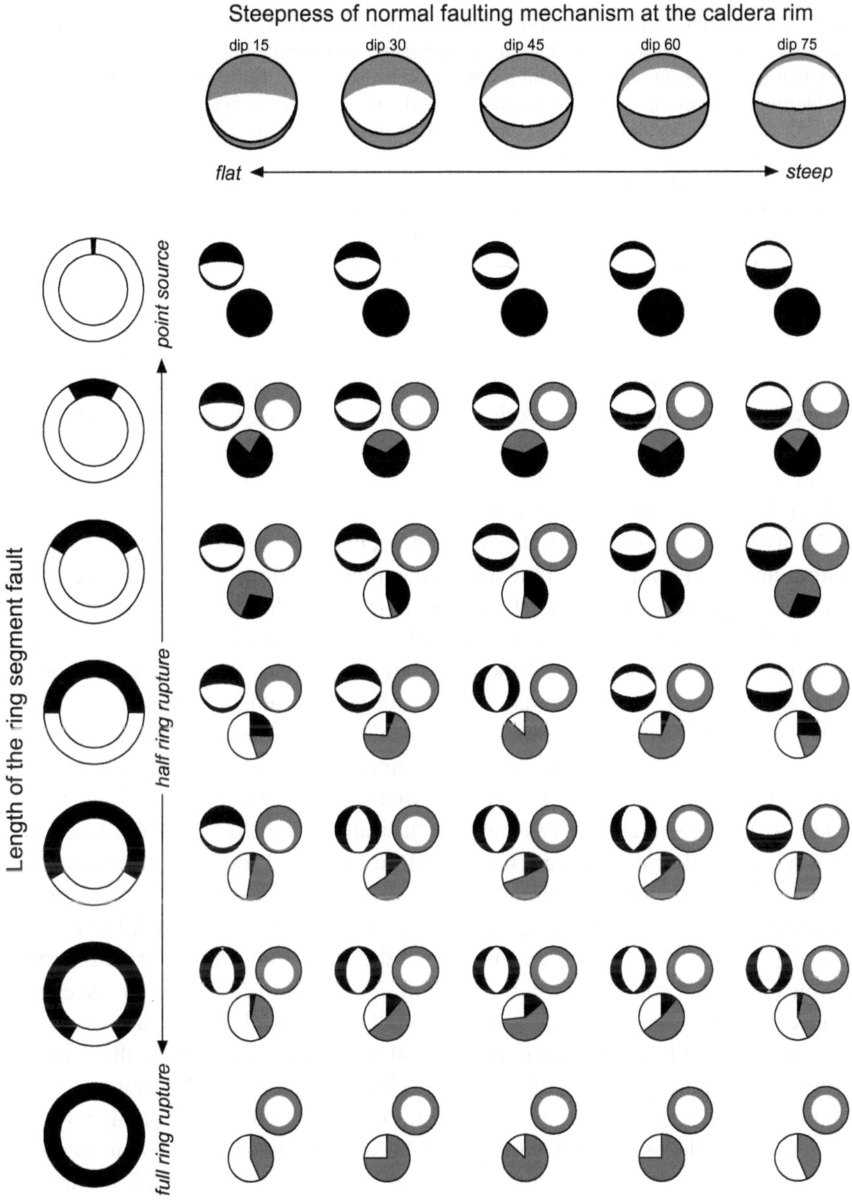
Steepness of normal faulting mechanism at the caldera rim
dip 15
dip 30
dip 45
dip 60
dip 75
flat
steep
Length of the ring segment fault
point source
half ring rupture
full ring rupture

◀**Fig. 7** Moment tensor decomposition results for a inward dipping ring fault segment rupture. A cumulative moment tensor is obtained by superposing the contribution of discretized source on the active ring segment, varying from a point source (top) to the full ring failure (bottom), and considering different dip angles of the caldera wall (increasingly steeper from left to right). The results of the decomposition for each combination of ring segment length and dip angle are shown by three plots: the upper left one represents the DC focal mechanism (black), the upper right one the CLVD (gray), and the bottom one the scalar moment contribution of each term, with respect to the cumulative moment of the original contributions (black for DC and gray for CLVD, whereas the white wedge denotes the undetected moment contribution, due to the fact that the sum of the scalar moment of DC and CLVD contribution underestimates the sum of the original DC contributions along the active ring segment)

which should be interpreted, as they can provide important information of the processes at the seismic source. Following the most general decomposition approach, i.e. decomposing the moment tensor into an isotropic and deviatoric term, and then further the deviatoric term into DC and CLVD, may not be always the best approach, and can lead to severe misinterpretations.

We assume we face a complex rupture problem, characterize by two seismic sources active within a short time and with similar locations. In this case, recorded seismic data will be characterize by the overlap of waveforms generated by the two sources. Considering sources of comparable size and duration, a waveform based inversion problem should reveal the cumulative (composite) source, e.g. the composition of the two original sources. We do not discuss here details of the inversion problem, but start from the composite source to infer the two contributing sources. We illustrate the problem considering the superposition of two competing source terms: a negative vertical CLVD and a 45° dipping reverse faulting mechanism. We choose these two source terms as they are illustrative for the selected real data cases, but obviously other deviatoric terms could be used to discuss the decomposition problem. The problem is illustrated in Fig. 8, where we combine the two sources assuming 5 different seismic moment ratios, from the case of a pure CLVD, to the case of a pure DC. In all five cases the sum of the scalar moments contributing to the complex source is constant. In the second, third and fourth cases, which correspond to all cases of incomplete ring segments modelled by multiple point sources, the decomposition fails to reproduce the original terms, both in terms of their mechanisms and in terms of their scalar moment. Even more, the scalar moment of the composite source is not necessarily equal to the sum of the original moments. The moment tensor decomposition simply fails because the two original sources do not share the pressure axis (equivalently, the tension axis for a positive CLVD), which is a constraint of the DC + CLVD decomposition. For strongly unbalanced contributions, where one of the two original terms has a much larger scalar moment, the dominant term is better resolved. So, in the case of a dominant negative vertical CLVD, this is resolved but accompanied with a DC term which shares its vertical pressure axis (thus, a wrong normal faulting), whereas for a dominant DC thrust, this is resolved but accompanied by a wrongly oriented CLVD.

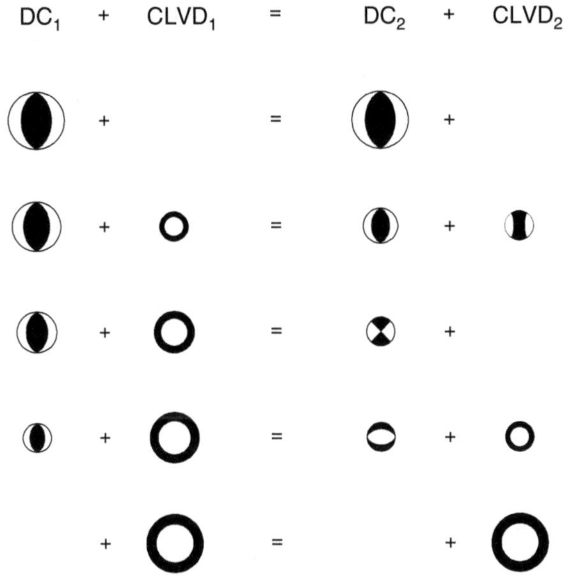

Fig. 8 Example of problematic decomposition of a moment tensor, illustrative for all the case studies discussed in this work. Left: superposition of a negative vertical CLVD and a thrust faulting with different relative sizes (from top to bottom 4:0, 3:1, 2:2, 1:3 and 0:4; the area of the focal spheres scales with their scalar moment). Right: the result of a standard decomposition (e.g. Jost and Hermann 1989). Since the original DC source does not share the pressure axis with the negative CLVD, the decomposition is unable to reconstruct the original mechanisms, nor the original scalar moment

One could question whether such unfavourable conditions are ever met. Complex rupture processes may not happen too often, but they can still affect an important number of seismological observations. They affected, indeed, all case studies discussed above. Let us reconsider them. At the Rudna mine, a small thrust mechanism is immediately followed by a large collapse. The regional moment tensor solution is controlled by the isotropic and CLVD terms associated to the collapse, which released most energy. A standard decomposition can retrieve the collapse signature, but would erroneously resolve a marginal normal faulting, hiding the true thrust mechanism term (Rudziński et al. 2015). At the Bardarbunga caldera, regional moment tensor inversions for events located at the northern rim are characterized by negative subvertical CLVD and normal faulting (Gudmundsson et al. 2016), although they can be better interpreted as the sum of a vertical CLVD and a thrust faulting. Finally, the North Korea case is further complicated by the role of the isotropic component, and its trade off with a negative vertical CLVD, discussed in a previous paragraph. The presence of a marginal DC term, possibly consequence of near source topographic interactions, has been proven on the base on the observation and modeling of weak amplitude Love waves. A recompilation of DC terms in regional moment tensor solutions for nuclear explosions in this region show important anomalies (Cesca et al. 2017). For some moment tensor solutions, characterized by a larger dominant isotropic source, the DC term is modeled as a sub-vertical thrust faulting. For other solutions, however, which favour a larger negative vertical CLVD, the DC term consists of a ~45° dipping normal faulting.

7 Conclusions

The moment tensor inversion is a powerful tool to investigate the seismic source of earthquakes at different spatial scales and for different magnitudes. As the result of any inversion problem, however, a moment tensor solution suffers of important uncertainties. Synthetic and real cases collected in this manuscript illustrate that moment tensor component uncertainties are heterogeneous. Their relative size may be affected by the network configuration, the quality of the velocity model, the inversion parametrization (e.g. frequency bandpass, waveform taper, time versus frequency inversion approach) and the source depth. The source depth, in particular has a very important role with respect to the characterization of moment tensor component uncertainties. We have discussed here extreme cases, with earthquake sources located at very shallow depth, close to the surface, where uncertainty and trade-offs become very large and problematic to handle. The problem, however, also affects crustal and deep earthquakes, with different uncertainties on different moment tensor components.

The MT uncertainties are not independent, but trade-offs exist and characterize the range of possible MT solutions. The application to the North Korea case is in this sense very illustrative, showing that a shallow explosion and a shallow vertical negative CLVD produce similar waveforms and are poorly distinguishable. This is likely only one example of possible trade-offs, which may affect a robust discrimination of the correct moment tensor.

Temporal and spatial complexity of the seismic source pose further challenges to moment tensor inversion. Among the temporal complexity, a complex source time function or the presence of multiple ruptures, as illustrated for the Rudna collapse case, may hinder the correct determination and interpretation of the seismic source parameters. Similarly, the spatial extent of the rupture, and the possible superposition of rupture processes with different mechanisms (here illustrated by the failure of curved rupture surface for a volcanic collapse), can be inferred only with an accurate interpretation of the resolved, composite moment tensor.

Finally, we illustrated the potential misinterpretation suggested by the adoption of improper moment tensor decomposition schemes, when dealing with complex rupture processes. Specifically in presence of multiple sources, or when DC, CLVD and isotropic contributions have different physical origins, the adoption of a standard decomposition into DC and CLVD with common pressure or tension axis may not resolve the true composition of rupture processes, and hide the rupture dynamic complexity.

The message of this paper is not to discourage the application of moment tensor inversion for the determination of seismic source parameters, which is and remains for us the most useful and robust tool for the determination of seismic source parameters. Uncertainties and limitations also affect other methods to determine earthquake focal mechanisms and moment tensors. Instead, we hope that our contribution helps to improve the way moment tensor inversion is used in the seismological community. A good practice should be to test any moment tensor inversion algorithm and application to realistic synthetic dataset cases, to judge the quality and robustness of solutions against a number of factors which can limit the inverse problem. Effects of seismic noise, network configurations, velocity model anomalies, waveform processing and filtering should be well assessed before coming to real data application. An accurate estimation of moment tensor uncertainty and trade offs may be further beneficial to improve the interpretation of moment tensor solutions.

References

Barth A (2014) Significant release of shear energy of the North Korean nuclear test on February 12, 2013. J Seismol 18:605–615. https://doi.org/10.1007/s10950-014-9431-6

Bernardi F, Braunmiller J, Kradolfer U, Giardini D (2004) Automatic regional moment tensor inversion in the Euro pean–Mediterranean region. Geophys J Int 157:703–716. https://doi.org/10.1111/j.1365-246X.2004.02215.x

Braunmiller J, Kradolfer U, Baer M, Giardini D (2002) Regional moment tensor determination in the European-Mediterranean area: initial results. Tectonophysics 356:5–22

Bukchin B, Clévédé E, Mostinskiy A (2010) Uncertainty of moment tensor determination from surface wave analysis for shallow earthquakes. J Seismol 14:601–614. https://doi.org/10.1007/s10950-009-9185-8

Cambaz MD, Mutlu AK (2016) Regional moment tensor inversion for earthquakes in turkey and its surroundings: 2008–2015. Seismol Res Lett. https://doi.org/10.1785/0220150276

Cesca S, Buforn E, Dahm T (2006) Moment tensor inversion of shallow earthquakes in Spain. Geophys J Int. https://doi.org/10.1111/j.1365-246X.2006.03073.x

Cesca S, Heimann S, Stammle K, Dahm T (2010) Automated procedure for point and kinematic source inversion at regional distances. J Geophys Res. https://doi.org/10.1029/2009JB006450

Cesca S, Dahm T, Juretzek C, Kühn D (2011) Rupture process of the 7 May 2001 Mw 4.3 Ekofisk induced earthquake. Geophys J. https://doi.org/10.1111/j.1365-246X.2011.05151.x

Cesca S, Rohr A, Dahm T (2013) Discrimination of induced seismicity by full moment tensor inversion and decomposition. J Seismol. https://doi.org/10.1007/s10950-012-9305-8

Cesca S, Heimann S, Kriegerowski M, Saul J, Dahm T (2017) Seismic analysis of nuclear explosions: what can we learn from the 2016 nuclear tests, North Korea? Seismol Res Lett. https://doi.org/10.1785/0220160139

Domingues A, Custodio S, Cesca S (2013) Waveform inversion of small to moderate earthquakes located offshore southwest iberia. Geophys J Int. https://doi.org/10.1093/gji/ggs010

Dreger DS, Helmberger DV (1993) Determination of source parameters at regional distances with single station or sparse network data. J Geophys Res 98:8107–8125

Dufumier H, Rivera L (1997) On the resolution of the isotropic component in moment tensor inversion. Geophys J Int 131(3):595–606. https://doi.org/10.1111/j.1365-246X.1997.tb06601.x

Ford AR, Dreger DS, Walter WR (2009) Source analysis of the memorial day explosion. Geophys Res Lett, Kimchaek, North Korea. https://doi.org/10.1029/2009GL040003

Ford SR, Dreger DS, Walter WR (2010) Network sensitivity solutions for regional moment-tensor inversions. Bull Seismol Soc Am 100(5A):1962–1970. https://doi.org/10.1785/0120090140

Gilbert F (1970) Excitation of the normal modes of the earth by earthquake source. Geophys J R Astron Soc 22:223–226

Gudmundsson MT, Jónsdóttir K, Hooper A, Holohan EP, Halldórsson SA, Ófeigsson BG, Cesca S, Vogfjörd KS, Sigmundsson F, Högnadóttir T, Einarsson P, Sigmarsson O, Jarosch AH, Jónasson K, Magnússon F, Hreinsdóttir S, Bagnardi M, Parks MM, Hjörleifsdóttir V, Pálsson F, Walter TR, Schöpfer MPJ, Heimann S, Reynolds HI, Dumont S, Bali E, Gudfinnsson GH, Dahm T, Roberts M, Hensch M, Belart JMC, Spaans K, Jakobsson S, Gudmundsson GB, Fridriksdóttir HM, Drouin V, Dürig T, Adalgeirsdóttir G, Riishuus MS, Pedersen GBM, van Boeckel T, Oddsson B, Pfeffer MA, Barsotti S, Bergsson B, Donovan A, Burton MR, Aiuppa A (2016) Gradual caldera collapse at Bárdarbunga volcano, Iceland. Science. https://doi.org/10.1126/science.aaf8988

Hudson JA, Pearce RG, Rogers RM (1989) Source type plot for inversion of the moment tensor. J Geophys Res 94:765–774

Ichinoise GA, Anderson JG, Smith KD, Zeng Y (2003) Source parameters of eastern California and western Nevada earthquakes from regional moment tensor inversion. Bull Seismol Soc Am 93(1):61–84. https://doi.org/10.1785/0120020063

Ide S, Beroza GC, Mcguire JJ (2005) Imaging earthquake source complexity, in seismic earth. In: Levander A, Nolet G (eds) Array analysis of broadband seismograms. American Geophysical Union, Washington, D. C. https://doi.org/10.1029/157gm08

Jost ML, Hermann RB (1989) A student's guide to and review of moment tensors. Seismol Res Lett 60(2):37–57

Kawakatsu H (1995) Automated near-realtime CMT inversion. Geophys Res Lett 22:2569–2572. https://doi.org/10.1029/95GL02341

Konstantinou KI, Melis NS, Boukouras K (2010) Routine regional moment tensor inversion for earthquakes in the Greek region: the national observatory of Athens (NOA) database (2001–2006). Seismol Res Lett. https://doi.org/10.1785/gssrl.81.5.750

Krizova D, Zahradnik J, Kiratzi A (2013) Resolvability of isotropic component in regional seismic moment tensor inversion. Bull Seismol Soc Am 103(4):247–2460. https://doi.org/10.1785/0120120097

Kubo A, Fukuyama E, Kawai H, Nonomura K (2002) NIED seismic moment tensor catalogue for regional earthquakes around Japan: quality test and application. Tectonophysics 356(1):23–48. https://doi.org/10.1016/S0040-1951(02)00375-X

Lizurek G, Rudziński Ł, Plesiewicz B (2014) Mining induced seismic event on inactive fault. Acta Geophys. https://doi.org/10.2478/s11600-014-0249-y

Minson SE, Dreger DS (2008) Stable inversion for complete moment tensors. Geophys J Int. https://doi.org/10.1111/j.1365-246X.2008.03797.x

Murphy JR, Stevens JL, Kohl BC, Bennett TJ (2013) Advanced seismic analyses of the source characteristics of the 2006 and 2009 North Korean nuclear tests. Bull Seismol Soc Am 103(3):1640–1661

Pasyanos ME, Dreger DS, Romanowicz B (1996) Toward real-time estimation of regional moment tensors. Bull Seismol Soc Am 86:1255–1269

Pondrelli S, Morelli A, Ekstrom G, Mazza S, Boschi E, Dziewonski AM (2002) European-Mediterranean regional centroid-moment tensors: 1997–2000. Phys Earth Planet Int 130:71–101

Romanowicz B, Dreger D, Pasyanos M, Uhrhammer R (1993) Monitoring of strain release in central and northern California using broadband data. Geophys Res Lett 20:1643–1646

Rudziński Ł, Cesca S, Lizurek G (2015) Complex rupture process of the March 19, 2013, Rudna mine (Poland) induced seismic event and collapse in the light of local and regional moment tensor inversion. Seismol. Res. Lett. https://doi.org/10.1785/0220150150

Rueda J, Mezcua J (2005) Near-real-time seismic moment-tensor determination in Spain. Seismol Res Lett 76(4):355–365. https://doi.org/10.1785/gssrl.76.4.455

Selby N, Eshun E, Patton H, Douglas A (2005) Unusual long-period Rayleigh wave from a vertical dip-slip source: the 7 May 2001. J geophys Res, North Sea earthquake. https://doi.org/10.1029/2005JB003721

Shuler A, Ekström G, Nettles M (2013a) Physical mechanisms for vertical-CLVD earthquakes at active volcanoes. J Geophys Res 118(B4):1569–1586. https://doi.org/10.1002/jgrb.50131

Shuler A, Nettles MK, Ekström G (2013b) Global observation of vertical-CLVD earthquakes at active volcanoes. J Geophys Res. https://doi.org/10.1029/2012JB009721

Š´ılený J, Campus P, Panza G (1996) Seismic moment tensor resolution by waveform inversion of a few local noisy records—I. Synthetic tests. Geophys J Int 126(3):605–619. https://doi.org/10.1111/j.1365-246x.1996.tb04693.x

Š´ılený J (2004) Regional moment tensor uncertainty due to mismodeling of the crust. Tectonophysics 383(3–4):133–147. https://doi.org/10.1016/j.tecto.2003.12.007

Sokos E, Zahradnik J (2013) Evaluating centroid moment tensor uncertainty in the new version of ISOLA software. Seismol Res Lett. https://doi.org/10.1785/0220130002

Stich D, Ammon CJ, Morales J (2003) Moment tensor solutions for small and moderate earthquakes in the Ibero-Maghreb region. J Geophys Res. https://doi.org/10.1029/2002JB002057

Valentine AP, Trampert J (2012) Assessing the uncertainties on seismic source parameters: Towards realistic error estimates for centroid-moment-tensor determinations. Phys Earth Planet Int 210–211:36–49. https://doi.org/10.1016/j.pepi.2012.08.003

Vavryčuk V, Kim SG (2014) Nonisotropic radiation of the 2013 North Korean nuclear explosion. Geophys Res Lett 41:7048–7056. https://doi.org/10.1002/2014GL061265

Vera Rodriguez I, Gu YJ, Sacchi MD (2011) Resolution of seismic-moment tensor inversions from a single array of receivers. Bull Seismol Soc Am 101(6):2634–2642. https://doi.org/10.1785/0120110016

Selby ND, Eshun E, Patton HJ, Douglas A (2005) Unusual long-period Rayleigh wave from a vertical dip-slip source: the 7 May 2001 North Sea earthquake. J Geophys Res. https://doi.org/10.1029/2005JB003721

Shuler A, Ekström G, Nettles M (2013a) Physical mechanisms for vertical-CLVD earthquakes at active volcanoes. J Geophys Res 118(B4):1569–1586. https://doi.org/10.1002/jgrb.50131

Shuler A, Nettles M, Ekström G (2013b) Global observation of vertical-CLVD earthquakes at active volcanoes. J Geophys Res. https://doi.org/10.1029/2012JB009721

Sileny J, Campus P, Panza GF (1996) Seismic moment tensor resolution by waveform inversion of a few local noisy records—I. Synthetic tests. Geophys J Int 126:605–619. https://doi.org/10.1111/j.1365-246X.1996.tb04693.x

Sileny J (2004) Regional moment tensor uncertainty due to mismodeling of the crust. Tectonophysics 383(3–4):133–147. https://doi.org/10.1016/j.tecto.2003.12.007

Sokos E, Zahradnik J (2013) Evaluating centroid-moment-tensor uncertainty in the new version of ISOLA software. Seismol Res Lett. https://doi.org/10.1785/0220130002

Stich D, Ammon CJ, Morales J (2003) Moment tensor solutions for small and moderate earthquakes in the Ibero-Maghreb region. J Geophys Res. https://doi.org/10.1029/2002JB002057

Valentine AP, Trampert J (2012) Assessing the uncertainties on seismic source parameters: Towards realistic error estimates for centroid-moment-tensor determinations. Phys Earth Planet Int 210–211:36–49. https://doi.org/10.1016/j.pepi.2012.08.003

Vavryčuk V, Kim SG (2014) Nonisotropic radiation of the 2013 North Korean nuclear explosion. Geophys Res Lett 41:7048–7056. https://doi.org/10.1002/2014GL061265

Vera Rodriguez I, Gu YJ, Sacchi MD (2011) Resolution of seismic-moment tensor inversions from a single array of receivers. Bull Seismol Soc Am 101(6):2634–2642. https://doi.org/10.1785/0120100246

The Role of Moment Tensors in the Characterization of Hydraulic Stimulations

Ismael Vera Rodriguez, James Rutledge and Sergey Stanchits

Keywords Moment tensor · Microseismic monitoring · Acoustic emission
Hydraulic stimulation

1 Introduction

Hydraulic stimulation is a technique aimed at increasing the effective permeability of rock reservoirs in connection to a producing/injecting well (Economides and Nolte 2003). It is applied in the oil and gas industry to increase production in unconventional reservoirs, and in geothermal systems to enhance fluid circulation. This technique (similar to others involving the injection of fluids such as CO_2 sequestration, natural gas storage and wastewater disposal) has the potential to induce small seismic events (generally with moment magnitude $M_w < 0$). The analysis of this induced microseismicity is of interest for purposes ranging from seismic hazard assessment to reservoir characterization.

In the oil and gas industry, hydraulic stimulations start with the high pressure injection of a water-based fluid. The process stimulates the reservoir by driving extension fractures from the injection point into the rock formations. This results in the creation and activation of new and pre-existing fractures. New and pre-existing fractures can be activated in combinations of tensile, compressive, and shear modes depending on the state of stress, the rock's fabric and strength, and pore-pressure coupling into the rock. Furthermore, as a side effect, some of these fractures also release

I. Vera Rodriguez (✉)
Schlumberger Cambridge Research, High Cross, Madingley Road, Cambridge CB3 0EL, UK
e-mail: verarodr@ualberta.ca

J. Rutledge
Schlumberger, 3750 Briarpark Drive, Houston, TX 77042, USA

S. Stanchits
Formerly at Schlumberger Research, 6350 West Sam Houston Pkwy N, Suite 200, Houston, TX 77041, USA

S. D'Amico (ed.), *Moment Tensor Solutions*, Springer Natural Hazards,
https://doi.org/10.1007/978-3-319-77359-9_8

seismic energy that we observe as microseismic activity. After fracture extension reaches a region away from the injection point, a propping agent is pumped down to pack-open the fracture planes that are connected with the wellbore to retain a high permeability path between the reservoir and the treated well after pumping has ceased (Fink 2013). Connectivity, asperities, tortuosity and settling effects, however, are expected to prevent the proppant from reaching every open fracture within the stimulated volume, especially in systems of fractures with planes that are connected at high angles (McLennan et al. 2008). At the end of the injection, when the pressure is reduced at the wellbore, the fracture(s) that are not supported by proppant close again. Therefore, the volume of rock containing proppant is normally smaller than the total stimulated volume (Mayerhofer et al. 2010). Some self-propping of the fracture system is also expected due to shear dislocations along irregular surfaces or at fracture discontinuities, but production enhancement obtained from the hydraulic stimulation is arguably coming primarily from the propped volume. For this reason, it is of great interest to obtain a picture of the stimulated fracture network, proppant transportation and proppant placement. Because the effective fracture network contributing to production is influenced by multiple variables including local, poorly known factors such as the rock's fabric and strength, the effectiveness of hydraulic stimulations can vary within a few tens of meters. This variability introduces the need for methodologies that permit a cost-efficient and reliable characterization of fracture growth and deformation for individual stimulation treatments.

Among the different diagnostic techniques available to characterize hydraulic stimulations, microseismic monitoring has been one of the most versatile and successfully applied (Cipolla and Wright 2002). Microseismic monitoring consists of the recording and analysis of the seismic energy released from the activation of some of the fractures stimulated by hydraulic injections. The monitoring is conducted with arrays of receivers deployed on the surface of the earth and/or inside shallow or deep boreholes (Duncan and Eisner 2010; Maxwell et al. 2010). From the analysis of the spatial locations of the microseismic events, information can be extracted regarding the geometry of the stimulated volume (e.g., Bennett et al. 2005), and the complexity of the stimulated fracture network (Cipolla et al. 2008). Combining spatial and temporal information, microseismic events have been used to obtain estimations of the permeability tensor of the fracture network (Shapiro et al. 1999), and to detect proppant bridging (Maxwell et al. 2004). The source spectrum of microseismic arrivals provides information about the seismic moment and corner frequency, which are useful for estimating event magnitudes, source dimensions, stress drops and radiated seismic energy. Graphs of cumulative seismic moment release can quantify seismic-measured deformation, uncovering interactions between small fracture networks and larger faults being activated by the hydraulic treatment (Rutledge et al. 2004; Maxwell et al. 2008). Moment-derived magnitudes (M_w) are also the basis of traffic-light systems used to control the size of events during injection operations (e.g., Department of Energy and Climate Change 2014). Stress drops derived from seismic moments and corner frequencies have been interpreted to reflect stress heterogeneities during fracture growth and help to discriminate between fluid-induced and stress-related activation of microseismic events (e.g., Urbancic and Rutledge 2000; Goertz-Allmann et al. 2011).

Moment tensor inversion, or MTI, describes the inelastic deformation of the source region that generates the microseismic signal. In earthquake seismology, the source mechanism is usually presented as a "fault-plane solution", providing the fault's orientation and the slip vector direction (rake). The fault-plane solution is also commonly referred to as the double-couple (DC) solution, which is a special case of the moment tensor (MT) that assumes that the signal is generated from shear dislocation along a planar surface in an isotropic medium. More generally, the MT can describe multiple other source types. For example, internal sources such as explosions/implosions, crack opening or closure, slip on a plane, or any combination and superposition of these are represented by symmetric MTs. The MT's general decomposition into displacement discontinuity sources, describing fracture deformation in anisotropic media, is described in Chapman and Leaney (2012), Leaney et al. (2014), and Vavryčuk (2015). Since hydraulic-stimulation microseismicity is generated by high-pressure injection of fluids, one of the primary interests in MTI is the distinction of fracture tensile/closure events from shear displacements.

MTI provides useful information on the geomechanical response of a reservoir to the hydraulic stimulation. In principle, MTI provides a partial description of how a fracture system, pre-existing, created and/or augmented through growth, is responding to some in situ stress field that is being altered by the injection. Additional information helpful in interpreting MTI results includes analyses of source locations with respect to the geology, target reservoir, and stress magnitudes and orientations. Also, natural fracture characterization, and the correlation between the temporal development of the seismicity and injection parameters provide useful context and input to MT interpretation. Generally, MTI results can be a reference to build and calibrate fracture models. Specifically, MTI can provide statistics on fracture orientations, indicate areas of volume creation, and by inference, the mapping of proppant placement (e.g., Neuhaus et al. 2014; Yu et al. 2015).

Events predominantly exhibiting shear mechanisms are common during hydraulic stimulations. Stress anisotropy in the earth's crust creates shear stresses. Since a rock's shear strength is typically greater than its tensile strength, shear stress is easier to sustain in the earth, and can be released when pore pressure during injection reaches some threshold value (e.g., as critical failure of the rock or slip along pre-existing fractures). Shear stresses are also generated near the tip of a tensile crack, which may also drive rock to shear failure just ahead of the hydraulic fracture. Conversely, tensile stresses are induced near the tips of shear cracks, so, in general, shear and tensile fracturing are closely coupled phenomena. Although microseismicity may under-represent the gross deformation occurring during hydraulic fracturing, MTI analysis provides an important measure of the geomechanical response to hydraulic stimulation.

In addition to the calibration of fracture networks, MTs have other applications in microseismic monitoring. For example, the seismic moment of a microseismic event can be predicted using source models developed in global seismology (e.g., Aki 1966) or from the normalization of its estimated MT. Source models in global

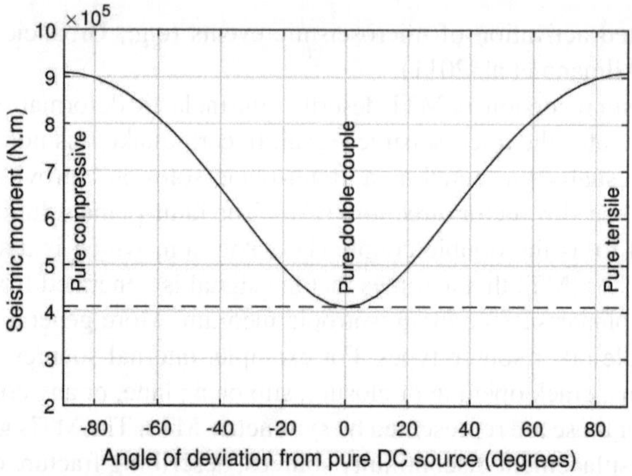

Fig. 1 Comparison of seismic moment (M_0) predictions from MT normalization (solid line) and a source model that assumes pure DC (dashed line). The medium is isotropic with Vp = 3633 m/s, Vs = 2142 m/s and $\rho = 2700$ kg/m^3. The fracture area is 1 m^2 and the stress drop $\Delta\sigma = 1$ MPa. The fracture plane is vertical, activated with a rake of 0° (strike-slip shear displacement in the pure DC case)

seismology often include the implicit assumption of a pure DC source in isotropic, homogeneous media (e.g., Brune 1970). In hydraulic stimulations, however, non-DC dislocations have been identified both at field and laboratory scales (e.g., Sileny et al. 2009; Vera Rodriguez et al. 2017), while most of the microseismicity of interest takes place in anisotropic formations such as shales. The seismic moment of a non-DC dislocation is systematically underestimated when computed with traditional source models that assume pure DC (Fig. 1). For this reason, seismic moment from MT normalization gives a more general approximation for the expected size of a microseismic event. The source types interpreted from MTs provide an insight into the local state of stress, where the variability of the source types can be directly linked to the complexity of the stress field at a specific snapshot in time and space (e.g., Aker et al. 2014). More recently, the evolution of fracture geometries and styles of activation obtained from MTs has received increased attention with the objective to gain a better understanding of the mechanics of hydraulic stimulations (e.g., Tan and Engelder 2016; Rutledge et al. 2016).

In this chapter, we illustrate two applications of MT analysis related to the study of hydraulic stimulations. Both applications study the mechanics of the problem at different scales. In the first case, field-scale data are analysed to develop an understanding of fracture propagation and its interaction with formation bedding. In the second application, the MTs of acoustic emissions observed in a laboratory experiment are analysed to develop an understanding of the source types that can be expected during hydraulic stimulations at field-scale.

2 Characterization of a Field-Scale Hydraulic Stimulation Through the Moment Tensors of Microseismicity

The growth of hydraulic fractures in layered sedimentary rocks is strongly controlled by the mechanical stratigraphy. The mechanical discontinuities and stress contrast across layer interfaces tend to impede vertical growth relative to lateral extension, allowing the creation of reservoir-contained fractures that extend laterally up to hundreds of meters (Chuprakov and Prioul 2015). Not surprisingly, the occurrence and mechanisms of microseismicity induced during hydraulic fracturing can also be controlled by mechanical stratigraphy. Precise microseismic source locations sometimes show fine-scale banding of events in depth (e.g., Rutledge et al. 2004, 2015), implying brittle deformation in some layers and stable, aseismic extensional growth in others. The elastic heterogeneity associated with the stratigraphy can explain the banding of events (e.g., Langenbruch and Shapiro 2014; Roche and van der Baan 2015). Depending on specific conditions, one could invoke a number of mechanisms to explain such correlations. For example, stiffer layers can support higher stress magnitudes and higher differential stresses, which can control the occurrence of natural fractures.

Dip-slip microseismic mechanisms are commonly observed during hydraulic fracture stimulations in shales in which vertical nodal planes tend to be aligned with the expected hydraulic-fracture orientation. Various investigators have suggested that the events represent shearing on bedding planes with the alternate, horizontal nodal plane corresponding to the slip surface (e.g., Rutledge et al. 2013; Stanek and Eisner 2013). We present a few cases of such observations and present a simple argument for bedding-plane slip rather than vertical dip-slip mechanisms. We then invoke a simple model of Chuprakov and Prioul (2015) to explain the persistent mechanisms. In their model, tensile crack vertical-growth can arrest at weak bedding interfaces. Fluid invasion and pressure can weaken the interface which can then shear from the added stress of adjacent crack opening. Under certain conditions, with subsequent net pressure recovery and gain at the weak interface, the parent tensile fracture propagation can continue across the interface. In one case presented here for a stimulation treatment in the Barnett shale, invoking this model qualitatively provides a consistent explanation of the continuity and time-space development of the microseismic structures, as well as explaining the observed patterns of shear-fracture first motions.

2.1 Observations

Figure 2 shows an example of the interpretation of surface microseismic data collected during a multi-well, multi-stage completion program in a shale reservoir. An extensive array of shallow-well geophone receivers allows the determination of the MTs. Within the encircled area, the mechanisms are fairly uniform dip-slip events with the near vertical nodal planes aligned close to the event trends and expected

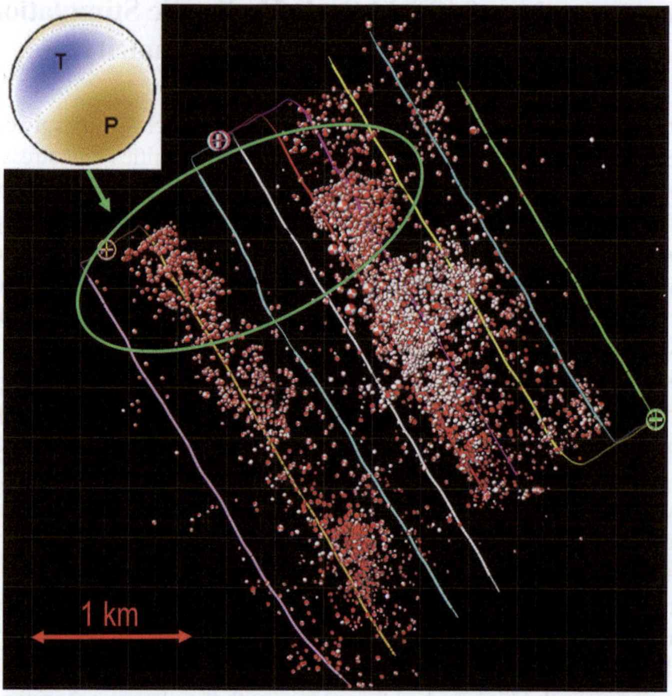

Fig. 2 Example of surface microseismic data from well completions in a shale reservoir. Events were detected on a grid array of 47 shallow-well receivers. Events are displayed showing upper hemisphere first motions. Within the green ellipse most events show dip-slip mechanisms on steep planes striking subparallel to the event trend and expected maximum horizontal stress (SHmax) direction. The inset solution is a high signal-to-noise ratio example from within the green ellipse. The alternate nodal plane is near horizontal

SHmax direction (Fig. 2). Similarly aligned and persistent mechanisms in shale reservoirs are also revealed in cases presented by Stanek and Eisner (2013) and Diller et al. (2015).

Figure 3 shows a similar population of mechanisms from downhole data in another shale reservoir. The data were acquired using multiple monitor wells so that the full MTs are well constrained. The events are primarily shear mechanisms showing predominantly dip-slip displacements on fractures aligned with the hypocenter trend and principal stress direction. Steeply dipping natural fractures over the same depth interval are sparse with orientations widespread and poorly correlated with the uniform vertical nodal plane orientations exhibited by the source mechanisms (Figs. 3 and 4).

Sometimes such aligned and repetitive mechanisms can be inferred from composite patterns of P- and S-wave first motions and amplitude ratios observed from a single vertical monitor well (Rutledge and Phillips 2003; Rutledge et al. 2013). An example of this is shown in Fig. 5 from the Barnett shale, where a persistent radiation

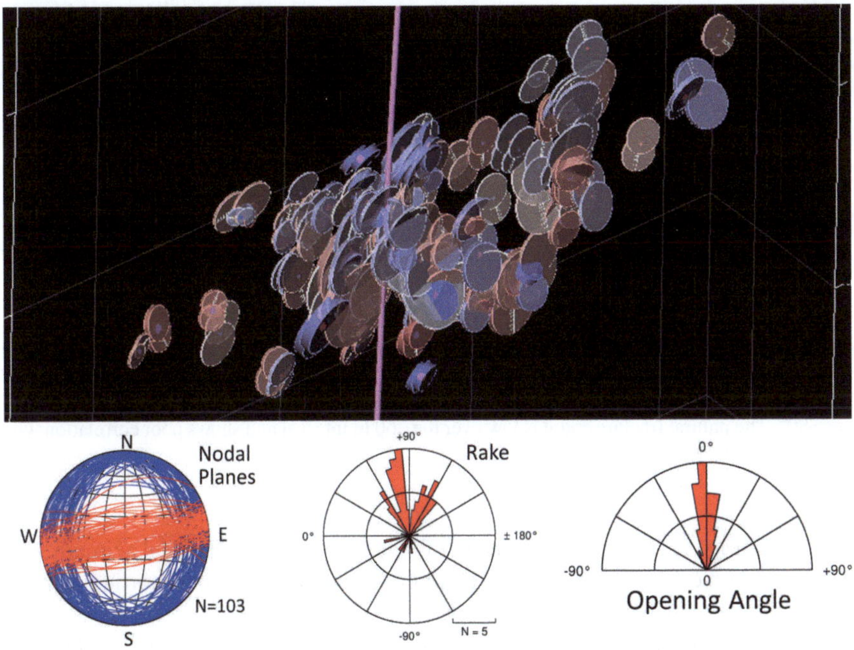

Fig. 3 Perspective view of a microseismic cloud (top). The cloud length is about 300 m, and height is about 150 m. The treatment well is vertical (purple line). The event glyphs represent the MT decompositions (see Leaney et al. 2014), where the displacement between the two disks portrays the component of slip and the disk thickness represents the tensile component, colored red for opening and blue for closing. For display purposes, a minimum disk thickness is required for a pure slip source. The radii of the disks scales with the moment scalar. The stereonet and rose diagrams summarize the geometry of the source mechanisms that show dip-slip displacements (73% of solutions shown in the perspective view), where the red nodal planes correspond to the glyph orientations. The vertical planes have a mean strike of $81° \pm 8°$, mean dip of $90° \pm 13°$ and are oriented subparallel to the hypocenter trend and principal stress direction (~92°). The mean opening angle is $-1.0° \pm 12°$ indicating they are primarily shear mechanisms. The opening angle would be 0° for a pure slip source and $\pm 90°$ for a pure opening (+) or closing (–) crack (Chapman and Leaney 2012). The remaining mechanisms represented in the perspective view (27%) are similarly oriented strike-slip events. Mean condition number is 5.4 ± 0.3

pattern is revealed by the pattern of P- to SH-wave amplitude ratios with respect to the monitor-well azimuthal position. The systematic radiation pattern corresponds to a dip-slip mechanism with a vertical nodal plane striking ~50°, close to the event trend and regional maximum horizontal stress (SHmax) direction.

The source locations shown in Fig. 5 were determined precisely using correlated picks and the method outlined by Rutledge and Phillips (2003). The mean rms residual is 0.2 ms. The gross geometry resolved are two vertical fractures, about 150 ft apart, growing simultaneously from the isolated perforation intervals. The depth locations form distinct bands separated by aseismic intervals. A gap of event detection is evident on the two depth projections near the origin of the horizontal distance-along-

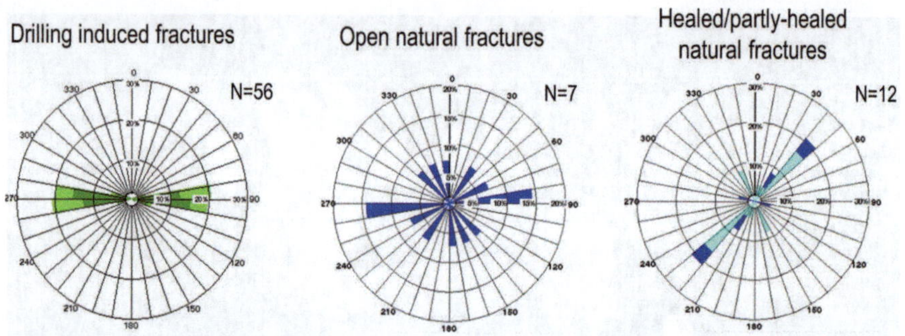

Fig. 4 Summary of borehole fracture image logs obtained in a well adjacent to the treatment well of Fig. 3, and covering a 260 m interval of the reservoir. SHmax inferred from drilling induced fractures is east-west parallel to the event trend. Natural fracture counts and strikes are for fracture dips >80°. The natural fracture count is low over the 260 m interval and shows poor correlation with the vertical nodal planes of Fig. 3

strike axis (Fig. 5). The gap is the result of signal loss at the SH-nodal azimuth with respect to the vertical receiver array, and our selection criteria of requiring clear P- and S-wave signals for precise locations.

The sequence of events and their general vertical alignment suggest through-going, vertical hydraulic fractures. Figure 6 shows the spatio-temporal development of the two resolved fracture zones. Most of the microseismicity is occurring at or near the leading edges, especially for the south fracture. The gap of events from about 800 to 1000 ft distance in Fig. 6 again corresponds to the SH-nodal azimuth. Fracture height comprising the lower two active intervals is established early and these intervals grow together laterally (red and blue symbols, Fig. 6). Upward growth to the shallowest interval occurs late (green symbols), but it also shows most of the seismicity is on the corresponding leading line moving laterally from the perforations.

2.2 *Bedding Plane Slip in Shale Reservoirs*

Injection-induced seismicity is often explained in terms of the Mohr diagram, where increased pore pressure reduces the effective normal stress and drives the rock closer to failure. For hydraulic-fracture microseismicity, this would correspond to driving the surrounding rock to failure via leak off and pressure coupling through the matrix or by direct connection into natural fracture networks. Our observations of nodal planes prevalently aligned close to the corresponding principal stress are difficult to reconcile with the Mohr diagram. In most cases, promoting shear preferentially at low angles to maximum stress is difficult without first crossing failure conditions for a much broader aperture of orientations (Rutledge et al. 2015). Further, matrix permeability of shales are typically extremely low, and natural fractures can be sparse and not necessarily aligned with principal stress (for example, Figs. 3 and 4). Thus,

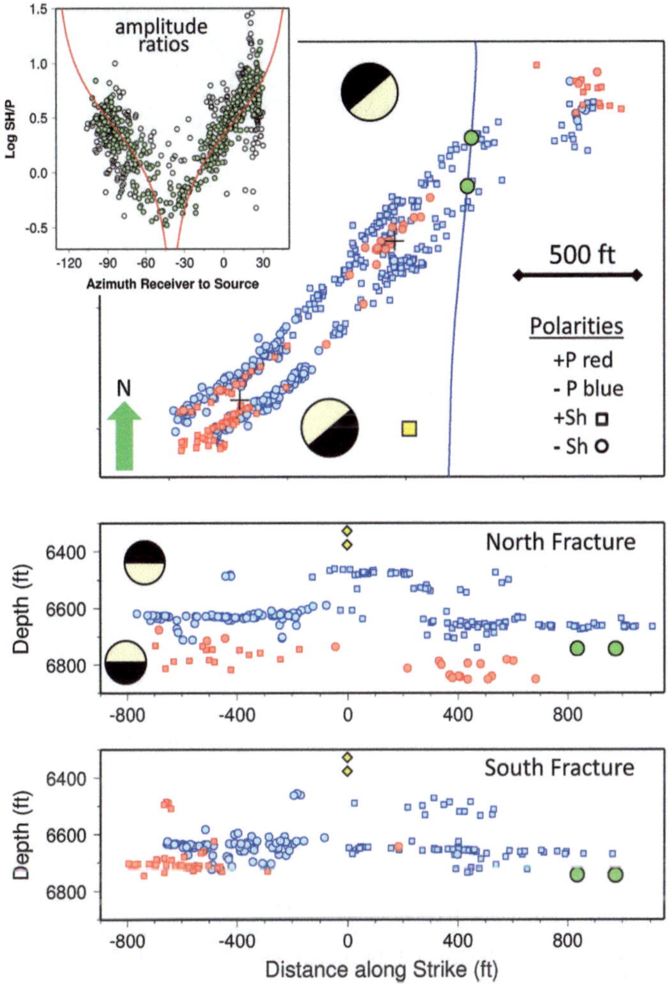

Fig. 5 High-precision microseismic source locations from one stage of a Barnett shale completion. Two mechanism families with opposite sense of first motions are distinguished by red and blue symbols. Two perforation intervals (green circles) spaced 200 ft apart were stimulated simultaneously. Two separate fractures growing from the isolated perforations intervals are resolved and are displayed separately in depth. The data were acquired on a 20-level vertical array of 3-component geophones spanning 950 ft. The monitor well is shown as the yellow square; the two deepest geophones are the yellow diamonds in the depth views. The SH- to P-wave amplitude ratios (SH/P) averaged over the array are shown with respect to the azimuth from the monitor well. The red curves correspond to the theoretical SH/P for dip-slip fault plane solutions on vertical planes striking 50°. SH/P values coloured green are within ±15° of the strike of the theoretical curve. The event P- and SH-phase polarities are distinguished by colour and symbol shape, respectively. The hemisphere projections for the P-wave first motions in depth view are towards the viewer and correspond to the red events (shown deeper) and the blue events (shallower). After Rutledge et al. (2015)

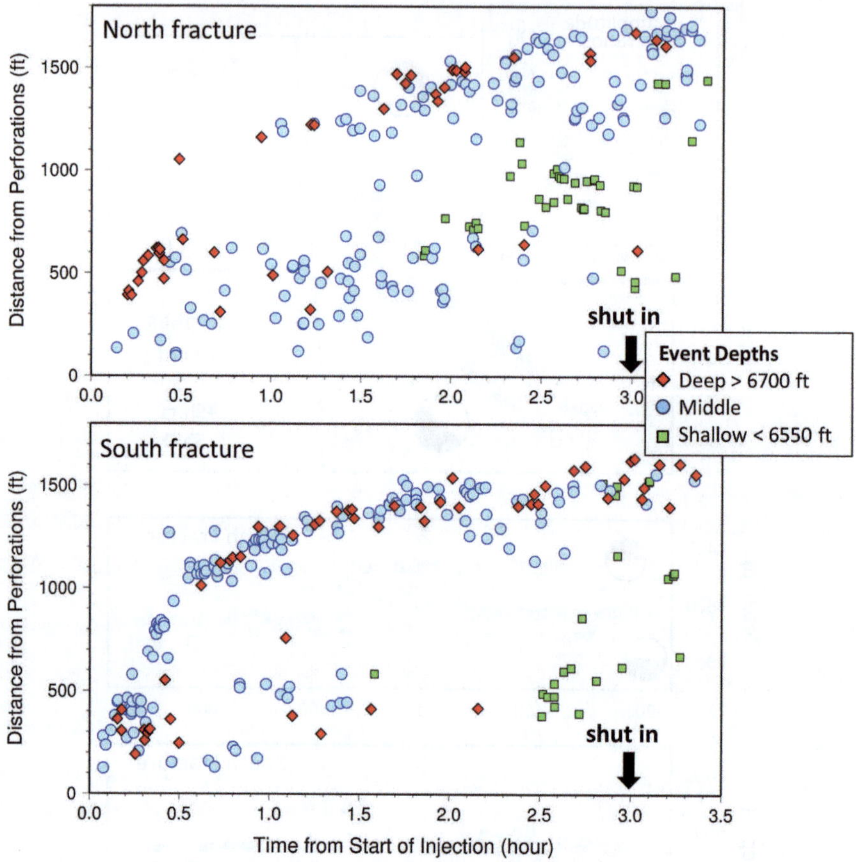

Fig. 6 Temporal development of the two separate fractures with respect to distance southwest from their respective perforation intervals for the Barnett completion shown in Fig. 5. Injection lasted 3 h at a rate of 80 bbl/min. Symbol color distinguishes three major depth intervals of events. After Rutledge et al. (2015)

shear failure driven by coupling of the hydraulic fracture pressure into the formation should often be weak.

Horizontal, or near-horizontal bedding planes are discontinuities common to most shale reservoirs. The common observation of dip-slip events in shale stimulations is probably easier to explain by considering the horizontal nodal plane as the slip plane, where the shearing occurs on weak bedding planes driven by the stresses/strains of hydraulic fracture opening. Shear at bedding planes can be driven by fracture tip stresses of a vertically-oriented extension crack approaching and interacting with the interface. The symmetry of stress changes should produce opposite senses of lateral displacements, left and right of the fracture tip (e.g., Cooke and Underwood 2001). From the time-space patterns and continuity of events and mechanisms displayed in Figs. 5 and 6, Rutledge et al. (2015) interpreted the microseismic shearing to be associated with step-overs formed as vertical hydraulic fractures jog at bedding interfaces, where the opening of a vertical crack is translated to a critical shear displacement. The development of step-overs is also described by crack tip stresses acting near a bedding interface. As a vertically-oriented extension crack approaches the mechanical discontinuity of bedding, the maximum tensile stresses offset symmetrically from the approaching crack (Helgeson and Aydin 1991; Cooke and Underwood 2001). The propensity for the crack to step one way is often controlled by presence of natural flaws, allowing continued vertical growth to reinitiate on one of the two maxima, left or right. Rutledge et al. (2015) proposed that the opposite senses of shear observed in the Barnett case are the result of opposite senses of step created at bedding interfaces.

As noted in Rutledge et al. (2016) the step-over interpretation is problematic. Modeling studies and field observations indicate that the offset of peak stresses from the parent tensile crack and the size of step-overs associated with crack tips stresses are typically on the order of a few centimeters (Cooke and Underwood 2001; Zhang and Jeffrey 2008; Chuprakov et al. 2011). Based on common magnitude scaling relationships, and for typical monitoring geometry, downhole detection of microseismic signals is limited to source lengths of about 0.5 m and generally larger for surface monitoring where dip-slip/horizontal-slip mechanisms are also commonly observed during shale stimulations. Thus, common step-over features should be too small to produce the microseismic observation unless the shear displacement extends considerably further along bedding. Further, larger jogs, if formed, will tend to elevate fracture pressures and arrest vertical fracture growth (Zhang and Jeffrey 2008).

Slip induced on bedding does not require a step-over or fracture termination. An extension crack can arrest, weaken and shear the interface and, under certain conditions with net pressure recovery and gain, then continue propagating across the interface. Chuprakov and Prioul (2015) present a model where vertical hydraulic fracture growth is slowed or terminated by contact with relatively weak bedding interfaces. A schematic of the model is shown in Fig. 7 where the fracture tips are blunted as they simultaneously contact interfaces above and below. During crack arrest, shearing can be promoted by fluid/pressure invasion along the bedding and the horizontal stress added by the fracture net pressure. For this model to be pertinent in describing microseismic observations, the scale length of the shearing along the interfaces would have to approach source dimensions on the order of a few meters.

2.3 Interpretation

The general pattern of P-wave first motions on both the north and south fractures show dilation first motions near the base of the fracture and compressional arrivals associated with the upward growth (Fig. 5). For the proposed geometry shown in Fig. 7, and the possible symmetries of mechanisms, the observations indicate that the bedding plane slip largely occurs on the northwest side of the fractures. This preference for shear to the northwest side may be due to the stress shadow effects of five stages previously completed to the southeast. The increased horizontal stresses associated

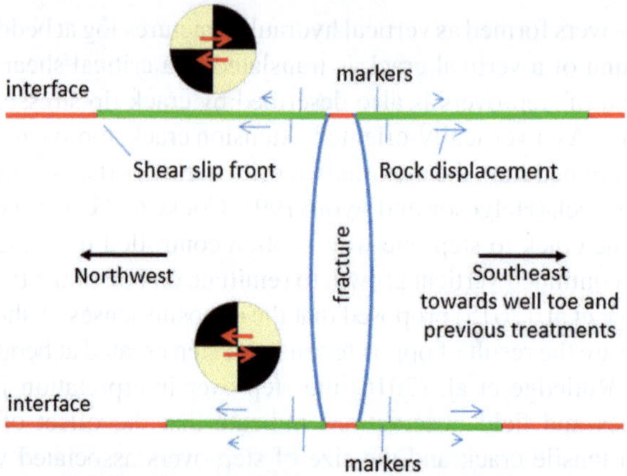

Fig. 7 Schematic of fracture arrest and shearing at bedding interface contacts. The rock displacements driven by fracture opening are shown with the blue arrows and markers. The markers would be aligned before fracturing, and here show the relative displacements of the rock on both sides of the interfaces with fracturing. Fluid invasion and pressure coupling weaken the interfaces. Opposite senses of shear would be generated both left and right of the fracture contacts, and at upper and lower interfaces. The P-wave first-motion diagrams show the relative sense of shear observed near the base of the Barnett fractures and that associated with upward growth at interfaces (Fig. 5). For this model, applying the first motion relationships observed in Fig. 5 indicates the bedding-plane shearing is preferentially occurring on the northwest side of the hydraulic fractures as highlighted by the two first-motion diagrams

with opened fractures to the southeast would tend to inhibit the induced displacements toe-ward of the treatment well and promote the shearing on the northwest side (Fig. 7). Similar symmetry in first motions of dip-slip/horizontal-slip mechanisms were observed in a series of Marcellus shale hydraulic-fracture completions, which may also be attributable to stress shadowing effects (Tan and Engelder 2016).

The general time-spatial sequence and continuity of events displayed in Figs. 5 and 6 show the shearing progressing down the horizontal structures. The effects not included in quasi-static models of interface slip would be the dynamic effects of critical shear failure and the mode III loading of the bedding surfaces ahead of the hydraulic fracture propagation direction by stress transfer. The dimensions of the shearing surfaces parallel to the fracture propagation direction may also be longer than the lengths orthogonal to the parent fracture, contributing to more detectable, larger magnitude signals. The events occurring behind the seismic front implied by Fig. 6 also indicate that bedding-plane slip can continue or repeat after the fracture tip propagates away from the active bedding surfaces.

The vertical development of the fractures appears to be more inhibited downward with the red events of Fig. 5 defining a dispersed lower boundary. The middle band of events near 6600 ft depth appear to define a clear bedding boundary where the fracture's upward growth may temporarily halt. But the eventual appearance of the top band of events indicates the upward growth continues through the mid-depth, seismically active interface.

3 Characterization of a Laboratory-Scale Hydraulic Stimulation Through the Moment Tensors of Acoustic Emissions

Laboratory experiments offer a controlled environment where one can gain insight into the mechanics of hydraulic stimulations. As with other smaller-scale simulations of earth phenomena, the main limitation of the understanding obtained from laboratory experiments is the extrapolation of results and interpretations to field-scale. Having that in mind, laboratory experiments are still an important tool in the study of hydraulic stimulations. In this section, we illustrate the last statement with the analysis of the MTs of acoustic emissions (AEs) observed during the hydraulic stimulation of a sandstone block under triaxial stress conditions.

AEs are seismic events that obey the same physics of elastodynamics. The name AE is used in fields such as geotechnics (Koerner et al. 1981; Hardy 2003), where the analysis is often focused upon compressional wave arrivals. For the experimental data considered in this section, we were able to isolate both direct P and Sv arrivals from the recordings. Therefore, the MTs described in the following paragraphs come from the fitting of both wave phases.

3.1 Description of the Experiment

The rock sample used was a Colton sandstone block. The dimensions of the block and other details of the experimental setup can be obtained from Table 1, Fig. 8 and Vera Rodriguez et al. (2017). The block was instrumented with 38 one-component, piezo-electric transducers distributed over its six surfaces. The monitoring array offered a coverage of the focal sphere never available, to the extent of our knowledge, in field-scale hydraulic stimulations. Further details regarding the acquisition system and characterization of the transfer function of the sensors can also be found in Vera Rodriguez et al. (2017). As part of the preparation of the block, vertical slots were sandblasted in the open hole section of the borehole to promote initiation of hydraulic fracturing in the direction of maximum horizontal stress. Before the start of the experiment, the block could be described as homogeneous in its fabric, without visible layering or noticeable pre-existing fracturing.

Once the block was subjected to the desired state of stress, the injection fluid of 2.5e6 cP viscosity was pumped at a rate of 5 mL/min until a few seconds after the breakdown point was detected in the borehole pressure measurements. The break-down corresponds to the point in time when the maximum borehole pressure is

Table 1 Completion details of the experiment

Rock type	Colton sandstone
Dimensions (mm)	228.6 × 279.4 × 381.0
Borehole diameter (mm)	25.4
Borehole depth (mm)	228.6
Cased hole section (mm)	165.1
Open hole section (mm)	50.8
Epoxy plug (mm)	12.7
Fracturing fluid	Silicon oil
Fluid viscosity (cP)	2.5e6
Injection rate (mL/min)	5
Vertical stress (σ_1) MPa	27.6
North-South stress (σ_2) MPa	13.8
East-West stress (σ_3) MPa	6.9

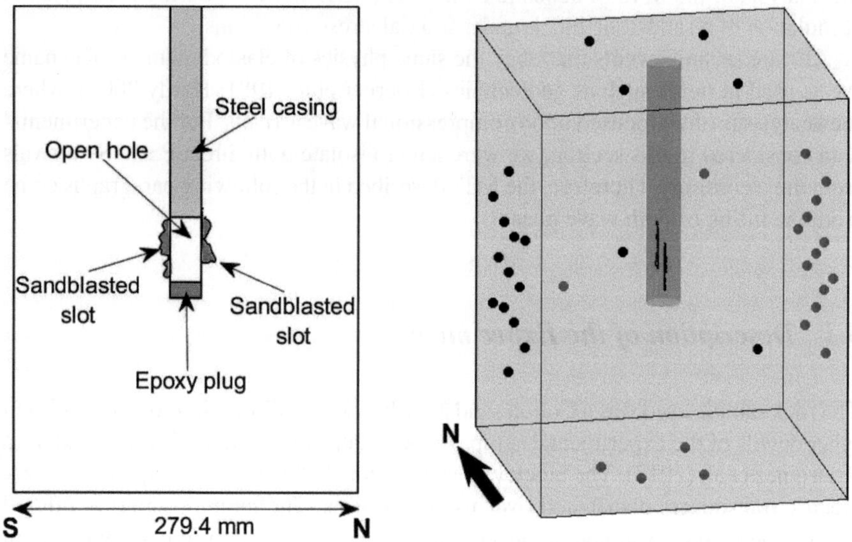

Fig. 8 Left: schematic diagram showing the completion details of the block and the approximate geometry of the sandblasted slots (as observed after the experiment over the open block). Right: distribution of receivers (black and grey dots) over the block boundaries

recorded. After the injection was stopped, the pumps were reversed to withdraw fluid from the borehole at a rate of 1000 mL/min. The objective was to stop the propagation of the hydraulic fracture before it reached the boundaries of the block. During the entire experiment, the stresses at the boundaries were controlled to be constant. After the block was removed from the presses, it was split in two along the surface of the created hydraulic fracture (Fig. 9).

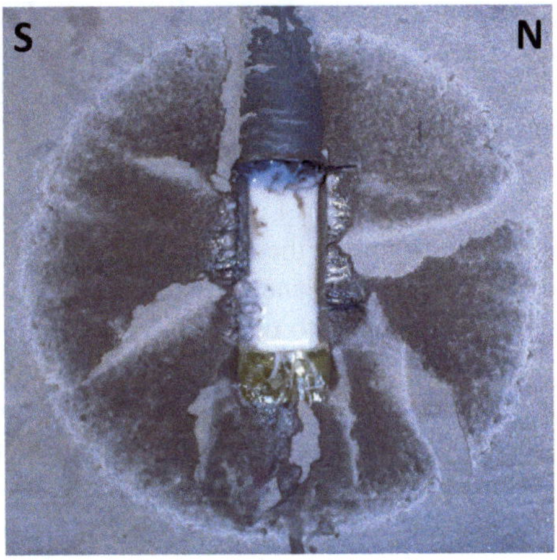

Fig. 9 Fracture plane observed after the block was split in two. The stripes displaying lighter colour correspond to regions of the fracture plane that were not invaded by the injection fluid. These are later referred to in the text as "dry stripes"

In total, 15,756 AEs were extracted from the monitoring records. To estimate their locations, a velocity model was constructed using measurements of compressional speed along ray paths connecting different pairs of transducers. For this purpose, the transducers were activated as ultrasonic transmitters one at a time while the rest of the array was listening. These measurements were taken at the start of the experiment with the block already compressed at the desired state of stress. The average shear wave velocity for the block was derived by matching the geometry of the cloud of located AEs to the surface of the fracture measured over the open block. The frequency spectra of all the AEs was also filtered to the range 90–200 kHz so the data would correspond to the far-field for the MTI. Additionally, the recordings were corrected for instrument response, coupling effects and the angular sensitivity of the receivers. Vera Rodriguez et al. (2017) present a detailed description of the processing workflow to obtain the MT solutions in this dataset.

More recently, we have conducted an analysis of the evolution of the attenuation properties of the block along the three coordinate directions and over time. From the results of this analysis, the recordings have been corrected for attenuation and new MT solutions have been estimated. The MTs presented in the following subsection correspond to our latest results incorporating spatio-temporally varying attenuation corrections.

3.2 *Analysis of the Estimated Moment Tensors*

The most reliable MT solutions were selected taking into account the following criteria: distance from the AE to each receiver (>50 mm to meet the far-field assumption), normalized misfit of the solution with the observations (<0.3, except for the interval 0–2300 s where we used <0.36), and the sensitivity of the MTI to random noise perturbations (measured as the dispersion of multiple, noise-perturbed inversions over a Hudson's plot<0.05). Using all of the above elements, the number of solutions left for analysis was reduced to 4,788. MTs were also decomposed, using the biaxial decomposition (Chapman and Leaney 2012), into general dislocations represented by the angles of strike, dip, rake, and deviation from pure shearing or α (Vavryčuk 2011; Chapman and Leaney 2012). Each MT can be associated with two different dislocations due to the ambiguity between the displacement vector and the normal to the dislocation plane. In every case, the dislocation with the largest dipping plane was categorized into one family of solutions and the remaining dislocation into a second family. Afterwards, we retained only those dislocations that could be associated with a stress regime. For example, normal faulting (negative rake or extensional stress regime) was linked to opening dislocations ($\alpha > 10°$). Similarly, reverse faulting (positive rake or compressional stress regime) was linked to closing dislocations ($\alpha < -10°$). Shear dislocations were accepted in all cases ($|\alpha| < 10°$). The cut-off angle of 10° is arbitrary but takes into consideration the uncertainty carried on from the MTI onto the decomposed dislocation angles. While in some instances this selection process allowed only one dislocation to be associated with an AE, in many cases, both dislocations were consistent with a stress regime and were retained. In the following, the source mechanisms are analysed focusing on their evolution during the experiment, for which we will present groups of AEs observed in progressive time intervals.

The first part of the experiment corresponds to the AE activity observed before the initiation of the hydraulic fracture (0–2300 s, Fig. 10). The initiation point is identified as an increase in the number of detected AEs and the deformation of the block in the direction of minimum stress (Zoback et al. 1977). The AE activity takes place north and south of the borehole (Fig. 11). The AEs located in the south appear next to the sandblasted slot and present predominantly opening components under 30°. These dislocations seem directly connected to the borehole; therefore, they are being consistently activated by the tensile push of the injection fluid. The initiation of the fracture could have started earlier on this side of the borehole. The AEs displayed in the north side took place farther away from the sandblasted slot and show a mix of low-angle opening and closing source mechanisms. If we added the locations of AEs for which MT solutions were discarded, we could see that these AEs could be connected to the borehole from its intersection with the sandblasted slot. In particular, the high-magnitude, closing event near the bottom, could be associated with the activation of a pre-existing crack, likely linked to the deformation (compression) of the block. During this part of the experiment, the fluid could not leak-off into the rock due to its high viscosity; therefore, the block was being internally compressed by the increase in fluid volume and pressure build up inside the borehole. Given the distribution of the principal stresses imposed over the outer boundaries (see Table 1), there is a preferential activation of highly dipping dislocations aligned closer to the north-south direction.

After initiation, the opening fracture can start accepting fluid depending on its width and the viscosity of the fluid. Lower viscosity fluids can flow more easily towards the opening fracture. In the case of our experiment, however, the high vis-

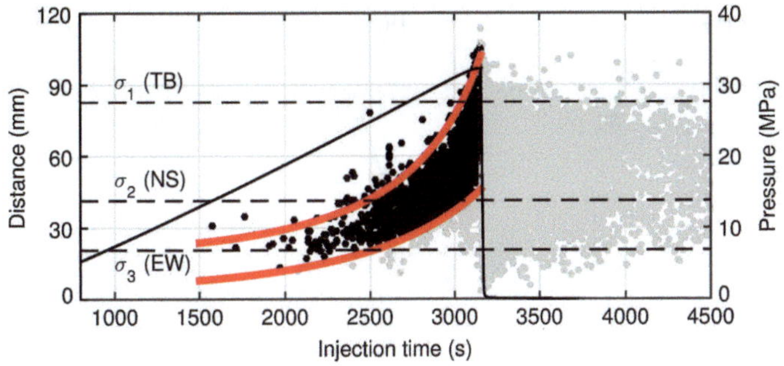

Fig. 10 AE distance from the injection point along the fracture plane. AEs between the two thick red lines (black dots) are interpreted to have taken place near the edge of the propagating fracture. The rest of the AEs (grey dots) are interpreted to have taken place at the back of the fracture's edge. The black solid line is the borehole pressure; the steep fall of this curve marks the moment when the pumps were reversed to withdraw fluid

cosity of the injection fluid resulted in a period of time during which the fracture propagated dry (Stanchits et al. 2015). Figure 12 shows a set of AEs observed at the start of this period of dry propagation. The orientation of the dislocations are now better defined close to the north-south direction with highly dipping angles.

The style of activation is also dominated by dip-slip components. The main fracture plane appears already well-defined in the south side of the borehole, while in the north side there is still activity in front but also growing from the sandblasted slot. It appears that the block had a network of weakness planes (e.g., micro-fractures) in this area that influenced the initial development of the north side of the main fracture plane. Most AEs reflect opening dislocations at angles predominantly under 30°. Up to this point there is no evidence of fracture growth below the bottom part of the borehole.

Non-opening events are observed in some cases separated from the main AE cluster. When inside the main cluster, non-opening events can often be associated with regions where the fracture found barriers restricting its growth (Fig. 13, bottom row). These regions are identified in the open block as dry stripes. We can also see an absence of opening events inside these stripes during some time intervals (e.g., Fig. 13, intervals 2650–2800 s and 2950–3010 s), and that some of the lower-angle opening AEs fall in their surroundings. We speculate that non-opening AEs that cannot be associated with dry zones in the open block could also correspond to regions where fracture propagation was initially restricted, but that were finally opened during the experiment, sometimes enough to permit fluid invasion. According to this interpretation, the lower part of the fracture plane was the most difficult to develop, since it contains a significant proportion of non-opening events, and also because multiple dry stripes can be observed in the open block. The north and south sides also display concentrations of non-opening events near dry stripes. Restrictions to fracture growth can be the result of variations in the rock strength,

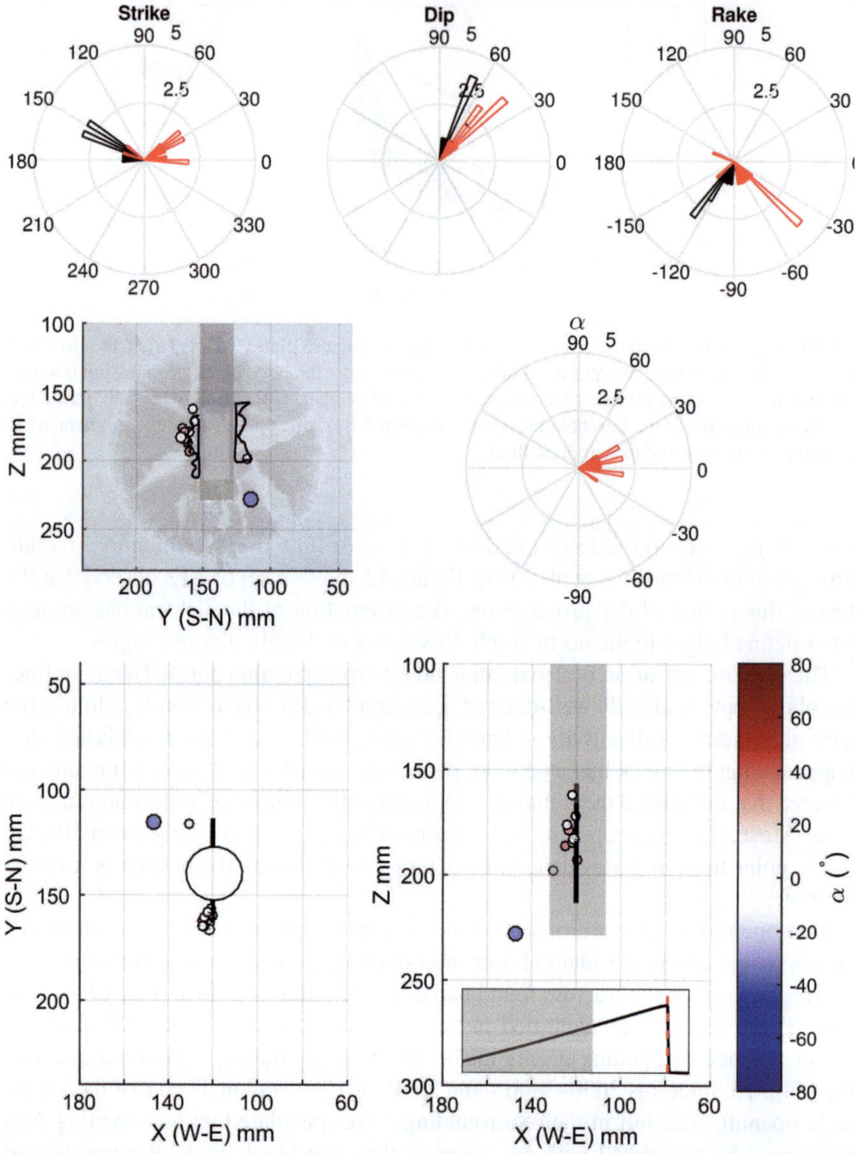

Fig. 11 Source mechanisms of AEs observed during the interval 0–2300 s. Black and red colours on the rose plots correspond to the two families of dislocations extracted from the AE MTs. Dots represent AEs coloured by their angle of deviation from the DC source (α) and sized by their relative magnitude. The picture of the actual fracture plane is also showed as a reference in the background of the front-view plot. Black-solid lines follow the approximate contour of the sandblasted slots. The inset plot displays the position of the time interval (grey rectangle) with respect to the borehole pressure (black line) and the breakdown point (red, vertical-dashed line)

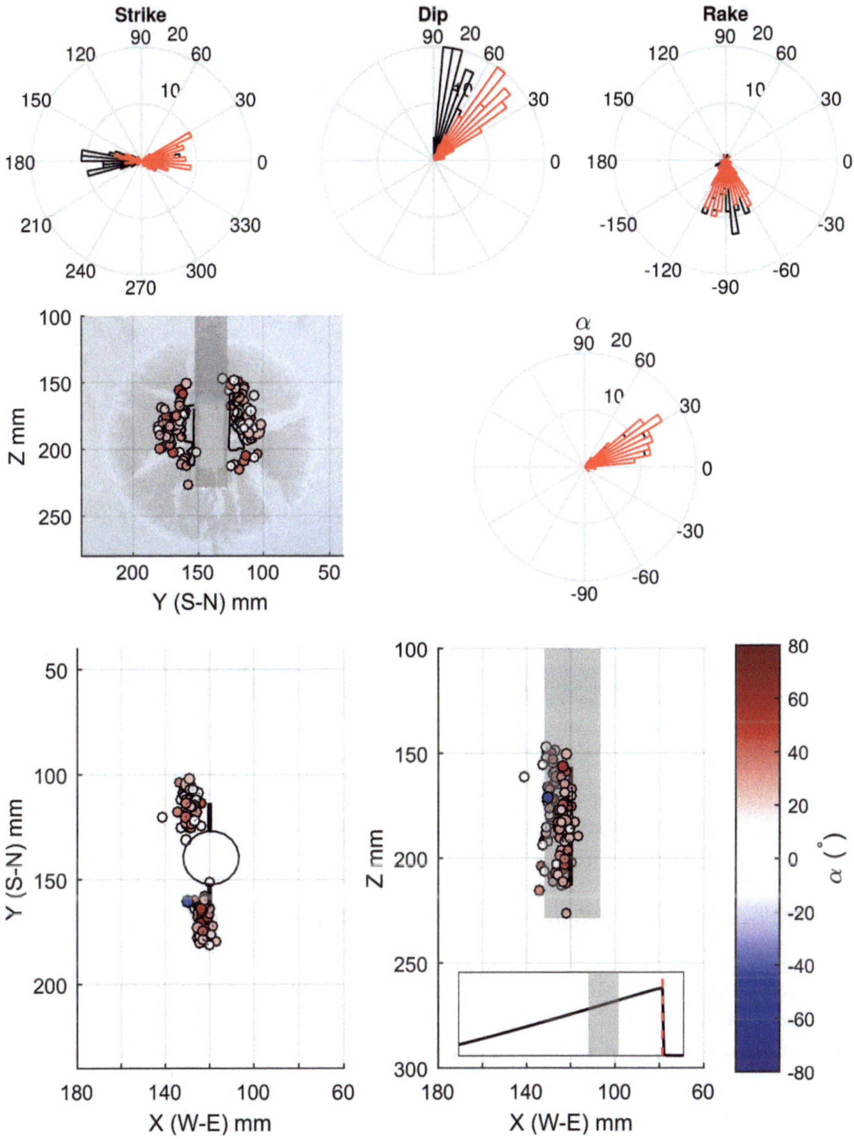

Fig. 12 As Fig. 11 except time interval is 2300–2650 s

fabric, distribution of stresses around the borehole and/or a combination of all these factors.

In experiments with low viscosity fluids, fracture initiation can coincide with breakdown, i.e., when the elastic energy stored in the system is suddenly released after the borehole pressure reaches its peak value (Zoback et al. 1977). With increasing

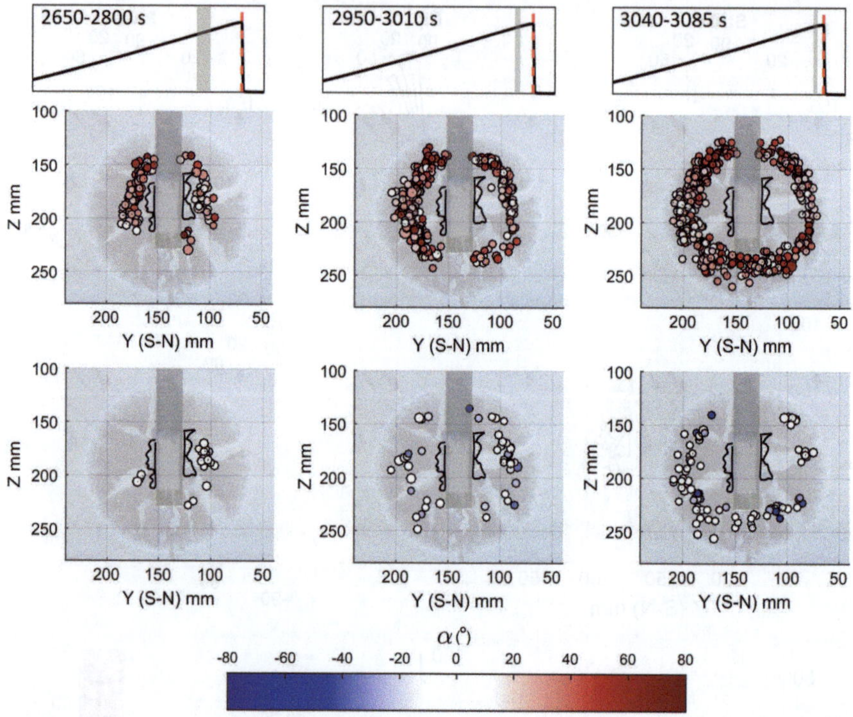

Fig. 13 AEs observed in successive time intervals during the dry fracture propagation stage. Top row are opening AEs ($\alpha > 10°$). Bottom row are shearing ($|\alpha| < 10°$) and closing ($\alpha < -10°$) AEs. The plots at the top display the position of the corresponding time intervals (grey rectangles) with respect to the borehole pressure (black line) and the breakdown point (red, vertical-dashed line)

fluid viscosity, fracture initiation takes place in advance of breakdown, introducing a period of controlled fracture growth. As mentioned before, part of this controlled growth can take place with fluid entering the fracture depending on fracture width and fluid viscosity. In our experiment, a period of lubricated fracture propagation is identified starting with a slowdown in the rate of increase of the borehole pressure (onset of leak-off towards the fracture) until the point where it reaches its maximum value at breakdown. This corresponds to the time interval from 3085 to 3155 s (see Fig. 10). Most of the fracture growth took place during the dry fracture propagation period as a result of the pressure build up at the borehole. With the flow of the fluid towards the fracture plane, this pressure build up was relaxed, which in turn slowed down fracture growth until it was finally stopped by the reversal of the pumps to withdraw fluid from the borehole (Fig. 14, left). The source mechanisms remain consistent during this period of fracture growth, with mainly sub-vertical dislocation planes aligned near a north-south orientation and being activated in a predominantly dip-slip direction with opening angles around 30°. Only a slight increase in the

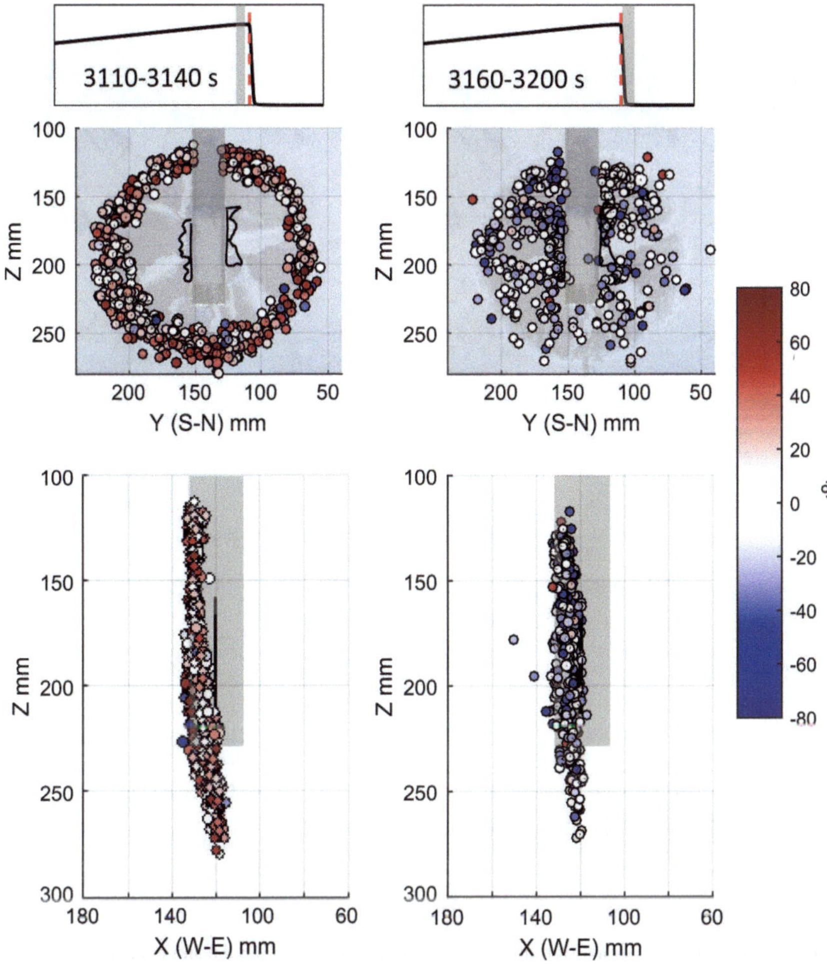

Fig. 14 AEs observed during two time intervals corresponding to the lubricated fracture propagation (left) and closing (right) periods of the experiment. The plots at the top display the position of the corresponding time intervals (grey rectangles) with respect to the borehole pressure (black line) and the breakdown point (red, vertical-dashed line)

dispersion of the dislocation angles with respect to the aforementioned directions of alignment could be observed.

With the reversing of the pumps to withdraw fluid from the borehole, the whole fracture plane started closing. This is reflected both by the locations and the source mechanisms of the AEs observed during this period of time (Fig. 14, right). Generally speaking, the amount of deviation of closing AEs from the DC case is smaller than in the case of opening AEs. Also, the maximum deviations from DC occur at the beginning of the closing period and reduce towards the end of the experiment. In other

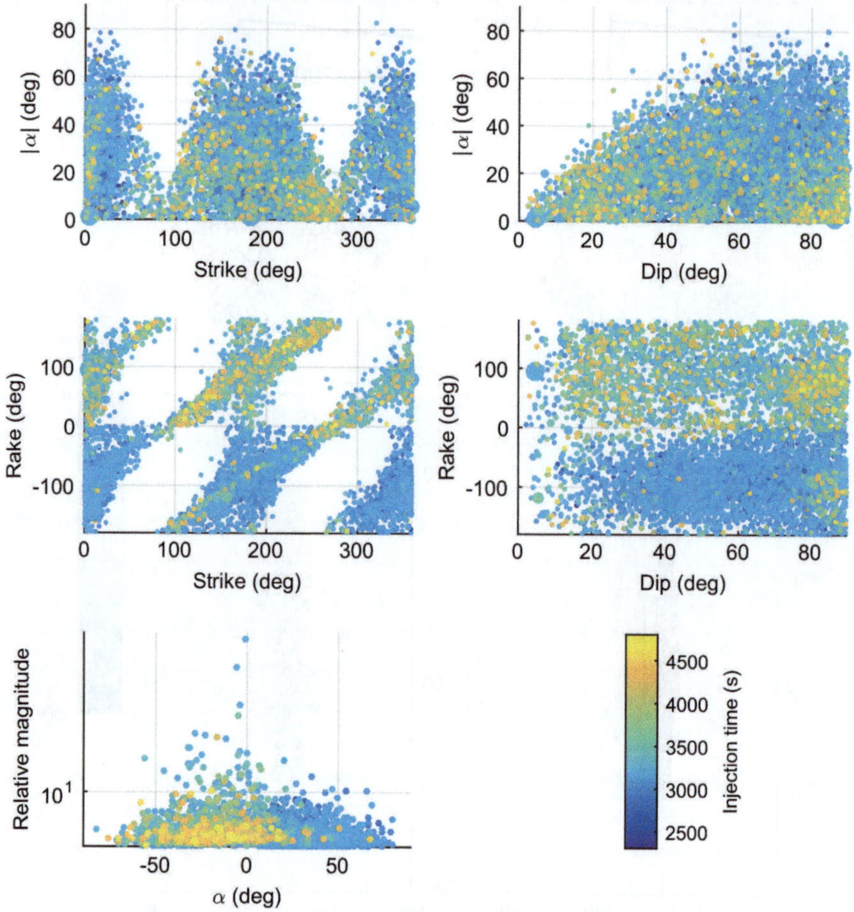

Fig. 15 Scatter plots between different dislocation angles and between the angle α and the relative size of the AEs. Only combinations that displayed some sort of correlation are presented. Both dislocation families, after filtering for stress regime consistency, are included in the plots

words, most of the last-observed AEs are predominantly of shearing type. Closing AEs are better explained by the collapse of asperities holding sections of the fracture open while the injection fluid was being withdrawn. Otherwise, the closing of flat dislocation surfaces would require a fast squeezing of the injection fluid, which is more difficult to support given its high viscosity.

Looking at correlations between dislocation angles during the complete duration of the experiment (Fig. 15) we can make the following observations:

- Dislocation planes lying parallel to the main fracture plane (north-south) present the largest components of opening/closing. As the strike of the dislocation deviates from the north-south direction towards the east-west direction the amount of permissible opening/closing is reduced, so the events are predominantly shearing. The same effect is observed when the dip of the dislocation reduces towards the horizontal plane.
- The maximum deviation from pure shearing is reached for dislocations dipping around 60°, this is 30° from the direction of the maximum principal stress. Above this dip angle, the maximum deviation from pure shearing remains stable around $|\alpha|$ ~70°.
- Dislocation planes lying parallel to the main fracture plane (north-south) are activated predominantly in dip-slip but their activation can also take place in practically any other direction from dip-slip to strike-slip. As the dislocation-plane strike deviates from the north-south direction towards the east-west direction, the activation mode becomes restricted towards strike-slip only.
- Opening AEs during the first part of the experiment (fracture propagation) are predominantly activated near dip-slip directions. Closing AEs during the second part of the experiment (fracture closure) show more deviations from dip-slip toward strike-slip activation.
- Relative AE magnitude reduces quickly for $|\alpha| > 50°$. AEs that are closer to the pure DC case display the largest relative magnitudes.

Shearing AEs

Shearing AEs (i.e., AEs with $|\alpha| < 10°$) account for 19% of the total number of MTs selected for analysis. If we assume that the criteria used for selection are independent of the angle α, we can consider this number is also representative of the general distribution of AE source mechanisms (α is unique for each AE). In a similar way, we estimate that shearing AEs account for 19% of the total seismic moment released during the experiment. The events with the larger individual magnitudes in the catalogue are also of shearing nature with dip-slip style of activation. The contribution from shearing AEs to the total number of observed events increased during the experiment from 11% during the dry fracture propagation period, to 17% during lubricated fracture propagation, and finally to 27% during the closing of the fracture. We speculate that the percentage of shearing AEs during the closing stage would have been larger if the pumps had not been reversed to control the growth of the fracture.

In field-scale hydraulic stimulations it is common to observe increases in the number of detected microseismic events after falls in borehole pressure. Similarly, source mechanisms derived from surface monitoring arrays (which usually provide better focal coverage compared to downhole monitoring arrays) tend to describe predominantly shearing events. Normally, only the events with higher signal-to-noise ratio are selected for MTI in field-scale applications. The results from this experiment suggest that, considering relative magnitudes, shearing events could be more likely to be detected and selected for MTI than opening and closing events in field-scale microseismic monitoring. The increase in shear activity observed during

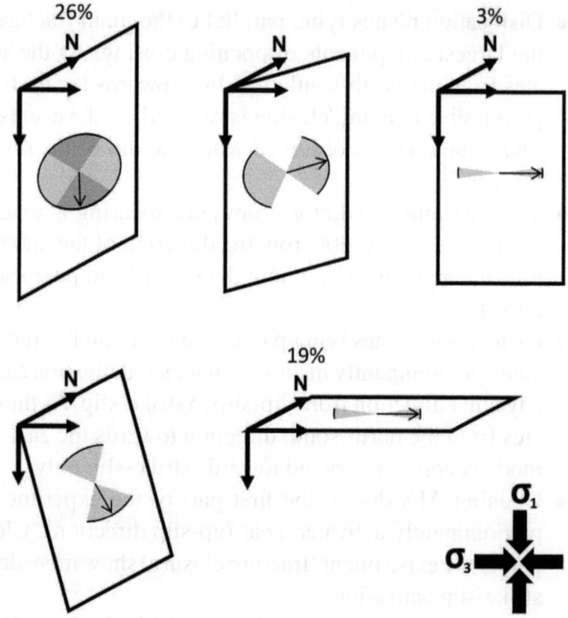

Fig. 16 Representation of the ranges of activation constrained by the orientation of the dislocation plane with respect to the stress field. The black arrow represents the direction of slip and the grey areas the permissible activation ranges. The darker areas in the top-left image denote that although activation can take place in all directions, dip-slip is dominant. The numbers are percentages (with respect to shear-only AEs) observed during the complete experiment. These percentages include both dislocation solutions from the MT decomposition

the closing stage of the experiment is also consistent with this hypothesis and the increase in microseismic event detections during pressure drops at field-scale.

The geometry of individual dislocations is controlled by the rock's strength and fabric, and the local state of stress. Similarly, the style of activation is constrained by the orientation of the dislocation plane with respect to the stress field. For instance, for dislocation planes aligned with the stress field (i.e., parallel to the main fracture plane) the rake or direction of slip can vary from dip-slip to strike-slip style of activation (although dip-slip is significantly dominant). Changing the strike while maintaining the dip constant near 90° reduces the range of permitted activation style towards strike-slip. On the other hand, changing the dip while maintaining the strike aligned with the main fracture plane reduces the range of activation style towards dip-slip. In both cases, the slip direction is constrained towards alignment with the direction of minimum principal stress (Fig. 16).

In this analysis of shear sources we have considered the two dislocation planes that are obtained from the decomposition of each MT. Additionally, we did not remove shear AEs with activation styles that contradicted the main stress conditions at a particular time during the experiment. For example, during fracture opening we retained shear sources denoting reverse faulting, and vice versa during the closing stage. These criteria accept the possibility of local variations in the stress field that include reversals. Further geomechanical modelling of the fracturing experiment is necessary to validate or discard this possibility.

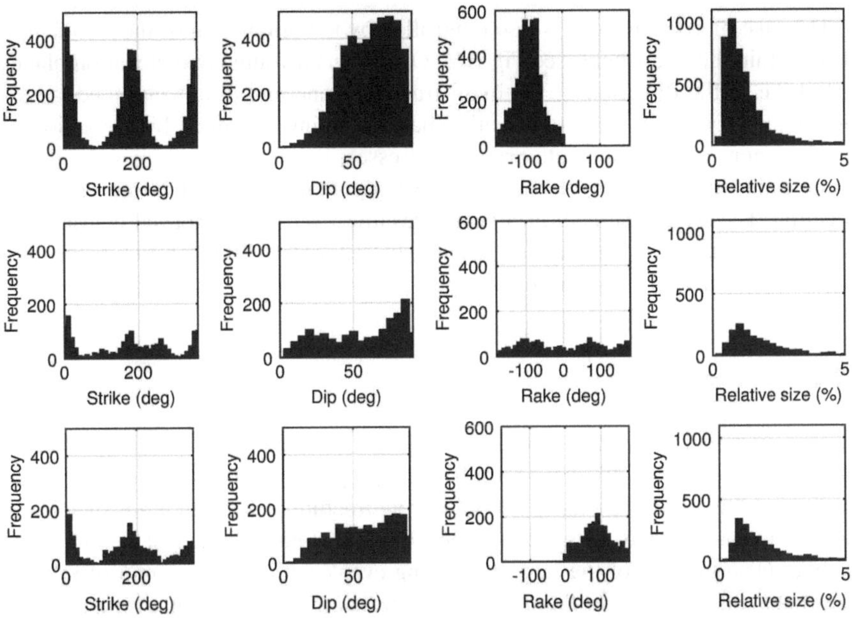

Fig. 17 Distribution of dislocation angles for opening (top), shearing (middle) and closing (bottom) AEs. Both dislocation solutions are represented in the histograms after filtering by consistency with stress regimes. Relative sizes are estimated with respect to the AE with the largest seismic moment in the complete catalogue, a dip-slip shear event

3.2.1 Opening AEs

Opening events are favoured by high viscosity injection fluids (Matsunaga et al. 1993; Bennour et al. 2015). In consequence, it is possible that with a lower viscosity injection fluid, the maximum AE opening angle observed in our dataset could have been smaller, and/or the frequency of opening AEs could have been less. Opening dislocations account for 52% of the catalogue of AEs. Their contribution in number decreases from 85% of the total during the dry opening period, to 74% during lubricated propagation, and down to 10% during the closing of the fracture. The seismic moment released from opening AEs corresponds to approximately 52% of the total during the complete duration of the experiment, most of it released during the opening stages. The largest magnitude opening event took place soon after the start of the fracture closure period. It was located near the border of a dry stripe on the north side of the fracture with $\alpha = 49°$. It is almost half the size of the largest shearing event and about 75% the size of the largest closing event. Most opening AEs are about 0.75% the size of the largest shearing event (Fig. 17, top). This is not too far from the 1% mode observed in the distribution of relative sizes of shearing events (Fig. 17, middle). The decay in the distribution of relative magnitudes is steeper for opening than for shearing AEs towards bigger-sized events.

The rake of opening AEs is also controlled by the orientation of the dislocation plane within the stress field (see Fig. 16). Opening is favoured in dislocation planes that strike in the direction of the main fracture plane and dip at intermediate-to-high angles (Fig. 17, top). On the other hand, opening is limited by the action of the maximum and intermediate principal stresses; therefore, there are virtually no opening dislocations with planes near the horizontal or aligned in the east-west direction. The style of activation concentrates towards dip-slip, normal faulting but, as mentioned above, rake can also deviate toward strike-slip when the dislocation plane is aligned with the main fracture plane.

3.2.2 Closing AEs

Closing AEs represent 29% of the total number of events in the catalogue. Their number increases from 4% during dry fracture propagation, to 9% during lubricated fracture propagation, and finally to 63% during fracture closure. In terms of seismic moment release, closing AEs account for 29% of the total. The largest closing event is about 60% the size of the largest shearing event, with the histogram of relative sizes behaving similarly to that of shearing events in its decay but with a mode at 0.75% as in the case of opening events (Fig. 17, bottom).

For a fluid-filled dislocation, closing accompanied by seismic energy release requires a fast compression and/or retreat of the fluid inside. The block was not artificially dried or saturated for the experiment; therefore, it contained a mix of interstitial fluids and air filling its pores. Closing AEs outside the fracture plane and before breakdown can be explained by the collapse of cracks in response to the compressive deformation of the block produced by the opening fracture. Cracks filled with air and/or connected to high permeability paths would be more likely to release seismic energy. Closing AEs located inside the fracture plane during the same period of time seem related to regions where fracture propagation was limited by the rock's strength and fabric. For AEs located within the fracture plane during closing, the low compressibility and high viscosity of the fluid used in the experiment makes the release of seismic energy while closing improbable without the external help of the pumps. Therefore, it is possible that a significant amount of closing AEs were the result of the reversal of the pumps producing sudden drops in pore pressure along high permeability paths across the fracture plane. The presence of asperities over the fracture plane could prevent a smooth closing as the pore pressure dropped, resulting in negative pressure build up and eventual, sudden collapses along the asperities. It is also possible that in the absence of the fluid withdrawal process, fewer closing and more shearing AEs could have been observed.

The distribution of dislocation geometries of closing AEs is similar to that of opening AEs, with the increase in intermediate-dipping planes not observed in shearing AEs (Fig. 17). The only noticeable difference is an increase in the proportion of dislocations dipping under 50° compared to opening AEs. The dislocation planes are also preferentially oriented in the north-south direction, although the decrease in the number of dislocation planes oriented in the east-west direction is not as smooth as with opening AEs. The activation of closing dislocations follows also a preferential dip-slip direction, although there is a higher proportion of deviations towards strike-slip compared to opening AEs.

4 Conclusions

The examples presented in this chapter illustrate the usefulness of MTs for the understanding of the mechanics of hydraulic stimulations both at different scales and in different lithologies. Furthermore, the examples highlight the benefits of incorporating additional, independent sources of information for the interpretation of dislocation planes extracted from MT solutions. For instance, in the field case study, we proposed that the observed microseismicity is generated by bedding-plane slip using a model where vertical fracture growth is halted or slowed at bedding interfaces. The general patterns of first motions observed on two distinct, parallel fractures propagating simultaneously suggest that events detected near the base of the fracture and those associated with upward growth tend to occur preferentially on the heelward side of the treatment well and away from the direction of previously completed hydraulic-fracture stages. We propose that stress-shadow effects with increased horizontal stresses generated southeastward by previously completed stages in the treatment well break the expected symmetry of opposite shearing left and right of the fracture contact at an interface, where the elevated stress inhibits displacements towards previously completed stages and further promotes the stress changes for bedding-plane towards the northwest.

The scale of the laboratory experiment permitted the recovery of a more complete catalogue of AEs, where the MTs describe general source types that correlate with the stress regime throughout the hydraulic stimulation process. On the other hand, the source types that are inconsistent can be associated to local effects modifying the general stress regime. For example, closing events inside the fracture plane observed during fracture opening could be linked to regions where fracture propagation was facing barriers; and those outside the fracture plane to the collapse of pre-existing cracks. Shear events and low-angle opening events could also be associated with regions of restricted fracture propagation over the fracture plane. Opening events during fracture closing were possibly the result of the injection fluid being drawn towards the borehole through regions of the fracture plane that had already closed. Notice that, despite the differences in scale and lithology, the laboratory data coincide with the field-scale example by ascribing at least part of shear activated AEs to some form of restriction to fracture growth.

In the laboratory experiment, the MT solutions describe a fracture plane that is not smooth, but rather the connection of small dislocation planes dipping at angles mostly above 30° with two possible distribution peaks, one around 45° and the other around 80°. The planes that are better aligned with the principal stresses (north-south, vertical) are those that can potentially open the most. As the planes deviate from this alignment, the action of the intermediate and maximum stress limits the maximum opening that a dislocation can experience. The direction of slip along the dislocation plane is also influenced by the principal stresses, with the slip being aligned toward the direction of the minimum principal stress as the dislocation plane deviates from the north-south, vertical case. As the dislocation plane aligns toward the horizontal plane, with strike defined near the maximum horizontal stress, the direction of slip is restricted toward dip-slip motion, which is also consistent with the observations from the field-scale example.

The results from the experiment can also be connected to other general observations in field-scale hydraulic stimulations. For instance, the larger number of event detections observed at field-scale after pressure drops and their shear nature is consistent with the larger relative magnitudes of shear events observed during the laboratory experiment and their increase in number after breakdown. In other words, the smaller number of publications reporting opening events (compared to shear events) in field-scale stimulations might be in part a consequence of a detectability limit issue. It has to be pointed out as well, however, that the very high viscosity of the injection fluid in the experiment likely promoted more opening AEs than might be expected in field-scale operations. The large number of closing events observed in our dataset during closure might also be a particular behaviour produced by the experimental procedure. Nevertheless, a similar increase in closing source types could be expected at least during flow-back periods in field-scale stimulations.

Acknowledgements The authors would like to thank Phil Christie, Chris Chapman, Gwenola Michaud and Michael Williams for useful comments that helped improve the presentation of this chapter. We are also grateful to Devon Energy Corporation for granting permission to publish the Barnett data results and Schlumberger for permission for the publication of the laboratory data.

References

Aker E, Kuhn D, Vavryčuk V, Soldal M, Oye V (2014) Experimental investigation of acoustic emissions and their moment tensors in rock failure. Int J Rock Mech Min Sci 70:286–295

Aki K (1966) Generation and propagation of G waves from the Niigata earthquake of June 16, 1964. Part 2. Estimation of earthquake moment, released energy, and stress-strain drop from the G wave spectrum. Bull EarthQ Res Inst 44:73–88

Bennett L, Le Calvez J, Sarver D, Tanner K, Birk W, Waters G, Drew J, Michaud G, Primiero P, Eisner L, Jones R, Leslie D, Williams M, Govenlock J, Klem R, Tezuka K (2005) The source for hydraulic fracture characterization. Oilfield Rev 17:42–57

Bennour Z, Ishida T, Nagaya Y, Chen Y, Nara Y, Chen Q, Sekine K, Nagano Y (2015) Crack extension in hydraulic fracturing of shale cores using viscous oil, water, and liquid carbon dioxide. Rock Mech Rock Eng 48:1463–1473

Brune J (1970) Tectonic stress and the spectra of seismic shear waves from earthquakes. J Geophys Res 75:4997–5009

Chapman C, Leaney S (2012) A new moment-tensor decomposition for seismic events in anisotropic media. Geophys J Int 188:343–370

Chuprakov D, Prioul R (2015) Hydraulic fracture height containment by weak horizontal interfaces. Paper presented at the SPE hydraulic fracture technology conference, The Woodlands, 3–5 Feb 2015

Chuprakov D, Akulich A, Siebrits E, Thiercelin M (2011) Hydraulic-fracture propagation in a naturally fractured reservoir. SPE Prod Oper 26:88–97
Cipolla C, Warpinski N, Mayerhofer M, Lolon E, Vincent M (2008) The relationship between fracture complexity, reservoir properties, and fracture treatment design. Paper presented at the SPE annual technical conference and exhibition, Denver, 21–24 Sept 2008
Cipolla C, Wright C (2002) Diagnostic techniques to understand hydraulic fracturing: what? why? and how? SPE Prod Facil 17:23–35
Cooke M, Underwood C (2001) Fracture termination and step-over at bedding interfaces due to frictional slip and interface opening. J Struct Geol 23:223–238
Department of Energy and Climate Change (2014) Fracking UK shale: understanding earthquake risk. http://m.northyorks.gov.uk/CHttpHandler.ashx?id=32806&p=0. Accessed 19 Jun 2017
Diller D, Shuck T, Fish B (2015) Estimation and interpretation of high-confidence microseismic source mechanisms. Lead Edge 34:918–924
Duncan P, Eisner L (2010) Reservoir characterization using surface microseismic monitoring. Geophysics 75:75A139–75A146
Economides M, Nolte K (eds) (2003) Reservoir stimulation. Wiley, New Jersey
Fink J (2013) Hydraulic fracturing chemicals and fluids technology. Gulf Publishing Company, Oxford
Goertz-Allmann B, Goertz A, Wiemer S (2011) Stress drop variations of induced earthquakes at the Basel geothermal site. Geophys Res Lett 38:L09308
Hardy H (2003) Acoustic emission microseismic activity, volume 1: principles, techniques and geotechnical applications. Taylor & Francis, Lisse
Helgeson D, Aydin A (1991) Characteristics of joint propagation across layer interfaces in sedimentary rocks. J Struct Geol 13:897–911
Koerner R, McCabe W, Lord A (1981) Overview of acoustic emission monitoring of rock structures. Rock Mech Rock Eng 14:27–35
Langenbruch C and Shapiro S (2014) Probability of brittle rock failure during hydraulic fracturing of conventional and unconventional reservoirs. In: SEG expanded abstracts, Denver, 2014
Leaney S, Yu X, Chapman C, Bennett L, Maxwell S, Rutledge J, Duhault J (2014) Anisotropic moment tensor inversion and visualization applied to a dual well monitoring survey. CSEG Recorder 39:48–54
Matsunaga I, Kobayashi H, Sasaki S and Ishida T (1993) Studying hydraulic fracturing mechanisms by laboratory experiments with acoustic emission monitoring. Paper presented at the 34th US symposium on rock mechanics, ARMA, Madison, 28–30 June 1993
Mayerhofer M, Lolon E, Warpinski N, Cipolla C, Walser D, Rightmire C (2010) What is stimulated reservoir volume? SPE Prod Oper 25:89–98
Maxwell S, Rutledge J, Jones R, Fehler M (2010) Petroleum reservoir characterization using downhole microseismic monitoring. Geophysics 75:75A129–75A137
Maxwell S, Shemeta J, Campbell E and Quirk D (2008) Microseismic deformation rate monitoring. Paper presented at the SPE annual technical conference and exhibition, Denver, 21–24 Sept 2008
Maxwell S, Urbancic T, Le Calvez J, Tanner K, Grant W (2004) Passive seismic imaging of hydraulic fracture proppant placement. In: SEG Expanded Abstracts, Denver, 2004
McLennan J, Green S, Bai M (2008) Proppant placement during tight gas shale stimulation: literature review and speculation. Paper presented at the 42nd US symposium on rock mechanics, ARMA, San Francisco, 29 June–2 July 2008
Neuhaus C, Ellison M, Telker C, Blair K (2014) Drainage estimation and proppant placement evaluation from microseismic data. Paper presented at the SPE/EAGE European unconventional resources conference and exhibition, Vienna, 25–27 Feb 2014
Roche V, van der Baan M (2015) The role of lithological layering and pore pressure on fluid-induced microseismicity. J Geophys Res 120:923–943
Rutledge J, Phillips W (2003) Hydraulic stimulations of natural fracture as revealed by induced microearthquakes, Carthage cotton valley gas field, east Texas. Geophysics 68:441–452

Rutledge J, Phillips W, Mayerhofer M (2004) Faulting induced by forced fluid injection and fluid flow forced by faulting: an interpretation of hydraulic-fracture microseismicity, Carthage cotton valley gas field, Texas. B Seismol Soc Am 94:1817–1830

Rutledge J, Downie R, Maxwell S, Drew J (2013) Geomechanics of hydraulic fracturing inferred from composite radiation patterns of microseismicity. Paper presented at the SPE annual technical conference and exhibition, New Orleans, 30 Sept–2 October 2013

Rutledge J, Yu X, Leaney S (2015) Microseismic shearing driven by hydraulic-fracture opening: an interpretation of source mechanism trends. Lead Edge 34:926–934

Rutledge J, Weng X, Yu X, Chapman C, Leaney S (2016) Bedding-plane slip as a microseismic source during hydraulic fracturing. In: SEG expanded abstracts, Dallas, 2016

Shapiro S, Audigane P, Royer J (1999) Large-scale in situ permeability tensor of rocks from induced microseismicity. Geophys J Int 137:207–213

Sileny J, Hill D, Eisner L, Cornet F (2009) Non-double-couple mechanisms of microearthquakes induced by hydraulic fracturing. J Geophys Res 114:B08307

Stanchits S, Burghardt J, Surdi A (2015) Hydraulic fracturing of heterogeneous rock monitored by acoustic emission. Rock Mech Rock Eng 48:2513–2527. https://doi.org/10.1007/s00603-015-0848-1

Stanek F, Eisner L (2013) New model explaining inverted source mechanisms of microseismic events induced by hydraulic fracturing. In: SEG expanded abstracts, Houston, 2013

Tan Y, Engelder T (2016) Further testing of the bedding-plane-slip model for hydraulic-fracture opening using moment-tensor inversions. Geophysics 81:KS159–KS168

Urbancic T, Rutledge J (2000) Using microseismicity to map Cotton Valley hydraulic fractures. In: SEG expanded abstracts, Calgary, 2000

Vavryčuk V (2011) Tensile earthquakes: theory, modelling and inversion. J Geophys Res 116:B12320

Vavryčuk V (2015) Moment tensor decompositions revisited. J Seismol 19:231–252

Vera Rodriguez I, Stanchits S, Burghardt J (2017) Data-driven, in-situ, relative sensor calibration based on waveform fitting moment tensor inversion. Rock Mech Rock Eng 50:891–911

Yu X, Rutledge J, Leaney S, Sun J, Pankaj P, Weng X, Onda H, Donovan M, Nielsen J, Duhault J (2015) Integration of microseismic data and an unconventional fracture modeling tool to generate the hydraulically induced fracture network: a case study from the cardium formation. West Central Alberta, Canada. Paper presented at the unconventional resources technology conference (URTeC), San Antonio, 20–22 July 2015

Zhang X, Jeffrey R (2008) Reinitiation or termination of fluid-driven fractures at frictional bedding interfaces. J Geophys Res 113:B08416

Zoback M, Rummel F, Jung R, Raleigh C (1977) Laboratory hydraulic fracturing experiments in intact and pre-fractured rock. Int J Rock Mech Min Sci 14:49–58

Constrained Moment Tensors: Source Models and Case Studies

Jan Šílený

1 Introduction—The Moment Tensor

In recent decades, the earthquake mechanism, regardless of scale, has commonly come to be described by the moment tensor (MT). The MT has been used when inverting local but, especially, teleseismic (Global CMT project: Dziewonski et al. 1981; Ekstrom et al. 2012; USGS: Sipkin 1982; Sipkin and Zirbes 2004; ERI: Kawakatsu 1995) and regional (e.g., Dreger and Helmberger 1993; Ritsema and Lay 1993; Nábělek and Xia 1995; Braunmiller et al. 2002; Pondrelli et al. 2002; Stich et al. 2003; Kubo et al. 2002) data. For local data, here, I cite events occurring within earthquake swarms (Jakobsdóttir et al. 2008; Horálek and Fischer 2008).

The unconstrained MT is the most comprehensive description of shear and non-shear sources and is the body force equivalent of a rupture (i.e. it consists of a system of forces, in fact, force couples) that generates the same wave field in a continuous medium as an actual rupture. In this way, it is not a physical source but substitutes for real processes occurring within the focus. As a system of body forces, the MT is not a priori convenient for offering a simple perception of an earthquake focus. Therefore, for the sake of interpretation, the MT is generally decomposed into simple sources. The method of decomposition is not unique. The most commonly and widely used procedure is one that splits the general MT into isotropic and deviatoric portions (unique), and then splits the deviatoric portion into a double couple (DC) and a compensated linear vector dipole (CLVD) with a common major tension or pressure axis (ambiguous). The somewhat painful procedure of searching for a reasonable decomposition amongst a theoretically infinite number of processes is described in Julian et al. (1998). The MT captures general combinations of dipoles and, as such, is able to approach a wide class of mechanisms that can occur within an earthquake source. An important advantage of the MT description is a linear relationship between

J. Šílený (✉)
Institute of Geophysics, Czech Academy of Sciences, Prague, Czech Republic
e-mail: jsi@ig.cas.cz

S. D'Amico (ed.), *Moment Tensor Solutions*, Springer Natural Hazards,
https://doi.org/10.1007/978-3-319-77359-9_9

source parameters and seismic observations (through the response of a medium, Green's function) that implies linearity for the inversion task. As such, the MT allows fast and unique retrieval of the six independent parameters—M_{11}, M_{12}, M_{13}, M_{22}, M_{23}, and M_{33}—without the need to specify an initial guess.

The MT is not only the most general description of a mechanism relevant to an earthquake source. In general, it is also relevant to the fracturing of a solid body, including all of the modes of fracturing recognized within fracture mechanics and their combinations. However, the MT also includes mechanisms that do not generally represent realistic physical sources because it does not describe the rupture itself but rather body force equivalents of actual rupturing. As such, the MT is needlessly general. In addition to rupture mode I (tensile fracturing), rupture modes II and III (plane and anti-plane shear slip), combinations of force systems that do not correspond to a physically feasible rupturing are also present. Therefore, for the simple rupturing expected within the foci of tectonic earthquakes, the MT is unnecessarily complex, leading to more parameters than those relevant to simple rupture models.

2 Constraining the Moment Tensor

From the viewpoint of solving the inverse problem of earthquake mechanism retrieval from seismic data, the tax of an advantage of linearity for the inverse task employing the MT as the source mechanism description is potentially small robustness in cases where the inverse problem is not well-posed (e.g. if the data are sparse or of low quality, or if the hypocenter localization or velocity/attenuation model of the medium are uncertain (e.g., Šílený et al. 1992; Dahm et al 2000; Stierle et al. 2014)). Therefore, in practice, it is advantageous to reduce generality within the MT (i.e., to decrease the number of parameters within the source mechanism model from six, for the MT, to five or four). The endpoint of this sequence is a DC that is described by three angles that depict the orientation of the corresponding shear slip and its size, capturing its scalar moment.

A widely used constraint is the deviatoric MT, where the MT trace preserves a zero value:

$$M_{11} + M_{22} + M_{33} = 0 \tag{1}$$

As a default, the deviatoric MT is applied within Global and Regional CMT procedures in order to simplify the MT to make an inversion more stable and to approach the shear slip as the foreseen mechanism for tectonic earthquakes. The result is a zero volume change within the focus. In spite of this, the constraint does not imply that the mechanism described is a desirable shear slip. The deviatoric MT contains both a double-couple (DC) (the equivalent of a pure shear slip along a plane fault embedded within an isotropic medium), as well as a compensated linear-vector dipole (CLVD). The CLVD is an artifact of the decomposition and, by itself, lacks clear physical interpretation. From this perspective, the deviatoric constraint is not ideal. Although the deviatoric constraint is widely used, its only advantage is that it maintains the linearity of the inverse task for unconstrained MT retrieval. Much of the use of Global CMT catalogs, as well as regional CMT catalogs, concern utilization of the DC portion of the CMT solution—its association to faults within the area under study, interpretation in terms of regional and local tectonics, etc. However, a DC obtained in this way, a traditional "best DC" (i.e. a component isolated from the MT obtained as a solution to the inverse problem), is not equal to a directly retrieved DC. Henry et al. (2002) demonstrated that the DC, decomposed from the MT, can be biased due to inaccurate velocity models and that this type of DC is only reliable when the non-DC component of the MT solution is small.

For completeness, the purely isotropic source itself, the complement to the deviatoric MT, should also be considered to be a constrained MT. The importance of this source model rests in the fact that explosions are processes that basically approach isotropic expansion. However, this may be modified by inhomogeneity and/or the anisotropy of the medium of the focal zone, the latter being caused by both intrinsic properties of the geologic material or by the state of pre-stress, as demonstrated by Vavryčuk and Kim (2014). Also, in industry, due to the technology of blasting, blasts may be far from isotropic expansion, especially blasts within a series of delayed shocks. Such a series may be perceived as a single event depending on the frequency window of the observation and possesses radiation directivity related to the geometry of the blast series.

The deviatoric constraint is a formal approach for reducing the number of parameters describing the source mechanism, as originated by a mathematical concept. A different approach is to constrain the MT by accepting some idea regarding the physics of rupturing during earthquake activity within the focus. An obvious procedure is assuming a fault within the focus and the earthquake process as a sudden slip on the fault. This idea leads to an equation relating the MT to fault and slip vectors (within an isotropic medium; Aki and Richards 2002, Eq. 3.21), as follows:

$$M_{ij} = \lambda n_k u_k \delta_{ij} + \mu \left(n_i u_j + n_j u_i\right) \tag{2}$$

where **n** and **u** are the fault normal and slip vector, respectively; and λ and μ are the Lamé constants. Although the setup of this source model is fairly generic and capable of describing many situations that researchers believe may occur within the earthquake focus, it is less general than the full MT. For example, it cannot describe a purely isotropic source (1).

Both implicit or explicit, the constraint (2) is the basis of many models for a seismic source. The former group consists of models intuitively designed to describe natural or man-made setups that are related to the occurrence of seismicity and are not formally treated using mathematics. For example, Hasegawa et al. (1989) suggested a series of models designed to describe seismoactive phenomena occurring in mines: (i) a single force for a cavity collapse, (ii) a vertical compressive single couple for a pillar burst, (iii) a vertical tensile crack for a cavity roof break, and (iv) combinations of implosion within a dip slip for a stope and cavity closure (Fig. 1, the b/w background). To be specific, model (i) is outside the framework of the moment

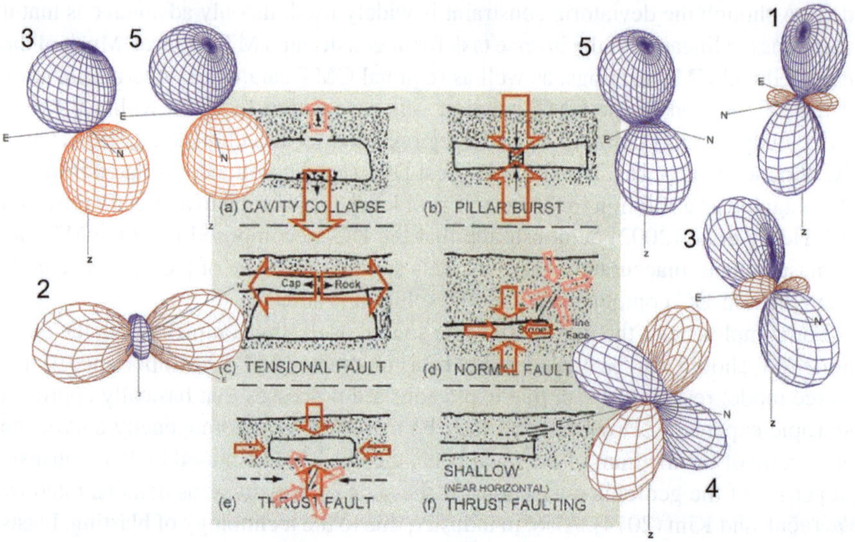

Fig. 1 Models (**a**–**f**) for seismoactive events occurring in mines by Hasegawa et al. (1989) (black/white background) and their body-force equivalents (red and rose arrows). 3D views of the P-radiation pattern for five Ridgeway mine events obtained using the inversion of P and S amplitudes are found as wireframe diagrams on the margin

tensor concept because a single force is applied as the equivalent for the break of the cavity roof (the MT consists of couples of forces). Additionally, for a more exact consideration, the equivalent for cavity roof collapse is a couple of single forces related to the detachment of rock from the roof and to its subsequent fall to the floor (Malovichko 2005). Below, in Fig. 1, body-force equivalents corresponding to individual physical phenomena (a–d) are marked using red/rose arrows.

Šílený and Milev (2008) applied the MT and single force models when inverting direct P and S amplitudes for the five seismic events recorded within the Driefontein gold mine in South Africa with the aim of interpreting their mechanisms in terms of the Hasegawa models. For two events (3 and 5), both a downward single force and nearly a vertical compressional single couple equally fit the data (i.e. a cavity collapse and a pilar burst were both plausible). A pillar burst may have also occurred for event 1, while event 2, with a nearly extensional single couple, may have corresponded to a break of the cavity roof. A comparison with the mine map documented a coincidence for the hypocentra within the mine works. Thus, the Hasegawa models seem to have been useful.

Explicitly, models related to (2) or to combinations of these models have appeared in numerous papers. Combinations are largely comprised of a traditional model for an earthquake mechanism, a pure shear slip along a fault within body-force equivalents corresponding to a DC, and an additional component designed to simulate a particular feature of an event to be investigated. As early as the 1980s, Teisseyre (1980) proposed several configurations for a shear-slip (with opening or closing tensile cracks at the

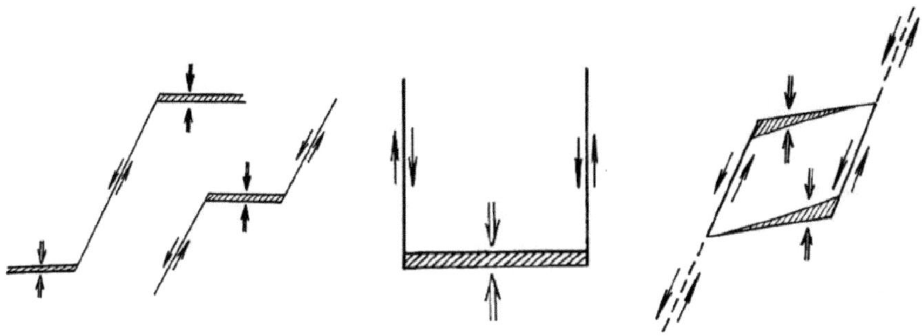

Fig. 2 Models of rockburst foci by Teisseyre (1980). From left to right: a simple shearing, shearing with compression, a complex opposite shearing, and a closed pattern for the rockburst process

tips of the fault) in order to simulate complex fracturing in either the extensional or compressional regime, the latter intended to describe a rockburst focus (Fig. 2). However, Teisseyre (1980) did not solve for the angle of the tensile crack with respect to fault direction, so the model remained conceptual.

The concept of Teisseyre (1980) was employed by Rudajev and Šílený (1985) and Šílený (1989) for inverting local seismic data related to mining tremors occurring within Kladno and Ostrava, Czechoslovakia, coal mines. Rudajev and Šílený (1985) embedded a model of a shear fault and a closing crack running perpendicular from its tip. In Šílený (1989), to avoid doubts regarding the precise angle of the tensile crack, the shear fault was complemented by an isotropic implosion. The combination of a shear fault with a tensile crack leaving the fault tip at various angles was studied by Julian et al. (1998) who investigated the radiation of such a source independent of the geometry of the fault versus the tensile crack and the direction of the opening.

A special case of geometry amongst the Julian et al. (1998) superpositions of a shear fault with a tensile crack is the geometry that occurs when the tensile crack formally coincides with a shear fault and opens perpendicularly to the fault. At the same time, this geometry is the genuine source described by (2). Previously described models were combinations of two or more sources compatible with (2). The one just described can be interpreted as a shear fault that simultaneously opens (or closes) with shear slipping. Although differently named by various authors, here, I refer to it as a shear-tensile crack (STC).

3 Shear-Tensile Crack (STC)

The concept of the shear slip and a simultaneous opening of the same fault originated in Dufumier and Rivera (1997), afterward appeared in Minson et al. (2007), and was revisited by Vavryčuk (2001, 2011). A big advantage in comparison to the MT is the fact that STC is a physically based concept for the source mechanism; simply, a

Fig. 3 The STC source model: a combination of shear-slip and tensile crack represented by the slip vector off the fault plane with the slope angle α (α = 0° is a pure shear slip and α = 90° is a tensile crack)

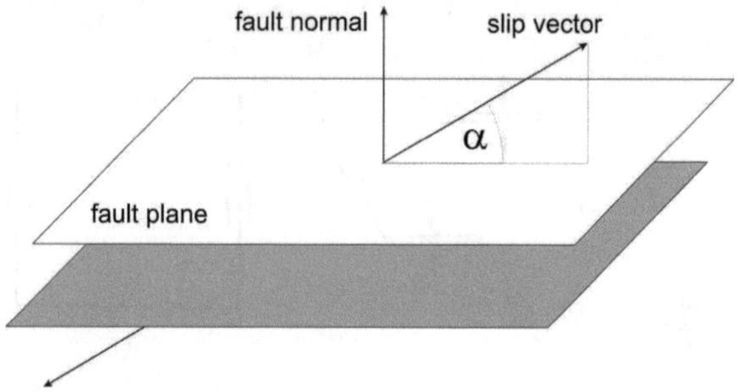

slip along a fault that may open or close at the same time (i.e. directly describing the seismoactive process in terms of the rupture occurring within the focus), Fig. 3.

At the same time, a source model representing a constrained MT is described by five parameters: (i) the dip of the fault, (ii) the strike of the fault, (iii) the rake angle specifying the shear component of the slip, (iv) angle, α, referred to as the slope, representing the off-plane component of the slip vector, and (v) the magnitude. The number of parameters is the same as that for the deviatoric MT. Contrarily, however, the model also has the capability of describing a change in volume within the focus. In fact, it is the simplest source model for incorporating a shear slip and a volume change. In this way, it combines a desirable amount of generality with maximum possible simplicity, ensuring robustness for the inverse task (Šílený 2009; Pesicek et al. 2012).

Despite the constraint in comparison to the full MT, the STC is a source model that is amazingly flexible with a varying slope angle, the endpoints being a traditional shear-slip (with a DC as its body force equivalent) and a tensile crack, Fig. 4. Thus, the properties of radiation change from the well-known quadrant pattern of the DC to the all-positive (or, alternatively, negative in the option of an implosion) onset of a P-wave. A similar feature, a change from a four lobbed pattern to a bimodal appearance, is also seen in the radiation of S waves. Illustrative behavior can be observed at the source planes (source lines within the projection onto the horizontal plane). Source planes within the STC include the plane of the fault that tangentially slips and the plane perpendicular to the slip vector (Vavryčuk 2001, 2011). For the pure shear-slip, the planes coincide with the traditional nodal planes of the fault-plane solution. With an increasing slope, they deviate from their originally perpendicular configuration. For the tensile crack, they merge into a single plane, namely, the plane of the crack.

In the design of their model, Dufumier and Rivera (1997) were motivated by thoughts regarding the relationship of a full moment tensor applied to tectonic earthquakes and rheological constraints, which may not always be realistic. Their solution was the proposal of an extended physical model that included tectonic and non-tectonic volumetric variations. Minson et al. (2007) essentially used the same model but referred to it as the CDC—the "crack plus double couple". They reminded readers that the model was used for geodetic inversions of seismic sources by Okada (1985)

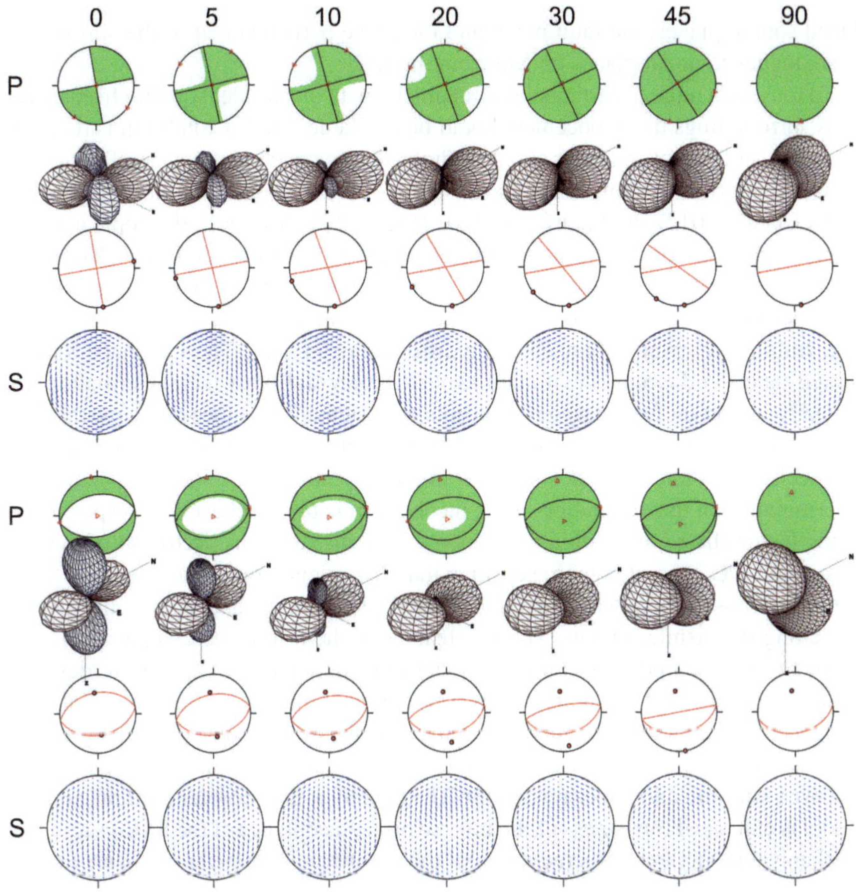

Fig. 4 Plots of the radiation pattern of P and S waves of the shear-tensile crack (STC) source dependent of the acquiring slope angle of 0, 5, 10, 20, 30, 45, and 90°. The upper half of the figure displays the vertical strike-slip mechanism. The lower half displays the dip-slip inclined 45° from the vertical. Row 1: traditional fault-plane solution plots (nodal lines, zones of compression for P-waves in green; principal axes, red triangles (T-axis, triangle up; P-axis, triangle right; N-axis, triangle left). Row 2: wireframe diagrams of P-radiation in 3-D (compressions, red; dilations, blue). Row 3: source lines (red lines) and poles of the source planes (red circles). Row 4: the S-radiation pattern (projection of polarization vectors of S-waves). In rows 1, 2, and 4, equal area projections of the lower hemisphere are used

and they applied it in order to interpret data from a volcanic earthquake swarm in Japan that had obvious volume changes within the focus. Features of the model were investigated, in detail, from the point of view of the theory by Vavryčuk (2001, 2011). Vavryčuk (2001, 2011) referred to events with both shear slip and tensile components as "tensile earthquakes". He derived various useful formulas for the decomposition of the equivalent moment tensor and the source tensor for the slope angle, and introduced source planes, the fault plane and the plane perpendicular to the slip vector, as analogues to nodal planes known from the DC.

To retrieve a reliable earthquake mechanism, data of sufficient quality from directions surroundings the hypocenter should be available. Such a condition is required for any seismic event. As far as small earthquakes are concerned, a generally low signal to noise ratio and limited data due to sparse observations are additional obstacles. Pesicek et al. (2012) and Šílený et al. (2014) recently investigated this topic in detail by performing a series of synthetic tests that explored: (i) various types of station distributions from very good to very poor, (ii) inaccurate locations for hypocenters, (iii) improper modeling for velocities, (iv) the noise contamination of data, and (v) the compatibility of results obtained using different techniques. These authors demonstrated that: (1) the number of stations and the regularity of their distribution are crucially important for the retrieval of both the orientation of the source mechanism and the DC versus non-DC content, (2) noise within the data distorts retrieval of the mechanism more than uncertainty within the location and velocity modeling, (3) determination of the orientation of the mechanism is fairly robust with respect to all of the factors listed, excluding very poor station distributions. Retrieval of the DC versus non-DC content is more sensitive than the orientation. Poor station distributions may cause a large bias and reliable determination is a challenge for the inverse modelling of seismic sources. The key feature of the task is the complexity versus simplicity of the model. A complex model obviously has the capacity to describe in-depth properties of the source. However, complex models need not be resolvable using a limited data set. By applying a constraint, the model exactly simulates source processes that are simpler and relevant to certain realistic features. In many cases, however, simplification does not matter. The benefit of the acceptance of the simpler model is enhanced robustness of the inverse task. Acceptance of the STC as a source mechanism model implies a reduction of the number of parameters from six to five.

4 Confidence Zones

To assess the reliability of the solution obtained, I evaluated the confidence zones of source model parameters or their combinations. Confidence zones are objects within the model space that specify the volume in which the solution of the inverse task is captured in an a priori specified probability if errors in the data are taken into account. In other words, confidence zones describe a distribution for model parameters that yield a good match to the data. The size and shape of confidence zones indicate the uncertainty of the determined parameters. Large confidence zones indicate a poor solution while small confidence zones indicate a good solution. Both reliability for determining geometry, the orientation of the mechanism, and its characteristics for the DC and non-DC content, must be estimated. To assess geometric reliability, I constructed confidence zones for the T, P, and N axes of the deviatoric portion of the MT solution and confidence zones for decomposition of the MT in order to determine the characteristics for both the DC and non-DC content. For the STC model, I evaluated the same simply by re-computing the MT that corresponded to a given STC and additionally computing the confidence zone for the slope angle. For the procedure, I scanned the model space within a regular grid and evaluated a match to the data and the posterior probability density (PPD) for grid points. I then integrated the PPD across a trial volume within the model space and looked for a patch in which the cumulative probability acquired certain values, say, 0.9, 0.95, and 0.99.

I considered the confidence zone of the model parameters, $\boldsymbol{m}$, using the probability content, p, to be a set of points, $\boldsymbol{m}$, satisfying the following:

$$\chi^2(\boldsymbol{m}) < \chi_p^2 \tag{3}$$

where $\chi^2(\boldsymbol{m})$ is the misfit function constructed from the residual least squares of data and synthetics, and the dispersion, χ_p^2, is determined from the following condition:

$$\frac{\int_{\chi^2(\mathbf{m})<\$^2_{\mathrm{p}}} \mathrm{PPD}\ \mathrm{d}\mathbf{m}}{\int_{\mathbf{m}} \mathrm{PPD}\ \mathrm{d}\mathbf{m}} = \mathrm{p} \tag{4}$$

that specifies the ratio of the cumulative probability within the region to the integral of the posterior probability density (PPD; e.g., Tarantola 1987) across the entire model space. The PPD is defined as follows:

$$\mathrm{PPD} = \exp\left(-\frac{1}{2}\chi^2\right). \tag{5}$$

Schematically, it can be depicted by the cartoon in Fig. 5.

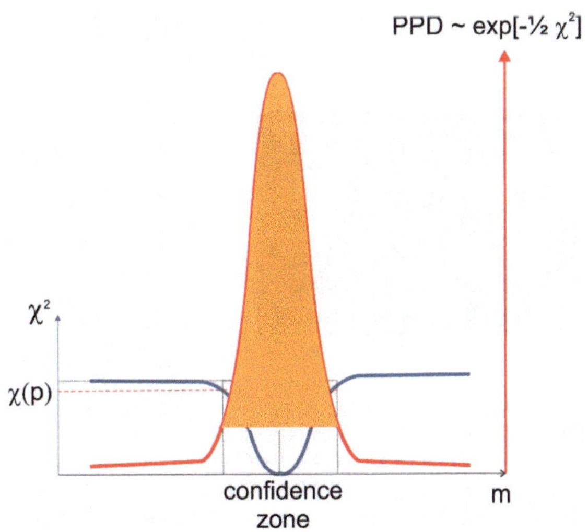

Fig. 5 A cartoon illustrating construction of a 1-D confidence interval. χ^2 is the residual function that is minimized during the course of the inversion, PPD is the posterior probability density function according to (5), and χ_p is the value satisfying (4)

Using the function $\chi^2(\boldsymbol{m})$, I determined particular confidence zones as regions limited by the contour χ_p^2 that possessed the probability content requested (Šílený 1998). In particular, I constructed confidence zones for the P, T, and N axes of the DC portion of the mechanism that provided information on the uncertainty for determining orientation and the confidence zone for the slope angle that yielded information on the uncertainty of the shear versus non-shear content. High content for a non-shear slip component was indicated by increased uncertainty in the P and N axes in cases of a fault opening, and the T and N axes if the fault closed. For a pure tensile crack (mode I of the fracturing), the P and N (or T, N) axes were undetermined and their confidence zones were merged into a single belt within the plane perpendicular to the crack opening (or closing). For the case of the MT source model, I again investigated the confidence zones of the P, T, and N axes, as well as the MT decomposition in terms of the percentage of the double-couple (DC), the isotropic component (ISO), and the compensated linear vector dipole (CLVD).

5 Case Studies

5.1 Tectonic Earthquakes: West Bohemia

Advantageous robustness of the STC as compared to the MT can be demonstrated by verifying non-shear content within the mechanism of the W-Bohemia/Vogtland earthquake swarm from 1997 (Dahm et al. 2000). The W-Bohemia/Vogtland is an intraplate area with extinct Quaternary Period volcanism that manifests present geo-

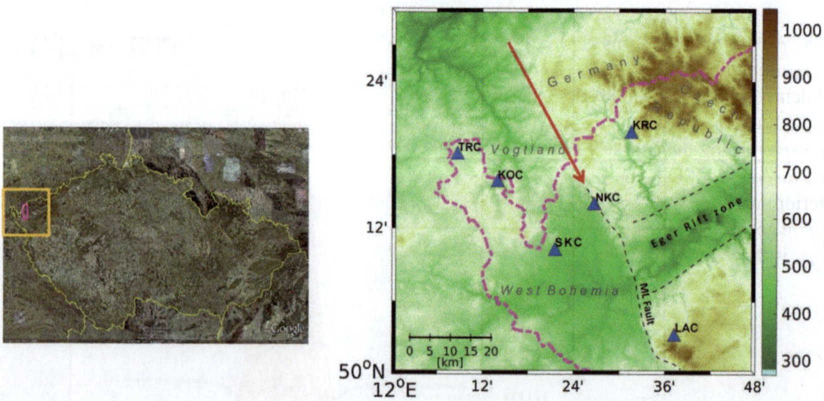

Fig. 6 The seismoactive area of west Bohemia/Vogtland. The figure on the right is a zoom-in of the yellow framed section within the Czech Republic (shown in the left figure) and displays stations (blue triangles) within the WEBNET seismic network that were active during 1997, as well as the principle tectonic features of the Mariánské Lázně (ML) Fault and the Eger Rift Zone (dashed black lines). The red arrow in the right figure marks the position of Nový Kostel, the most active portion of the zone. The purple line indicates the border between the Czech Rep. and Germany

dynamic unrest by persistent swarm seismicity and the degassing of a large amount of CO_2 of upper-mantle origin, Fig. 6. The local magnitudes of swarm events are most often lower than 4.0; larger earthquakes are rather exceptional.

During the 1997 swarm, two types of prevailing mechanisms were observed, one characterized by a notable non-DC component. Dahm et al. (2000) reviewed whether or not these mechanisms were real by changing the attenuation model during inversion and by changing the inversion approach. These authors inverted each single event into the MT, and also determined relative MTs for the entire group. The result was stability of the solutions and, thus, the non-DC component was declared to be reliable. By employing a constrained source model possessing the ability to resolve a volume change and by exhibiting better resolution than inversion into the full MT, the test conducted with the STC represents a good opportunity for checking this conclusion on a different basis. The conclusion has been confirmed by performing

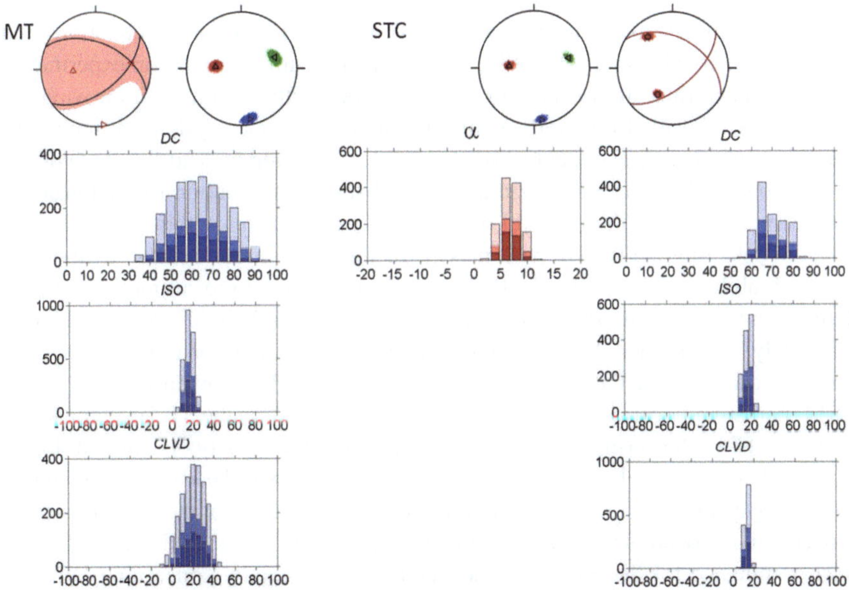

Fig. 7 Mechanism for the earthquake on January 26, 1997, 12:31:27, at a depth 9.2 km, M_0 1.95 $\times 10^{12}$ Nm, ML = 1.1 in relation to the MT (left) and the STC (right) source models. The MT portion is shown as follows: Top left, the fault-plane solution (solid lines are the nodal lines of the DC; triangle up, T-axis; triangle right, P-axis; triangle left, N-axis; and red patch, zone of P compressions) in an equal-area projection of the lower focal hemisphere. Top right, the T, P, and N axes and their confidence regions at the 90% (dark color), 95% (medium), and 99% (pale color) probability levels. From middle to bottom, the histograms, in turn, for the DC, ISO, and CLVD components displaying confidence regions for the 90% (dark color), 95% (medium), and 99% (pale color) probability levels. The STC: Top left, T, P, and N axes and their confidence regions at the 90, 95, and 99% probability levels. Top right, source lines (red) and the confidence regions of the poles for the source planes. Middle left, a histogram of the slope angle, α. Right from middle to bottom, histograms, in turn, for the DC, ISO, and CLVD components

the parallel MT and STC inversion using a confidence region estimate for events previously marked as non-DC phenomena. However, for several events, the non-DC component was determined to not be significant in terms of the confidence region evaluation. Here, I present an example of the former group (Fig. 7), whereas the non-DC characteristics of the event have been confirmed both by the MT and by the STC inversion (the ISO and CLVD histograms are out of the zero value for both the MT and STC solutions). However, thanks to markedly narrower confidence regions, the STC solution is preferred against the MT.

5.2 *Earthquakes Induced by Fluid Injection into Geothermal Wells: Soultz-sous-Forêts, Alsace*

The geothermal reservoir at Soultz-sous-Forêts is a unique natural laboratory for studying the mechanisms of micro-earthquakes generated during stimulations and circulation tests (Genter et al. 2009). Numerous hydraulic stimulation experiments have been performed in this area in recent decades with the aim of improving the connectivity of drilled boreholes within the natural fracture system. Stimulations were accompanied by abundant seismic activity, monitored by a surface network of 20+ stations with a very good azimuthal and distance distribution located near the site. Efforts regarding the retrieval of mechanisms occurring within this area were aimed at detecting the type of rock massif fracturing required to aid estimations of the permeability of the reservoir. Recent studies indicate a prevailingly shear slip. Rarely, a non-shear pattern has also been observed. As a default, a MT description has largely been applied. With the goal of assessing the resolution of the mechanism (both concerning orientation and, potentially, the non-DC content), Šílený et al. (2014) performed a series of synthetic experiments for testing the STC in comparison with the MT in the set-up of a seismic activity monitoring network. In addition to the complete configuration (23 stations surrounding the well-head) and four subsets mimicking deteriorating coverage, they assumed mismodeling of the velocity profile, hypocenter mislocation, and random noise contamination. The experiments exhibited clear dominance for the STC over the MT model, especially concerning DC/non-DC resolution. Then, the expectation of a better resolution was demonstrated on earthquakes from the reservoir (Fig. 8).

Here, there is obviously good similarity for the MT and STC solutions for monitoring geometry. Concerning orientation, both approaches yield practically the same results. However, as for the non-DC content, there is a difference. The STC solutions are closer to pure DCs and in several cases markedly closer (events 8, 12, 19, 23, and 25). Thus, in comparison to the MT, the STC behaves conservatively, reduces the non-DC suggested by the MT, and suggests that they may be spurious.

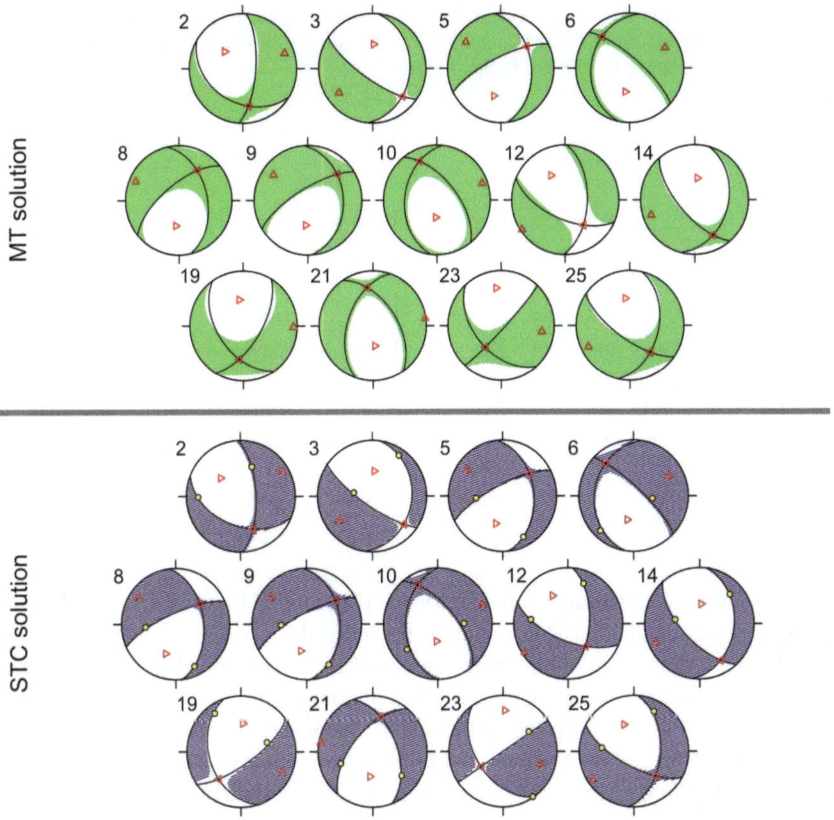

Fig. 8 A comparison of moment tensor (MT) solutions and shear-tensile crack (STC) solutions for 13 earthquakes occurring within the geothermal reservoir at Soultz-sous-Forêts. Source mechanisms are represented using traditional fault-plane solution plots. The nodal lines of the DC portion of retrieved MTs and source lines of the STC are shown as gray lines. The corresponding principal axes T, P, and N are shown with a triangle apex up, right, and left, respectively, within an equal-area, lower-hemisphere projection

5.3 *Mining Associated Seismic Events Recorded at the Driefontein-Sibanye Gold Mine in South Africa*

Dominance of the STC model over the traditional MT in resolving the mode of fracturing can be well-demonstrated in several mining-associated seismic events occurring at the Driefontein-Sibanye gold mine in South Africa. In the Introduction, I discussed the single-force and MT inversion of these events in relation to the Hasegawa models (Šílený and Milev 2008). In a follow-up paper, Šílený and Milev (2017) also determined their mechanisms based on the STC model, and complemented both inverse tasks by evaluating confidence regions in order to assess the uncertainty of solutions. The message of this upgrade for processing was twofold:

Table 1 A qualitative assessment of resolution for orientation of the mechanism and its decomposition using the MT and STC source models for five, mining-associated earthquakes recorded at the Driefontein-Sibanye Gold Mine in South Africa

Event			1	2	3	4	5
Resolution of orientation	MT	Single axis	Good	Fair	Fair	Excellent	Fair
		All 3 axes	Good	Poor	Poor	Poor	Very poor
	STC	Single axis	Excellent	Good	Good	Excellent	Fair
		All 3 axes	Excellent	Good	Good	Poor	Very poor
Resolution of decomposition: MT			Fair	Poor	Very poor	Very poor	Very poor
MT: DC versus non-DC			None	None	None	None	None
Resolution of decomposition: STC			Good	Good	fair	Fair	Excellent
STC: shear versus non-shear			**Implosion**	**Tensile**	**None**	**Tensile**	**Implosion**

(1) the orientation of the mechanism retrieved using the MT and STC is equal; and (2) the type of mechanism basically remains unchanged (i.e. relevance to the individual Hasegawa models mentioned above remains valid). The added value of upgraded processing (construction of the confidence regions), however, largely changed the view on the MT solution. Based on the results, it appeared that confidence of the DC versus non-DC assessment within the MT approach was very low, resulting in a non-DC percentage that could not be considered significant. Contrary to the MT, the STC offered a much more valuable outcome: the DC versus non-DC issue was resolved for all five events (Table 1) and allowed an interpretation of their mechanisms in terms of geomechanics.

5.4 Seismicity Due to the Hydrofracturing of Oil/Gas Wells: Cotton Valley, Texas

Hydrofracturing, the treatment aimed to enhance production, is routine within the oil and gas industry. Related seismicity provides valuable information on the success of the operation. Determination of the mechanisms yields information on the mode of rock-mass fracturing in terms of fracture mechanics: mode I (tensile crack) versus mode II (shear fracturing). Such knowledge is vital for mapping the permeability of the reservoir. Unfortunately, since observations are only performed from a few monitoring wells, the monitoring geometry is extremely unfavorable in this area. Even more problematic, the most frequent setup (monitoring within a single monitoring well) is a worst case scenario since an observation from only one azimuth is available. Then, far-field data are not sufficient for resolving the full MT (Nolen-Hoeksema

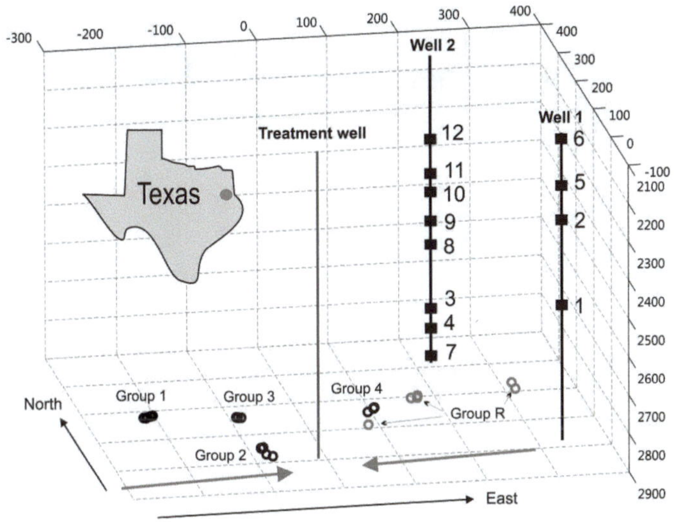

Fig. 9 An illustration of the hydraulic fracture imaging geometry for the Cotton Valley experiment (inset, gray dot on the sketch of Texas). Black heavy lines indicate monitoring wells 1 and 2 (full squares mark the positions of the geophones used in this study). The gray heavy line indicates the treatment well. The hypocenters of selected microearthquakes within groups G1–G4 and R are marked by circles. Gray arrows at the bottom of the model indicate the direction of maximum horizontal stress. Distance is provided in meters

and Ruff 2001; Vavryčuk 2007). The solution is adding more data (e.g., observations from a near-field (Song and Toksöz 2011), a priori constraints of the mechanism (Jechumtálová and Eisner 2008), or applying a constrained MT. Such a solution was performed by Grechka et al. (2016), who used the "tensile model", essentially the STC, and designed an inversion algorithm and demonstrated its performance on single-well data from the oil industry.

Here, I demonstrate improvement for the determination of a mechanism from data gathered within two monitoring wells. In theory, data from two wells is enough to obtain a full MT. Nevertheless, its resolution is still very poor due to extremely sparse coverage of the focal sphere. Thus, employment of a constrained MT is also reasonable here. As an example, I reinterpreted microearthquakes obtained from the Cotton Valley, E. Texas data set. The data were processed within the complete MT description by Šílený et al. (2009), in terms of the STC source model, and the MT and STC solutions were compared (Fig. 9).

Hydrofracturing of the treatment well resulted in abundant seismicity; classified by Rutledge et al. (2004) into several groups, G1-G4. Šílený et al. (2009) processed selected events from each group using a full MT analysis and determined notable non-DC mechanisms. However, some of the mechanisms seemed to be rather unrealistic, especially events from the G4 group. Šílený (2012) reprocessed them in terms of the STC source model and documented a similar fit for the data. The solutions were largely similar, in several cases (for the G4 group, especially) simpler mecha-

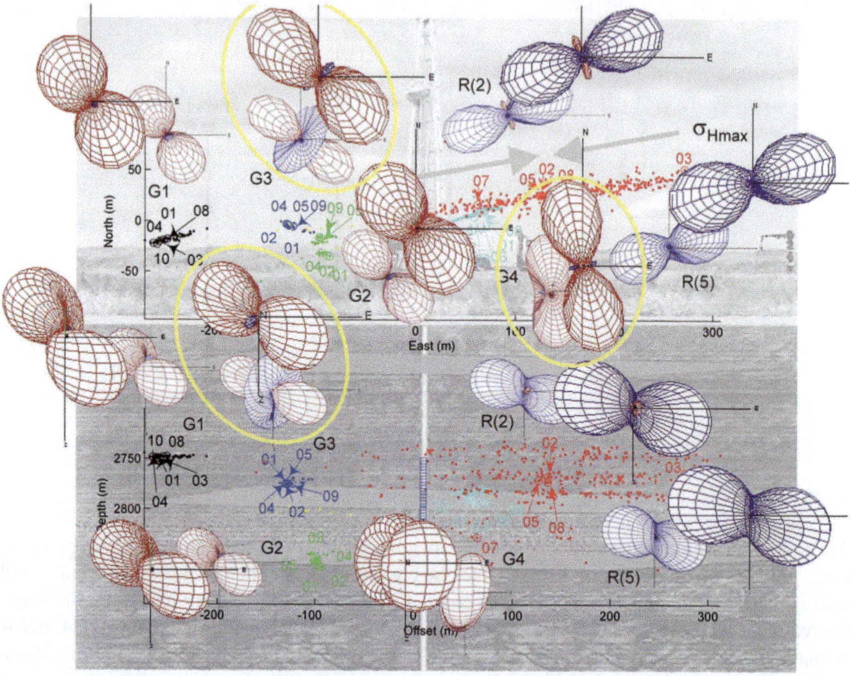

Fig. 10 Cotton Valley mechanisms for G1–G4 and R groups within the MT (smaller-size plots are provided in the background) and the STC (large-size plots are provided in the foreground) source description, plotted as 3-D, wire-frame diagrams of P radiation (red—expansion, blue—compression). The top panel provides a map view, while the bottom panel provides a depth section viewed from S. The MT solutions are reprinted from Šílený et al. (2009). MT versus STC couples that differed the most are provided in yellow ovals

nisms were obtained, allowing a more feasible interpretation from the viewpoint of geomechanics (Fig. 10).

6 Concluding Remarks

The moment tensor (MT) has become a standard for describing seismic sources, both in earthquake seismology and for various types of induced seismicity. In general, the MT is a dipole source but, in practice, may be too general because its generality causes trouble during its reconstruction from noisy data during the inverse process that may also be ill-conditioned due to an inexact hypocenter location and/or to the availability of a rough velocity/attenuation model. Then, the retrieved source may be biased. Thus, in practice, a constrained MT is frequently used. The traditional deviatoric constraint, routinely employed within Centroid Moment Tensor (CMT) solutions of moderate to strong earthquakes, may not be adequate for earthquakes without a purely tectonic origin (e.g., for volcanic earthquakes), among them events induced by industrial activities such as mining or fluid injection into boreholes. Therefore, it seems reasonable to use a simpler source model for directly describing the physical phenomena anticipated within a particular focus. The simple combination of a shear slip with a tensile crack or a 1D implosion (a Shear-Tensile Crack, STC) may be a good model both for natural earthquakes and induced events. In many cases, constraining the full MT helps one obtain a better resolution for the mechanism: sometimes the determination of its orientation and, very frequently, its double-couple (DC) versus non-DC content. Urgency for constraining the full MT increases with deteriorating quality of the geometry of the observation. With dense observations, well-distributed surrounding a focus, the full MT can yield good resolution for both the geometry and the non-DC content. Resolution especially falls due to notable gaps in station distribution. In extreme cases of sensor configuration depletion, during single azimuth monitoring, the MT completely fails and model simplification in terms of the STC is crucial.

Acknowledgements This work was supported by the following grant from the Czech Science Foundation: "Solid body fracturing mode by shear-tensile source model: acoustic emission laboratory study", Grant Agreement No. P108 16-03950S.

References

Aki K, Richards PG (2002) Quantitative seismology. Freeman, San Francisco

Braunmiller J, Kradolfer U, Baer M, Giardini D (2002) Regional moment tensor determination in the European–Mediterranean area—initial results. Tectonophysics 356:5–22

Dahm T, Horálek J, Šílený J (2000) Comparison of absolute and relative moment tensor solutions for the January 1997 West Bohemia earthquake swarm. Studia Geophys et Geodaet 44:233–250

Dreger DS, Helmberger DV (1993) Determination of source parameters at regional distances with three-component sparse network data. J Geophys Res 98:8107–8125

Dufumier H, Rivera L (1997) On the resolution of the isotropic component in moment tensor inversion. Geophys J Int 131:595–606. https://doi.org/10.1111/j.1365-246X.1997.tb06601.x

Dziewonski AM, Chou TA, Woodhouse JH (1981) Determination of earthquake source parameters from waveform data for studies of global and regional seismicity. J Geophys Res 86:2825–2852

Ekstrom G, Nettles M, Dziewonski AM (2012) Centroid-moment tensors for 13,017 earthquakes. Phys Earth Planet Int 200:1–9

Genter A, Fritsch D, Cuénot N, Baumgärtner J, Graff JJ (2009) Overview of the current activities of the European EGS Soultz project: from exploration to electricity production. In: Proceedings thirty-fourth workshop on geothermal reservoir engineering, Stanford, California, 9–11 Feb 2009, SGP-TR-187

Grechka V, Li Z, Howell B, Vavryčuk V (2016) Single-well moment tensor inversion of tensile microseismic events. Geophysics 208:1724–1739

Hasegawa HS, Wetmiller RJ, Gendzwill DJ (1989) Induced seismicity in mines in Canada—an overview. Pure Appl Geophys 129:423–453

Henry C, Woodhouse JH, Das S (2002) Stability of earthquake moment tensor inversions: effect of the double-couple constraint. Tectonophysics 356:115–124

Horálek J, Fischer T (2008) Role of crustal fluids in triggering the west Bohemia/Vogtland earthquake swarms: just what we know (a review). Stud Geophys Geodet 52:455–478

Jakobsdóttir SS, Roberts MJ, Gudmundsson GB, Geirsson H, Slunga R (2008) Earthquake swarms at Upptyppingar, north-east Iceland: a sign of magma intrusion? Stud Geophys Geodet 52:513–528
Jechumtálová Z, Eisner L (2008) Seismic source mechanism inversion from a linear array of receivers reveals non-double-couple seismic events induced by hydraulic fracturing in sedimentary formation. Tectonophysics 460:124–133. https://doi.org/10.1016/j.tecto.2008.07.011
Julian BR, Miller AD, Foulger GR (1998) Non-double-couple earthquakes 1. Theory. Rev Geophys 36:525–549
Kawakatsu H (1995) Automated near-realtime CMT inversion. Geophys Res Lett 22:2569–2572
Kubo A, Fukuyama E, Kawai H, Nonomura K (2002) NIED seismic moment tensor catalog for regional earthquakes around Japan: quality test and application. Tectonophysics 356:23–48
Malovichko DA (2005) Personal communication
Minson SE, Dreger DS, Bürgmann R, Kanamori H, Larson KM (2007) Seismically and geodetically determined nondouble-couple source mechanisms from the 2000 Miyakejima volcanic earthquake swarm. J Geophys Res 112:B10308. https://doi.org/10.1029/2006JB004847
Nolen-Hoeksema RC, Ruff LJ (2001) Moment tensor inversion of microseisms from the B-sand propped hydrofracture, M-site, Colorado. Tectonophysics 336:163–181
Nábělek J, Xia G (1995) Moment-tensor analysis using regional data: application to the 25 March, 1993, Scotts Mills, Oregon, earthquake. Geophys Res Lett 22:13–16
Okada Y (1985) Surface deformation due to shear and tensile faults in a half-space. Bull Seismol Soc Am 75(4):1135–1154
Pesicek JD, Šílený J, Prejean SG, Thurber CH (2012) Determination and uncertainty of moment tensors for microearthquakes at Okmok Caldera, Alaska. Geophys J Int 190:1689–1709. https://doi.org/10.1111/j.1365-246X.2012.05574.x
Pondrelli S, Morelli A, Ekström G, Mazza S, Boschi E, Dziewonski AM (2002) European-Mediterranean regional centroid-moment tensors: 1997–2000. Phys Earth Planet Int 130:71–101
Ritsema J, Lay T (1993) Rapid source mechanism determination of large (Mw $\geq$ 4.5) earthquakes in western United States. Geophys Res Lett 20:1611–1614
Rudajev V, Šílený J (1985) Seismic events with non-shear component II. Rock bursts with implosive source component. Pure Appl Geophys 123:17–25
Rutledge JT, Phillips WS, Mayerhofer MJ (2004) Faulting induced by forced injection and fluid flow forced by faulting: an interpretation of hydraulic-fracture microseismicity, Carthage Cotton Valley Gas Field. Bull Seism Soc Am 94:1817–1830
Stierle E, Vavryčuk V, Šílený J, Bohnhoff M (2014) Resolution of non-double-couple components in the seismic moment tensor using regional networks—I: a synthetic case study. Geophys J Int 196(3):1869–1877. https://doi.org/10.1093/gji/ggt502
Šílený J (1989) The mechanism of small mining tremors from amplitude inversion. PAGEOPH 129:309–324
Šílený J (2009) Resolution of non-double-couple-mechanisms: simulation of hypocenter mislocation and velocity structure mismodeling. Bull Seismol Soc Am 99. https://doi.org/10.1785/0120080335
Šílený J (2012) Shear-tensile/implosion source model vs. moment tensor: benefit in single-azimuth monitoring. Cotton Valley set-up. In: 74th EAGE conference & exhibition incorporating SPE EUROPEC 2012. Houten, EAGE, 2012, pp 1–5. ISBN 978-90-73834-27-9
Šílený J, Hill DP, Eisner L, Cornet FH (2009) Non–double-couple mechanisms of icroearthquakes induced by hydraulic fracturing. J Geophys Res 114:B08307. https://doi.org/10.1029/2008JB005987
Šílený J, Milev A (2008) Source mechanism of mining induced seismic events—resolution of double couple and non double couple models. Tectonophysics 456:3–15. https://doi.org/10.1016/j.tecto.2006.09.021
Šílený J, Milev A (2017) Mechanism of mining associated seismic events recorded at Driefontein-Sibanye gold mine in South Africa. In: Feng X-T (eds) Rock mechanics and engineering, vol 5, © CRC Press/Balkema, Taylor & Francis Group

Šílený J, Jechumtálová Z, Dorbath C (2014) Small scale earthquake mechanisms induced by fluid injection at the enhanced geothermal system reservoir Soultz (Alsace) in 2003 using alternative source models. Pure Appl Geophys 171:2783–2804. https://doi.org/10.1007/s00024-013-0750-2

Šílený J, Panza GF, Campus P (1992) Waveform inversion for point source moment tensor retrieval with variable hypocentral depth and structural model. Geophys J Int 109:259–274

Sipkin SA (1982) Estimation of earthquake source parameters by the inversion of waveform data: synthetic waveforms. Phys Earth Planet Int 30:242–259

Sipkin SA, Zirbes MD (2004) Moment-tensor solutions estimated using optimal filter theory: global seismicity, 2002. Phys Earth Planet Int 145:203–217

Song F, Toksöz MN (2011) Full-waveform based complete moment tensor inversion and source parameter estimation from downhole microseismic data for hydrofracture monitoring. Geophysics 76(6):WC103–WC116

Stich D, Ammon CJ, Morales J (2003) Moment tensor solutions for small and moderate earthquakes in the Ibero-Maghreb region. J Geophys Res 108, Art. No. 2148

Tarantola A (1987) Inverse problem theory. methods for data fitting and model parameter estimation. Elsevier, Amsterdam

Teisseyre R (1980) Some remarks on the source mechanism of rockhursts in mines and on the possible source extension. Acta Mont CSAV Praha 58:7–13

Vavryčuk V (2001) Inversion for parameters of tensile earthquakes. J Geophys Res 106:16339–16355. https://doi.org/10.1029/2001JB000372

Vavryčuk V (2007) On the retrieval of moment tensors from borehole data. Geophys Prospect 55(3):381–391

Vavryčuk V (2011) Tensile earthquakes: theory, modeling and inversion. J Geophys Res 116:B12320. https://doi.org/10.1029/2011JB008770

Vavryčuk V, Kim SG (2014) Nonisotropic radiation of the 2013 North Korean nuclear explosion. Geophys Res Lett 41:7048–7056. https://doi.org/10.1002/2014GL061265

Šílený J, Jechumtálová Z, Dorbath C (2014) Small scale earthquake mechanisms induced by fluid injection at the enhanced geothermal system reservoir Soultz (Alsace) in 2003 using alternative source models. Pure Appl Geophys 171:2783–2804. https://doi.org/10.1007/s00024-014-0780-2

Šílený J, Panza GF, Campus P (1992) Waveform inversion for point source moment tensor retrieval with variable hypocentral depth and structural model. Geophys J Int 109:259–274

Sipkin SA (1982) Estimation of earthquake source parameters by the inversion of waveform data: synthetic waveforms. Phys Earth Planet Inter 30:242–259

Sipkin SA, Zirbes MD (2004) Moment-tensor solutions estimated using optimal filter theory: global seismicity, 2002. Phys Earth Planet Inter 145:203–217

Song F, Toksöz MN (2011) Full-waveform based complete moment tensor inversion and source parameter estimation from downhole microseismic data for hydrofracture monitoring. Geophysics 76:WC103–WC116

Stich D, Ammon CJ, Morales J (2003) Moment tensor solutions for small and moderate earthquakes in the Ibero-Maghreb region. J Geophys Res 108, Art. No. 2148

Tarantola A (1987) Inverse problem theory: methods for data fitting and model parameter estimation. Elsevier, Amsterdam

Teisseyre R (1980) Some remarks on the source mechanism of rockbursts in mines and on the possible source extension. Acta Mont ČSAV Praha 58:61

Vavryčuk V (2001) Inversion for parameters of tensile earthquakes. J Geophys Res 106:16339–16355. https://doi.org/10.1029/2001JB000372

Vavryčuk V (2007) On the retrieval of moment tensors from borehole data. Geophys Prospect 55:381–391

Vavryčuk V (2011) Tensile earthquakes: theory, modeling, and inversion. J Geophys Res 116:B12320. https://doi.org/10.1029/2011JB008770

Vavryčuk V, Kim SG (2014) Nonisotropic radiation of the 2013 North Korean nuclear explosion. Geophys Res Lett 41:7048–7056. https://doi.org/10.1002/2014GL061265

Seismic Deformation Derived from Moment Tensor Summation: Application Along the Hellenic Trench

Anastasia Kiratzi, Christoforos Benetatos and Filippos Vallianatos

1 Introduction

How continental areas deform has been a subject of intense research, especially in the prism of plate tectonics. The distribution of seismicity in continental areas is not confined to a single fault, but earthquakes are distributed over wide zones, hundreds or even thousands wide, which contain many faults. Within these deforming zones, as for example the Alpine-Himalayan Belt, there are blocks, which are relative aseismic, as the region of central Turkey, and are commonly referred to as microplates. Continental deformation cannot be described by the relative motion of the bounding plates. For example, the frequent earthquakes in the southern Mediterranean countries (Spain, Italy, Greece) and Middle East are caused by the northward motion of the African and Arabian plate, relative to Eurasia, but knowing the relative motions of these plates is not enough, in order to describe or forecast how the deformation will be taken up by the whole family of active faults. In such a framework, it is more useful to describe the overall deformation by a velocity field, rather than by the relative motions of the rigid plates that bound the deforming region.

Kostrov (1974) was the first to set the theoretical formulations in order to quantify the distributed deformation (average strain) due to the seismic activity, from the analysis of moment tensors. Jackson and McKenzie (1988) deduced a relationship between average strain rate and the relative velocity between two plates. Papazachos and Kiratzi (1992) introduced a slightly modified approach to determine the strain

A. Kiratzi
Department of Geophysics, Aristotle University of Thessaloniki, Thessaloniki, Greece

C. Benetatos
Petroleum Engineering Group, DIATI, Politecnico di Torino, Turin, Italy

F. Vallianatos (✉)
UNESCO Chair on Solid Earth Physics & Geohazards Risk Reduction, Laboratory of Geophysics & Seismology, Technological Educational Institute of Crete, Heraklion, Greece
e-mail: fvallian@chania.teicrete.gr

S. D'Amico (ed.), *Moment Tensor Solutions*, Springer Natural Hazards,
https://doi.org/10.1007/978-3-319-77359-9_10

rate field. All the above formulations, became very popular and a significant number of publications are available in the literature.

Here, we provide the theoretical background, discuss the parameters which are important for the analysis, and apply this technique, as formulated in Papazachos and Kiratzi (1992), in order to obtain the deformation field, along the Hellenic Trench. To do so, we use an updated zonation (Vamvakaris et al. 2016) and a recent catalogue of focal mechanisms. The aim of this work is to provide the background that is required to apply the method, and not the results per se, as previous studies are already available (e.g. Eyidogan 1988; Jackson and McKenzie 1988; Ekstrom and England 1989; Ambraseys and Jackson 1990; Kiratzi 1991, 1993; Papazachos and Kiratzi 1996 among many others).

2 Theoretical Background

2.1 Focal Mechanisms—Moment Tensors

A very useful tool for seismologists to investigate the geometry of the faults and the orientation of the stress field in an active region, is the focal mechanism of the seismic source, and in particular, its moment tensor representation. From the classic work of Jost and Hermann (1989), the displacement on any point of the Earth's surface can be expressed by a linear combination of time dependent moment tensor elements. At large distances from the fault (dimension of fault small in respect to the wavelengths of the seismic waves) the displacement, $u_n(\vec{x}, t)$, can be expressed as:

$$u_n(\vec{x}, t) = M_{pq} * G_{np,q} \tag{1}$$

where M_{pq} is the seismic moment tensor, $G_{np,q}$ are the derivative of the Green's functions and $u_n(\vec{x}, t)$ the displacement at the Earth's surface. The seismic tensor can be calculated as:

$$M = G^{-1}u \tag{2}$$

The equivalent forces for the seismic sources of different geometries, can be represented by the seismic moment tensor, M_{pq}, the components of which are the nine couples of the equivalent forces:

$$M_{pq} = \begin{Bmatrix} M_{xx} & M_{xy} & M_{xz} \\ M_{yx} & M_{yy} & M_{yz} \\ M_{zx} & M_{zy} & M_{zz} \end{Bmatrix} \tag{3}$$

A moment tensor can be decomposed to elementary tensors following two different approaches: A mathematical approach and an analysis at source components with larger physical meaning (see Jost and Herrmann 1989 for more details).

In most seismological applications, the moment tensor is decomposed to an isotropic part and a deviatoric part:

$$M = M_{ISO} + M_{DEV} \tag{4}$$

The first term (M_{ISO}) of the Eq. (4) expresses the isotropic part of the tensor, that involves volumetric change. The second term (M_{DEV} =deviatoric) can be further analysed using various approaches. Most commonly, the deviatoric part is analysed in a component that describes the double couple of forces applied to the source (DC, double couple component) and to a component, known as CLVD (Compensated Linear Vector Dipole), (Knopoff and Randall 1970).

$$M_{DEV} = M_{DC} + M_{CLVD} \tag{5}$$

The seismic moment tensor, if we assume that no volumetric change occurs at the source, has 5 independent components due to its symmetric nature since we accept that a double-couple is acting at the source. The components of the moment tensor can be expressed in relation to the scalar moment (M_o), strike (φ), dip (δ) and rake (λ) of a shear dislocation (Aki and Richards 1980) using the following relations:

$$\begin{aligned}
M_{xx} &= -M_o\,(\sin\delta\,\cos\lambda\,\sin 2\varphi + \sin 2\delta\,\sin\lambda\,\sin 2\varphi)\\
M_{yy} &= M_o\,(\sin\delta\cos\lambda\,\sin 2\varphi - \sin 2\delta\,\sin\lambda\cos 2\varphi)\\
M_{zz} &= M_o\,(\sin 2\delta\,\sin\lambda)\\
M_{xy} &= M_o\,(\sin\delta\,\cos\lambda\,\cos 2\varphi + 0.5\,\sin 2\delta\,\sin\lambda\,\sin 2\varphi)\\
M_{xz} &= -M_o\,(\cos\delta\,\cos\lambda\cos\varphi + \cos 2\delta\,\sin\lambda\,\sin\varphi)\\
M_{yz} &= -M_o\,(\cos\delta\,\cos\lambda\,\sin\varphi - \cos 2\delta\,\sin\lambda\,\cos\varphi)
\end{aligned} \tag{6}$$

In the following sections, we will use the moment tensor formulation to derive the average strain rate within a seismically deforming volume.

2.2 *Moment Tensor Summation to Derive the Strain Rate Field*

2.2.1 Method of Analysis

The analysis is based on the work of Papazachos and Kiratzi (1992) which takes into account the formulation of Kostrov (1974), Jackson and McKenzie (1988) and Molnar (1979). More specifically, let us assume we want to study the seismic deformation of a seismic zone which has the shape of a parallelepiped, of volume, V, with known dimensions l_1, l_2, l_3 (length, width, and thickness) and that the deformation occurring in that zone is seismic. In the following the different steps of the analysis are discussed.

Step 1: Calculation of the Annual (Scalar) Moment Rate of the Zone

The Gutenberg–Richter law:

$$logN\,(>M) = a - bM \tag{7}$$

that relates the cumulative frequency of earthquake occurrence $N(>M)$ to different magnitude (M), in principle, does not have upper and lower values for the magnitude M, even though there is a finite breaking strength of the earth's crust and a finite extent of the source zone. The *a* value in (7) is related to the seismicity rate and the *b* value, or the negative slope of the regression straight line, indicates the relative frequency of occurrence of large and small earthquakes. To apply upper and lower magnitude limits the truncated Gutenberg–Richter expression was introduced. Based on the seismic moment and the Gutenberg–Richter law, Molnar (1979) and Papastamatiou (1980) developed a number of relations between the earthquake occurrence parameters and the rates of crustal deformation. Using an additional empirical relation between earthquake magnitude and the seismic moment M_o

$$\log Mo = cM + d \tag{8}$$

where *c* and *d* are the regression constants, Molnar (1979) developed a formula that relates the cumulative number of earthquake occurrences per year $N(>M_o)$ to the seismic moment (M_o) and the rate of seismic moment release:

$$N\,(>Mo) = \frac{(1-\beta)\,\dot{M}_0 Mo^{-\beta}}{M_{omax}^{1-\beta}} \tag{9}$$

where $M_{o\,\max}$ is the maximum seismic moment which likely occurs in the region of interest, $\dot{M}_0$ is the seismic moment rate and β equals (*b*/*c*). The *b* and *c* values are the slopes of Eqs. (7) and (8), respectively. Here, we will generalize the Molnar (1979) model to produce an expression based on truncated Gutenberg–Richter expression, which can be applied in similar analyses, as the one that follows in the subsequent sections, if so desired.

Although the Gutenberg–Richter frequency–magnitude expression is an inadequate description of the seismicity, its density distribution may be retained when restricted to a range bounded by a maximum magnitude ($M_{\max}$)). The density distribution then becomes

$$n\,(M) = n_o e^{-b'M}\,for\ M < Mmax\ \text{and}\ n\,(M) = 0\ for\ M > Mmax \tag{10}$$

The cumulative frequency $N(\geq M)$ is equal to the integral of n(M) over the magnitude range (M, Max), (Main 1984) and is given as:

$$N\,(>M) = \frac{e^{-b'M} - e^{-b'M_{max}}}{e^{-b'M_{min}} - e^{-b'M_{max}}} \tag{11}$$

where $b' = b\ \ln 10$ and M_{max} and M_{min} the upper and lower limits in the truncated Gutenberg–Richter expression. If the truncated Gutenberg–Richter expression is combined with the empirical moment–magnitude relations (8), substituting $d = \log$ K, we are led to $M = \frac{loge}{c} \ln\left(\frac{Mo}{K}\right)$ and after simple algebra, Eq. (11) becomes:

$$N\,(>M_o) = \frac{M_o^{-b/c}}{\left(e^{-b'M_{min}} - e^{-b'M}\text{max}_{)}\right) K^{-b/c}} - \frac{e^{-b'M_{max}}}{e^{-b'M_{min}} - e^{-b'M_{max}}} \tag{12a}$$

and follows the form

$$\text{N}\,(>M_o) = A^* M_o^{-\beta} - \Delta \tag{12b}$$

In a region undergoing seismic deformation, there is a maximum seismic moment above which the number of earthquakes is equal to zero in this region i.e., for $M_o > M_{omax}$, $N(>M_{omax}) = 0$. Equation (12b) can be rewritten as $\text{N}\,(>M_o) = \left(A^* M_o^{-\beta} - \Delta\right)[1 - H\,(M_o - M_{omax})]$ where H(x) is a Heaviside function. Using the aforementioned expression, the density function $n(M_o)$ defined as $n(M_o) = -dN/dMo$ could be calculated

$$\begin{aligned} n\,(M_o) = -\big[&\left[-\left(A^* M_o^{-\beta} - \Delta\right)\delta\,(M_o - M_{omax})\right] \\ &+ \left[[1 - H\,(M_o - M_{omax})]\left(-\beta \text{A}^* \text{M}_o^{-\beta-1}\right)\right]\big] \end{aligned} \tag{13}$$

where δ(x) is the Dirac δ function. For a long time period T, the seismic moment rate could be approximated as $\dot{M}_o = \frac{\sum M_o}{T}$ where $\sum M_o\, is$ the accumulated moment. The rate of occurrence of seismic moment should also be equal to (Main and Burton 1984) $\dot{M}o = \int_0^{\infty} M_o n\,(M_o)\, dM_o$. Introducing it to Eq. (12a, 12b) and after integrating we lead to:

$$\dot{M}_o = A^* M_{omax}^{1-\beta}\left(\frac{1}{1-\beta}\right) - \Delta\, M_{omax}, \quad \text{where}\, \beta \neq 1 \tag{14}$$

Thus the parameter A* could be expressed as $A^* = (1-\beta)\,\frac{\dot{M}o + \Delta M_{omax}}{M_{omax}^{1-\beta}}$ which could be substituted in (e.g. 12b) leading to

$$\text{N}\,(>M_o) = \frac{\left(\dot{M}_o + \Delta M_{omax}\right)(1-\beta)\, M_o^{-\beta}}{M_{omax}^{1-\beta}} - \Delta = C M_o^{-\beta} - \Delta \tag{15}$$

If the parameter $\Delta=0$, then we obtain the Molnar (1979) expression. The physical meaning of parameter Δ is to introduce a limit that above a certain value of maximum seismic moment M_{omax} there is no earthquake.

In most of the cases, Δ is assumed equal to zero and the model of Molnar (1979) is applied. If this is adopted in an analysis, as in our case here, we assume that the amount of deformation, let us call it the 'size of deformation', within the zone, is expressed by the *annual* (scalar) seismic moment rate, $\dot{M}_0$ and can be calculated by the formula (Molnar 1979) (an approximation of Eq. (14)),

$$\dot{M}_0 = \frac{A}{1-\beta} M_{0,max}^{(1-\beta)} \tag{16}$$

where $M_{0,max}$ is the scalar moment of the largest event ever occurred in the zone. Obviously, this event can be either instrumentally recorded or historically documented. The constants A and B are calculated by:

$$A = 10^{\left(a+\frac{bd}{c}\right)} \text{and}\, \beta = \frac{b}{c} \tag{17}$$

Step 2: Calculation of the Annual Moment Rate Tensor of the Zone

If in the zone under study, we have available focal mechanisms for N earthquakes, then the *ij*th components of the moment tensor, **M**, of the nth event, are represented as:

$$\boldsymbol{M}_{ij}^{\boldsymbol{n}} = M_0^n\,(\bar{u}\,\bar{n} + \bar{n}\,\bar{u}) = M_0^n F_{ij}^n\,(\varphi, \delta, \lambda) \tag{18}$$

where M_0^n is the scalar seismic moment of this event, and $\bar{u}$ and $\bar{n}$ are unit vectors parallel to the slip vector and normal to the fault plane, and the components of F^n are easily calculated, from the strike (φ), dip (δ) and rake (λ) of each focal mechanism of the earthquakes (Aki and Richards 1980), as shown in Eq. (12a, 12b). The scalar seismic moment for each earthquake, if not available by other means, can be calculated from an empirical scaling relation from the magnitudes (e.g. Eq. 8).

Rewriting Eq. (18) in terms of the annual rates for the zone, we get that the annual moment rate tensor, $\dot{\boldsymbol{M}}$, is the product of the scalar moment rate, $\dot{M}_0$ and a "shape" tensor, $\bar{F}$:

$$\dot{M}_{ij} = \dot{M}_0 \overline{F_{ij}} \tag{19}$$

The tensor, $\bar{F}$, can be calculated as simple average (divided by N) or a normalized average (divided for example by the scalar moment, as in Eq. 20), applicable to the zone, as follows:

$$\overline{F_{ij}} = \frac{\sum_{n=1}^{N} M_0^n F^n}{\sum_{n=1}^{N} M_0^n} \tag{20}$$

Step 3: Calculation of the Strain Rate and Velocity Tensor of the Zone
In a coordinate system with the horizontal axes, x_1 and x_2 parallel and normal to the trend of the zone, respectively and the vertical x_3 axis pointing downwards, following Kostrov (1974), the average strain rate tensor, $\dot{e}_{ij}$, in the volume V, can be calculated, from the following relations:

$$\dot{\varepsilon_{ij}} = \frac{1}{2\mu V}\dot{M}_0\overline{F_{ij}}\, i,\ j = 1,\ 2,\ 3 \tag{21}$$

where μ is the shear modulus.

In the same reference frame, the components of the velocity tensor, U_{ij} are calculated as follows (Jackson and McKenzie 1988):

$$\begin{aligned} U_{11} &= \frac{1}{2\mu l_2 l_3}\dot{M}_0\overline{F_{11}},\ U_{22} = \frac{1}{2\mu l_1 l_3}\dot{M}_0\overline{F_{22}},\quad U_{33} = \frac{1}{2\mu l_1 l_2}\dot{M}_0\overline{F_{33}} \\ U_{12} &= \frac{1}{\mu l_1 l_3}\dot{M}_0\overline{F_{12}},\ U_{13} = \frac{1}{\mu l_1 l_2}\dot{M}_0\overline{F_{13}},\quad U_{23} = \frac{1}{\mu l_1 l_2}\dot{M}_0\overline{F_{23}} \end{aligned} \tag{22}$$

It is worth noting, that tensor $\bar{F}$, is usually calculated in a reference system 1-North, 2-East, 3-down, so before using it in Eqs. 21 and 22, a rotation of $\bar{F}$, in the zone's reference system is necessary. The diagonal components of the velocity tensor show the component of motion normal to the zone, its sideways expulsion of material and its thickening (or thinning) rate, respectively. The off-diagonal components of the velocity tensor are less significant for the interpretation.

2.2.2 Summarising the Steps

Summarizing the above, a rule of thumb in terms of application steps, includes the following:

- We divide the area, where we seek to calculate seismic strain rates and velocity rates, in zones which exhibit the same type of deformation style. To do so, we care to include in each zone, an adequate number of earthquake focal mechanism solutions.
- For each zone we calculate the annual moment rate as described in step 1. To do so, we use a catalogue of historical and present day seismicity, we carefully examine the periods and the threshold magnitudes of completeness of the catalogue. Using the complete catalogue, we determine the parameters of the Gutenberg–Richter relation, and we adopt a scaling relation between the magnitude of the catalogue we use and seismic moment, for example (Mo and Mw or Mo and Ms) and then we calculate the annual moment rate.
- We subsequently calculate the components of the average tensor, $\bar{F}$, as described in step 2. In this case, and in order to increase the number of the available data, we may calculate its components, by including focal mechanisms of similar type, from a broader area, that may include several seismic zones, undergoing the same pattern

of deformation. As mentioned previously, the determination of $\bar{F}$ (using Eq. 20) for each belt is in the North-East-Down coordinate system, and then a rotation is performed into the zone system Ox_1,x_2,x_3. In practice, the eigenvalues of tensor $\bar{F}$ correspond to the average P, T and null axis of the typical focal mechanism of the region. Their deviation from the typical double couple values $(-1, +1, 0)$ is a measure of the variability of the mechanisms used.

- Then, we calculate the components of the strain rate tensor and of the velocity tensor, as described in step 3. A rotation to the North-East-Down coordinate frame, if desired, can be performed in this stage.

3 Seismic Strain Rates Along the Hellenic Trench

As an application example we apply the previously described formulation to calculate the deformation rates for the southern part of Aegean Sea, along the Hellenic Trench in particular, using seismological input.

3.1 Tectonic Setting

The Aegean Sea and the surrounding lands (Fig. 1), usually, are described as a "*natural geophysical laboratory*". This stems from the fact that, within a small geographical area, a variety of deformation mechanisms take place, as for example: subduction of the front part of the Eastern Mediterranean lithosphere beneath the Aegean plate, continental collision in the west, alongside shear motions imposed by the North Anatolian Fault entering into the Aegean Sea, and back-arc extension.

The Hellenic subduction zone extends from approximately south of the Ionian Islands in the west, up to Rhodes Island in the east. The Mediterranean Ridge is the accretionary prism formed by the pile of sediments, up to 10 km thick, that overlies the oceanic crust. The Hellenic Trench is a scarp in the bathymetry 2–3 km high, which is more clearly developed in the western part of the subduction zone, between Peloponnese and Crete, while it splits into branches towards east, the best developed of which are known as the Ptolemy, Pliny and Strabo Trenches (Kiratzi 2014 and references therein). The Hellenic Trench does not define the plate boundary, as it is common in other ocean trenches; it is not the surface projection of the subduction interface. The African-Aegean plate boundary itself, lies south of the Mediterranean Ridge and most likely is buried beneath it. The degree of coupling on the Hellenic subduction interface, i.e. the fraction of the motion across the plate boundary accommodated by elastic strain accumulation, is estimated to be low (10–20% of the full African-Aegean convergence rate).

Here we focus on the deformation related mainly to the low-angle thrusts, underlying the Hellenic Trench at depths up to 40–50 km, which they are considered to represent the interface between the subducting African lithosphere and the overriding

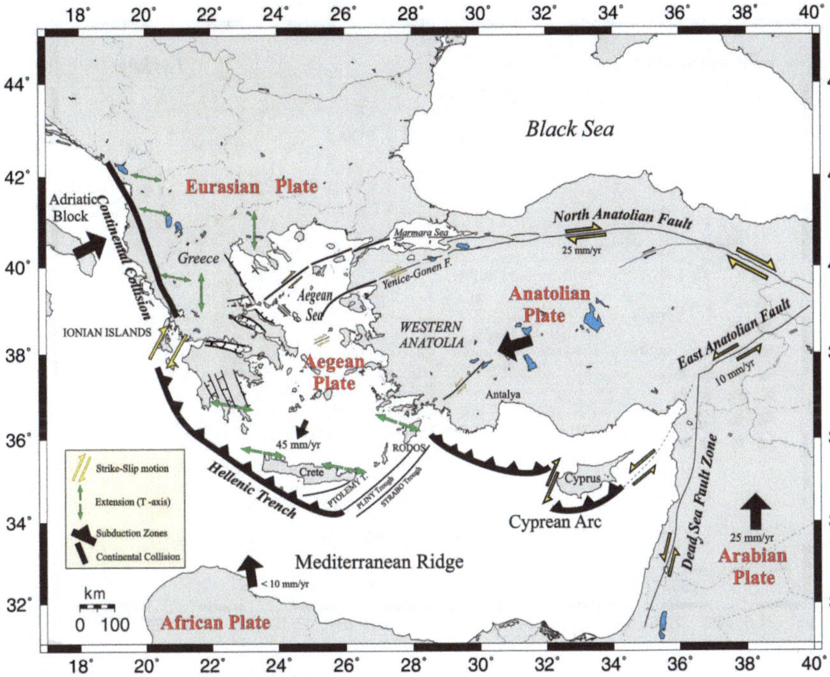

Fig. 1 Simplified map of the main tectonic features of the broader Aegean Sea and the surrounding lands

Aegean plate. Taking this into consideration it is easy to understand that their surface projection, which would mark the African-Aegean plate boundary itself, should be sought at least 150 km south of Crete, at the Mediterranean Ridge.

3.2 Data and Analysis

We adopted the most recent zonation of the region and the seismicity parameters (Fig. 2 and Table 1) as proposed by Vamvakaris et al. (2016). In order to define the geometry of each zone, e.g. spatial dimensions (length, width) as well the azimuth, we applied a least square analysis to the actual seismicity distribution (Papazachos and Kiratzi 1992), using the seismicity catalogue compiled by the Geophysical Department of Aristotle University of Thessaloniki. We adopted a constant thickness of the seismogenic layer of 15 km for all zones, based on previous work (Kiratzi 2014 and references therein). When the seismicity within a zone was not sufficient to sample adequately the spatial dimensions (e.g. zones NL3, NL4, NL5 in Fig. 2) we visually calculated the area and the azimuth.

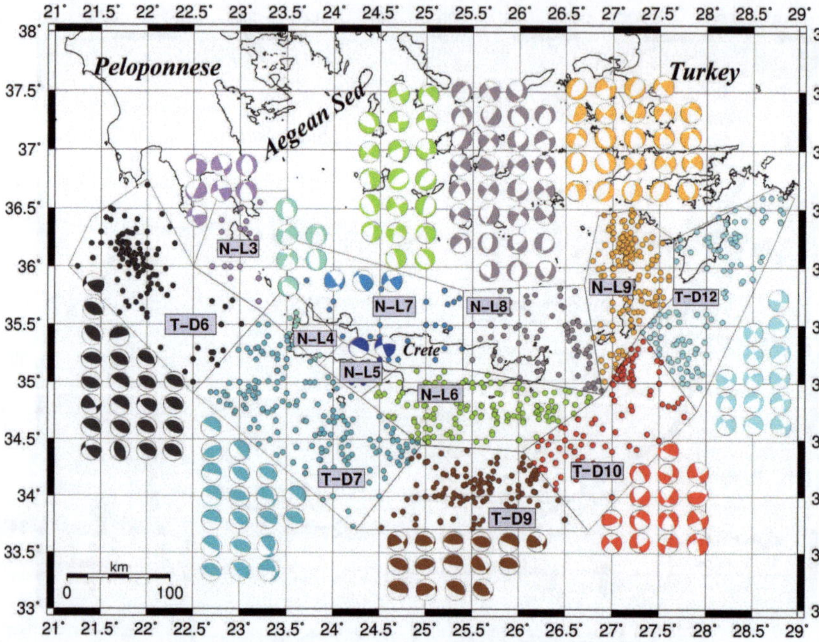

Fig. 2 Zonation adopted in the analysis (from Vamvakaris et al. 2016) and the focal mechanisms used, colour coded as the background shallow (h < 20 km) seismicity, for each zone

To compile a catalogue of focal mechanisms for each zone (Table 2), we chose to use the GCMT catalogue, which is easily accessible. We retrieved the focal mechanisms for all earthquakes from 1977–2016 with magnitude M > 4.5 and focal depths less than 40 km. To enrich the catalogue we obtained additional focal mechanisms from moderate magnitude events, retrieved from the databases of the National Observatory of Athens (NOA) and the one of the Istituto Nazionale di Geofisica e Vulcanologia (INGV) of Italy, all available at the site of the European Mediterranean Seismological Centre (http://emsc-csem.org).

3.3 *Application Results*

The strain rates and velocity tensors were calculated for each zone, in the zone's coordination system alongside the north/east/down reference system (Table 3). In Fig. 3 we chose to plot the eigensystem of the velocity tensor, that is we plot the maximum horizontal component, and in the cases we had two horizontal components of comparable size, we plot both of them. Variability is observed in the velocities for the outer ring of seismogenic sources (T-D type), both in terms of amplitude and orientation, moving from west towards east. Obviously, the deformation is taken

Table 1 Zonation and seismicity parameters used in the analysis. (Adopted from Vamvakaris et al. 2016)

Zone	Area vertices		Azimuth	M_{max}	Length	Width	G-R parameters	
	N (°)	E (°)	ξ (°)		l_1 (km)	l_2 (km)	a	b
NL3	36.65	22.78	177	6.6	64	89	4.28	1.03
	36.00	22.50						
	35.60	23.25						
	35.85	23.50						
	36.25	23.50						
	36.65	23.50						
NL4	35.60	23.25	135	8.2	78	32	4.06	1.05
	35.15	23.85						
	35.35	24.10						
	35.85	23.50						
NL5	35.15	23.85	127	8.2	51	47	4.19	1.08
	34.87	24.30						
	35.13	24.70						
	35.35	24.10						
NL6	35.13	24.70	96	8.2	202	84	5.4	1.13
	34.87	24.30						
	34.45	24.95						
	34.40	26.05						
	34.90	26.90						
	35.12	25.40						
NL7	36.25	23.50	105	6.5	178	44	4.23	1.06
	35.85	23.50						
	35.35	24.10						
	35.13	24.70						
	35.12	25.40						
	35.80	25.40						
NL8	35.80	25.40	86	6.6	117	89	5.08	1.1
	35.12	25.40						
	34.90	26.90						
	35.85	26.70						
NL9	36.45	26.85	1	6.7	49	83	5.16	1.09
	35.85	26.70						
	34.90	26.90						
	35.45	27.30						
	35.75	27.65						
	36.25	27.65						
	36.50	27.20						
TD6	36.00	21.15	135	8.0	173	103	5.15	1.05
	34.90	22.50						
	35.60	23.25						
	36.00	22.50						
	36.75	22.00						
	36.42	21.40						
TD7	34.90	22.50	130	8.2	208	105	5.48	1.1
	33.70	24.25						

(continued)

Table 1 (continued)

Zone	Area vertices		Azimuth	Mmax	Length	Width	G-R parameters	
	N (°)	E (°)	ξ (°)		l_1 (km)	l_2 (km)	a	b
	34.45	24.95						
	34.87	24.30						
	35.15	23.85						
	36.60	23.25						
TD9	34.45	24.95	90	8.2	165	77	5.42	1.14
	33.70	24.25						
	33.70	26.75						
	34.40	26.05						
TD10	34.40	26.05	41	8.2	160	100	5.28	1.13
	33.70	26.75						
	34.75	27.90						
	35.45	27.30						
	34.90	26.90						
TD12	36.25	27.65	31	8.2	379	95	5.27	1.09
	35.75	27.65						
	35.45	27.30						
	34.75	27.90						
	36.05	28.65						
	36.60	28.95						
	36.70	28.75						

up by compression along the Hellenic Trench, south of Crete. The velocities are significantly larger at the western and eastern part of the trench (TD-6 and TD-12) compared to its central part. The maximum velocity is 1.23 cm/yr (TD-6) and the minimum is 0.7 cm/yr (TD-9). The deformation of the southern part of the Aegean is taken up by earthquakes with mostly low angle thrusting located at the southern and south-western part of the Hellenic trench, whereas towards east a significant strike slip component is introduced. This occurs probably due to the presence in this area of strike-slip faults bounding the basins, with tenths of kilometres of length. However, during the instrumental era, no strong earthquake has ever occurred in the shallow part of the easternmost part of the Hellenic trench that could reveal in a clearer way the strike slip characteristics of the rupture at that area.

In the inner part of the Hellenic Trench and in the zones coded as NL the deformation is mainly taken up by ~E-W extension. The seismogenic source belonging to the inner part of the Aegean Arc (N-L type) exhibit very small deformation compared to the rest of the zones. Small variations from the pure E-W orientation are mostly due to the fact that also in these areas (in particular sources N-L3 to N-L8) the focal mechanism database we used mainly consists of small and a very few moderate magnitude events that occur on small faults that locally can have various orientations.

Table 2 Catalogue of focal mechanisms used for each zone. All data were retrieved from public and easily accessible sources, as referenced

Event n.	Date (dd/mm/yyyy)	Lon (°E)	Lat (°N)	Mw	h (km)	Strike (°)	Dip (°)	Rake (°)	Institute	Zone
1	18/04/2015	23.150	36.480	4.9	19	163	56	164	CMT	NL3
2	27/12/2016	22.965	36.555	4.6	22	348	41	−104	NOA	NL3
3	13/03/2015	23.218	36.430	4.4	23	69	56	1	NOA	NL3
4	10/04/2013	22.986	36.508	3.5	17	328	59	−122	NOA	NL3
5	09/04/2013	23.026	36.526	4.6	21	359	89	−129	NOA	NL3
6	09/06/2012	23.416	36.313	3.5	9	110	30	174	NOA	NL3
7	05/01/2012	23.225	36.331	3.9	31	252	74	18	NOA	NL3
8	24/07/2004	23.590	35.370	4.8	11	181	39	−38	CMT	NL4
9	12/03/2016	23.671	35.415	4.5	14	206	57	−11	NOA	NL4
10	06/03/2016	23.586	35.397	3.9	8	328	36	−139	NOA	NL4
11	09/02/2016	23.570	35.367	3.6	6	325	42	−123	NOA	NL4
12	06/07/2012	23.450	35.532	3.8	22	211	80	−35	NOA	NL4
13	27/05/2011	23.513	35.783	4.3	6	162	26	−103	NOA	NL4
14	15/06/1979	24.240	34.960	5.4	33	216	11	10	CMT	NL5
15	02/01/2014	24.366	35.056	4.1	32	164	76	159	NOA	NL5
16	09/07/1990	26.750	34.920	5.4	33	129	27	−106	CMT	NL6
17	19/03/1991	26.280	34.820	5.5	18	245	36	−33	CMT	NL6
18	12/10/2002	26.370	34.770	5.4	10	250	29	2	CMT	NL6
19	02/05/2015	25.780	34.450	5.0	13	295	70	−159	CMT	NL6
20	25/05/2016	26.200	34.740	5.5	25	149	9	−103	INGV	NL6
21	28/05/2015	26.767	34.894	4.0	14	127	73	174	NOA	NL6
22	05/02/2015	25.433	34.908	4.2	10	159	84	−173	NOA	NL6
23	18/05/2014	26.163	34.904	4.1	26	230	63	−104	NOA	NL6
24	06/09/2013	24.901	34.761	3.8	6	174	89	−175	NOA	NL6
25	06/09/2013	24.904	34.843	4.1	2	258	70	10	NOA	NL6
26	09/01/2013	25.486	34.934	3.9	3	264	20	5	NOA	NL6
27	23/04/2012	25.352	34.902	4.7	16	336	47	−48	NOA	NL6
28	06/04/2012	26.325	34.983	3.5	14	64	40	−65	NOA	NL6
29	08/03/2012	24.891	34.765	3.7	30	331	87	180	NOA	NL6
30	02/03/2012	25.797	34.759	4.1	16	140	22	−102	NOA	NL6
31	16/12/2011	24.800	34.928	3.5	20	349	55	−160	NOA	NL6
32	14/11/2011	24.492	34.905	3.7	15	111	64	−151	NOA	NL6
33	25/10/2011	26.228	34.978	3.9	11	194	73	−99	NOA	NL6
34	11/10/2011	24.843	34.822	3.5	26	216	44	−10	NOA	NL6
35	27/10/2015	24.000	35.877	4.1	30	324	87	135	NOA	NL7
36	24/03/2012	24.048	35.923	3.7	6	283	39	42	NOA	NL7
37	11/12/2011	23.721	35.773	3.8	4	241	71	14	NOA	NL7
38	23/07/1979	26.320	35.480	5.6	33	61	35	−40	CMT	NL8
39	30/04/1992	26.710	35.070	5.7	17	172	38	−106	CMT	NL8
40	12/06/2008	26.190	35.110	5.1	29	117	40	−151	CMT	NL8
41	06/11/2011	25.660	35.800	4.8	30	193	40	−79	CMT	NL8
42	16/04/2015	26.830	34.920	5.0	25	30	53	−39	CMT	NL8
43	17/04/2015	26.780	35.030	5.5	15	332	51	−163	CMT	NL8
44	09/06/2015	26.790	35.040	5.4	16	158	53	−150	CMT	NL8
45	03/03/2017	25.922	35.498	4.5	20	67	58	36	NOA	NL8

(continued)

Table 2 (continued)

Event n.	Date (dd/mm/yyyy)	Lon (°E)	Lat (°N)	Mw	h (km)	Strike (°)	Dip (°)	Rake (°)	Institute	Zone
46	18/02/2016	26.319	35.178	4.5	26	18	45	−29	NOA	NL8
47	29/05/2015	26.761	34.949	4.1	20	124	82	176	NOA	NL8
48	22/04/2015	26.710	35.292	4.2	28	129	81	160	NOA	NL8
49	21/04/2015	26.757	35.228	4.4	30	315	60	−155	NOA	NL8
50	21/04/2015	26.803	35.194	4.1	28	348	30	−143	NOA	NL8
51	17/04/2015	26.671	35.197	4.2	28	149	88	161	NOA	NL8
52	17/04/2015	26.740	35.174	4.1	25	325	66	−153	NOA	NL8
53	17/04/2015	26.739	34.956	4.5	24	127	89	174	NOA	NL8
54	16/04/2015	26.786	35.154	4.4	28	63	47	−32	NOA	NL8
55	16/04/2015	26.786	35.154	4.1	25	324	83	−176	NOA	NL8
56	10/01/2015	26.204	35.309	4.0	10	328	63	−143	NOA	NL8
57	06/10/2014	26.546	35.118	4.3	4	35	81	−66	NOA	NL8
58	02/07/2014	26.433	35.219	3.9	26	123	76	161	NOA	NL8
59	24/12/2013	26.234	35.055	4.3	18	26	78	−70	NOA	NL8
60	20/05/2013	26.615	35.438	3.8	30	56	64	7	NOA	NL8
61	03/09/2012	25.899	35.769	3.8	13	212	89	−21	NOA	NL8
62	20/11/2011	26.581	35.115	4.0	32	41	88	14	NOA	NL8
63	07/11/2011	25.648	35.788	3.5	9	26	74	−100	NOA	NL8
64	15/09/2011	25.655	35.798	4.2	19	337	45	−143	NOA	NL8
65	15/09/2011	25.655	35.798	3.9	9	198	38	−34	NOA	NL8
66	13/01/2009	26.430	35.740	5.2	35	56	50	−31	NOA	NL8
67	26/06/2006	26.490	34.970	4.4	25	247	83	−32	NOA	NL8
68	20/07/1996	27.100	36.150	6.2	33	196	38	−102	CMT	NL9
69	22/07/1996	27.220	36.150	5.0	33	223	36	−78	CMT	NL9
70	01/05/2001	27.500	35.700	5.2	33	176	25	−105	CMT	NL9
71	07/02/2004	26.910	36.040	5.1	25	211	27	−80	CMT	NL9
72	25/05/2004	27.180	35.920	4.8	5	25	39	−38	CMT	NL9
73	23/09/2007	27.120	35.270	5.3	24	234	41	−23	CMT	NL9
74	15/03/2017	27.589	35.945	4.3	22	12	41	−79	NOA	NL9
75	03/12/2016	27.279	35.883	3.9	10	353	60	−63	NOA	NL9
76	11/05/2015	26.900	35.015	4.5	32	315	88	172	NOA	NL9
77	16/04/2015	26.896	35.175	4.8	25	321	77	179	NOA	NL9
78	16/04/2015	26.858	35.222	4.0	21	218	68	1	NOA	NL9
79	02/10/2014	27.035	36.030	4.2	5	17	65	−87	NOA	NL9
80	30/06/2013	27.510	35.943	3.8	10	343	54	−81	NOA	NL9
81	26/06/2013	27.260	35.808	3.9	4	317	47	−134	NOA	NL9
82	19/06/2013	27.228	35.826	3.8	6	304	59	−142	NOA	NL9
83	19/06/2013	27.218	35.805	4.1	5	191	72	−47	NOA	NL9
84	29/01/2013	27.065	35.483	4.1	25	238	55	10	NOA	NL9
85	19/06/2012	27.142	35.289	3.4	35	295	63	−172	NOA	NL9
86	17/05/2012	27.023	35.813	3.7	18	130	84	174	NOA	NL9
87	07/01/2012	27.226	35.876	4.0	23	76	29	147	NOA	NL9
88	21/11/2011	27.328	35.596	3.7	20	250	84	−28	NOA	NL9
89	09/11/2011	27.561	35.767	3.8	30	20	84	−37	NOA	NL9
90	05/11/2011	26.908	35.569	3.6	5	215	80	−21	NOA	NL9
91	01/11/2011	27.146	36.423	3.6	20	185	62	−36	NOA	NL9

(continued)

Table 2 (continued)

Event n.	Date (dd/mm/yyyy)	Lon (°E)	Lat (°N)	Mw	h (km)	Strike (°)	Dip (°)	Rake (°)	Institute	Zone
92	05/03/2008	27.410	36.160	4.5	14	55	33	−60	NOA	NL9
93	05/03/2008	27.410	36.160	4.4	13	48	39	−70	NOA	NL9
94	21/04/1985	22.200	35.680	5.3	36	269	36	71	CMT	TD6
95	08/06/1986	21.510	36.070	5.1	29	109	34	86	CMT	TD6
96	11/01/1994	21.910	35.950	5.5	33	77	61	30	CMT	TD6
97	13/10/1997	22.070	36.380	6.4	24	298	20	89	CMT	TD6
98	29/04/1998	21.940	36.140	5.5	33	260	53	36	CMT	TD6
99	17/04/1999	21.680	36.040	5.3	40	160	43	95	CMT	TD6
100	24/05/2000	22.100	35.920	5.2	33	263	12	93	CMT	TD6
101	14/02/2008	21.670	36.500	6.9	29	332	6	120	CMT	TD6
102	14/02/2008	21.860	36.350	6.5	28	337	5	127	CMT	TD6
103	26/02/2008	21.870	35.910	5.4	5	286	34	74	CMT	TD6
104	26/02/2008	21.890	36.090	5.0	38	321	35	114	CMT	TD6
105	04/03/2008	21.810	36.060	4.9	11	293	40	80	CMT	TD6
106	07/03/2008	21.760	36.140	4.9	13	262	40	45	CMT	TD6
107	14/03/2008	21.760	36.010	5.1	5	304	35	101	CMT	TD6
108	23/03/2008	21.860	36.150	5.0	35	279	31	69	CMT	TD6
109	08/05/2008	21.950	36.110	4.9	10	304	37	97	CMT	TD6
110	21/06/2008	21.930	36.100	5.2	17	299	33	91	CMT	TD6
111	21/06/2008	21.820	36.060	5.6	5	290	33	80	CMT	TD6
112	21/06/2008	21.700	35.990	4.9	5	292	34	83	CMT	TD6
113	22/06/2008	21.980	36.170	4.8	22	297	33	91	CMT	TD6
114	04/11/2012	21.678	36.191	3.7	31	315	33	91	NOA	TD6
115	28/02/2008	21.780	36.020	4.6	11	254	64	45	NOA	TD6
116	14/02/2008	21.700	36.270	4.5	22	115	47	85	NOA	TD6
117	11/09/1977	23.030	35.050	5.8	33	74	28	100	CMT	TD7
118	15/05/1979	24.440	34.530	6.1	33	172	4	−20	CMT	TD7
119	21/06/1984	23.290	35.350	6.2	27	79	7	−128	CMT	TD7
120	07/12/1995	24.150	34.790	5.6	0	319	6	123	CMT	TD7
121	10/12/1995	24.250	34.830	5.2	19	289	22	75	CMT	TD7
122	05/11/1997	23.980	34.770	5.4	33	350	34	154	CMT	TD7
123	13/03/2000	23.380	35.360	5.0	50	296	4	68	INGV	TD7
124	25/11/2005	23.320	35.020	5.2	32	289	34	67	CMT	TD7
125	18/01/2007	22.670	34.840	5.0	39	300	40	111	INGV	TD7
126	19/03/2009	23.440	35.100	4.9	37	299	33	79	CMT	TD7
127	12/09/2012	24.110	34.780	5.4	33	278	26	69	CMT	TD7
128	06/04/2013	24.090	34.800	5.3	34	298	32	93	CMT	TD7
129	12/10/2013	23.250	35.510	6.8	40	339	3	130	CMT	TD7
130	30/07/2016	22.800	35.130	5.2	10	311	1	91	INGV	TD7
131	15/01/2017	23.064	35.063	4.7	36	30	37	159	NOA	TD7
132	10/03/2016	23.228	34.875	4.4	30	355	21	145	NOA	TD7
133	24/10/2015	23.250	35.316	4.1	21	115	54	92	NOA	TD7
134	29/04/2014	23.155	35.331	4.0	14	359	36	173	NOA	TD7
135	26/01/2013	23.961	34.532	4.1	2	280	36	75	NOA	TD7
136	16/11/2012	23.132	35.257	3.7	30	138	64	101	NOA	TD7
137	28/03/1989	24.680	34.060	5.5	33	67	53	29	CMT	TD9
138	07/10/1998	25.620	33.800	5.1	10	318	60	150	CMT	TD9

(continued)

Table 2 (continued)

Event n.	Date (dd/mm/yyyy)	Lon (°E)	Lat (°N)	Mw	h (km)	Strike (°)	Dip (°)	Rake (°)	Institute	Zone
139	25/01/1999	25.750	34.120	5.0	33	305	44	124	INGV	TD9
140	10/03/2000	26.030	34.360	5.2	10	256	22	72	CMT	TD9
141	05/04/2000	25.690	34.220	5.5	38	276	43	80	CMT	TD9
142	01/07/2009	25.470	34.160	6.4	19	295	32	108	CMT	TD9
143	02/07/2009	25.220	34.250	4.8	22	95	40	43	CMT	TD9
144	01/02/2010	25.390	34.000	5.4	14	282	32	77	CMT	TD9
145	15/06/2013	25.020	34.400	6.3	10	235	7	45	CMT	TD9
146	16/06/2013	25.160	34.350	6.1	19	222	7	20	CMT	TD9
147	03/04/2016	25.620	34.220	4.9	10	265	32	98	INGV	TD9
148	28/01/2015	25.089	34.361	4.9	32	108	70	94	NOA	TD9
149	24/11/2013	25.597	33.950	4.5	2	302	22	94	NOA	TD9
150	07/07/2009	25.450	34.030	4.7	8	61	59	59	NOA	TD9
151	07/07/2009	25.200	33.920	4.6	3	99	41	59	NOA	TD9
152	22/05/1986	26.590	34.510	5.5	30	227	37	24	CMT	TD10
153	05/09/1988	26.650	34.510	5.2	12	15	55	−11	CMT	TD10
154	09/05/1997	27.280	35.380	5.0	20	133	52	−168	INGV	TD10
155	22/11/1999	27.480	35.000	5.0	33	214	60	−34	INGV	TD10
156	13/06/2000	27.120	35.150	5.4	10	147	45	−166	CMT	TD10
157	30/12/2001	27.400	34.800	5.0	10	34	67	157	INGV	TD10
158	08/02/2005	26.500	34.440	5.0	36	189	45	−27	INGV	TD10
159	03/07/2005	27.260	35.110	4.7	21	62	58	−26	CMT	TD10
160	09/12/2005	27.180	35.240	4.8	24	216	39	−27	CMT	TD10
161	09/04/2006	27.240	35.170	5.3	33	144	36	−117	CMT	TD10
162	13/08/2006	26.570	34.420	5.3	33	107	76	169	CMT	TD10
163	04/08/2008	26.580	34.100	5.3	31	117	46	−18	CMT	TD10
164	03/10/2014	26.300	34.440	5.0	16	303	51	136	CMT	TD10
165	12/10/2015	26.380	34.500	4.9	14	230	70	−3	CMT	TD10
166	28/07/2012	27.547	35.126	4.1	40	160	86	−155	NOA	TD10
167	20/05/2012	27.170	35.248	3.5	20	127	87	−11	NOA	TD10
168	16/02/2012	26.648	34.663	4.2	30	301	36	−103	NOA	TD10
169	21/01/2012	26.786	34.554	4.1	20	332	74	173	NOA	TD10
170	23/09/2007	27.170	35.140	5.1	11	254	49	−7	NOA	TD10
171	22/08/2006	27.210	35.090	4.8	40	56	35	59	NOA	TD10
172	29/05/2001	27.780	35.410	5.1	21	42	66	13	CMT	TD12
173	21/09/2004	27.790	35.280	4.9	29	311	79	180	CMT	TD12
174	11/10/2005	27.470	35.580	4.9	21	215	61	−27	CMT	TD12
175	21/05/2007	27.810	35.120	5.0	10	1	39	−146	CMT	TD12
176	17/01/2010	27.860	35.260	4.9	29	229	72	−16	CMT	TD12
177	21/08/2014	27.690	35.650	4.7	1	241	55	−1	CMT	TD12
178	16/09/2015	27.790	35.350	5.0	11	216	21	3	CMT	TD12
179	26/12/2015	27.390	35.520	5.2	37	232	63	11	CMT	TD12
180	12/03/2015	27.821	35.483	4.4	8	316	75	155	NOA	TD12
181	16/12/2012	27.663	35.459	4.2	4	323	27	−178	NOA	TD12
182	09/10/2012	27.720	35.372	4.4	20	331	46	160	NOA	TD12
183	06/10/2012	27.798	35.433	3.9	10	73	68	22	NOA	TD12
184	24/10/2011	28.744	36.461	3.7	10	9	68	−8	NOA	TD12
185	15/07/2008	28.080	35.740	6.4	44	252	77	−24	NOA	TD12

Table 3 Components of the strain rate tensor $\dot{\varepsilon}$ ($\times 10^{-8}$/yr) and of the velocity tensor U (cm/yr) for the seismogenic sources used in this chapter. Coordinate system of the tensors is North-East-Down

	Strain rate tensor $\dot{\varepsilon}$ ($\times 10^{-8}$/yr)						Velocity tensor U (cm/yr)					
Zone	ε_{11}	ε_{22}	ε_{33}	ε_{12}	ε_{13}	ε_{23}	U_{11}	U_{22}	U_{33}	U_{12}	U_{13}	U_{23}
NL3	−0.51	0.78	−0.26	−0.46	−0.12	0.36	−0.04	0.07	0.00	−0.09	0.00	0.01
NL4	−2.22	6.78	−4.55	2.43	3.14	−0.35	−0.21	0.35	−0.07	0.08	0.09	−0.01
NL5	2.03	−0.82	−1.21	−1.28	1.44	4.75	0.18	−0.11	−0.02	−0.09	0.04	0.14
NL6	−0.51	1.43	−0.93	−0.47	0.52	0.72	−0.04	0.29	−0.01	−0.08	0.02	0.02
NL7	−0.40	0.23	0.18	−0.25	−0.08	0.08	−0.01	0.05	0.00	−0.02	0.00	0.00
NL8	−1.03	1.68	−0.65	−0.16	0.07	0.04	−0.10	0.20	0.01	−0.04	0.00	0.00
NL9	−2.22	5.95	−3.66	0.41	0.51	0.73	−0.11	0.49	−0.05	0.08	0.02	0.02
TD6	−7.53	−1.52	9.05	−4.06	3.49	2.28	−1.10	0.14	0.14	−0.40	0.10	0.07
TD7	−3.83	−1.71	5.54	−2.53	6.01	3.64	−0.47	−0.12	0.08	−0.25	0.18	0.11
TD9	−6.21	0.45	5.77	−1.52	3.36	−0.75	−0.48	0.07	0.09	−0.23	0.10	−0.02
TD10	−3.23	3.77	−0.53	−0.03	0.10	0.65	−0.67	0.71	0.00	0.04	0.00	0.02
TD12	−3.08	3.17	−0.09	−0.74	1.42	−0.23	−0.93	0.35	0.00	−0.22	0.04	0.00

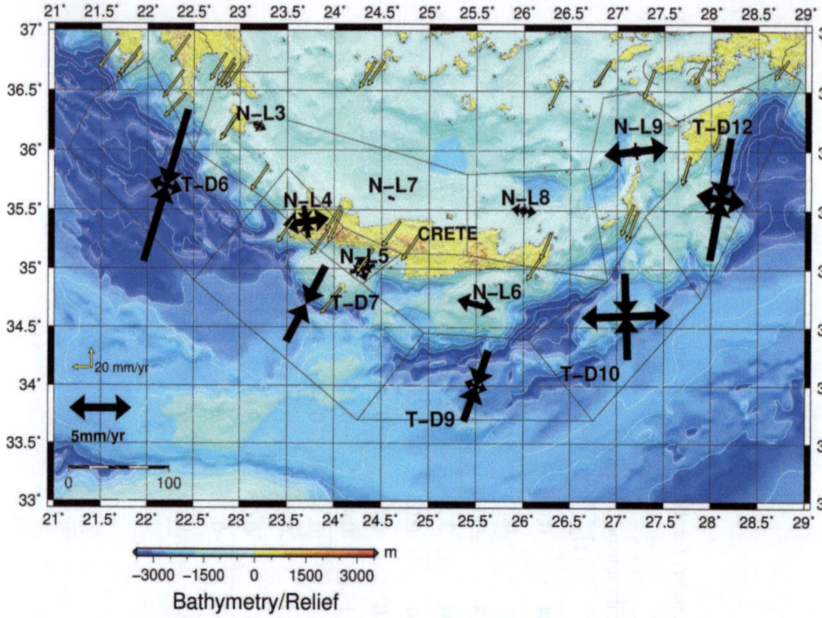

Fig. 3 Spatial distribution of deformation velocities (in mm/yr) along the Hellenic Trench. Converging and diverging arrows denote compressional and extensional velocities, respectively. These velocities were derived from the eigensystem of the velocity tensor and here we plot the maximum horizontal components. The yellow arrows are GPS obtained velocities and are shown for comparison

4 Data Sources

The data used in this work have been all retrieved from published sources, as referenced in the text. Most of the figures were produced using the GMT software (Wessel and Smith 1998).

References

Aki K, Richards P (1980) Quantitative seismology: theory and methods. Freeman, San Francisco, California, p 557

Ambraseys NN, Jackson J (1990) Seismicity and associated strain of central Greece between 1890 and 1988. Geophys J Int 101(3):663–708. https://doi.org/10.1111/j.1365-246X.1990.tb05577.x

Ekström G, England P (1989) Seismic strain rates in regions of distributed continental deformation. J Geophys Res 94:10231–10257

Eyidogan H (1988) Rates of crustal deformation in western Turkey as deduced from major earthquakes. Tectonophysics 148:83–92

Jackson J, McKenzie D (1988) The relationship between plate motions and seismic moment tensors, and the rates of active deformation in the Mediterranean and Middle East. Geophys J R Astron Soc 93:45–73
Jost ML, Herrmann RB (1989) A student's guide to and review of moment tensors seismological research letters, vol 60, pp 37–57. https://doi.org/10.1785/gssrl.60.2.37
Kiratzi A (1991) Rates of crustal deformation in the north Aegean trough-north Anatolian fault deduced from seismicity. Pageoph 136:421–432
Kiratzi A (1993) A study on the active crustal deformation of the North and East Anatolian fault zones. Tectonophysics 225:191–203
Kiratzi A (2014) Mechanisms of earthquakes in Aegean. In: Encyclopedia of earthquake engineering, pp 1439–1460. https://doi.org/10.1007/978-3-642-35344-4_299
Knopoff L, Randall MJ (1970) The compensated linear-vector dipole: a possible mechanism for deep earthquakes. J Geophys Res 75:4957–4963
Kostrov V (1974) Seismic moment and energy of earthquakes, and seismic flow on rock. Izv Acad Sci USSR Phys Solid Earth 1:23–44
Main I, Burton P (1984) Physical link between crustal deformation, seismic moment and seismic hazard for regions of varying seismicity. Geophys J R Astr Soc 79:469–488
Molnar P (1979) Earthquake recurrence intervals and plate tectonics. Bull Seism Soc Am 69:115–133
Papastamatiou D (1980) Incorporation of crustal deformation to seismic hazard analysis Bull. Seism Soc Am 72:1–14
Papazachos C, Kiratzi A (1992) A formulation for reliable estimation of active crustal deformation and its application to central Greece. Geophys J Int 111:424–432
Papazachos CB, Kiratzi AA (1996) A detailed study of active crustal in the Aegean and surrounding region, 1996. Tectonophysics 253:129–153
Vamvakaris DA, Papazachos CB, Papaioannou ChA, Scordilis EM, Karakaisis GF (2016) A detailed seismic zonation model for shallow earthquakes in the broader Aegean area. Nat Hazards Earth Syst Sci 16:55–84. https://doi.org/10.5194/nhess-16-55-2016
Wessel P, Smith WHF (1998) New improved version of the generic mapping tools released. EOS Trans AGU 79:579

Estimation of Empirical Green's Tensor Spatial Derivative Elements: A Preliminary Study Using Strong Motion Records in Southern Fukui Prefecture, Japan

Michihiro Ohori

1 Introduction

The empirical Green's tensor spatial derivative (EGTD) method, proposed by Plicka and Zahradnik (1998), has the potential to deal with differences in focal mechanisms between a targeted event and other small events, and to predict the ground motion for an event with an arbitrary focal mechanism. The EGTD elements are estimated through a form of single-station inversion using waveform data from several small events whose focal mechanisms and source time functions have been well determined. Although this technique is expected to provide results of considerable accuracy and prediction stability, discussion of its application has been limited in the literature (Ito et al. 2001; Ito 2005; Pulido et al. 2006; Ohori and Hisada 2006, 2011; Ohori 2014, 2016). Further data accumulation and investigation will enhance the applicability of the EGTD method. I estimated the EGTD using seven small events (M_J 3.7–4.2) in the southern part of Fukui Prefecture, Japan. Fukui Prefecture is an area with relatively low seismicity compared with other parts of Japan and strong motion data for EGTD estimation is less abundant. Most of the events used in this study occurred within or near the Mikata Fault Zone, where earthquakes of magnitude 7.2 are expected with a recurrence period of 3,800–6,300 years (Ishibe and Shimazaki 2012). In my recent work (Ohori 2016), I targeted 0.25–1 Hz bandpass filtered velocity waveforms obtained at FKI007, one of the K-NET stations operated by the National Research Institute for Earth Science and Disaster Prevention (NIED). In this article, I target velocity waveforms bandpass-filtered between 0.25 and 1 Hz and 0.25 and 10 Hz obtained at three stations, including FKI007. There are nuclear power stations of various types along the coastal area close to the fault.

M. Ohori (✉)
Research Institute of Nuclear Engineering, University of Fukui, 1-3-33 Kanawa, Tsuruga 914-0055, Japan
e-mail: ohorim@u-fukui.ac.jp

S. D'Amico (ed.), *Moment Tensor Solutions*, Springer Natural Hazards,
https://doi.org/10.1007/978-3-319-77359-9_11

This preliminary study in such an area is significant in its evaluation of the value of strong motion prediction using EGTD, which demonstrates the potential for greater use of EGTD estimation.

2 Method

The estimation method of EGTD has been explained fully by Ohori and Hisada (2006, 2011). It is applicable to simulation of strong motion in a frequency range below the corner frequency. I briefly summarize the method below. Ground motion displacement $u_i(x_o, t)$ $(i = x, y, z)$, excited by a double-couple point source, is theoretically expressed as the convolution of moment tensor elements $M_{pq}(x_s, \tau)$ $(p, q = x, y, z)$ and Green's tensor spatial derivative elements $G_{ip,q}(x_o, t|x_s, \tau)$.

$$u_i(x_o, t) = M_{pq}(x_s, \tau) * G_{ip,q}(x_o, t|x_s, \tau) \tag{1}$$

Hereafter, I abbreviate $u_i(x_o, t)$, $M_{pq}(x_s, \tau)$, and $G_{ip,q}(x_o, t|x_s, \tau)$ as u_i, M_{pq}, and $G_{ip,q}$. Explicit expressions of M_{pq} for a double-couple point source are available in the literature (e.g. Aki and Richards (1980)). Considering symmetrical conditions ($M_{pq} = M_{qp}$) and no volume change [$M_{xx} = -(M_{yy} + M_{zz})$] of the moment tensor elements, we can rewrite Eq. (1) as

$$u_i = \sum_{j=1}^{5} M_j * G_{ij} \tag{2}$$

where M_j $(j = 1, 2, \ldots, 5)$ is defined by $M_1 = M_{xy}$, $M_2 = M_{yy}$, $M_3 = M_{yz}$, $M_4 = M_{xz}$, $M_5 = M_{zz}$, and G_{ij} $(j = 1, 2, \ldots, 5)$ is defined by $G_{i1} = G_{ix,y} + G_{iy,x}$, $G_{i2} = G_{iy,y} - G_{x,x}$, $G_{i3} = G_{iy,z} + G_{iz,y}$, $G_{i4} = G_{ix,z} + G_{iz,x}$, $G_{i5} = G_{iz,z} - G_{ix,x}$. In a moment tensor inversion, u_i and G_{ij} are given and M_j are the unknowns to be solved in a least-squares sense. Conversely, in the EGTD inversion, u_i and M_j are given and G_{ij} are the unknowns to be solved. Note that the EGTD inversion is carried out for each component at each station using data from several events simultaneously, whereas the moment tensor inversion is computed for a particular event using data of all possible components at all possible stations simultaneously. It should also be emphasized that the moment tensor elements are determined by the source parameters while the Green's tensor spatial derivative elements are dependent on the underground structure of the area surrounding both the source and the station.

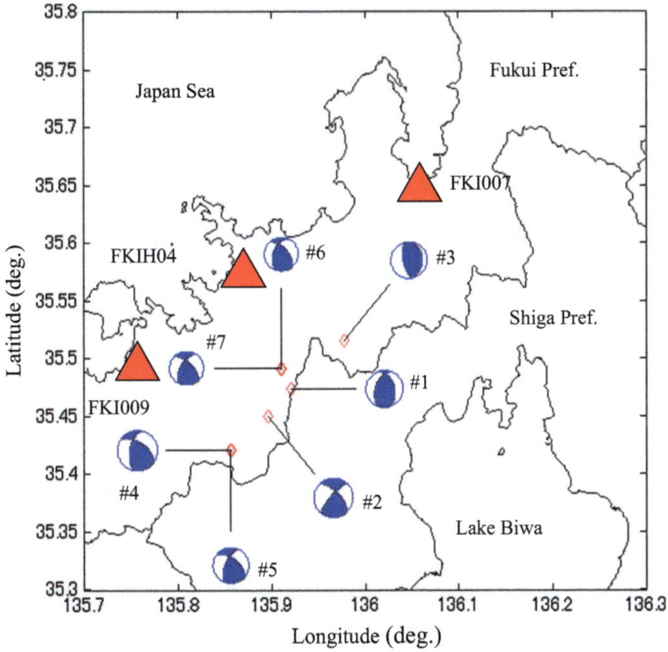

Fig. 1 Map showing epicenter locations of small events (M_J 3.7–4.2) and target stations, FKI007, FKIH04, FKI009. Focal mechanisms, as determined by the F-net, are also shown. Numerals after # show correspondence to the events listed in Table 1. This figure is modified from Ohori (2016)

3 Targeted Events and Stations

Strong motion records at FKI007 and FKI009 of the K-NET stations and FKIH04 of the KiK-net stations were used, all of which are operated by NIED. The map in Fig. 1 shows the stations and the epicenter locations of the target events. The source locations determined by the Japan Meteorological Agency (JMA) are summarized in Table 1. The focal mechanisms determined by the F-net (NIED) are also shown in Fig. 1. They all are classified as reverse-faulting mechanisms, but strike-slip faulting components are found in most events. Their epicentral distance was 16.1–30.8 km for FKI007, 8.2–15.3 km for FKIH04, and 12.6–20.6 km for FKI009. I estimated G_{ij}, the EGTD elements of the target stations, from these small events. To treat each event as a point source at the same location with different source mechanisms, I conducted, as described below, some corrections to the focal mechanisms and the waveform data prior to EGTD estimation.

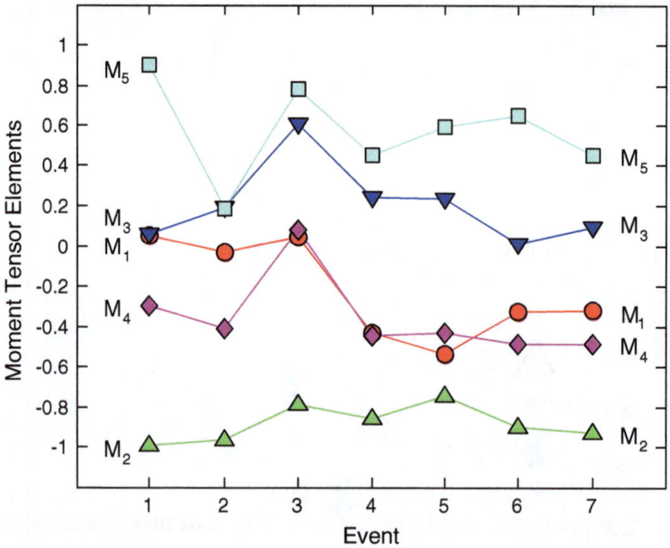

Fig. 2 Distributions of moment tensor elements of seven small events for FKI007. This figure is modified from Ohori (2016)

3.1 Correction for Focal Mechanisms

As seen in Fig. 1, the sources of the target events are almost located in a line from north-east to south-west. The station azimuth, with respect to the epicenters, is between 26.2° and 38.3° for FKI007, measured clockwise from North. It more widely varies between −63.7° and 4.5° for FKIH04 and between −95.7° and −48.6° for FKI009. To compensate for this discrepancy, I selected event 2 as a reference event and horizontally rotated the focal mechanisms of the other six events so that the station azimuth of each event can be treated as the same as that of event 2. Figure 2 shows the

Table 1 The source information of target events determined by the united hypocenter catalog of the JMA (Japan Meteorological Agency)

Event	Date (y/m/d)	Clock (h:m:s)	Latitude (°)	Longitude (°)	Depth (km)	M_J
1	2001/04/16	19:05:18.19	35.4737	135.9208	14.30	4.0
2	2001/12/28	3:28:02.73	35.4497	135.8957	6.74	4.2
3	2003/03/13	21:04:56.02	35.5152	135.9773	14.22	4.1
4	2008/08/08	4:35:16.24	35.4213	135.8563	14.56	4.2
5	2008/08/30	18:28:30.11	35.4207	135.8573	14.51	4.2
6	2009/11/22	23:49:22.73	35.4907	135.9105	11.79	3.7
7	2009/11/22	23:52:29.04	35.7920	135.9097	11.83	3.7

distributions of moment tensor elements for seven events after the horizontal rotation with respect for FKI007. The take-off angle from each source is also different. For example, assuming the source depth as in Table 1, it varies from 100.6° to 130.3° for FKI007. In our previous study (e.g. Ohori and Hisada 2011), we examined the effect of vertical rotation of focal mechanisms to correct for differences in take-off angles and found that this is not significant when epicentral distances are larger than source depths. This seems to be effective for all source-station pairs of FKI007 and most pairs of FKIH03 and FKI009. Therefore, for simplicity, I disregarded the discrepancy of the take-off angles.

3.2 Correction Applied to the Waveform Data

The observed acceleration records at three stations for seven small events were integrated into 0.25–1 Hz bandpass-filtered velocity waveform data and 0.25–10 Hz bandpass-filtered velocity waveform data. Two horizontal components were rotated into transverse and radial ones. According to Ohori (2014), corner frequencies of these small events are expected to be higher than 1 Hz, so I first targeted the frequency components lower than 1 Hz. To adjust the timing among events, I applied a time shift to the observed data to match the S-wave arrival time with that of the reference event 2. Differences in the seismic moments among events were corrected by normalization to 1.0×10^{15} Nm, approximately equal to $M_w 4.0$. To estimate the EGTD for simulation of broadband strong motion, next I targeted the 0.25–10 Hz bandpass-filtered velocity waveform data. To simulate broadband strong motion based on the empirical Green's spatial derivative method, the discrepancy among the corner frequencies of events should be removed. Ohori (2014) assumed the scaling law based on the ω^{-2} model (Aki 1967) and compensated the spectral amplitude decay beyond the corner frequency of each event so as to have the same corner frequency each other. In this study, I neglected this procedure for simplicity.

4 Results

Through the above procedure, I estimated the EGTD elements. In Fig. 3, I show example of empirical Green's tensor spatial derivative (EGTD) elements for FKI007 from bandpass-filtered velocity waveforms of 0.25–1 Hz (3a) and 0.25–10 Hz (3b). They both show very long duration characteristics in each EGTD element, after the S-wave main portion. These are difficult to produce by wave propagation theory for a stratified underground structure and reflect the complicated underground structure of the area surrounding both the source and the station. As suggested by Ohori and Hisada (2011), the EGTD elements could be useful for the structural study. In Figs. 4 and 5, I compare the observed velocity waveforms (in red) and corresponding syntheses calculated using EGTD (in blue). In Fig. 4, velocity waveform data with

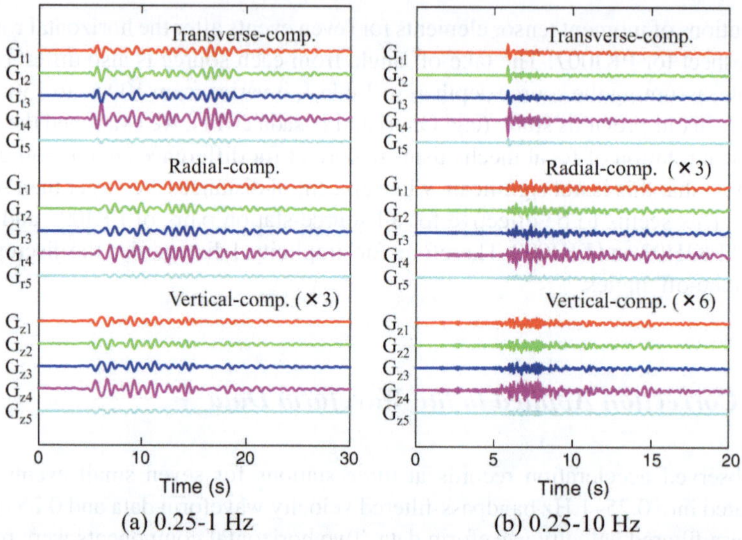

Fig. 3 Example of empirical Green's tensor spatial derivative (EGTD) elements for FKI007 estimated from 0.25–1 and 0.25–10 Hz bandpass-filtered velocity waveforms. For graphical purposes, the amplitudes of radial components in (**b**) and vertical components in (**a**) and (**b**) are three times or six times exaggerated compared with transverse components

0.25–1 Hz bandpass filtering are targeted. From Fig. 4, an acceptable match for the whole waveforms of all events can be found in three components, except for the underestimation found in transverse and vertical components of FKIH04 and transverse and radial components of FKI009 for event 4. In Fig. 5, velocity waveform data with 0.25–10 Hz bandpass filtering are targeted. From Fig. 5, an acceptable match for the whole waveforms of all events can be found in three components, except for the underestimation found in three components of FKIH04 and FKI009 for event 4. In Fig. 6, the ratio of the maximum amplitude between the synthesized and observed waveforms (hereafter "ratio") are summarized. As mentioned above, except for event 4, the ratio agrees well, no more than 1.5 for most of events. To improve the underestimation for event 4 found in most cases, the slight modification of source mechanism and source time function as well as the appropriate seismic moment should be considered for this event.

5 Conclusions

I applied the EGTD method to the velocity waveforms bandpass-filtered between 0.25–1 Hz and 0.25–10 Hz, respectively. On the whole, I consider that the agreement of the simulation results obtained using the EGTD method with observation data is acceptable. To improve the waveform match, I will reconsider the appropriate source

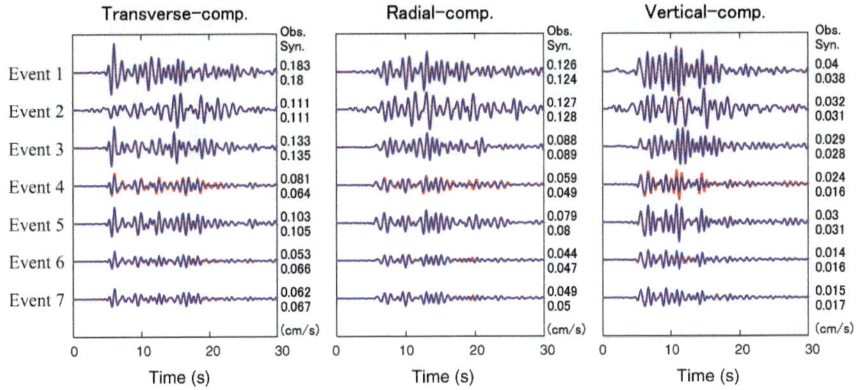

(a) FKI007

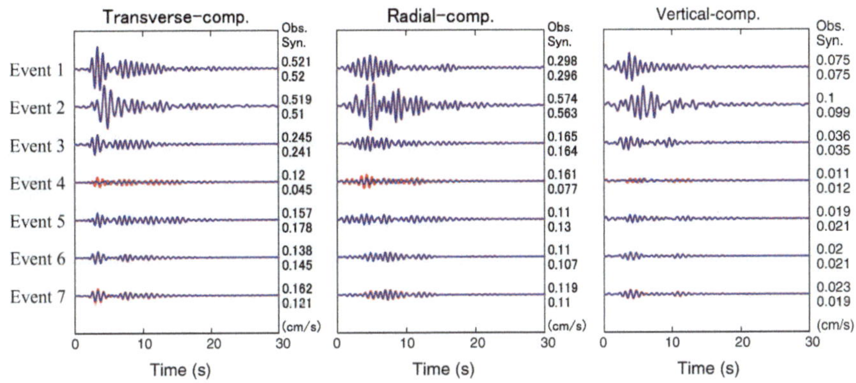

(b) FKIH04

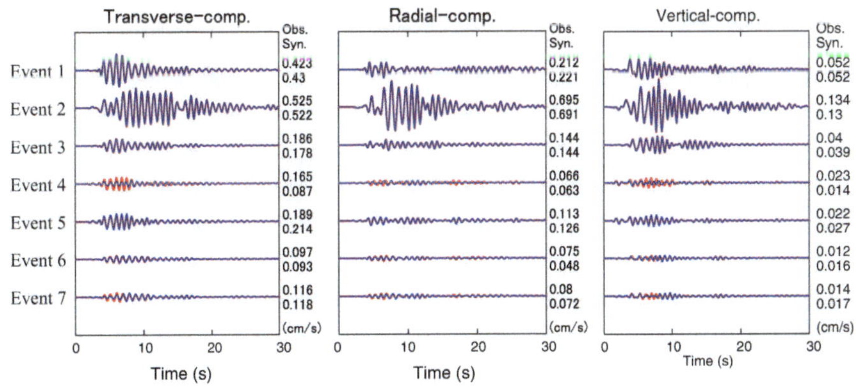

(c) FKI009

Fig. 4 Comparison of the 0.25–1 Hz bandpass-filtered observed velocity waveforms used in the EGTD estimation (in red) and the corresponding syntheses calculated from the EGTD (in blue). The absolute peak amplitude (in cm/s) is given at the end of each trace

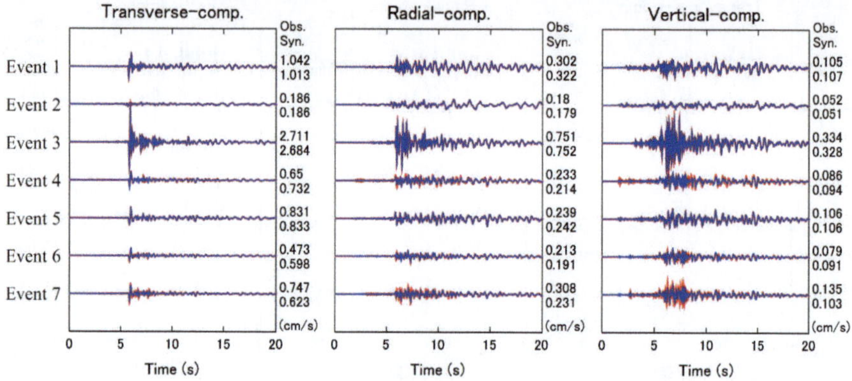

(a) FKI007

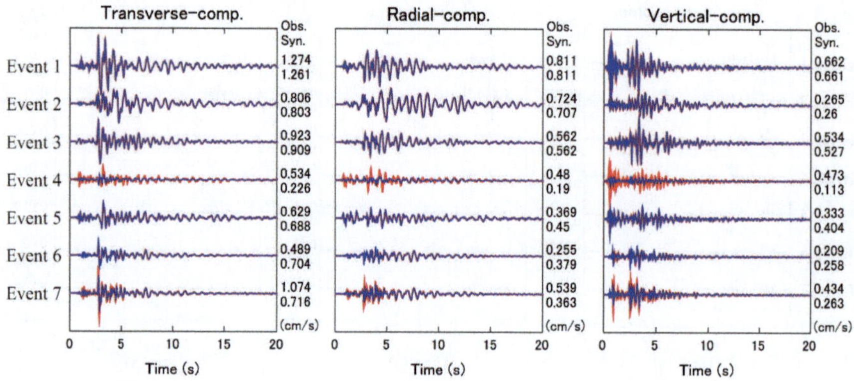

(b) FKIH04

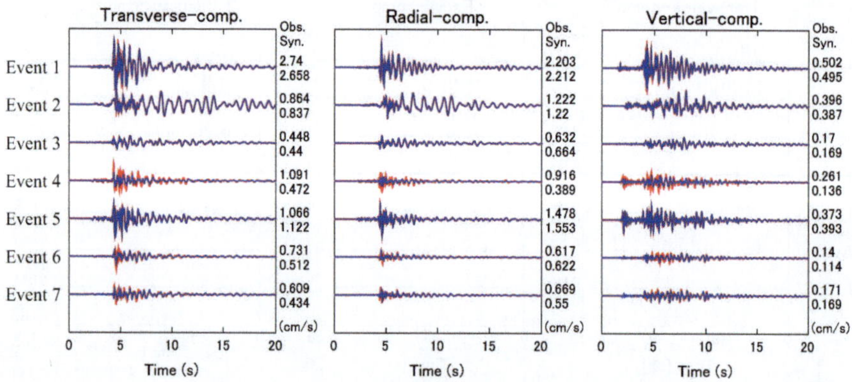

(c) FKI009

Fig. 5 Comparison of the 0.25–10 Hz bandpass-filtered observed velocity waveforms used in the EGTD estimation (in red) and the corresponding syntheses calculated from the EGTD (in blue). The absolute peak amplitude (in cm/s) is given at the end of each trace

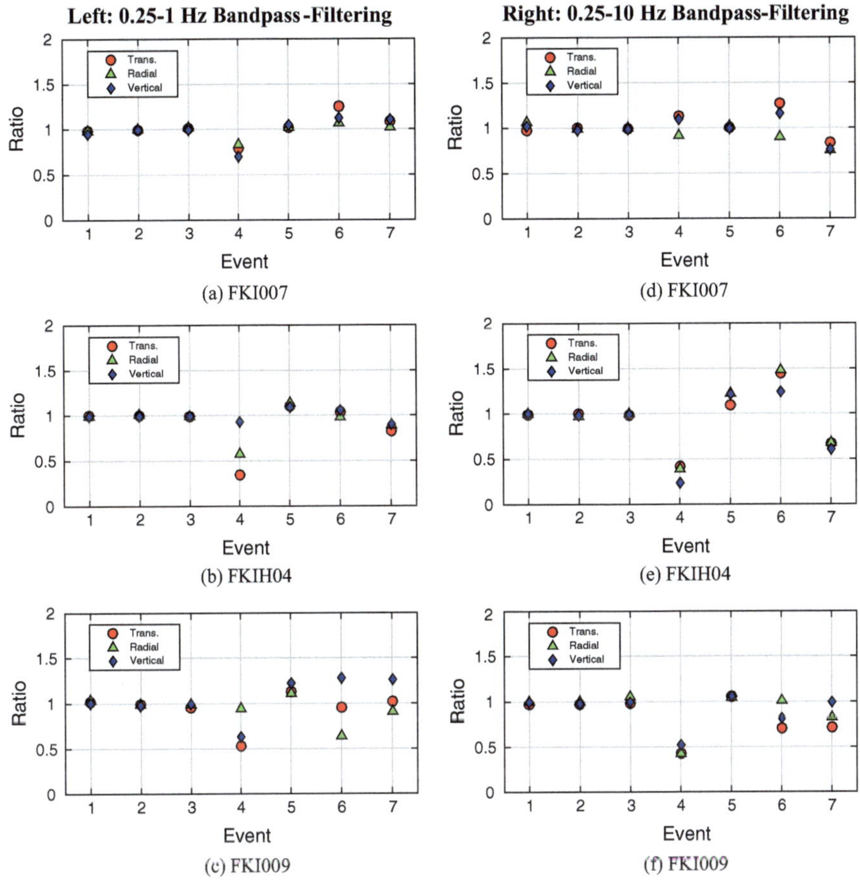

Fig. 6 Comparison of the maximum amplitude ratio between the synthesis and observatory data. Results from 0.25–1 Hz bandpass-filterd velocity waveforms are shown in (**a**, **b**, **c**) on a left panel, and those from 0.25–10 Hz are in (**d**, **e**, **f**) on a right panel

parameters (e.g. seismic moments, source mechanisms, source time functions, and corner frequencies) in a future study. The EGTD estimates in this report should be confirmed when future earthquakes occur around the same source area. I hope that this preliminary study in an area of relatively low seismicity is encouraging for the further investigation of EGTD estimation and evaluation of strong motion predictions using EGTD.

Acknowledgements The strong ground motion data used in this study were recorded at the K-NET and KiK-net stations provided by the National Research Institute for Earth Science and Disaster Prevention (NIED) on their Websites (http://www.kyoshin.bosai.go.jp/kyoshin/, last accessed April 2015). The Japan Meteorological Agency (JMA) unified hypocenter catalogue and the F-net source parameters were also provided by the NIED on their Website (http://www.fnet.bosai.go.jp/freesia/

top.php, last accessed April 2015). This study was partially supported by Grants-in-Aid for Scientific Research (C) (16K01316).

References

Aki K (1967) Scaling law of seismic spectrum. J Geophys Res 72(4):1217–1231

Aki K, Richards PG (1980) Quantitative seismology, theory and methods. W. H, Freeman, San Francisco, p 932

Ishibe T, Shimazaki K (2012) Characteristic earthquake model and seismicity around late quarternary active faults Japan. Bull Seism Soc Am 102:1041–1058

Ito Y (2005) A study on focal mechanisms of aftershocks. Report National Res Inst Earth Sci Disaster Prev 68:27–89 (in Japanese)

Ito Y, Okada T, Matsuzawa T, Umino N, Hasegawa A (2001) Estimation of stress tensor using aftershocks of 15 September 1998 M5.0 Sendai, NE Japan, earthquake. Bull Earthq Res Inst 76:51–59 (in Japanese)

Ohori M (2014) Simulation of broadband strong motion based on the empirical Green's spatial derivative method. In: Second conference on european earthquake engineering and seismology. Istanbul, Aug 25–29

Ohori M (2016) Estimation of empirical green's tensor spatial derivative elements: a preliminary study using strong ground motion records in southern Fukui prefecture, Japan. J Earth Sci 2:11–18

Ohori M, Hisada Y (2006) Estimation of empirical green's tensor spatial derivatives using aftershocks of the 2001 Hyogo-ken Hokubu earthquake and simulation of mainshock (M_J5.4) strong motion. Zisin (J Seismol Soc Japan) 59:133–146. (in Japanese with English abstract)

Ohori M, Hisada Y (2011) Comparison of the empirical Green's spatial derivative method empirical Green's function method. Bull Seism. Soc Am 101:2872–2886

Plicka V, Zahradnik J (1998) Inverting seismograms of weak events for empirical Green's tensor derivatives. Geophys J Int 132(3):471–478

Pulido N, Dalguer L, Fujiwara H (2006) Strong motion simulation on a dynamic fault rupture process and empirical Green's tensor derivatives. In: Fall meeting of the seismologic society of Japan, D018

Retrieval of the Seismic Moment Tensor from Joint Measurements of Translational and Rotational Ground Motions: Sparse Networks and Single Stations

Stefanie Donner, Heiner Igel, Céline Hadziioannou and the Romy group

Abstract Seismic moment tensors help us to increase our understanding about e.g. earthquake processes, tectonics, Earth or planetary structure. Based on ground motion measurements of seismic networks their determination is in general standard for all distance ranges, provided the velocity model of the target region is known well enough. For sparse networks in inaccessible terrain and planetary seismology, the waveform inversion for the moment tensor often fails. Rotational ground motions are on the verge of becoming routinely observable with the potential of providing additional constraints for seismic inverse problems. In this study, we test their benefit for the waveform inversion for seismic moment tensors under the condition of sparse networks. We compare the results of (1) inverting only traditional translational data with (2) inverting translational plus rotational data for the cases of only one, two, and three stations. Even for the single station case the inversion results can be improved when including rotational ground motions. However, from data of a single station only, the probability of determining the correct full seismic moment tensor is still low. When using data of two or three stations, the information gain due to rotational ground motions almost doubles. The probability of deriving the correct full moment tensor here is very high.

S. Donner (✉)
Bundesanstalt für Geowissenschaften und Rohstoffe (BGR), Hannover, Germany
e-mail: stefanie.donner@lmu.de; stefanie.donner@bgr.de
URL: http://romy-erc.eu

S. Donner · H. Igel · the Romy group
Ludwig-Maximilians-University Munich, Munich, Germany

C. Hadziioannou
University of Hamburg, Hamburg, Germany

S. D'Amico (ed.), *Moment Tensor Solutions*, Springer Natural Hazards,
https://doi.org/10.1007/978-3-319-77359-9_12

1 Introduction

1.1 Why Single Station/Sparse Network?

Regional and local seismic moment tensors are an important observation for various applications, such as spatio-temporal analysis of earthquake sequences, volcanic studies, monitoring of geothermal facilities, and tomographic studies. In order to avoid errors, the moment tensors must be determined precisely. However, this objective can be challenging, especially in the regional/local distance range, where waveforms are more complex than in the teleseismic distance range.

Usually, translational ground motion observations from a network of stations are used to determine the seismic moment tensor. The more data are available and the better the stations are distributed around the source in distance and azimuth (thus sampling the entire radiation pattern), the better is the resolution of the determined moment tensor. However, in inaccessible environments such as deserts or arctic regions it is not always possible to install a large network. Even worse, in planetary seismology often only single observation sites can be instrumented (e.g. InSight: Banerdt et al. 2013). The examination and understanding of the structure of other planets such as Mars and Moon are hampered because most seismological methods are based on station networks. Establishing entire networks on other planets is logistically and economically impossible, at least for now. Therefore, the question remains if and to what extent the full seismic moment tensor is resolvable from data of just one single station or a sparse network of maximum three stations.

1.2 Difficulties in the Single Station Approach

Theoretically, the three components of translational motion recorded at a single location should be sufficient to unambiguously determine the full seismic moment tensor in the teleseismic distance range (Ekstöm et al. 1986). Converted into an event-centric coordinate system (ZRT) the different wave types P, SV, and SH can be separated on the three components. Their amplitude ratios then provide the necessary information about the mechanism. However, in the regional distance range waveforms are more complex and more strongly influenced by noise. In addition, the inversion is performed on surface waves rather than on body waves. Especially for shallow sources this approach results in difficulties for the resolution of specific moment tensor components (e.g. Bukchin 2006; Bukchin et al. 2010). Therefore, the single station approach almost always fails in the regional distance range.

1.3 Rotational Ground Motions and Outline

Rotational ground motions are a relatively new observable, providing an additional three out of twelve components that completely describe the motion of a measurement point. So far, this observable has often been ignored, since compared to the traditional three components of translation it is much smaller (up to a factor 10^{-9}) and suitable measurement devices were not available. Only recently, sensitive portable broadband measurement devices based on fibre-optic techniques are available (Bernauer et al. 2018).

Rotation $\vec{\omega}$ is the curl of the wave field defined through a linear combination of the space derivatives of the translation vector $\vec{u}$:

$$\begin{pmatrix} \omega_x \\ \omega_y \\ \omega_z \end{pmatrix} = \frac{1}{2} \begin{pmatrix} \partial_x \\ \partial_y \\ \partial_z \end{pmatrix} \times \begin{pmatrix} u_x \\ u_y \\ u_z \end{pmatrix} = \frac{1}{2} \begin{pmatrix} \partial_y u_z - \partial_z u_y \\ \partial_z u_x - \partial_x u_z \\ \partial_x u_y - \partial_y u_x \end{pmatrix} \tag{1}$$

where $\times$ denotes a vector product. Measured at the Earth's surface, the horizontal components of rotation are commonly known as tilt. Rotational ground motions provide us with information about the vertical displacement gradient which is not available from conventional seismic stations on the Earth's surface.

In an earlier study, we could numerically show that data of a network based on six component measurement devices (3 translations + 3 rotations) considerably improves the resolution of the full moment tensor when performing a Bayesian (i.e. probabilistic) waveform inversion while keeping the total number of components constant (Donner et al. 2016). Motivated by the promising results of this earlier study, here, we evaluate the benefits of rotational ground motions for single or sparse station configurations.

In addition to the Bayesian inversion using the set-up of the former study, we also show the analytical solution for a double-couple moment tensor in a homogeneous space for translational and rotational ground motion. We will use the analytical solution to discuss under which geometries the reliable retrieval of the seismic moment tensor has the highest probability.

2 The Test Case Scenario and Applied Method

In an earlier study, we designed a synthetic earthquake scenario in Iran to analyse the performance of an entire network for waveform inversion to retrieve seismic moment tensors (Donner et al. 2016). The motivation was to examine the benefits of rotational ground motion measurements for the moment tensor retrieval. For this purpose, we have calculated Greens functions for two scenarios: one based on the real station distribution of the Iranian broadband network and one based on a fictional grid of stations covering the whole country. In this study, we continue with the fictive grid of stations and evaluate the cases when only one, two, or three stations are available. In the following, we briefly review the set-up of the experiment and the Bayesian, i.e. probabilistic, inversion approach.

2.1 Set-Up of the Synthetic Test Case

In Fig. 1 the blue star marks the location of the synthetic earthquake in north-central Iran. It is an almost vertical strike-slip event with a magnitude M_w4.0 at a depth of 6 km. Its double-couple (DC) and isotropic (ISO) parts have 80% and 15%, respectively. The full moment tensor is given by

$$M_{kj} = \begin{pmatrix} -0.75 & -0.16 & 0.41 \\ & 1.17 & 0.26 \\ & & 0.02 \end{pmatrix} \quad (2)$$

The theoretically observed waveforms as well as the Greens functions are calculated using a spectral-element solver based on a 1-D structural model for the Alborz mountains (Fichtner et al. 2009; Donner et al. 2013). Gaussian noise was added to the synthetic seismograms to render them more realistic. For this study, since only the relative noise levels are relevant, we applied a noise of 10% of the maximum amplitude and evaluate translation and rotation data separately.

The fictional stations are distributed equally covering almost the whole surface of Iran (Fig. 1). There is a grid of 21 by 31 stations along latitude and longitude, respectively. The Greens functions are calculated for depths between 2 and 20 km in 2 km steps for frequencies up to 0.1 Hz to include surface waves, which are usually used for inversion in the regional distance range.

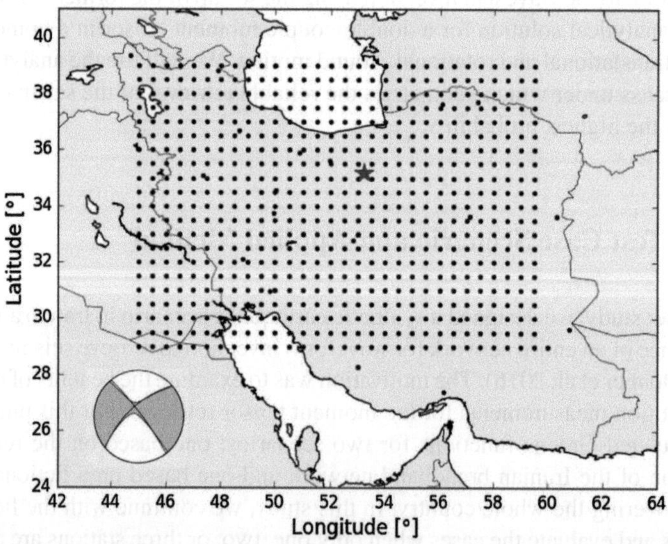

Fig. 1 Study area in Iran. The blue star marks the position of the synthetic M_w4.0 earthquake. The mechanism is shown in the lower left with black lines and grey shaded areas being the nodal planes of the double-couple part and the full moment tensor solution, respectively. Black dots show the grid of theoretical stations

2.2 *Probabilistic Waveform Inversion for Seismic Moment Tensors*

Probabilistic, i.e. Bayesian, inversion has the advantage that it results in a solution class instead of one single solution. A selection of "the best" solution according to a more or less subjective criterion is omitted. Thus, Bayesian inversion provides an unbiased measure of resolution and possible trade-offs.

As in our former studies, we use the Bayesian inversion approach based on equations from Tarantola (2005):

$$\sigma(\mathbf{m}) = k\rho(\mathbf{m})L(\mathbf{m}) \tag{3}$$

where $\sigma(\mathbf{m})$ and $\rho(\mathbf{m})$ are the *posterior* and *prior probability density functions* (pdf) of the model parameters, respectively, k is a normalization constant, and $L(\mathbf{m})$ is the likelihood function providing a measure of how well a model $\mathbf{m}$ fits the data.

The inversion is performed for seven parameters, i.e. the six moment tensor components and centroid depth, in the time domain according the forward problem for the seismic point-source moment tensor (Aki and Richards 2002):

$$u_n(\mathbf{x}, t) = M_{kj} \cdot G_{nk,j}(\mathbf{x}, t - \tau) \tag{4}$$

with u_n is the ground motion at observation point $\mathbf{x}$ and time t, M_{kj} the components of the seismic moment tensor, and $G_{nk,j}$ the spatial derivatives of the Green's function components. τ is the origin time of the source. Einstein summation convention was applied to facilitate readability of the equation.

The *posterior pdf* is approximated from the homogeneous *prior pdf* by testing one million start models by applying the Metropolis-Hastings algorithm (e.g. Metropolis et al. 1953; Hastings 1970). Intuitively spoken, this algorithm performs a random walk through the model space. Based on Eq. 4 waveforms for the model candidate are calculated and compared to the theoretical observations. Depending on whether or not the model candidate improves the data fit, it will be added to the solution class or not.

For an objective estimate on the information content of the *posterior* relative to the *prior pdf* we use Shannon's measure of information gain:

$$I(\rho, \sigma) = \int \rho(x) \log\left[\frac{\rho(x)}{\sigma(x)}\right] dx. \tag{5}$$

Due to the logarithm base 2 in Eq. 5, the unit of the information gain is termed a *bit*. For further details on the set-up, on the method and Shannon's measure of information gain, we refer to our earlier studies (Bernauer et al. 2014; Donner et al. 2016; Reinwald et al. 2016) and to (Tarantola 2005).

3 Results

In the following, we show the results of the inversion of the theoretically observed waveforms for the test case shown in Fig. 1 and described in Sect. 2. We tested three scenarios with only one, two and three stations. For each scenario, we performed 670 inversions where we first inverted only translational ground motions and then repeated the 670 inversions inverting translational and rotational ground motions. For each inversion, the station(s) are randomly selected from the grid of possible stations. For each scenario, we show the generalised result over all 670 inversions along with the result of the inversion with the best outcome.

3.1 Single Station

Figure 2 shows the generalised result of the 670 inversions based on data of only one station. Each sub-plot represents one of the seven parameters inverted for. The x-axes show the value of Shannon's measure of information gain, while the y-axes show the Gaussian kernel density estimations (kde), i.e. the marginal probability density function over the information gain. The higher the information gain, the better is the resolution of the parameters inverted for. Coral and green functions give the results for inversion of translational data only and translational with rotational data, respectively. The same colour-coding is used for all following figures.

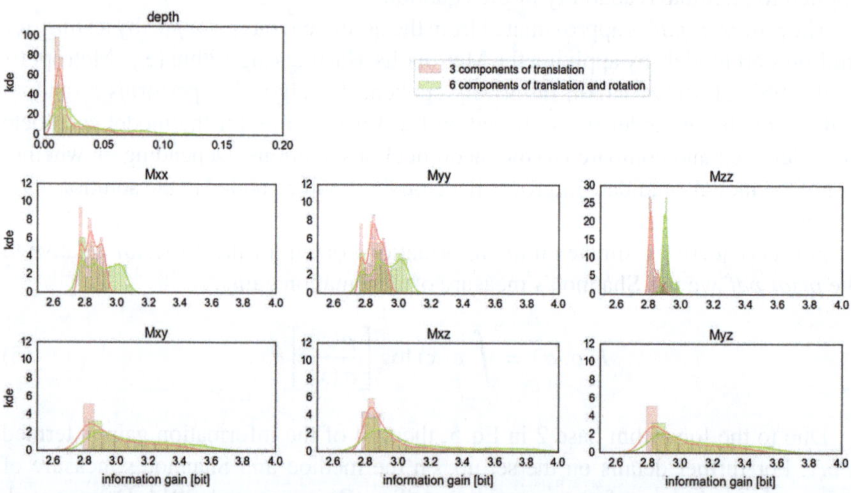

Fig. 2 Gaussian kernel density estimations (kde) for the information gain corresponding to 670 inversions for the **single station case**. Coral and green distributions show the results for inverting translational ground motions only and translational with rotational ground motions, respectively. The higher the information gain, the better is the resolution of the parameters

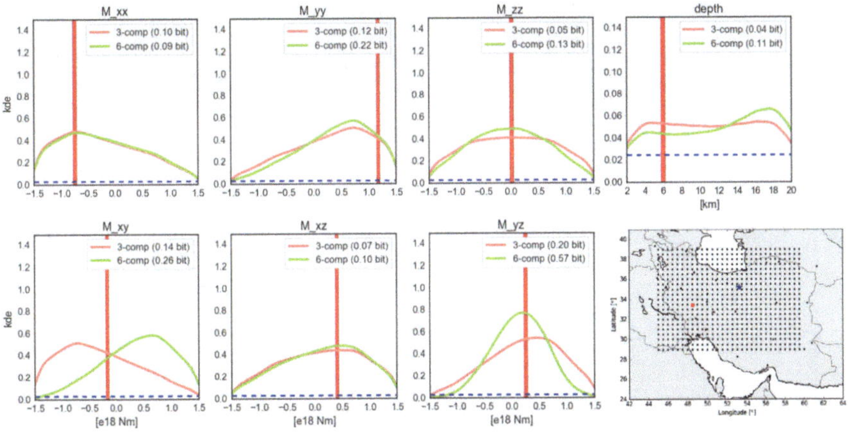

Fig. 3 Inversion result for the **single station case** for the inversion with highest information gain (0.72/1.48 bit for 3/6 components). The station used for this inversion is marked by a red dot on the map. The red bars show the real value of the target solution. Blue dotted lines are the homogeneous *prior pdf*. Coral and green curves show the *posterior pdf* for inversion of translations only and translations plus rotations, respectively. Numbers in the upper right corner give increase of information content relative to the prior according to Shannon's measure

No matter which of the stations within the grid is used, four out of the seven parameters almost never benefit from the inclusion of rotational ground motions. These are the centroid depth and the moment tensor components M_{xy}, M_{xz}, and M_{yz}. In contrast, the other three components M_{xx}, M_{yy}, and M_{zz} can benefit remarkably. Especially the component M_{zz} almost always benefits from the inclusion of rotational ground motions, no matter which station is used for the inversion.

The question remains whether the real moment tensor solution was determined. Figure 3 shows the result of only one of the 670 inversions, the one with the highest resulting information gain. For each of the parameters inverted for the *posterior pdf* kde's are plotted over the parameter space. In the upper right corners the information gains are given. The vertical red bars show the real values of the target solution. Most of the parameters benefit only slightly from the rotational ground motions, except for the moment tensor component M_{yz} which shows a substantial improvement. For the components M_{yy}, M_{xy}, and the centroid depth the highest probability is not at the real value of the target model. When considering the rotational ground motions, the highest probability for centroid depth is different from the true value. Despite this, the overall benefit of the rotational ground motion for inversion based on a single station, i.e. the increase in information gain, is 105% (Fig. 3, $(1.48 - 0.72) * 100/0.72$).

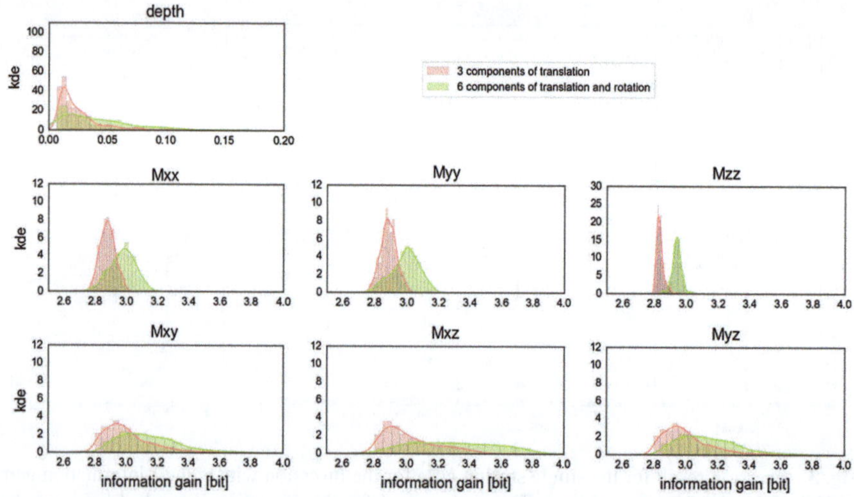

Fig. 4 Generalized inversion results for the **two station case**. Colour-coding is the same as in Fig. 2

3.2 *Two Stations*

Figure 4 shows the generalised result of the 670 inversions based on data from two stations. Now, compared to the single station case, the green kde functions (6-component data) of the components M_{xy}, M_{xz}, and M_{yz} are slightly shifted to higher values for information gain compared to the coral kde functions. These shifts indicate that these components have a higher probability to be determined reliably when including rotational ground motions to the inversion based on two stations. In addition, the probability to estimate the correct centroid depth is increased. For the remaining three moment tensor components the benefits due to rotations are also more pronounced.

The result of the inversion with the highest resulting information gain (0.98/2.73 bit for 3/6 components) based on data from two stations is shown in Fig. 5. Almost all moment tensor components show a much higher probability to be determined correctly when including rotational ground motions, except the component M_{xy}. By far the biggest improvement can be seen for the moment tensor component M_{xz}. The most critical parameter is still the centroid depth, which again shows a low probability to be resolved. The overall increase in information gain is 178%, almost 75% more than for the single station case.

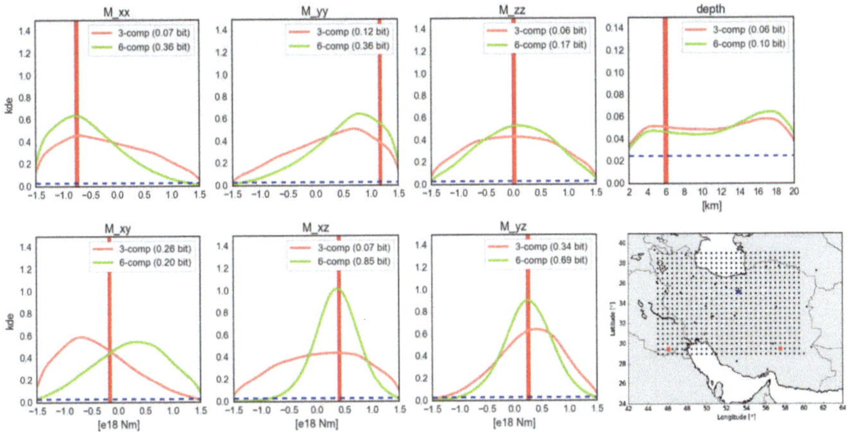

Fig. 5 Inversion result for the **two station case** for the inversion with highest information gain (0.98/2.73 bit for 3/6 components). The stations used for this inversion are marked by red dots on the map. Colour-coding is the same as in Fig. 3

3.3 Three Stations

Figure 6 shows the generalised result of the 670 inversions based on data from three stations. As expected, the shift to values of higher information gain of the green kde's (6-component data) compared to the coral kde's (3-component data) is much more

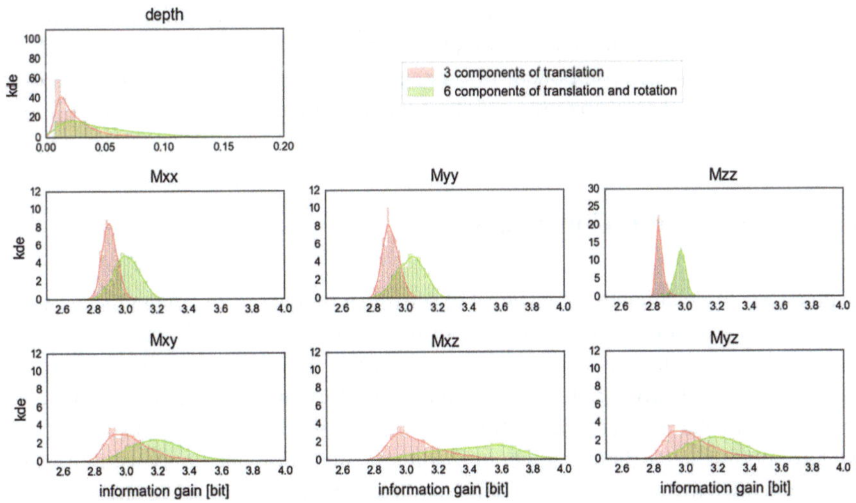

Fig. 6 Generalized inversion results for the **three station case**. Colour-coding is the same as in Figs. 2 and 4

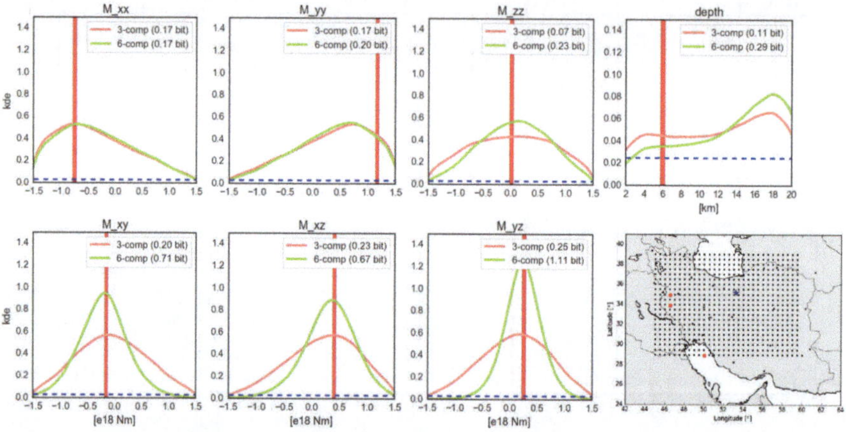

Fig. 7 Inversion result for the **three station case** for the inversion with highest information gain (1.20/3.38 bit for 3/6 components). The stations used for this inversion are marked by red dots on the map. Colour-coding is the same as in Figs. 3 and 5

distinct for all moment tensor components compared to the two station case. The overlap between coral and green curves is greatly reduced, implying that for almost all station distributions, the inclusion of the rotational ground motion data results in a much higher probability of producing in the correct solution.

The result of only one inversion showing the best result (i.e. highest information gain) based on three stations is shown in Fig. 7. Now, there is even a significant probability to retrieve the correct moment tensor from translational data alone (coral functions). However, the probability is 181% higher when including the rotational ground motions. Especially the components including depth dependencies (M_{xz}, M_{yz}, and M_{zz}) and the component M_{xy} benefit most from including rotational data into the inversion. The probability of retrieving the correct centroid depth is still low.

4 Discussion and Conclusion

In this study, we evaluated the retrieval of seismic moment tensors under the assumption of having data from only one, two, or three stations with both translational and rotational ground motions. The study is based on a theoretical scenario in Iran already used in an earlier study, where we evaluated the performance of a network under consideration of rotational ground motions (Donner et al. 2016). For each station case, we have performed a Bayesian, i.e. probabilistic, inversion once only using translational and once using translational plus rotational ground motions. We have repeated these inversions 670 times, always randomly selecting station(s) out of a grid of theoretical station locations (Fig. 1). Next to the generalized results based on these

670 inversions (Figs. 2, 4 and 6), we also showed the results of the inversion with the highest information gain (Figs. 3, 5 and 7).

The results show that it might be difficult to fully retrieve the moment tensor from one station alone. Figure 3 shows that even when including the rotations, the probability to retrieve the true values for the inversion parameters is not very high. Nevertheless, the increase of information gain due to the rotations is still 105%. Here, we tested the inversion for the full moment tensor. There might be a chance to retrieve at least the mechanism of an earthquake from data of a single station when applying constraints to the inversion. Finally, when including data of a second and third station the probability to retrieve the correct full moment tensor is very high. The overall increase of information gain is 178% for two and 181% for three 6-component stations. The generalisations of the two and three station case in the Figs. 4 and 6 show that this is the case for a wide range of station distributions. Especially for planetary seismology, this is an important result. It might be logistically and economically possible to bring two or three instruments to another planet such as Mars or Moon, even though the locations for these instruments would be determined according to practical rather than scientific criteria.

Centroid depth is the only parameter which in all station cases has a very low probability of being resolved. This effect is a general problem in waveform inversion for moment tensors. Especially, in the regional distance range, where inversion is done on surface waves, and for shallow earthquakes depth is a very critical parameter (e.g. Bukchin 2006; Bukchin et al. 2010). Our test case scenario includes exactly these difficulties. Therefore, the unsatisfying results for centroid depths are to be expected.

The question remains what the cause for the benefit from rotational data is. As can be seen from Eq. 1, rotational ground motions are linear combinations of the space derivatives of the translational ground motions. Measured at the Earth surface, their horizontal components correspond to tilt. Therefore, rotational ground motions include information about the velocity gradient with depth which is not available from translational ground motions. However, the comparison is not completely fair. When including the rotational ground motions, we have doubled the number of data (3 versus 6 traces for the single station case). Thus, the improved inversion results may be partly due to higher amount of information. However, in our former study based on an entire network, we kept the number of data in both cases the same (by cutting the number of stations in half when including rotational ground motions) and still received greatly improved inversion results. Therefore we argue that the increased amount of data is responsible only for a small portion of the improvement, if at all.

Already in 1986, when waveform inversion for moment tensors was newly established as a standard in the teleseismic distance range, Ekstöm et al. successfully inverted waveforms of the 1985 M 8.1 Michoacan, Mexico, earthquake from only three components of translational data for the deviatoric moment tensor (i.e. isotropical part is constrained; $ISO = \mathrm{tr}\, M_{ij} = 0$). They emphasised the necessity of a very good signal-to-noise ratio (SNR), a recording system with a broad frequency response and large dynamic range (not standard at that time), a well known velocity model, and exclusion of antipodal event-station geometry. Nowadays, three out of these four criteria are fulfilled in the teleseismic distance range and strong enough earthquakes result in a good SNR. Due to the well developed global seismic network it is now also possible to find an appropriate station, which meets the geometry criterion. However, in the regional distance range this approach is hampered due to more complex waveforms, less well known velocity models, and a limited number of stations. Jiménez et al. (1989) have extended the approach of Ekstöm et al. (1986) to regional and moderate-sized (i.e. $M_S \approx 4.5 - 5.5$) earthquakes. They further limited the inversion to the double-couple part of the moment tensor (i.e. tectonic part; $DC = \det M_{ij}^{dev} = 0$) and emphasized the need for path corrections using the dispersion properties of Love and Rayleigh waves. Fan and Wallace (1991) used a similar approach and found a trade-off between structure and depth in that a reliable mechanism might come with a "wrong" depth due to inaccuracies in the assumed structural model. Additionally, they stated that the noise level should be 10% as a maximum to retrieve a reliable mechanism. The results of these studies based on only three components of translation combined with the results of the Bayesian inversion in this study are promising. We are confident that having six components available there are chances to at least determine the deviatoric moment tensor reliably from a single station measurement, if not the full moment tensor.

The velocity model needed to calculate the Greens functions used for inversion is the more crucial point in the entire process. In the regional distance range, a larger proportion of the seismic energy travels through the crust than in the teleseismic distance range (where the more homogeneous deeper structure of the Earth is more important). Thus, the structural details of the crust have a stronger influence on the waveforms and therefore cause the waveform complexities. In planetary seismology, where the structures are less well known than on Earth, the same difficulties will be faced. Sampling an additional part of the seismic wave field by including rotational ground motions into the inversion can compensate for these issues, at least partly.

Next to the velocity model, the event-station geometry is of great importance when restricted to only a few stations. In Fig. 8 we show the radiation pattern of the theoretical earthquake for the six components of translation and rotation. Each sub-plot shows the maximum energy for this specific earthquake mechanism and for each of the theoretical stations of the grid for the component mentioned above the sub-plot. Blue, green, and turquoise dots show the station(s) used for the best inversions for the single, two, and three station case shown in Figs. 3, 5, and 7. The energy is calculated as the square root of the sum of squared amplitudes divided by the maximum amplitude of the trace: $E = \sqrt{\sum tr_i^2 / A_{max}(tr)}$.

When comparing the locations of the stations used within the radiation pattern of the different components, it is clearly visible that they lead to significant amplitude ratios between the different components. Inverting only waveforms of the three translational components results in only three amplitude ratios. The additional three components of rotation provide further three amplitude ratios of rotation plus further nine amplitude ratios of translation versus rotation. Therefore, the position within the radiation pattern at which the station(s) are located is better constrained. Not the single traces themselves but rather their amplitude ratios contribute to the

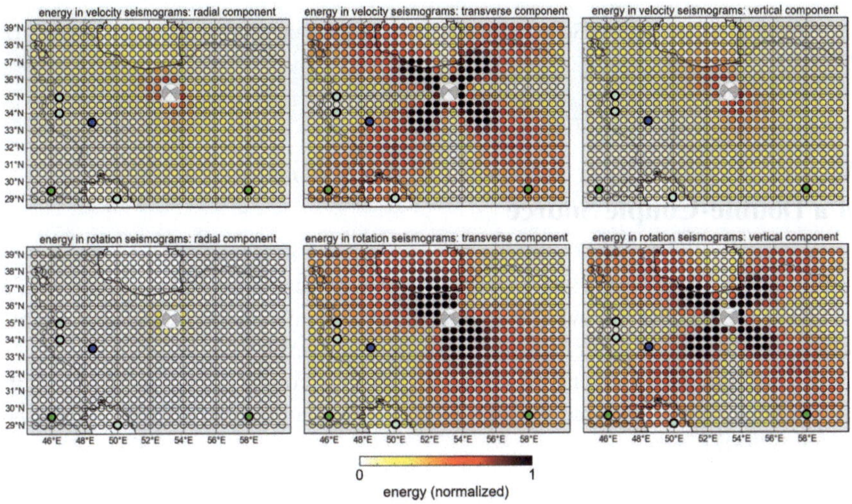

Fig. 8 Radiation pattern for the synthetic earthquake scenario in Iran (Fig. 1). The plots are in the event-centric cartesian coordinate system (ZRT). Top row shows radial, transverse, and vertical component of the velocity energy, respectively, while the bottom row shows the same for rotation. The mechanism of the theoretical earthquake is shown at its location. Blue, green, and turquoise dots show the station(s) used for the inversions based on one, two, and three stations, respectively; inversion results are shown in Figs. 3, 5 and 7

resolution of the moment tensor solution. Therefore, we think that it might be possible to at least derive the pure deviatoric mechanism from only one station, i.e. adding constraints to the waveform inversion. This argumentation is supported by the evaluation of the analytical solution for the radiation pattern of a double-couple source in a homogeneous medium in Appendix A.

In summary, there are two main crucial points when inverting waveforms for moment tensors from sparse networks: a limited knowledge of the underlying velocity model and the question where in the radiation pattern the station(s) is/are located. Including rotational ground motion data can at least partly compensate for these limitations. First, by sampling another property of the wave field, and second, by adding more input data, in the form of more components or local wavefield gradients. We see potential to greatly improve the performance of existing seismic networks with respect to waveform inversion for moment tensors by adding just one single rotational sensor to the network. In regions where the same mechanism occurs regularly, it is possible to find the most advantageous station position with respect to the radiation pattern.

Acknowledgements Parts of this study have been done using ObsPy (ObsPy Development Team 2015; Beyreuther et al. 2010; Megies et al 2011; Krischer et al. 2015) and we are very thankful to the developers. The research presented in this article was funded by the European Research Council (Advanced grant: ROMY, number: 339991). The numerical computations were performed on the National Supercomputer SuperMUC maintained by the Leibniz-Rechenzentrum (Project ID: pr63qo).

Appendix: Analytical Solution for the Radiation Pattern of a Double-Couple Source

To further analyse the event station geometry with respect to the radiation pattern of a seismic source, in the following, we show the analytical solution of a double-couple source, i.e. the tectonic part of the seismic moment tensor.

The far-field displacement $u(\vec{x}, t)$ due to a double-couple point-source in an infinite, homogeneous, isotropic medium in a spherical coordinate system is (Aki and Richards 2002):

$$\mathbf{u}(\mathbf{x}, t) = \frac{1}{4\pi\rho\alpha^3}\mathbf{A}^{FP}\frac{1}{r}\dot{M}_0\left(t - \frac{r}{\alpha}\right) + \frac{1}{4\pi\rho\beta^3}\mathbf{A}^{FS}\frac{1}{r}\dot{M}_0\left(t - \frac{r}{\beta}\right) \tag{6}$$

where α, β, and ρ are the velocities for P- and S-wave and the density of the medium, respectively. r is the source-receiver distance and $M_0(t)$ the seismic moment function. A^{FP} and A^{FS} are the far-field radiation patters for the P- and S-wave in spherical coordinates:

$$\begin{aligned} \boldsymbol{A}^{FP} &= \sin 2\theta \cos\phi \hat{\boldsymbol{r}} \\ \boldsymbol{A}^{FS} &= \cos 2\theta \cos\phi \hat{\boldsymbol{\theta}} - \cos\theta \sin\phi \hat{\boldsymbol{\phi}} \end{aligned} \tag{7}$$

Similarly, the far-field rotation can be described as

$$\boldsymbol{\omega}(\mathbf{x},t) = \frac{-1}{8\pi\rho\beta^4}\boldsymbol{A}^{R}\frac{1}{r}\ddot{M}_0\left(t - \frac{r}{\beta}\right) \tag{8}$$

with

$$\boldsymbol{A}^{R} = \cos\theta \sin\phi \hat{\boldsymbol{\theta}} + \cos 2\theta \cos\phi \hat{\boldsymbol{\phi}} \tag{9}$$

as the rotational radiation pattern (Cochard et al. 2006). Figure 9 shows the (a) translational and (b) rotational radiation patterns for a source in the XY-plane. Blue, red, and pink are the radiation patterns for the P-wave, S-wave, and rotation, respectively. The black, red, and blue circles mark positions of equally spaced stations in the XY-, XZ-, and YZ-plane around the source, respectively. It is easy to see that the biggest amplitude variations with azimuth are to be expect for stations along the black ring in the XY-plane, i.e. in plane with the source.

The Figs. 10 and 11 show the theoretical waveforms at the stations around the source shown in Fig. 9 in cartesian (ZXY) and spherical (r θ ϕ) coordinate systems,

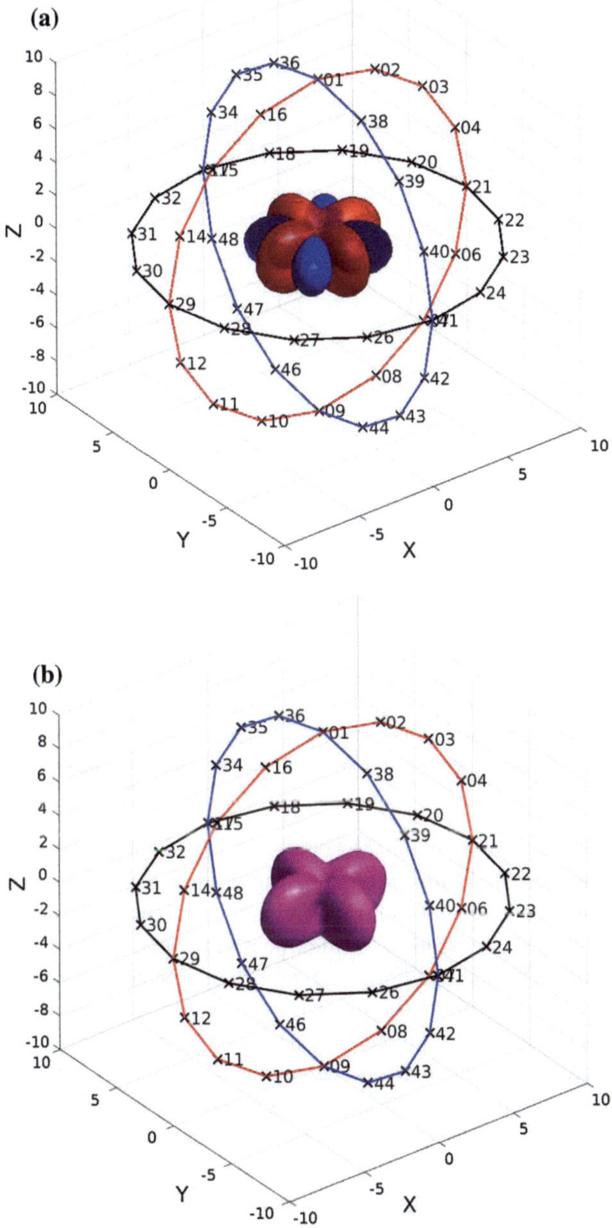

Fig. 9 Radiation patterns for slip in the XY-plane with slip in X direction in a homogeneous half-space. **a**, **b** are the radiation patterns for translational and rotational ground motions, respectively. In **a**, the blue and red radiation patterns are for P- and S-waves, respectively. Black, red, and blue circles mark positions of equally spaced stations in the XY-, XZ-, and YZ-plane, respectively. The colours and numbers correspond to the waveform colours and trace numbers in Figs. 10 and 11

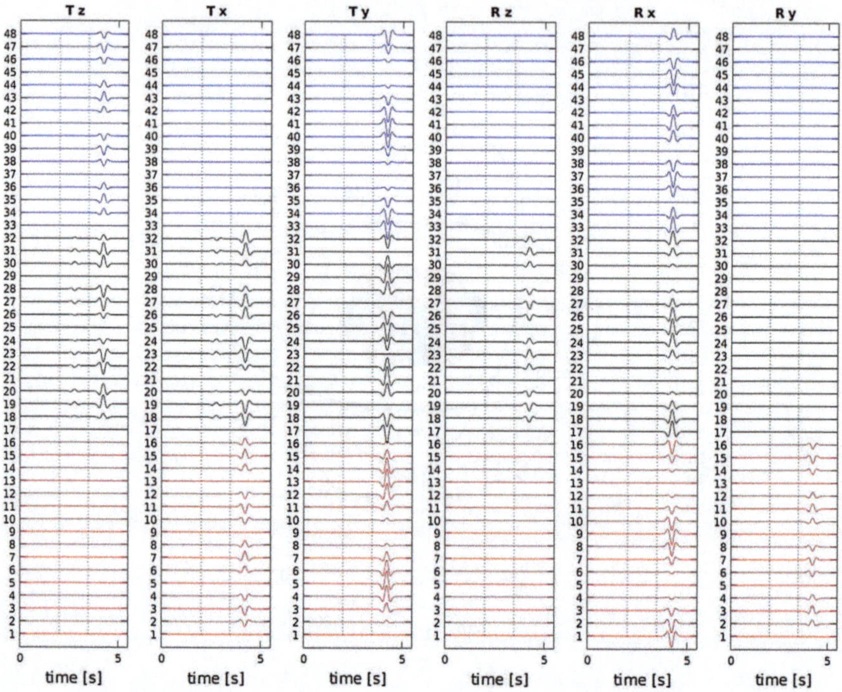

Fig. 10 Waveforms for the double-couple source of Fig. 9 in the ZXY coordinate system. Vertical numbers and waveform colours correspond to the three station rings around the source in Fig. 9. First three columns show translational displacement, while last three columns show rotational ground motions. Black dotted lines mark the time of the P-and S-wave arrivals. Waveforms are normalized preserving the ratios between the components. Rotational waveforms are additionally amplified by a factor 1.5×10^4

respectively. The colour-coding and numbering of the waveforms correspond to the station rings. As expected, the black waveforms for stations in-plane with the source show a clear signal on the majority of the traces compared to the waveforms of the other ring stations. The stations 17, 21, 25, and 29 are located on the nodal planes of the mechanism. In Fig. 11 there is only energy on their θ component, corresponding to the SH-wave energy of the source. The stations 19, 23, 27, and 31, rotated by 45° to the nodal planes, show the opposite behaviour with energy on the radial (P-wave energy) and the ϕ component (SV-wave energy) but no energy on the θ component.

Similar patterns can be found for the rotational waveforms. In the spherical coordinate system, the stations 21 and 29 have no rotation energy at all on all three components (nodal planes along rupture direction), while they show a maximum of rotation energy for the ϕ component at the stations 17 and 25 (nodal planes perpendicular to rupture direction). Slightly different, but still azimuth dependant energy patters are visible for the other two station rings.

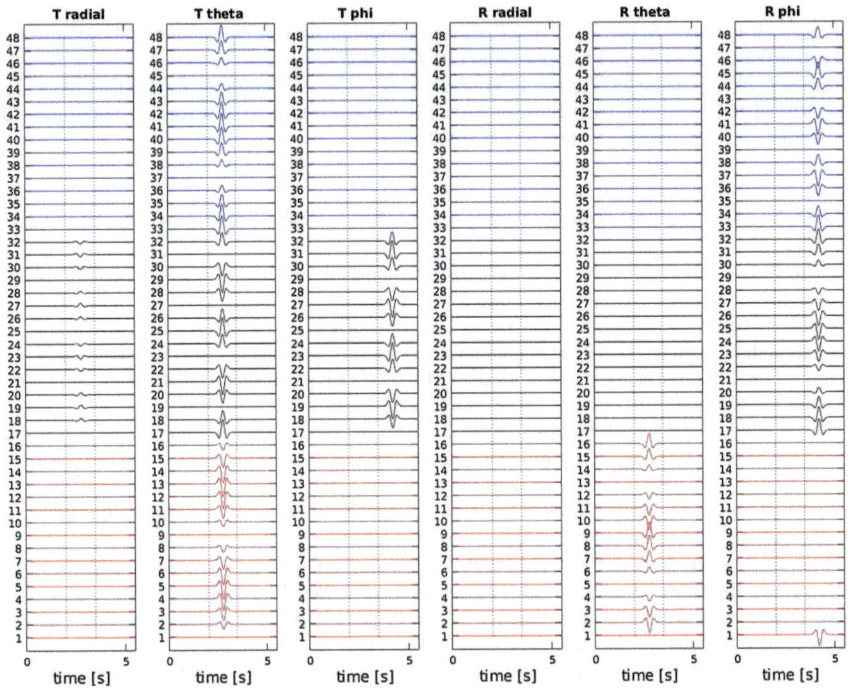

Fig. 11 Same as Fig. 10, the waveforms are now shown in the spherical coordinate system (r, θ, and ϕ) as used by e.g. Aki and Richards (2002)

These amplitude patterns in the form of ratios between the different components are the information needed to resolve the mechanism of an earthquake during inversion. When including rotational ground motion data to the inversion, 15 instead of only 3 amplitude ratios can be determined. Therefore, the position of the station within the radiation pattern can be much better constrained. As a consequence, the mechanism can be determined much more reliably, at least in cases of tectonic events where the double-couple part of the moment tensor should dominate.

References

Aki K, Richards PG (2002) Quantitative seismology, 2nd edn. University Science Book, Sausalito, CA

Banerdt WB, Smrekar S, Lognonné P, Spohn T, Asmar SW, Banfield D, Boschi L, Christensen U, Dehant V, Folkner W, Giardini D, Goetze W, Golombek M, Grott M, Hudson T, Johnson C, Kargl G, Kobayashi N, Maki J, Mimoun D, Mocquet A, Morgan P, Panning M, Pike WT, Tromp J, van Zoest T, Weber R, Wieczorek MA, Garcia R, Hurst K (2013) InSight: a discovery mission to explore the interior of Mars. In: Lunar and planetary science conference, vol 44 of Lunar and Planetary Inst. Technical Report

Bernauer M, Fichtner A, Igel H (2014) Reducing nonuniqueness in finite source inversion using rotational ground motions. J Geophys Res 119:4860–4875

Bernauer F, Wassermann J, Guattari F, Frenois A, Bigueur A, Gaillot A, de Toldi E, Ponceau D, Schreiber U, Igel H (2018) BlueSeis3A: full characterization of a 3C broadband rotational seismometer, SRL, https://doi.org/10.1785/0220170143

Beyreuther M, Barsch R, Krischer L, Megies T, Behr Y, Wassermann J (2010) ObsPy: a python toolbox for seismology. Seismol Res Lett 81:530–533

Bukchin B (2006) Specific features of surface wave radiation by shallow sources. Phys Solid Earth 42:712–717

Bukchin B, Clévédé E, Mostinskiy A (2010) Uncertainty of moment tensor determination from surface wave analysis for shallow earthquakes. J Seismol 14:601–614

Cochard A, Igel H, Schuberth B, Suryanto W, Velikoseltsev A, Schreiber U, Wassermann J, Scherbaum F, Vollmer D (2006) Rotational motions in seismology: theory, observation, simulation. In: Teisseyre R, Majewski E, Takeo M (eds) Earthquake source asymmetry. Structural Media and Rotation Effects, Springer, Berlin Heidelberg

Donner S, Rößler D, Krüger F, Ghods A, Strecker MR (2013) Segmented seismicity of the Mw 6.2 Baladeh earthquake sequence (Alborz mountains, Iran) revealed from regional moment tensors. J Seismol 17:925–959

Donner S, Bernauer M, Igel H (2016) Inversion for seismic moment tensors combining translational and rotational ground motions. Geophys J Int 207:562–570

Ekström G, Dziewonski AM, Steim JM (1986) Single station CMT: application to the Michoacan, Mexico, earthquake of September 19, 1985. Geophys Res Lett 13:173–176

Fan G, Wallace T (1991) The determination of source parameters for smaller earthquakes from a single, very broadband seismic station. Geophys Res Lett 18:1385–1388

Fichtner A, Kennett BLN, Igel H, Bunge H-P (2009) Spectral-element simulation and inversion of seismic waves in a spherical section of the Earth. J Numer Anal Ind Appl Math 4:11–22

Hastings WK (1970) Monte Carlo sampling methods using Markov chains and their applications. Biometrika 57:97–109

Jiménez E, Cara M, Rouland D (1989) Focal mechanisms of moderate-size earthquakes from teh analysis of single-station three-component surface wave records. Bull Seismol Soc Am 79:955–972

Krischer L, Megies T, Barsch R, Beyreuther M, Lecocq T, Caudron C, Wassermann J (2015) ObsPy: a bridge for seismology into the scientific Python ecosystem. Comput Sci Discov 8:014003

Megies T, Beyreuther M, Barsch R, Krischer L, Wassermann J (2011) ObsPy—What can it do for data centers and observatories. Ann Geophys 54:47–58

Metropolis N, Rosenbluth MN, Teller AH, Teller E (1953) Equation of state calculations by fast computing machines. J Chem Phys 21:1087–1092

ObsPy Development Team (2015). ObsPy 0.10.1. https://doi.org/10.5281/zenodo.16248

Reinwald M, Bernauer M, Igel H, Donner S (2016) Improved finite-source inversion through joint measurements of rotational and translational ground motions: a numerical study. Solid Earth 7:1467–1477

Tarantola A (2005) Inverse problem theory and methods for model parameter estimation, 2nd edn. Society for Industrial and Applied Mathematics, Philadelphia, Penn

Overview of Moment Tensor Analysis in New Zealand

John Ristau

1 Introduction

The determination of earthquake source parameters is of fundamental importance in seismological research. Moment tensor analysis involves fitting theoretical waveforms to observed broadband waveforms and inverting for the moment tensor elements, and allows for the calculation of focal mechanism (strike, dip, and rake), seismic moment (M_0), moment magnitude (M_w) which is calculated directly from M_0, and centroid depth of an earthquake. A comprehensive catalogue of moment tensor solutions is of great importance in seismic hazard analysis and tectonic studies. For example, seismic hazard estimates typically use M_w in earthquake forecasts and risk analysis, and moment release rates along plate boundaries are important in calculating predicted plate motions in tectonic studies. Routine regional moment tensor (RMT) analysis was implemented in New Zealand in early-2007 (Ristau 2008), and since that time more than 2000 RMT solutions have been calculated for New Zealand earthquakes with the catalogue beginning August 2003. This article will provide an overview of RMT analysis in New Zealand with specific focus on five earthquake sequences which significantly impacted New Zealand since July 2009.

2 Tectonic Setting of New Zealand

New Zealand straddles the boundary of the Pacific and Australian plates, and the active tectonics are dominated by three main features (Fig. 1). North Island seismicity ranges from shallow activity with surface ruptures to events with hypocentres

J. Ristau (✉)
GNS Science, Lower Hutt, New Zealand
e-mail: j.ristau@gns.cri.nz

S. D'Amico (ed.), *Moment Tensor Solutions*, Springer Natural Hazards,
https://doi.org/10.1007/978-3-319-77359-9_13

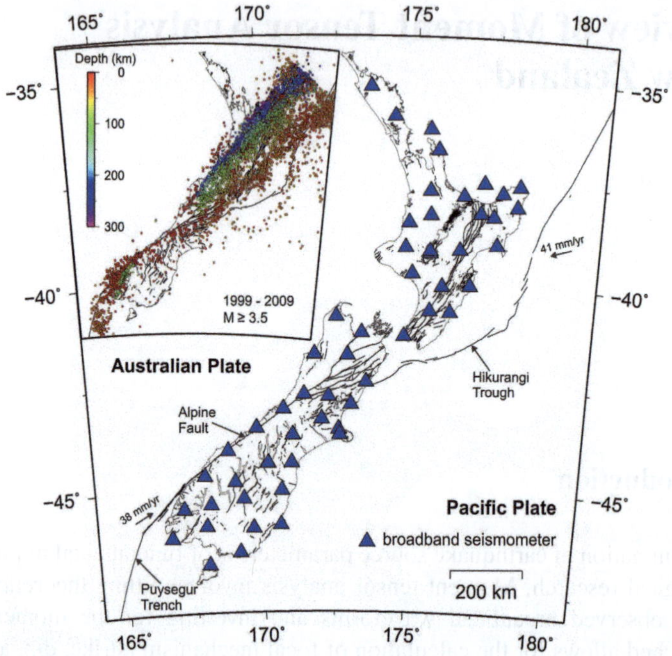

Fig. 1 New Zealand broadband seismometer network with more than 50 broadband seismometers available for calculating RMT solutions. New Zealand sits atop the boundary between the Pacific and Australian plates. (inset) 10 years of seismicity in New Zealand from (shown is $M \geq 3.5$). New Zealand seismicity ranges from earthquakes with surface ruptures to earthquakes several hundred kilometres deep beneath North Island. To represent New Zealand seismicity prior to the earthquake sequences discussed in this article, the 10 years shown spans the decade before the 15 July 2009 M_w 7.8 Dusky Sound earthquake

several hundred kilometres deep. There are also several active volcanoes present in central North Island and north of North Island. Beneath North Island, there is active subduction of the Pacific plate beneath the Australian plate at the Hikurangi trough which is capable of producing great ($M > 8.0$) megathrust earthquakes (e.g. Wallace et al. 2014; Clark et al. 2015). The Wellington region, the capital of New Zealand in southern North Island, is also at risk from shallow strike-slip earthquakes, notably the Wellington Fault which is capable of producing $M > 7.5$ earthquakes with recurrence intervals of 610–1100 years (e.g. Little et al. 2010; Langridge et al. 2011), and the Wairarapa Fault which last ruptured in 1855 with M 8.2 (e.g. Van Dissen and Berryman 1996; Rodgers and Little 2006).

At the southern end of South Island subduction polarity is reversed with the Australian plate subducting beneath the Pacific plate at the Puysegur Trench, which also produces earthquakes with a wide range of focal depths. The largest recorded earthquake at the Puysegur Trench was the 15 July 2009 M 7.8 Dusky Sound earthquake (Fry et al. 2010; Beavan et al. 2010). Between the two subduction zones is the Alpine

Fault, a 650 km long, right-lateral strike-slip fault forming the Pacific/Australian plate boundary. The Alpine Fault has ruptured in major earthquakes (M > 7.5) with recurrence intervals of ~300 years, with the last major earthquake occurring in 1717 (e.g. Cooper and Norris 1990; Berryman et al. 2012; Cochran et al. 2017). Earthquakes in central South Island are primarily within the overlying crust with focal depths < 40 km.

3 New Zealand Seismic Network

In 2001 the New Zealand Earthquake Commission (EQC) provided funding to GNS Science to launch the GeoNet project. The purpose of GeoNet is to upgrade the existing seismic network, provide data communication links, modernize data management practices, and introduce new initiatives for volcano surveillance, landslide response, earth deformation monitoring, and tsunami warning systems. GeoNet manages a large network of three-component broadband seismometers, strong-motion accelerometers, GPS sites, volcano surveillance equipment, and tsunami tide gauges. Figure 1 shows the broadband seismograph network in New Zealand with more than 50 three-component broadband sites across the country which are used for moment tensor analysis. Petersen et al. (2011) provides a detailed overview of the New Zealand seismic network.

4 Regional Moment Tensor Analysis

4.1 Moment Tensor Theory

Moment tensor analysis involves fitting theoretical waveforms with observed broadband waveforms and inverting for the moment tensor elements. A brief description of moment tensor theory is included here, and detailed descriptions of the theory and methodology are provided in a number of sources (e.g. Aki and Richards 1980; Jost and Herrmann 1989; Frohlich and Apperson 1992; other chapters in this volume). The general representation of a seismic source is given in Aki and Richards (1980) as

$$d_n(x,t) = \int_{-\infty}^{\infty} \int_{V} G_{nk}\left(x,t;r,\bar{t}\right) f_k\left(r,\bar{t}\right) dV(r)\, d\bar{t}, \tag{1}$$

where d_n is the observed displacement at an arbitrary position x at time t due to a distribution of equivalent body force densities f_k. G_{nk} are the components of the

Green's function containing the effects of the Earth structure, and V is the source volume. The problem can be discretised as

$$d_n(x, t) = \sum_{i=1}^{5} m_i * G_{in}, \tag{2}$$

where d_n is the vertical, radial, or transverse displacement, and m_i are the moment tensor elements with $m_1 = M_{11}$, $m_2 = M_{22}$, $m_3 = M_{12}$, $m_4 = M_{13}$, and $m_5 = M_{23}$ (note $M_{33} = -(M_{11} + M_{22})$). G_{in} are the Green's functions corresponding to each of the respective moment tensor elements. The Green's functions are usually calculated using simple 1-D velocity models appropriate to the source region; however, 2-D or 3-D models can also be used. Using the observed displacements d_n, and Green's functions G_n, the moment tensor elements can be inverted for, most commonly using a least-squares inversion that can be written as

$$d = G\overline{m}, \tag{3}$$

where d consists of n waveforms, G is an $n \times 6$ matrix consisting of the Green's functions, and $\overline{m}$ is the vector containing the moment tensor elements to be determined.

Moment tensor analysis of larger magnitude earthquakes ($M_w > 5.0$–5.5) at teleseismic distances ($\gtrsim$1000 km) has been ongoing since the late-1970s by the Global Centroid Moment Tensor (GCMT) Project (formerly Harvard CMT solutions; e.g. Dziewonski et al. 1981), and the United States Geological Survey (USGS; e.g. Sipkin 1986). Regional moment tensor (RMT) solutions differ from teleseismic methods in that they use regional data (source-receiver distances $\lesssim$ 1000 km) and velocity models specific to the source region to calculate Green's functions. RMT analysis for smaller magnitude earthquakes ($M_w < 5.0$–5.5) has become a routine technique with a number of agencies providing catalogues of RMT solutions (e.g. Bent 2015; Braunmiller and Nábelek 2002; Braunmiller et al. 2002; Clinton et al. 2006; Kao et al. 1998; Kao et al. 2012; Kubo et al. 2002; Ristau et al. 2007; Ristau 2013; Ruppert and Hansen 2010; Scognamiglio et al. 2009; Whidden and Pankow 2012).

5 Regional Moment Tensor Analysis in New Zealand

RMT solutions at GeoNet are calculated using the method developed by Doug Dreger at the University of California, Berkeley Seismological Laboratory (e.g. Dreger and Helmberger 1993; Pasyanos et al. 1996; Dreger 2003) for use on smaller magnitude earthquakes. Ristau (2008) introduced RMT analysis in New Zealand, provided the velocity models that were developed to calculate Green's functions, and outlined the procedure for calculating a RMT solution. Ristau (2008) showed examples of RMT solutions and compared them with GCMT (see Sect. 9) and first-motion solutions, and discussed some the issues encountered in implementing the procedure. Prior to August 2003 the New Zealand broadband seismograph network was not sufficiently dense to calculate RMT solutions (Matcham et al. 2006).

GeoNet locates over 15 000 earthquakes each year in New Zealand and adjacent offshore regions (Fig. 1 *inset*). Detailed overviews of New Zealand seismicity can be found in Robinson (1986), Reyners (1989), Anderson and Webb (1994), and Ristau (2008). Typically, more than 1000 earthquakes each year have local magnitude (M_L) ≥ 3.5 which make them candidates for RMT analysis. In September 2012 GeoNet began using SeisComP3 (SC3) for earthquake analysis, replacing the CalTech-USGS seismic processor (CUSP) system which had been in use since 1986. Ristau (2009) showed there is a significant discrepancy between M_L calculated with CUSP and M_w for New Zealand earthquakes, with M_L normally being larger than M_w and the discrepancy being particularly large for deep focus (>33 km) earthquakes. Therefore, CUSP M_L often needs to be ≥~4.0 to calculate a moment tensor solution; however, it has been possible to calculate RMT solutions for some smaller events depending on the signal-to-noise ratio. SC3 M_L correlates better with M_w, and earthquakes with smaller SC3 M_L's are able to have moment tensor solutions calculated (Ristau et al. 2016).

Since RMT analysis was implemented by GeoNet in early-2007, more than 2000 RMT solutions have been calculated for earthquakes in New Zealand and adjacent offshore regions with M_w 3.2–7.4 and depth 2–375 km, with the catalogue beginning August 2003. The vast majority are for earthquakes in the range M_w 3.5–5.0; however, there are 34 earthquakes in the RMT catalogue with M_w ≥ 6.0, with five having M_w ≥ 7.0 (Fig. 2). This does not include two M_w 7.8 earthquakes that were too large to calculate reliable RMT solutions (the reasons are discussed later in this article).

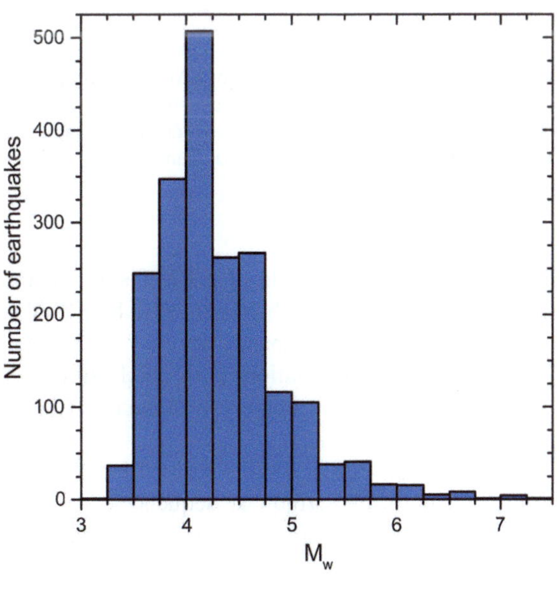

Fig. 2 M_w distribution of earthquakes in the RMT catalogue. Most earthquakes are in the range M_w 3.5–5.0; however, 34 have M_w ≥ 6.0. The 15 July 2009 M_w 7.8 Dusky Sound and 13 November 2016 M_w 7.8 Kaikoura earthquakes are not included in this plot as they were too large to calculate reliable RMT solutions

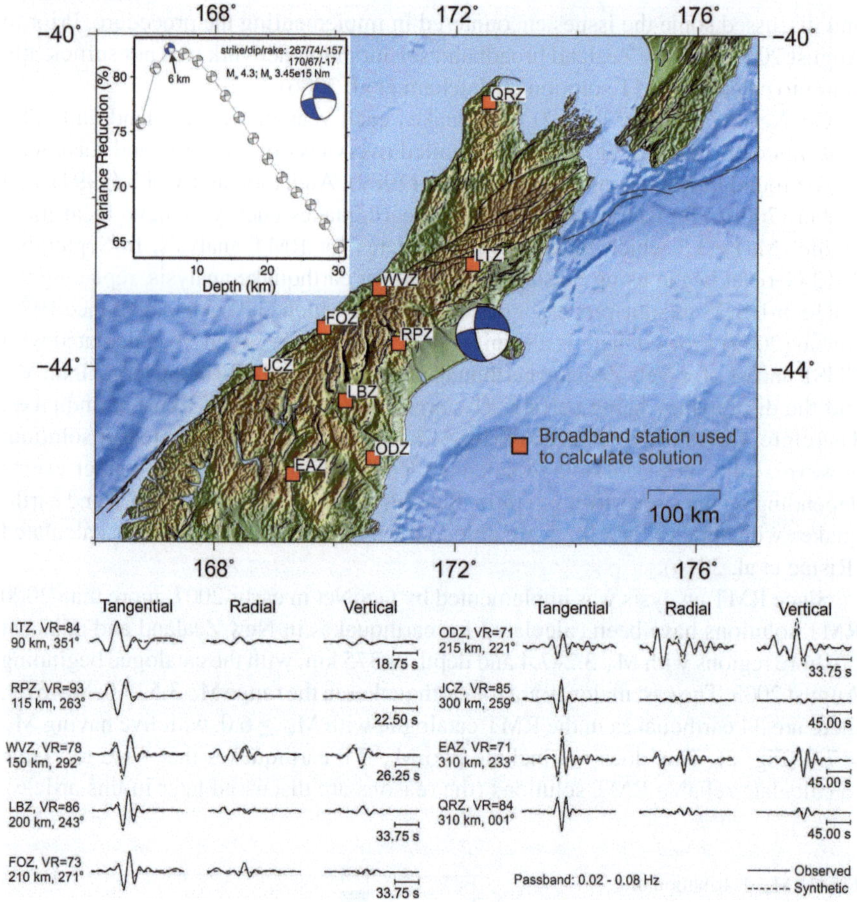

Fig. 3 Example of a RMT solution for a M_W 4.3 earthquake in the Canterbury region. (top) Map showing the focal mechanism and the nine broadband seismometers used to calculate the solution. (inset) Variance reduction versus centroid depth showing the best-fit at a depth of 6 km. (bottom) The waveforms were filtered using a passband of 0.02–0.08 Hz and there is a very good fit between the synthetic and observed waveforms

Figure 3 shows an example of a RMT solution for a M_W 4.3 earthquake in the Canterbury region of South Island. Reliable RMT solutions have been calculated using as few as three broadband stations; however, the average is around six stations. It is preferable to have stations with good azimuthal coverage, but in a region like New Zealand—a narrow country with many offshore earthquakes—this is often not possible.

The more than 2000 RMT solutions for New Zealand include RMT solutions from five significant earthquake sequences that have impacted New Zealand since July 2009 (Fig. 4). In chronological order these earthquake sequences are:

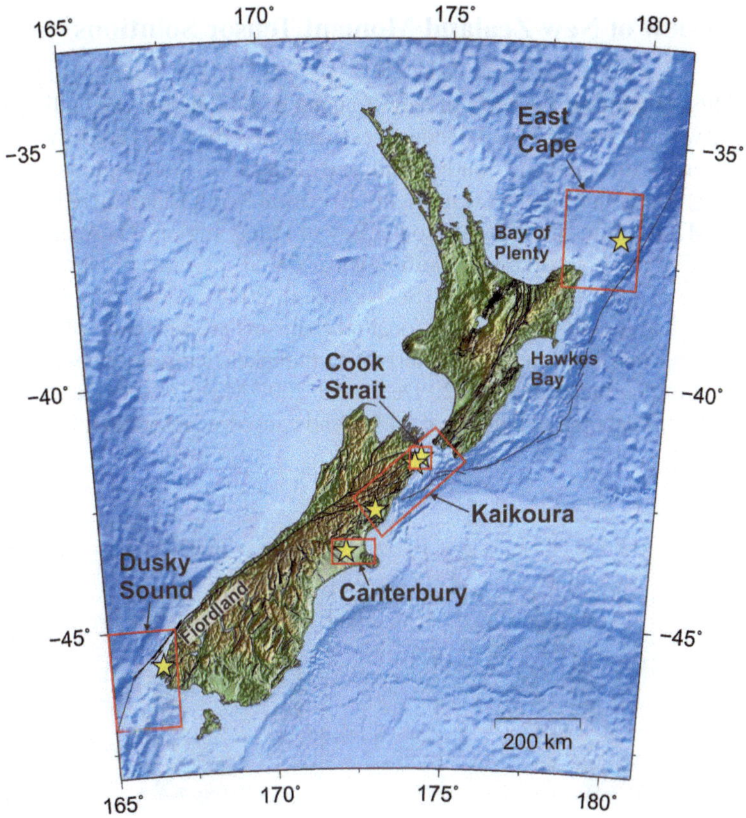

Fig. 4 Locations of the five significant earthquakes in New Zealand since July 2009 which are discussed in this article (yellow stars) and aftershock zones where RMT solutions were calculated (red boxes)

1. 15 July 2009 M_W 7.8, Dusky Sound, Fiordland (Fry et al. 2010; Beavan et al. 2010);
2. 3 September 2010 M_W 7.1, Darfield, Canterbury (Gledhill et al. 2011);
3. 21 July 2013 M_W 6.6 and 16 August 2013 M_W 6.6, Cook Strait (Hamling et al. 2014);
4. 1 September 2016 M_W 7.1, East Cape (Te Araroa) (Ristau 2017);
5. 13 November 2016 M_W 7.8, Kaikoura (Kaiser et al. 2017).

At the time this overview chapter was written the M_W 7.1 East Cape and M_W 7.8 Kaikoura earthquakes had recently occurred; therefore, much work is still needed to fully characterise these earthquakes and aftershock sequences.

6 Overview of New Zealand Moment Tensor Solutions

I will first provide an overview of New Zealand RMT solutions, excluding events associated with the five earthquake sequences mentioned above. Figure 5 shows RMT solutions for North Island and South Island, and adjacent offshore regions, separated into depth ≤ 33 km (left) and >33 km (right). In central North Island there is a band of shallow RMT solutions that extends to the Bay of Plenty region. These earthquakes are typically normal or strike-slip faulting with NW-SE oriented T-axes and reflect the influence of the Taupo rift zone (Seeback et al. 2014). Along the east coast of North Island and the offshore region, focal mechanisms are a mixture of strike-slip, reverse, and normal faulting. The two largest earthquakes in this region were in-slab, normal faulting earthquakes. The 20 December 2007 M_w 6.7 earthquake occurred near the city of Gisborne at a depth of 24 km and caused widespread minor damage

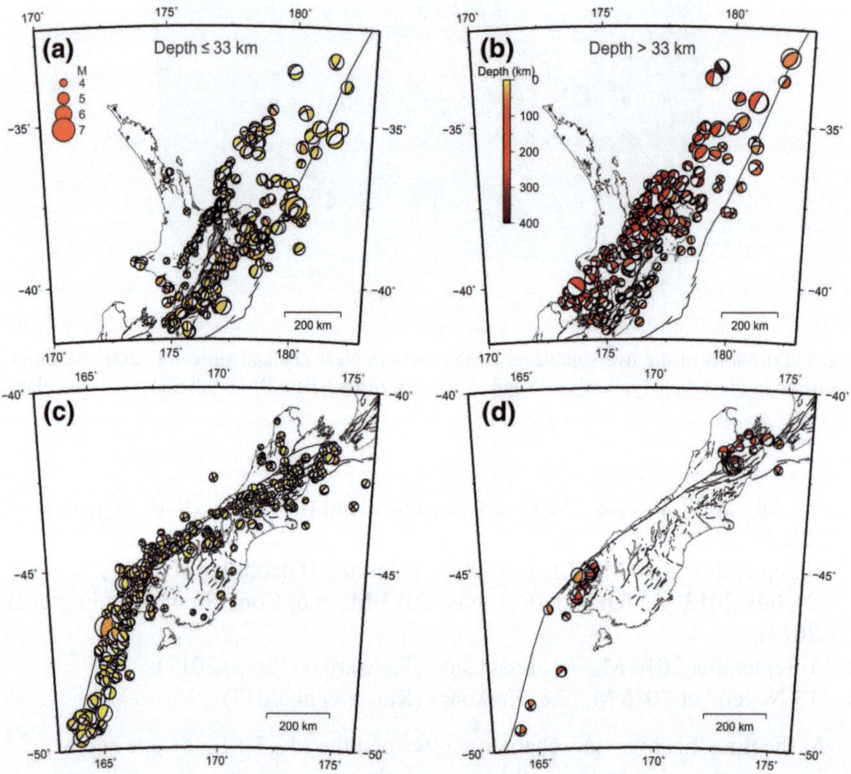

Fig. 5 RMT solutions for New Zealand, excluding earthquakes associated with the five earthquake sequences shown in Fig. 4. **a** RMT solutions for North Island for earthquakes with centroid depth ≤ 33 km. **b** RMT solutions for North Island earthquakes with centroid depth > 33 km. **c** RMT solutions for South Island for earthquakes with centroid depth ≤ 33 km. **d** RMT solutions for South Island earthquakes with centroid depth > 33 km

along with some structural damage (Holden et al. 2008). A similar earthquake struck on 16 November 2014 with M_w 6.8, but was located further offshore and resulted in less human impact. Two other significant earthquakes were the 20 January 2014 M_w 6.3 Eketahuna earthquake (e.g. Abercrombie et al. 2017), and the 20 January 2005 M_w 5.3 Upper Hutt earthquake (e.g. Reyners and Bannister 2007), with epicentres in southern North Island. These earthquakes were widely felt throughout southern North Island and caused widespread minor damage. Both earthquakes had normal faulting mechanisms at depths of ~20–30 km, consistent with being in-slab events. The deeper North Island earthquakes are also a mixture of normal, reverse, and strike-slip faulting; however, many of the deeper earthquakes are reverse faulting and related to subduction of the Pacific plate. The deep earthquakes can have $M_w \geq 6.0$ and be widely felt throughout North Island, but they typically do little damage due to their depth.

South Island has a different pattern of seismicity than North Island, with the majority of the seismic activity being dominated by the plate boundary between the Australian and Pacific plates, and the Alpine Fault delineating the boundary. The onshore shallow seismicity is almost all strike-slip or reverse faulting with P-axes consistently oriented ~115°. Further south, in the Fiordland region of the southwest South Island and south of South Island, the Puysegur subduction zone dominates with many reverse faulting focal mechanisms. The Fiordland region is the most seismically active region of New Zealand with 11 earthquakes $M_w \geq 6.0$ in the RMT catalogue; however, due to the sparse population in Fiordland the damage is typically minor. The 21 August 2003 M_w 7.1 Fiordland earthquake was the largest shallow earthquake in New Zealand in 35 years at the time, and was a reverse faulting event on the shallow part of the subduction interface (Reyners et al. 2003). The 15 October 2007 M_w 6.7 George Sound earthquake was centred northeast of the M_w 7.1 Fiordland earthquake, and was also a reverse faulting event on the subduction interface (Petersen et al. 2009). South of South Island a M_w 7.4 earthquake occurred on 30 September 2007 with a reverse faulting mechanism and may also have been on the subduction interface. This earthquake was followed by a M_w 6.6 aftershock. Central South Island is largely devoid of deep seismicity; however, there is some deeper seismic activity at the north end which marks the southern boundary of the Hikurangi subduction zone. In particular, near the town of Hanmer Springs at the northern end of South Island, there is a cluster of oblique faulting events, several with $M_w \geq 5.0$, at depths of ~60–70 km. There is deep seismic activity in the Fiordland region associated with the Puysegur subduction zone and the faulting style varies between strike-slip and reverse, but the vast majority have depths $\lesssim 150$ km.

7 Discussion of Earthquake Sequences

7.1 Dusky Sound, Fiordland: 15 July 2009, M_w 7.8

The M_W 7.8 Dusky Sound earthquake was the largest earthquake in New Zealand since the 1931 M 7.8 Hawkes Bay earthquake, east coast of North Island (Fry et al. 2010; Beavan et al. 2010). The earthquake was a thrust faulting event on the subduction interface of the Puysegur subduction zone. Most of the energy release was at very low frequencies and, combined with the remote location of the epicentre, resulted in mainly minor damage. Using the focal mechanism of the mainshock, rupture area, and location of nearby faults, Coulomb stress models show a stress increase of ~0.2 MPa on the offshore extension of the Alpine Fault (Fry et al. 2010; Beavan et al. 2010).

The RMT method used at GeoNet assumes a point-source (i.e. a source with no dimensions), which is adequate when the wavelengths analysed are much greater than the source dimensions. For earthquakes less than M_W ~ 7.0–7.5 a point-source assumption is sufficient; however, for larger earthquakes the point-source assumption using the frequency bands and source-receiver distances normally used for RMT analysis no longer applies, and either more complex source modelling or very long wavelengths are needed. For the Dusky Sound earthquake most of the broadband seismometers in New Zealand saturated which made the data unusable for RMT analysis, and even the furthest stations (source-receiver distance ~1000 km), were not distant enough to utilise long enough wavelengths to calculate a reliable RMT solution for the mainshock. Therefore, the W-phase teleseismic moment tensor solution calculated by the USGS is used for the mainshock (see Sect. 9).

132 RMT solutions for aftershocks were calculated, with the largest being M_W 6.1 (Fig. 6). The shallowest solutions (Fig. 6a) are mainly normal faulting mechanisms with one near-vertical fault plane and one near-horizontal fault plane. In the 7–12 km depth range (Fig. 6b) the mechanisms are mainly reverse or strike-slip faulting with a P-axis oriented approximately E-W, similar to the mainshock. Many of the strike-slip faulting mechanisms are located near the offshore extension of the Alpine Fault. However, given the uncertainty in epicentres in the Fiordland region it is not clear whether these aftershocks are on the Alpine Fault and induced by the Dusky Sound earthquake, or on structures near the Alpine Fault. At 13–19 km (Fig. 6c) the mechanisms are mainly reverse faulting and very similar to the mainshock; therefore, they may have also occurred on the plate interface. At 20–30 km (Fig. 6d) the mechanisms are a mixture of reverse, normal, and strike-slip faulting suggesting that the tectonic setting at greater depths becomes more complicated. Some aftershocks may be occurring in the crust above the plate interface and others within the subducting plate.

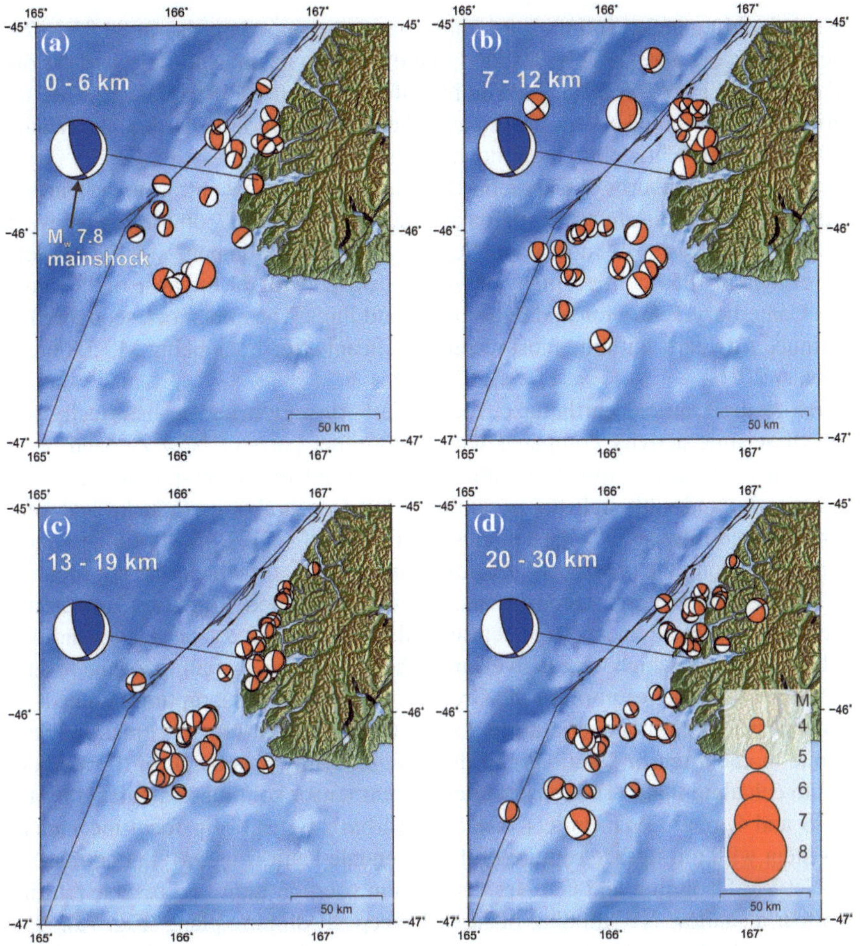

Fig. 6 RMT solutions for aftershocks of the Dusky Sound earthquake. The mainshock, taken from the USGS W-phase solution, is the blue focal mechanism, and aftershocks are orange focal mechanisms. **a** Focal mechanisms for aftershocks with depth 0–6 km. **b** Focal mechanisms for aftershocks with depth 7–12 km. **c** Focal mechanisms for aftershocks with depth 13–19 km. **d** Focal mechanisms for aftershocks with depth 20–30 km

7.2 *Darfield, Canterbury: 3 September 2010, M_w 7.1*

The M_w 7.1 Darfield earthquake occurred in the Canterbury region of New Zealand, ~10 southeast of the town of Darfield and ~40 km west of Christchurch, New Zealand's second largest city (population ~377 000) (Gledhill et al. 2011). The earthquake was widely felt throughout South Island and southern North Island, with over 7300 felt reports received, and caused significant damage in Christchurch, with maximum intensity MM 9 in the epicentral region. Through a fortunate combination of strict building codes and the earthquake occurring on a Saturday at 04:35 NZST when the streets were largely deserted, there were no deaths and only two serious injuries. On 21 February 2011 (~5½ months after the Darfield earthquake) the M_w 6.2 Christchurch earthquake struck as an aftershock to the Darfield earthquake (Kaiser et al. 2012). The Christchurch earthquake struck ~6 km southeast of the city centre beneath the outer suburbs of Christchurch at shallow depth. In contrast to the Darfield earthquake, the Christchurch earthquake struck at 12:51 NZST on a weekday when the city centre was at its most populated. Severe building damage was widespread and resulted in 185 fatalities, many of which occurred with the collapse of the multi-story Canterbury Television and Pyne Gould buildings. The aftershock sequence has continued for years, with the most recent significant aftershock being M_w 5.8 on 14 February 2016.

Before the Darfield earthquake, the Canterbury region had a historically low level of seismic activity compared with many other parts of New Zealand. As a result, an earthquake similar to Darfield was considered a low-probability event for New Zealand, although the possibility of a comparable earthquake had been acknowledged in seismic hazard models (Stirling et al. 2012). Most of the moment release was along the previously unknown Greendale Fault which had not ruptured for 18 000–20 000 years. Geodetic and strong-motion modelling (e.g. Holden and Beavan Holden and Beavan 2012; Beavan et al. 2012) identified at least seven individual fault segments which were active during the earthquake, including a blind reverse fault several kilometres north of the Greendale Fault where the rupture initiated.

The Darfield earthquake is an important case study on how regional and teleseismic moment tensor solutions can differ significantly as a result of the methods used to calculate them. The teleseismic moment tensor solutions calculated by the USGS and Global CMT Project (see Sect. 9) have a strike-slip focal mechanism consistent with the surface rupture of the Greendale Fault (Fig. 7a). In contrast, the GeoNet RMT solution and GeoNet first-motion solution indicated reverse faulting on either a shallow NW-dipping plane or a steep SE-dipping plane (Fig. 7a). As a result of the high density of strong-motion stations in the vicinity of the mainshock, the hypocentre was well constrained about 4 ± 0.5 km north of the surface trace of the Greendale Fault (Gledhill et al. 2011). Due to the well-constrained hypocentre with an estimated depth of about 11 km, the discrepancy between the hypocentre location and the trace of the Greendale Fault cannot be explained by the location uncertainty. A shallow-dipping fault plane could account for the discrepancy, but there should be near co-incidence of the epicentre with the trace of the Greendale Fault for any near-vertical strike-slip mechanism as indicated in the teleseismic moment tensor solutions.

Teleseismic moment tensor methods use lower frequency bands and much greater source-receiver distances than RMT methods. As a result, they are not able to resolve distinct mechanisms but instead provide an average over the whole event, which is dominated in the case of the Darfield earthquake by strike-slip movement along the Greendale Fault. The RMT solution uses higher frequency bands and smaller source-receiver distances, ~300–860 km for the GeoNet Darfield RMT solution, which makes the solution more sensitive to small-scale features. As a result, the

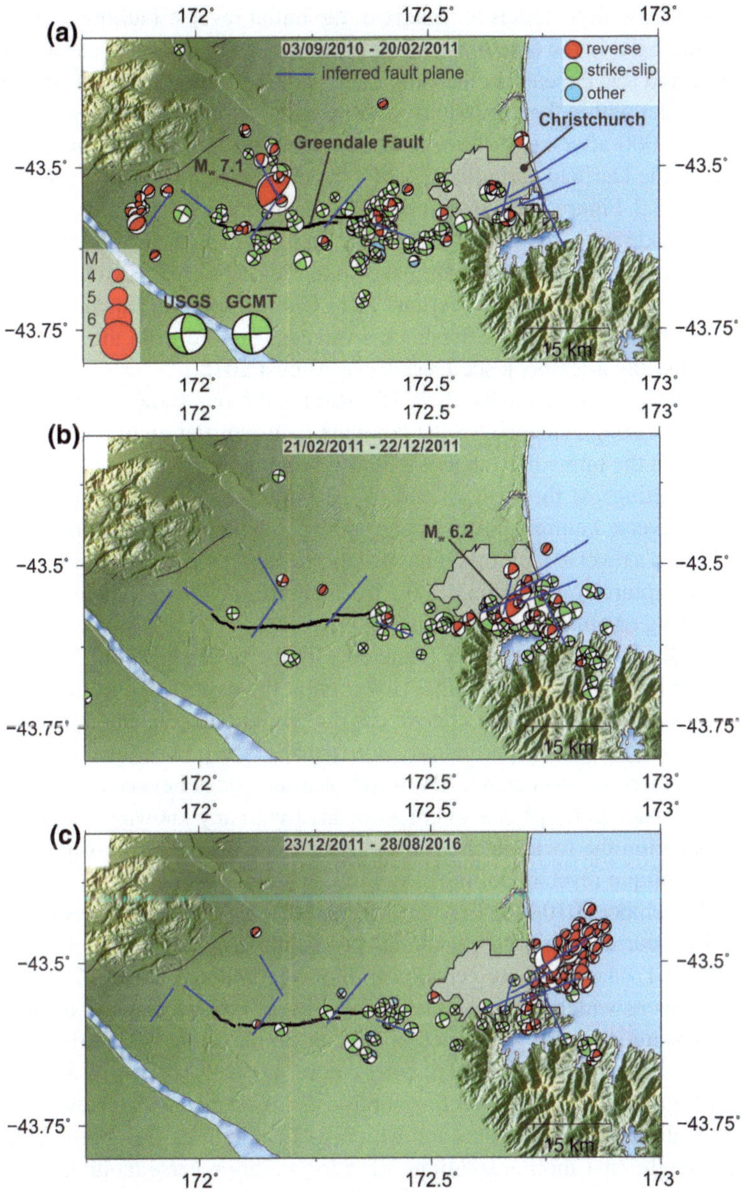

Fig. 7 RMT solutions for the Canterbury earthquake sequence. **a** The M_W 7.1 Darfield earthquake and aftershocks up to 20 February 2011. Also shown are the USGS and GCMT focal mechanisms for the mainshock showing strike-slip faulting as opposed to reverse faulting for the RMT solution. Focal mechanisms are colour-coded according to faulting style. Also shown in the surface fault trace of the Greendale Fault and inferred fault planes from geodetic studies. **b** The M_W 6.2 Christchurch earthquake and aftershocks up to 22 December 2011. **c** The 23 December 2011 Pegasus Bay aftershock sequence and aftershocks up to 28 August 2016, including the 14 February 2016 M_W 5.8 aftershock

GeoNet RMT solution models the nature of the initial reverse-faulting rupture. This demonstrates how, in the case of a complex rupture like the Darfield earthquake, the teleseismic and RMT focal mechanisms can be quite different from one another, but both are correct within the limitations of the methods used to calculate them.

Nearly 15 000 aftershocks have been recorded in the Canterbury/Christchurch region since the Darfield earthquake, and 378 RMT solutions have been calculated with M_w 3.4–6.2. Figure 7 separates the RMT solutions into three groups based on the main earthquakes in the Canterbury sequence. In Fig. 7a are the RMT solutions from the 3 September 2010 Darfield earthquake up to the 21 February 2011 Christchurch earthquake; Fig. 7b from the 21 February 2011 Christchurch earthquake up to the 23 December 2011 Pegasus Bay aftershock sequence; Fig. 7c from the 23 December 2011 Pegasus Bay aftershock sequence to 28 August 2016.

In Fig. 7a, focal mechanisms from 153 RMT solutions show a variety of faulting styles, providing evidence for the complex nature of the rupture process. Focal mechanisms in the immediate area of the initial rupture are a mixture of reverse and strike-slip faulting. At the western end of the fault zone the mechanisms are predominantly reverse faulting, consistent with the geodetic model of the main rupture, which includes a reverse faulting segment at the western end of the rupture zone. East of the main rupture zone, leading into Christchurch, focal mechanisms are mainly reverse faulting or oblique-reverse faulting.

The aftershock locations mostly coincide with the Greendale Fault trace and the location of inferred subsurface faults. However, at the eastern end of the Greendale Fault there is a NE-SW trend of aftershocks that are not associated with any known subsurface fault, and this is particularly noticeable in the plot of focal mechanisms. Further east there is another NE-SW trend of aftershocks between the Greendale Fault and Christchurch that is also not associated with any known subsurface fault, and in this region the focal mechanisms change from mainly strike-slip faulting in the west to oblique-reverse faulting closer to Christchurch.

On 26 December 2010 NZST (25 December 2010 UTC) a cluster of very shallow aftershocks occurred near the Christchurch city centre. The largest, M_w 4.7, occurred at 12:30 NZST when the city centre was highly populated (Ristau 2011). These aftershocks were widely felt and the M_w 4.7 event caused damage to brick and masonry structures already weakened in the city centre. Three RMT solutions were calculated for events in the series of aftershocks, all with strike-slip mechanisms. Ristau (2011) also calculated 16 first-motion focal mechanism for events in this series, including the three events for which RMT solutions had been calculated, and although the first-motion solutions were for mainly reverse faulting, the P-axis orientation is consistent with those in the RMT solutions.

Figure 7b shows 113 focal mechanisms beginning with the 21 February 2011 M_w 6.2 Christchurch earthquake. The Christchurch earthquake occurred on a previously unmapped NE-SW striking fault in the Port Hills area of the outer suburbs of Christchurch (Fig. 7b), in an area of small positive stress increase resulting from the Darfield earthquake (Kaiser et al. 2012). The RMT solution shows primarily reverse faulting with a strike-slip component, consistent with the complex nature of the faulting as indicated by the geodetic modelling which requires two, and possibly three faults to be active (Beavan et al. 2012). The earthquake triggered a rejuvenated aftershock sequence centred around Christchurch and into Pegasus Bay east of Christchurch, most notably a M_w 6.0 aftershock on 13 June 2011. The 13 June 2011 aftershock was a strike-slip event that occurred ~4 km east of the Christchurch earthquake epicentre (Fig. 7b) and also involved at least two fault ruptures (Beavan et al. 2012). Following the 13 June 2011 aftershock many of the aftershocks extended SE into Banks Peninsula where little aftershock activity had occurred previously. Whereas the Christchurch earthquake had mainly reverse faulting, focal mechanism derived from RMT solutions for aftershocks to the Christchurch earthquake indicated mainly strike-slip faulting, although there were some with reverse or oblique-reverse faulting (Fig. 7b).

Three earthquakes on 23 December 2011 UTC (M_w 5.4–5.9) centred near Pegasus Bay, east of Christchurch, triggered a NE-SW series of aftershocks that extended offshore (Fig. 7c). These earthquakes were widely felt in Christchurch but damage was minimal due to their offshore location (Ristau et al. 2013). 53 RMT solutions were determined for events in Pegasus Bay with a majority (45 of 53) indicating reverse or oblique-reverse faulting. This is in contrast with the rest of the Canterbury aftershock sequence where ~74% of the focal mechanism indicated strike-slip faulting (Ristau et al. 2013). Pegasus Bay was also the location of the most recent significant aftershock in the Canterbury sequence, a M_w 5.8 oblique-reverse faulting event on 14 February 2016 (Herman and Furlong 2016).

7.3 Cook Strait/Lake Grassmere: 21 July 2013, M_w 6.6 and 16 August 2013, M_w 6.6

The Cook Strait and Lake Grassmere earthquakes of July and August 2013 was an earthquake doublet—two earthquakes with similar epicentre, magnitude, and source properties that occur closely spaced in time (Hamling et al. 2014; Holden et al. 2013). Cook Strait is a narrow strait that separates North Island and South Island, and Lake Grassmere is near the town of Seddon at the northern end of South Island. The earthquakes were ~50–70 km southwest of New Zealand's capital, Wellington. The earthquakes were widely felt throughout northern South Island and southern North Island, and caused significant structural damage in northern South Island communities and significant non-structural damage in the greater Wellington region.

The sequence began with two foreshocks of M_w 5.5 on 18 July 2013 and M_w 5.8 on 20 July 2013 (Fig. 8). The foreshocks were located in Cook Strait and were both reverse faulting mechanisms with depths ~20 km. Most of the RMT solutions for aftershocks to the foreshocks were also reverse faulting mechanisms with similar depth. On 21 July 2013 the M_w 6.6 Cook Strait earthquake struck with over 5500 felt reports, mostly from North Island and northern South Island. The Cook Strait earthquake was a right-lateral strike-slip earthquake at a depth of ~20 km along a

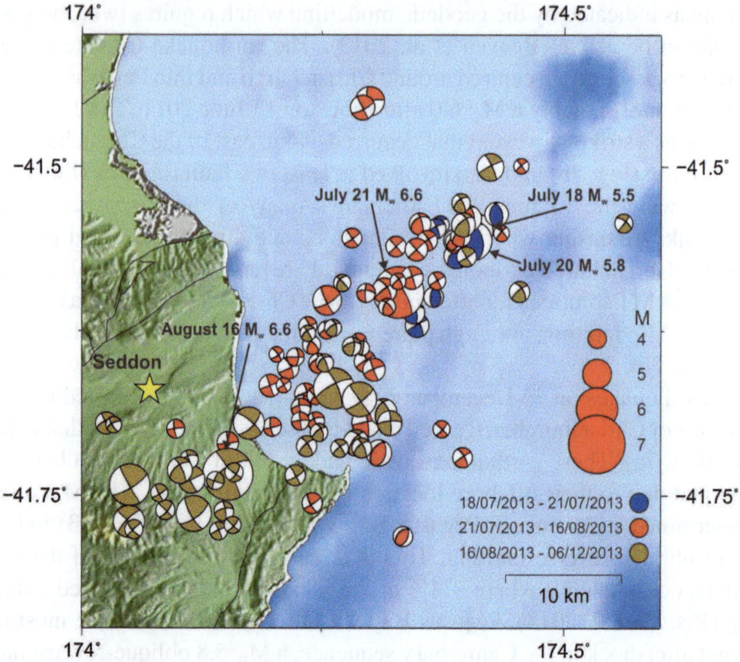

Fig. 8 RMT solutions for the Cook Strait/Lake Grassmere earthquake sequence. Focal mechanisms are colour-coded by time with blue mechanisms corresponding to foreshocks to the Cook Strait earthquake, orange mechanisms for the M_W 6.6 Cook Strait earthquake and aftershocks, and gold mechanisms for the M_W 6.6 Lake Grassmere earthquake and aftershocks. The town of Seddon is shown by the yellow star

previously unmapped offshore fault. RMT solutions for aftershocks were predominantly strike-slip with most located SW of the mainshock towards South Island. The depths of the RMT solutions trend towards shallower depths as they get closer to South Island.

On 16 August 2013 the M_W 6.6 Lake Grassmere earthquake occurred with nearly 6200 felt reports from North Island and northern South Island. This earthquake was also a right-lateral strike-slip earthquake, very similar to the earlier Cook Strait earthquake. As a result of their similar source properties, magnitudes, epicentres, and occurring less than a month apart they can be considered an earthquake doublet. Geodetic modelling and Coulomb stress analysis suggest that the Lake Grassmere earthquake was triggered by the Cook Strait earthquake, which in turn was triggered by the earlier foreshocks (Hamling et al. 2014). As with the Cook Strait earthquake, most of the RMT solutions for Lake Grassmere aftershocks are strike-slip faulting but with shallower depths.

Northern South Island, Cook Strait, and southern North Island have a large number of active faults, and the region sit atop the Hikurangi subduction zone. Slip models derived using focal mechanism information can be used to resolve Coulomb stress transfer onto nearby mapped faults. One of the most important faults with a positive stress change following the Cook Strait/Lake Grassmere sequence is the southern extension of the Wellington Fault which showed a positive stress change of ~0.1 MPa (Hamling et al. 2014). The Wellington Fault extends through the heart of the Wellington city centre and north to the populated Hutt Valley, and is capable of rupturing in M > 7.5 earthquakes with a recurrence interval of 610–1100 years (Little et al. 2010; Langridge et al. 2011). The second area of concern is the Hikurangi subduction zone which has the potential to rupture in M > 8.0 earthquakes (Wallace et al. 2014; Clark et al. 2015). Hamling et al. (2014) concluded that it is likely that the Cook Strait and Lake Grassmere earthquakes resulted in a positive stress change approaching 0.3 MPa on the Hikurangi plate interface.

7.4 East Cape: 1 September 2016, M_w 7.1

The M_w 7.1 East Cape (Te Araroa) earthquake occurred ~125 km NE of East Cape, at the northeast end of North Island (Fig. 9). The offshore epicentre resulted in mainly minor damage; however, GeoNet received more than 4700 felt reports from across North Island and northern South Island. A minor tsunami of ~30 cm was recorded at the East Cape and Great Barrier Island tide gauges, and many local residents self-evacuated to higher ground when they felt strong shaking that lasted for more than 30 s.

The earthquake was preceded by a M_w 5.8 foreshock on 31 August 2016, which was a normal faulting mechanism with one high angle and one low angle fault plane, at a depth of 15 km (Fig. 9). The mainshock was an oblique-normal faulting mechanism with one high angle and one low angle fault plane, but deeper at 23 km. Initially a tsunami was not anticipated due to the magnitude, depth, and faulting style, and it was not until it was detected at tide gauges that it was known a tsunami had been generated. RMT solutions have been calculated for 27 aftershocks with M_w 4.4–6.0 and depths 6–24 km. Due to the complex nature of the local crustal structure and the offshore location, it was not possible to calculate reliable RMT solutions for smaller aftershocks. The aftershocks have similar mechanisms to the foreshock and mainshock, and almost all are located southwest of the mainshock and at shallower depth (Fig. 9). One notable exception is the largest aftershock (M_w 6.0) which was located east of the mainshock. With the uncertainty in the hypocentre of the mainshock it is not clear how the East Cape earthquake relates to the plate interface. The centroid depth of 23 km is consistent with the depth of the plate interface in the hypocentral region, but there is currently no evidence of rupture on the interface.

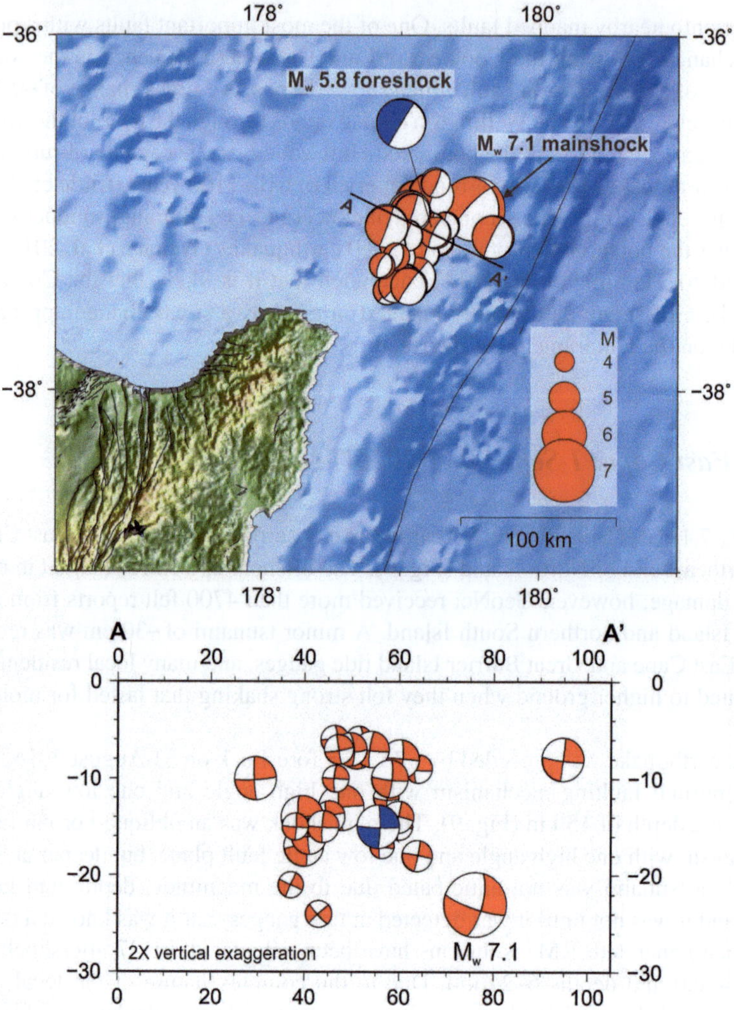

Fig. 9 (top) RMT solutions for the East Cape earthquake and aftershocks. The blue mechanism is the M_W 5.8 foreshock which occurred the day before the mainshock. (bottom) Cross-section showing aftershocks located shallower and west of the mainshock. The exception is the largest aftershock (M_W 6.0) which was located east of the mainshock

7.5 *Kaikoura: 13 November 2016, M_W 7.8*

The M_W 7.8 Kaikoura earthquake occurred just past midnight NZST with an epicentre ~30 km from the northeast coast of South Island, near the town of Waiau. This was the largest earthquake in New Zealand since the 15 July 2009 M_W 7.8 Dusky Sound earthquake, and GeoNet received more than 15 800 felt reports from almost all parts of the country. Heavy damage was experienced in the epicentral region, and widespread structural damage was reported in Wellington and Christchurch. One death resulted as a result of a collapsed house, and another from a heart attack. More than 100 000 landslides have been documented, and a tsunami of ~2 m was recorded on the closest tide gauge (Kaikoura); however, field observations show >4 m tsunami in some areas.

Almost every broadband seismometer in New Zealand saturated making it impossible to calculate a RMT solution for the mainshock. Teleseismic moment tensor solutions (USGS; GCMT; GFZ) showed mainly reverse faulting at a depth of ~10 km, and the rupture propagated north of the epicentre (Fig. 10). Geodetic and strong-motion modelling, along with field observations, show surface rupture on at least 12 different faults (Duputel and Rivera 2017; Hamling et al. 2017), and the earthquake involved at least 21 faults with the Kekerengu Fault showing up to 10–12 m of dextral slip (Kaiser et al. 2017). Although the teleseismic moment tensor solutions have a reverse faulting mechanism, they all have a significant non-double-couple component which could suggest a complex source. Fault motion from GPS and strong-motion modelling is a mixture of strike-slip and reverse faulting which may explain the large non-double-couple component in the moment tensor solutions. The rupture propagated above the southern end of the Hikurangi subduction zone and it is currently unclear how much, if any, of the plate interface was involved.

RMT solutions have been calculated for 213 aftershocks, including four with $M_W \geq 6.0$ (Fig. 10). Similar to the Canterbury aftershock sequence, the focal mechanisms show a variety of faulting styles providing evidence for the complex nature of the rupture process. Three of the $M_W \geq 6.0$ aftershocks have strike-slip mechanisms, with the largest (M_W 6.5) possibly associated with the Hundalee Fault. The second largest aftershock (M_W 6.3) has a thrust mechanism and may be associated with the Jordan Thrust Fault. The aftershock zone extends for ~230 km, with the northern end falling in the same region as the 2013 Cook Strait/Lake Grassmere earthquake sequence. The northern end of the aftershock zone approaches to within ~15 km of the south coast of the Wellington region. Over the entire aftershock zone, the RMT solutions have approximately equal numbers of reverse and strike-slip focal mechanisms. However, there appear to be two groups of aftershocks based on focal mechanism and depth. In the south (Fig. 10, orange) there are more reverse than strike-slip mechanisms, while to the north (Fig. 10, gold) there are more strike-slip than reverse mechanisms. The northern focal mechanisms are ~5 km deeper on average than the southern focal mechanisms.

8 Summary and Conclusions

Since routine RMT analysis for New Zealand earthquakes was implemented in early-2007, more than 2000 RMT solutions have been calculated with the catalogue beginning August 2003. The RMT solutions have made important contributions to

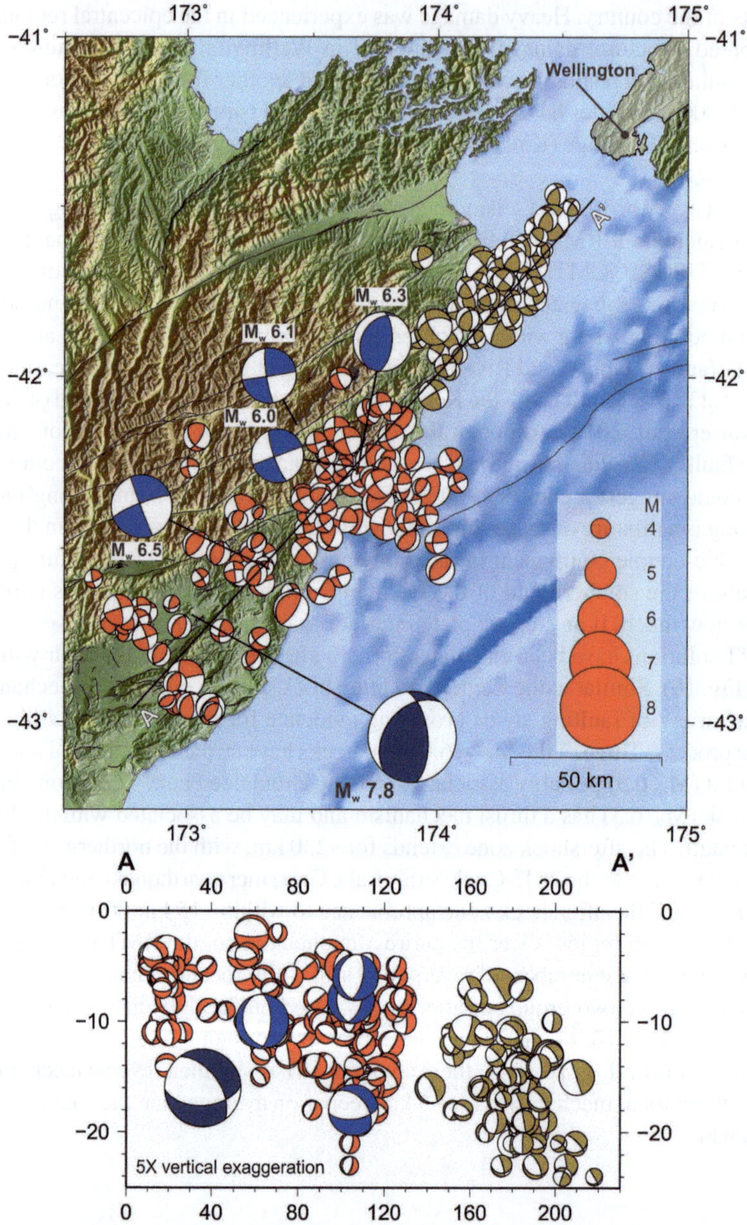

Fig. 10 (top) RMT solutions for aftershocks of the Kaikoura earthquake. The mainshock, taken from the USGS W-phase solution, is shown in dark blue, and the $M_W \geq 6.0$ aftershocks are shown in blue. The orange mechanisms are the southern group which have more reverse faulting than strike-slip faulting, and the gold mechanisms are the northern group which have more strike-slip faulting than reverse faulting. The northern end of the aftershock zone is ~15 km SW of the south coast of Wellington. (bottom) Cross-section showing the centroid depths becoming deeper from south-to-north Focal mechanism colours are the same as in map view

the understanding of the active tectonics and seismic hazard of New Zealand. The RMT catalogue includes solutions from five earthquake sequences that significantly impacted New Zealand since July 2009 and account for ~40% of the catalogue. The RMT solutions have helped to characterise the aftershock sequences and understand how they evolve over time.

RMT solutions are valuable for a variety of seismological and tectonic research topics. The improved magnitude estimates from using M_w are important for seismic hazard and risk studies, including calculating aftershock forecasts (e.g. Gerstenberger et al. 2012). The focal mechanisms are used to better understand stress transfer resulting from large earthquakes (e.g. Ellis et al. 2016), and for improved models of estimated shaking which can be used to direct resources for emergency response (e.g. Horspool et al. 2015). RMT solutions are also used to study regional scale tectonic stress and strain orientations (e.g. Ristau et al. 2007; Holt et al. 2013; Townend et al. 2012). RMT solutions are often needed to perform simulations for 3D seismic wave propagation, which can be used to validate and improve a local 3D velocity model (e.g. Loubet et al. 2016). RMT solutions can also be used to constrain the source geometry of a large earthquake in the case of complex ruptures where finite-fault inversion initially struggles to find the correct fault geometry (e.g. Gledhill et al. 2011).

9 Data and Resources

The regional moment tensor solutions used are calculated as part of the GNS Science GeoNet project (http://www.geonet.org.nz/). A complete list is available at http://info.geonet.org.nz/display/appdata/Earthquake+Catalogue (last updated 1 March 2016). The Global Centroid Moment Tensor Project (GCMT) database was searched using http://www.globalcmt.org/CMTsearch.html (last accessed 27 March 2017). The United States Geological Survey (USGS) earthquake database was searched using http://earthquake.usgs.gov/earthquakes/search/ (last accessed 27 March 2017). TheGeoForschungsZentrum(GFZ) moment tensor database was searched using https://eqarchives.wordpress.com/gfz-moment-tensors/ (last accessed 27 March 2017).

Acknowledgements Caroline Holden and Yoshihiro Kaneko provided many valuable comments which helped improve this article. Many of the figures were created using Generic Mapping Tools (GMT; Wessel and Smith 1998). Regional moment tensor solutions were computed using the mtpackagev1.1 package developed by Doug Dreger of the Berkeley Seismological Laboratory, and Green's functions were computed using the FKRPROG software developed by Chandan Saikia of URS.

References

Abercrombie RE, Bannister S, Ristau J, Doser D (2017) Variability of earthquake stress drop in a subduction setting, the Hikurangi margin, New Zealand. Geophys J Int 208:306–320. https://doi.org/10.1093/gji/ggw393

Aki K, Richards PG (1980) Quantitative seismology. W.H. Freeman and Co., San Francisco, p 932

Anderson H, Webb T (1994) New Zealand seismicity: patterns revealed by the upgraded national seismograph network. NZ J Geol Geophys 37:477–493

Beavan J, Samsonov S, Denys P, Sutherland R, Palmer N, Denham M (2010) Oblique slip on the Puysegur subduction interface in the 2009 July M_W 7.8 Dusky Sound earthquake from GPS and InSAR observations: implications for the tectonics of southwestern New Zealand. Geophys J Int 183:1265–1286. https://doi.org/10.1111/j.1365-246X.2010.04798.x

Beavan J, Motagh M, Fielding E, Donnelly N, Collett D (2012) Fault slip models of the 2010–2011 Canterbury, New Zealand, earthquakes from geodetic data, and observations of post-seismic ground deformation. NZ J Geol Geophys 55:207–221. https://doi.org/10.1080/00288306.2012.697472

Bent AL (2015) Regional centroid moment tensor solutions for eastern Canadian earthquakes: 2014. Geological Survey of Canada, Open File 7834, 35 p. https://doi.org/10.4095/296795

Berryman KR, Cochran UA, Clark KJ, Biasi GP, Langridge RM, Villamor P (2012) Major earthquakes occur regularly on an isolated plate boundary. Science 336:1690–1693. https://doi.org/10.1126/science.1218959

Braunmiller J, Nábělek J (2002) Seismotectonics of the explorer region. J Geophys Res 107:ETG1-1–ETG1-25. https://doi.org/10.1029/2001jb000220

Braunmiller J, Kradolfer U, Baer M, Giardini D (2002) Regional moment tensor determination in the European-Mediterranean area—initial results. Tectonophysics 356:5–22

Clark KJ, Hayward BW, Cochran UA, Wallace LM, Power WL, Sabaa AT (2015) Evidence for past subduction earthquakes at a plate boundary with widespread upper plate faulting: southern Hikurangi margin, New Zealand. Bull Seismol Soc Am 105:1661–1690. https://doi.org/10.1785/0120140291

Clinton JF, Hauksson E, Solanki K (2006) An evaluation of the SCSN moment tensor solutions: robustness of the M_W magnitude scale, style of faulting, and automation of the method. Bull Seismol Soc Am 96:1689–1705. https://doi.org/10.1785/0120050241

Cochran UA, Clark KJ, Howarth JD, Biasi GP, Langridge RM, Villamor P, Berryman KR, Vandergoes MJ (2017) A plate boundary record from a wetland adjacent to the Alpine Fault in New Zealand refines hazard estimates. Earth Planet Sci Lett 464:175–188. https://doi.org/10.1016/j.epsl.2017.02.026

Cooper AF, Norris RJ (1990) Estimates for the timing of the last coseismic displacement on the Alpine Fault, northern Fiordland, New Zealand. NZ J Geol Geophys 33:303–307

Dreger D, Helmberger DV (1993) Determination of source parameters at regional distances with single station or sparse network data. J Geophys Res 98:8107–8125

Dreger DS (2003) TDMT_INV: time domain seismic moment tensor INVersion. In: Lee WK (ed) International Handbook of earthquake and engineering seismology, vol 81B. Academic Press, Boston, p 1627

Duputel Z, Rivera L (2017) Long-period analysis of the 2016 Kaikoura earthquake. Phys Earth Planet Inter 265:62–66. https://doi.org/10.1016/j.pepi.2017.02.004

Dziewonski AM, Chou T-A, Woodhouse JH (1981) Determination of earthquake source parameters from waveform data for studies of global and regional seismicity. J Geophys Res 86:2825–2852

Ellis SM, Williams CA, Ristau J, Reyners ME, Eberhart-Phillips D, Wallace LM (2016) Calculating regional stresses for northern Canterbury: the effect of the 2010 Darfield earthquake. NZ J Geol Geophys 59:202–212. https://doi.org/10.1080/00288306.2015.1123740

Frohlich C, Apperson KD (1992) Earthquake focal mechanisms, moment tensors, and the consistency of seismic activity near plate boundaries. Tectonics 11:279–296

Fry B, Bannister SC, Beavan RJ, Bland L, Bradley BA, Cox SC, Cousins WJ, Gale NH, Hancox GT, Holden C, Jongens R, Power WL, Prasetya G, Reyners ME, Ristau J, Robinson R, Samsonov S, Wilson KJ, The GeoNet team (2010) The M_w 7.6 Dusky Sound earthquake of 2009: preliminary report. Bull NZ Soc Earthq Eng 43:24–40

Gerstenberger MC, Fry B, Abercrombie R, Doser D, Ristau J (2012) On the relation of stresses to aftershock decay. p 365. In: SSA 2012 annual meeting announcement Albany, California: Seismological Society of America. Seismol Res Lett *83*

Gledhill K, Ristau J, Reyners M, Fry B, Holden C (2011) The Darfield (Canterbury, New Zealand) M_w 7.1 earthquake of September 2010: a preliminary seismological report. Seismol Res Lett 82:378–386. https://doi.org/10.1785/gssrl.82.3.378

Hamling IJ, D'Anastasio E, Wallace LM, Ellis SM, Motagh M, Samsonov S, Palmer NG, Hreinsdottir S (2014) Crustal deformation and stress transfer during a propagating earthquake sequence: the 2013 Cook Strait sequence, central New Zealand. J Geophys Res 119:6080–6092. https://doi.org/10.1002/2014JB011084

Hamling IJ, Hreinsdóttir S, Clark K, Elliot J, Liang C, Fielding E, Litchfield N, Villamor P, Wallace L, Wright TJ, D'Anastasio E, Bannister S, Burbridge D, Denys P, Gentle P, Howarth J, Mueller C, Palmer N, Pearson C, Power W, Barnes P, Barrell DJA, Van Dissen R, Langridge R, Little T, Nicol A, Pettinga J, Rowland J, Stirling M (2017) Complex multifault rupture during the 2016 M_w 7.8 Kaikoura earthquake, New Zealand. Science. https://doi.org/10.1126/science.aam7194

Herman MW, Furlong KP (2016) Revisiting the Canterbury earthquake sequence after the 14 February 2016 M_w 5.7 event. Geophys Res Lett 43:7503–7510. https://doi.org/10.1002/2016GL069528

Holden C, Beavan J (2012) Kinematic source studies of the ongoing (2010–2011) sequence of recent large earthquakes in Canterbury. Paper 061 (8 p). In: Implementing lessons learnt: 2012 Conference, 13–15 April, Christchurch, New Zealand. New Zealand Society for Earthquake Engineering, Christchurch

Holden C, Bannister SC, Beavan RJ, Cousins WJ, Field BJ, McCaffrey R, Reyners ME, Ristau J, Samsonov S, Wallace LM (2008) The M_w 6.6 Gisborne earthquake of 2007: preliminary records and general source characterisation. Bull NZ Soc Earthq Eng 41:266–277

Holden C, Kaiser A, Van Dissen R, Jury R (2013) Sources, ground motion and structural response characteristics in Wellington of the 2013 Cook Strait earthquakes. Bull NZ Soc Earthq Eng 46:188–195

Holt RA, Savage MK, Townend J, Syracuse EM, Thurber CH (2013) Crustal stress and fault strength in the Canterbury Plains, New Zealand. Earth and Planet Sci Lett 383:173–181. https://doi.org/10.1016/j.epsl.2013.09.041

Horspool NA, Chadwick MP, Ristau J, Salichon J, Gerstenberger MC (2015) ShakeMapNZ: informing post-event decision making. Paper no. O-40. In: New dimensions in earthquake resilience: 2015 New Zealand Society for Earthquake Engineering technical conference and AGM, 10–12 Apr 2015, Energy Events Centre, Rotorua. New Zealand Society for Earthquake Engineering

Jost ML, Herrmann R (1989) A student's guide to and review of moment tensors. Seismol Res Lett 60:37–57

Kaiser A, Holden C, Beavan J, Beetham D, Benites R, Celentano A, Collet D, Cousins J, Cubrinovski M, Dellow G, Denys P, Fielding E, Fry B, Gerstenberger M, Langridge R, Massey C, Motagh M, Pondard N, McVerry G, Ristau J, Stirling M, Thomas J, Uma SR, Zhao J (2012) The M_w 6.2 Christchurch earthquake of February 2011: preliminary report. NZ J Geol Geophys 55:67–90. https://doi.org/10.1080/00288306.2011.641182

Kaiser A, Balfour N, Fry B, Holden C, Litchfield N, Gerstenberger M, D'Anastasio E, Horspool N, McVerry G, Ristau J, Bannister S, Christophersen A, Clark K, Power W, Rhoades D, Massey C, Hamling I, Wallace L, Mountjoy J, Kaneko Y, Benites R, Van Houtte C, Dellow S, Wotherspoon L, Elwood K, Gledhill K (2017) The Kaikoura (New Zealand) earthquake: preliminary seismological report. Seismol Res Lett 88. https://doi.org/10.1785/0220170018

Kao H, Jian P-R, Ma K-F, Huang B-S, Liu C-C (1998) Moment-tensor inversion for offshore earthquakes east of Taiwan and their implications to regional collision. Geophys Res Lett 25:3619–3622

Kao H, Shan S-J, Bent A, Woodgold C, Rogers G, Cassidy JF, Ristau J (2012) Regional centroid-moment-tensor analysis for earthquakes in Canada and adjacent regions: an update. Seismol Res Lett 83:505–515. https://doi.org/10.1785/gssrl.83.3.505

Kubo A, Fukuyama E, Kawai H, Nonomura K (2002) NIED seismic moment tensor catalogue for regional earthquakes around Japan: quality test and application. Tectonophysics 356:23–48

Langridge R, Van Dissen R, Rhoades D, Villamor P, Little T, Litchfield N, Clark K, Clark D (2011) Five thousand years of surface ruptures on the Wellington fault, New Zealand: implications for recurrence and fault segmentation. Bull Seismol Soc Am 101:2088–2107. https://doi.org/10.1785/0120100340

Little TA, Van Dissen R, Rieser U, Smith EGC, Langridge RM (2010) Coseismic strike slip at a point during the last four earthquakes on the Wellington fault near Wellington, New Zealand. J Geophys Res 115:B05403. https://doi.org/10.1029/2009JB006589

Loubet A, Kaneko Y, Tape C (2016) Validation of 3D velocity and attenuation models in the North Island of New Zealand with full-wavefield simulations. Abstract S41A-2739. In: AGU fall meeting, San Francisco, 12–16 Dec 2016: 2016 fall meeting program. American Geophysical Union, USA

Matcham I, Savage MK, Taber JJ, Reyners ME (2006) Earthquake source mechanism analysis for events between 1992 and 1997 using sparse New Zealand broadband data. NZ J Geol Geophys 49:75–89

Pasyanos ME, Dreger DS, Romanowicz B (1996) Towards real-time determination of regional moment tensors. Bull Seismol Soc Am 86:1255–1269

Petersen T, Ristau J, Beavan J, Denys P, Denham M, Field B, Francois-Holden C, McCaffrey R, Palmer N, Reyners M, Samsonov S, The GeoNet team (2009) The M_W 6.7 George Sound earthquake of October 15, 2007: response and preliminary results. Bull NZ Soc Earthq Eng 42:129–141

Petersen T, Gledhill K, Chadwick M, Gale N, Ristau J (2011) The New Zealand seismograph network. Seismol Res Lett 82:9–20. https://doi.org/10.1785/gssrl.82.1.9

Reyners M (1989) New Zealand seismicity 1964–87: an interpretation. NZ J Geol Geophys 32:307–315

Reyners M, McGinty P, Cox S, Turnbull I, O'Neill T, Gledhill K, Hancox G, Beavan J, Matheson D, McVerry G, Cousins J, Zhao J, Cowen H, Caldwell G, Bennie S, The GeoNet team (2003) The M_W 7.2 Fiordland earthquake of August 21, 2003: background and preliminary results. Bull NZ Soc Earthq Eng 36:233–248

Reyners M, Bannister S (2007) Earthquakes triggered by slow slip at the plate interface in the Hikurangi subduction zone, New Zealand. Geophys Res Lett 34:L14305. https://doi.org/10.1029/2007GL030511

Ristau J (2008) Implementation of routine regional moment tensor analysis in New Zealand. Seismol Res Lett 79:400–415. https://doi.org/10.1785/gssrl.79.3.400

Ristau J (2009) Comparison of magnitude estimates for New Zealand earthquakes: moment magnitude, local magnitude, and teleseismic body-wave magnitude. Bull Seismol Soc Am 99:1841–1852. https://doi.org/10.1785/0120080237

Ristau J (2011) Focal mechanism analysis of Christchurch Boxing Day aftershocks. GNS Science Consultancy Report 2011/43, 7 p

Ristau J (2013) Update of regional moment tensor analysis for earthquakes in New Zealand and adjacent offshore regions. Bull Seismol Soc Am 103:2520–2533. https://doi.org/10.1785/0120120339

Ristau J (2017) Source characteristics of the 2016 Kaikoura and Te Araroa, New Zealand, earthquake sequences from regional moment tensor analysis. Seismol Res Lett 87:654. https://doi.org/10.1785/0220170035

Ristau J, Rogers GC, Cassidy JF (2007) Stress in western Canada from regional moment tensor analysis. Can J Earth Sci 44:127–148. https://doi.org/10.1139/E06-057

Ristau J, Holden C, Kaiser A, Williams C, Bannister S, Fry B (2013) The Pegasus Bay aftershock sequence of the M_W 7.1 Darfield (Canterbury), New Zealand earthquake. Geophys J Int 195:444–459. https://doi.org/10.1093/gji/ggt222

Ristau J, Harte D, Salichon J (2016) A revised local magnitude (M_L) scale for New Zealand earthquakes. Bull Seismol Soc Am 106:398–407. https://doi.org/10.1785/0120150293

Robinson R (1986) Seismicity, structure and tectonics of the Wellington region, New Zealand. Geophys J R Astron Soc 87:379–409

Rodgers DW, Little TA (2006) World's largest coseismic strike-slip offset: the 1855 rupture of the Wairarapa fault, New Zealand, and implications for displacement/length scaling of continental earthquakes. J Geophys Res 111(B12408). https://doi.org/10.1029/2005jb004065

Ruppert NA, Hansen RA (2010) Temporal and spatial variations of local magnitudes in Alaska and Aleutians and comparison with body-wave and moment magnitudes. Bull Seismol Soc Am 100:1174–1183. https://doi.org/10.1785/0120090172

Scognamiglio L, Tinti E, Michelini A (2009) Real-time determination of seismic moment tensor for the Italian region. Bull Seismol Soc Am 99:2223–2242. https://doi.org/10.1785/0120080104

Seeback H, Nicol A, Villamor P, Ristau J, Pettinga J (2014) Structure and kinematics of the Taupo rift, New Zealand. Tectonics 33:1178–1199. https://doi.org/10.1002/2014TC003569

Sipkin SA (1986) Estimation of earthquake source parameters by the inversion of waveform data: global seismicity. Bull Seismol Soc Am 76:1515–1541

Stirling M, McVerry G, Gerstenberger M, Litchfield N, Van Dissen R, Berryman K, Barnes P, Wallace L, Villamor P, Langridge R, Lamarche G, Nodder S, Reyners M, Bradley B, Rhoades D, Smith W, Nicol A, Pettinga J, Clark K, Jacobs K (2012) National seismic hazard model for New Zealand: 2010 update. Bull Seismol Soc Am 102:1514–1542. https://doi.org/10.1785/0120110170

Townend J, Sherburn S, Arnold R, Boese C, Woods L (2012) Three-dimensional variations in present-day tectonic stress along the Australia-Pacific plate boundary in New Zealand. Earth Planet Sci Lett 353–354:47–59. https://doi.org/10.1016/j.epsl.2012.08.003

Van Dissen RJ, Berryman KR (1996) Surface rupture earthquakes over the last ~1000 years in the Wellington region, New Zealand, and implications for ground shaking hazard. J Geophys Res 101(B3), 5999–6019

Wallace LM, Cochran UA, Power WL, Clark KJ (2014) Earthquake and tsunami potential of the Hikurangi subduction thrust, New Zealand: insights from paleoseismology, GPS, and tsunami modelling. Oceanography 27:104–117. https://doi.org/10.5670/oceanog.2014.46

Wessel P, Smith WHF (1998) New, improved version of generic mapping tools released, *EOS, Transactions*. Am Geophys Union 79:579

Whidden KM, Pankow KL (2012) A catalog of regional moment tensors in Utah from 1998 to 2011. Seismol Res Lett 83:775–783. https://doi.org/10.1785/0220120046

Applications of Moment Tensor Solutions to the Assessment of Earthquake Hazard in Canada

J. F. Cassidy, H. Kao, John Ristau and A. Bent

1 Introduction

Centroid Moment Tensor solutions (CMT's) provide valuable information on the physics of an earthquake source, focal depth, and seismic moment. The earthquake rupture is described in terms of nine generalised force couples (a 3×3 matrix) that represent shear dislocation and volume change (see Jost and Herrmann 1989). Seismic moment (Mo) is directly related to the area of the fault that broke during the earthquake and the average displacement along the fault. Moment magnitude (Mw) is calculated directly from Mo and does not saturate unlike other magnitude scales. These moment tensor solutions, in turn, provide critical information on seismotectonics, the regional stress field, earthquake hazards, and tsunami hazards. It is noteworthy that moment tensor solutions are not restricted to earthquake sources, but also can be used for studies of landslides, explosions, implosions, and induced events.

The development of the moment tensor theory, including Green's functions and inversion, is outlined in Jost and Herrman (1989). Moment tensor inversions began in the early-mid 1970s (Gilbert 1973; Gilbert and Dziewonski 1975). Their popularity increased rapidly in with the initiation of Harvard's Global Centroid-Moment-Tensor (CMT) Project (Dziewonski et al. 1981; Ekström et al. 2012). Since 1976 more than 25,000 global CMT's (M > 5) have been computed and are available at the website http://www.globalcmt.org/CMTsearch.html. Currently most major seismological organisations around the world compute (often in near-real-time) moment tensor solutions.

J. F. Cassidy (✉) · H. Kao
Natural Resources Canada, Geological Survey of Canada, Sidney, BC, Canada
e-mail: john.cassidy@canada.ca

J. Ristau
GNS Science, Wellington, New Zealand

A. Bent
Natural Resources Canada, Geological Survey of Canada, Ottawa, ON, Canada

S. D'Amico (ed.), *Moment Tensor Solutions*, Springer Natural Hazards,
https://doi.org/10.1007/978-3-319-77359-9_14

The goal of this article is to highlight the numerous applications of Regional Moment Tensor solutions (RMT's) to understanding seismotectonics, and earthquake and tsunami hazards across Canada (sampling a wide variety of tectonic settings).

2 Tectonic Setting and Seismicity

The Canadian landmass spans a wide variety of tectonic settings from a stable continental craton with the oldest rocks on earth to an active subduction zone off the west coast (Fig. 1). The active plate margin along the west coast includes the Cascadia subduction zone of southwest BC where the oceanic Juan de Fuca and Explorer plates are subducting beneath the North American Plate at a rate of 2–4 cm/year (Riddihough and Hyndman 1991); the mainly transcurrent Queen Charlotte Fault (just west of Haida Gwaii) where the Pacific and North America plates slide past one another; and, further north, the Yakutat collision zone near the Alaska-British Columbia-Yukon border region. The Canadian cordillera is comprised of numerous terranes and microterranes that have collided with North America over the past ~80 Ma (e.g., Gabrielse et al. 1991). To the east of the Cordillera lies the Plains and stable North American craton—including some of the oldest rocks on earth. To the east of the stable craton lies the Grenville Province and Appalachian Orogen (Fig. 1).

The pattern of seismicity across the country (Fig. 2) follows the tectonics. The largest and most frequent earthquakes occur along the active plate boundary off the west coast and through the northern Canadian Cordillera. There are few earthquakes through the stable craton in the continental interior. Moderate seismicity (including larger M 6-7 earthquakes) occurs throughout eastern Ontario, the St. Lawrence Valley region, and through the Appalachians (Fig. 2). One of the challenges in assessing earthquake hazards in Canada is in understanding the linkage between seismicity and tectonics. Moment tensor solutions provide a means to directly examine those links, to help identify active faults, and to better assess earthquake hazards.

3 Methodologies

There are several key steps involved in determining moment tensor solutions. These are: (1) data acquisition and pre-processing (good signal-to-noise ratio waveforms, removing instrument response); (2) calculation of synthetic Green functions for local earth models, earthquake-receiver combinations (this is the most computationally expensive part, but can be done in advance and archived); and (3) inversion and interpretation (including uncertainties and resolution).

The primary moment tensor solution methodologies used in Canada are described in detail in Ristau et al. (2007), Kao et al. (1998, 2012). In both cases, these techniques utilise regional broadband waveforms (typically source-receiver distances of 1500 km or less, rather than teleseismic data). The advantage of using regional data

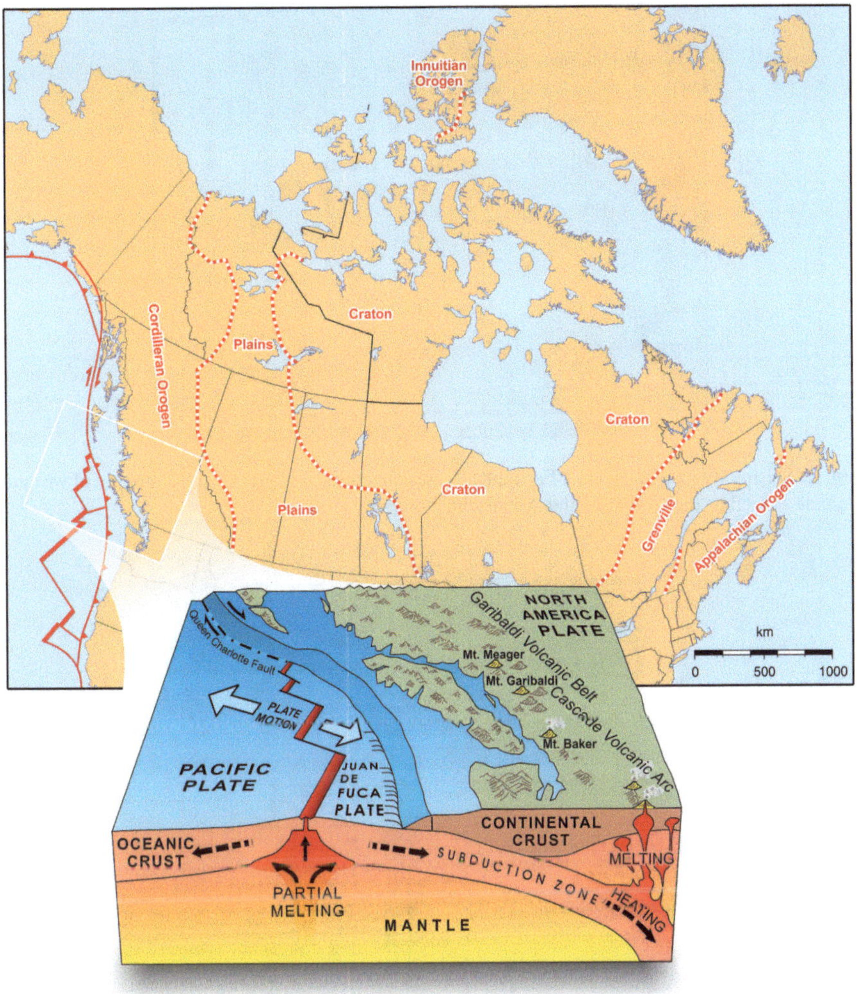

Fig. 1 Tectonic environments of Canada. Inset shows the active tectonics along Canada's west coast, including the Queen Charlotte Fault and the Cascadia subduction zone (from Cassidy et al. 2010)

is that lower magnitude earthquakes can be modelled, one disadvantage is that relatively detailed local earth models are required for the waveform modelling. The period range used for the waveform modelling ranges from about 12.5–50 s (Ristau et al. 2007) to ~16–100 s (Kao et al. 2012). Typically these "regional moment tensor" techniques (RMT's) can determine focal mechanisms for earthquakes of ~Mw 4.0, and sometimes as small as Mw ~ 3.5. An example of a typical regional moment tensor solution in Canada is shown in Fig. 3.

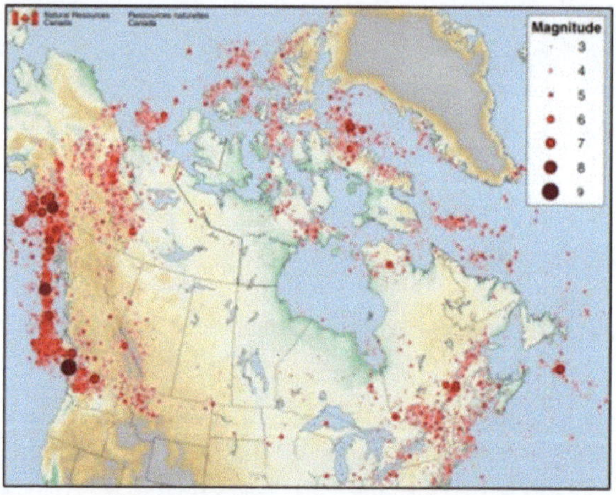

Fig. 2 Seismicity map of Canada. Earthquakes are concentrated along the active margin of western Canada as well as the Grenville and Appalachian orogeny

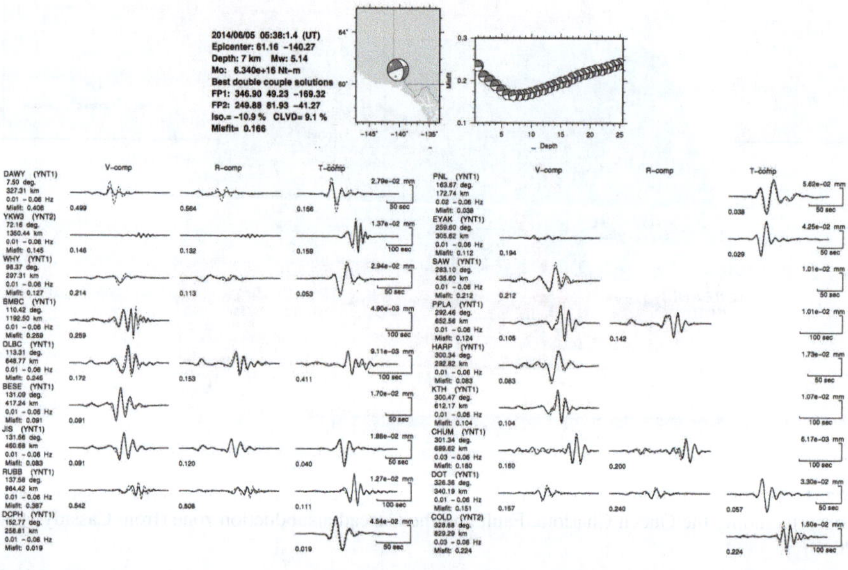

Fig. 3 Example of a typical regional CMT solution for a Canadian earthquake. A summary of source parameters is given in the text block at top with a map showing the event's focal mechanism placed at the epicenter. The misfit function with respect to focal depth is plotted to the right. In the lower panel, a station-by-station plot of the three-component observed (solid lines) and synthetic (dashed lines) waveforms is given with a brief description of model parameters used for each station and the plotting scale on the left and right sides, respectively. The number at the lower-left corner of each component corresponds to the individual misfit. The source time function used in the inversion (rise time, flat duration, and fall time) is given at the bottom of the waveform plot

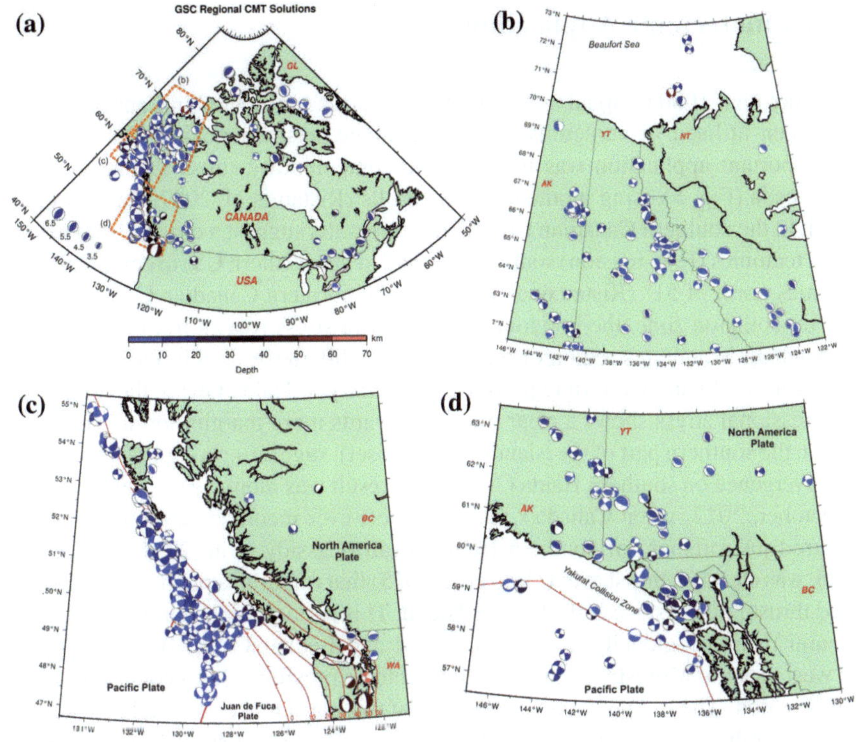

Fig. 4 A composite plot showing all GSC regional CMT solutions (1995–2017) for earthquakes in and adjacent to Canada (**a**) and zoom-in plots for events in Yukon and Northwest territories (**b**), the southwest Yukon and Alaska Panhandle (**c**), and southwestern British Columbia (**d**). Each beach ball shows the best double-couple solution in lower-hemisphere projection for one event. The colored areas correspond to quadrants with dilatational strain (i.e., upward first motions). Different colors represent different ranges of the focal depth, and the size of the beach ball is proportional to the corresponding Mw, as shown by the legends. The iso-depth contours of the subducted Juan de Fuca plate are adopted from McCrory et al. (2004)

Since 1995, more than 1000 (as of Jan. 2017) regional moment tensor solutions have been computed for Canadian earthquakes (Fig. 4), including 934 for western Canada and 72 for eastern Canada. These have provided new opportunities (as described below) to explore the seismotectonics of the Canadian landmass and have been utilised in developing earthquake hazard models for the 2015 National Building Code of Canada.

4 Seismotectonic Applications

The more than 1000 regional moment tensor solutions computed in Canada (Fig. 4) have been utilised in numerous ways to evaluate seismotectonics. For example, one important application was mapping the contemporary stress regime in western Canada (Fig. 5) using moment tensor results (Ristau et al. 2007; Cassidy et al. 2008). In the southern Canadian Cordillera, the results suggest a change from extensional tectonics (E-W tension) south of 51° (and in the northern US) to compressional tectonics north of 51° (Ristau et al. 2007). In the northern Canadian Cordillera and Yakutat collision zone, the horizontal compressive stress rotates from near E-W in the Nahanni Mountains region of the eastern Cordillera to ~N-S in the Richardson Mountains (YT) and western regions. In the vicinity of Haida Gwaii, the horizontal compressional stress shows a clear rotation towards more margin-normal compression on the southern part of the islands (Fig. 5, inset), suggesting a significant amount of convergence on southern Haida Gwaii. This result was more clearly demonstrated in October, 2012, when Canada's second-largest-ever recorded earthquake (M7.8) occurred just offshore of southern Haida Gwaii (Cassidy et al. 2014). The mainshock was clearly a thrust event (Kao et al. 2015) that ruptured a previously unrecognised thrust fault, just west of Moresby Island. This earthquake generated the largest tsunami recorded globally in 2012 (Leonard and Bednarski 2014) that impacted the west coast of Moresby Island and highlighted a previously-unrecognised hazard in this region. The moment tensor solutions of this earthquake and the largest aftershocks (Farahbod and Kao 2015) were applied to evaluate changes in Coulomb stress caused by the mainshock and linkages with locations of large aftershocks (Hobbs et al. 2015).

In eastern and northern Canada, CMT solutions provide focal mechanisms (Bent 2015a, b, 2017) in areas of sparse station coverage, where P-nodal solutions could not previously be determined. These CMT solutions, in turn, provide valuable new information on faulting styles, the stress regime, and seismotectonics (e.g., Bent et al. 2016).

Another key piece of information from CMT solutions across Canada is focal depth estimates. These are particularly valuable in areas of sparse seismic coverage, or in offshore regions. Focal depths provide information on seismogenic thickness, which in turn is a critical parameter for modelling crustal deformation (e.g., Hyndman et al. 2005).

5 Hazard Applications

Earthquake hazard assessments require a complete and uniform earthquake catalogue. Currently, the global standard for magnitude is M_w. This magnitude has a physical meaning, in that it is calculated from Mo which is directly related to the area of a fault that ruptures during an earthquake and the average displacement along

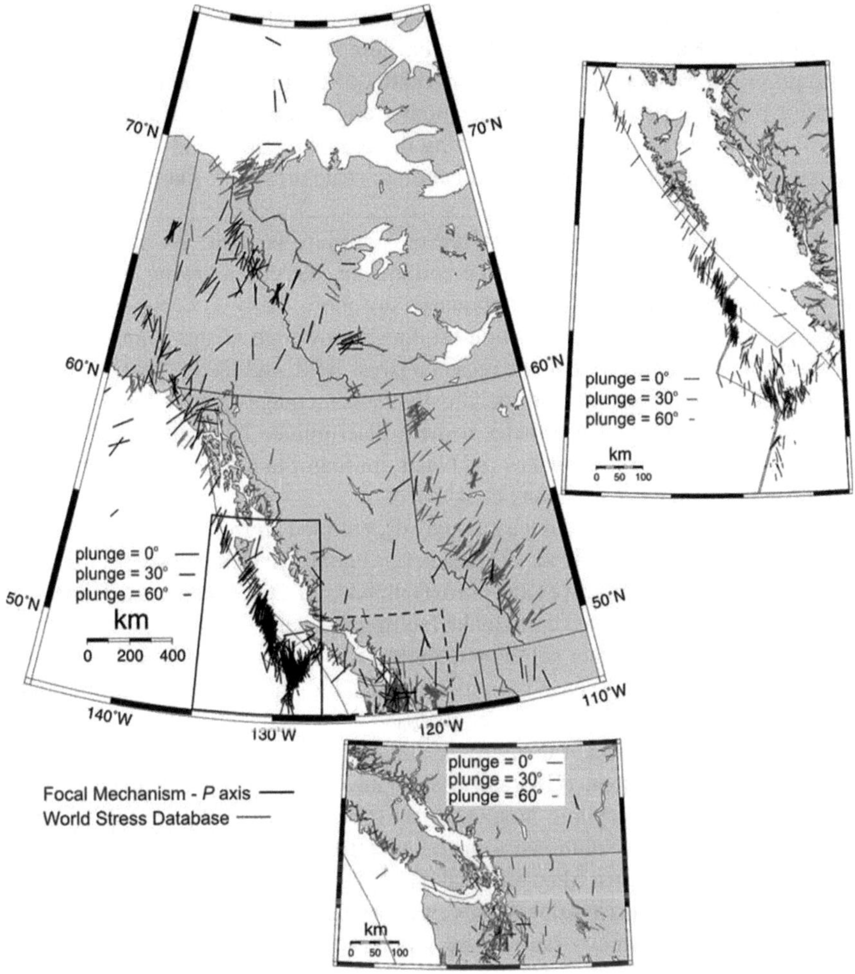

Fig. 5 P-axis azimuths from focal mechanisms in Canada with compressive stress directions from the world stress database. The arrow lengths are scaled according to the plunge of the axis, with 0° plunge being the longest and 90° having zero length. The compressive stress direction is generally oriented NE-SW for most of western Canada, with the exception of southwest British Columbia and northwest Washington where the compressive stress direction is north-south

the fault. M_w is available directly from centroid moment tensor solutions. Through a series of studies (e.g., Ristau et al. 2003, 2005; Kao et al. 2012; Bent 2011, 2016) M_w estimates for recent earthquakes have been used to develop relationships with other magnitude scales that have been traditionally used in Canada (e.g., M_L and M_N) thus allowing for the magnitude of older earthquakes to be converted to M_w and thereby allowing for the development of a uniform M_w catalog (Halchuk et al. 2015).

Identification of active faults using moment tensor analysis (e.g., the Haida Gwaii Thrust Fault—Kao et al. (2015)) allows for these structures to be incorporated into earthquake hazard maps, and for scenario earthquakes (and tsunamis) to be considered for planning purposes. This was another direct application of moment tensor solutions in Canada. In addition, knowing the regional stress field and the orientation of surface faults allows for identification of structures more likely to experience earthquakes (Ristau et al. 2007).

Another application of moment tensor solutions is their use in calculating Coulomb stress changes after a large earthquake, thereby allowing for estimates of where aftershocks or future earthquakes are more likely to occur (e.g., Hobbs et al. 2015). This type of calculation requires knowledge of the mainshock focal mechanism and rupture extent, the regional stress field, and likely "receiver faults".

Moment tensor solutions, and specifically the percentage of "non-double-couple" solution, is an important parameter to help discriminate between typical "shear-dislocation" (tectonic) earthquakes and blasts, mining-related events, and other types of induced seismicity (e.g., Zhang et al. 2016).

The applications of moment tensor analysis, outlined above, have all been used in the 2015 National Building Code of Canada Earthquake Hazard maps. Specifically: a uniform M_W catalog was derived and used (Halchuk et al. 2015); improved delineation of seismic source zones used earthquake faulting style and crustal stress maps (both derived from moment tensor solutions); and several new subduction faults (e.g., the Haida Gwaii Thrust Fault) have been incorporated into the earthquake hazard model for the first time in Canada (Adams et al. 2015).

Finally, an important and evolving seismic hazard application is incorporating deformation measurements (GPS and GNSS) into hazard maps. One method being explored is using the seismic moment rates (obtained from the seismic moment determined from CMT solutions) to estimate deformation rates and compare with deformation observations. Mazzotti et al. (2011) compared deformation rates from seismicity with GPS observations across western Canada. In some areas (especially regions of higher seismicity rates) there was reasonable agreement, while in other areas (e.g., Cordillera and east) GPS deformation rates exceeded the seismic rates by factor of 6–150. This may reflect the short (and incomplete) seismic catalogue, or it may represent aseismic deformation, or, more likely, a combination of the two. Studies such as this are possible, in large part, because of centroid moment tensor solutions and estimates of seismic moment. Applications of CMT solutions to earthquake hazard estimation will continue to increase as more seismic stations re deployed, and more data becomes available.

6 Summary and Conclusions

During the past 22 years, more than 1000 earthquake moment tensor solutions have been computed across Canada. These have contributed directly to our understanding of the seismotectonics and seismicity in all tectonic environments of the country, and have been utilised in mapping earthquake and tsunami hazards. A few specific examples include developing a uniform Mw catalogue for Canada, mapping the crustal stress field, allowing for improved delineation of earthquake source zones in hazard models, identification of a new active subduction fault (the Haida Gwaii Thrust Fault) to the west of Moresby Island, and an improved understanding of associated tsunami hazards. Moment tensor solutions also contribute to the important and rapidly evolving field of understanding induced seismicity. Seismic depths estimated from these solutions allow for improved estimates of seismogenic strength, and modelling of crustal deformation data. Undoubtedly there will be many more applications in the coming years, as the database of moment tensor solutions continues to grow.

Acknowledgements We thank the reviewers for their comments that have improved this manuscript. We gratefully acknowledge the many organisations that have provided waveform data, including CNSN, POLARIS, ARSN, PNSN, ANSS, and IRIS-DMC. This is ESS Contribution Number 20170017.

References

Adams J, Halchuk S, Allen TI, Rogers GC (2015) Canada's 5th generation seismic hazard model, as prepared for the 2015 national building code of Canada. In: Proceedings of the 11th Canadian conference on earthquake engineering, Victoria, BC., paper 93782

Bent AL (2011) Moment magnitude (Mw) conversion relations for use in hazard assessment in eastern Canada. Seismol Res Lett 82(6):984–990. https://doi.org/10.1785/gssrl.83.3.984

Bent AL (2016) Moment magnitude (MW) conversion relations for use in hazard assessment in offshore eastern Canada. In: Bent AL (ed) Geological Survey of Canada, Open File 8027, 12 pp. https://doi.org/10.4095/297965

Bent AL (2015a) Regional centroid moment tensor solutions for eastern Canadian earthquakes: 2011–2013, Geological Survey of Canada, Open File 7726, 71 pp. https://doi.org/10.4095/296795

Bent AL (2015b) Regional centroid moment tensor solutions for eastern Canadian earthquakes: 2014, Geological Survey of Canada, Open File 7834, 35 pp. https://doi.org/10.4095/296795

Bent AL (2017) Regional centroid moment tensor solutions for eastern Canadian earthquakes: 2015, Geological Survey of Canada, Open File 5080, 26 pp. https://doi.org/10.4095/299816

Bent AL, Peci V, Halchuk S (2016) The 2015 Canada day, Mw 3.8, earthquake in Nova Scotia. Seismol Res Lett 87(5):1224–1231. https://doi.org/10.1785/0220160074

Cassidy JF, Rogers GC, Ristau J, Adams J, Bent A (2008) The Canadian crustal stress database: update to 2008, 3rd world stress map conference, frontiers of stress research: observation, integration and application, p. 92

Cassidy JF, Rogers GC, Lamontagne M, Halchuk S, Adams G (2010) Canada's earthquakes: the Good, the bad, and the ugly. Geosci Can 37:1–16

Cassidy JF, Rogers GC, Hyndman RD (2014) An overview of the October 28, 2012 Mw 7.7 earthquake in Haida Gwaii, Canada: a tsunamigenic thrust event along a predominantly strike-slip margin. Pure Appl Geophys, Feb. 2014, ISSN: 0033-4553 (Print) 1420-9136 (Online), 171, 283–292. https://doi.org/10.1007/s00024-014-0775-1

Dziewonski AM, Chou T-A, Woodhouse JH (1981) Determination of earthquake source parameters from waveform data for studies of global and regional seismicity. J Geophys Res 86(2825–2852):1981. https://doi.org/10.1029/JB086iB04p02825

Ekström G, Nettles M, Dziewonski AM (2012) The global CMT project 2004–2010: centroid-moment tensors for 13,017 earthquakes. Phys Earth Planet Inter 200–201:1–9. https://doi.org/10.1016/j.pepi.2012.04.002
Farahbod AM, Kao H (2015) Spatiotemporal distribution of events during the first week of the 2012 Haida Gwaii aftershock sequence. Bull Seismol Soc Am 105(2B):1231–1240. https://doi.org/10.1785/0120140173
Gabrielse H, Monger JWH, Wheeler JO, Yorath CJ (1991) Tectonic framework part A. Morphological belts, tectonic assemblages and terranes, Chap. 2. In: Gabrielse H, Yorath CJ (eds) Geology of the Cordilleran orogen in Canada, geological society of America, Denver, CO, pp 15–28
Gilbert E (1973) Derivation of source parameters from low-frequency spectra. Philos Trans R Soc 274:369–371
Gilbert F, Dziewonski DM (1975) An application of normal mode theory to the retrieval of structural parameters and source mechanisms from seismic spectra. Philos Trans R Soc Lond A 278, 187–269. https://doi.org/10.1098/rsta.1975.0025
Halchuk S, Allen TI, Rogers GC, Adams J (2015) Seismic hazard earthquake epicentre file (SHEEF2010) used in the fifth generation seismic hazard maps of Canada, Geological Survey of Canada, Open File 7724, 21 pp. https://doi.org/10.4095/296908
Hobbs TE, Cassidy JF, Dosso SE, Brillon C (2015) Coulomb stress changes following the Mw 7.8 2012 Haida Gwaii, Canada earthquake: implications for seismic hazard. Bull Seismol Soc Am 105(2B):1253–1264. https://doi.org/10.1785/0120140158
Hyndman RD, Flück P, Mazzotti S, Lewis TJ, Ristau J, Leonard L (2005) The lithoprobe slave-northern cordillera lithospheric evolution (SNORCLE) transect/current tectonics of the northern Canadian Cordillera; La géotraverse de l'évolution lithosphérique de la Cordillère du Nord et de la province du lac des Esclaves (SNORCLE). In: Cook FA, Erdmer P (eds) Can J Earth Sci 42, 1117–1136. https://doi.org/10.1139/e05-023
Jost ML, Herrmann RB (1989) A student's guide to and review of moment tensors. Seismol Res Lett Apr 60(2):37–57. https://doi.org/10.1785/gssrl.60.2.37
Kao H, Juan P-R, Ma K-F, Huang B-S, Liu C-C (1998) Moment-tensor inversion for offshore earthquakes east of Taiwan and their implications to regional collision. Geophys Res Lett 25:3619–3622
Kao H, Shan S-J, Bent A, Woodgold C, Rogers GC, Cassidy JF, Ristau J (2012) Regional centroid-moment-tensor inversion for earthquakes in Canada and adjacent regions: an update. Seismol Res Lett 83(3):505–515
Kao H, Shan S-J, Farahbod AM (2015) Source characteristics of the 2012 Haida Gwaii earthquake sequence. Bull Seismol Soc Am 105(2B):1206–1218. https://doi.org/10.1785/0120140165
Leonard LJ, Bednarski JM (2014) Field survey following the 28 October 2012 Haida Gwaii tsunami. Pure and Appl Geophys, 16 pp. https://doi.org/10.1007/s00024-014-0792-0
McCrory PA, Blair JL, Oppenheimer DH, Walter SR (2004) Depth to the Juan de Fuca Slab beneath the Cascadia Subduction Margin: a 3-D Model for sorting earthquakes. USGS Data Series 91, 1 CD-ROM. http://pubs.usgs.gov/ds/91/
Mazzotti S, Leonard LJ, Cassidy JF, Rogers GC, Halchuk S (2011) Seismic hazard in Western Canada from GPS strain rates versus earthquake catalog. J Geophys Res 116:B12310. https://doi.org/10.1029/2011JB008213
Riddihough RP, RD Hyndman (1991) Modern plate tectonic regime of the continental margin of western Canada, in Geology of North America, vol G-2, geology of the Cordilleran orogen in Canada. In: Gabrielse H, Yorath CJ (eds) Geological Survey of Canada, Ottawa, pp 435–455
Ristau J, Rogers GC, Cassidy JF (2007) Stress in western Canada from regional moment tensor analysis. Can J Earth Sci 44:127–148

Ristau J, Rogers GC, Cassidy JF (2003) Moment magnitude—local magnitude calibration for earthquakes off Canada's west coast. Bull Seismol Soc Am 93:2296–2300

Ristau J, Rogers GC, Cassidy JF (2005) Moment magnitude—local magnitude calibration for earthquakes in western Canada. Bull Seismol Soc Am 95:1994–2000

Zhang H, Eaton DW, Li G, Liu Y, Harrington RM (2016) Discriminating induced seismicity from natural earthquakes using moment tensors and source spectra. J Geophys Res Solid Earth 121:972–993. https://doi.org/10.1002/2015JB012603

Intraplate Earthquakes in Europe—Source Parameters from Regional Moment Tensor Analysis

Jochen Braunmiller

1 Introduction

Plate tectonics provides a highly successful framework to describe a wide range of geological observations invoking the motion of lithospheric plates. In its simplest form the plates are rigid and earthquakes are confined to boundaries where plates move relative to each other. Over 90% of all earthquakes (Johnston 1989) follow this tenet and, considering that subduction zone earthquakes are the largest seismic events, an even higher percentage of the global seismic moment is released along plate boundaries. However, plate interiors are not entirely aseismic and experience rare, large earthquakes such as the 1811–1812 New Madrid earthquakes (e.g., Johnston 1996). Causes for intraplate earthquakes are debated but probably involve transfer of stresses exerted at plate boundaries to plate interiors with seismicity seemingly concentrated along zones of preexisting weakness (e.g., Stein and Mazzotti (2007) and Talwani (2014) for overviews on intraplate earthquakes).

Seismicity in the European-Mediterranean region follows the plate boundary zones (Fig. 1) related to the convergence of the Africa (Nubia), Adria, and Arabia plates relative to Eurasia. A significant number of earthquakes, however, are located inside the Eurasian plate. The earthquakes considered here include events in the stable continental region (SCR) of northern and central-western Europe (blue and brown shading in Fig. 1) as well as events in broad, actively deforming regions such as the Hellenic backarc and the Caucasus Arabia-Eurasia collision zone (yellow in Fig. 1). SCRs, as defined in Johnston et al. (1994), are areas of continental crust with no major tectonic, magmatic, basement metamorphic or anorogenic intrusive event since the early Cretaceous and no other major extension, rifting or transtension event since the Paleogene. Shading in Fig. 1 indicates regions experiencing the last major tectonic event in the Precambrian (brown), Paleozoic (blue), and Mesozoic (yellow)

J. Braunmiller (✉)
School of Geosciences, University of South Florida, Tampa, FL 33620, USA
e-mail: jbraunmiller@usf.edu

S. D'Amico (ed.), *Moment Tensor Solutions*, Springer Natural Hazards,
https://doi.org/10.1007/978-3-319-77359-9_15

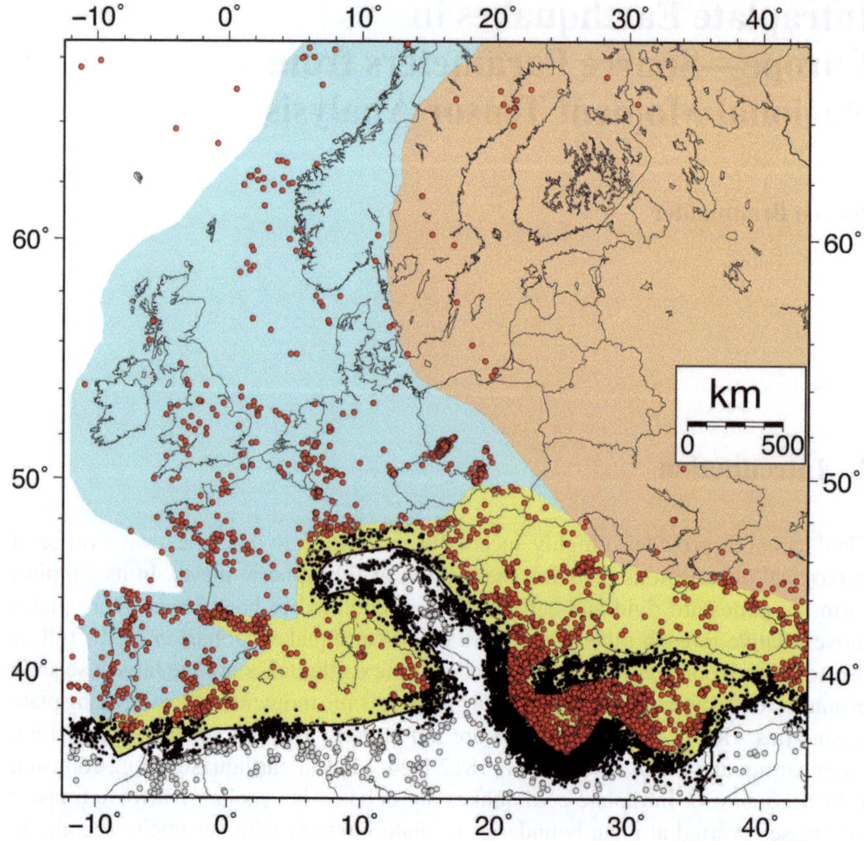

Fig. 1 Seismicity in Europe. Shown are 23,100+ magnitude M ≥ 3.5 earthquakes from the USGS PDE catalog from 1989 to June 2016 (27.5 years). Red, gray: Intraplate earthquakes (taken as ≥ 125 km from plate boundary) in Eurasia and other plates, respectively. Black: plate boundary events. Solid black lines are plate boundaries from the PLATES Project (Coffin et al. 1998) and extended to include the Adria microplate. Brown, blue, and yellow shading for regions experiencing last major tectonic phase in the Precambrian, Paleozoic, and Mesozoic era, respectively, after Johnston et al. (1994)

eras, respectively, following maps in Johnston et al. (1994). In Western Europe, I include earthquakes within ~50 km of the Paleozoic-Mesozoic boundary to the SCR dataset, but also show results excluding these events.

The systematic retrieval of earthquake source parameters may improve the understanding of causes for intraplate seismicity, for example, in relation to acting stresses and existing geologic structures. However, seismicity away from broad deforming plate boundary zones is low (Fig. 1) and larger events (moment magnitude M_w 5.0 and more) that can be analyzed with teleseismic waveforms occur infrequently. The broadband seismic networks in Europe, which started to be installed in the 1990s, offer an opportunity to obtain moment tensors for the more frequent moderate-sized events ($M_w \geq 4.0$–4.5). Here, I combine robust and reliable regional moment tensor (for 1995–2007) and Global CMT (for 1976–2015) databases to investigate deformation style, stress field orientation, and depth distribution of crustal intraplate earthquakes in Europe with focus on its SCR part.

2 Seismicity

Seismicity inside the European plate is considerably lower than at its plate boundaries and large intraplate events are rare. Figure 1 shows 27.5 years (1989-June 2016) of seismicity with magnitude $M \geq 3.5$ from the USGS PDE catalog. More than 68% of the 23,100+ events occurred along plate boundaries (black in Fig. 1) and another 20% occurred in the broadly deforming Aegean and western Anatolia regions, the Caucasus and the Vrancea zone, Romania (red circles in Fig. 1 east of 20° E and south of 46° N). Less than 6% of seismicity occurred in the rest of Europe (remaining red circles in Fig. 1). For simplicity, I will refer to earthquakes in this latter region as "intraplate" in contrast to events in the broad plate boundary zone in the larger Aegean-Anatolian region. (The rest are events in northern Africa, Arabia, and Adriatic—2%—and mining related events in southwest Poland—4%).

The Precambrian shield region (brown shading) with only about 1 $M \geq 3.5$ event per year is essentially aseismic away from its exterior boundaries. Seismicity rate in the Paleozoic (blue) and the western Mesozoic (yellow, west of 20° E) parts of Europe are similar and combine to about 50 events per year. Seismicity shows clear areas of concentration that are separated from relatively aseismic regions (such as northern France or southwestern France away from the Pyrenees).

The time span shown in Fig. 1 is too short to characterize the distribution and expected maximum size of intraplate earthquakes properly. The largest intraplate earthquake since 1989 in Europe was the 1992 Roermond, Netherlands $M_w = 5.3$ event in the Lower Rhine embayment (Van Eck and Davenport 1994), which is small compared to the largest event shown on Fig. 1, the $M_w = 7.6$ 1999 Izmit, Turkey earthquake, but caused considerable losses of about US$150 Million (Berz 1994). Large intraplate earthquakes are rare but a repeat of the 1356 Basel, Switzerland earthquake, the largest historic earthquake north of the Alps, estimated as M_w 6.7–7.1 (Fäh et al. 2009), could cause billions of US$ in damage.

3 Data and Method

The bulk of the moment tensor solutions shown in this paper are obtained by regional moment tensor (RMT) inversion. The method is presented in Nabelek and Xia (1995) and has been successfully applied in the European-Mediterranean region (Braunmiller et al. 2002). I refer to the two papers for details of the method and its application. Long-period, three-component waveform modeling provides robust results for the seismic moment tensor, seismic moment, and centroid depth. The method can be applied routinely to M_w 4.0–4.5 and larger earthquakes and in regions of dense broadband seismic networks (such as Switzerland and parts of Germany) to events as small as $M_w \approx 3.0$.

Regional broadband data were obtained from several international (IRIS, Geofon, Geoscope, ORFEUS) and national (Austria, Germany, Israel, and Switzerland) data centers. The network of broadband seismic stations in the European-Mediterranean region has expanded rapidly in the last 25 years and for an overview of currently openly available broadband stations visit the European Integrated Data Archive hosted by the ORFEUS data center (www.orfeus-eu.org). The stations available for this study are roughly similar to the station map shown in Braunmiller et al. (2002).

The RMT database covers the 1995–2007 period and includes the 1978 $M_w = 5.2$ Swabian Alb, Germany and other selected 1989–1994 events for which sufficient broadband data exist. Some solutions shown here have been published elsewhere (Braunmiller et al. 2002; Braunmiller 2002; Bernardi et al. 2005; and yearly reports of the Swiss Seismological Service: Baer et al. (2007) and references therein). The European intraplate and plate boundary zone events discussed here are a subset of the complete RMT database; its publication is in preparation.

4 Results

The database of European intraplate source parameters considers only events that are at least 125 km from an active plate boundary (defined by thick solid lines in Fig. 2). Analysis of regional data reduces the magnitude threshold compared to teleseismic methods. The database includes 451 RMT solutions (1995–2007: 433) compared to 225 Global CMT solutions (Dziewonski et al. 1981; Ekström et al. 2012) for 1976–2015, or 33 compared to fewer than 6 events per year. The 87 common events are compared (see Appendix) and high similarity of results attests to robustness and reliability of waveform modeling derived source parameters.

For further analysis, I excluded sub-crustal earthquakes (depth $z > 40$ km) associated with current subduction zones (Tyrrhenian Sea, Hellenic arc) and the slab remnant of the Carpathian arc resulting in 504 unique crustal seismic events. Strong events ($M \geq 3.5$) induced by mining and hydrocarbon extraction are relatively common in parts of Poland, Germany, the Netherlands, and in the North Sea and 29 such events are identified in Fig. 2; while not focus of this paper, it is noteworthy that their source parameters overall agree well with Cesca et al. (2013).

The color scheme in Fig. 2 and subsequent maps follows the World Stress Map project (www.world-stress-map.org) with red for normal faulting, green for strike-slip, blue for thrust faulting, and black for "unknown". The categories of the stress regime are defined by principal axes orientations and given in Zoback (1992). The results in terms of deformation style and orientation of the maximum horizontal stress direction S_{Hmax}, as expected, agree extremely well with the existing stress data shown for the European-Mediterranean region, e.g., in Heidbach et al. (2016). An

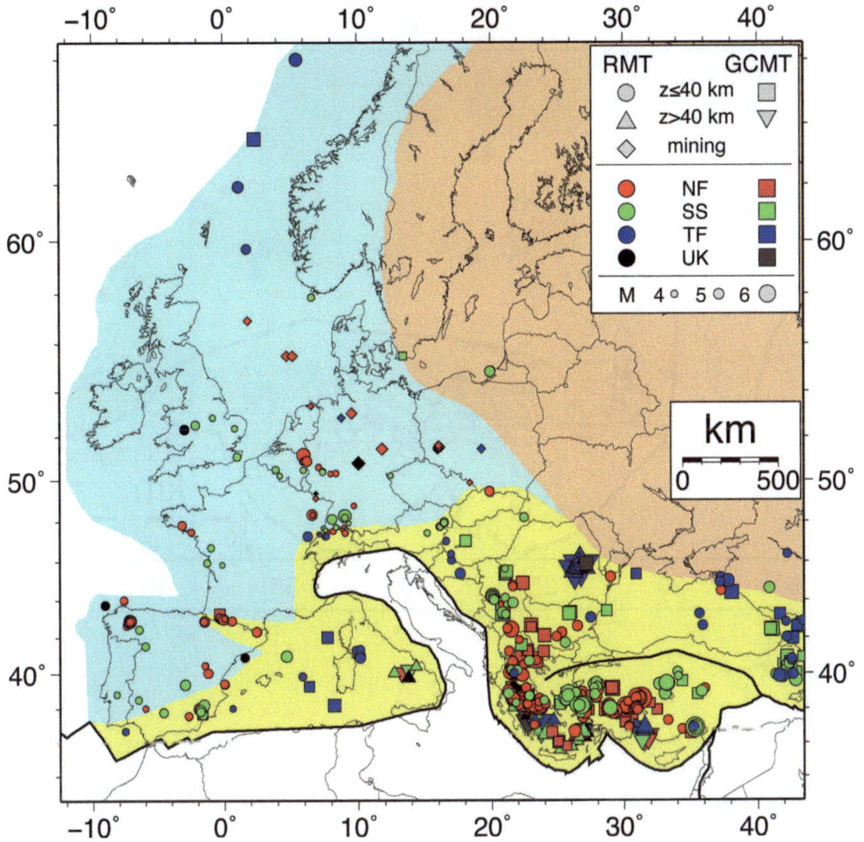

Fig. 2 Distribution of seismic events with regional moment tensor (RMT) and/or Global CMT solution in Europe. The 475 shallow crustal earthquakes are shown as circles (RMT) and squares (GCMT), respectively, and color-coded to indicate different stress regimes (following World Stress Map classification). Symbol size scales with M_w. The 29 suspected mining and hydrocarbon extraction events are marked as diamonds. In addition, 85 sub-crustal (z > 40 km) events are shown as triangles (RMT) and inverted triangles (GCMT), respectively, but are not discussed further. Otherwise as Fig. 1

important point, though, is that all source parameters shown here are derived from long-period waveform modeling and, as shown in Figs. 3 and 4, provide internally highly consistent results for different sub-regions.

4.1 Southeastern Europe, Anatolia and Caucasus

The focus of this contribution is on Western Europe and results for other areas are only summarized. Figure 3 shows fault plane solutions and S_{Hmax} directions for

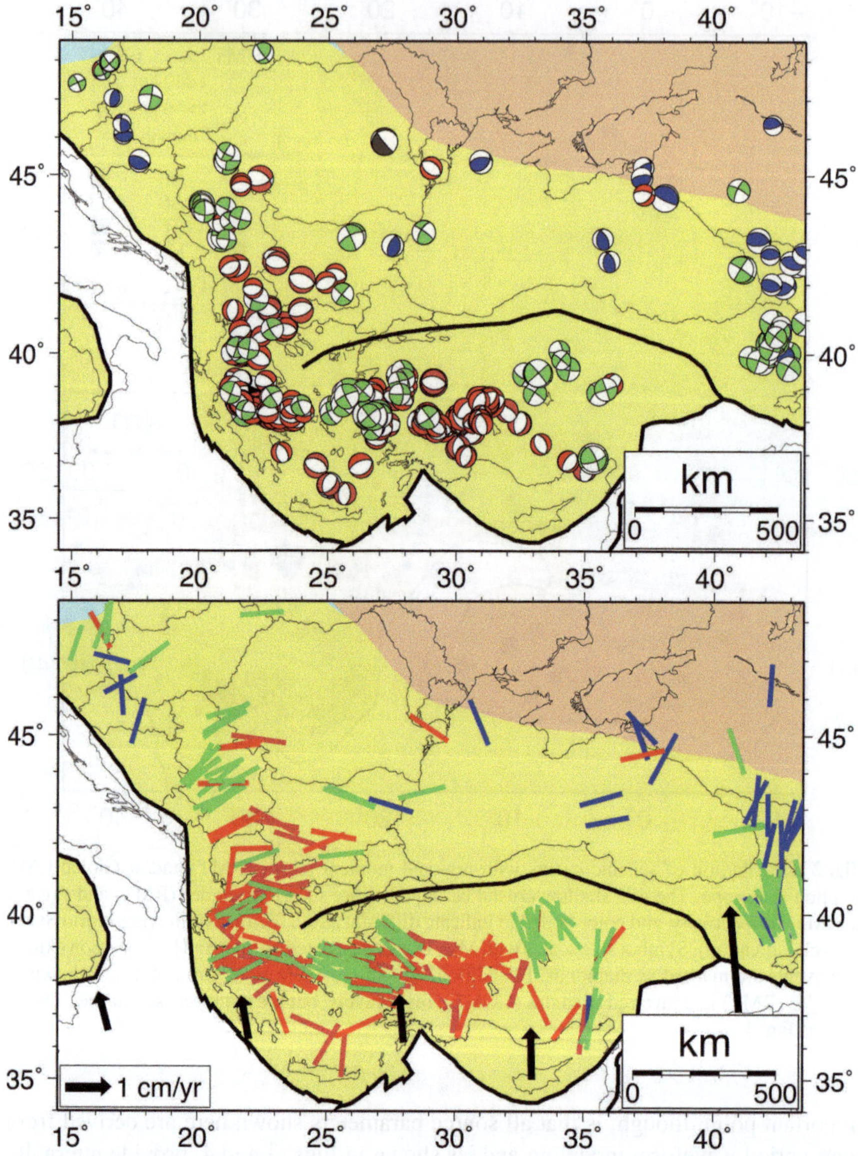

Fig. 3 Top: Fault plane solutions in southeast Europe, Anatolia and the Caucasus. Shown are 356 crustal events with depth z $\leq$ 40 km and a minimum double couple contribution of 20% to exclude mechanisms with ill-constrained principal axes. Event color-coding as in Fig. 2 and background color scheme as in Fig. 1. Note high consistency of mechanisms within regions. Plot order: "unknown", thrust, normal and strike-slip on top. Bottom: S_{Hmax} directions for same earthquakes ("unknown" excluded). Black arrows show MORVEL (DeMets et al. 2010) Nubia-Eurasia and Arabia-Eurasia plate motion vectors, respectively

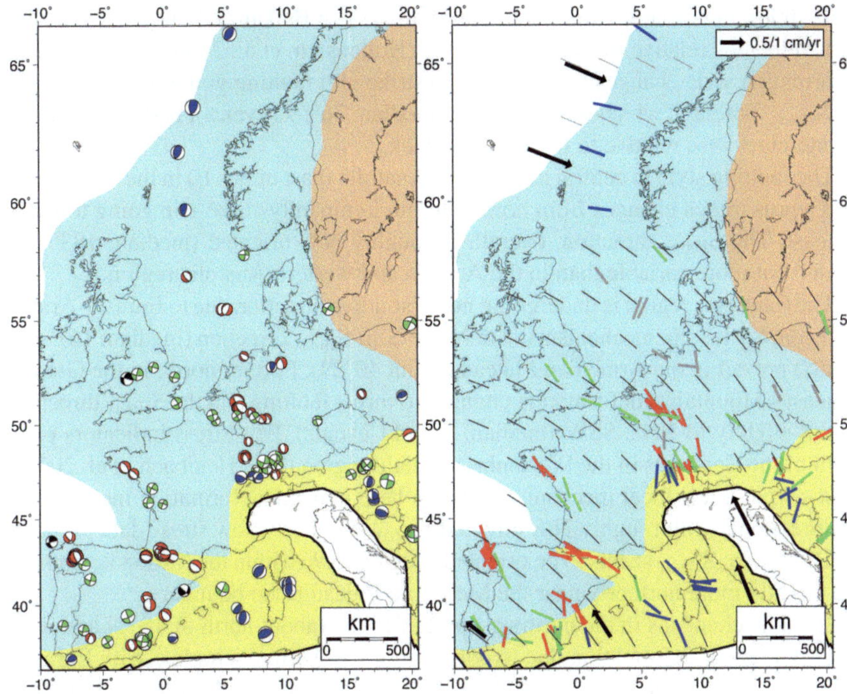

Fig. 4 Left: Fault plane solutions in Western Europe. Shown are 139 crustal seismic events ($z \leq 40$ km) with a minimum double couple and deviatoric moment tensor contribution of 20% and 60%, respectively, to exclude mechanisms with ill-constrained axes orientations. Non-deviatoric sources are considered only for the 29 mining or hydrocarbon-extraction related seismic events; 10 have a significant deviatoric contribution and are shown. Otherwise as Fig. 3. Right: S_{Hmax} directions for same earthquakes ("unknown" excluded and gray bars for non-tectonic events). Black arrows show MORVEL Nubia-Eurasia (scale is 0.5 cm/yr) and North America-Eurasia (scale is 1.0 cm/yr) plate motion vectors, respectively. Thin black lines are trajectories of MORVEL Nubia plate motion relative to Europe; thin gray lines (north of 60° N) show same for North America

events in southeastern Europe, Anatolia and the western Caucasus. The majority of events in the combined (RMT and GCMT) dataset occurred in the Aegean region and in western-central Anatolia in the back arc of the Hellenic and Cyprus arcs where Africa (Nubia) slowly subducts beneath the Aegean and Anatolian blocks.

In the Aegean Sea and western Anatolia region (about south of 39.5° N and west of 30° E), N-S extension and westward escape of Anatolia are accommodated by E-W striking normal faults and by right-lateral strike slip faulting; the S_{Hmax} direction is close to WNW-ESE (median: 105°; 153 events); within uncertainties (~21°), this is consistent with Müller et al. (1992). A few normal faulting mechanism in the southern Aegean Sea suggest arc parallel strikes of active normal faults in the southern Cyclades and southern Peloponnesus (three events near 37° N, 23°–26.5° E), which further south (~36° N) seems to give way to E-W extension in the overriding crust perpendicular to the arc consistent with observed GPS vectors (Hollenstein et al. 2008). In northern Greece and the southwestern Balkans (about 39.5°–45° N), GPS velocities relative to Eurasia decrease (Hollenstein et al. 2008). The dominant deformation style changes from normal to strike-slip faulting going northward, and the S_{Hmax} direction, which is WSW-ENE (median: 76°; 92 events), is slightly rotated counterclockwise relative to the Aegean block.

The faulting style in central and eastern Anatolia (east of 30° E) in the back arc of the Cyprus trench changes from normal to predominantly strike-slip going towards the east. The S_{Hmax} direction, overall, is roughly N-S oriented (median: 183°; 47 events) but more variable than in the Aegean and western Anatolia region.

The Caucasus region is undergoing north-south contraction due to Eurasia-Arabia collision. Strike-slip mechanisms with NNW-SSE S_{Hmax} direction (median: 167°; 24 events) prevail in the Armenian Plateau (about 40° N). Further north, in the Greater Caucasus Mountains, deformation changes to thrust faulting and the S_{Hmax} direction rotates slightly to NNE-SSW (median: 13°; 11 events). Few stress indicators exist north of the Caucasus in the Precambrian shield region (e.g., Heidbach et al. 2016); results from a few focal mechanisms shown here indicate deformation may remain compressional with roughly N-S directed S_{Hmax}. Similarly, few stress data exist for the Black Sea; S_{Hmax} seems to be oriented roughly E-W within the Mesozoic domain and slightly more northerly near the Mesozoic-Precambrian boundary.

Focal mechanisms in the northwest part of Fig. 3 (about north of 45° N and west to 25° E) reflect deformation due to collision of the Adria microplate with Eurasia. GPS data (e.g., Grenerczy et al. 2005) indicate a few mm/yr NNE-directed motion, which is taken up over a broad region from the coast across the Dinarides and into the Pannonian basin. This is consistent with the thrust mechanisms with roughly NNE-SSW S_{Hmax} direction between 45°–47° N and left-lateral strike-slip motion further north reflecting eastward Alpine extrusion as suggested, e.g., by Grenerczy et al. (2005) and Grenerczy and Kenyeres (2006).

4.2 *Northern and Western-Central Europe*

Focal mechanisms and corresponding S_{Hmax} directions for intraplate earthquakes in northern and central-western Europe are shown in Fig. 4. I separate the events into three broad groups. The first consists of earthquakes in the Tyrrhenian Sea and the western Mediterranean Sea within the Mesozoic domain. Fault plane solutions in the western Mediterranean (west of the Tyrrhenian Sea) and south of Portugal show a mix of normal, strike-slip, and thrust faulting but have consistent NNW-SSE S_{Hmax} orientations, which is in broad agreement with the Nubia-Eurasia plate motion direction suggesting seismic activity is governed by plate interactions. The events probably reflect deformation in a broad plate boundary zone. The S_{Hmax} orientations are different at the western boundary (near 10° E) of the Tyrrhenian basin an extensional feature associated with the Apennine and Calabrian arcs (e.g., Malinverno 2012) where several N-S trending thrust mechanisms indicate E-W directed compression.

This suggests that stress conditions in the Tyrrhenian Sea might be dominated by local rather than larger-scale plate boundary forces.

The small second group consists of four thrust events between 59°–66° N along the continental margin offshore Norway. Their S_{Hmax} orientations are sub-parallel to the North America-Eurasia relative plate motion direction indicating the stress field is likely due to plate boundary forces ("ridge push") transmitted through the oceanic plate (e.g., Stein et al. 1989). A significant percentage of the larger SCR earthquakes globally are located at passive rifted margins (e.g., Johnston 1989). The largest SCR event ($M_w = 5.6$) during the last 40+ years in the study area is the margin earthquake near ~63° N (Fig. 4) and two additional margin events reached $M_w = 5$.

The third group includes events in the Precambrian (brown) and Paleozoic (blue, south of 59° N) domains, as well as events within ~50 km of the Mesozoic-Paleozoic domain boundary (Fig. 4). This group represents seismicity in stable continental crust (SCR) and includes 122 seismic events of which 29 are non-tectonic; of the 122 solutions, 120 are regional moment tensors. This group will be discussed below.

5 Discussion

The seismic events over the broad SCR region of western-central Europe show remarkably consistent source mechanisms and S_{Hmax} orientations. Away from the vicinity of the Alps in northwest Switzerland and excluding non tectonic events, all focal mechanisms are either normal or strike-slip (Fig. 4 left). This is consistent with compilations of first motion focal mechanisms for Switzerland (Kastrup et al. 2004), the Rhine Graben area (Bonjer 1997; Plenefisch and Bonjer 1997), and the Armorican Massif and Massif Central regions of France (Nicolas et al. 1990; Delouis et al. 1993) that show predominance of normal and strike-slip faulting. Mechanisms within small sub-regions are generally similar (such as the western Pyrenees) and change smoothly (e.g., from normal faulting in Pyrenees to strike-slip and then back to normal faulting along the French Atlantic coast in the western Armorican Massif). In northwestern Switzerland and the southern Rhine Graben region (near 47° N and 8° E), RMTs show a range of deformation styles within a small source region consistent with, e.g., Kastrup et al. (2004) and the mechanisms agree well with first motion solutions from the dense Swiss seismic network (Baer et al. (2007) and references therein for comparison of 1999–2006 events).

The orientation of the maximum compressive horizontal stress S_{Hmax} in Western Europe is predominantly NW-SE directed (Fig. 4, right). The orientation changes systematically and remains roughly parallel to the relative motion of Nubia (Africa) with respect to the Eurasian plate. This suggests that plate boundary forces exerted by the Nubia-Eurasia collision are the primary control on the European intraplate stress field. The average S_{Hmax} orientation is 145° ± 26° (Fig. 5 left) and the S_{Hmax} orientations differ from the local trajectories of the Nubia-Eurasia plate motion by only 12° ± 26° (Fig. 5 right). As expected, the results are very similar to the World Stress Map estimates (Müller et al. 1992; Heidbach et al. 2007, 2016) and to formal

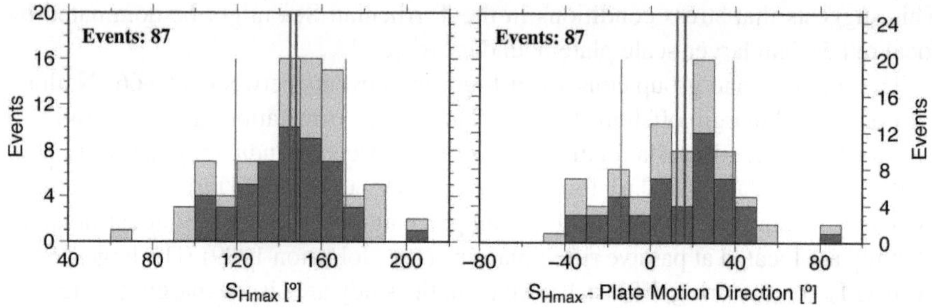

Fig. 5 Left: Histogram of S_{Hmax} orientations in central-western Europe for 87 (of 93) earthquakes with defined stress regime (following Zoback 1992) south of 59° N (Fig. 4). Light gray shows all events; dark gray only earthquakes in the Paleozoic domain. Average orientation is 145° (bold red line) with standard deviation of 26° (thin red lines); median value is 148° (bold blue line). Values for the Paleozoic domain only are 144° ± 21° and 146°, respectively. Right: Histogram of orientation differences between S_{Hmax} and the Nubia-Eurasia plate motion trajectory (calculated at the event location using MORVEL rotation pole). Gray and color scheme as on left. Average and median for all events are: 12° ± 26° and 16°, and for the Paleozoic domain only: 14° ± 25° and 18°

stress tensor inversions for central Europe north of the Alps (e.g., Plenefisch and Bonjer 1997; Kastrup et al. 2004).

Outliers in S_{Hmax} orientations (Fig. 5, left) are associated with anomalous events and almost all are located in the Mesozoic-Paleozoic boundary region (light gray shading in Fig. 5). A single Paleozoic outlier with $S_{Hmax} \geq 200°$ comes from an earthquake in northwestern Spain, which is part of a 1995–1998 sequence of 11 events where all other events have "normal" orientations between 146° and 171°. The second event with $S_{Hmax} \geq 200°$, and three others with $S_{Hmax} \geq 180°$, occurred in eastern Austria possibly related to eastward extrusion of the Eastern Alps and are thus not part of stable central-western Europe. The two other $S_{Hmax} \geq 180°$ events, one in the French Jura and one in southern Spain, are likely affected by stress perturbations caused by the Alps and Betic Mountains within the Nubia-Eurasia boundary zone, respectively.

The single $S_{Hmax} \approx 60°$ observation in Fig. 5 is from an unusual normal faulting earthquake in southern Poland (near 50° N) (Wiejacz and Debski 2009); it is not clear how the event's stress indicators relate to the rest of central-western Europe. It is interesting to note that S_{Hmax} from mining related events with significant deviatoric components in southwestern Poland (gray lines, Fig. 4) are consistent with earthquake observations in central-western Europe and with local Nubia-Eurasia plate motion trajectories. However, mining and hydrocarbon extraction related events often show significant S_{Hmax} deviations (e.g., northern Germany and Central Graben, North Sea) probably caused by local stress field perturbations.

The other three events with $S_{Hmax} \leq 100°$ are in the western Pyrenees, where a group of 8 events with a median $S_{Hmax} = 107°$ indicates a counterclockwise rotation of the stress field relative to the Nubia-Eurasia direction that could be due to a crustal root as suggested for the Alps (Kastrup et al. 2004). Most outliers come from the Mesozoic-Paleozoic boundary region suggesting larger local stress field variations occur in the more mobile Mesozoic domain. Without the outliers, the S_{Hmax} distribution tightens to a ~100°–170° range similar to the World Stress Map data (e.g., Müller et al. 1992). A certain amount of S_{Hmax} spread is expected because earthquakes usually occur on pre-existing faults that rupture under the prevailing stress field even if a fault is not optimally aligned leading to differences between expected and actual S_{Hmax} directions.

The region north of the Alps to the Lower Rhine Embayment (Fig. 6) is one of the most seismically active regions in central Europe and the RMT dataset includes the strong 1978 Swabian Jura, Germany (M_w = 5.2) and 1992 Roermond, Netherlands (M_w = 5.3) earthquakes. Like Plenefisch and Bonjer (1997), I find the S_{Hmax} direction rotates by about 30° counterclockwise from the northern Alpine Foreland and broader Southern Rhine Graben region south of 49° N (median: 159°) to events in the Lower Rhine Embayment and Ardennes Massif north of 50° N (median: 130°). The rotation is possibly due to a regional stress field perturbation caused by the lateral density contrast of a crustal root beneath the Alps as proposed by Kastrup et al. (2004) for the Alps proper.

Subdividing events south of 49° N according to latitude or depth does not reveal significant differences between a northern and southern group, or between shallow and deeper events, respectively. The deformation regime, however, changes with depth as strike-slip events dominate in the upper crust ($z < 15$ km) while normal faulting events are more common at depth $z > 15$ km (Plenefisch and Bonjer 1997).

Figure 7 (left) shows the earthquake depth distribution for western-central Europe. Only events with well-constrained RMT depth are shown and considering their size (Fig. 7 right), centroids obtained from long-period waveforms are equivalent to hypocenters. Earthquakes are observed throughout most of the crust from about 0–30 km depth. Most events occur at 10–15 km depth but shallow events ($z < 10$ km) are common while activity overall decays with depth from 15 to 30 km. The small, but well-constrained dataset does not support the idea of a bimodal distribution of SCR event depth (cf. Klose and Seeber 2007).

Consistent with earlier observations of lower crustal earthquakes east of the Southern Rhine Graben and beneath the Alpine Foreland (e.g., Deichmann 1992; Bonjer 1997), I find 8 of 13 events south of 48° N (Fig. 6) at mid-to-lower crustal depths of 18-27 km. Brittle failure in the lower crust probably requires the presence of fluids (Deichmann 1992).

Earthquakes in the Lower Rhine Embayment (Fig. 6) occur primarily from shallow-to-mid crustal depths of ~4–20 km (Reamer and Hinzen 2004). The RMT database includes 4 km shallow events associated with the 2001 Voerendaal, Netherlands swarm (Goutbeek and Dost 2004) but the largest events (1992 M_w = 5.3 Roermond and 2002 M_w = 4.6 Alsdorf, Germany) seem to be concentrated at 15–18 km close to the bottom of the seismogenic zone; these depths are consistent with traveltime based locations and other modeling (e.g., Van Eck and Davenport 1994 and references therein; Braunmiller et al. 1994; Hinzen and Reamer 2007).

A main advantage of using regional waveform data is that smaller seismic events can be analyzed. Large earthquakes that can be analyzed with teleseismic data occur

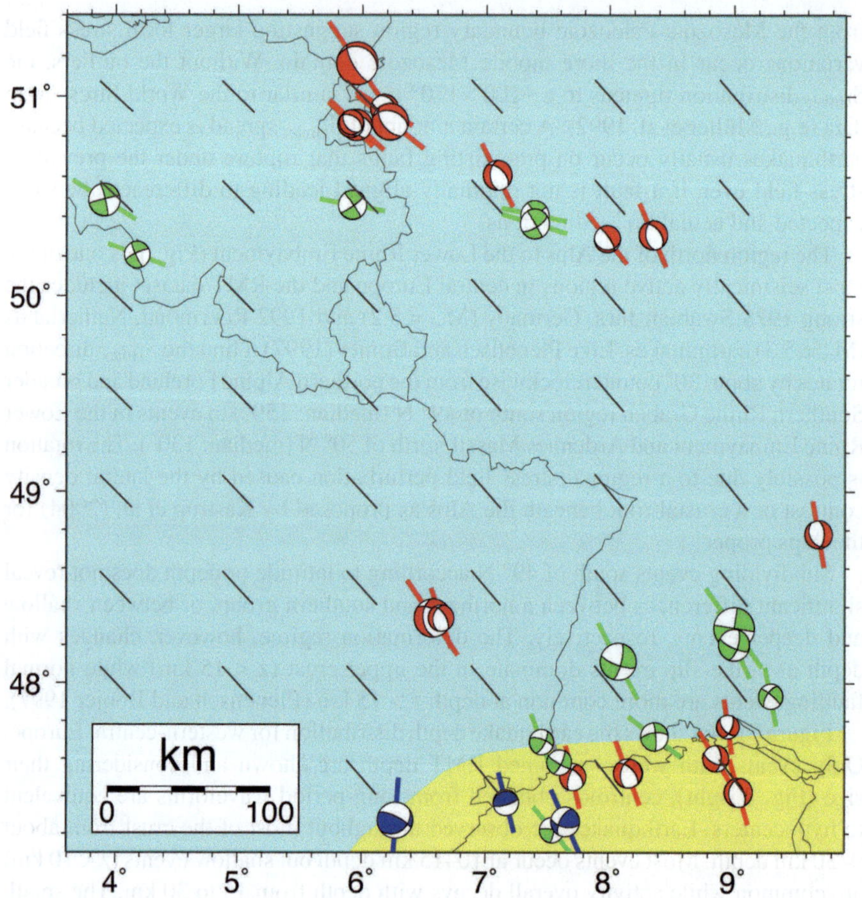

Fig. 6 Close-up from northern Alpine foreland to Lower Rhine Embayment with fault plane solutions, S_{Hmax} orientations, and Nubia-Eurasia motion trajectories. Symbols and colors as in Fig. 4. Note south to north counterclockwise S_{Hmax} rotation. Mining related event near 49° N (Fig. 4) is omitted

rarely in SCRs making it difficult to study SCR seismic behavior. The 40-year GCMT catalog (1976–2015) contains 13 SCR events in central-western and northern Europe (events away from yellow "Mesozoic" region of Fig. 4) compared to 94 earthquakes in the RMT database. Figure 7 (right) shows the size distribution of the combined set (94 RMTs plus 3 unique GCMTs). The dark gray shading shows 85 RMTs for 1995–2007 for which the number of solutions increases down to M_w =4.2–4.3 providing a rough estimate for RMT catalog completeness. More than half of the RMTs are M_w < 4.2 from events in regions with relatively dense broadband seismic networks already during 1995–2007. The threshold is probably lower now. The recent and ongoing expansion of broadband networks in Europe and elsewhere provides an unprecedented opportunity to obtain reliable earthquake source parameters for stable

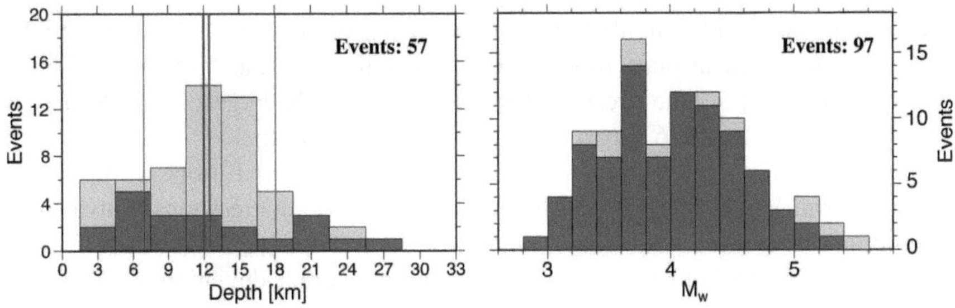

Fig. 7 Left: Centroid depths from RMT analysis in central-western Europe for 57 earthquakes with well-constrained depth (variance increase ≥5% for depth change of up to ±10 km). Light gray are all events, dark gray only earthquakes in the Paleozoic-Mesozoic boundary domain to highlight the Alpine Foreland where most of the deeper events occurred. Average depth: 12.5 ± 5.6 km (red), median: 12 km (blue). Right: Size distribution of 97 earthquakes (mining events excluded) in Precambrian, Paleozoic, and Paleozoic-Mesozoic boundary domains of Fig. 4; 94 RMTs and 3 GCMTs (events with no RMT). Dark gray are 85 RMTs for 1995–2007 when station distribution allowed systematic analysis of entire region

continental regions globally from regional waveform analysis with direct applications such as to the World Stress Map project (Heidbach et al. 2007) and to gain understanding of seismogenesis in stable continental regions and other intraplate settings towards improving hazard assessments (e.g., Johnston et al. 1994; Stein and Mazzotti 2007).

6 Conclusions

Analysis of regional broadband seismic data significantly increases the European source parameter database (moment tensor, depth, M_w, S_{Hmax}) for intraplate earthquakes and earthquakes within plate boundary deformation zones compared to teleseismic analysis. The combined set of 504 unique crustal seismic events within Europe and at least 125 km away from an active plate boundary includes 390 RMT solutions, primarily for 1995–2007, and 114 GCMT solutions for 1976–2015.

Most solutions (378) are from earthquakes in the broadly deforming plate boundary zones associated with the Nubia-Eurasia subduction in the western Mediterranean and the Aegean and Anatolian backarc extension and deformation regions, as well as with Arabia-Eurasia collision in the Caucasus. Deformation styles and orientations of the maximum horizontal compressive stress direction S_{Hmax} are consistent with earlier findings and indicate a dominance of plate boundary forces on deformation and stress orientations in actively deforming regions.

For the stable continental regions of northern and central-western Europe, the 123 RMT solutions constitute the bulk of available source parameters (total: 126—plus 10 "doubles" see Appendix) derived from stable, reliable long-period three-component waveform modeling. Earthquakes near the continental margin offshore Norway have thrust mechanisms with roughly E-W oriented S_{Hmax} orientations sub-parallel to the North America-Eurasia plate motion direction indicating that "ridge-push" plate boundary forces control the stress field and deformation; the largest ($M_w = 5.6$) SCR event in Europe since 1976 is part of this small 4-event group.

The average S_{Hmax} orientation in the SCR of central-western Europe is $145° \pm 26°$ consistent with World Stress Map observations and formal stress tensor inversions of compilations of first motion fault plane solutions. The S_{Hmax} orientation is sub-parallel to the Nubia-Eurasia plate motion direction (average deviation is $12° \pm 26°$) indicating that Nubia-Eurasia plate boundary forces control the large-scale stress field in stable Europe. Local deviations observed in the Pyrenees and in the northern Alpine Foreland are probably due to stress field perturbations caused by the crustal roots of the mountain ranges. Event depths do not show a general bi-modal depth distribution in the European SCR. However, earthquake depths in the northern Alpine Foreland reach into the lower crust (to ~30 km) and likely require the presence of fluids to allow brittle deformation. In the Lower Rhine Embayment, earthquakes are slightly shallower (to ~20+ km) but the largest recent events seem to concentrate towards the bottom of the seismogenic zone ($\geq$15 km).

Regional waveforms allow systematic source parameter retrieval of earthquakes that are significantly smaller than events that can be analyzed with teleseismic data. In regions of low seismicity and infrequent larger events, such as most SCRs and intraplate regions, deformation style, event depth, and stress orientations can be estimated from the more frequent moderate-sized earthquakes ($M_w \geq 3.5$ depending on broadband seismic network density) through regional moment tensor inversion.

Appendix

Of the earthquake shown in Fig. 2, including the sub-crustal events with depth $z > 40$ km, 87 have a RMT and a Global CMT solution. This appendix compares results from the two methods to illustrate overall robustness and reliability of waveform modeling derived source parameter estimates.

Figure 8 shows the double couple (DC) fault plane solutions for the common events. With few exceptions, the agreement between the RMT (black and white) and GCMT (gray) solutions is high. Most events with large focal mechanism differences occurred before 2004 when GCMT analysis started to include intermediate-period surface waves (Ekström et al. 2012), which lowered GCMT analysis threshold and stabilized results for $M \approx 5$ and smaller events common in the European intraplate dataset. Expansion of broadband seismic networks in the European-Mediterranean region also improved RMT solution quality.

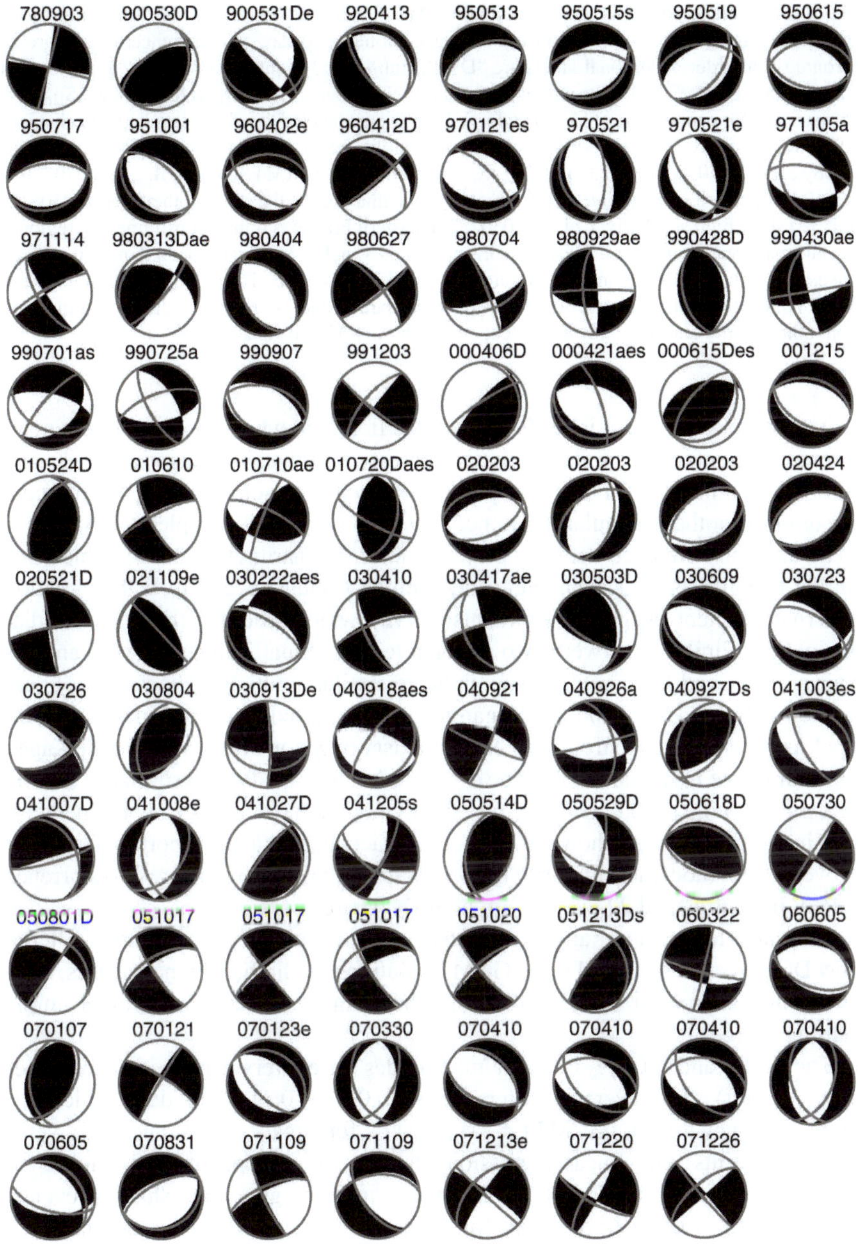
780903
900530D
900531De
920413
950513
950515s
950519
950615
950717
951001
960402e
960412D
970121es
970521
970521e
971105a
971114
980313Dae
980404
980627
980704
980929ae
990428D
990430ae
990701as
990725a
990907
991203
000406D
000421aes
000615Des
001215
010524D
010610
010710ae
010720Daes
020203
020203
020203
020424
020521D
021109e
030222aes
030410
030417ae
030503D
030609
030723
030726
030804
030913De
040918aes
040921
040926a
040927Ds
041003es
041007D
041008e
041027D
041205s
050514D
050529D
050618D
050730
050801D
051017
051017
051017
051020
051213Ds
060322
060605
070107
070121
070123e
070330
070410
070410
070410
070410
070605
070831
071109
071109
071213e
071220
071226

◀**Fig. 8** Fault plane solutions for 87 common RMT (black and white) and GCMT (gray lines) earthquakes in Europe. Numbers on top are year, month, and day of occurrence; events are in chronological order. Additional labels are "D": 20 sub-crustal events (z > 40 km); "a": 13 events with $|\Delta Ax| \geq 30°$; "e": 20 events with $\eta_p \leq 0.8$; and "s": 12 events with S_{Hmax} difference $\Delta S_{Hmax} \geq 15°$ (see text for explanation of $|\Delta Ax|$ and η_p). Four events labeled "aes" indicate large mechanism differences, they are: **A**. Event 000421 in western Anatolia, which is characterized by N-S extension; nearby events (970121, 030723, 030726) are consistent with the RMT solution. **B**. Event 010720 is one of the common intermediate-depth Vrancea, Romania events; the closest other events (050618, 051213) show high similarity. **C**. Event 030222 in eastern France is tightly constrained by regional waveforms (Deichmann et al. 2004); the event predates inclusion of intermediate-period surface waves and with $M_w = 4.8$ is small for traditional GCMT analysis. **D**. The GCMT solution for event 040918 in the Pyrenees has a 6% double-couple contribution and its orientation differs significantly from other nearby and internally consistent events in the RMTs database

The overall agreement is further illustrated in Fig. 9, which shows different measures of moment tensor similarity plotted against each other. The Kagan angle (Kagan 2007) is the minimum 3-D angle required to rotate the principal axes of one moment tensor onto another ('angular difference' between two double couples), $|\Delta Ax|$ is the mean angular difference between principal axes (Bernardi et al. 2004), which is a similar quantity as the Kagan angle and simple to calculate, and η_p is the radiation pattern coefficient (Kuge and Kawakatsu 1993), which describes the P-wave radiation pattern similarity between two moment tensors. Small values of $|\Delta Ax|$ and the Kagan angle, and large values of η_p indicate similar solutions. The roughly linear relation between $|\Delta Ax|$ and the Kagan angle is expected since both describe principal axes rotations. A strong correlation exists only for events with small Kagan angle and large η_p. About 74% of the events (64 of 87) satisfy high similarity criteria $|\Delta Ax| \leq 30°$ and $\eta_p \geq 0.8$ while the rest shows moderate similarity. Low DC percentages in moment tensors, shown as dark circles, account for a disproportional number of events with moderate similarity (23%) compared to their frequency of occurrence (14%). In contrast, all 22 events with high DC source contributions, shown as open circles, have low rotation angles and 86% of them satisfy high similarity criteria. Low DC percentages in RMT and GCMT solutions might thus, in many cases, indicate limited source parameter resolution rather than actual deviations from a simple faulting source.

The RMT and GCMT moment magnitudes agree very well with each other (Fig. 10 left). The average difference is 0.00 ± 0.11. The data indicate a slight tendency of M_w(GCMT) > M_w(RMT) for small events and vice versa for large events with linear regression resulting in a slope of 1.08. Figure 10 (right) shows centroid depths of all events in light gray. For shallow events, GCMT depth is often fixed while for deep events, depth is sometimes not well resolved by long-period regional data. The subset of dark gray circles consists of events with free (GCMT) and resolved (RMT) depth. For these 35 events, the average depth difference GCMT-RMT is 2.8 ± 10.1 km, and, for 19 crustal events ($z \leq 40$ km) is 1.3 ± 3.4 km indicating excellent agreement considering that RMT depth analysis is performed as grid search with 3 km steps.

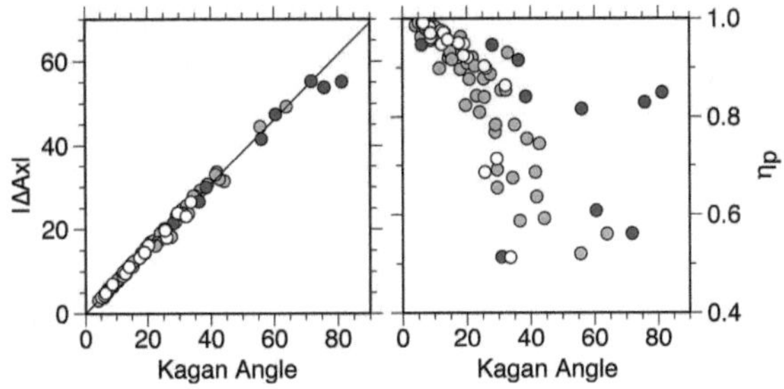

Fig. 9 Comparison between RMT and GCMT solutions. Kagan angle and |ΔAx| measure angular difference between principal axes systems while η_p measures P-wave radiation pattern similarity (normalized to [−1, 1]). An interchange of two axes results in |ΔAx| = 60°. Left: Kagan angle versus |ΔAx|. Dark circles indicate solutions with double-couple percentage of DC ≤ 35% in either RMT or GCMT; these solutions on average show larger differences. Open circles are events with DC ≥ 85% in both RMT and GCMT; all have |ΔAx| ≤ 30°, which Bernardi et al. (2004) considered to indicate high similarity. Gray circles are all other events. Kagan angle and |ΔAx| roughly show a 4/3 scaling (solid line) Right: Kagan angle versus η_p. Symbols are as on left. High correlation exists only for low Kagan angle and high η_p events (upper left of diagram); the larger scatter at lower similarities reflects that the two indicators measure separate aspects of radiation differences

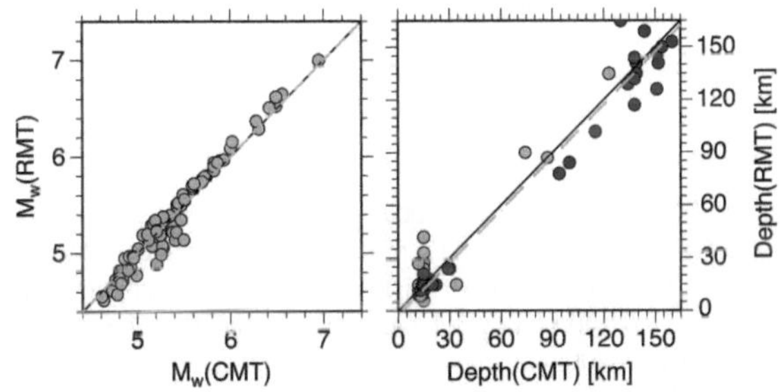

Fig. 10 Left: Comparison of M_w(GCMT) and M_w(RMT). Solid black line is a 1:1 relation and gray dashed line shows the average difference (0.00 ± 0.11). Right: Comparison of GCMT and RMT centroid depth. Light gray shows all 87 events. Dark gray shows 35 events with free (GCMT) and resolved (RMT) depth; their average difference, shown as gray dashed line, is 2.8 ± 10.1 km and for crustal events (z ≤ 40 km) is 1.3 ± 3.4 km

A histogram of the differences in the maximum horizontal stress orientation S_{Hmax} is shown in Fig. 11. Excluding low DC events (≤20% for either GCMT or RMT) and events with "unknown" stress regime (using Zoback (1992) assignments) results in 70 events. The average difference between RMT and GCMT is −0.7° ± 10.2°, while the median is −1° with 75% of all observations between −7° and 9° indicating S_{Hmax} orientations are stable and well resolved. The three outliers in Fig. 11 correspond to the suspicious 000421 (−48°), 030222 (24°), and 010720 (28°) events discussed

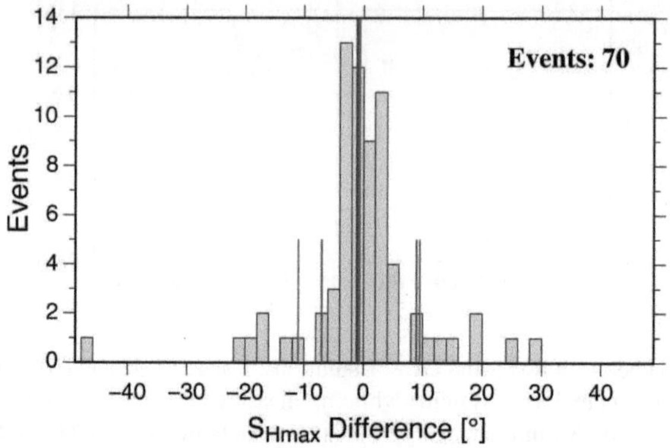

Fig. 11 Histogram of S_{Hmax} differences between RMT and GCMT solutions. Events with low DC (≤20%) or of "unknown" stress regime are excluded resulting in 70 common events. Red lines show average and standard deviation ($-0.7° \pm 10.2°$) and blue lines median ($-1°$) and range containing 75% of observations ($-7°$–$9°$) indicating a tight distribution

in the context of Fig. 8 and should be disregarded. This leaves a remarkably small largest difference of $-22°$ (normal faulting event 041003 at northwestern edge of Black Sea).

References

Baer M, Deichmann N, Braunmiller J et al (2007) Earthquakes in Switzerland and surrounding regions during 2006. Swiss J Geosci 100:517–528. https://doi.org/10.1007/s00015-007-1242-0

Bernardi F, Braunmiller J, Giardini D (2005) Seismic moment from regional surface-wave amplitudes: applications to digital and analog seismograms. Bull Seismol Soc Am 95(2):408–418. https://doi.org/10.1785/0120040048

Bernardi F, Braunmiller J, Kradolfer U et al (2004) Automatic regional moment tensor inversion in the European-Mediterranean region. Geophys J Int 157:703–716. https://doi.org/10.1111/j.1365-246X.2004.02215.x

Berz G (1994) Assessment of the losses caused by the 1992 Roermond earthquake, the Netherlands (extended abstract). In: van Eck T, Davenport CA (eds) Seismotectonics and seismic hazard in the Roer Valley graben; with emphasis on the Roermond earthquake of April 13, 1993. Geologie en Mijnbouw 73(2–4):281

Bonjer K-P (1997) Seismicity pattern and style of seismic faulting at the eastern border fault of the southern Rhine Graben. Tectonophysics 275:41–69

Braunmiller J (2002) Moment tensor solutions of stronger earthquakes in Germany with GRSN data. In: Korn M (ed) Ten years of german regional seismic network (GRSN). DFG Senate Commission for Geosciences, Report 25, pp 227–235

Braunmiller J, Kradolfer U, Baer M et al (2002) Regional moment tensor determination in the European-Mediterranean area—initial results. Tectonophysics 356(22):5–22

Braunmiller J, Dahm T, Bonjer K-P (1994) Source mechanism of the 1992 Roermond earthquake from surface-wave inversion of regional data. Geophys J Int 116:663–672

Cesca S, Rohr A, Dahm T (2013) Discrimination of induced seismicity by full moment tensor inversion and decomposition. J Seismol 17:147–163. https://doi.org/10.1007/s10950-012-9305-8
Coffin MF, Gahagan LM, Lawver LA (1998) Present-day plate boundary digital compilation. UTIG Technical Report No. 174, Progress Report No. 204-0598
Deichmann N (1992) Structural and rheological implications of lower-crustal earthquakes below northern Switzerland. Phys Earth Plant Int 69:270–280
Deichmann N, Baer M, Braunmiller J et al (2004) Earthquakes in Switzerland and surrounding regions during 2003. Eclogae Geol Helv 97:447–458
Delouis B, Haessler H, Cisternas A et al (1993) Stress tensor determination in France and neighboring regions. Tectonophysics 22:413–438
DeMets C, Gordon RG, Argus DF (2010) Geologically current plate motions. Geophys J Int 181:1–80. https://doi.org/10.1111/j.1365-246X.2009.04491.x
Dziewonski AM, Chou T-A, Woodhouse JH (1981) Determination of earthquake source parameters from waveform data for studies of global and regional seismicity. J Geophys Res 86:2825–2852. https://doi.org/10.1029/JB086iB04p02825
Ekström G, Nettles M, Dziewonski AM (2012) The global CMT project 2004–2010: centroid moment tensors for 13,017 earthquakes. Phys Earth Planet Inter 200–201:1–9
Fäh D, Gisler M, Jaggi B et al (2009) The 1356 Basel earthquake: an interdisciplinary revision. Geophys J Int 178:351–374
Goutbeek FH, Dost B (2004) First results on the analysis of the Voerendaal, the Netherlands, earthquake swarm of 2000–2002. In: Paper presented at the 29th general assembly of the European seismological commission, Potsdam, 2004
Grenerczy G, Kenyeres A (2006) Crustal deformation between Adria and the European platform from space geodesy. In: Pinter N et al (eds) The Adria microplate: GPS geodesy, tectonics and hazards. Springer, pp 321–334
Grenerczy G, Sella G, Stein S et al (2005) Tectonic implications of the GPS velocity field in the northern Adriatic region. Geophys Res Lett 32:L16311. https://doi.org/10.1029/2005GL022947
Heidbach O, Custodio S, Kingdon A et al (2016) Stress map of the Mediterranean and central Europe 2016. GFZ data service. https://doi.org/10.5880/WSM.Europe2016
Heidbach O, Reinecker J, Tingay M et al (2007) Plate boundary forces are not enough: second- and third-order stress patterns highlighted in the world stress map database. Tectonics 26:TC6014. https://doi.org/10.1029/2007tc002133
Hinzen K-G, Reamer SK (2007) Seismicity, seismotectonics, and seismic hazard in the northern Rhine area. In: Stein S, Mazzotti S (eds) Continental intraplate earthquakes: science, hazard, and policy issues. Geological Society of America Special Paper 425, pp 225–242. https://doi.org/10.1130/2007.2425(15)
Hollenstein C, Müller MD, Geiger A et al (2008) Crustal motion and deformation in Greece from a decade of GPS measurements, 1993–2003. Tectonophysics 449:17–40. https://doi.org/10.1016/j.tecto.2007.12.006
Johnston AC (1996) Seismic moment assessment of earthquakes in stable continental regions—III. New Madrid 1811–1812, Charleston, 1886 and Lisbon, 1755. Geophys J Int 126:314–344
Johnston AC (1989) The seismicity of 'stable continental interiors'. In: Gregersen S, Basham PW (eds) Earthquakes at North-Atlantic passive margins: neotectonics and postglacial rebound. Kluwer Academic Press, pp 299–327
Johnston AC, Coppersmith KJ, Kanter LR et al (1994) The earthquakes of stable continental regions: assessment of large earthquake potential. In: Schneider JF (ed) TR-102261, vol 1–5, Electric Power Research Institute (EPRI), Palo Alto, CA
Kagan YY (2007) Simplified algorithm for calculating double-couple rotation. Geophys J Int 171:411–418. https://doi.org/10.1111/j.1365-246X-2007.03538.x
Kastrup U, Zoback ML, Deichmann N et al (2004) Stress field variations in the Swiss Alps and the northern Alpine foreland derived from inversion of fault plane solutions. J Geophys Res 109:B01402. https://doi.org/10.1029/2003JB002550

Klose CD, Seeber L (2007) Shallow seismicity in stable continental regions. Seismol Res Lett 78(5):554–562
Kuge K, Kawakatsu H (1993) Significant non-double couple source components in deep and intermediate-depth earthquakes: Implications from moment tensor inversion of long-period seismic waves. Phys Earth Planet Inter 75:243–266
Malinverno A (2012) Evolution of the Tyrrhenian Sea-Calabria Arc system: the past and the present. Rend Online Soc Geol It 21:11–15
Müller B, Zoback ML, Fuchs K et al (1992) Regional patterns of tectonic stress in Europe. J Geophys Res 97(B8):11,783–11,803
Nabelek J, Xia G (1995) Moment-tensor analysis using regional data: application to the 25 March, 1993, Scotts Mills, Oregon earthquake. Geophys Res Lett 22:13–16
Nicolas M, Santoire JP, Delpech PY (1990) Intraplate seismicity: new seismotectonic data in Western Europe. Tectonophysics 179:27–53
Plenefisch T, Bonjer K-P (1997) The stress field in the Rhine Graben area inferred from earthquake focal mechanisms and estimation of frictional parameters. Tectonophysics 275:71–97
Reamer SK, Hinzen K-G (2004) An earthquake catalog for the Northern Rhine area, central Europe (1975–2002). Seismol Res Lett 75(6):713–725
Stein S, Mazzotti S (eds) (2007) Continental intraplate earthquakes: science, hazard, and policy issues. Geological Society of America Special Paper 425
Stein S, Cloetingh S, Sleep NH et al (1989) Passive margin earthquakes, stresses and rheology. In: Gregersen S, Basham PW (eds) Earthquakes at North-Atlantic passive margins: neotectonics and postglacial rebound. Kluwer Academic Press, pp 231–259
Talwani P (ed) (2014) Intraplate earthquakes. Cambridge University Press
Van Eck T, Davenport CA (1994) Seismotectonics and seismic hazard in the Roer Valley Graben; with emphasis on the Roermond earthquake of April 13, 1992. Geologie en Mijnbouw 73(2–4):91–92
Wiejacz P, Debski W (2009) Podhale, Poland earthquake of November 30, 2004. Acta Geophys 57:346–366. https://doi.org/10.2478/s11600-009-0007-8
Zoback ML (1992) First- and second-order patterns of stress in the lithosphere: the world stress map project. J Geophys Res 97:11,703–11,728

Source Characteristics of the January 8, 2013 ($M_W = 5.7$) and May 24, 2014 ($M_W = 6.8$) North Aegean Earthquakes Sequence

Doğan Kalafat, Kıvanç Kekovalı and Ali Pınar

Keywords North Aegean Region · Seismicity · Active faulting · Aftershock
CMT solution

1 Introductıon

The Aegean Sea, is one of the most seismically active areas of the Eastern Mediterranean region (Fig. 1). Generally, North Aegean Sea region has been tectonically developed after the collision of Arabian plate with the Eurasian in the Late Miocene time and the subsequent westward escape of the Anatolian Plate relative to the Eurasian Plate, during the Early Pliocene (McKenzie 1978; McKenzie and Jackson 1983; Taymaz et al. 1991; Barka 1992; Barka and Gülen 1988; Papazachos et al. 2000; Tranos 2009; Kiratzi and Svigkas 2013). The North Anatolia Fault Zone (NAFZ) accommodates much of the right-lateral, strike-slip motion between the Anatolian Block and the Eurasia Plate (Allen 1969; McKenzie 1972; Dewey 1976; Dewey and Şengör 1979; Jackson and McKenzie 1984; Şengör et al. 1985; Barka et al. 1987).

NAFZ is one of the most active and important fault zones in the world. Especially, faults within the North Aegean Trough (NAT) represent the northern branch of the North Anatolian Fault Zone (NAFZ), the major transform faulting structure in northern Turkey accommodating the westward motion of the Anatolian plate with respect

D. Kalafat (✉) · K. Kekovalı · A. Pınar
Regional Earthquake-Tsunami Monitoring Center, Boğaziçi
University Kandilli Observatory & Earthquake Research Institute, 34684 Çengelköy, Istanbul, Turkey
e-mail: kalafato@boun.edu.tr

S. D'Amico (ed.), *Moment Tensor Solutions*, Springer Natural Hazards,
https://doi.org/10.1007/978-3-319-77359-9_16

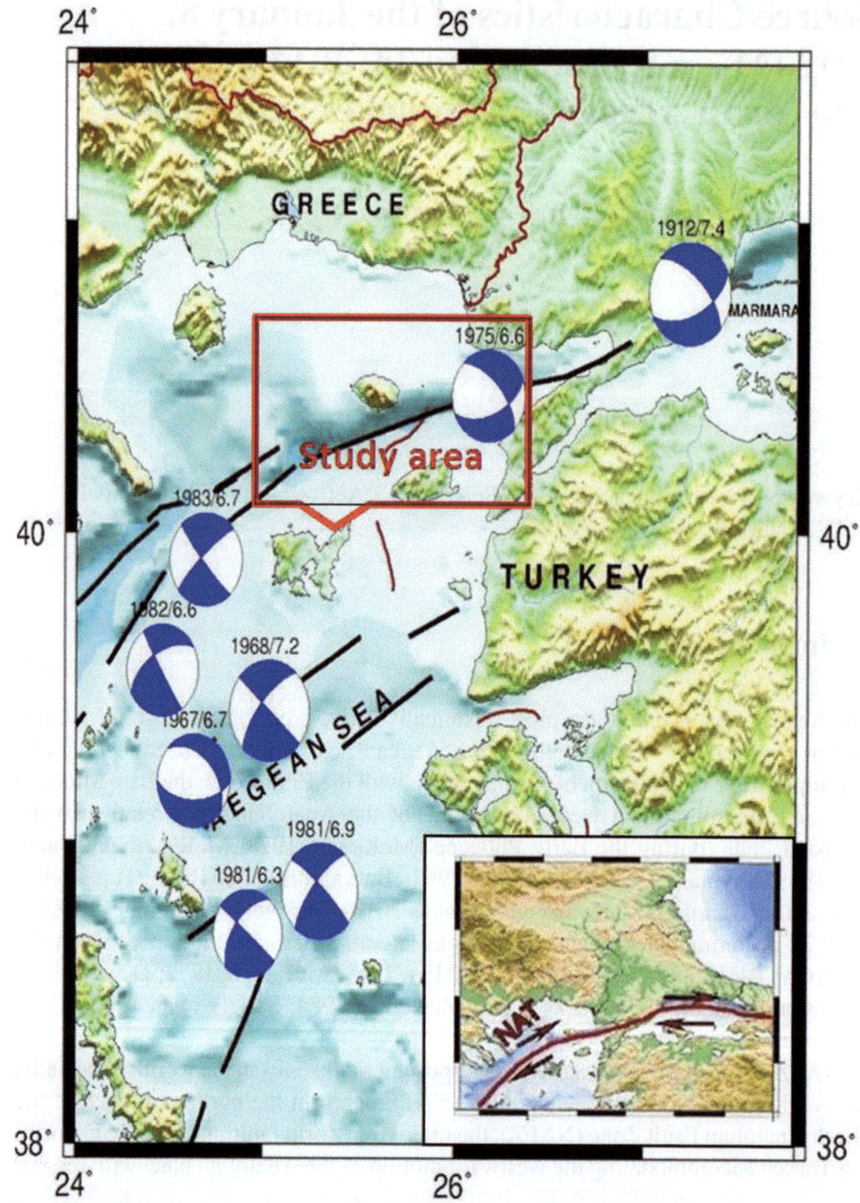

Fig. 1 Active faults in the study area and faulting mechanisms of major earthquakes in the last century (faults are taken from Şaroğlu et al. 1992, MTA; NAT North Aegean Trough)

to Eurasia, at a rate of approximately 25 mm/year (Reilinger et al. 1997; Papazachos 1999; McClusky et al. 2000).

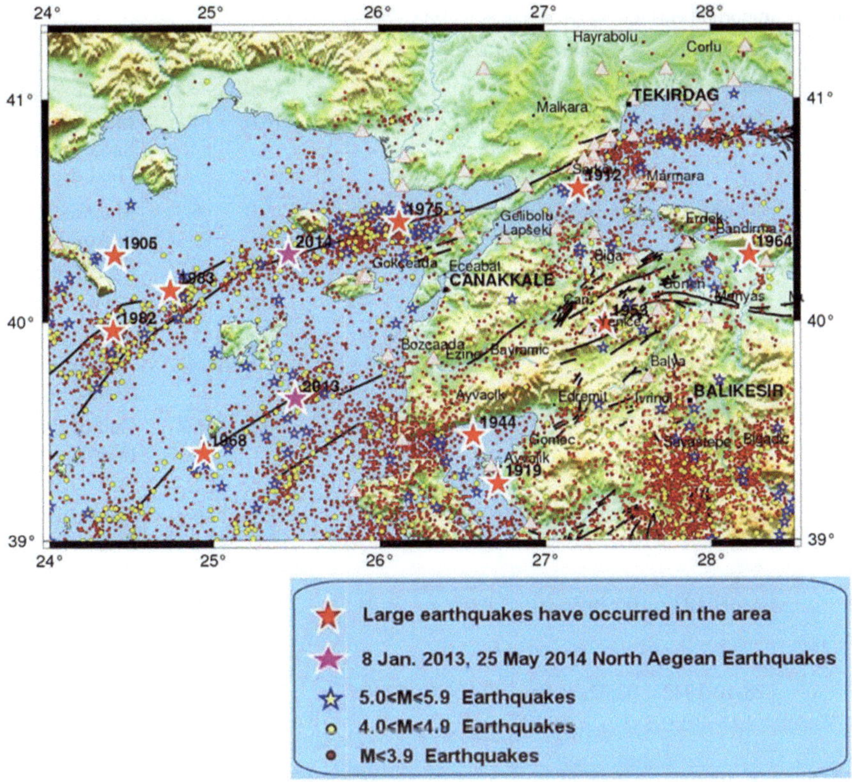

Fig. 2 Strong earthquakes, recent seismicity, active faults in the region and the triangles show the seismic stations in the region from Turkish and Greece Seismological Networks

The NAT corresponds to an approximately 300 km long system of tectonically active marine basins, representing the extension of the North Anatolian Fault Zone (NAFZ) into the Aegean Sea (Le Pichon et al. 1987; Kreemer et al. 2004). The study area is dominated by dextral strike-slip faulting and is characterized by frequently occurring strong earthquakes (Fig. 1).

For example, during the last century many destructive earthquakes occurred in this region (Fig. 2; Table 1). The 1905 Greece (Io = IX; many casualties, M = 7.5), 1912 Saros-Mürefte-Şarköy (Io = X; 2836 casualties; Ms = 7.4), 1919 Soma-Ayvalık (Io = IX; many casualties; Ms = 6.9), 1968 North Aegean Sea (Io = IX; 20 casualties; Ms = 7.1), 1975 Gulf of Saros (Io = VII; Ms = 6.6), 1982 (M = 7.0), 1983 (M = 6.8) are the most important earthquakes in the region (Drakopoulos and Ekonomides 1972; Eyidoğan et al. 1991; Papazachos et al. 1998; Papadimitriou and Sykes 2001; Vannucci and Gasperini 2003; Kalafat et al. 2011).

On 8th January 2013 at 14:16 UTC (16:16 local time) a moderate earthquake (M_w = 5.7) occurred between the south of Gökçeada and southwest of Bozcaada Islands.

Table 1 Important earthquakes in the region (1900–2014)

Eq.no.	Date	O. Time	Latitude	Longitude	Depth	Ms	Location
1	08-11-1905	22:06:00.00	40.30	24.40	14.0	7.4	Greece
2	09-08-1912	01:29:00.00	40.60	27.20	16.0	7.3	Murefte-Sarkoy (Tekirdag)
3	10-08-1912	09:23:00.00	40.60	27.10	15.0	6.3	Murefte-Sarkoy (Tekirdag)
4	20-08-1917	23:02:09.60	40.30	25.43	40.0	6.0	Aegean Sea
5	18-11-1919	21:54:50.30	39.26	26.71	10.0	7.0	Kucukkoy-Ayvalik (Balikesir)
6	04-01-1935	14:41:30.40	40.40	27.49	30.0	6.4	Gulf of Erdek (Marmara Sea)
7	04-01-1935	16:20:04.60	40.30	27.45	20.0	6.3	Guvemalani-Biga (Çanakkale)
8	22-09-1939	00:36:36.60	39.07	26.94	10.0	6.6	Dikili-Bergama (Izmir)
9	28-10-1942	02:22:53.10	39.10	27.80	50.0	6.0	Karakurt-Kirkagac (Manisa)
10	06-10-1944	02:34:48.70	39.48	26.56	40.0	6.8	Gulf of Edremit (Ege Denizi)
11	18-03-1953	19:06:16.10	39.99	27.36	10.0	7.2	Yenice-Gönen (Çanakkale)
12	06-10-1964	14:31:23.00	40.30	28.23	34.0	7.0	Manyas-Karacabey (Bursa)
13	04-03-1967	17:58:09.00	39.25	24.60	60.0	6.5	Aegean Sea
14	19-02-1968	22:45:42.40	39.40	24.94	7.0	7.1	Aegean Sea
15	27-03-1975	05:15:07.90	40.45	26.12	15.0	5.7	Gulf of Saros (Aegean Sea)
16	19-12-1981	14:10:51.10	39.22	25.25	10.0	7.2	Aegean Sea
17	18-01-1982	19:27:25.00	39.96	24.39	10.0	6.9	Greece
18	18-01-1982	19:31:07.90	40.03	24.56	10.0	5.4	Aegean Sea
19	06-08-1983	15:43:51.90	40.14	24.75	2.0	6.9	Aegean Sea
20	08-01-2013	14:16:07.19	39.65	25.50	8.0	5.8	Aegean Sea
21	24-05-2014	09:25:01.59	40.30	25.46	21.2	6.9	Aegean Sea

On 24th May 2014, at 09:25 UTC (12:25 local time), a powerful Ml = 6.7 (M_w = 6.8) earthquake occurred 87 km west of Çanakkale in the Agean Sea.

This study covers seismicity, aftershock distribution, moment tensor inversion for important aftershocks of 2013–2014 sequences. This is important in order to understand more accurate tectonic regime of the study region. Also we calculated stress tensor inversion, b values for 2013–2014 North Aegean Sea Earthquakes. We found source features of the May 24, 2014 North Aegean Earthquake. According to our solution the source rupture area extends. The direct relationship wasn't found between the 2014 earthquake and the NAFZ by Saltogianni et al. (2015).

Also we found long period pulses associated with the 2014 North Aegean earthquake recorded at Gökçeada strong motion station. We have applied teleseismic bodywave modelling for 2014 Earthquake. In addition, we investigated Coulomb Stress Changes in this area in order to asses areas of subsequent off fault aftershocks.

2014 North Agean Sea Earthquakes have studied using the seismological and geodetic inversions of teleseismic waveforms by Saltogianni et al. (2015). They found that the faulting of the earthquake shows shallow strike slip, likely to a fault consisting of two segments.

2 Data and Methodology

2.1 *The Turkish and Greek Seismic Networks*

The 2013 and the 2014 North Aegean Earthquake sequences were recorded by the stations of the Turkish and Greek Seismological Networks (Fig. 2). The Turkish seismic stations are located to the east of the source regions of both the earthquakes resulting in large seismic gap. Similarly, the Greek stations are located mostly to the west of the source regions. For this reason, we need to merge data from the Turkish and Greek seismic stations to provide better station coverage for the location of the mainshock and also for the aftershocks. Moreover, a good azimuthal coverage is essential for CMT solution algorithms we applied. Generally, both the networks are equipped with broadband and strong motion sensors such as Guralp CMG 3T, CMG 5TD, CMG-3ESP.

For the location of the aftershocks we used HYPO71 (Lee and Lahr 1975) and HYPOINV (Klein 2002) to obtain initial locations. For the locations we used 1-D local crustal and velocity model (Fig. 3; Kalafat et al. 1987) and Vp/Vs ratio of 1.73 which we calculated from our database using the Wadati technique.

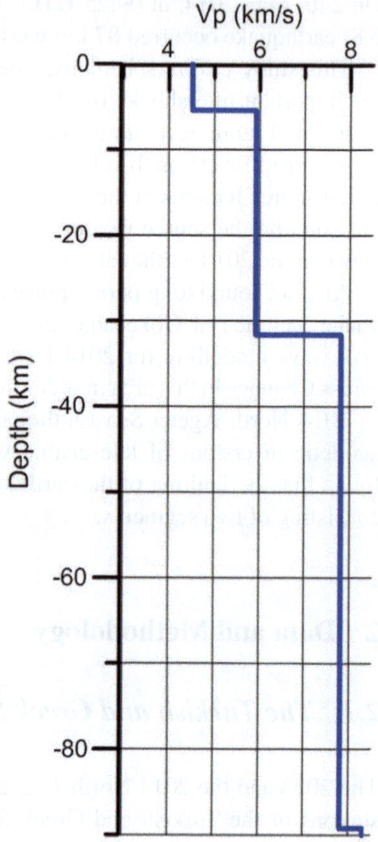

Fig. 3 1-D velocity model used to locate the earthquakes of Aegean Sea

3 Source Features of the May 24, 2014 North Aegean Earthquake

3.1 *The Source Rupture Area Extent*

The most prominent feature of the earthquake is the widespread distribution of the aftershocks. The routine fast locations carried out by the National Earthquake Monitoring Center (NEMC) of KOERI portrays a lateral variation of longitudes between 25.0° E and 26.2° E. This corresponds to approximately 120 km fault rupture length if all the aftershocks take place along the ruptured fault plane. Using the USGS estimated seismic moment magnitude of $M_w = 6.9$ and the relation between fault rupture length and moment magnitude of Wells and Coppersmith (1994), ($Log(L) = (M_w\text{-}5.16 \pm 0.13)/1.12 \pm 0.08$) yields a rupture length between 35 and 60 km. Thus estimated rupture length is a few times shorter than the rupture derived from the aftershock distribution (Fig. 4).

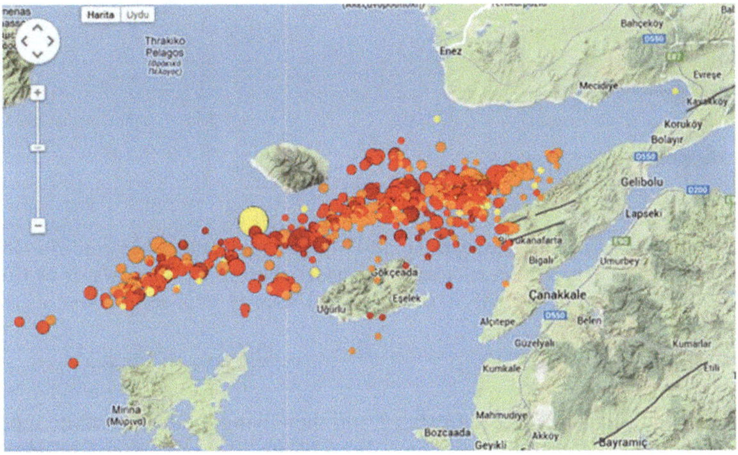

Fig. 4 The aftershock distribution of events between May 24th and 30th May 2014 (quick solution by KOERI)

One of the prerequisites for such long rupture area is a shallow seismogenic source zone. Indeed, the Aegean crustal structure possesses quite a thin seismogenic crustal thickness manifested by the unusually long duration small amplitude Pn phases observed at the land stations at NW Anatolia followed by large amplitude Pg phases. Even such Pn phases are recorded at short epicentral distances being a strong evidence for such a thin crust overlaying the Mantle.

Yet another, distinct characteristic feature associated with the North Aegean Sea Earthquake is the long period seismic waves recorded at the stations deployed in eastern Marmara region. The causatives for such long period waves is a challenge both for the Earthquake Early Warning and Earthquake engineering studies since the long periods waves are effective even at locations several hundreds km away from the source region. The Gökçeada strong motion station located about 50 km to the south of the source zone acquired the long period source pulses associated with the rupture of the fault plane. Considering the strike-slip faulting mechanism of the earthquake and the location of the Gökçeada station the long pulses portrays the fault normal motions acquired because of the fact that the station is located close to the nodal plane for the shear waves (Fig. 5).

4 Teleseismic Body-Wave Modeling

One of the most effective tools to infer details on the source rupture process of large earthquakes is modeling of the teleseismic bodywaves. The large aftershock area may reflect the co-seismic rupture zone and the events triggered by the mainshock as a result of Coulomb failure stress changes. To test the hypotheses first we obtain a

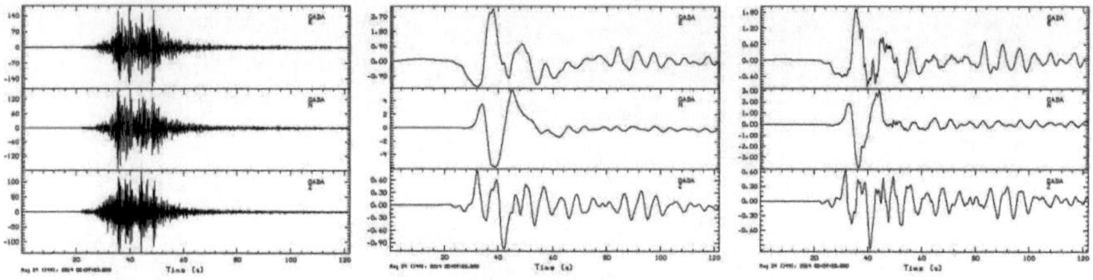

Fig. 5 Long period pulses associated with the 2014 North Aegean earthquake recorded at Gökçeada strong motion station. The station is close to the nodal plane of the shear waves (**a** Gokceada station acceleration record (from AFAD); **b** calculated velocity from acceleration; **c** calculated displacement)

slip distribution model for the mainshock and then using the finite source model we estimate the Coulomb failure stress changes with the aim to understand which part of the aftershock zone are associated as triggered events.

Using the complex teleseismic bodywave records generated by the earthquake and the method developed by Kikuchi and Kanamori (2003) we estimated the seismic moment release on each subfault grid distributed along the strike and dip of the ruptured fault plane (Fig. 6).

The grid size of 10 × 5 km was chosen as 10 km along the strike and 5 km along the dip of the fault plane. The inversion results yield a seismic moment of Mo = 2.9 × 10^{19} Nm (M_w = 6.9) and approximately 30 s source rupture duration. That is consistent with the result of Saltogianni et al. (2015), obtained relatively long source duration (~30 s). The size and arrow of the vectors (rake) shown on the fault plane (Fig. 6) characterize the seismic moment tensor derived for each grid point. The rakes illustrated in Fig. 6 suggests that the region to the west of the epicenter experienced mainly strike-slip motion while to the east considerable dip-slip component contributed to the motion on the fault plane.

5 Coulomb Failure Stress Changes

The slip distribution model portrayed in Fig. 6 was used to estimate the co-seismic static stress changes associated with the mainshock. In our calculations we used a frictional coefficient of 0.3 which is one of the parameters affecting the spatial distribution of the Coulomb stress changes for the optimally oriented fault planes. Considering the predominantly strike slip mechanism for most of the events in North Aegean we used a regional stress tensor appropriate for strike-slip tectonic regimes. The azimuth of the maximum principle axis was fixed at 290°. The results with the fixed parameters are shown in Fig. 7. Here, the red colour indicates the areas of increased stress changes and the blue regions show the region were the stress changes are negative.

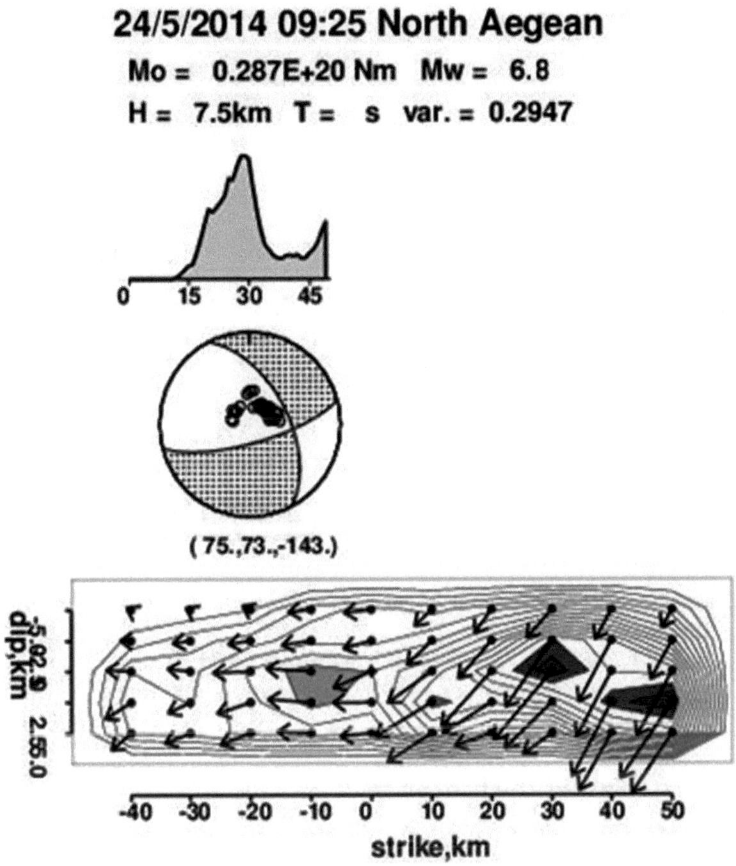

Fig. 6 2D slip distribution along the ruptured fault plane

The source rupture process of the North Aegean earthquake is rather important from the view point of understanding whether the increased static stress changes are high enough to trigger the expected large Marmara earthquake. To explore the hypothesis, we constructed an east-west cross-section of the Coulomb stress changes based on the results presented in Fig. 7 so as to explore the eastward extension of the stress changes toward the Saros bay (Fig. 8).

Such a cross section reveals that the Coulomb stresses exponentially decreases starting from the eastern termination of the rupture toward the east. Besides, it is obvious from Fig. 8 that the aftershocks taking to the east of the mainshock area are the events triggered by the static stress increase rather than events taking place on a ruptured fault plane. Most of the aftershocks concentrate at region where the stress increases is between 0.5 and 3.0 bars. This result has been shown to be consistent with the previous study (Görgün and Görgün 2015). The region where the stress is less than 0.5 bar the aftershock activity diminishes.

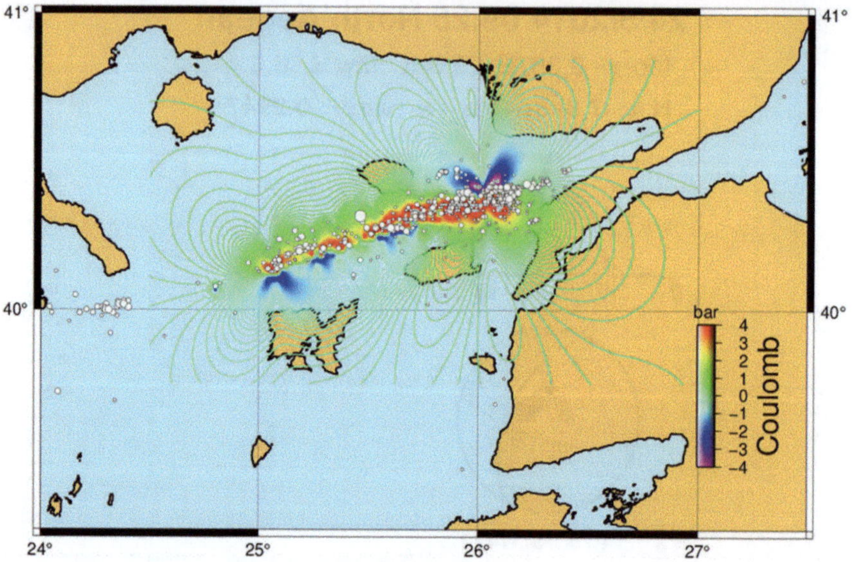

Fig. 7 Coulomb stress changes associated with the mainshock

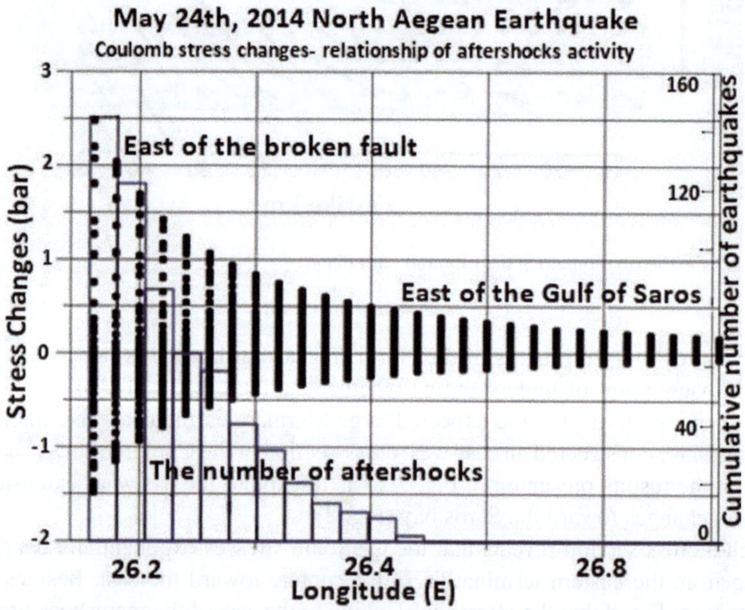

Fig. 8 A relation between the number of aftershocks and the stress changes to the east of the ruptured plane (the black dots are the stresses and the histogram show the number of the aftershocks)

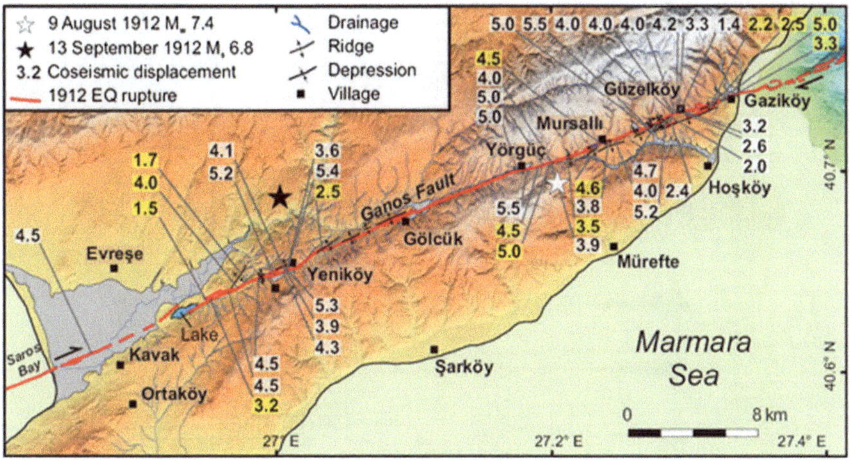

Fig. 9 Lateral displacements measured from the surface ruptures associated with the 1912 Şarköy-Mürefte Earthquake (Aksoy et al. 2010)

The fault segments to the east of the ruptured area were broken by 1912 Şarköy-Mürefte ($M_W = 7.4$) and the 1975 Saroz bay ($M_W = 6.3$) earthquakes. The surface ruptures on the Ganos fault segment, extending from Saros bay towards Marmara sea, associated with the 1912 earthquakes reveal that the coseismic maximum displacements were in the range between 4 and 5 m (Fig. 9; Aksoy et al. 2010; Görgün and Görgün 2015). This in turn imply that the Ganos fault segment is a strong fault to bear stresses capable to generate 5 m slip.

On the other hand, the GPS study carried out by Ergintav et al. (2007) shows that the slip rate along the Ganos fault segment is about 17 mm/year. Thus, the level of strain already accumulated on that fault is far below the maximum bearable stress range of the Ganos segment. Thus, considering all these facts and the stress increases on the Ganos fault caused by the last North Aegean earthquake one may claim that the increased seismic risk is within the range already predicted by the seismic hazard maps.

By virtue of the fact that, the fault segments expected to be ruptured by the impending Marmara earthquake occur further east of the Ganos fault, the coseismic static stress loading caused by the last North Aegean earthquake on those fault segments should be negligible in the order of millibar (Figs. 7 and 8).

6 Moment Tensor Inversion

In this study we calculated 25 CMT parameters for the strongest events of the 2013 and 25 CMT solutions for the 2014 sequences. We used regionally recorded broad-band velocity waveforms for CMT solutions. We calculated moment tensors

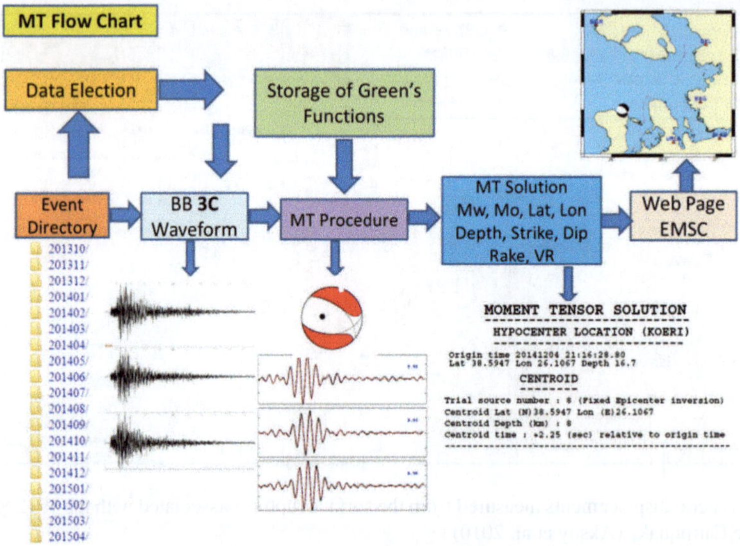

Fig. 10 Fault mechanism/CMT solution flowchart (Kalafat et al. 2009)

of the mainshocks and their strong aftershocks. Moment Tensor Inversion Technique (TDMT_INV time-domain inversion code Dreger 2002) was used for the earthquakes recorded by at least 4 digital broadband seismic stations of KOERI and other Greek seismological networks with 3-component recordings filtered by a band-pass filters described below. An example for a well-constrained CMT solution is illustrated in Figs. 10 and 11. The quality (good signal-to-noise ratio) of the available data allowed the computation of 50 earthquake focal mechanisms and source depths (Table 2). Green's functions were calculated using the frequency-wave number integration code (FKRPROG) developed by Saikia (1994).

Once we have the location and magnitude of an event we can proceed on CMT estimation following the two stages given below:

(1) Data preparation for inversion
(2) CMT Inversion process

In the first stage, the three component digital broadband waveforms are cut according to the origin time, station response is removed, the horizontal components are rotated to get the radial and transversal components, integration or derivation is applied depending on the data type to get the displacements in cm. The modeling process is carried out using the long periods of the seismograms where the frequency range depends on the magnitude of the event. We apply the following frequency ranges for different magnitude ranges:

Table 2 January 8, 2013 North Aegean Sea earthquake source parameters (Main Shock+ Big Aftershocks)

Moment tensor solutions of the 8 January 2013 North Aegean Sea Earthquake and its important aftershocks

Source parameters

EQ	Date	Time	Latitude	Longitude	Depth	Magnitude	S. Moment	Fault parameters							Region
Number	D/M/Y	UTC	Degrees	Degrees	km	Mw	Nm	Strike	Dip	Sli p	P axis	P axis	T axis	T axis	
											Azim	Pl unge	Azimutl	Plunge	
1	08-01-2013	14:16	39.65	25.50	14.0	5.7	4.15E+24	138.7	88.4	−12.8	93.5	10.6	−175.0	7.7	Aegean Sea
2	08-01-2013	14:45	39.63	25.45	6.0	3.6	3.03E+21	144.2	85.6	11.7	278	5.6	9.1	11.3	Aegean Sea
3	09-01-2013	15:36	39.65	25.54	6.0	3.3	5.01E+21	215.1	89.7	−135.8	89.3	29.4	−19.3	29.4	Aegean Sea
4	09-01-2013	15:41	39.68	25.67	4.0	4.6	7.78E+22	246.0	89.3	−119.1	129.8	39.1	1.5	37.3	Aegean Sea
5	09-01-2013	15:56	39.70	25.61	10.0	3.6	2.77E+22	78.9	89.8	150.5	128.1	20.7	29.9	20.7	Aegean Sea
6	09-11-2013	17:28	39.70	25.60	10.0	3.5	1.82E+21	24.3	83.0	−159.7	249.3	19.0	156.2	8.9	Aegean Sea
7	10-01-2013	05:49	39.66	25.52	8.0	4.1	1.57E+22	19.2	84.3	−163.9	244.1	15.5	152.2	6.9	Aegean Sea
8	11-01-2013	00:30	39.68	25.51	10.0	4.2	2.59E+22	132.9	76.6	−27.2	88.6	28.2	−176.6	8.8	Aegean Sea
9	11-01-2013	15:07	39.69	25.60	10.0	3.8	4.95E+21	235.8	67.5	−143.9	97.2	40.6	1.2	6.9	Aegean Sea

(continued)

Table 2 (continued)

Moment tensor solutions of the 8 January 2013 North Aegean Sea Earthquake and its important aftershocks

Source parameters

EQ	Date	Time	Latitude	Longitude	Depth	Magnitude	S. Moment	Fault parameters							Region
Number	D/M/Y	UTC	Degrees	Degrees	km	Mw	Nm	Strike	Dip	Sli p	P axis	P axis	T axis	T axis	
											Azim	Pl unge	Azimutl	Plunge	
10	11-01-2013	21:56	40.41	25.95	14.0	4.1	1.78E+22	208.6	82.2	−161.0	73.9	19.0	−18.6	7.4	Gulf of Saros-Aegean Sea
11	12-01-2013	13:47	39.67	25.55	6.0	3.7	4.13E+21	202.5	76.3	−159.6	65.4	23.9	−26.3	3.6	Aegean Sea
12	12-01-2013	15:00	39.69	25.62	8.0	3.6	2.76E+21	119.4	85.7	−12.2	74.2	11.3	165.4	5.6	Aegean Sea
13	12-01-2013	20:50	38.86	26.00	12.0	3.8	5.16E+21	48	74.4	−162.2	270.6	23.9	−179.7	0.9	Aegean Sea
14	13-01-2013	08:55	39.68	25.55	12.0	4.4	4.91E+22	66.5	80.2	−150.1	292.5	28.1	−164.6	13.0	Aegean Sea
15	13-01-2013	17:54	39.64	25.62	6.0	4.0	1.12E+22	65.8	74.0	−145.2	290.4	35.8	−167.8	11.2	Aegean Sea
16	16-01-2013	03:17	39.71	25.58	10.0	3.9	7.99E+21	242.6	89.1	−140.4	114.2	27.8	9.1	26.2	Aegean Sea
17	19-01-2013	15:19	39.72	25.40	8.0	3.8	6.03E+21	24.1	83.2	129.2	84.0	27.3	329.1	39.2	Aegean Sea

(continued)

Table 2 (continued)

Moment tensor solutions of the 8 January 2013 North Aegean Sea Earthquake and its important aftershocks

Source parameters

EQ	Date	Time	Latitude	Longitude	Depth	Magnitude	S. Moment	Fault parameters							Region
Number	D/M/Y	UTC	Degrees	Degrees	km	Mw	Nm	Strike	Dip	Sli p	P axis	P axis	T axis	T axis	
											Azim	Pl unge	Azimutl	Plunge	
18	19-01-2013	19:26	39.68	25.57	4.0	3.7	3.65E+21	293.8	87.9	−8.5	248.9	7.1	−20.6	4.2	Aegean Sea
19	30-01-2013	22:30	39.69	25.62	6.0	4.0	1.21E+22	112.1	83.1	−29.6	64.9	25.9	162.6	15.4	Aegean Sea
20	10-02-2013	18:10	39.73	25.59	6.0	3.8	6.39E+21	301.5	85.8	−13.4	256.8	12.0	−11.9	6.3	Aegean Sea
21	05-03-2013	09:44	39.69	25.54	8.0	4.1	1.66E+22	248.9	87.4	−113.2	136.6	43.4	−0.5	37.8	Aegean Sea
22	01-04-2013	16:54	38.85	25.98	14.0	3.9	8.16E+21	324.6	78.7	−18.7	280.8	21.2	12.8	5.2	Aegean Sea
23	30-07-2013	05:33	40.30	25.82	15.0	4.9	2.79E+23	150.2	85.6	−9.5	105.2	9.9	−164.0	4.2	Aegean Sea
24	15-11-2013	00:10	39.62	25.61	8.0	3.8	6.11E+21	55.3	82.1	−131.9	287.9	38.2	176.4	24.9	Aegean Sea
25	18-12-2013	21:20	39.83	25.39	8.0	4.0	1.23E+22	308.2	87.9	−15.9	262.4	12.7	−5.4	9.8	Aegean Sea

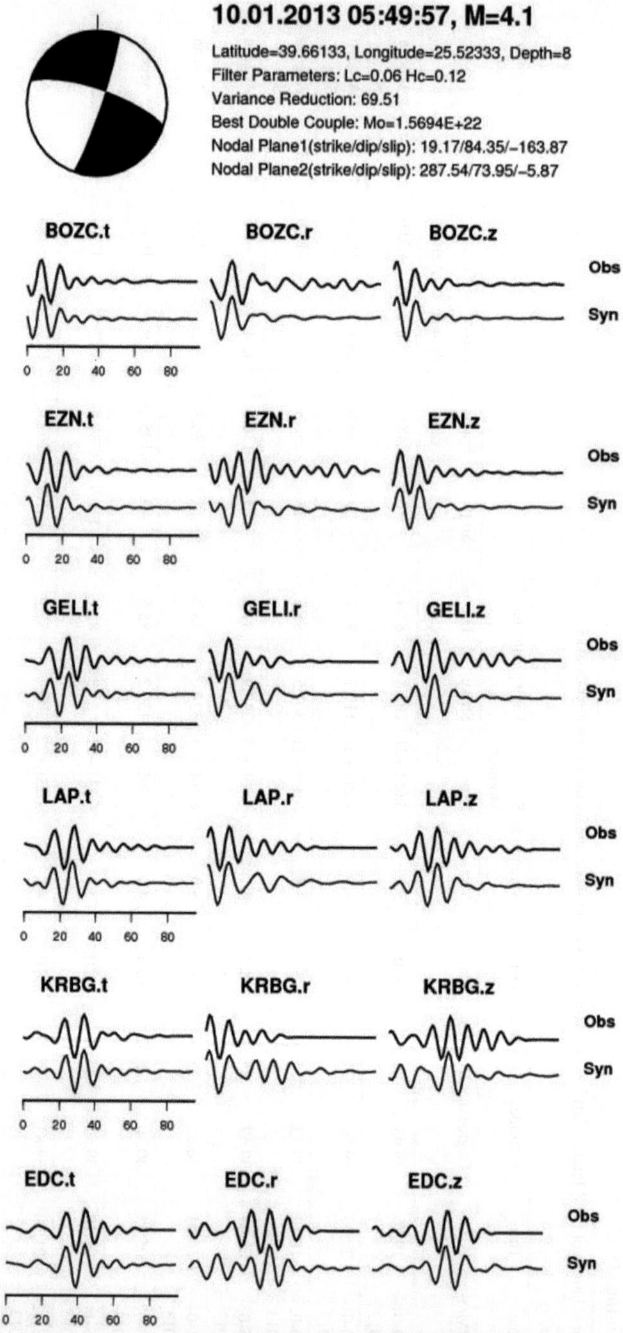

Fig. 11 Examples of inversion

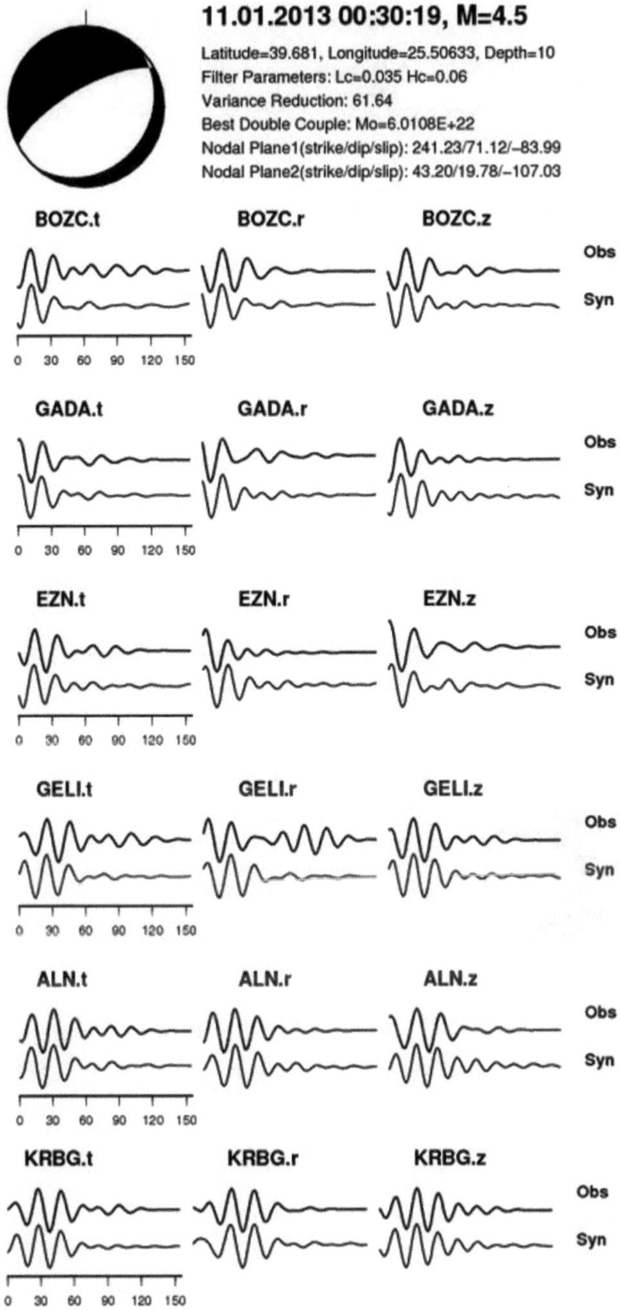

Fig. 11 (continued)

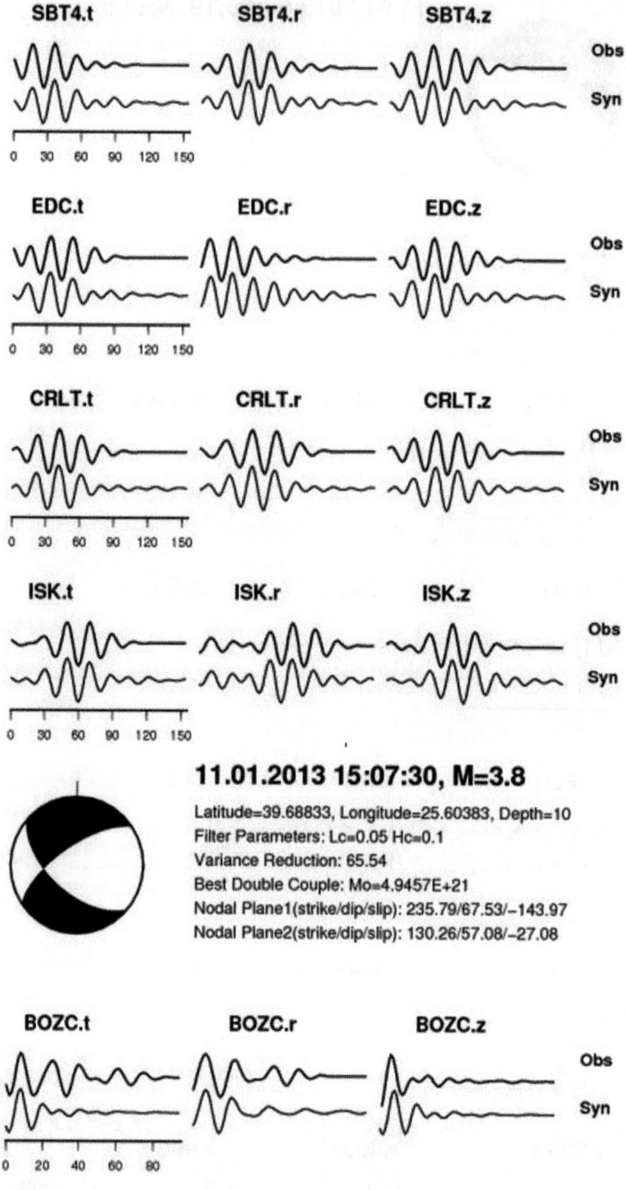

Fig. 11 (continued)

$M < 4$	0.02–0.1 Hz,
$4.0 \leq M < 5.0$	0.02–0.05Hz,
$5.0 < M \leq 7.5$	0.01–0.05Hz
$M > 7.5$	0.005–0.05Hz.

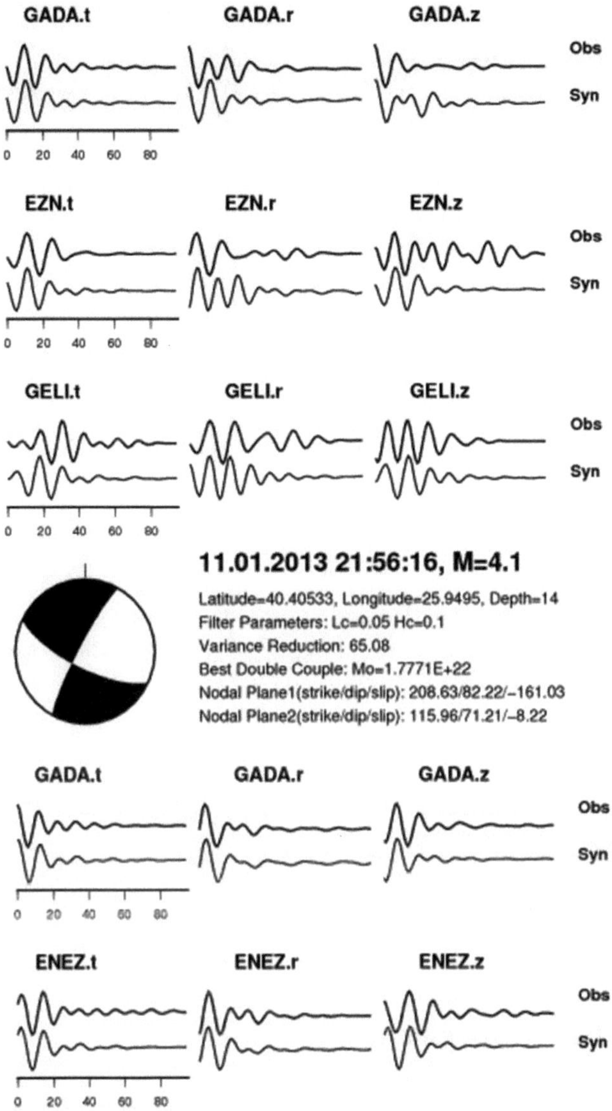

Fig. 11 (continued)

The sampling rate of the observed and the calculated seismograms are modified to have the same sampling rate.

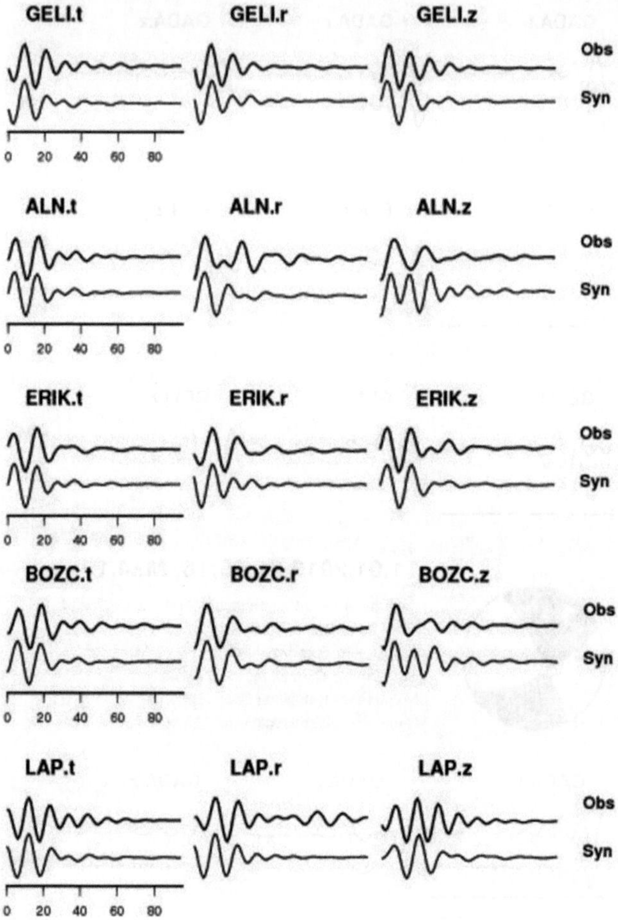

Fig. 11 (continued)

7 Stress Tensor Inversion

The method we use to derive the stress tensor acting on the faults in the North Aegean Sea is described by Gephart (1990) following the same approach we used to derive stress tensor from the focal mechanisms of the events in Marmara region (Pınar et al. 2003). Our data are the orientation of the P- and T-axes of the fault plane solutions we determined. In the method, the earthquakes are assumed to have occurred in a region with no spatial or temporal changes in the stress field, and the associated slip direction is the shear stress direction on the fault plane. The method yields a stress tensor defined by the three principal stress components, namely, maximum compression, ($\sigma 1$), intermediate compression, ($\sigma 2$), minimum compression, ($\sigma 3$), and the stress magnitude ratio defined as $R = (\sigma 2 - \sigma 1)/(\sigma 3 - \sigma 1)$. The value of R is

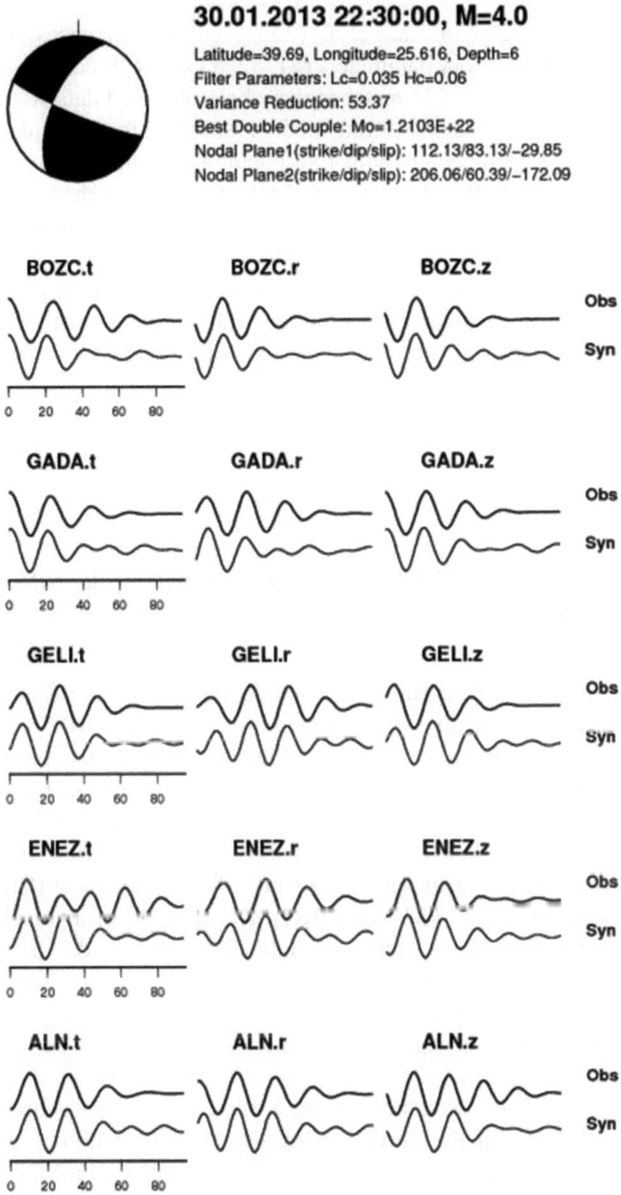

Fig. 11 (continued)

an indicator of the dominant stress regime acting in the region under investigation; R = 0 when σ1 ≈ σ2 (biaxial deviatoric compression or state of confined extension), R = 1 when σ2 ≈ σ3 (uniaxial deviatoric compression or state of confined compression) and R = 0.5 when σ1 ≈ σ2 ≈ σ3 (uniform triaxial compression). For more information on the subject please refer to Pınar et al. (2003) and the references given therein. The combination of these four parameters (σ1, σ2, σ3 and R) is called a stress model and the model that most closely matches the whole observed data set is called the best-fitting stress model. The best-fitting model is searched for in a grid over the four model parameters, systematically adjusting one at a time through a wide range of possibilities (Gephart 1990). The measure of misfit is given by the smallest rotation about an axis of any orientation that brings one of the nodal planes and its slip direction into an orientation consistent with the stress model.

Thus, for each stress model, the misfits between the orientation of the observed data and prediction are estimated and summed. The minimum misfit is the one that yields the smallest sum of misfits and is selected as the regional stress tensor for the region.

8 Results

8.1 *The January 8, 2013 North Aegean Earthquake and Its Aftershocks Sequence*

The sequence included nine events of magnitude $M_w \geq 4.0$ and greater. During the 8 January–31 March 2014 time period (approximately 2 months) 385 aftershocks were relocated in the rupture region. The calculation of epicenters have been done by P and S readings of the seismograms recorded by the Turkish and Greece stations.

In the study, events with at least 5 P- and 1 S-wave arrivals, having azimuthal gap lower than 200, location RMS lower than 0.30 s were chosen for processing. The earthquakes were obtained with horizontal (ERH) and vertical (ERZ) location errors lesser than 2.0 and 1.0 km, respectively. Based on the estimated hypocentral depths, most of the aftershocks occur within the upper crust, within the first 15 km. We relocated 385 events which were well recorded by at least five stations for two month period following the mainshock. Aftershocks are complete for $Ml \geq 2.3$ and from the Gutenberg–Richter curve the b-value was calculated to be b = 0.703 and the a-value = 4.74 (Fig. 12).

8 January 2013 seismic sequence occurred between Lemnos Island, Greece and Bozcaada Island, Turkey in the North Aegean Sea. The distribution of relocated epicentres and the focal mechanisms clearly show the activation of a NE-SW trending right lateral strike-slip fault. This study provided CMT solutions for the mainshock and 24 important aftershocks. The results of this study supports the work done previously which showed that the main shock has a right-lateral faulting mechanism (Kiratzi and Svigkas 2013; Ganas et al. 2014; Kürçer et al. 2015). The source parameters of the earthquakes are given Table 2 and Figs. 13 and 14 and the stress tensor inversion results using the focal mechanisms given in Fig. 14 is illustrated in Fig. 15.

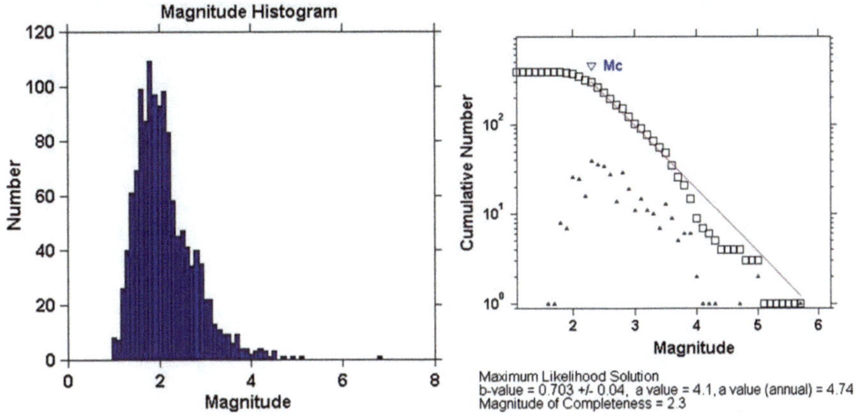

Fig. 12 Cumulative number of earthquakes versus magnitude (completeness magnitude *Mc*)

8.2 *The May 24, 2014 North Aegean Earthquake and Its Aftershocks Sequence*

The second earthquake in this region, 24 May 2014 North Aegean sequence was recorded by the stations of the Turkish and Greek Seismological Networks. In the study, events with at least 7 P- and 1 S-wave arrivals, having azimuthal gap lower than 200, location RMS lower than 0.40 s were chosen for processing. The earthquakes were obtained with horizontal (ERH) and vertical (ERZ) location errors lesser than 2.0 and 1.0 km, respectively. Based on the estimated hypocentral depths, most of the aftershocks occurs within the upper crust, in the first 20 km. 24 May–31 July 2014 time period (approximately 2 months) 1305 aftershocks were relocated in this region. Most of the aftershocks activity took place within the first two months of the earthquake sequence. Aftershocks are complete for Ml ≥ 2.1 and from the Gutenberg–Richter curve the b-value was calculated to be b = 0.598 and the a-value = 4.03 (Fig. 16).

24 May 2014 earthquake occurred also in the North Aegean Sea. The earthquake was felt very strongly in the North Aegean Sea area between Greece and Turkey (Io = VIII, MM). About 300 houses were damaged in Turkey (Fig. 17), 11 houses collapsed in Greece and totally 350 people injured in the earthquake.

The aftershocks continued about 2 months in this area. Along strike dimension of the aftershock zone is approximately ~120 km long, and its width is ~10 km, in accordance to what is expected from the magnitude of the largest event of the sequence (M_W = 6.8) assuming 50 cm average displacement (Fig. 18). The aftershock zone of the 2014 earthquake was defined a distance of 180–250 km long by Saltogianni et al. (2015). Most of the aftershock focal mechanisms show right-lateral strike-slip faulting (Fig. 19; Table 3), as is the case of the mainshock (No. 1). After the mainshock the seismic activity in NW Turkey considerably increased (Fig. 20).

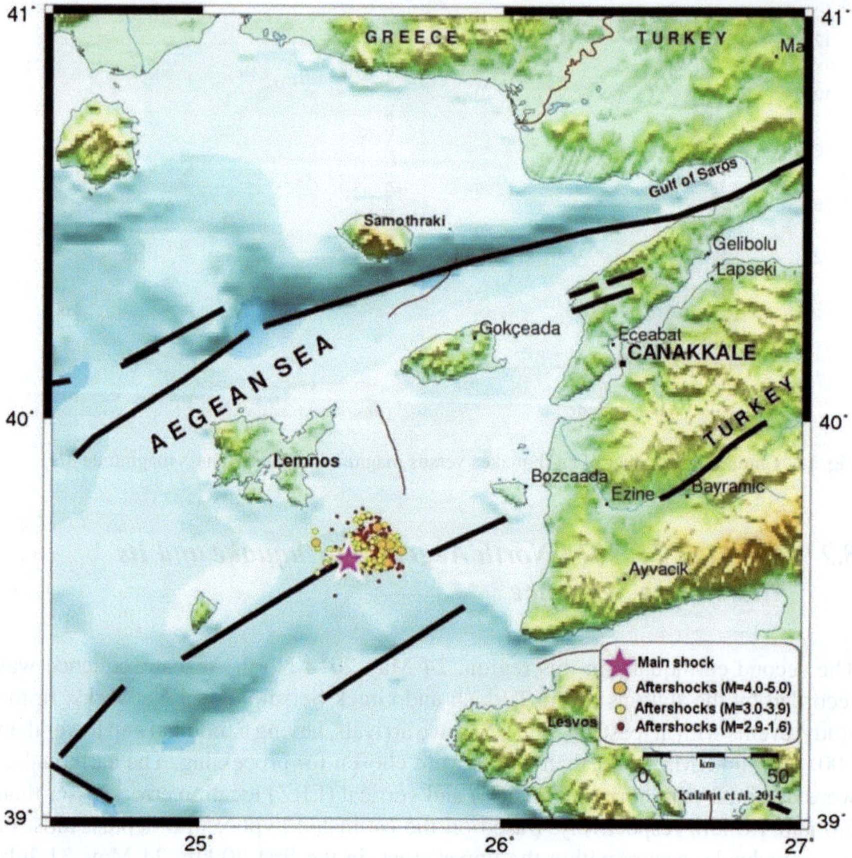

Fig. 13 Aftershocks distribution of 2013 North Aegean Sea earthquake

Based on the distribution of the aftershocks and the focal mechanisms, the ENE–WSW trending planes are the fault planes. Also, similar results have been found by Görgün and Görgün (2015).

Stress Tensor Analysis (STA) was done within the study area for understanding the tectonic properties. The stress tensor parameters are calculated using the azimuth and dip of P and T axis couples of 50 earthquakes (Fig. 21) occurring in the aftershock region and the Stress Tensor Inversion software developed by Gephart (1990).

For the region, the best fit was attained for $R = 0.5$ and for the azimuth and plunge pair of (107°, 31°) for σ1, (252°, 53°) for σ2 and (6°, 17°) for σ3, respectively. For more explanation see the caption for Fig. 15.

The most prominent feature of the earthquake is the widespread distribution of the aftershocks. Whether with, all aftershocks took place along the 120 km length, rupture length was a few times shorter than the rupture derived from the aftershock

Table 3 May 24, 2014 North Aegean Sea earthquake source parameters

Moment tensor solutions of the 24 May 2014 North Aegean Sea Earthquake and its important aftershocks

Source parameters														
EQ	Date	Time	Latitude	Longitude	Depth	Magnitude	S. moment	Fault parameters						Region
Number	D/M/Y	UTC	Degrees	Degrees	km	Mw	Nm	Strike	Dip	Slip	P axis	P axis	T axis	T axis	
											Azimuth	Plunge	Azimuth	Plunge	
1	24-05-2014	09:25	40.31	25.45	20.0	6.8	2.03E+26	43.2	89.0	−127.0	281.9	35.2	163.9	33.5	Aegean Sea
2	24-05-2014	09:31	40.43	26.25	8.0	5.3	1.12E+24	251.3	85.0	−139.1	122.5	31.6	16.9	23.6	Gulf of Saros-Aegean Sea
3	24-05-2014	10:11	40.40	26.13	12.0	4.4	4.47E+22	198.0	61.0	−141.9	54.2	46.3	−38.1	2.2	Gulf of Saros-Aegean Sea
4	24-05-2014	10:35	40.42	26.14	10.0	4.4	4.65E+22	240.6	89.2	−162.6	107.1	12.6	14.6	11.2	Gulf of Saros-Aegean Sea
5	24-05-2014	11:18	40.40	26.19	12.0	4.2	2.09E+22	215.2	80.3	−147.6	81.5	29.4	−16.8	14.2	Gulf of Saros-Aegean Sea
6	24-05-2014	11:33	40.29	25.62	10.0	4.7	1.11E+23	116.5	66.3	−64.9	63.9	60.6	−171.8	17.6	Aegean Sea
7	24-05-2014	12:30	40.29	25.60	12.0	4.0	1.24E+22	317.6	85.8	50.8	78.9	29.9	194.3	36.7	Aegean Sea

(continued)

Table 3 (continued)

Moment tensor solutions of the 24 May 2014 North Aegean Sea Earthquake and its important aftershocks

Source parameters

EQ	Date	Time	Latitude	Longitude	Depth	Magnitude	S. moment	Fault parameters							Region
Number	D/M/Y	UTC	Degrees	Degrees	km	Mw	Nm	Strike	Dip	Slip	P axis	P axis	T axis	T axis	
											Azimuth	Plunge	Azimuth	Plunge	
8	24-05-2014	12:45	40.24	25.18	14.0	4.5	6.26E+22	219.7	89.7	−137.6	93.4	28.2	−13.4	28.2	Aegean Sea
9	24-05-2014	14:49	40.41	25.99	12.0	4.8	2.10E+23	110.6	81.3	−11.7	66.7	14.8	157.2	1.9	Gulf of Saros-Aegean Sea
10	24-05-2014	15:01	40.39	26.13	12.0	4.0	9.46E+21	325.9	82.3	49.2	86.9	25.5	199.4	38.8	Gulf of Saros-Aegean Sea
11	24-05-2014	16:34	40.29	25.62	6.0	4.1	1.69E+22	60.0	64.9	−71.3	2.7	64.8	136.0	17.9	Aegean Sea
12	24-05-2014	20:28	40.28	25.50	12.0	3.7	3.84E+21	62.1	89.4	172.3	107.2	4.9	16.8	6.4	Aegean Sea
13	24-05-2014	23:00	40.41	25.93	12.0	4.0	1.19E+22	197.1	75.0	−145.6	61.7	34.4	−36.3	11.5	Gulf of Saros-Aegean Sea

(continued)

Table 3 (continued)

Moment tensor solutions of the 24 May 2014 North Aegean Sea Earthquake and its important aftershocks

Source parameters

EQ	Date	Time	Latitude	Longitude	Depth	Magnitude	S. moment	Fault parameters							Region
Number	D/M/Y	UTC	Degrees	Degrees	km	Mw	Nm	Strike	Dip	Slip	P axis Azimuth	P axis Plunge	T axis Azimuth	T axis Plunge	
14	25-05-2014	05:44	40.42	26.07	12.0	3.9	7.31E+21	125.2	87.9	−36.9	74.2	26.7	176.9	23.6	Gulf of Saros-Aegean Sea
15	25-05-2014	11:38	40.42	26.15	14.0	5.0	3.31E+23	71.2	83.2	−139.6	303.5	27.0	−161.5	27.0	Gulf of Saros-Aegean Sea
16	25-05-2014	11:47	40.41	26.10	28.0	4.6	1.00E+23	89.7	75.5	−41.2	42.9	38.4	146.2	16.2	Gulf of Saros-Aegean Sea
17	25-05-2014	14:55	40.25	25.25	14.0	4.0	9.52E+21	92.9	47.9	−114.7	290.9	71.7	−159.6	52.7	Aegean Sea
18	26-05-2014	21:28	40.26	25.19	16.0	4.4	3.88E+22	25.3	56.4	−80.2	326.0	76.2	107.8	10.9	Aegean Sea

(continued)

Table 3 (continued)

Moment tensor solutions of the 24 May 2014 North Aegean Sea Earthquake and its important aftershocks

Source parameters

EQ	Date	Time	Latitude	Longitude	Depth	Magnitude	S. moment	Fault parameters							Region
Number	D/M/Y	UTC	Degrees	Degrees	km	Mw	Nm	Strike	Dip	Slip	P axis Azimuth	P axis Plunge	T axis Azimuth	T axis Plunge	
19	28-05-2014	03:59	40.42	26.13	14.0	4.8	2.00E+23	251.5	58.3	−109.0	118.8	70.3	−4.4	11.1	Gulf of Saros-Aegean Sea
20	30-05-2014	04:06	40.18	25.56	14.0	4.4	4.97E+22	260.5	89.2	154.1	128.8	18.8	32.7	17.3	Aegean Sea
21	15-06-2014	19:46	40.26	25.17	18.0	4.4	5.13E+22	257.9	88.2	−153.2	125.8	20.2	29.3	17.2	Aegean Sea
22	26-06-2014	16:22	39.70	25.56	8.0	4.2	1.94E+22	115.4	80.1	−38.0	66.9	33.5	169.2	17.8	Aegean Sea
23	09-07-2014	23:45	40.43	26.26	16.0	4.3	2.86E+22	99.9	56.1	−80.2	41.9	76.5	−177.2	10.5	Gulf of Saros-Aegean Sea
24	15-07-2014	10:33	40.22	25.18	12.0	4.2	1.95E+22	133.8	79.9	−11.0	89.9	14.8	−179.9	0.6	Aegean Sea
25	04-09-2014	17:43	40.13	24.87	8.0	4.8	1.51E+23	236.0	85.1	139.8	289.9	23.0	185.2	30.9	Aegean Sea

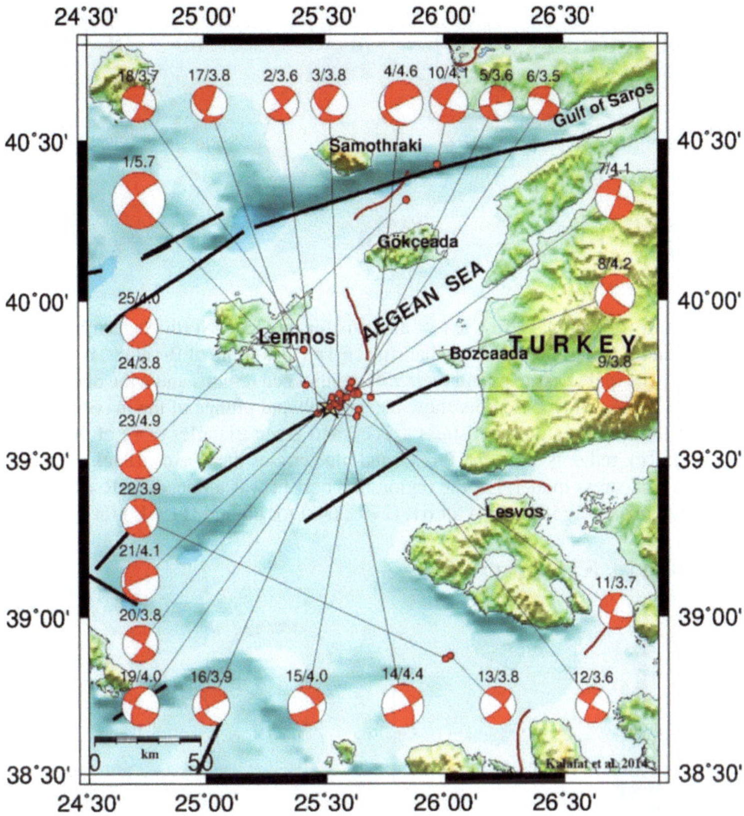

Fig. 14 CMT solutions of 2013 North Aegean Sea earthquakes

distribution. 2014 North Aegean Earthquake showed that, the long period seismic waves recorded at the stations deployed in eastern Marmara region. The causatives for such long period waves is a challenge both for the Earthquake Early Warning and Earthquake engineering studies since the long periods waves are effective even at locations several hundreds km away from the source region. The Gökçeada strong motion station located about 50 km to the south of the source zone acquired the long period source pulses associated with the rupture of the fault plane. Considering the strike-slip faulting mechanism of the earthquake and the location of the Gökçeada station the long pulses portrays the fault normal motions acquired because of the fact that the station is located close to the nodal plane for the shear waves.

On the other hand, One of the most effective tools to infer details on the source rupture process of large earthquakes is modeling of the teleseismic bodywaves. We obtain a slip distribution model for the mainshock and then using the finite source model we estimate the Coulomb failure stress changes with the aim to understand which part of the aftershock zone are associated as triggered events. Coulomb stress changes associated with the mainshock. The aftershocks tend to take place at regions

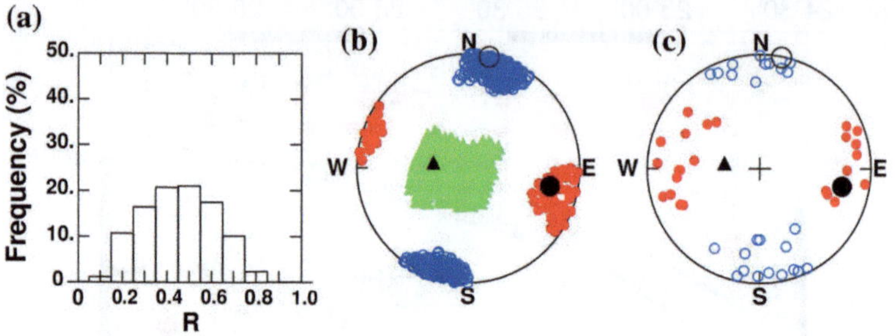

Fig. 15 The results of regional stress tensor analysis for the 2013 North Aegean Sea earthquake, based on the P- and T-axes of the focal mechanisms. **a** The histogram of R-values, **b** the distribution of the predicted principal stress axes and their 95% confidence regions and **c** the distribution of the observed P- and T-axes. In (**b**), red dots show the azimuth and plunge of the maximum stress axis σ1, blue circles those of the minimum stress axis σ3 and green triangles those of the intermediate stress axis σ2. In (**c**), red dots show the P-axes and blue circles the T-axes. Black symbols denote the axes for the best stress model. For the region, the best fit was attained for R = 0.5 and for the azimuth and plunge pair of (102°, 26°) for σ1, (279°, 64°) for σ2 and (11°, 1°) for σ3, respectively

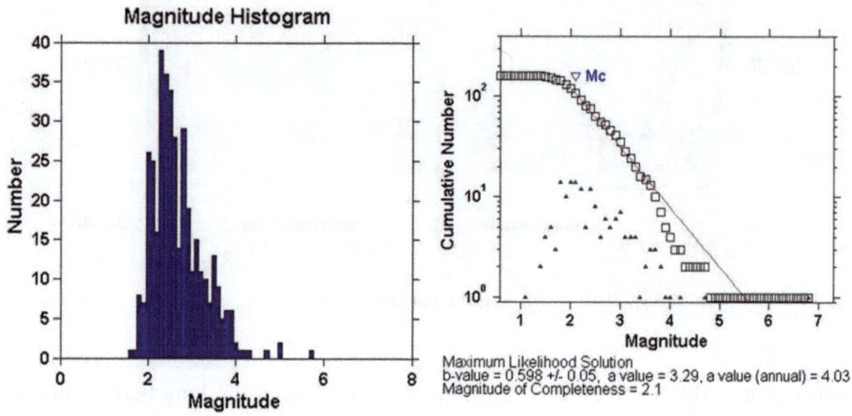

Fig. 16 Cumulative number of earthquakes versus magnitude (completeness magnitude, *Mc*)

of increased stress. The southward extension of the aftershock area and the area of increased stress to the east of the epicenter where dip-slip component was derived.

The source rupture process of the North Aegean earthquake is rather important from the view point of understanding whether the increased static stress changes are high enough to trigger the expected large Marmara earthquake. Coulomb stresses exponentially decreases starting from the eastern termination of the rupture toward the east. Most of the aftershocks concentrate at region where the stress increases is between 0.5 and 3.0 bars.

Fig. 17 Some examples of heavily damaged house from Gökçeada Island and Çanakkale (Turkey; by AA)

9 Discussion

The source regions of the North Aegean earthquakes are influenced by both the Aegean extensional regime and the strike-slip regime in the western part of the North Anatolian Fault Zone. Strike-slip faulting changes to oblique-slip faulting, with significant component of extension, as one goes from the Aegean to the coastal area of Western Turkey. Evidence from the distribution of large earthquakes, and geodetic measurements suggests that the active faulting in mainland Greece and the North Aegean Sea is concentrated into a small number of discrete, linear zones that bound relatively rigid blocks (Goldsworthy et al. 2002).

The prevailing N-S extension in the whole back arc Aegean region is then the key motion that results in dextral strike-slip movement on NE-SW trending faults and sinistral strike-slip movement on NW-SE trending faults. This interpretation relies on the assumption that the faults occupying the western Aegean coast are

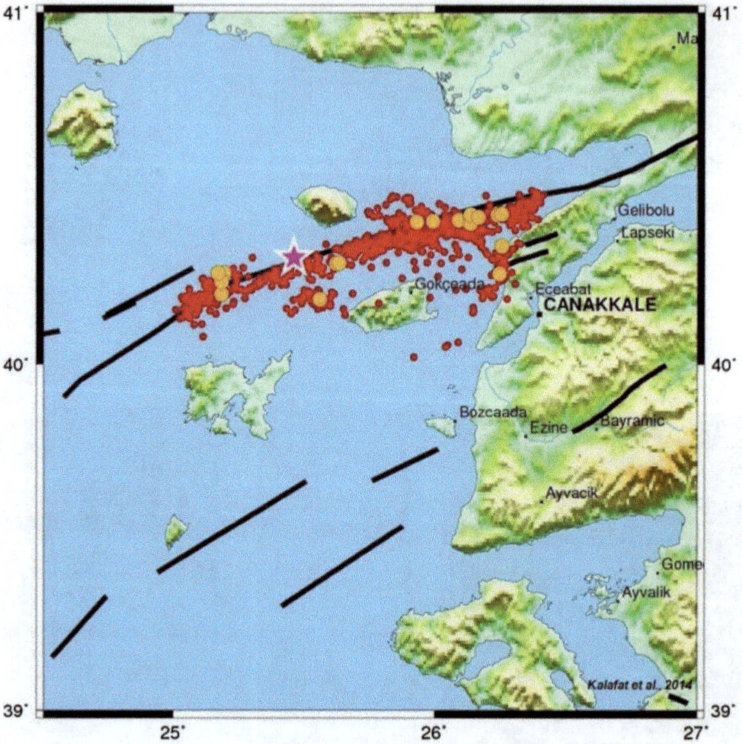

Fig. 18 Aftershocks distribution of 2014 North Aegean Sea earthquake

orthogonal to the NE-SW dextral strike-slip faults and mark the boundary between them and E-W normal faults in the mainland of Greece and Western Turkey (Genç et al. 2001; Koukouvelas and Aydin 2002; Karakostas et al. 2003; Yaltırak et al. 2012; Chatzipetros et al. 2013; Görgün and Görgün 2015).

The 2013 and 2014 North Aegean Sea Earthquakes caused a regional stress change and triggered earthquakes on nearby active fault segments in the region. For this reason, seismicity increased very abruptly in the region (Fig. 20).

The rupture of the main fault oriented west to east direction caused stress increase and triggered related seismic activity in the North Aegean Sea region. Therefore high seismic activity is observed at northeast-southwest oriented directions of the main NE-SW trending fault which was broken during the 24th May 2014 Earthquake.

Also according to Görgün and Görgün (2015), the May 24th, 2014 Earthquake clearly indicated that active NE–SW trending right-lateral strike–slip faulting systems are wide-spread in the Aegean region.

The distribution of the aftershocks of the 2014 earthquake support the presence of a rupture of approximately 120 ± 10 km. Aftershocks occurred within an area of approximately 1200 km^2 (Fig. 18). The a and b-values for these sequences were estimated to be equal to 4.03–4.74 and 0.598–0.703, respectively with 90% goodness

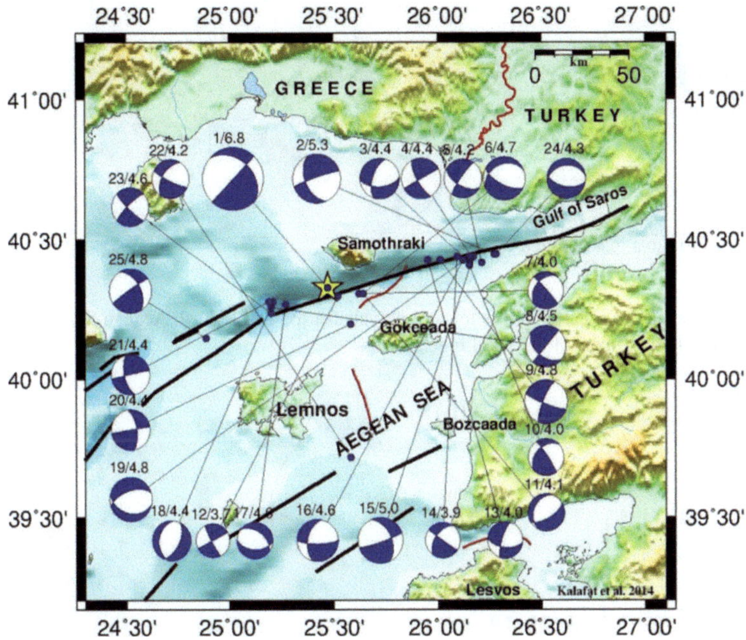

Fig. 19 CMT solutions of 2014 North Aegean Sea earthquakes

Fig. 20 Cumulative distribution of January 8, 2013 and May 24, 2014 North Aegean Sea earthquakes in the region

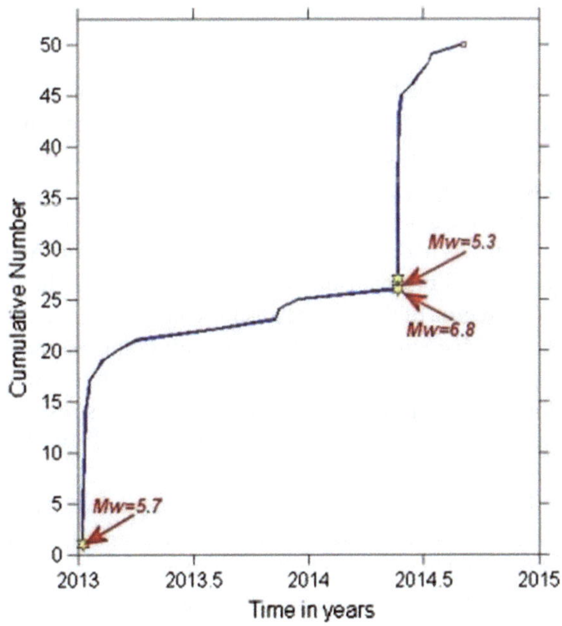

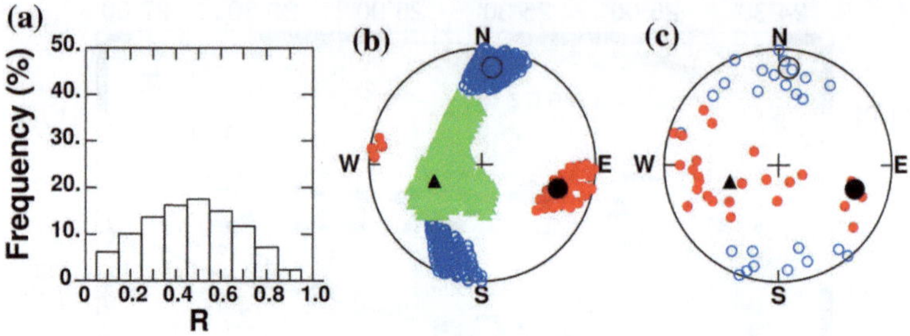

Fig. 21 The results of regional stress tensor analysis for the 2014 North Aegean Sea earthquakes, based on the P- and T-axes of the focal mechanisms

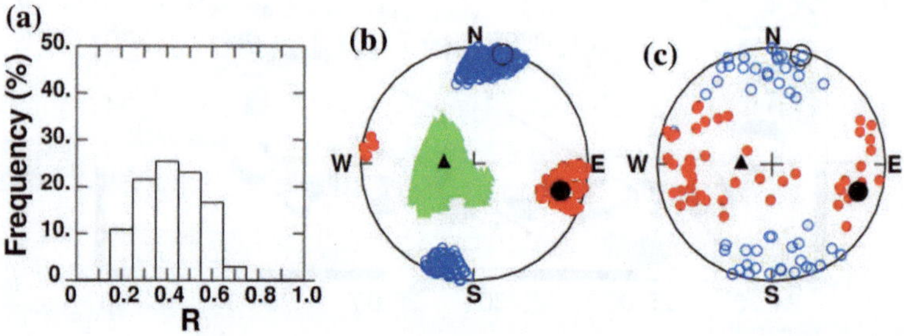

Fig. 22 The results of regional stress tensor analysis for the 2013 and 2014 North Aegean Sea earthquakes, based on the all the P- and T-axes of the focal mechanisms

of fit level. The b-value is lower than the global mean value of 1.0, which indicates that the North Aegean sequence consists of larger magnitude aftershocks and high differential crustal stress in the regime (Wiemer and Katsumata 1999; Wiemer and Wyss 2002).

We have obtained local stress tensors for the source regions of the 2013 and 2014 earthquakes and a regional stress tensor for North Aegean Sea region by combining the all focal mechanisms (Figs. 15, 21 and 22). The local stress tensor acting in the source region of the 2013 mainshock and the regional stress tensor exhibits maximum compressive stress axis and minimum compressive stress axis close to horizontal implying pure strike-slip tectonic regime (Fig. 9). Same result given by Görgün and Görgün (2015). According to their study, the stress tensor inversion results indicate a predominant strike–slip stress regime with a NW–SE oriented maximum principal compressive stress (σ1).

However, the events in the source region of the 2014 event yield stress tensor inversion results where significant deviation of the intermediate stress axis from the vertical and bias of the minimum compressive stress axis from horizontal suggesting transtential tectonic regime. The local stress tensor derived from the 2013 events is close to the parameters of the regional stress tensor suggesting no stress perturbation caused by the 2013 $M_w = 5.8$ mainshock.

The orientation of the maximum compressive stress axis estimated for the three stress tensors is ESE-WNW and the orientation of the minimum compressive stress axis is NNE-SSW. The plunge of both the axes is close to horizontal. Comparing these stress tensor inversion results with the ones obtained using the events located further east along the North Anatolian fault in Marmara Sea region (Kiratzi 2002 and Pınar et al. 2003) point out significant counterclockwise rotation of the stress field going from east to west.

10 Conclusions

The focal mechanism solutions of the important aftershocks of the North Aegean Sea earthquakes show the region is undertaking deformation where right-lateral strike-slip and oblique normal faulting occur (Figs. 14 and 19).

The distribution of the important earthquakes and aftershocks also provided that E-W and NE-SW trending faults caused the seismic activity in the region (Figs. 13 and 18). 2013 and 2014 North Aegean earthquakes are good example of the right-lateral fault zones terminate in well-defined extensional basins. The focal depth solutions show that the seismogenic zone producing earthquakes is in 8–15 km depth range which is not very deep.

Likewise, this result is supported by earlier studies. The thickness of brittle seismogenic crust in NAT area is about 25 km (Güngör and Güngör 2015; Karabulut et al. 2006).

The results of the stress analysis show that the (**P**-compressional) direction of the stress axes is in **WNW-ESE** direction and (**Tmax**-extensional) direction is in **NNE-SSW** direction (Fig. 22).

For the combined region, the best fit was attained for R = 0.4 and for the azimuth and plunge pair of (107°, 22°) for $\sigma 1$, (275°, 68°) for $\sigma 2$ and (15°, 4°) for $\sigma 3$, respectively.

The stress field along the North Anatolian fault zone in NW Turkey and North Aegean Sea region rotates counterclockwise moving from east to west.

Coulomb stress changes associated with the mainshock. The aftershocks tend to take place at regions of increased stress. The southward extension of the aftershock area and the area of increased stress to the east of the epicenter where dip-slip component was derived.

One of the characteristics of the North Aegean Earthquake has been shown large amplitude seismic waves. The causatives for such long period waves is a challenge both for the Earthquake Early Warning and Earthquake engineering studies since the long periods waves are effective even at locations several hundreds km away from the source region.

According to the GPS study, the slip rate along the Ganos fault segment is about 17 mm/year (Ergintav et al. 2007). Thus, the level of strain already accumulated on that fault is far below the maximum bearable stress range of the Ganos segment. Thus, considering all these facts and the stress increases on the Ganos fault caused by the last North Aegean earthquake one may claim that the increased seismic risk is within the range already predicted by the seismic hazard maps.

Acknowledgements The first author was supported by the Department of Science Fellowship and Grant programs (2014-2219) of TUBITAK (The Scientific and Technological Research Council of Turkey). Also this study was supported by Bogazici University Research Projects Commission under SRP/BAP project No. 6040.

References

Aksoy ME, Meghraoui M, Vallée M, Çakır Z (2010) Rupture characteristics of the A.D. 1912 Mürefte (Ganos) earthquake segment of the North Anatolian fault (western Turkey). Geology 1(38):991–994

Allen CR (1969) Active faulting in northern Turkey, Contr. No. 1577. Div Geol Sci Calf Inst Tech 32

Barka AA (1992) The North Anatolian Fault zone. Annales Tectonicae 6:164–195

Barka A, Gülen L (1988) New constrains on age and total offset on the North Anatolian Fault Zone: implications for tectonics of the Eastern Mediterranean Region. METU J Pure Appl Sci 21:39–63

Barka AA, Toksöz MN, Gülen L, Kadinsky-Cade K (1987) Kuzey Anadolu Fayının Doğu Kesiminin Segmentasyonu, Sismisitesi ve Deprem Potansiyeli, Yerbilimleri C14:337–352

Chatzipetros A, Kiratzi A, Sboras S, Zouros N, Pavlides S (2013) Active faulting in the north-eastern Aegean Sea Islands. Tectonophysics 597–598:106–122

Dewey JF (1976) Seismicity of Northern Anatolia. Bull Seis Soc Am 3:843–868

Dewey JF, Şengör AMC (1979) Aegean and surrounding regions: Complex multiplate and continuum tectonics in a convergent zone. Geol Soc Am Bull 90:84–92. https://doi.org/10.1130/0016-7606

Drakopoulos JC, Ekonomides AC (1972) Aftershocks of February 19, 1968 earthquake in Northern Aegean Sea and Related Problems. Pure Appl Geophys 95(1):100–115

Dreger D (2002) Manual of the time-domain moment tensor inverse code (TDMT_INVC), Release 1.1. Berkeley Seismology Laboratory, p 18

Ergintav S, Doğan U, Gerstenecker C, Çakmak R, Belgen A, Demirel H, Aydın C, Reilinger R (2007) A snapshot (2003–2005) of the 3D postseismic deformation for the 1999, $M_W = 7.4$ İzmit Earthquake in the Marmara Region, Turkey, by first results of joint gravity and GPS monitoring. J Geodyn 44(1):1–18

Eyidoğan H, Güçlü U, Utku Z, Değirmenci E (1991) Türkiye Büyük Depremleri Makro -Sismik Rehberi (1900–1988). İTÜ Maden Fak. Jeofizik Müh. Böl, İstanbul

Ganas A, Roumelioti Z, Karastathis V, Chousianitis K, Moshou A, Mouzakiotis E (2014) The Lemnos January 2013 (Mw = 5.7) earthquake: fault slip, aftershock properties and static stress transfer modeling in the north Aegean Sea. J Seismol 2014(18):433–455

Genç CŞ, Altunkaynak Ş, Karacık Z, Yazman M, Yılmaz Y (2001) The Çubukludağ graben, south of İzmir: its tectonic significance in the Neogene geological evolution of the western Anatolia. Geodin Acta 14(1–3):45–55

Gephart JW (1990) FMSI: A FORTRAN program for inverting fault/slickenside and earthquake focal mechanism data to obtain the regional stress tensor. Com Geosci 16:953–989

Goldsworthy M, Jackson J, Haines J (2002) The continuity of active fault systems in Greece. Geophys J Int 148:596–618

Görgün E, Görgün B (2015) Seismicity of the 24 May 2014 Mw 7.0 Aegean Sea earthquake sequence along the North Aegean Trough. J Asian Earth Sci 111:459–469

Jackson J, McKenzie DP (1984) Active tectonics of the Alpine-Himalayan Belt between Western Turkey and Pakistan. Geophys J R Astr Soc 77:185–246

Kalafat D, Gürbüz C, Üçer SB (1987) Batı Türkiye'de Kabuk ve Üst Manto Yapısının Araştırılması. Deprem Araştırma Bülteni, Sayı 59:43–64 (in Turkish)
Kalafat D, Kekovalı K, Güneş Y, Yılmazer M, Kara M, Deniz P, Berberoğlu M (2009) Türkiye ve Çevresi Faylanma-Kaynak Parametreleri (MT) Kataloğu (1938–2008): A catalogue of source parameters of moderate and strong earthquakes for Turkey and its surrounding area (1938–2008). Boğaziçi University Publication,Bebek-İstanbul, no 1026, 43 pp
Kalafat D, Güneş Y, Kekovalı K, Kara M, Deniz P, Yılmazer M (2011) Bütünleştirilmiş Homojen Deprem Kataloğu (1900–2010; M ≥4.0)—a revised and extended earthquake catalogue for Turkey since 1900 (1900–2010; M ≥4.0). Boğaziçi University Library Cataloging-in Publication Data, İstanbul, 640 pp
Karakostas VG, Papadimitriou EE, Karakaisis GF, Papazachos CB, Scordilis EM, Vargemezis G, Aidona E (2003) The 2001 Skyros, Northern Aegean, Greece, earthquake sequence: off-fault aftershocks, tectonic implications, and seismicity triggering. Geophys Res Lett 30(1):1012. https://doi.org/10.1029/2002gl015814
Kikuchi M, Kanamori H (2003) Note on teleseismic body-wave inversion program. http://www.eri.u-tokyo.ac.jp/ETAL/KIKUCHI
Kiratzi AA (2002) Stress tensor inversions along the westernmost North Anatolian Fault Zone and its continuation into the North Aegean Sea. Geophys J Int 151:360–376
Kiratzi AA, Svigkas N (2013) A study of the 8 January 2013 Mw5.8 earthquake sequence (Lemnos Island, East Aegean Sea). Tectonophysics 608:452–460. http://dx.doi.org/10.1016/j.tecto.2013.09.002
Klein FW (2002) User's Guide to HYPOINVERSE–2000, a Fortran program to solve for earthquake locations and magnitudes, USGS Open-File Report 02-171
Koukouvelas I, Aydin A (2002) Fault structure and related basins of the North Aegean Sea and its surroundings. Tectonics 21(5)
Kreemer C, Chamot-Rooke N, Le Pichon X (2004) Constraints on the evolution and vertical coherency of deformation in the Northern Aegean from a comparison of geodetic, geologic and seismologic data. Earth Planet Sci Lett 225(3–4):346. https://doi.org/10.1016/j.epsl.2004.06.018
Kürçer A, Yalçın H, Gülen L, Kalafat D (2015) 8 January 2013 Mw = 5.7 North Aegean Sea Earthquake and its seismotectonic significance. Geodin Acta. https://doi.org/10.1080/09853111.2014.957503
Le Pichon X, Lybéris N, Alvarez F (1987) Discussion on the subsidence of the North Aegean Trough: an alternative view. J Geol Soc 144(2):349–351. https://doi.org/10.1144/gsjgs.144.2.0349
Lee WHK, Lahr JC (1975) HYPO71: a computer program for determining hypocenter, magnitude and first motion pattern of local earthquakes. USGS Open File Rep 75–311:1–116
McClusky S, Balassanian S, Barka A, Demir C, Ergintav S, Georgiev I, Gurkan O, Hamburger M, Hurst K, Kahle H, Kastens K, Kekelidze G, King R, Kotzev V, Lenk O, Mahmoud S, Mishin A, Nadariya M, Ouzounis A, Paradisis D, Peter Y, Prilepin M, Reilinger R, Sanli I, Seeger H, Tealeb A, Toksöz MN, Veis G (2000) Global Positioning System constraints on plate kinematics and dynamics in the eastern Mediterranean and Caucasus. J Geophys Res 105(B3):5695–5719
McKenzie D (1972) Active tectonics of the Mediterranean region. Geophys J R Astr Soc 30(2):109–185. https://doi.org/10.1111/j.1365-246x.1972.tb02351.x
McKenzie D (1978) Active tectonics of the Alpine-Himalayan belt, the Aegean Sea and surrounding regions. Geophys J Roy Astron Soc 55:217–254
McKenzie D, Jackson J (1983) The relationship between strain rates, crustal thickening, paleomagnetism, finite strain and fault movements within a deforming zone. Earth Planet Sci Lett 65:182–202
Papadimitriou EE, Sykes LR (2001) Evolution of the stress field in the northern Aegean Sea (Greece). Geophys J Int 146(3):747–759
Papazachos CB (1999) Seismological and GPS evidence for the Aegean–Anatolia interaction. Geophys Res Lett 26(17):2653–2656. https://doi.org/10.1029/1999gl.900411

Papazachos BC, Papadimitriou EE, Kiratzi AA, Papazachos CB, Louvari EK (1998) Fault plane solutions in the Aegean Sea and the surrounding area and their tectonic implication. Boll Geof Teor Appl 39(3):199–218

Papazachos BC, Karakaisis GF, Papazachos CB, Scordilis EM (2000) Earthquake triggering in the North and East Aegean Plate Boundaries due to the Anatolia Westward Motion. Geophys Res Lett 27. https://doi.org/10.1029/2000GL011425. issn: 0094–8276.

Pınar A, Kuge K, Honkura Y (2003) Moment tensor inversion of recent small to moderate sized earthquakes: implications for seismic hazard and active tectonics beneath the Sea of Marmara. Geophys J Int 153:133–145

Reilinger R, McClusky S, Oral B, King R, Toksoz M, Barka A, Kinik I, Lenk O, Sanli I (1997) GPS measurements of present-day crustal movements in the Arabia Africa-Eurasia Plate collision zone. J Geophys Res 102(9983–9999):1997

Saikia C (1994) Modified frequency-wave number algorithm for regional seismo-grams using Filon's quadrature; modelling of Lg waves in eastern North America. Geophys J Int 118:142–158

Saltogianni V, Gianniou M, Taymaz T, Yolsal-Çevikbilen S, Stiros S (2015) Fault slip source models for the 2014 M_w 6.9 Samothraki-Gökçeada earthquake (North Aegean Trough) combining geodetic and seismological observations. J Geophys Res Solid Earth 120(12):8610–8622

Şaroğlu F, Emre Ö, Herece E (1992) MTA Genel Müdürlüğü web site 1.500 000 scale geological map

Şengör AMC, Görür N, Şaroğlu F (1985) Strike-slip deformation, basin formation and sedimentation. Soc Econ Paleontol Mineral Spc Publ 37:227–264

Taymaz T, Jackson J, McKenzie D (1991) Active tectonics of the north and central Aegean Sea. Geophys J Int 106:433–490

Vannucci G, Gasperini P (2003) The earthquake mechanisms of the Mediterranean Area (EMMA). http://www.emsc-csem.org/Earthquake/emma.php

Wells DL, Coppersmith KJ (1994) New empirical relationships among magnitude, rupture length, rupture width, rupture area, and surface displacement. Bull Seismol Soc Am 84(4):974–1002

Wiemer S, Katsumata K (1999) Spatial variability of seismicity parameters in aftershock zones. J Geophys Res 104:13135–13151

Wiemer S, Wyss M (2002) Spatial and temporal variability of the b-value in seismogenic volumes: an overview. Adv Geophys 45:259–302

Yaltırak C, İşler EB, Aksu AE, Hiscott RN (2012) Evolution of the Bababurnu Basin and shelf of the Biga Peninsula: western extension of the middle strand of the North Anatolian Fault Zone, Northeast Aegean Sea, Turkey. J Asian Earth Sci 57(2012):103–119

Investigating the Focal Mechanisms of the August 4th, 2003, M_w 7.6, South Orkney Islands Earthquake and its Aftershocks Sequence

M. P. Plasencia Linares, M. Guidarelli, M. Russi and G. F. Panza

Abstract The $M_w = 7.6$ earthquake, known as *Centenary Earthquake*, occurred in the Scotia Sea region near the South Orkney Islands, Laurie Island, where is located the permanent Argentinean Antarctic Base Orcadas, here from 1997 operates a seismographic station ORCD, which has recorded several thousands of aftershocks, the most energetic ones recorded by all the instruments of the Antarctic Seismographic Argentinean-Italian Network (ASAIN). The aftershocks data available at ORCD station, till 60 days following the main shock were compiled. The plot of aftershocks rate with time was found to be oscillatory decay. Then, we inverted regional waveforms from ASAIN and International Federation of Digital Seismograph Networks (FSDN) stations to determine source parameters and source time functions for a set of aftershocks with magnitudes in the range 4.3–5.6 m_b. For the regional inversion we applied a methodology for the determination of the seismic moment tensor by means of full waveform inversion. The results obtained reflect the normal character of the main system fault, characterizing the study area.

1 Introduction

The South Orkney Microcontinent (SOM), is the largest (250 × 350 Km) continental element of the South Scotia Ridge which located between the Antarctic and the Scotia plates, and represents a remnant of the original continuous ridge linking the tip of South America with the Antarctic Peninsula (Dalziel and Elliot 1973; De Wit 1977; Dalziel 1984).

The South Scotia Ridge is located along the Antarctic—Scotia Plate boundary, and is characterized by a geodynamic setting reflecting the complicated evolution of the Scotia Sea area which included several spreading episodes (Barker et al. 1991).

M. P. Plasencia Linares (✉) · M. Guidarelli · M. Russi
Istituto Nazionale di Oceanografia e Geofisica Sperimentale - OGS,
Borgo Grotta Gigante 42/C, 34010 Sgonico, Trieste, Italy
e-mail: mplasencia@inogs.it

G. F. Panza
Department of Earth Sciences, University of Trieste, Via E. Weiss 4, 34127 Trieste, Italy

S. D'Amico (ed.), *Moment Tensor Solutions*, Springer Natural Hazards,
https://doi.org/10.1007/978-3-319-77359-9_17

The main tectonic feature is the left-lateral transcurrent fault running all along the plate boundary located at the northern shelf edge of the SOM (Forsyth 1975). A broad area of complex deformation accomplishes the strike-slip movements, with both extension and compression tectonics along it.

The western part of the Antarctica Plate-Scotia Plate boundary, north of Powell Basin, has been interpreted by Acosta and Uchupi (1996) as characterised by transtensional tectonic; while evidences of accretionary prism at the northern side were considered as indication of compression, due to plate convergence, in the westernmost part of the plate boundary (Lodolo et al. 1997) and along the northern side of the SOM (Kavoun and Vinnikovskaya 1994; Maldonado et al. 1998). Busetti et al. (2000) suggested that the northern edge of the SOM is characterised by strain partitioning with strike-slip zone accommodating the transcurrent motion and oceanward a trust zone accommodating the convergent component, that according to Braun and Beaumont (1995) model, correspond to an obliquely convergent plate boundary dominated by transcurrent condition. Eastward, at the plate boundary between the SOM and Discovery Bank, Pelayo and Wiens (1989), from seismological studies, suggested a transtensional tectonic, characterised by a combination of extension and transcurrent motion, by zones of oblique extension as well as discrete transform faults and extensional segments. The southeast and southwest borders of the SOM are passive margins connecting the SOM to the proto-oceanic Powell Basin on the west and to the oceanic Jane Basin on the east sides, relative to the rifting that separated the SOM from the Antarctic Peninsula.

2 Data

Even if the SOM is a small continental fragment of the Scotia Sea region, the study of its tectonic characteristics benefited from the integration between seismological data and other geophysical data. A large amount of seismological data have been made available after the deployment of broadband instruments in the Scotia Sea region since the early 90 as an extension of the global seismological network.

In the last 15 years several at least seven permanent seismographic stations were deployed in the Scotia Sea regions and Antarctica. Esperanza, Hope Bay (ESPZ), Jubany—King George Islands (JUBA), Despedida—Tierra del Fuego (DSPA), Orcadas—South Orkney Islands (ORCD), San Martín—Marguerite Bay (SMAI), Belgrano II—Nunatak Bertrab (BELA) and Marambio—Seymour Island (MBIO) stations constitute the Antarctic Seismographic Argentinean Italian Network (ASAIN), a cooperation between the Italian Programma Nazionale di Ricerche in Antartide (PNRA) and the Argentinean Dirección Nacional del Antártico (DNA); while East Falkland Islands (EFI), Palmer, Anvers Islands (PMSA) and South Georgia Island (HOPE) are also distributed by the Incorporated Research Institutions for seismology (IRIS) consortium. In spite of the development of a permanent seismographic network in the area, the application of seismological methodologies to the study of the South Orkney region was not possible until the end of 2003, because the low level

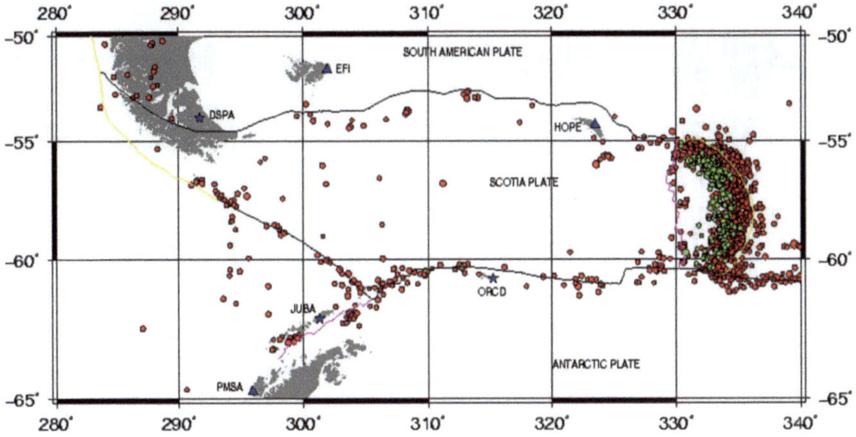

Fig. 1 Seismicity of the Scotia Sea region, 1973–2007. NEIC catalogue. Stars represents ASAIN stations and triangles global network stations

of seismicity. Figure 1 shows the distribution of earthquakes in the Scotia Sea region until 2007. This gap in the seismicity of the South Orkney area was filled after the Centenary Earthquake sequence which started on August 4, 2003 with a $M_w = 7.6$ earthquake, located offshore of the South Orkney Island and 2 years later a low level seismicity had yet remained.

This study provides an analysis of the *Centenary Earthquake* sequence centered on the determination of the source parameters for seven aftershocks. The determination of source mechanisms for the earthquakes in the South Orkney region can provide information about the stress conditions and the tectonic setting in the area and help to remove the limitations posed by the scarcity of events recorded in the past.

3 The Centenary Earthquake and Its Aftershock Sequence

On August 4, 2003 a major earthquake occurred in the south-eastern Scotia Sea, along the South Scotia Ridge and about 70 Km East from the Argentinean Base Orcadas (Laurie Island, South Orkney Islands), where the ASAIN station ORCD has been operating since 1997 (Russi and Febrer 2001). The moment magnitude was estimated as $M_w = 7.6$ according to CMT-Harvard. The August 4, 2003 earthquake was nicknamed Orcadas *Centenary Earthquake* because the Orcadas Base was opened exactly 100 years before and has been continuously operating since then. The *Centenary Earthquake* is the strongest event recorded in the SOM region. Recordings of the main shock from the ASAIN broadband digital stations ORCD and JUBA are shown in Fig. 2. The top trace corresponds to the clipped ORCD station records at a distance of about 100 Km from the epicenter, the bottom trace shows records at JUBA station at a distance of about 700 Km from the epicenter. All ASAIN stations

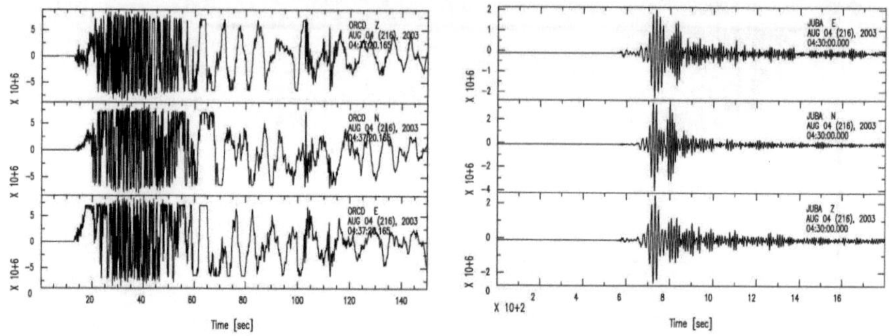

Fig. 2 Centenary Earthquake seismograms recorded by ORCD and JUBA, ASAIN stations

are equipped with Güralp broad-band seismographs: CMG-3T seismometer 0.01 s–50 Hz, CMG-DM24 24-bit Datalogger, GPS timing, great capacity storage media, allowing continuous recording of more than 2 years of the three component seismic channels at 100, 20 and 2 samples/s and connected to Internet.

Immediately after the *Centenary Earthquake* a noticeable aftershock sequence started. Within three hours from the main shock about twenty events, with magnitude larger than 4.0 occurred. In the following 24 hours, more than 50 events with magnitude within the range 3.6–5.6 m_b were recorded. The stress energy following a main shock is slowly released as a sequence of aftershocks. According to Omori law, the rate of aftershocks decays with time as $1/t$, where t is the time from the main shock. A generalization of Omori's Law was proposed by (Utsu 1961): the rate of aftershocks decays with time as $1/(t+t_0)^n$, where t_0 is a constant and n is the exponent. From this generalized Omori's relation, the number of aftershocks, N, occurring per unit time is proportional to $1/t^n$, when $t \gg t_0$. The aftershocks activity of the California earthquakes gave the value of n lying between 0.5 and 1.5 (Reasenberg and Jones 1989), which is also observed in most of the aftershock activity in the rest of the world. The plot of the rate of aftershocks against time for 60 days, following the main shock, is shown in Fig. 3.

In the Fig. 4 dash and dash-dotted lines indicate the hyperbolic fit, $77.58\,t^{-1.1}$, and the exponential fit, $41.41\,e^{-0.17\,t}$, respectively.

The superimposed oscillations present in the rate of decay of aftershocks are analytically extracted by subtracting the exponential and hyperbolic fits from the observed decay curve. Then we calculated the amplitude and phase spectra of the resulting curves. In order to find the nature of aftershocks it is necessary to examine the phase spectra, whether these oscillations correspond to random time series or deterministic series.

The result of this test is called z score or test statistic value. If $|z|$ score is greater than 1.96, then the data under consideration are not random at a confidence level of 95%. Using the test, the z score of the phase spectra of hyperbolic and exponential cases are found to be -7.52 and -6.97 respectively. Hence the oscillations present in the aftershocks activity are not random.

The ORCD station recorded several hundreds of aftershocks. The geographical distribution of aftershocks with magnitude larger than 3.6 m_b, occurred in the period August-December 2003, covers an ellipsoidal area, with the major axis extending for more than 150 Km in the E-W direction, along the main trend of the border between the Antarctica and the Scotia plates.

4 Determination of Source Mechanisms

To study the source mechanism we considered events occurred between August 4, 2003 and December 3, 2003. To investigate the source characteristic of the main shock we applied an amplitude spectra inversion methodology (Bukchin 1995) to teleseismic records from FSDN stations. We were able to determine source characteristics for the larger aftershocks using a regional waveform inversion methodology (Šilený et al. 1992; Šilený 1998). We selected 7 events, among those with magnitude greater than 4.3 m_b, which have sufficient signal to noise ratio and were recorded of FDSN stations that operates in the Antarctic area.

The moment tensor inversion is performed through the following processing steps. After data acquisition and pre-processing (we need data that have a good coverage of the focal sphere), linear trends are identified and removed. High frequency noise in

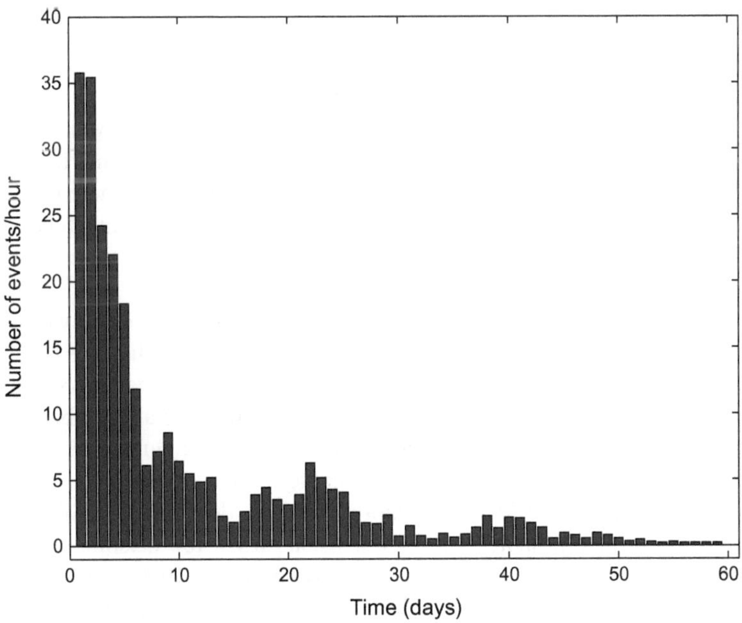

Fig. 3 Number of aftershocks per hour after the Centenary Earthquake, observed at ORCD station, for sixty days following the main shock

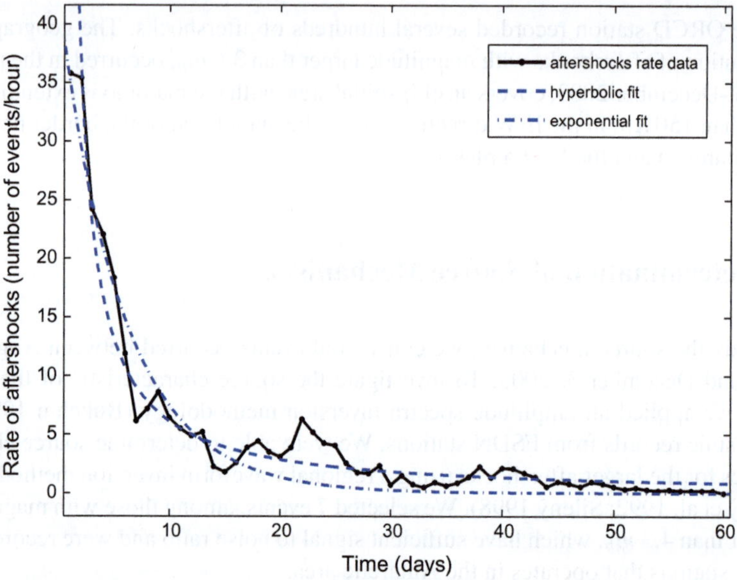

Fig. 4 Continuous trace shows the plot of rate of aftershocks of Centenary Earthquake, 2003, against time, dash trace shows the hyperbolic fit and dash-dotted shows the exponential fit

the data is removed by low-pass filtering. We low-pass filter the data at the frequency of 0.07 Hz for all the events. This filtering frequency was chosen because it enables us to make use of the point source approximation when studying the source in that frequency range. To consider the instrument effect we apply the instrument response to the synthetic Green's functions and compare them with the observed data. We do not deconvolve the instrument response from the data. After mean removing, tapering and filtering, we select the temporal window of the seismograms to be inverted for the retrieval of the moment tensor components. The synthetic Green's functions are calculated using the structural models obtained by Vuan et al. (2000) from group velocity tomography.

To study the lower magnitude events of the aftershock sequence we used the regional waveforms recorded by the broadband instruments of the ASAIN network and the IRIS consortium located in the Scotia Sea region. For the source characteristics determination we decided to use a methodology that had already been applied for earthquakes in the Scotia Sea region (Vuan et al. 2001; Guidarelli et al. 2003; Guidarelli and Panza 2006) and produced reliable results.

The INPAR (INdirect PARameterization) method, developed by Šílený (1992, 1998), proved to give reliable results even when only a few seismograms from a limited number of stations are available (Kravanja et al. 1999); such possibility is particularly important in regions where logistics is a major problem like in the Scotia Sea region. Kravanja et al. (1999) demonstrated that the INPAR method can absorb part of the inconsistency caused by poor forward modeling, which makes the availability of average models sufficient for our purposes. Limits and possibilities of this approach have been discussed in several papers (e.g., Panza 2000; Sarao et al. 2001; Šilený et al. 1996).

The method works in the point source approximation and consists of two main steps. In the first step (linear) the six Moment Tensor Rate functions (MTRF), obtained deconvolving from the data the Green's functions computed by the modal summation (e.g., Panza 1985; Panza et al. 2000) are inverted, without any constraints, with a damped least squares algorithm. In the second step (non-linear) the retrieved MTRFs, describing a source mechanism varying in time, are reduced to a constant moment tensor and the corresponding source time function taking only the correlated part from each MTRF. This is a basic point of the Šilený's method since, when taking only the coherent part at different stations, the influences on the solutions of strong non-modelled structural heterogeneities and of scattering can be reduced. The problem is non-linear and is solved iteratively by imposing constraints such as positivity of the source time function and, when clear readings of first arrivals are available, consistency with polarities. The genetic algorithm is employed in the search of the solutions and in the estimate of the error areas (Šilený 1998).

5 Aftershocks Source Mechanisms

Here we will describe in detail the solution obtained for each event analysed. The map with the fault plane solution of events is reported in Figs. 5, 6.

The resulting mechanisms for events 1 and 2 are normal faulting mechanisms, and our solution agrees with the results given by Harvard Centroid Moment Tensor (CMT). Both of the events are shallow. Event 3 present a normal source mechanism with a small strike slip component, and also agrees with the results of the Harvard-CMT. Event 4 show a normal fault solution while the Harvard-CMT present a dominantly strike slip mechanism. Event 5 is characterized by a normal source mechanism with strike slip component. For events 2–5 a Harvard-CMT solution is available, the main differences exist for the event number 4. The possibility of relocation of hypocentral coordinates even for shallow events is one of the main features of the INPAR method, since it permits a refinement of focal depth mainly when depth values are fixed (15.0 Km) a priori in Harvard-CMT inversion scheme.

6 Conclusions

The source characteristics of the most relevant aftershocks have been investigated by means of the INPAR method, which is a suitable tool to study medium energy sources in areas where a small number of stations is available to record the regional seismicity and the standard procedures employing first arrival data are not effective to analyze situations where a relevant noise level is present (Kravanja et al. 1999).

DATE	Depth (km)	Half duration	Scalar moment (dyn cm)	Mw	Fault Plane Solution	
04/08/2003 12:53:51.0	4	1.2	4.35E+25	6.3	19 58 -145 / 268 61 -38	
04/08/2003 18:18:29.6	5	1.2	8.31E+24	5.8	160 11 -76 / 326 80 -93	
	15	1.1	1.12E+24	5.3	6 23 -81 / 176 68 -94 CMT	
06/08/2003 17:01:54.4	49	1.1	1.36E+24	5.4	119 18 -155 / 5 83 -74	
	15	1.0	6.88E+23	5.2	268 18 -128 / 128 76 -78 CMT	
14/08/2003 08:41:31.6	2	1.0	1.11E+24	5.3	61 12 -179 / 330 90 -78	
	15	1.0	7.45E+23	5.2	30 47 -171 / 295 84 -43 CMT	
29/08/2003 14:50:34.8	24	1.1	4.44E+24	5.7	253 24 98 / 65 66 86	
	15	0.8	3.29E+23	5.0	138 36 -92 / 320 54 -89 CMT	
09/10/2003 18:49:33.6	9	1.1	8.06E+23	5.2	226 19 95 / 41 71 88	
03/12/2003 03:17:24.8	2	1.1	3.90E+24	5.7	17 54 24 / 272 71 141	

Fig. 5 Aftershocks focal mechanisms determined with constrained epicenter location

DATE	Hypoc. Coord.	Half duration	Scalar moment (dyn cm)	Mw	Fault Plane Solution	
04/08/2003 12:53:51.0	7 -60.54 -42.46	1.2	5.28E+25	6.4	84 27 -99 / 275 63 -85	
04/08/2003 18:18:29.6	2 -61.12 -43.49	1.2	2.93E+24	5.5	42 20 -65 / 196 72 -99	
	15 -61.23 -43.59	1.1	1.12E+24	5.3	6 23 -81 / 176 68 -94 CMT	
06/08/2003 17:01:54.4	18 -60.23 -44.73	1.1	1.46E+24	5.3	292 22 -152 / 176 80 -70	
	15 -60.34 -45.03	1.0	6.88E+23	5.2	268 18 -128 / 128 76 -78 CMT	
14/08/2003 08:41:31.6	22 -60.84 -43.59	1.2	2.90E+24	5.5	310 25 -93 / 134 65 -88	
	15 -60.61 -42.37	1.0	7.45E+23	5.2	30 47 -171 / 295 84 -43 CMT	
29/08/2003 14:50:34.8	17 -60.52 -43.29	0.9	2.98E+24	5.5	129 32 -103 / 324 59 -82	
	15 -60.56 -43.21	0.8	3.29E+23	5.0	138 36 -92 / 320 54 -89 CMT	
09/10/2003 18:49:33.6	6 -59.94 -44.55	1.1	5.87E+23	5.1	62 19 -161 / 315 84 -72	
03/12/2003 03:17:24.8	6 -61.05 -42.27	1.2	9.28E+23	5.2	190 40 -143 / 70 67 -56	

Fig. 6 Aftershocks focal mechanisms determined with not constrained epicenter location

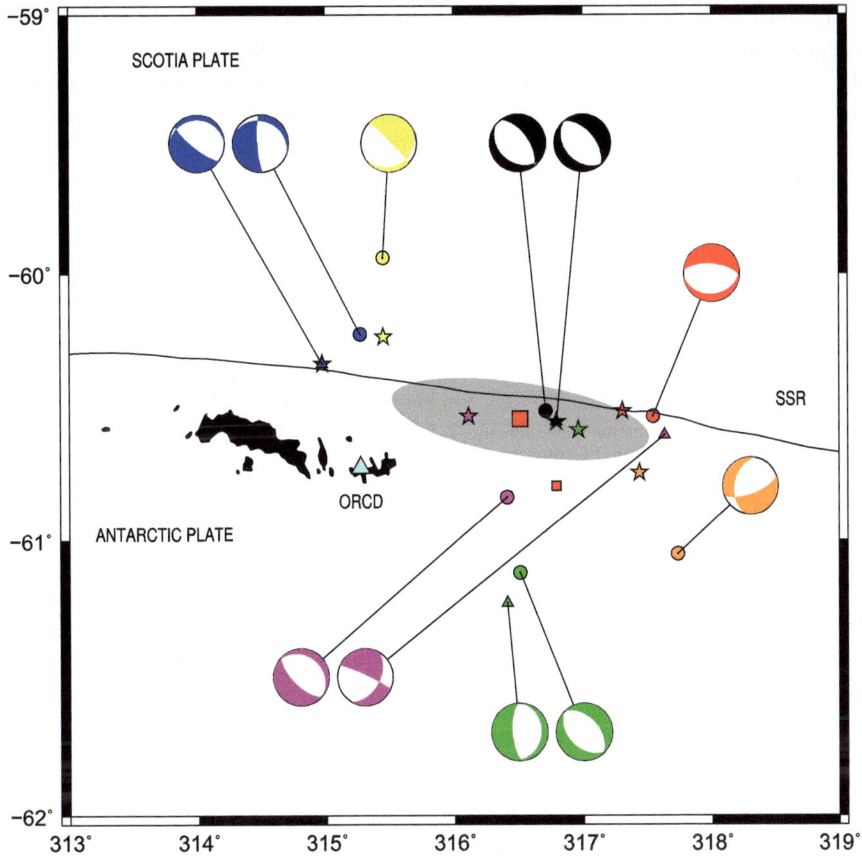

Fig. 7 Map with the location of events analysed in this study, stars: USGS locations, circles: INPAR locations and triangles: Harvard-CMT locations, in grey the rupture area

Still few information is available on the structural of the South Orkneys micro-continent and its tectonical setting. Most of the existing information on the structural and geodynamics setting comes from some seismic profiles performed by Italian, Spanish and Russian vessels during the late decade of the twentieth century, and the of average structural parameters for the lithosphere of the Scotia Sea region by surface waves tomography (Vuan et al. 2001).

According to the interpretation of this information plus gravity and bathymetry data a structural map for the SOM area has been constructed, this map is displayed in Fig. 7 together with the focal mechanisms of the events analised. The strike slips kinematics of the fault system on the northern border, which is also part of the Scotia Plate–Antarctic Plate margin along the South Scotia Ridge, and the existence of a convergent component responsible for the subduction of the Scotia Plate below the Antarctic Plate are put into in evidence.

Besides allowing us to fill a relevant gap in the seismicity map of the South Scotia Ridge the analysis of *Centenary Earthquake* and its aftershock sequence waveform data resulted in an increased seismological information about the characteristics of the seismic sources acting in the area which confirm the structural evidences collected by seismic experiments and gravimetry and the hypotheses of the existence both of transpresive and transtensive areas along the northern border of the SOM found in the literature.

Acknowledgements The project "Broad Band seismology in the Scotia Sea Region" is financially supported by the Italian Programma Nazionale di Ricerche in Antartide in the framework of the PNRA research area "Geodesia e Osservatori". Thanks are due to the Argentinean institutions cooperating to the ASAIN: Dirección Nacional del Antártico–Instituto Antártico Argentino. We are also extremely grateful to the military and civilian personnel operating in the Argentinean Bases Esperanza, Jubany (Now Carlini), Orcadas, San Martín, Belgrano II and Marambio for their help in maintaining the field installations and in the seismographic stations operation. We also thank the Abdus Salam International Centre for Theoretical Physics (ICTP) and the OGS for financing Milton P. Plasencia Linares's TRIL fellowship. Thanks are due to M. Busetti for her help in the geological interpretation of the results.

References

Acosta J, Uchupi E (1996) Transtensional Tectonics along the South Scotia Ridge; Antarctica. Tectonophysics 267:31–56

Barker PF, Dalziel IWD, Storey BC (1991) Tectonic development of the Scotia Arc region. In: Tingey RJ (eds) The geology of Antarctica. Oxford University Press, Oxford, pp 215–248

Braun J, Beaumont C (1995) Three-dimensional numerical experiments of strain partitioning at oblique boundaries: Implications for contrasting tectonic styles in the southern Coast Ranges, California, and central South Island. New Zealand. J Geophys Res 100(B9):18059–18074

Bukchin BG (1995) Determination of stress glut moments of total degree 2 from teleseismic surface wave amplitude spectra. Tectonophysics 248(3–4):185–191

Busetti M, Zanolla C, Marchetti A (2000) Geological structure of the South Orkney Microcontinent. Terra Antarct 8(2):71–78

Dalziel IWD (1984) Tectonic evolution of a forearc terrane, southern Scotia Ridge. Antarctica. Geol Soc Am Spec pap 200:32

Dalziel IWD, Elliot DH (1973) The Scotia Arc and Antarctic margin. In: Stehli FG, Nairm AEM (eds) The ocean basins and continental margins, vol I. The South Atlantic Plenum Press, New York, pp 171–246

De Wit MJ (1977) The evolution of the Scotia Arc as a key to the reconstruction of the southwestern Gondwanaland. Tectonophysics 37:53–81

Forsyth DW (1975) Fault plane solutions and tectonics of the South Atlantic and Scotia Sea. J Geophys Res 80:1429–1443

Guidarelli M, Russi M, Plasencia Linares MP, Panza GF (2003) The Antarctic Seismographic Argentinean-Italian Network and the progress in the study of structural properties and stress conditions in the Scotia Sea region. Terra Antartica Rep 9:25–34

Guidarelli M, Panza GF (2006) Determination of the seismic moment tensor for local events in the South Shetland Islands and Bransfield Strait. Geophys J Int 167:684–692

Kavoun M, Vinnikovskaya O (1994) Seismic stratigraphy and tectonic of the Northwestern Weddel Sea (Antarctica) inferred from marine geophysical surveys. Tectonophysics 240:299–341

Kravanja S, Panza GF, Šilený J, (1999) Robust retrieval of seismic point source time function. Geophys J Int 136:385–394
Lodolo E, Coren F, Schreider AA, Ceccone G (1997) Geophysical evidence of a Relict Oceanic Crust in the Southwestern Scotia Sea. Mar Geophys Res 19:439–450
Maldonado A, Zitellini N, Leitchenkov G, Balanya JC, Coren F, Galindo Zaldívar J, Lodolo E, Jabaloy A, Zanolla C, Rodrigues Fernandez L, Vinnikovskaya O (1998) Small ocean basin development along the Scotia-Antarctica plate boundary and in the northern Weddell Sea. Tectonophysics 236:371–402
Panza GF (1985) Synthetic Seismograms: the rayleigh waves modal summation. J Geophys 58:125–145
Panza GF, Romanelli F, Vaccari F (2000) Seismic wave propagation in laterally heterogeneous anelastic media: theory and applications to seismic zonation. Adv Geophys 43:1–95
Panza GF, Saraó A (2000) Monitoring volcanic and geothermal areas by full seismic moment tensor inversion: are non-double couple components always artifacts of modelling? Geophys J Int 143:353–364
Pelayo AM, Wiens DA (1989) Seismotectonics and relative plate motions in the Scotia Sea Region. J Geophys Res 94(B6):7293–7320
Reasenberg PA, Jones LM (1989) Earthquake hazard after a mainshock in California. Science 243:1173–1176
Russi M and Febrer JM (2001) Broad-band seismology in the Scotia Sea region, Antarctica. Italian and Argentinean contributions to the Scotia Sea Broad-band Network. In: Brancolini G, Panza GF, Russi M, Wu F (eds) Results from the workshop broadband seismic observations and the geodynamics of the Scotia Sea Region, Trieste, Italy, 25-26 ottobre 1999. Terra Antartica 82:29–34
Saraò A, Panza GF, Privitera E, Cocina O (2001) Non double couple mechanisms in the seismicity preceding 1991–1993 Etna volcano eruption. Geophys J Int 145:1657–1667
Šilený J, Panza GF, Campus P, (1992) Waveform inversion for point source moment tensor retrieval with optimization of hypocentral depth and structural model. Geophys J Int 108:259–274
Šilený J, Campus P, Panza GF, (1996) Seismic moment tensor resolution by waveform inversion of a few local noisy records-I. Synthetic test. Geophys J Int 126:605–619
Šilený J, (1998) Earthquake source parameters and their confidence regions by a genetic algorithm with a "memory". Geophys J Int 134:228–242
Utsu T (1961) A statistical study on the occurrence of aftershocks. Geophys Mag 3(3):710–724
Vuan A, Russi M, Panza GF (2000) Group velocity tomography in the subantarctic Scotia Sea region. Pure Appl Geophys 157:1337–1357
Vuan A, Russi M, Costa G, Panza GF (2001) Moment tensor waveform inversion in the sub-Antarctic Scotia Sea: feasibility tests and preliminary results. Terra Antartica 8(2):55–62
Wessel P, Smith WHF (1998) New, improved version of generic mapping tools released. EOS Trans Amer Geophys U 79(47):579

Waveform Modelling of 2009 Bhutan Earthquake of Magnitude 6.1 (Mw) Using Local Network Data of North East India

Santanu Baruah and Midusmita Boruah

Keywords Bhutan earthquake · Waveform inversion · Transverse tectonics

1 Introduction

A strong earthquake of Mw 6.1 struck Bhutan on September 21, 2009 with casualties of several people. The epicentre of the event was given at latitude 27.34° N and longitude 91.41° E, and depth ~ 10 km (USGS report; http://earthquake.usgs.gov). Shaking from the earthquake was felt in the Bhutan, Tibet and in the adjoining North East region of India including Bangladesh.

The objective of this work is to carry out the numerical modeling of the 21st September 2009 Bhutan earthquake and its aftershocks and the identification of physical mechanism based on 1D velocity model. To accomplish this, modeling of main shock and its aftershocks is carried out through waveform inversion which in turn help in understanding the physics of the earthquake process by characterizing the associated fault. The reason for using the waveform inversion tool is that it enables estimation of reliable focal mechanism solutions that estimates seismic moment and moment magnitude, quantifying the earthquake to the extent. Fault-plane solution thus obtained are tried to correlate with the local geology in order to identify the causative fault of the Bhutan Himalaya 2009 earthquake. Additionally, to identify the true fault plane orientation, the H (Hypocentre)—C (Centroid) method (Zahradnik et al. 2008) was used with combined knowledge of the centroid position, MT solution (nodal planes) and hypocentre parameter.

S. Baruah (✉)
Geosciences & Technology Division, CSIR-North East Institute of Science and Technology, Jorhat 785006, Assam, India
e-mail: santanub27@gmail.com

M. Boruah
Department of Applied Geology,
Dibrugarh University, Dibrugarh 786004, Assam, India

S. D'Amico (ed.), *Moment Tensor Solutions*, Springer Natural Hazards,
https://doi.org/10.1007/978-3-319-77359-9_18

2 Tectonic Settings of Bhutan Himalaya

The Himalaya is the largest orogenic belt of the world form by the collision between Indian subcontinent and Tibet about 50 Ma ago (Gansser 1964; McKenzie and Sclater 1971; Molnar and Tapponnier 1975; Mitchell 1981; Curry et al. 1982; Zhao et al. 1993; Nandy 2001). Main Boundary Thrust (MBT) and the Main Central Thrust (MCT) are the major tectonic features that are traversing Bhutan Himalaya to the north, albeit a circular overturned of the later (Ravi Kumar et al. 2012; Pradhan et al. 2013). The MCT in Bhutan Himalaya has taken a curvilinear loop shape indicating thicker sediments (Fig. 1). Based on the known geology, geophysics and teleseismic hypocentral data Seeber and Armbruster (1981) first proposed a conceptual tectonic model of the Himalaya known as Steady-state model. The model consists of the underthrusted Indian plate beneath the Himalaya (known as plane of detachment), overriding Tethyan slab and the Himalayan sedimentary wedge, which is decoupled from the two converging slabs. In this model, the MBT and the MCT are two active thrusts and are contemporaneous features. Further north, below the MCT lies the *ramp,* named *basement thrust front* that accumulates the tectonic stress due to northward movement of the Indian plate, and abrupt release is believed to be the main cause of earthquakes on the plane of detachment at a shallower depth (0–20 km).

The Bhutan Himalaya is different from the western section of the Himalayas with the presence of the Kakhtang thrust (KT) which is an out of sequence thrust located between the MCT and the STDS (South Tibetan Detachment System) (Gansser 1993; Grujic et al. 1996, 2002, 2006; Drukpa et al. 2006). Seismic pattern in relationship to the major tectonic faults and lineaments of the region have been examined by several researchers in the past (e.g. Seeber and Armbruster 1981; Ni and Barazangi 1984; Fitch 1970; Baranowski et al. 1984; Searle 1996; Chen and Kao 1996). It is found that most of the earthquakes in this region lie between the MCT and the MBT in the 50 km wide relatively narrow belt and are mostly concentrated south of the MCT (Ni and Barazangi 1984). The fault plane solutions of the past studies as well as from Harvard G-CMT, indicate that most of the earthquakes along the Himalayan region are thrust faulting (Fitch 1970; Baranowski et al. 1984; Searle 1996; Chen and Kao 1996). The slip vectors of these solutions are found to be perpendicular to the mountain ranges with steeper plunges (~25°) in the western Himalaya, in contrast to shallow plunges of (~10°) in the Eastern Himalaya (Baranowski et al. 1984; Drukpa et al. 2006). Chen and Kao (1996) also reported the presence of strike-slip and normal faulting in the Himalayan region. However, strike-slip faulting becomes more frequent near the eastern and western Himalayan syntaxes (Drukpa et al. 2006).

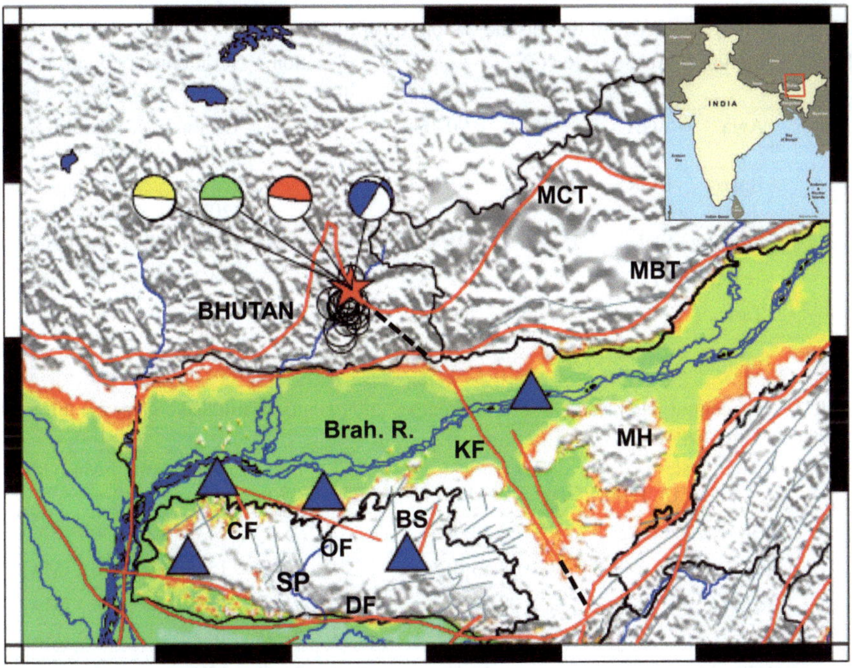

Fig. 1 Tectonic map of Bhutan Himalaya and its adjoining areas (after, GSI 2000) showing main tectonic features of the region. The major tectonic features are MBT: Main Boundary Thrust; MCT: Main Central Thrust; Brah. R.: Brahmaputra River, KF: Kopili fault, MH: Mikir Hills, SP: Shillong plateau; OF: Oldham fault; BS: Barapani Shear zone; DF: Dauki fault; CF: Chedrang fault. The black dotted lines indicate the extended part of Kopili fault. The 21st September 2009 Bhutan Himalaya main shock and 17 significant aftershocks ($M \geq 4.0$) are shown by red stars and open black circle respectively. The blue line indicates the river system. Five stations namely AGA, BOK, SIIL, TZR and TRA are also shown as blue triangles. The yellow beach ball indicates the NEIC solution, green beach ball the USGS solution, red beach ball the GCMT solution and the blue beach ball the CSIR-NEIST solution respectively. *Inset* Map of India

3 Databases Used

In order to find out the hypocentral parameters of the 2009 Bhutan earthquake and its aftershocks, digital seismic waveforms from different seismic stations operated by CSIR-NEIST-Jorht are obtained (Table 1). Several locations have been reported for the main shock of the 2009 Bhutan earthquake by different national and international agencies (Table 2). The reported source depth was the most uncertain parameter, ranging from 10 to 15 km. We have reanalyzed the hypocentral parameters using the HYPOCENTER programme of Lienert et al. (1986) and a velocity model proposed by Bhattacharya et al. (2005) for the region. The estimated location errors are ±2.1 km and ±5.1 km along the horizontal and depth respectively, and the root mean square error 1.3 s. The same programme is used for locating the aftershocks

Table 1 Station parameters of the seismic stations of North East India

No.	Station	Abb.	Lat (°N)	Long (°E)	Elevation (m)	Seismometer used	Digitizer used
1	Agia	AGA	26.066	90.464	75	Trilium 120P	Guralp DCM
2	Boko	BOK	25.969	91.244	50	Trilium 120P	Guralp DCM
3	Shillong	SHL	25.566	91.859	1590	Trilium 240	TAURAS
4	Tezpur	TZR	26.617	92.783	140	CMG 3ESP	REFTEK
5	Tura	TRA	25.546	90.242	305	Trilium 120P	Guralp DCM

Table 2 Hypocentral parameters of 21st September 2009 Bhutan earthquake given by different national and international agencies

No	Source	Date (YYYY/MM/DD)	OT (HH:MM:SS)	Lat (°N)	Lon (°E)	Depth (km)	Mag (Mw)
1	NEIC	2009/09/21	08:53:06	27.33	91.43	14.0	6.1 (Mw)
	USGS		08:53:06	27.31	91.41	15	6.2 (Mw)
	GCMT		08:53:10	27.2	91.63	12.0	6.1 (Mw)
	ISC		08:53:06	27.36	91.45	16.1	6.2 (Mb)
	CSIR-NEIST		08:53:10	27.32	91.45	15.4	6 (Mw)
	IMD		08:53:04	27.30	91.50	8	6.3 (Mw)

of the 2009 Bhutan earthquake. However, towards retrieval and utilization of waveforms, emphasis should be given to higher signal to noise ratio (clear phases and free from noise) and good azimuthal coverage of the recording stations. Out of these waveforms, inversions are carried out for those waveforms which have undergone very low frequency band pass filtering towards generation of synthetic seismogram.

4 Method of Waveform Inversion

The velocity seismograms selected for waveform inversion are prepared by removal of mean, tapering and band pass filtering. The lower and higher cut off values for the band pass filter depends on the background noise level that exists in the signal as the three component seismograms are filtered using 8-pole Butterworth band pass filter (Scherbaum 1994). The seismograms are filtered within the range 1–2, 2–4 Hz that give smooth noise free waveform to be used for inversion. The observed waveforms are corrected for instrument response so as to get the corresponding ground velocity. Using instrument response, the amplitude in counts are converted into cm/sec unit.

The angle of incidence and the back azimuths for each event are calculated from the hypocentral parameters. The back—azimuth is the angle measured between the vector pointing from the station to the source and the vector pointing from station to the north while the incidence angle is defined as the angle measured between the ray vector at the station (from the source to the station) and the vector pointing from the station straight up. Using these two parameters the observed waveforms are rotated to get vertical, radial and transverse components. After this, the vertical, radial and transverse components are windowed to an appropriate length to include the P and S onsets for each component excluding the later part of the S wave i.e., coda wave, because coda waves are considered to be scattered waves with little relationship to the source process (Aki and Chouet 1975).

4.1 Velocity Model

The first step in inversion procedure is to define a proper crustal velocity model. Use of average crustal velocity model in waveform modeling of local events is difficult because of high frequency contents in the seismograms. A local velocity model can generate the recorded characteristics of near earthquake phases, is very much essential for retrieval of source parameter of small events. Hence, a local crustal velocity model is the prime requirement for waveform modeling. The crustal velocity model along with attenuation parameter helps to determine most important parameter i.e., the Green's function: the seismic impulse response of the Earth. Green's functions when convolved with source time function and observed seismograms produce synthetic seismograms.

For successful applications of full waveform inversion, it is critical to generate a sufficiently accurate starting velocity model in order to allow the misfit function to converge to the acceptable limit. In this study, an initial model is applied to waveform inversion considering the requirement of waveform inversion tool [ISOLA-GUI; (https://github.com/esokos/isola)]. In the waveform inversion tool, there is a fixed option for a maximum of 15 layers [ISOLA-GUI]. For each layer the depth of the top layer, the V_p, V_s (velocities of P and S waves) in km/sec., density in gm/cm^3, Q_P, Q_s (attenuation parameters of P and S waves) are given. As a result the tool generates a graph of V_p and V_s versus depth (km). Many researchers have proposed different velocity model for North East, India using variety of seismological and geophysical methods (e.g., Gupta et al. 1984; Mukhopadhayay et al. 1997; Parvez et al. 2003; Kumar et al. 2004; Mitra et al. 2005; Bhattacharya et al. 2005, 2008). The application of suitable velocity model always plays an important role in waveform inversion. Sensitivity of different velocity models to the focal mechanism solutions of 2009 Bhutan earthquake and its aftershocks are tested first. We observed that the models are sensitive mostly in terms of the DC% and correlation value. The mechanisms obtained using the Bhattacharya et al. (2005) model show high DC% and high correlation values among all other proposed velocity model of North East

Table 3 Crustal velocity model derived by Bhattacharya et al. (2005). The corresponding values of density and attenuation parameters of P and S waves (Q_P and Q_S) are also depicted

Depth (km)	Vp (km/s)	Vs (km/s)	ρ (g/cm^3)	Q_P	Q_S
0.0	5.56	3.19	2.56	300	300
10.0	6.10	3.50	2.94	300	300
20.0	6.45	3.70	3.00	400	200
30.0	6.90	3.96	3.00	400	200
40.0	7.60	4.36	3.30	2000	1000
50.0	8.40	4.82	3.40	2000	1000

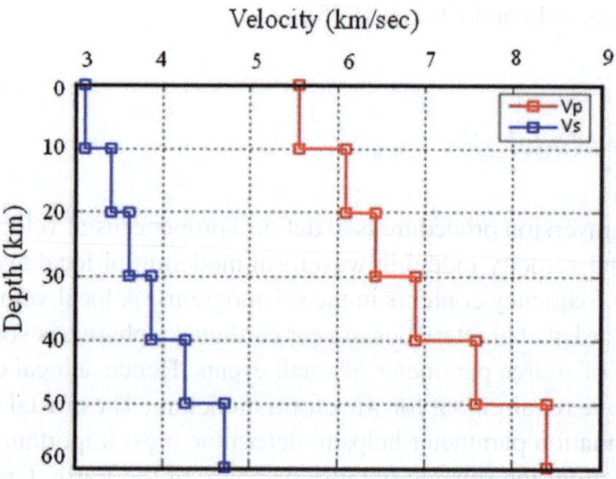

Fig. 2 Velocity model of Bhattacharya et al. (2005) for North East India used in waveform inversion

India region. Hence, in this study also we have selected Bhattacharya et al. (2005) (Table 3; Fig. 2) as the best suitable velocity model of for waveform inversion.

4.2 *Steps Involved in Waveform Inversion*

The procedure of waveform inversion technique adopted in the present study can be summarized as below. It is assumed that the crustal velocity model used is realistic one and hypocentral parameters of the studied events are accurate. In the process, complete velocity records are used without selecting any particular phases.

(i) The earthquake events recorded by the digital broadband stations are selected on the basis of high signal to noise ratio and clear records of P and S arrival.
(ii) The 3-component digital records are first converted into displacement waveforms.
(iii) The displacement waveforms are then low pass filtered below the corner frequency in order to remove the any offset. The high frequency components

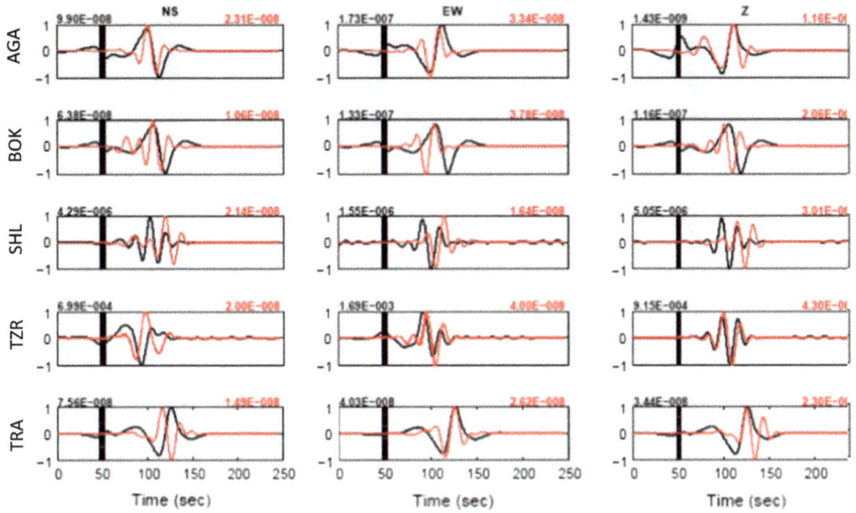

Fig. 3 Waveform inversion for the 2009 Bhutan event; a comparison of synthetic (red lines) and observed (black lines) waveforms at stations AGA, BOK, SHL, TZR and TRA for the preferred focal mechanism solution. X-axis indicates time scale in sec and Y-axis represents amplitude scale in meter. Maximum amplitudes are shown at traces

are excluded because it is difficult to model high frequency components as it requires a precise knowledge of detailed sub-surface crustal velocity model.

(iv) Subsequently resampling of records from a frequency of 100–33 Hz is carried out having with transfer function of the seismometers. Necessary DC removal and Trend line removal is performed.

(v) Green's functions are then computed in the complex spectral domain using the suitable crustal velocity model pertinent to the region at a point source by Discrete Wavenumber (DW) method (Bouchan 1981) and using the program called AXITRA by Coutant (1989). The Green's functions are then convolved with appropriate instrument response and source time function. While calculating the Green's functions, the position of the epicenters are kept fixed.

As we know that the depths of the foci are the most uncertain parameter among all the hypocentral parameters, the depth is allowed to vary in the inversion program from the optimum depth found in the location procedure by up to say ±5 km in step of 0.5–2 km. Similarly the origin (centroid) time the earthquake was also allowed to vary up to say ±(1–2) s from that found by location procedure. The time step in the grid search was identical with the sampling interval of 0.01 s (e.g., Fojtikova et al. 2010). The inversion was performed in a frequency band of 0.03–0.05 Hz varying for each event at each depths and times. The final validation of the best fitting solutions was accomplished by comparing the observed and synthetic amplitude waveform.

Table 4 Focal mechanism solutions of 21st September 2009 Bhutan earthquake and its 17 numbers of aftershocks determined through waveform inversion using the broadband data of CSIR-NEIST, Jorhat

No	Source	Date (YYYY/MM/DD)	OT (HH:MM:SS)	Lat (°N)	Long (°E)	Depth (km)	Mag (Mw)	First plane			Second plane			P axis		T axis	
								Strike (°)	Dip (°)	Rake (°)	Strike (°)	Dip (°)	Rake (°)	Azm (°)	Plunge (°)	Azm (°)	Plunge (°)
Main shock parameters																	
1	NEIC	2009/09/21	08:53:06	27.33	91.43	14	6.1 (Mw)	12	4	−173							
	USGS		08:53:06	27.31	91.41	15	6.2 (Mw)	89	86	88							
	GCMT		08:53:10	27.2	91.63	12	6.1 (Mw)	97	84	90							
	ISC		08:53:06	27.36	91.45	16.1	6.2 (Mb)										
	CSIR-NEIST		08:53:10	27.32	91.45	15.4	6 (Mw)	32	86	58	296	31	173	148	35	273	41

(continued)

Table 4 (continued)

No	Source	Date (YYYY/MM/DD)	OT (HH:MM:SS)	Lat (°N)	Long (°E)	Depth (km)	Mag (Mw)	First plane			Second plane			P axis		T axis	
								Strike (°)	Dip (°)	Rake (°)	Strike (°)	Dip (°)	Rake (°)	Azm (°)	Plunge (°)	Azm (°)	Plunge (°)
After shock parameters																	
1	IDC	2009/09/21	09:31:57	27.3	91.45	10											
	ISC		09:31:58	27.34	91.48	15											
	CSIR-NEIST		09:32:01	27	91.38	16.32	3.42	251	77	−161	157	71	−13	115	23	23	4
2	CSIR-NEIST	2009/09/21	09:41:15	27.05	91.37	16.09	3.44	198	85	130	293	40	6	256	30	142	37
3	CSIR-NEIST	2009/09/21	11:07:54	27.2	91.39	14.18	3.4	293	71	4	201	85	161	248	10	156	16
4	CSIR-NEIST	2009/09/21	11:11:37	27.19	91.41	13.72	3.62	180	79	133	281	44	15	238	23	129	40
5	CSIR-NEIST	2009/09/21	11:51:00	27.38	91.45	7.78	3.07	157	84	178	247	88	5	22	3	112	5
6	CSIR-NEIST	2009/09/21	12:31:19	27.1	91.42	16.04	3.15	2	82	−172	272	82	−7	227	11	317	1
7	CSIR-NEIST	2009/09/21	12:37:38	27.15	91.48	8.7	3.29	182	54	31	306	51	47	245	2	152	58
8	IDC	2009/09/21	02:34:02	27.24	91.51	10											

(continued)

Table 4 (continued)

No	Source	Date (YYYY/MM/DD)	OT (HH:MM:SS)	Lat (°N)	Long (°E)	Depth (km)	Mag (Mw)	First plane			Second plane			P axis		T axis	
								Strike (°)	Dip (°)	Rake (°)	Strike (°)	Dip (°)	Rake (°)	Azm (°)	Plunge (°)	Azm (°)	Plunge (°)
	ISC		02:34:09	27.49	91.44	44.4											
	CSIR-NEIST		02:34:07	27.23	91.41	14.64	3.01	234	45	141	353	63	51	109	11	214	55
9	CSIR-NEIST	2009/09/21	04:17:54	27.24	91.46	13.29	3.26	81	88	−175	351	85	−1	306	5	216	2
10	CSIR-NEIST	2009/09/21	04:26:01	27.27	91.36	11.79	2.76	252	54	−8	347	82	−144	215	30	114	19
11	CSIR-NEIST	2009/09/21	05:43:16	27.09	91.5	27.49	2.86	43	66	4	311	85	156	359	14	264	20
12	CSIR-NEIST	2009/09/21	07:58:59	27.16	91.42	11.31	3.02	303	82	−171	212	81	−7	168	11	78	1
13	CSIR-NEIST	2009/09/21	09:45:48	27.27	91.44	11.66	3.07	121	84	174	212	84	5	167	1	77	8
14	CSIR-NEIST	2009/09/21	09:55:47	27.22	91.32	11.92	3.44	212	69	−10	306	79	−158	171	23	78	8
15	CSIR-NEIST	2009/09/21	11:19:52	27.22	91.39	14.25	3.24	351	87	38	259	51	176	118	25	222	29
16	CSIR-NEIST	2009/09/21	11:27:57	27.2	91.4	12.18	3.39	271	64	1	180	88	154	238	18	132	19
17	CSIR-NEIST	2009/09/21	08:27:58	27.19	91.47	10.23	3.5	299	86	3	209	86	176	254	1	164	6

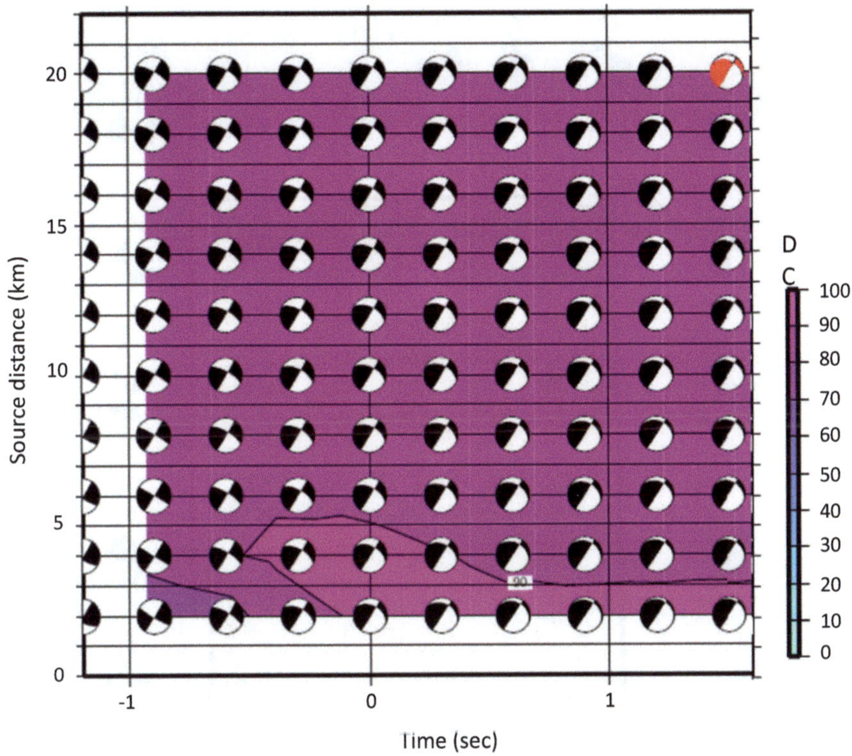

Fig. 4 Plot of source position versus time-shift and focal mechanism for the multiple source inversion. Largest correlation was obtained for source 20 and 1.4 s time shift. Preferred solution is depicted by a red-beachball. Scales for DC% is shown in the right

The inversion procedure inverts the point source moment tensor using Green's functions computed for 1D velocity model. The 1D velocity model may not be perfect and small time shift is required for maximum correlation between the observed and synthetics (Marzooqi et al. 2008). In the inversion technique, the shift is not known and it becomes a non-linear model parameters which has to be inverted along with the moment tensor and optimal source depth. To account for the horizontal mislocation, the synthetics are shifted relative to the observed by changing the origin time few seconds before/after location origin time during the inversion. In this study, the inversion is carried out for full moment tensor. The preferred solution is obtained by a simple grid search over the focal depth with certain steps and also over the origin time between (1–2) s. before/after the location origin time, since the origin time trades off with focal depth. The solution that has a large percentage of variance and double couple component is selected. To accomplish the above procedure the software package by Sokos and Zahradnik (2008) has been used.

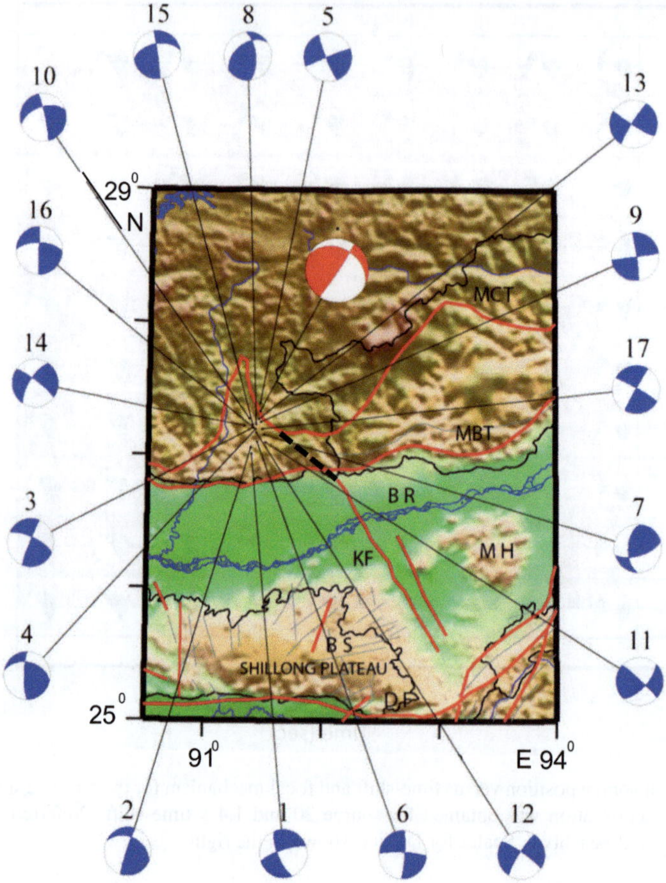

Fig. 5 Focal mechanism solutions of the 2009 Bhutan earthquake (Red beach-ball) and its 17 aftershocks (Blue beach-balls) obtained through waveform inversions. The blue-beach balls are numbered as per Table 4 (Aftershocks Parameters). All other explanations as in Fig. 1

5 Results and Discussions

Focal mechanism solutions of 21st September 2009 Bhutan earthquake along with its 17 aftershocks are determined using waveform inversion technique. There is no doubt that the usage of large station data with good azimuthal coverage yield a reliable focal mechanism solution. But in wave-form inversion technique single station data (less azimuthal coverage) may also become sufficient to get an accurate solution. Several authors made comprehensive studies on the focal mechanism solutions derived from single or double station data e.g. Rao (2009), Delouis and Legrand (1999), Fojtíkova and Zahradnik (2014), Zahradník et al. (2015) etc. In this

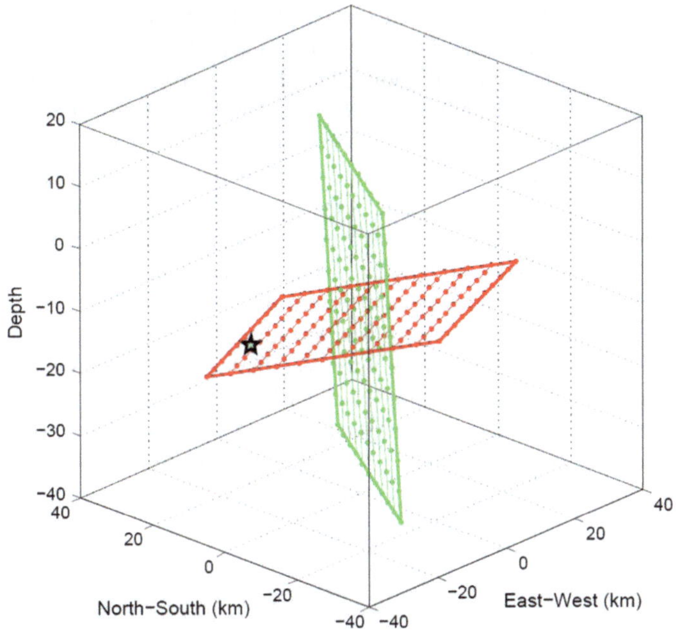

Fig. 6 Fault plane identification of 2009 Bhutan event. Two nodal planes (red and green planes) passing through the centroid. Hypocentre exists on one of the nodal planes (red planes), and thus indicates the fault plane

study, Seismic waveform data from five stations namely AGA, BOK, SHL, TZR and TRA are used in the search of focal mechanism.

The inversion for 2009 Bhutan earthquake (Date: 21-09-2009, OT: 08 h:53 min:10.4 s, Lat. 27.329°, Long. 91.459°, Depth. 15.4 km, $M_D = 3.5$) is performed at a starting depth of 5 km (15.4 km being the actual depth) having a spacing of 2 km with 10 trial source positions. The used frequency band for the inversion is fixed at (0.03–0.05 Hz) with cosine tapering applied in order to obtain the detailed information about the source rupture process. The best waveform match between observed and synthetic (Fig. 3) is obtained with the calculation of Green's function at the epicentral distance of 155, 153, 171, 232 and 200 km respectively for stations namely BOK, TZR, AGA, TRA, and SHL. Ultimately a preferred solution is obtained (red beach ball) based on maximum correlation between source position and time-shift. Figure 4 shows that the preferred solution (red beach ball) differs from the nearby solutions in terms of DC percentage and correlation value. The focal parameters are illustrated in Table 4.

As illustrated in Fig. 4, it is ascertained from correlation value that the preferred solution is well correlated at a depth of 20 km for a time-shift of 1.4 s having a DC% of 82. Thus, we can conclude from the observation that the 2009 Bhutan event is characterized by thrust with strike-slip faulting; all of the 17 aftershocks are also consistent with the main event (Fig. 5).

Earthquake source parameters were used to identify the orientation of event's fault plane. To identify the true fault plane orientation, the H (Hypocentre)—C (Centroid) method was used with combined knowledge of the centroid position, MT solution (nodal planes), and hypocenter key to identify this fault plane. The nodal planes pass through C, with the fault plane corresponding to the plane that also passes through H. If the hypocentre, centroid and nodal planes are consistent, they enable the identification of which nodal plane is the fault plane (Zahradnik et al. 2008).

The hypocenter coordinate for Bhutan event is obtained using CSIR-NEIST data whereas centroid coordinate is obtained using GCMT data. Based on the analysis results, the Bhutan event has hypocentre-centroid distance of 22.42 km. The fault plane 1 (green plane) is characterized by strike = 32°; dip = 86°; rake = 58° with distance of 22.27 km from the hypocentre. The fault plane 2 (red plane) is characterized by strike = 296°; dip = 31° and rake = 173° with a distance of 2.05 km from the hypocentre. The distance of fault plane 2 to the hypocentre is closer than fault plane 1. Therefore, the real fault plane is fault plane 2 as shown in Fig. 6.

6 Conclusion

The Mw 6.1 2009 Bhutan earthquake is used for the study of seismotectonics of the Bhutan Himalaya through a state of art technique called waveform inversion of seismic waves. Focal mechanism solutions of 2009 Bhutan earthquake is found to be thrust with strike slip faulting which clearly indicates the existence of transverse faults in the region. The same observation was also reported by Kayal et al. (2010). The geologically mapped curvilinear structure of the Main Central Thrust (MCT) in the Himalaya, where the epicentre of the Bhutan earthquake is located, is possibly caused by the transverse Kopili fault beneath the MCT. Further, identification of the true fault plane orientation is made by using H-C method. The distance of fault plane 2 (strike = 296°; dip = 31° and rake = 173°) to the hypocentre is closer than the fault plane 1 (strike = 32°; dip = 86°; rake = 58°). Therefore, the real fault plane is fault plane 2. The plane 2 coincides with the orientation of the extended Kopili fault in the Bhutan Himalaya region. The P-axis orientations of the 2009 Bhutan earthquake and most of its aftershocks are found to be nearly along N-S direction which display the northward underthrusting of the Indian plate to the Eurasian plate along Bhutan Himalaya.

Acknowledgements We thank Dr. D. Ramaiah FNASc, FNA, Director, CSIR-North East Institute of Science and Technology (CSIR-NEIST), Jorhat, for his kind support and encouragement in carrying out this work. We thank Prof. Harsh K. Gupta, Chairman, Research Council, CSIR-NEIST-J for his encouragement. We also thank Prof. J. R. Kayal, Visiting Professor, ISR, Gujarat, India and Dr. Saurabh Baruah, Head, GSTD, CSIR-NEIST-Jorhat, Assam, India for their help. Dr. Sebastiano D'Amico, Editor and the reviewers are thankfully acknowledged for their constructive comments. The Science and Engineering Research Board (SERB), Government of India provided the financial support for this under Young Scientist Scheme, vide Sanction No. SR/FTP/ES-121/2014.

References

Aki K, Chouet B (1975) Origin of coda waves: source, attenuation, and scattering effects. J Geophys Res (Solid Earth and Planets) 80:3322–3342. https://doi.org/10.1029/jb080i023p03322

Baranowski J, Armbruster J, Seeber L, Molner P (1984) Focal depths and fault plane solutions of earthquakes and active tectonics of the Himalaya. J Geophys Res 89(B8):6918–6928

Bhattacharya PM, Pujol J, Mazumdar RK, Kayal JR (2005) Relocation of earthquakes in the North-east India region using joint Hypocenter determination method. Curr Sci 89(8):1404–1413

Bhattacharya PM, Mukhopadhyay S, Mazumdar RK, Kayal JR (2008) 3-D seismic structure of the northeast India region and its implications for local and regional tectonics. J Asian Earth Sci 33:25–41

Bouchon M (1981) A simple method to calculate Green's functions for elastic layered media. Bull Seism Soc Am 71:959–971

Chen WP, Kao H (1996) Seismotectonics of Asia: some recent progress. In: Yin A, Harrison M (eds) The tectonic evolution of Asia. Cambridge University Press, Cambridge, pp 37–62

Coutant O (1989) Program of numerical simulation AXITRA, Research Report LGIT, Grenoble in French

Curray JR, Emmel FJ, Moore DG, Raitt RW (1982) Structure tectonics and geological history of the northeastern Indian Ocean. In: Nairn AEM, Stehli FG (eds) The ocean basins and margins, The India Ocean, vol VI. Plenum, New York, pp 399–450

Delouis B, Legrand D (1999) Focal mechanism determination and identification of the fault plane of earthquakes using only one or two near-source seismic recordings. Seismol Soc Am Bull 89:1558–1574

Drukpa D, Velasco AA, Doser DI (2006) Seismicity in the Kingdom of Bhutan (1937–2003): evidence for crustal transcurrent deformation. J Geophys Res 111:B06301. https://doi.org/10.1029/2004jb003087

Fitch TJ (1970) Earthquake mechanisms in the Himalaya, Burmese and Andaman Regions and continental tectonics in Central Asia. J Geophys Res 75:2699–2709

Fojtikova L, Vavrycuk V, Cipciar A, Madaras J (2010) Focal mechanisms of micro-earthquakes in the Dobra Voda seismoactive area in the Male Karpaty Mts. (Little Carpathians), Slovakia. Tectonophysics 492:213–229. https://doi.org/10.1016/j.tecto.2010.06.007

Fojtíkova L, Zahradnik J (2014) A new strategy for weak events in sparse networks: the first-motion polarity solutions constrained by single-station wave-form inversion. Seismol Res Lett 85(6):1265–1274. http://dx.doi.org/10.1785/0220140072

Gansser A (1964) Geology of the Himalaya. London (Inter Science), p 289

Gansser A (1993) The Himalayas seen from Bhutan. J Geol B-A Band 136:335–346. ISSN: 0016-7800

Grujic D, Casey M, Davidson C, Hollister LS, Kundig R, Pavlis T, Schmid S (1996) Ductile extrusion of the higher Himalayan crystalline in Bhutan: evidence from quartz microfabrics. Tectonophysics 260:21–43

Grujic D, Hollister L, Parrish R (2002) Himalayan metamorphic sequence as an orogenic channel: insight from Bhutan. Earth Planet Sci Lett 198:177–191

Grujic D, Coutand I, Bookhagen B, Bonnet S, Blythe A, Duncan C (2006) Climatic forcing of erosion, landscape, and tectonics in the Bhutan Himalayas. Geology 34:801–804

GSI (2000) Seismotectonic atlas of India and its environs, geological survey. India Pub. 86 pp

Gupta HK, Singh SC, Dutta TK, Saikia MM (1984) Recent investigations of North East seismicity. In: Gongxu G, Xing-Yuan M (eds) Proceedings of the international symposium continental seismicity and earthquake prediction. Seismological Press. Beijing, pp 63–71

Kayal JR, Arefiev S, Baruah S, Tatevossian R, Gogoi N, Sanoujam M, Gautam JL, Hazarika D, Bora D (2010) The 2009 Bhutan and Assam felt earthquakes (Mw 6.3 and 5.1) at the Kopili fault in the northeast Himalaya region. Geomat Nat Hazards Risk 1:273–281

Kumar MR, Raju PS, Devi EU, Saul J, Ramesh DS (2004) Crustal structure variations in northeast India from converted phases. Geophys Res Lett 31:1–4. https://doi.org/10.1029/2004GL020576

Lienert BR, Berg BE, Frazer LN (1986) Hypocenter: an earthquake location method using centered, scaled and adaptively damped least squares. Bull Seism Soc Am 76:771–783
Marzooqi YAl , Abou Elenean KM, Megahed AS, El-Hussain I, Rodgers A, Al Khatibi E (2008) Source parameters of March 10 and September 13, 2007, united Arab Emiratus earthuqkae. Tectonophysics, 460:237–247
Mckenzie DP, Sclater JG (1971) The evolution of the Indian Ocean since the late cretaceous. Geophys J 24:437–528
Mitchell AHG (1981) Phanerozoic plate boundaries in mainland SE Asia, the Himalayas and Tibet. Geol Soc London J 138:109–122
Mitra S, Pristley K, Bhattacharya A, Gaur VK (2005) Crustal structure and earthquake focal depths beneath northeastern India and southern Tibet. Geophys J Int 160:227–248
Molnar P, Tapponnier P (1975) Cenozoic tectonics of Asia: effects of a continental collision. Science 189:419–425
Mukhopadhyay S, Chander R, Khattri KN (1997) Crustal properties in the epicentral tract of the Great 1897 Assam Earthquake, northeastern India. Tectonophysics 283:311–330
Nandy DR (2001) Geodynamics of Northeastern India and the adjoining region. ACB publications, Calcutta
Ni J, Barazangi M (1984) Seismotectonics of the Himalayan collision zone: geometry of the underthrusting Indian plate beneath the Himalaya. J Geophys Res 89:1147–1163
Parvez IA, Vaccari F, Panza GF (2003) A deterministic seismic hazard map of India and adjacent areas. Geophys J Int 155:489–508
Pradhan R, Prajapati S, Chopra S, Kumar A, Bansal BK, Reddy CD (2013) Causative source of Mw 6.9 Sikkim-Nepal border earthquake of September 2011: GPS baseline observations and strain analysis. J Asian Earth Sci 70–71:179–192
Rao NP (2009) Single station moment tensor inversion for focal mechanisms of Indian intra-plate earthquakes. Curr Sci 77(9):1184–1188
Ravi Kumar M, Hazarika P, Prasad GS, Singh A, Saha S (2012) Tectonic implications of the September 2011 Sikkim earthquake and its aftershocks. Curr Sci 102:788–792
Scherbaum F (1994) Modelling the Roermond earthquake of 1992 April 13 by stochastic simulation of its high-frequency strong ground motion. Geophys J Inr 119:31–43
Seeber L, Armbruster JG (1981) Great detachment earthquakes along the Himalayan arc and long-term forecasting. In: Earthquake prediction—An international review. Manrice Ewing Serires, American Geophysical Union, vol 4, pp 259–277
Searle MP (1996) Cooling history, erosion, exhumation, and kinematics of the Himalayan-Karakoram-Tibet orogenic belt. In: Yin A, Harrison M (eds) The tectonic evolution of Asia. Cambridge University Press, New York, pp 110–137
Sokos E, Zahradnik J (2008) ISOLA—A Fortran code and Matlab GUI to perform multiple-point source inversion of seismic data. Comput Geosci 34:967–977
Zahradnik J, Gallovic F, Sokos E, Serpetsidaki A, Tselentis G-A (2008) Quick fault-plane identification by a geometrical method: application to the Mw 6.2 Leonidio earthquake, 6 January 2008, Greece. Seismol Res Lett 79(5):653–662
Zahradník J, Fojtíkova L, Carvalho J, Barros LV, Sokos E, Janský J (2015) Compromising polarity and waveform constraints in focal-mechanism solutions; the Mara Rosa 2010 Mw 4 central Brazil earthquake revisited. J Sci Am Earth Sci 63:323–333. https://doi.org/10.1016/j.jsames.2015.08.011
Zhao W, Nelson KD, Project INDEPTH Team (1993) Deep seismic reflection evidence for continental underthrusting beneath southern Tibet. Nature 366:557–559

Resolving the Tectonic Stress by the Inversion of Earthquake Focal Mechanisms. Application in the Region of Greece. A Tutorial

Ioannis G. Kassaras and Vasilis Kapetanidis

Keywords Earthquakes · Active tectonics · Stress tensor · Strain tensor
Moment tensor · Focal mechanism · Inversion · Seismotectonics
Wallace-Bott hypothesis

1 Introduction

This work describes the derivation of the tectonic stress from the inversion of focal mechanisms of double-couple earthquakes. The presented material is based, in large part, on several review papers, lecture notes and practices on the matter, developed by the authors during the last years.

The rupture of geological faults produces the so-called "double-couple" or "tectonic earthquakes" that constitute the major cause of deformation of the Earth's lithosphere. The causes of tectonic earthquakes are forces applied on surfaces (stresses) that drive the earthquake cycle, during which energy is accumulated and released abruptly, generating elastic waves which propagate through the Earth. The analysis of earthquake driving forces is complicated, and their accurate knowledge is a challenging milestone, with numerous applications in geosciences.

Although present-day kinematics of the Earth's surface is efficiently monitored by geodetic remote sensing (GNSS), an accumulation of evidence regarding the earthquake cycle is required, involving the scientific fields of structural geology and seismology. Both provide observations regarding the deformation of the crust, while seismology sets robust constraints on the coseismic stress release within the entire volume of the crust. The most efficient tool in seismology for studying the

I. G. Kassaras (✉) · V. Kapetanidis
Department of Geophysics and Geothermics, National and Kapodistrian University of Athens, Panepistimiopolis, 15784 Zografou, Greece
e-mail: kassaras@geol.uoa.gr

V. Kapetanidis
e-mail: vkapetan@geol.uoa.gr

S. D'Amico (ed.), *Moment Tensor Solutions*, Springer Natural Hazards,
https://doi.org/10.1007/978-3-319-77359-9_19

relative magnitudes and directions of the stresses released during earthquakes is the determination of focal mechanisms.

However, the use of individual focal mechanisms for the assessment of the tectonic stress is not a straightforward solution, since the degree that they sample intrinsic tectonic episodes varies. This is due to several uncertainties related to inherent ambiguities in the definition of the fault plane and slip direction or to the accuracy of the individual focal mechanisms and the respective inversion methods' assumptions. Moreover, the stress field can be heterogeneous, complicated by processes that operate at different 2D/3D spatial scales, which may be a) large, related to plate motion, b) intermediate, reflecting lithospheric flexure/lateral density contrasts (Zoback 1992; Richardson 1992), or c) small, associated with geothermal or hydrocarbon reservoirs (Heidbach et al. 2007). These sorts of processes may be obscured or overestimated by the individual solutions, depending on the available data.

Consequently, individual focal mechanisms should be used with caution and only as a proxy for resolving the tectonic stress (McKenzie 1969; Célérier 2010). As a countermeasure to the above, several methodologies for the determination of the stress field from the inversion of sets of focal mechanism data have been proposed (e.g. Gephart and Forsyth 1984; Gephart 1990; Michael 1984, 1987; Angelier 2002), the application of which has been proven effective for resolving stress at various spatial scales, proportionally to the data multitude and quality (e.g. Zoback et al. 1989; Zoback 1992; Sperner et al. 2003; Heidbach et al. 2010). A largely acknowledged outcome of these concepts and methods is the World Stress Map Project (WSMP, Heidbach et al. 2016), a global compilation of information on the crustal present-day stress field maintained since 2009 at the Helmholtz Centre Potsdam GFZ German Research Centre for Geosciences (http://www.world-stress-map.org).

Apart from the seismic hazard assessment, which is a particular concern, applications of the stress inversion method include: calculation of deviations from the uniform fault motion at a range of spatial and temporal scales related to fault coupling variance, identification of fault zones or segments with significant strain accumulation that have not ruptured in a recent earthquake, migration of stresses, strain deficit etc. Moreover, the method favorably applies to the current needs of the oil industry, such as the exploitation of new, naturally fractured reservoirs. The latter requires knowledge of the present and past stress field and is important for the drilling and borehole stability through the prediction of the orientation of fractures induced by hydraulic fracturing, or "fracking" (Lejri 2015).

The first part of this work invokes the basic theory regarding the fundamentals of stress inversion, while the second presents the application of the method to a comprehensive dataset for Greece (Kassaras et al. 2016a), the most seismically active and tectonically complex region in Europe (SHARE project; Stucchi et al. 2013).

2 Basic Concepts

In seismically active regions, tectonic stress drives the "interseismic" phase, during which energy is accumulated within the rock volume, and the "post-seismic" phase, related to the viscoelastic relaxation of the crust after the abrupt stress release or "coseismic" phase (Bock and Melgar 2016). Tectonic stress is a long-term entity described by a second-order "stress tensor", distinct from the **moment** or **stress release**, represented by the "moment tensor".

2.1 The Stress Tensor

Typical models of tectonic stress are described in terms of forces acting at some distance from the area where a tectonic structure might develop (e.g. Muller and Pollard 1977). In the simplest cases, this distance is assumed to be large with respect to the dimensions of the structure and hence these 'regional' stresses are considered to be homogeneous (Pollard and Saltzer 1993), specified by a single force.

The generalized case of the applied stresses inside a material can be described by a second-order stress tensor σ_{ij}.

$$\sigma_{ij} = \begin{bmatrix} \sigma_{11} & \sigma_{12} & \sigma_{13} \\ \upsilon_{21} & \upsilon_{22} & \upsilon_{23} \\ \sigma_{31} & \sigma_{32} & \sigma_{33} \end{bmatrix} \quad (1)$$

The stresses are applied as coupled forces at opposite sides of an infinitesimal cube and towards opposite directions, cancelling one another to preserve equilibrium, but producing strain. The elements of the diagonal correspond to vectors which are normal to the surface of the cube that is indicated by the first index, *i*, and responsible for strain that changes the cube's volume. The other, non-diagonal stress elements refer to vectors which are coplanar to the applied surface and can cause shape-changing shear strain to the cube. In the literature, the latter are sometimes represented by τ_{ij} $(i \neq j)$, to distinguish between normal and shear stresses.

The elements of the stress tensor depend on the coordinate system that is chosen; the orientation of the infinitesimal cube with respect to the stress field. The stress tensor is a diagonalizable matrix, which means it can be rotated to a specific coordinate system where its non-diagonal elements (shear stresses) become zero. In that system, the only stresses acting on the cube are oriented in directions indicated by the cube's normal vectors, called principal stress axes. The directions, $\mathbf{S}_1$, $\mathbf{S}_2$, $\mathbf{S}_3$, and magnitudes, σ_1, σ_2 and σ_3, of the principal stresses can be derived from the eigenvectors and eigenvalues of the stress tensor, respectively. Eigenvalues are real numbers that represent the magnitude of the principal stresses while eigenvectors define their geometry. In geological applications where the stresses are all nega-

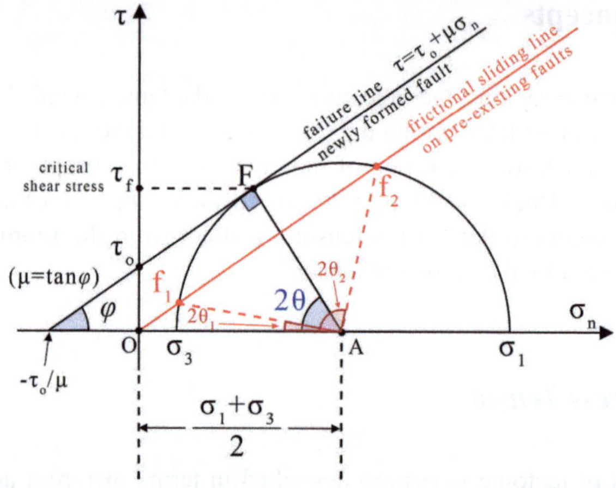

Fig. 1 Mohr's circle for the definition of frictional slide on pre-existing faults (red line) or failure angles on newly formed faults (black line parallel to the red one), where φ is the angle of internal friction, θ is the angle between the $\mathbf{S}_1$ axis and the failure plane, σ_n is the applied normal stress, τ is the applied shear stress, τ_o the cohesive strength and $(-\tau_o/\mu)$ the tensile strength of the material. The circle's radius is AF = $(\sigma_1 - \sigma_3)/2$, that is also the value of maximum shear stress for $\theta = 45°$ (if $\varphi = 0$) (Figure after Kapetanidis 2017)

tive (compressive) they are usually sorted by order of decreasing absolute values $|\sigma_1| \geq |\sigma_2| \geq |\sigma_3|$ (Stein and Wysession 2003).

In the general case, however, the shear stresses are non-zero. Shear stresses inside a material can produce rupture when the applied forces surpass the material's strength. Intuitively, for a given stress field, the material is more likely to break at a surface that is oriented in such a way that the shear stresses maximize. This condition occurs when the principal axes $\mathbf{S}_1$ and $\mathbf{S}_3$ are at a 45° angle from the weak plane (Stein and Wysession 2003).

However, as Reches (1987) notes, slip occurs when Coulomb's failure criterion is fulfilled, i.e. when the shear stress, τ, exceeds frictional resistance:

$$\tau = \tau_o + \mu \cdot \sigma_n \tag{2}$$

where τ_o is the cohesive strength of the rock, or simply cohesion, $\mu = \tan\Phi$ (Fig. 1) is the coefficient of internal friction of the material and σ_n in this formula is considered positive for compression (Scholz 2002). In this more realistic model, the angle θ between the $\mathbf{S}_1$ axis and the failure plane also depends on the angle of internal friction, φ, of the material:

$$\theta = 45° - \varphi/2 \tag{3}$$

When the material contains weak discontinuities, e.g. pre-existing fractures that have been cemented, where the cohesive strength of the material is $\tau_o \approx 0$, frictional

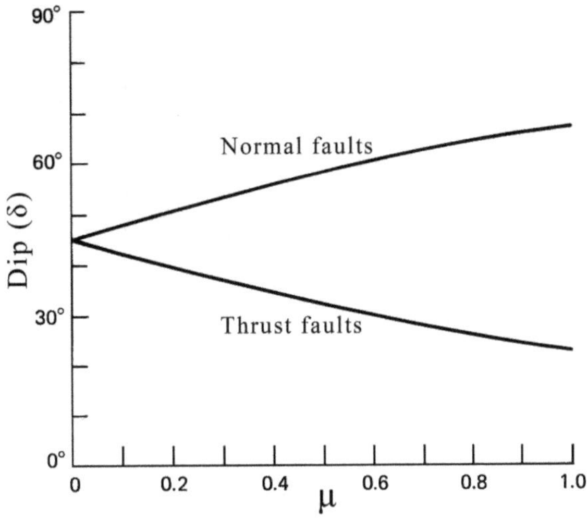

Fig. 2 Relation between the expected fault dip angle, δ, for a given coefficient of internal friction, μ, for normal and thrust faults (Figure after Turcotte and Schubert 2002)

slide can occur for a given stress state if discontinuities are found at either of two preferred orientations, θ_1 and θ_2 (Fig. 1).

The orientation of the maximum and minimum principal stresses, $\mathbf{S}_1$ and $\mathbf{S}_3$, in a regional scale largely defines the expected tectonic regime. When $\mathbf{S}_1$ is sub-vertical and $\mathbf{S}_3$ is sub-horizontal, extension occurs, creating normal fault systems under the vertical lithostatic pressure ($\mathbf{S}_1$). When the contrary is true, thrust faulting occurs under compression imposed by the sub-horizontal $\mathbf{S}_1$. Lastly, when both $\mathbf{S}_1$ and $\mathbf{S}_3$ are subhorizontal, strike-slip faulting takes place. Furthermore, this has implications on the expected dip angles, δ, that normal and thrust faults are likely to occur, depending on the coefficient of internal friction, μ, as calculated in a simple model by Turcotte and Schubert (2002) (Fig. 2).

2.2 *The Moment Tensor*

Contrary to the tectonic stress, the earthquake rupture cannot be described by a single force, only applicable to certain types of sources, i.e. those related to the collapse of a cave's ceiling. The model of "single-couple" source, with a pair of opposite forces applied at the two walls, parallel to the fault plane, although intuitively valid and correctly reproducing the P-wave radiation pattern, fails to produce the observed polarities and amplitudes for the generated shear waves (Aki and Richards 2002).

To describe the earthquake rupture, Vvedenskaya (1956) introduced the concept of the double-couple, a system of exerting forces at a point of a fault surface, the so-called "*point-source double-couple*", consisting of two opposing torques with equal moment (M_0) imposing volumetric and shear deformation of the elastic medium, with the assumption that translation and rotation are negligible. Thereafter, the *double-*

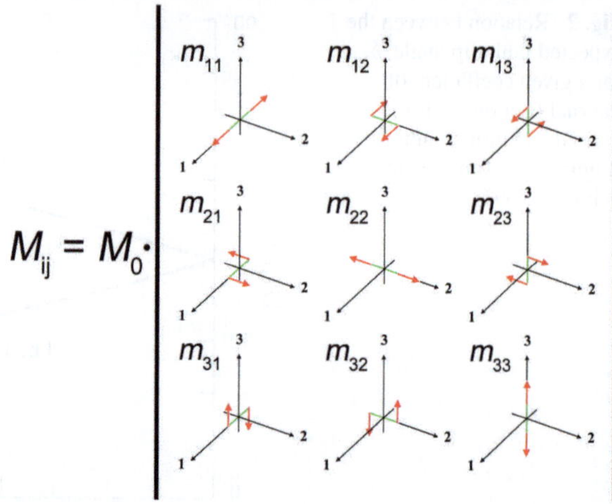

Fig. 3 The components of a Cartesian moment tensor

couple (DC) source model has been widely adopted as adequately explaining sources of tectonic origin, known as double-couple earthquakes. It also correctly predicts the radiation pattern of the shear-waves, as opposed to the single-couple model.

The **Moment Tensor** (MT) is a physical entity that defines the magnitude of a seismic source in terms of its seismic moment, M_0, and the radiation pattern of the recorded seismic waves, i.e. the characteristics of the seismic energy's distribution. A force couple m_{ij} in a Cartesian coordinate system is defined as a pair of opposing forces pointing in the direction of the component indicated by i and separated in the respective j direction (Fig. 3). The acting pairs of forces and torques applied on a point-source are represented by a second-order tensor, i.e. a 3×3 matrix, M_{ij}. The magnitude of M_{ij} is given by the scalar seismic moment M_0, which is defined as the largest singular value of the moment tensor. The diagonal elements in the matrix represent linear vector dipoles; the off diagonal elements represent shear force couples with non-zero torque.

As a second order 3×3 tensor, the MT (Fig. 3) has several similarities to the stress tensor, although the forces described in the former are not acting on infinitesimal surfaces, but rather on points at an infinitesimal distance from the point-source. The elements on its diagonal act as linear dipoles and produce volumetric changes (explosion, implosion and/or tensile fracturing) when the trace is non-zero. Removing the trace from the diagonal leaves behind the deviatoric part of the MT. The latter can still be divided into a pure DC component and another that is called a Compensated Linear Vector Dipole (CLVD). The CLVD component could be significant in earthquakes of volcano-tectonic origin, related to magma inflation or propagation through dykes, fissure eruptions, crack opening or closing without volume changes etc. DC earthquakes, on the other hand, are related to purely tectonic movements. Figure 4 illustrates the representation of a rupture (Fig. 4a) by the DC source model (Fig. 4b). The one couple of forces is acting parallel to the fault plane, defined by its

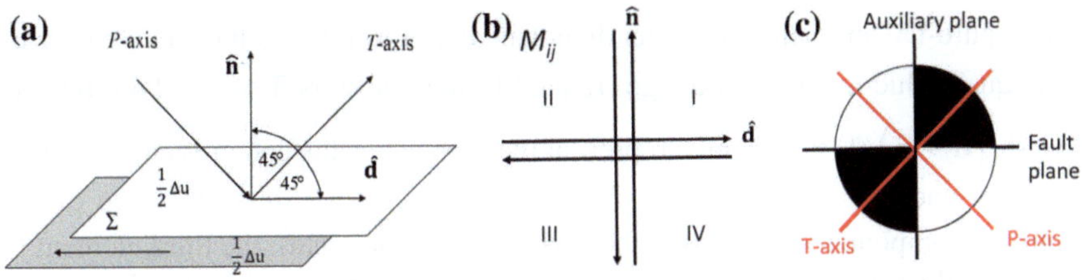

Fig. 4 **a** Dislocation on a fault surface (Σ). The total slip Δu is the difference between the displacements at the hanging-wall and the footwall (modified after Aki and Richards 2002). **b** "Double-couple" force system, equivalent to the dislocation model of panel (**a**) when in equilibrium state. **c** Beachball representation of the P-wave radiation pattern of the models shown in panels (**a**) and (**b**). Black and white quadrants correspond to compressional and dilatational first motion while they also contain the tension (**T**) and compression (**P**) axes, respectively, at 45° from the fault and auxiliary planes

normal vector $\hat{\mathbf{n}}$, and the other parallel to an "auxiliary plane" that is normal both to the fault plane and to the slip vector, **d**.

The unitary vectors $\hat{\mathbf{n}}$ and **d** are related in a way similar to how the stress tensor produces the traction $\hat{\mathbf{t}}$ from the normal vector $\hat{\mathbf{n}}$:

$$\mathbf{d}\left(\hat{\mathbf{n}}\right) = (M_{ij})\hat{\mathbf{n}} \tag{4}$$

The condition that angular momentum be conserved in equilibrium requires that the moment tensor is symmetric ($M_{ij} = M_{ji}$) to prevent rotation, thus leaving 6 independent elements (or 5, if the trace is zero):

$$M_{ij} = M_0\left(\hat{d}_i\hat{\mathbf{n}}_j + \hat{\mathbf{n}}_i\hat{d}_j\right) \tag{5}$$

Interchangeability of $\hat{\mathbf{d}}$ and $\hat{\mathbf{n}}$ in the DC model makes the tensor symmetric ($M_{ij} = M_{ji}$). Practically, this shows that slip on either the fault or the auxiliary plane in the directions indicated by $\hat{\mathbf{d}}$ or $\hat{\mathbf{n}}$, respectively, yields the same seismic radiation pattern, which leads to an inherent ambiguity, contrary to the single-couple model where the fault plane is unambiguously defined. In terms of the radiation pattern, this can be seen in Fig. 4c, where in quadrants I and III (black) the radiated (P-wave) first motion is compressional, or "thrusting away" from the source, while in quadrants II and IV (white) the first motion is dilatational, or "pulling towards" the source.

Another common characteristic between moment- and stress-tensors is that they are both diagonizable. They can be rotated to a common coordinate system which can be expressed as a product with a transformation matrix **A** or **B**:

$$M'_{ij} = A(M_{ij})\mathrm{A}^T, \quad \sigma'_{ij} = B\left(\sigma_{ij}\right)B^T \tag{6}$$

In a pure-DC moment tensor, its diagonalizable form would have two elements with equal value and opposite sign, related to the bisectors $\mathbf{T} = \frac{1}{\sqrt{2}}\left(\hat{\mathbf{d}} + \hat{\mathbf{n}}\right)$ and $\mathbf{P} = \frac{1}{\sqrt{2}}\left(\hat{\mathbf{d}} - \hat{\mathbf{n}}\right)$ (Fig. 4), which represent the axes of tension and pressure, or rather minimum and maximum compressive stress, respectively. Since the trace of M_{ij} (isotropic component) is zero (i.e. the $\hat{\mathbf{d}}$ and $\hat{\mathbf{n}}$ are orthogonal), the third element of the diagonal (or third eigenvalue) would be zero and correspond to the eigenvector in the direction of the intersection of the nodal planes $\mathbf{N} = \left(\hat{\mathbf{d}} \times \hat{\mathbf{n}}\right) = \mathbf{T} \times \mathbf{P}$, where no motion occurs (null axis; sometimes also symbolized as **B**).

The principal axes **P, N, T** of the moment tensor fully describe the dynamics and kinematics of a DC point source and are independent of the choice of the two nodal (fault and auxiliary) planes. They can be determined through the eigenvalue and eigenvector analysis of the moment tensor (M_{ij}), similar to the principal stresses $\mathbf{S}_1$, $\mathbf{S}_2$, $\mathbf{S}_3$ and their magnitudes σ_1, σ_2 and σ_3 for the stress tensor, which are related to **P**, **N** and **T**, respectively.

However, the **P**- and **T**-axes are always at a 45° angle from both the fault and the auxiliary planes, while the same is true for $\mathbf{S}_1$ and $\mathbf{S}_3$ only in case of new fracture generation in a homogeneous isotropic medium (e.g. Vavryčuk 2015). Hence, the moment tensor's principal axes **P**, **N**, and **T** do not always correspond to the respective principal stress axes $\mathbf{S}_1$, $\mathbf{S}_2$, and $\mathbf{S}_3$, because earthquakes typically occur on pre-existing faults in a heterogeneous anisotropic medium such as the crust (e.g. Scholtz 2002; Stein and Wysession 2003), due to directional differences caused by the fault shape, fault maturity and frictional anisotropy (Pollard and Saltzer 1993). However, $\mathbf{S}_1$ and $\mathbf{S}_3$ are expected to lie within the same quadrants as **P** and **T**, respectively (Fig. 4).

Inversely, in some methodologies (Vavryčuk 2015) the principal fault planes are considered those which are at an angle of less than 45° from the $\mathbf{S}_1$ axis. In Western Greece, it has indeed been observed that the $\mathbf{S}_1$ axis is usually at a ~30° angle from the principal fault planes (Kassaras et al. 2016b). Various methods have been devised for automatically choosing the fault plane (Lund and Slunga 1999), including the "Slip Angle Method" (e.g. Gephart and Forsyth 1984), which considers the nodal plane that has the smallest misfit between the predicted and observed slip vector orientation and the "Instability Method", which considers the most unstable plane as the one that slips (Lund and Slunga 1999).

In a fault-related earthquake, the DC component of the moment tensor can describe the orientation of the causative fault (its strike, φ_f, and dip angle, δ), the direction of the slip on its plane (rake angle, λ) and the magnitude of the rupture (expressed in the form of the released seismic moment, M_o), albeit with an ambiguity between the two nodal planes (fault and auxiliary). The elements of the moment tensor in a DC source can be directly calculated from the φ_f, δ, λ parameters and vice versa (Aki and Richards 2002; Gasperini and Vannucci 2003). The type of faulting for a certain set of φ_f, δ, λ parameters, when they correspond to the fault plane rather than the auxiliary one, can be determined by the value of λ, with $\lambda \approx -90°$ corresponding to normal faulting, $\lambda \approx 90°$ related to reverse faulting and $\lambda \approx 0°$ or $\pm 180°$ associated

Table 1 Stress regime characterization based on the plunge, γ_i, of **P**, **N** and **T** axes (on the basis of them being an approximation of the principal stress axes $\mathbf{S}_1$, $\mathbf{S}_2$ and $\mathbf{S}_3$), where NF, SS and TF refer to normal, strike-slip and thrust faulting, while NS and TS to oblique normal and thrust faulting, respectively. Symbols γ and α refer to the plunge and trend, respectively. The trend of the maximum horizontal stress axis, $\mathbf{S}_{Hmax}$, is also indicated (after Zoback 1992). The last two lines ("Odd", or "Unknown" faulting type) have been added for completeness

P (σ_1)	N (σ_2)	T (σ_3)	Regime or faulting type	S_{Hmax} trend
$\gamma_P \geq 52°$		$\gamma_T \leq 35°$	NF	α_N
$40° \leq \gamma_P < 52°$		$\gamma_T \leq 20°$	NS	$\alpha_T + 90°$
$\gamma_P < 40°$	$\gamma_N \geq 45°$	$\gamma_T \leq 20°$	SS	$\alpha_T + 90°$
$\gamma_P \leq 20°$	$\gamma_N \geq 45°$	$\gamma_T < 40°$	SS	α_P
$\gamma_P \leq 20°$		$40° \leq \gamma_T < 52°$	TS	α_P
$\gamma_P \leq 35°$		$\gamma_T \geq 52°$	TF	α_P
$40 \leq \gamma_P \leq 50°$	$\gamma_N \approx 0°$	$40 \leq \gamma_T \leq 50°$	"Odd"	–
$25 \leq \gamma_P \leq 45°$	$25 \leq \gamma_N \leq 45°$	$25 \leq \gamma_T \leq 45°$	"Odd"	–

with sinistral or dextral strike-slip faulting, respectively. However, the faulting type represented by a certain DC moment tensor can also be determined by the orientation of its eigenvectors, **P**, **N** and **T**, in a sense which may apply to the characterization of the stress regime in an area. Zoback (1992) determined a set of rules for the classification of faulting type depending mainly on the plunge angles, γ, of the **P** and **T** axes, complemented, in some cases, by the respective plunge of the **N** axis (Table 1). These also determine the trend of the maximum horizontal stress axis, $\mathbf{S}_{Hmax}$. There is a region of **P** and **T** plunges (γ_P and γ_T) in which focal mechanisms are not classified in a typical faulting type, but rather as "odd" or "unknown". This is the case of either sub-horizontal faults with horizontal slip or sub-vertical faults with (nearly vertical) dip-slip (both γ_P and $\gamma_T \approx 45°$ or all three axes with $25° < \gamma < 45°$). Such cases are generally rare and may characterize very low-angle normal faults or thrusts where the principal stress field is tilted out of horizontal and vertical planes (Zoback 1992). An additional set of criteria for faulting classification (e.g. Milano and Di Giovambattista 2011) can be used to further limit the zone of the "Odd" type.

2.3 *Methods to Determine Focal Mechanism Solutions (FMS)*

Nowadays, Focal Mechanism Solutions (FMS) are routinely obtained through the analysis of waveform recordings from modern deployments. Methods for determining FMS are based on the observed radiation pattern of seismic waves that depends on the location and orientation of the active fault and slip direction with respect to the receivers. The most widely applied methods include the determination of FMS through the distribution of first motion polarities of the P-waves (e.g. Kassaras et al. 2014a), polarization and amplitude of the S-waves (e.g. Bernard and Zollo 1989), S/P (or other types of waves) amplitude ratio (e.g. Godano et al. 2014) and moment tensor inversion using body-wave modeling (e.g. Kapetanidis et al. 2015). The solution determined by the latter is known as Moment Tensor Solution (MTS) and it is considered as the most reliable seismological method for defining the rupture dynamics and kinematics.

A variation of MTS, the Centroid Moment Tensor (CMT; http://www.globalcmt.org), includes additional inversion for the source origin time and location, in terms of the centroid of moment release (or slip) as opposed to the hypocenter, which is defined as the location where rupture began (nucleation point). The centroid location for very large earthquakes may significantly differ from the hypocenter; the offset of the centroid relative to the hypocentral location gives a first indication on the fault's extent and directivity of the rupture (Dahm and Krüger 2014). A sound example is the case of the August 17th, 1999 Izmit (Turkey) earthquake, where the centroid (quick CMT solution) was resolved about 30 km east of the epicenter, as determined by the arrival times of seismic waves (Tibi et al. 2001), and coincided with the area where the maximum surface ruptures were observed.

The FMS of tectonic earthquakes can be easily reconstructed from the analysis of the polarities and amplitudes of the P-wave onsets, which can be complemented by the amplitudes of the first arrival of S-waves. However, this must be performed at an adequate number of seismological stations distributed along a wide range of azimuths and for seismic rays at various angles of emergence. Such methods are sufficient to describe the mechanics of tectonic processes within the Earth's crust, mainly at small (local) epicentral distances, using up-going seismic rays, and assuming net slip on the fault (i.e. pure DC source).

Typically, FMS are displayed graphically by the so-called "*beachball*" diagram (Figs. 4c and 5), which gives a sense of the orientation of the nodal planes and the respective direction of the slip vector (for the nodal plane that corresponds to the fault), quadrants of alternating P-wave polarities and orientation and plunge of the **P**, **N** and **T** axes.

For a more detailed analysis of intrinsic geological, mineralogical and other processes, additional information can be derived from the Moment Tensor Solution (MTS) by inversion of seismograms at local, regional or teleseismic distances. The MTS uses the radiation pattern of body waves (e.g. Herrmann and Ammon 1997), surface waves (e.g. Aki and Patton 1978) and normal modes of the free oscillations of the Earth for very large earthquakes (e.g. Bogiatzis and Ishii 2014). MTS involves finding the moment tensor M_{ij} that produces synthetic waveforms which best match the observed seismograms in a least squares sense (e.g. Dahlen and Tromp 1998). Inversion for a full (unrestricted) moment tensor may result in various types of seismic sources (i.e. isotropic, tensile, CLVD, double-couple and their combinations), which is useful when a complex wavefield is involved, related to deep earthquakes, volcanic earthquakes, landslides, nuclear tests, etc. However, the non-DC component (e.g. isotropic and/or CLVD) of earthquake FMS, especially in

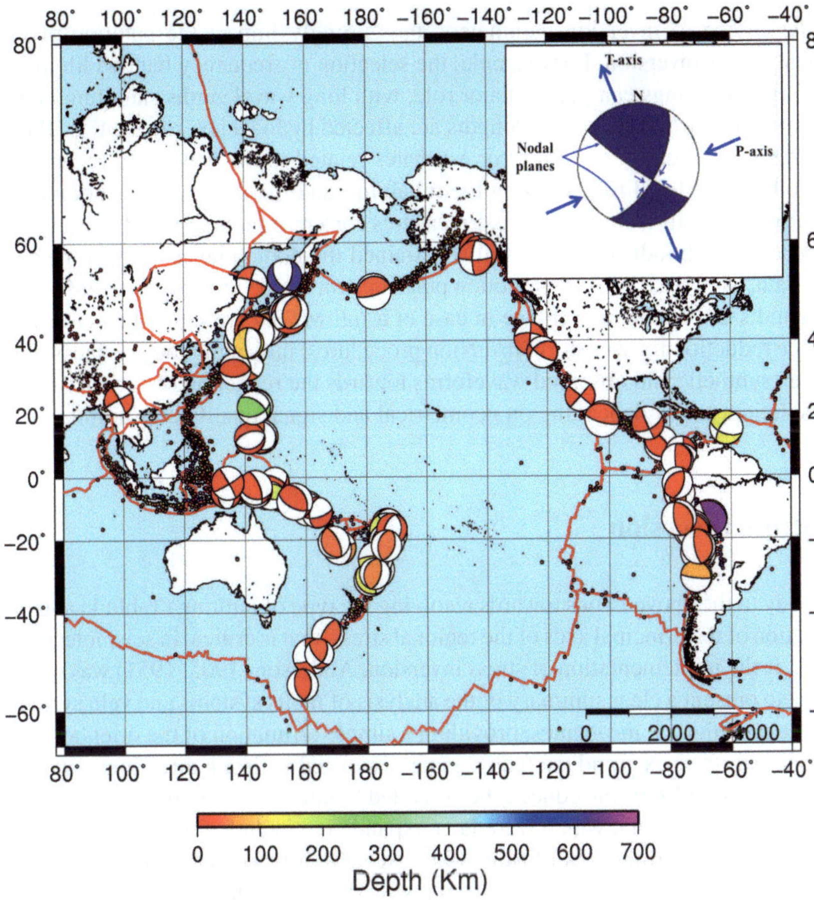

Fig. 5 CMT solutions of earthquakes with $M_w \geq 6.5$ (http://www.globalcmt.org). The "*beachball*" diagram in the upper right shows the stereographic projection of the lower hemisphere of a focal mechanism solution. White quadrants represent dilatational first motion and contain the maximum compression axis, **P**, while dilatational quadrants containing the maximum extension axis (**T**) are presented in color

non-volcanic regions, is usually attributed to noise and/or unmodeled structures in the Earth model employed for calculating Green's functions, in order to produce synthetic seismograms at selected frequency bands.

The quality of either solution, as determined by moment tensor inversion or first-motion analysis, depends not only on the datasets and a typical error estimation but, also, on several additional factors. Dufumier and Cara (1995) and Dufumier (1996) provide a systematic overview of how FMS are affected by the geographical coverage of stations, the type and quantity of the seismic waves involved (P, S, SH, SV), the signal-to-noise-ratio of the waveforms, site-effects and the knowledge of the Earth's structure. Moreover, methodological limitations due to different approaches and the choice of inversion parameters may strongly impact the solution of the moment tensor inversion. For example, the selection of frequency bandwidth in the waveform recordings can play a major role, with long wavelengths sampling large-scale structures while short wavelengths are affected by local heterogeneities (Barth et al. 2008). Therefore, several error estimates related to the waveform recordings have to be considered for assessing the reliability and the degree of uncertainty of a focal mechanism solution. Typical measures for error include the RMS angular difference, which indicates how well constrained the nodal planes are with respect to the data, as in the case of first motion polarities for single or composite FMS (e.g. Kapetanidis 2017), DC percentage in case of a full MTS, correlation percentage or variance reduction for waveform inversion procedures, indicating the goodness of fit between synthetic and observed waveforms towards the reduction of variance in the final solution, partly depending on geometrical and signal quality constraints.

3 Stress Inversion

A remedy to the above issues towards resolving the type of faulting (Table 1) and the orientation of the principal axes of the regional stress, that is critical in seismotectonic studies, is the implementation of stress inversion. Anderson (1905, 1951) was one of the first to present a clear summary of the analysis of faults systems and vein systems by analogy with rock mechanics, providing a simple estimation of the orientation of the principal stress axes and the fault regime associated with slip directions. Wallace (1951) and Bott (1959) introduced the so-called "faultless" methods in the analysis of fault slickenside data, which were later expanded to include focal mechanisms.

Since the pioneering work of Bott (1959), several different methods for inferring certain elements of the stress tensor from populations of faults have been proposed. These can be grouped into graphical (Compton 1966; Arthaud 1969; Angelier and Mechler 1977; Aleksandrowski 1985; Lisle 1987; Marrett and Allmendinger 1990) and numerical techniques (Carey and Brunier 1974; Etchecopar et al. 1981; Armijo et al. 1982; Angelier 1984, 1989, 2002; Gephart and Forsyth 1984; Michael 1984, 1987; Reches 1987; Gephart 1988, 1990; Huang 1988; Fry 1999; Shan et al. 2004).

A variety of graphical analyses methods have been proposed for fault-slip analysis. Among these, the "P- and T-dihedra" method (Angelier and Mechler 1977) and the kinematic P- and T-axes (Marrett and Allmendinger 1990) have been proven the most robust, providing similar results with the more sophisticated numerical methods.

The inverse problem that is encountered in the numerical "faultless" methods is to reconstruct the regional stress field for given groups of observed slip vectors and fault plane orientations. The procedure identifies the principal stress directions of the local stress tensor and a factor (stress ratio) that characterizes the relative differences between the principal stresses' magnitudes. All methodologies rely on the following four hypotheses:

H_1 *Faults slip in the direction indicated by the resolved shear stress*: Seismic slip is assumed to be parallel to and in the same direction as the one indicated by the shear stress resolved on the fault plane. This is also known as the Wallace-Bott hypothesis (Wallace 1951; Bott 1959).

H_2 *The rocks are physically homogeneous*: This practically assumes that stress is uniform within the rock volume, which can usually be achieved by dividing a large study region into smaller subareas (Maury et al. 2013).

H_3 *Stress is invariant over time*: This assumption holds for paleo-stress determination, since the age of slickenside data typically cannot be specified. This limitation does not apply when using focal mechanism data, in which case it is possible to perform an analysis of the temporal variations of stress (Hardebeck and Michael 2006).

H_4 *The input focal mechanisms are independent from each other*: This is required so that the input data actually sample manifestations of slip on several faults due to a common stress regime without local stress perturbations that could have been caused by a large earthquake that occurred previously (Maury et al. 2013). This suggests that some sort of data filtering, such as declustering, might be required before the inversion takes place.

Even though the inverse techniques may be used for either fault/slickenside or earthquake focal mechanism data, the above assumptions apply more accurately to the latter (Allmendinger et al. 1989). Among the inverse methods, there are two major types of algorithms to determine the stress tensor: the **grid search** and the **least-squares** methods.

Both approaches use the minimization of the angle between the calculated shear stress and the observed slip vector as a criterion for the best fit. Their difference concerns the data regression approach. The grid search method involves a large number of regression lines, for each of which their correlation with the data distribution is calculated. The least-squares approach uses regression to directly determine the best fitting line, in terms of an equation that expresses the slope and intercept of the line that minimizes the squares of the deviations between data and model.

The inversion procedure involves a system of linear equations for a population of focal mechanisms, which can be written in a matrix form (Bott 1959):

$$A\mathbf{x} = \mathbf{y} \tag{7}$$

where A is a matrix containing nodal plane orientations, $\mathbf{y}$ is a vector containing shear stress information inferred from the rake of the focal mechanisms and $\mathbf{x}$ is the vector of unknowns related to the regional stress tensor. Since the stress field is presumed to be homogeneous (H_2), the fault location, its size and shape are unimportant. A least-squares inversion of Eq. 7 would be (Menke 1984):

$$\mathbf{x} = \left[A^{\mathrm{T}}A\right]^{-1}A^{\mathrm{T}}\mathbf{y} \tag{8}$$

The stress inversion problem is nonlinear, but can be linearized by assuming that the magnitude of the shear traction on each fault plane is approximately the same. The procedure aims at finding the stress state which minimizes the discrepancy between the resolved shear stress direction and the slip direction for all earthquakes in the data set. In practice, when only the fault orientation and the slip direction are provided by the FMS, but not the magnitude Δu of the relative displacement, only part of the regional stress tensor can be estimated regarding the direction of the principal axes and their relative magnitudes. More specifically, the inversion yields a "reduced" deviatoric stress tensor, T, characterized by four parameters, three of which define the orientations of the principal stress axes $\mathbf{S}_1, \mathbf{S}_2, \mathbf{S}_3$, and another the stress ratio (also called shape factor/ratio, stress shape, aspect ratio or stress magnitude parameter), R (Gephart and Forsyth 1984):

$$R = \frac{\sigma_2 - \sigma_1}{\sigma_3 - \sigma_1} \tag{9}$$

where σ_1, σ_2, and σ_3 are the magnitudes of the maximum, intermediate, and minimum principal stress axes of the resolved stress tensor, respectively. As R increases from 0 to 1, σ_2 decreases in the range between σ_1 and σ_3, with a value of R near 0 indicating that $\sigma_2 \approx \sigma_1$ and a value of R near 1 meaning $\sigma_2 \approx \sigma_3$ (Fig. 6). These two limits represent the cases of uniaxial extension ($R=0$) and uniaxial compression ($R = 1$), respectively, while $R = 0.5$ means that $\sigma_2 = (\sigma_1 + \sigma_3)/2$. Any combination of principal stresses which produces $R > 1$ or $R < 0$ is invalid (Gephart 1985). An equivalent parameter Φ was proposed (Angelier et al. 1982; Angelier 1984, 1989; Michael 1984, 1987):

$$\Phi = \frac{\sigma_2 - \sigma_3}{\sigma_1 - \sigma_3} \tag{10}$$

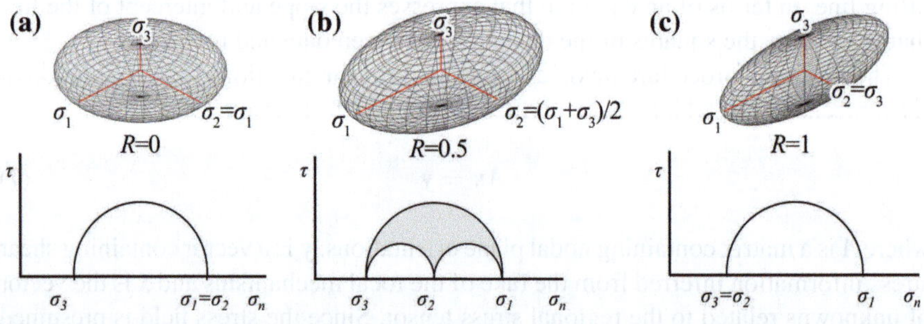

Fig. 6 Stress ellipsoids and associated Mohr diagrams for different stress states indicated by the stress ratio, R, for: **a** Uniaxial extension (R = 0, oblate ellipsoid), **b** Triaxial stress (R = 0.5), **c** uniaxial compression (R = 1, prolate ellipsoid). Note that in this example the stress magnitudes appear as all having positive values. Figure based on Lejri (2015)

The combination of (9) and (10) gives the relation $\Phi = 1 - R$. A geometrical representation of R, that is the inter-relation of the principal stress magnitudes, can be given by Lamé's stress ellipsoid and Mohr diagrams, which describe a certain state of stress (Fig. 6).

To acquire a full stress tensor, rather than a "reduced" one, further assumptions must be made about lithostatic stresses or magnitudes of shear stresses (Etchecopar et al. 1981). It is generally accepted (Byerlee 1978), that the magnitudes of shear and normal stress on the fault plane are linearly related (Amonton's Law of friction) (Eq. 2). By accepting this condition it is possible to estimate a fifth parameter, which relates the magnitudes of normal (σ_m) and shear stresses (τ_m) of the stress tensor (Allmendinger 1989):

$$\sigma_m \equiv (\sigma_1 + \sigma_3)/2 \quad \text{and} \quad \tau_m \equiv (\sigma_1 - \sigma_3)/2 \tag{11}$$

where σ_m and τ_m are the coordinate of the center of Mohr's circle and its radius, respectively (Fig. 1). A sixth parameter of the stress tensor provides a scaling factor, and thus the magnitude of all stress elements. Since this number is scaled, it cannot be estimated from orientations, which are inherently dimensionless (Allmendinger 1989). The magnitude of the displacement vector is required to estimate the scaling factor, which can be provided by surface and/or geodetic data.

Faultless methodologies are simple and fast, easy to understand and use, they have been widely implemented and provide consistent results. Despite the various conceptual differences, the most commonly used methodologies often derive comparable results for similar data sets.

3.1 Limitations of the Faultless Methods

The basic hypotheses (H_1, H_2, H_3, H_4) of the faultless methods have been questioned in cases of local and/or temporal stress perturbations; hence, one needs to verify a method's reliability on resolving the stress regime. More specifically, the main limitations of these methods are mostly attributed to: (a) the assumption that shear stress is parallel to the slip direction (H_1), (b) the assumption of stress invariance over space (H_2) and time (H_3); and an issue that can be reckoned as a limitation (c) the ambiguity of the nodal planes, which introduces an additional uncertainty when employing DC focal mechanisms instead of fault-slip data.

3.1.1 Parallelism Between Shear Stress and Slip Vector (H_1)

Faultless models do not take into account mechanical coupling of segments across the fault surface and/or mechanical interaction between adjacent fault systems, thus empirical relations are assumed to correlate slip with stress or strain. Among others, Twiss and Unruh (1998) argue on the assumption that infinitesimal strain is parallel to shear stress caused by rotations of the finite strain. The latter may occur either due to local complications at the edges of fault blocks, or because of complex kinematic conditions at regional boundaries, invalidating inversions for stress (Gapais et al. 2000). Another parameter that can have a strong effect on the mechanical interactions on the rupture surface, hence producing heterogeneous slip vector distribution, is the friction coefficient, which affects the angle of internal friction and, consequently, the angle between fault plane and $\mathbf{S}_1$ to fulfill Coulomb's failure criterion. Reches et al. (1992) suggest the selection of an 'optimum' model among different solutions yielded by varying the friction coefficient that minimizes the sum of the angular difference between the computed regional stress tensor orientations and an 'ideal' stress tensor orientation for each dataset.

3.1.2 Stress Invariance Over Space and Time ($H_{2\text{-}3}$)

The main limitations of the method come mostly from the basic hypotheses about the stress invariance both in space and time. Thus, when processes that operate at different spatial and/or temporal scales are involved, the inversion may lead to erroneous results. Any variations within the space-time window of the dataset that is considered in the faultless models are treated as noise (Angelier 1979, 1984; Angelier et al. 1982; Gephart and Forsyth 1984; Michael 1984; Reches 1987) and, hence, the validity of the inversion should be in terms of the model-data misfit. A high average misfit may indicate that the stress field is highly heterogeneous and that the stress inversion's results may not be meaningful. The problem of heterogeneous stress can be addressed by dividing the focal mechanism dataset into subsets, depending on their spatiotemporal distribution. Declustering algorithms (e.g. Wiemer 2001) can also be employed, to ensure uniform sampling over time and space or remove events below a threshold magnitude, so that the FMS used for the inversion represent ruptures caused by regional stresses rather than due to local stress perturbations.

3.1.3 Ambiguity of the Nodal Planes

A main uncertainty in defining stress using fault data is that it is impossible to identify the time of slippage. Focal mechanisms are free of this drawback (Maury et al. 2013). On the other hand, a complication of using earthquake focal mechanisms for stress inversion is that the fault plane must be distinguished between the two nodal planes, as its choice influences the derived stress tensor.

Unless a principal stress axis is parallel to the "null axis", **N**, or the considered stress tensor presents axial symmetry, with *R* having an extreme value, the theoretical shear stress and the slip vector cannot be aligned to both nodal planes and one of them must be chosen to represent the fault plane (Gephart 1985).

Resolving the ambiguity between the fault and auxiliary planes becomes significant when studying small variations in the stress field, since wrong selections may lead to inconclusive results and additional information should be considered, e.g., the distribution of seismicity, surface faulting or the kinematics of the region. The most common approach is to calculate the difference between the slip vector and the resolved shear stress orientation for both nodal planes, and choose the one that will produce the smallest misfit (Gephart and Forsyth 1984; Gephart 1990; Julien and Cornet 1987; Mercier and Gailhardis-Carey 1989). Lund and Slunga (1999) suggest the usage of an instability method based on the Mohr-Coulomb failure criterion, but only for nodal plane selection and not to be used as a misfit criterion. Others (Michael 1987; Hardebeck and Michael 2006) propose to randomly select one nodal plane when a sufficient number of focal mechanisms are available. Angelier (2002) searches for the stress tensor that maximizes the so called Shear Stress Slip Component (SSSC), τ_S, defined as the orthogonal projection of the shear stress $\boldsymbol{\tau}$ on the direction of the slip vector $\mathbf{d}$:

$$\tau_s = \boldsymbol{\tau} \cdot \mathbf{d} = \left(T\hat{\mathbf{n}}\right) \cdot \mathbf{d} \tag{12}$$

where $\hat{\mathbf{n}}$ is the unitary normal vector, T is the reduced stress tensor and $T\,\hat{\mathbf{n}}$ produces $\boldsymbol{\tau}$, the shear stress on the selected nodal plane, coplanar with the slip vector $\mathbf{d}$. The method uses grid search to find the maximum value of τ_s for k focal mechanisms:

$$S = \sum_{i=1}^{k} \tau_{s_i} \tag{13}$$

The maximum value S implies for parallelism between shear stress and slip vector (H_1), while, in addition, the magnitude of the shear stress is taken into account. Given that $\widehat{\mathbf{n}_1}.\mathbf{d}_1 = \widehat{\mathbf{n}_2}.\mathbf{d}_2 = 0$, by its definition, the SSSC criterion is independent of the choice between nodal planes. However, this method requires an empirical threshold on the number of minimum acceptable data to be defined (Angelier et al. 2004).

3.2 *Data-Model Misfit*

Validation of the acquired model is achieved through the data-model misfit. This is derived by solving the forward problem, i.e. by determining the difference between the observed slip directions and the predicted maximum shear stress directions of the model (e.g. Carey and Brunier 1974; Angelier 1984, 1984; Michael 1984, 1987). The magnitude of misfit reflects both the observational error and the degree of heterogeneity in stress orientations as determined by the focal mechanisms that have been used as input data.

Michael (1987) computes the model confidence regions by a bootstrap technique. The data set is resampled hundreds or thousands of times to simulate repeated samples of the population and each subset is inverted for the stress tensor. For a certain confidence level, a percentage of the inversion results which are closer to the preferred solution define the confidence region (Hardebeck and Hauksson 2001). Julien and

Cornet (1987) use an iterative numerical stress inversion procedure by employing the fixed point method. Hardebeck and Michael (2006) suggest a damped grid-search inversion method (SATSI), where damping is used to decrease data singularities which are treated as "random noise". Gephart and Forsyth (1984), and Gephart (1990) use a grid search, in terms of searching the best fit through a set of models, to find the best stress tensor that minimizes the misfit between model and data, suggesting equal uncertainty for fault plane orientations and slip directions. The total misfit for a given stress state is found by applying the L1- rather than the L2-norm (least absolute deviations instead of least squared ones), to avoid problems due to inconsistent or erratic data which may cause errors that do not follow a Gaussian distribution (Gephart and Forsyth 1984). With respect to Michael's (1987) method, the grid search approach yields a noticeable difference in the confidence regions only for small data sets (Hardebeck and Hauksson 2001).

Hardebeck and Hauksson (2001) by studying the stress field in southern California, using the methods of Gephart and Forsyth (1984) and Michael (1984, 1987), propose that the former provides more accurate estimates of stress orientation, especially for high-quality data sets, but the confidence regions are generally large. The method of Michael (1984, 1987) is more accurate for noisy data sets and provides a more appropriate estimate of uncertainty. Kassaras et al. (2016a), in a study of the western part of the Hellenic Arc, arrived at a similar conclusion, suggesting that the method of Michael (1984, 1987) is more suitable for investigating spatially heterogeneous regimes. Figure 7 illustrates the results from the application of the two methods in two focal mechanism datasets in Greece (Kassaras et al. 2016a) using the ZMAP software (Wiemer 2001).

The abovementioned approaches assume that all faults in the studied rock volume are activated by a uniform stress state (H_2). If that is not the case, the stress distribution is called heterogeneous. One way to handle heterogeneity is to divide the area's volume into sub-volumes and invert for stress separately for each sub-volume. However, even though this approach is frequently effective, it is characterized as subjective.

Several numerical techniques have been proposed to cope with separating stresses from spatially or temporarily varying states of stress (e.g. Yamaji 2000). Separating stresses is particularly useful in complex regimes, involving different tectonic states, which likely enhance the activation of both new and old fault systems that produce vague focal mechanisms patterns (e.g. Kassaras et al. 2014a). Moreover, the issue favorably meets the needs in the oil industry, which requires the discrimination of tectonic phases for forecasting the orientation of fractures induced by hydraulic fracturing towards ensuring drilling and borehole stability (Lejri 2015).

Most known computer programs available for the faultless method are TectonicVB (Ortner et al. 2002), FaultKin (Allmendinger et al. 2012), Faille (Etchecopar et al. 1981), FSA (Célérier 1988), Tensor (Delvaux and Sperner 2003), Stress (Michael 1984), SoftStructure (Reches 1987), Tectonics FP (Ortner et al. 2002), Tector XXI (Angelier 2002), SATSI (Hardebeck and Michael 2006), ZMAP (Wiemer 2001), MSATSI (Martínez-Garzón et al. 2014), MIM (Yamaji 2000), STRESSINVERSE (Vavryčuk 2014).

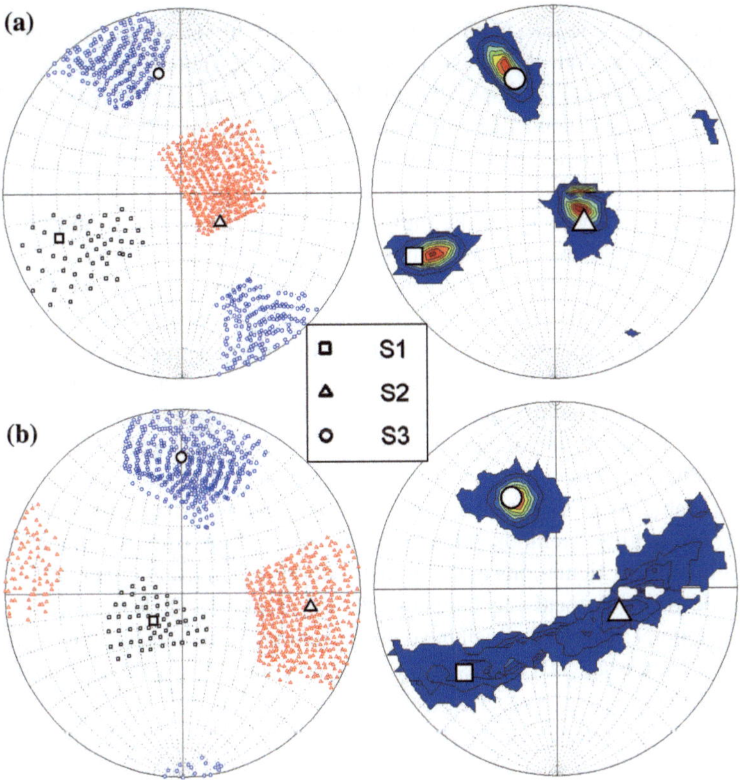

Fig. 7 Application of the methods of Gephart and Forsyth (1984) (Left) and Michael (1984) (Right) for two datasets (A, B) of focal mechanisms in W. Greece illustrating (A) similar results for both methods but larger confidence (colored regions) of the former and (B) better performance (compliant with established tectonics, Kassaras et al. 2016a) of the latter method in case heterogeneous data are involved

3.3 Geomechanical Models

The hypotheses H_1–H_3, which most of stress inversion schemes rely upon (i.e. the stress field is spatially homogeneous, temporally invariable and the direction of slip coincides with the direction of the resolved maximum shear stress), neglect the mechanical properties of the deformed region that may lead into heterogeneous slip distribution, and hence invalidate inversions for regional/remote stress (e.g. Gapais et al. 2000). Moreover, the ambiguity between the two nodal planes of the FMS introduces noise in the data when the wrong plane is selected as the fault, leading to erroneous slip vectors and interfering with the first basic assumption (H_1) (e.g. Molnar 1983; Jackson and McKenzie 1988). Another disadvantage of the abovementioned stress inversion methods concerns the data types that can be used, which are limited to fault-slip vectors and focal mechanisms.

More recently, "Geomechanical" stress inversion models have been proposed, which take into account the physical properties of the rupture surface, instead of assuming empirical relationships between slip and stress or strain (Kaven et al. 2011). The method requires knowledge of the fault geometry and information on at least one slip vector component along portions of the known fault surface(s). The concepts of fracture mechanics, such as the effects of elastic medium properties (Lamé constants), the friction and morphology of the fault surfaces and the effects of other proximal segments of the fault are considered in this method.

Both analytical (e.g. Pollard and Segall 1987; Bürgmann et al. 1994; Martel and Shacat 2006) and numerical (e.g. Segall and Pollard 1980; Maerten et al. 2005) methods have been proposed to solve the three-dimensional problem (multiple three-dimensional faults that interact mechanically) and acquire the slip distribution at every element on the fault. When displacement data are available (GPS, InSAR) the method also allows for the determination of regional stress magnitudes.

Geomechanical inversions for stress and slip distributions provide results which are comparable to principal stress axes orientations and stress magnitude ratios obtained by faultless methods (Maerten et al. 2010). Moreover, the method yields meaningful stress orientations and slip distributions when using either surface fault slickenside data, or focal mechanisms (Kaven et al. 2011). Some of the better known computer programs which have implemented the geomechanical method are Coulomb (Toda et al. 2011), based on Okada's code (Okada 1992), and Poly3D (Thomas 1993).

4 Application of Stress-Inversion on a Massive Dataset for Greece

In this section we exploit a massive dataset of individual focal mechanisms for Greece, towards investigating robustness and validity of different stress inversion methods. More specifically, our purpose is to define the degree of discrepancy/goodness of fit of each applied methodology, i.e. their consistency relative to the used FMS data and, also, in terms of interpretation, by comparing the results to quantitative strain parameters obtained by GPS measurements. Also, conclusions will be drawn on the data configuration required by each of the examined methods to reliably resolve the stress field for a given spatial scale.

The focal mechanisms dataset used is the one compiled by Kassaras et al. (2013, 2016a), which has been updated with new FMS for the purposes of the current work. The database is the outcome of thorough investigation in the literature including research papers, theses and catalogues provided by international seismological institutions. It includes 3590 individual focal mechanisms, spanning from 1861 until the end of 2016. Few focal mechanisms prior to 1962 have been derived from macroseismic observations (Papazachos and Papazachou 2003). Thereafter, the dataset follows the worldwide advances in instrumentation and computational methods. In brief, for the period between 1962 and 1980, the era of analogue records, FMS have been determined from P-wave first motion polarities. After 1980, the beginning of the digital era, focal mechanisms of moderate and strong earthquakes were mainly obtained by moment tensor inversion using teleseismic and regional waveform modeling. A significant number of focal mechanisms in the dataset concern P-wave first motion polarities solutions for earthquakes of small magnitude ($M < 4$), recorded by temporary networks installed for the purposes of research projects (e.g. Papadimitriou et al. 1994; Hatzfeld et al. 1995; Rigo et al. 1996; Haslinger et al. 1999) or monitoring aftershock sequences (e.g. Lyon-Caen et al. 1988; Makropoulos et al. 1996; Kassaras et al. 2014a, b; Kapetanidis et al. 2015).

The dominant tectonic processes in the region (Fig. 8) include the subduction of the African lithosphere beneath the Aegean plate occurring along the Hellenic arc, the extrusion of the North Anatolian Fault (NAF) and the collision of the system with the Apulian plate and the Eurasian margin, with most of strain accommodated by the Cephalonia Transform Fault Zone (CTFZ) (Laigle et al 2002). Figure 9a presents the spatial distribution of selected focal mechanisms from the comprehensive catalogue, the vast majority of which samples the crust with an average thickness of ~35 km (Kassaras et al. 2009), while sparse events are located beneath.

By evaluating the characteristics of the focal mechanism distribution (Fig. 9) and principal stresses orientations (Figs. 10 and 11), one can ascertain that the data exhibit (a) variability, with certain areas presenting more than one type of faulting, (b) spatial sampling inhomogeneity, with some areas having more data than others, (c) significant spatial heterogeneity, implying for processes that operate at distinctly different spatial scales, with focal mechanisms of small events varying more than those of the larger ones. The inversion is expected to correctly reconstruct the stress tensor as long as the magnitude of its uniform part is larger than the one of its variable part (Michael 1991). In order for the hypotheses $H_{1\text{-}4}$ to be plausible, the data will have to be divided into subsets, mainly according to criteria concerning their spatial distribution.

Because of the sparsity of the subcrustal earthquakes, we chose to restrict the inversion within the crust by removing deep events. After removing 342 events located deeper than 35 km, the catalogue employed in the inversions is composed of 3248 individual focal mechanisms. For the presented application, the study area was divided in sub-regions of $1° \times 1°$ dimensions (Fig. 12), assuming that significant variations of the regional stress occur at a larger scale than the one chosen. This is a rather crude approximation; a better division would require the employment of seismic zones, as defined recently by state-of-the-art methodologies (e.g. Stucchi et al. 2013), but this is out of the scope of the present study, although the conclusions drawn from the analyses that follow will be useful towards this future task. Thereafter, the $1° \times 1°$ sub-regions approximation is considered feasible for the educational purposes of the current application, on the one hand, and the restrictions of our dataset to sufficiently sample the entire area of study, on the other.

The data were analyzed by implementing two numerical inversion schemes and one graphical analysis method. The graphical scheme is based on P- and T-axes analysis as it was implemented in the FaultKin algorithm of Allmendinger et al.

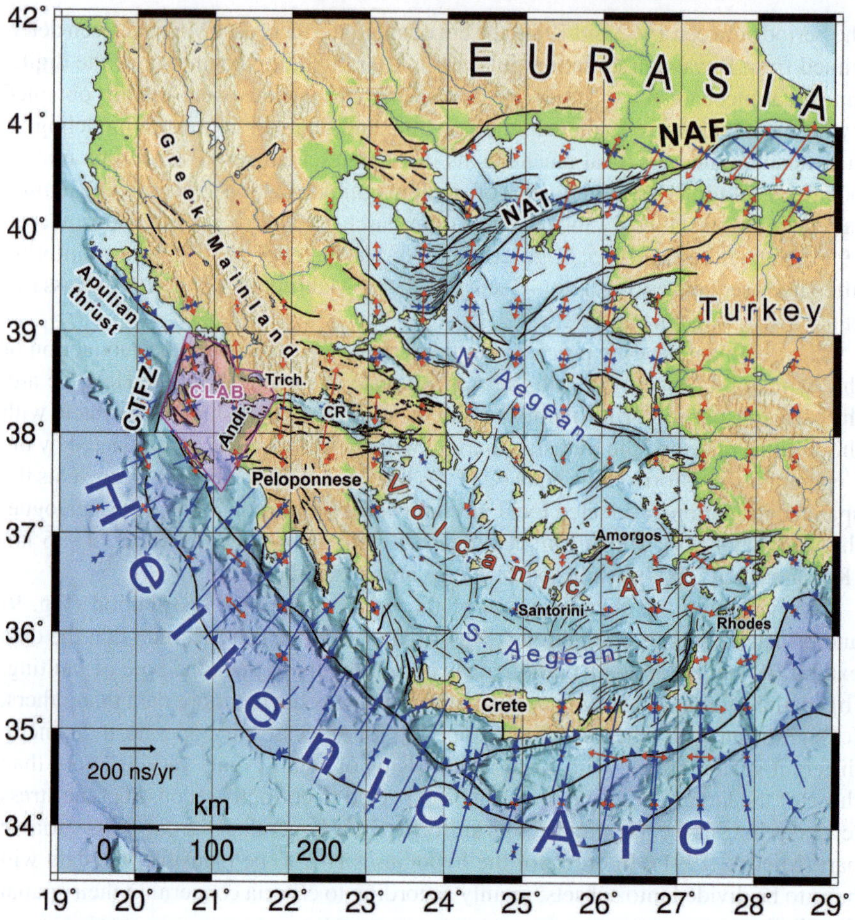

Fig. 8 Map of Greece showing the most prominent tectonic structures (Karakonstantis and Papadimitriou 2010; Sakellariou and Tsampouraki-Kraounaki 2016) and the direction of the compressional (e_{H1}-blue arrows) and extensional (e_{H3}-red arrows) principal components of horizontal strain (Kreemer et al. 2014), measured in units of nanostrain (10^{-9}) per year (ns/yr). Dextral strike-slip zones include: CTFZ: Cephalonia Transform Fault Zone, Andr.: Andravida fault, NAT: North Aegean Trough and NAF: North Anatolian Fault. Other labeled zones: CR: Corinth Rift and Trich.: Trichonis Lake while CLAB (purple polygon) marks the Cephalonia-Lefkada-Aitoloakarnania Block

(2012). The first numerical scheme is based on Michael's method (Michael 1984), using the SATSI algorithm (Hardebeck and Michael 2006) and the second one is the Multiple Inverse Method (MIM) of Yamaji (2000). Although for several regions in Greece the faulting geometry is well known, the input fault plane for the faultless methods was randomly chosen; hence a "noisy" dataset was employed aiming at investigating the resolving power of the applied methods. For the MIM, both nodal

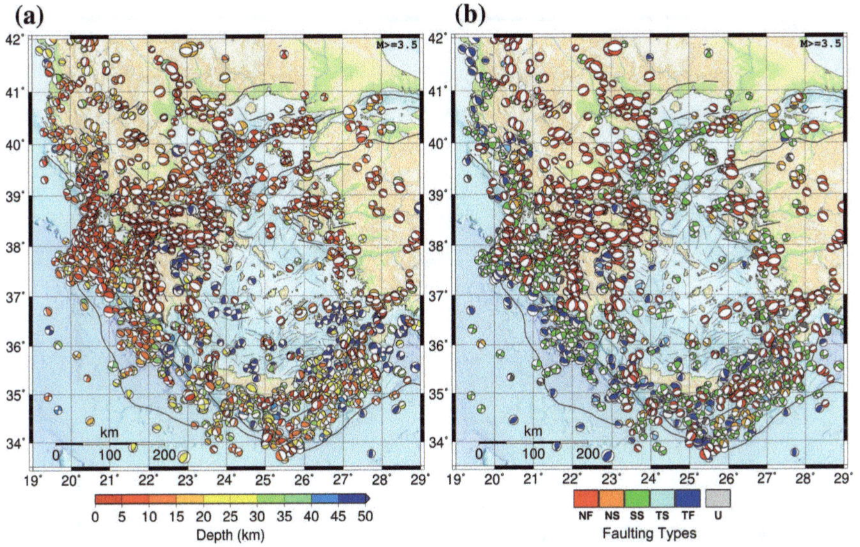

Fig. 9 Illustration of selected ($M \geq 3.5$) focal mechanism data contained in the comprehensive catalogue for Greece, **a** sorted by focal depth and **b** sorted by faulting type, according to the classification rules of Zoback (1992)

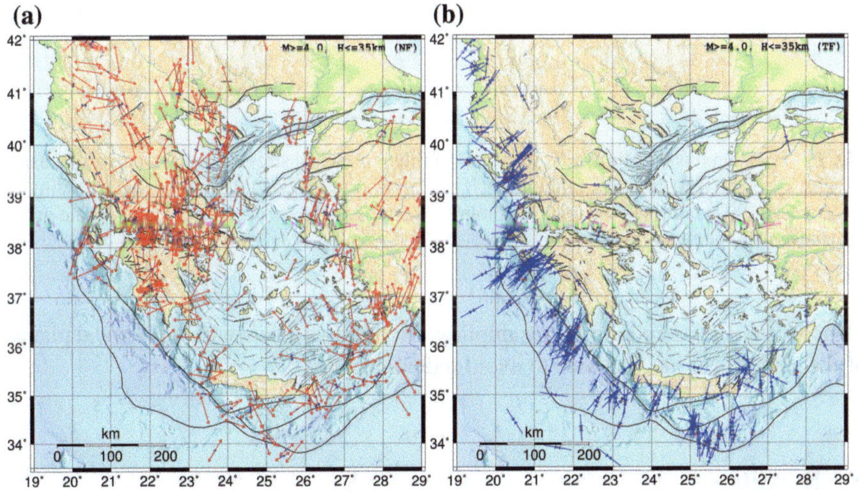

Fig. 10 Horizontal orientation of $\mathbf{S}_1$ (red) and $\mathbf{S}_3$ (blue) principal stress axes for focal mechanisms of selected ($M \geq 4.0$) crustal earthquakes ($H \leq 35$ km) in Greece, **a** for events classified as NF, **b** for events classified as TF, according to the Zoback (1992) faulting type criteria (Table 1). Vector length is proportional to the cosine of the respective axis' plunge

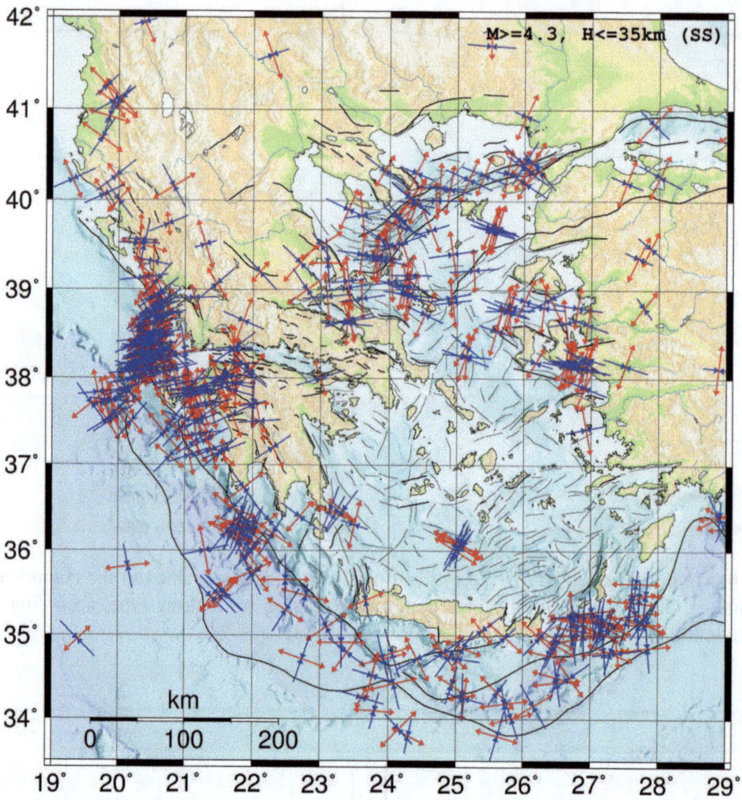

Fig. 11 Horizontal orientation of $\mathbf{S}_1$ (red) and $\mathbf{S}_3$ (blue) principal stress axes for focal mechanisms of selected ($M \geq 4.3$) crustal earthquakes ($H \leq 35$ km) in Greece classified as SS, according to the Zoback (1992) faulting type criteria (Table 1). Vector length is proportional to the cosine of the respective axis' plunge

planes were considered during resampling, excluding subsets with both nodal planes of the same focal mechanisms to avoid redundancy.

5 Results

5.1 The "FaultKin" Analysis

The methods that were described in Sect. 3 are based on the calculation of the regional stress field for given groups of observed slip vectors and fault plane orientations, which in the case of focal mechanisms are uncertain due to the nodal planes ambiguity. Besides the techniques developed for selecting the most relevant

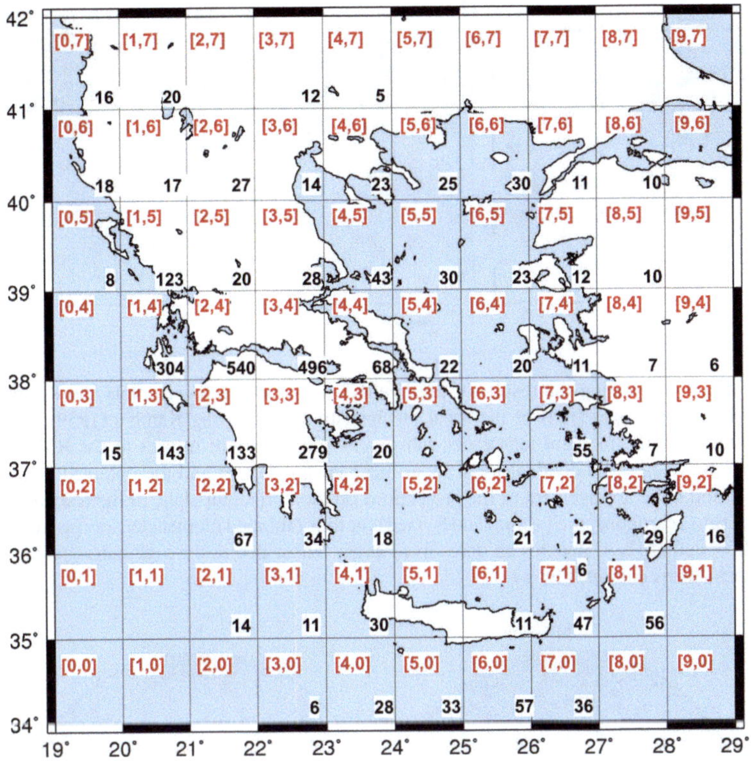

Fig. 12 The grid configuration applied for the inversions. Numbering indicates coordinates (red, upper-left) and available focal mechanisms for each sub-area (black, lower-right)

information by minimizing the data-model misfit, the degree up to which focal mechanism data are capable to reproduce the regional strain field is rather questionable. This problem arises with associating P- and T-axes either with the principal strains (Marrett and Allmendinger 1990; Twiss and Unruh 1998), or with principal stresses (e.g. Angelier and Mechler 1977; Gephart and Forsyth 1984). The concept behind the implementation of "FaultKin" analysis in our dataset is the direct association of P- and T-axes with fault geometry and sense of slip (Andersonian faulting type; Anderson 1905). Thus, the demand of defining the fault out of the two nodal planes, which is prerequisite in other methods, is eliminated.

"FaultKin", a computer program for the analysis of fault slip data (Marret and Allmendinger, 1990; Allmendinger et al. 2012), is classified among the "graphical" stress inversion methods, a variety of which has been proposed for fault slip analysis. Among these, the "P- and T-dihedra" method (Angelier and Mechler 1977) and the kinematic P- and T-axes, having been proven the most robust, seldom differing substantially from numerical analyses (Allmendinger 1989), are implemented in the Faultkin code. FaultKin uses Kamb's method (1959) for presenting orientation data, P- and T-axes are displayed as contoured plots that allow statistical

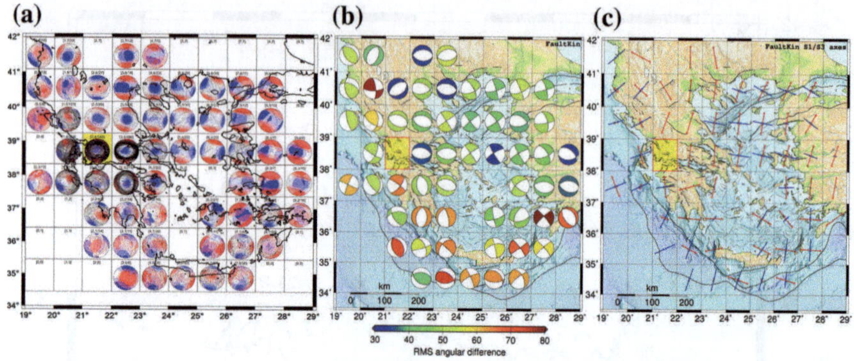

Fig. 13 Maps illustrating the results of the FaultKin inversion: **a** Stereo plots show contouring of **P**- (blue) and **T**-axes (red) of the focal mechanism subsets using Kamb's (1959) method, **b** unweighted summed moment tensor per sub-region (Eq. 14) with respect to the RMS angular difference between the individual input focal mechanisms dataset and the obtained solution for each cell, and **c** the arrangement of the projections on the horizontal plane of the resolved best-fit principal stress components $\mathbf{S}_1$ (blue) and $\mathbf{S}_3$ (red) for the combined moment tensors per sub-region. The yellow cell at [2, 4] denotes an unresolved stress tensor due to the high heterogeneity of the focal mechanisms dataset (Fig. 14)

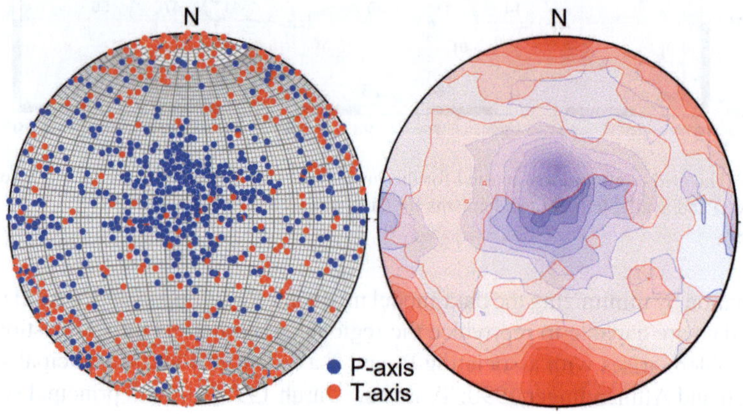

Fig. 14 Stereoplots produced by FaultKin, indicating the heterogeneous distribution of the input focal mechanisms P- and T-axes in the NE part of the CLAB (grid [2, 4] in Fig. 12; see also Figs. 8 and 13a): **a** scatter plot, **b** contour plot (Kamb 1959)

inferences to be drawn directly from the stereonet diagrams (Figs. 13a and 14b). An unweighted moment tensor summation scheme is used by applying the Bingham distribution (Binham, 1964; 1974), which can successfully describe circular, elliptical or bar/axes distributions and has been often used for delineating structural orientations (e.g. Yamaji 2016). The unweighted moment tensor is composed of the sums of the products and the squares of the direction cosines of the individual P- and T-axes (Allmendinger 1989):

$$M = \begin{vmatrix} \sum(CN(P))^2 - (CN(T))^2 & \sum CN(P) * CE(P) - CN(T) * CE(T) & \sum CN(P) * CD(P) - CN(T) * CD(T) \\ \sum CE(P) * CN(P) - CE(T) * CN(T) & \sum(CE(P))^2 - (CE(T))^2 & \sum CE(P) * CD(P) - CE(T) * CD(T) \\ \sum CD(P) * CN(P) - CD(T) * CN(T) & \sum CD(P) * CE(P) - CD(T) * CE(T) & \sum(CD(P))^2 - (CD(T))^2 \end{vmatrix} \quad (14)$$

where CN, CE, CD denote the North, East and Down direction cosines (NED) of the P- and T-axes, respectively. The eigenvalues and eigenvectors of M give the relative magnitudes and orientations of the three kinematic axes $\mathbf{S}_1$, $\mathbf{S}_2$, $\mathbf{S}_3$, respectively. A detailed description of the theoretical background of the method can be found in Marrett and Allmendinger (1990), and Allmendinger et al. (2012).

A user-friendly interface is provided by the "FaultKin" software, while data and results are displayed as graphical plots. The input from individual focal mechanisms includes strike, dip and rake of one nodal plane, following the Aki and Richards (2002) convention. In this format, with the slip vector representing the relative displacement of the hanging wall with respect to the footwall, the rake is measured counterclockwise from the strike direction to the slip vector, so that faults with a thrust component have a positive rake and those with a normal component have a negative rake. Equal uncertainty is assumed for fault plane orientations and slip directions. The output is the orientation (trend) and dipping (plunge) angles of the three principal axes of the stress tensor, the faulting parameters (strike, dip, rake). Details on the program's usage and its capabilities can be found in Allmendinger (2016).

Application of the Method

For each dataset corresponding to a sub-region of $1° \times 1°$ dimensions, the kinematic P- and T-axes analysis was applied. The results of the FaultKin procedure are presented in Fig. 13. Contoured plots in Fig. 13a indicate large heterogeneity for the majority of sub-regions, especially in the NE area of the CLAB (Fig. 8), where the reduced stress tensor could not be resolved by the method (rectangle with yellow fill in Fig. 13a, cell [2, 4] in Fig. 12). The heterogeneity of strain release in this area is evidenced by the scatter and Kamb plots of Fig. 14. The results are generally in close accordance to the expected stresses from the breakdown of data to different faulting types (Figs. 10 and 11).

The misfit between the optimal stress tensor, as derived by the inversion, and the individual FMS input data is measured in terms of the RMS angular difference between the optimal $\mathbf{S}_1$, $\mathbf{S}_2$, $\mathbf{S}_3$ and the respective orientations of the FMS eigenvectors **P**, **N**, **T**. The angular difference, $\delta\theta$, between two tensors with normalized eigenvectors represented by matrices **A** and **B**, respectively, is defined as the minimum angle of rotation, also known as "Kagan angle" (Kagan 1991; Tape and Tape 2012), that must be applied relative to a special axis so that the rotated **A** becomes equal to **B** (Kapetanidis 2017). The optimal 3D rotation matrix, **R**, is calculated using a method called "Kabsch algorithm", which provides the transformation that minimizes the RMS deviation between two sets of vectors in their exact order (Kabsch 1978). The rotation matrix, **R**, can then be converted to an axis/angle representation by a procedure known in the literature as the inverse of Rodrigues' rotation formula (e.g. Shuster 1993). The procedure has to be repeated for all combinations with each of the eigenvectors pointing to either direction, e.g. $\mathbf{S}_1$ and $-\mathbf{S}_1$, in order to find the minimum angle $\delta\theta$ between **A** and **B**.

5.2 *The SATSI Stress Inversion*

Faultless methods rely on the assumption of stress uniformity within space and time (H_2–H_3). Accordingly, inversions are conducted within volumes of a small enough size for the assumption of stress uniformity to remain valid (Plateaux et al. 2010). The fit of a stress tensor is performed independently to the focal mechanisms of each sub-area, regardless of noisy or non-uniquely fitting data that does not completely constrain the stress tensor (Hardebeck and Michael 2006).

SATSI (Spatial And Temporal Stress Inversion, Hardebeck and Michael 2006) is a modified version of Michael's code that inverts focal mechanism data for a spatially and/or temporally varying stress field. The stress field in SATSI is resolved by dividing a region into sub-regions and inverting the focal mechanisms simultaneously for all sub-regions, searching for the optimal stress within each cell while minimizing the stress differences between neighboring regions (Hardebeck and Michael 2006). Hence, apart from the noisy data issue, it also smoothens out anomalous stress perturbations that may be unimportant for the regional stress state configuration.

A damped grid-search is used in the method to optimize the fit between the observed and the calculated stress parameters by minimizing the weighted sum of two values: the data misfit and the model length. The model length is represented by the L2 norm of the vector containing the differences between each stress tensor component for each pair of adjacent grid points. The damped inversion helps in removing stress artifacts exhibited by isolated data singularities and resolving sharper true stress rotations compared to a simple smoothed model or a moving-window inversion (Hardebeck and Michael 2006).

Since appropriate damping should produce a regional stress field model that contains only those variations of the stress field that are strongly related to the data (Hardebeck and Michael 2006), in the first stage of the procedure, the optimum damping parameter, e, is searched. This parameter controls the amplitude of the solutions' variations between neighboring sub-regions. It is usually chosen from a range of damping values using the trade-off curve between the data misfit and the model length (e.g. Eberhart-Phillips 1986) to produce a balance between excessive misfit and unnecessary model length (Hardebeck and Michael 2006). In the second step, the stress inversion is performed and the best stress tensors are found for each grid point. Finally, the input dataset is resampled using a bootstrap procedure and the confidence intervals of the stress tensor axes are computed. The uncertainty for each grid node is determined as the range of stress orientations for this set of models (Hardebeck and Michael 2006).

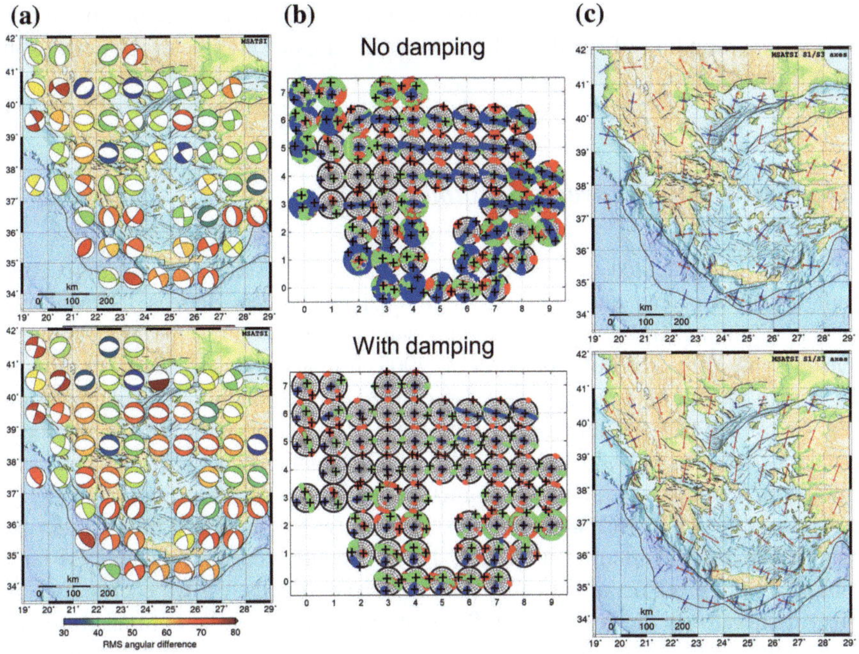

Fig. 15 Maps illustrating the results of the MSATSI inversion for 57 sub-regions. (Top) without damping; (Bottom) with auto damping ($e=2$); **a** the inverted focal mechanisms with respect to the RMS angular difference between the focal mechanism dataset and the obtained solution for each grid point. **b** Stereo plots showing the 95% confidence regions of 2000 bootstrap resamples for the $\mathbf{S}_1$ (blue), $\mathbf{S}_2$ (green) and $\mathbf{S}_3$ (red) axes, and **c** the arrangement of the projections on the horizontal plane of the resolved best-fit principal stress components $\mathbf{S}_1$ (blue) and $\mathbf{S}_3$ (red)

Application of the Method

We applied the SATSI method using the MSATSI software (Martínez-Garzón et al. 2014). MSATSI facilitates an automated procedure and MATLAB based visualization tools to represent the inversion results using a variety of plots. The inversion of the individual focal mechanisms for the subset of 3248 shallow events was performed over the $1° \times 1°$ grid of Fig. 12, using at least 5 focal mechanisms for each rectangle, by adopting solutions lying within the 95% confidence interval of 2000 resamples as determined by the bootstrapping method (Hardebeck and Michael 2006). For comparison, the SATSI inversion was performed with and without damping (Fig. 15). In the first case, the damping parameter was determined at $e=2$ by the aforementioned 'knee' of the trade-off curve between the data misfit and the model length.

The RMS angular difference between the input FMS datasets and the respective stress tensor's orientation, as obtained for each sub-region (Fig. 15a), appears relatively high, likely related to the ambiguity of the input nodal planes and the heterogeneity of the data. Considering the un-damped/damped solutions, the former exhibits smaller RMS angular difference values and broader confidence regions for the geometry of the principal stress axes (Fig. 15b), typically observed in sub-regions with sparse sampling of focal mechanisms or in cases of high heterogeneity, as in the NE part of the CLAB (Fig. 14).

5.3 *The Multiple Inverse Method (MIM)*

A broadly implemented method aiming to separate stresses from heterogeneous fault slickenside data and later on from focal mechanism data is the Multiple Inverse Method (MIM) of Yamaji (2000), which employs a resampling technique (Otsubo and Yamaji 2006; Otsubo et al. 2008; Yamaji and Sato 2011). It utilizes the Hough transform, an image processing method based on hierarchical clustering (e.g. Leavers 1992). In this scheme, the fitness of a stress state to a fault-slip/focal mechanism datum is evaluated using the angular misfit, d, between the observed and theoretical slip directions. A symmetric function, $\rho(d)$, of arbitrary form (e.g. $\rho(d) = (\cos|d| + 1)/2$), with a unique maximum at $d = 0$, is used as a fitness criterion (Choi et al. 1996; Yamaji et al. 2006). The procedure samples repeatedly k-fault subsets from a given set of data and calculates the optimal stress for each subset by applying iteratively the Angelier's (1979) inversion scheme. The optimal stress for each sample subset is determined by maximizing a function called "M-estimator" (Yamaji et al. 2006; similar to Eq. 14):

$$M(\sigma) = \frac{1}{\mathrm{N}} \sum_{k=1}^{N} \rho(d(k, \sigma)) \qquad (15)$$

where N is the number of fault-slip (or FMS nodal planes assumed as fault planes) in the subset and $d(k, \sigma)$ is the misfit angle of the kth fault datum for the respective trial tensor, σ. $M(\sigma)$ presents a peak when σ is compatible with one or more subsets. Namely, the match of each datum to various reduced tensors is represented by a spike in $M(\sigma)$; thereafter, significant stresses are separable according to the prominent peaks of $M(\sigma)$. Each kth point is projected on the surface of a unit hypersphere in a five-dimensional Euclidean space, representing five parameters that define the reduced stress tensor (orientation of $\mathbf{S}_1$, $\mathbf{S}_2$, $\mathbf{S}_3$, stress-ratio R or Φ) and the fault-slip generated through the application of hypothesis H_1 (Otsubo et al. 2006). Individual observations are plotted as tadpole symbols (5-elements reduced stress tensors) on the 5D hypersphere (i.e. panels b and c in Fig. 16). Distinct stress states compatible with a dataset are identified as groups of reduced stress tensors on the hypersphere (similar characteristics) by k-means clustering. The latter method needs a well-defined distance parameter to quantify the uncertainty of the resolved stresses. The required metric is provided by the great-circle distance between points on the hypersphere, termed "angular stress distance", Θ; the optimum partitioning into clusters (stress states) is determined by minimizing the sum of squared distances (stress angles) between each point and the respective cluster's center, also referred to as "spread"

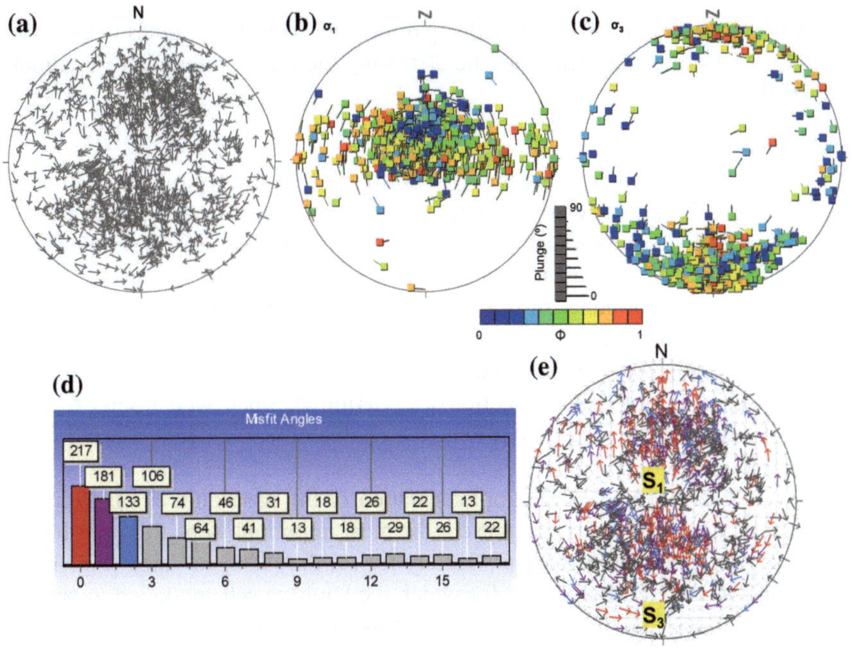

Fig. 16 Stereograms presenting the projection of: **a** tangent-lineation diagram of the input data without filtering, **b**, **c** optimized diagrams of tadpole symbols for (**b**) σ_1 ($\mathbf{S}_1$) and (**c**) σ_3 ($\mathbf{S}_3$) after declustering (Screening #1) and removal of erroneous data by using the EF option (Screening #2). The $\mathbf{S}_1$ tadpole tails indicate the azimuth and plunge of the corresponding $\mathbf{S}_3$ vector and vice versa for the $\mathbf{S}_3$ tail. **d** Histogram presenting the data-model misfit angles in terms of the slip vector orientations. The horizontal axis is in tens of degrees. The vertical axis denotes the multitude of reduced stress tensor pairs with the corresponding misfit value. **e** Diagram showing tangential-lineation of the input data, color-coded with respect to the misfit angles shown in panel (**d**). Thin gray arrows denote fault movements predicted by the Wallace-Bott hypothesis for the resolved stress state

(Otsubo et al. 2008). Significant stresses in the FMS dataset can be identified graphically as concentrations of tadpoles with similar colour, tail direction and length (see Fig. 16).

The MIM considers both the nodal planes of DC focal mechanisms during the resampling procedure to account for the fault/auxiliary plane ambiguity. The resolution can be improved by discarding subsets which contain both nodal planes of the same focal mechanism (the data need to be independent) or those not compatible with any single state of stress. The latter is considered when the theoretical slip direction makes a sufficiently large angle with the observed one; a subset is left out if no tensor is compatible with any of its members (Otsubo and Yamaji 2006; Otsubo et al. 2008).

The identification of significant stresses and the grouping of faults follow the classical scheme of Wallace-Bott, that is to minimize the sum of misfit angles

(d_T) between observed and calculated slip directions. Threshold values in the range $15° \leq d_T \leq 30°$ are suggested, with the 20° being the most preferable (Yamaji and Sato 2011).

Application of MIM

This section presents the application of MIM to selected sub-regions (1° × 1° cells; Fig. 12) that exhibit a particularly high degree of heterogeneity. An integer in the range $2 \leq k \leq 8$, called "data combination number" (Yamaji and Sato 2011), is allowed in the inversion, whilst for focal mechanism data $k = 5$ is suggested (Yamaji et al. 2011). However, due to performance restraints of the code, a limitation exists regarding the number of iterations, which depends on the multitude N of the dataset (maximum N=96 focal mechanisms are allowed in the current 32-bits version of the code). If larger datasets are employed, a reduction of k is required. Since our datasets exceeded in several cases the allowed dimensions, the choice of k was decided accordingly between the values of 3 and 5.

A detailed description of the implementation of the procedure can be found in Yamaji et al. (2011). Hereby, we present the most important steps followed for the MIM analysis of the Greek dataset, namely for (a) the [2, 4] sub-area (Fig. 12) at the NE part of the CLAB (Fig. 8), for which the Faultkin scheme failed to provide any solution due to the high heterogeneity of the involved data (Figs. 13a and 14), and (b) the [5, 5] sub-area related with NAT (Fig. 8) for which FaultKin and un-damped SATSI solutions provided controversial results compared to the damped SATSI solution (Figs. 13 and 15).

Sub-region [2, 4]—NE part of the CLAB

This area is characterized by complex tectonics related to different types of faulting, possibly acting at different scales (Kassaras et al. 2014a). More specifically, it contains (a) NNW–SSE structures parallel to the modern Hellenic Arc (McKenzie 1972; LePichon and Angelier 1979) and following the geometry of the fold-and-thrust belt of the External Hellenides (e.g. Philippson 1998), (b) Holocene ~E-W trending normal faults (Kapetanidis 2017) developed on top of the thrust sheets of the Hellenides (Mariolakos and Papanikolaou 1987) and (c) NE-SW dextral strike-slip structures, parallel to the CTFZ, suggested to be related to the Central Hellenic Shear Zone (Papanikolaou et al. 2009) and the North Anatolian Fault Zone (Serpetsidaki et al. 2014).

Figure 16 presents the stress field resolved for each k set of data produced by continuous resampling of the focal mechanisms dataset. The diagrams of Fig. 16b, c are the result of an optimization procedure of the MIM including: (a) the removal of subsets with obsolete nodal planes that correspond to similar focal mechanisms and do not contribute to the solution (termed Screening #1 in MIM; Otsubo et al. 2008), which has a similar effect to the declustering of aftershock sequences, thus fulfilling the assumption H_4, (b) rejection of subsets with a spread that surpasses the misfit threshold (d_T = 20°) for any possible stress configuration (termed Screening #2 in MIM; Otsubo et al. 2008), and (c) a procedure that employs a user-defined "enhance factor" (EF), combined with the kth stresses standard deviation with respect to the

Table 2 The average stress state resolved for cell [2, 4]

Node	Lat (°N)	Lon (°E)	Stress state A				
			S_1 (°)		S_3 (°)		Φ
			Trend	Plunge	Trend	Plunge	
[2, 4]	38.5	21.5	352	77	183	12	0.5

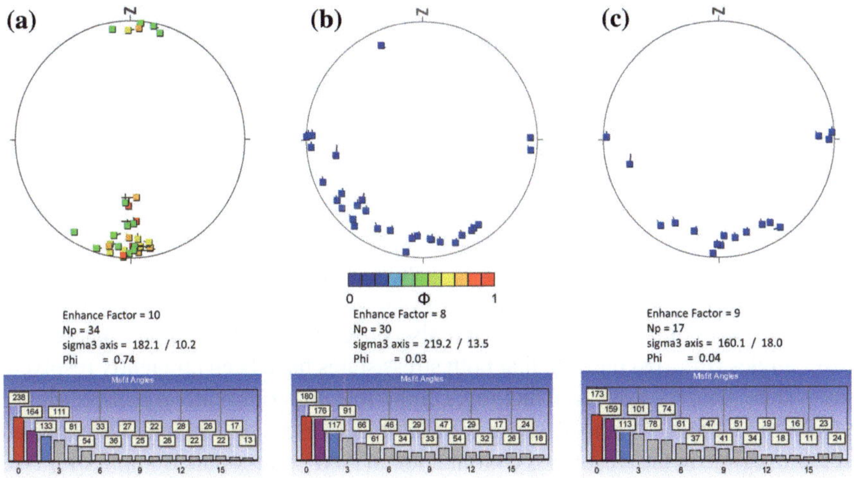

Fig. 17 (Top panels) Stereograms presenting the $\mathbf{S}_3$ stress axes distribution of k-subsets across the hypersphere. "Filtering" of k-means data points is based on the aforementioned optimization procedure with: **a** EF = 10, $\Phi = 0.74$, **b** EF = 8, $\Phi = 0.03$ and **c** EF = 9, $\Phi = 0.04$. Np is the number of tadpole symbols (reduced stress tensors) plotted on each stereogram. Bottom panels present the distribution of the respective angular misfit between the calculated and observed slip directions. The horizontal axes are in tens of degrees. The vertical axes denote the multitude of k-means solutions for each bin

global solution, to thin out erroneous solutions and enhance correct ones (Yamaji 2000). Although a complicated pattern is derived in terms of the stress shape (Φ), the filtered distribution of individual reduced stress tensors (Fig. 16b, c), is more indicative of a physical process (i.e. a distinct stress state) than the initial noisy distribution of the tangential slip on the nodal planes (Fig. 16a).

The average stress state resolved by the MIM (Table 2) implies for sub-vertical $\mathbf{S}_1$ and sub-horizontal (approximately N-S) $\mathbf{S}_3$, compatible with normal E-W faulting in the area. The obtained pattern is very similar to the one obtained by the SATSI procedure for this sub-region (Fig. 15). However, stress is highly variable as the data-model misfit angles distribution shows (Fig. 16d, e), consistently with Kassaras et al. (2012), who found predominant extension axes at different directions throughout this area. Indeed, by using different EFs and by applying selection of stresses according to Φ values, it has been made possible to distinguish between different $\mathbf{S}_3$ orientations, found to range between ~N160° and ~N220° in the region (Fig. 17).

The histograms in Fig. 17, presenting the misfit angles between the inverted and the observed slip directions, indicate that an $\mathbf{S_3}$ orientation of ~N180° explains well the majority of the observations. The ~N220° and ~N160° $\mathbf{S_3}$ arrangements, appear compatible with a smaller number of tadpole symbols, which present large dispersion across the hypersphere (blue tadpoles in Fig. 17), likely related to the large heterogeneity of the input dataset.

Sub-region [5, 5]—North Aegean
The MIM analysis was applied (for sub-area [5, 5], Fig. 12) in north Aegean, where damped SATSI yielded ~NS extension, compatible with EW normal faulting, whereas FaultKin and un-damped SATSI yielded almost horizontal $\mathbf{S_1}$ and $\mathbf{S_3}$ axes, corresponding to strike-slip style of faulting. The latter is consistent with the prominent branches of the dextral strike-slip North Anatolian Fault (NAF) in the area (Fig. 8). Figure 18 presents the distribution of *k*-FMS resampling inversion solutions as tadpole symbols, color coded according to the value of the shape ratio (Φ), while Table 3 summarizes the outcome. The results of the analysis exhibit two distinct stress clusters A and B, with different positions within the hypersphere separated by an angular stress distance $\Theta = 57°$ and also different Φ values. Small data-model misfit angles (Fig. 18) indicate that both stress states have been efficiently resolved. Stress state A infers strike-slip faulting, compatible with right-lateral motion when a NE-SW nodal plane is considered as the fault plane, in agreement with the expected tectonic regime in the westwards continuation of NAF. Stress state B implies for N-S extension on ~E-W conjugate faults, compatible with the orientation of pull-apart basins in north Aegean, i.e. extensional basins at the tips of conjugate strike-slip faults (e.g. Ganas et al. 2005). As a conclusion, the obtained results indicate an efficient performance of the MIM scheme on resolving patterns where more than one stress states are distinctly involved.

6 Discussion—Conclusions

In this work we overview state-of-the-art stress inversion methodologies using earthquake focal mechanism data. We have applied three methods based on the Wallace-Bott "faultless" approximation for determining the regional stress in Greece by employing a comprehensive focal mechanisms dataset (Kassaras et al. 2016a). The regional stress field was determined in terms of the reduced stress tensor that includes the orientations of the principal stress axes $\mathbf{S_1}$, $\mathbf{S_2}$, $\mathbf{S_3}$ and the shape ratio (*R* or Φ). The inversion schemes applied are: FaultKin (Allmendinger et al. 2012), SATSI (Hardebeck and Michael 2006), and MIM (Yamaji 2000). FaultKin belongs to the graphical methods for P- and T-axes analysis, SATSI applies the Michael's (1984) method combined with spatial and temporal smoothing and MIM (Yamaji 2000) applies Angelier's (1979) technique and hierarchical clustering (Leavers 1992).

In general, the three methods provided comparable results in most regions regarding the eigenvectors of the principal stress axes $\mathbf{S_1}$, $\mathbf{S_2}$, $\mathbf{S_3}$. However, the resolving

Table 3 The two stress states resolved for cell [5, 5] in north Aegean. Θ denotes the angular stress distance between the two stress states, A and B, on the hypersphere

Node	Lat (°N)	Lon (°E)	Stress state A					Stress state B					Θ (°)
			S_1 (°)		S_3 (°)		Φ	S_1 (°)		S_3 (°)		Φ	
			Trend	Plunge	Trend	Plunge		Trend	Plunge	Trend	Plunge		
[5, 5]	39.5	24.5	277	9	9	12	0.5	243	90	7	0	0.75	57

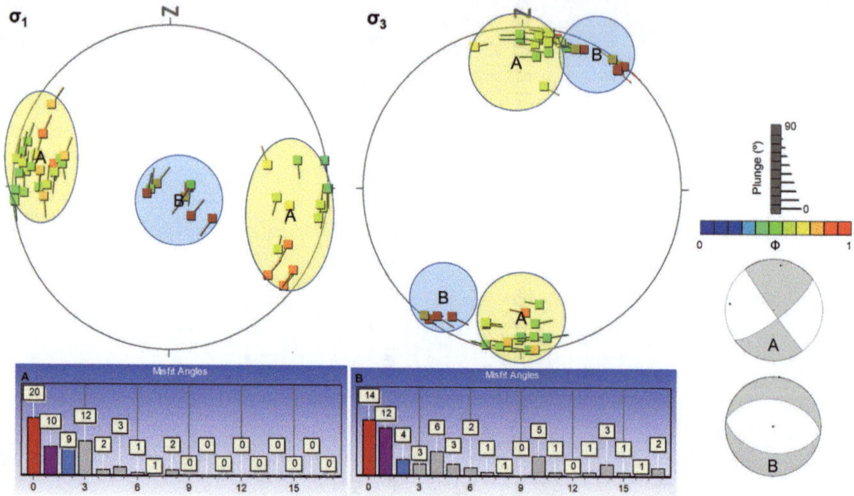

Fig. 18 Stereograms presenting the projection of $\mathbf{S}_1$ and $\mathbf{S}_3$ (with corresponding magnitudes σ_1, and σ_3). The notation is similar to Fig. 16. Colored areas indicate *k*-means clusters characterized as stress states A (yellow) and B (blue). The beachballs (bottom-right) correspond to the Andersonian (Anderson 1905) faulting types that can be produced by stress states A and B (assuming $\mathbf{P} \equiv \mathbf{S}_1$ and $\mathbf{T} \equiv \mathbf{S}_3$). Bottom panels present the data-model misfit angles for the two stress states (A, B). The horizontal axes are in tens of degrees. The vertical axes denote the multitude of *k*-means solutions for each bin

power of the different methods was found to vary. FaultKin resolved 56 out of 57 datasets corresponding to $1^\circ \times 1^\circ$ sub-regions, failing to invert cell [2, 4], a largely heterogeneous dataset at the NE part of the complex CLAB area (Fig. 8). SATSI provided results for all the sets of data. The un-damped SATSI procedure yielded similar results with those of FaultKin, but also managed to resolve cell [2, 4]. Both models exhibit quite large uncertainties for tectonically complex and sparsely sampled sub-regions. Damped SATSI yielded a consistent, smoothly varying model; however discrepancy occurs compared to the un-damped SATSI and FaultKin solutions, especially for the north Aegean (NAT). Judging by the resolved strain field (Fig. 8) and the distribution of **P** and **T** axes for individual FMS of different faulting types (Figs. 10 and 11), we suggest this discrepancy to be the effect of over-smoothing of the damped model that leads into artifacts for this region by treating real stress perturbations as "noise". In this respect, damping improves the model by smoothing only over adjacent regions where an individual stress tensor is efficiently resolved and where there is lack of strong contrast in the stresses. Otherwise, when the stress field is expected to exhibit strong variations, sub-regions should be further subdivided, likely including some degree of overlapping between neighboring cells, or more appropriate techniques for resolving heterogeneous datasets should be employed.

MIM, a technique suggested to sufficiently resolve heterogeneous stresses, was applied for the two datasets (NE part of CLAB and NAT), for which SATSI and FaultKin failed due to the aforementioned issues. Thanks to a sophisticated procedure, including redundant/erroneous data removal, hierarchical stress clustering and Φ filtering, MIM enabled the distinction of different stress states within each dataset, consistent with known active tectonics in both sub-regions.

The outcome of the FaultKin and un-damped/damped SATSI inversions shown in Figs. 13 and 15 infers for predominant extensional and strike-slip tectonics. Extensional tectonics, resolved by both inversion schemes, is widely observed across the continental part of the Greek territory. Along CTFZ and the Hellenic Arc mainly strike-slip tectonics with a reverse component is observed. Reverse faulting has been resolved only in NW Greece and offshore west/south Peloponnese and Crete, with the principal axis of compression being perpendicular to the Hellenic Arc (clearly observed from the axes of individual focal mechanisms in Fig. 10b). These results are in good agreement with Kassaras et al. (2016a) and Konstantinou et al. (2017), demonstrating that the applied methods succeeded, in general, to provide a meaningful stress state in this area.

The region of CLAB is well known for complex deformation (e.g. Chousianitis et al. 2015; Perouse et al. 2016), as it is situated between the Apulia-NW Greece continental collision and the Hellenic oceanic subduction and is possibly related to a large scale shear zone which bridges NAT with CTFZ (e.g. Royden and Papanikolaou 2011). Despite the data heterogeneity, SATSI inversions resolved a ~N-S extensional regime, compatible with regional kinematics (e.g. Chousianitis et al. 2015). The application of MIM in this area highlights distinctly variable orientations of the $\mathbf{S_3}$ stress axes between ~N160°–220°. Similar results were obtained by Kassaras et al. (2012). This is likely the effect of conjugate planes' interaction producing all types of faulting (Kassaras et al. 2014a).

To the east, the FaultKin and the un-damped SATSI inversions converge into strike-slip tectonics, related to dextral motion in north Aegean (NAT) and sinistral motion in the south Aegean and in the troughs south of Crete and Rhodes islands (Sakellariou and Tsampouraki-Kraounaki 2016). In the area of NAT though, FaultKin resolved prevailing maximum horizontal compression in ~E-W direction and N-S extension (producing SW-NE trending, dextral strike-slip faulting), while un-damped SATSI indicates a less pronounced horizontal E-W compression, with dominant N-S extension in some cells. For the same region, the damped SATSI inversions yielded pure N-S extension, which is considered as an artifact due to lateral over-smoothing of the model. This effect is likely connected with the heterogeneity of the input data, as measured by the RMS angular differences in Figs. 13 and 15. Nonetheless, the regime in the area of NAT is not straightforward and further analysis is required, since the application of MIM highlights two distinct stress states capable for producing strike-slip and normal faulting, attributed to the dextral motion of NAT and tensile motion between conjugate strike-slip faults (e.g. Ganas et al. 2005), respectively. Eastwards, in continental Turkey, all methods converge in ~N-S extensional tectonics, similar to what is observed in continental Greece north of Peloponnese (Figs. 13 and 15).

In addition, towards evaluating the outcome, we compare the resolved stress field to qualitative displacement data as represented by the strain-field (Fig. 8). Figure 19, presents the minimum 3D misfit angle, or "Kagan angle" (Kagan 1991; Tape and Tape 2012), between the resolved stress tensor and the respective strain tensor available

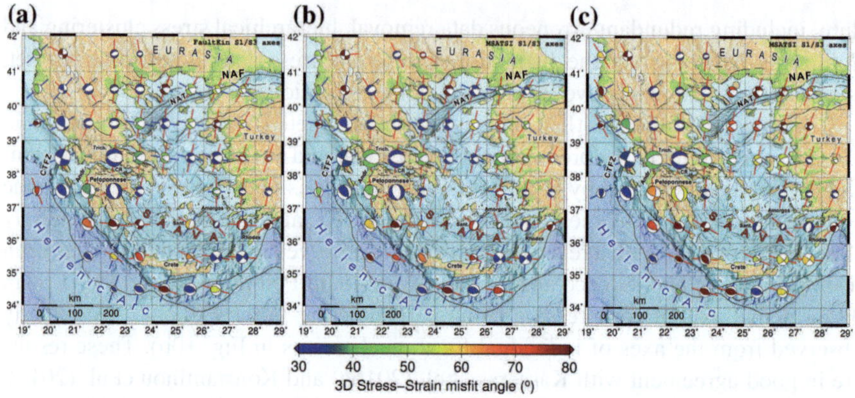

Fig. 19 Map of the resolved stress field using **a** FaultKin, **b** SATSI without damping and **c** SATSI with damping ($e = 2$), where the blue and red lines represent the orientation of the compressional, $\mathbf{S}_1$, and extensional, $\mathbf{S}_3$, principal stress axes, respectively, with the length being reversely proportional to the corresponding axis' plunge. Beach-balls represent the expected focal mechanisms according to the strain field, as resolved by GPS measurements (Kreemer et al. 2014), interpolated to the point at the middle of each cell of Fig. 12. The color of the compressional P-wave first motion quadrant corresponds to the 3D misfit angle between stress and strain, using the method described in Sect. 5.1 for the minimum angular difference between two tensors. The beachballs' size is proportional to the logarithm of the number of available FMS data in each cell. SAAVA: South Aegean Active Volcanic Arc, Sant.: Santorini

from GSRM (http://gsrm2.unavco.org/model/model.html; Kreemer et al. 2014) for each cell and for different inversion methods or parameterizations. From this figure, one can note significant differences existing in numerous sub-regions, even in those cells for which robust results are thought to have been obtained. Those are mainly related to the issues discussed above and in the previous sections of the manuscript.

The stress ratio indicates uniaxial stress when it takes an extreme value such as $\Phi = 1$ ($\sigma_2 \approx \sigma_1$) when $\mathbf{S}_3$ is generally stable but the same is not true for $\mathbf{S}_2$ and $\mathbf{S}_1$, with their values being comparable so that it isn't clear whether they should interchange. This also occurs for $\Phi = 0$ ($\sigma_2 \approx \sigma_3$) where the maximum principal (compressive) stress, $\mathbf{S}_1$, is stable but $\mathbf{S}_3$ and $\mathbf{S}_2$ are not, resulting in comparable magnitudes. Even without stress inversion, this can be demonstrated by a distribution of **P** and **T** axis which indicates that e.g. despite a large RMS angular difference between FMS data and their average solution (such as the one derived by FaultKin; Fig. 13b), the **T** axis trend/plunge distribution has a generally small range, but this is not the case for the **P** axis, which is less stable.

The distribution of stress ratios ($\Phi = 1 - R$) are shown in Fig. 20. A good agreement is observed between FaultKin and un-damped SATSI, indicating that stresses tending towards uniaxial compression ($\Phi \to 0$, blue colors) are distributed mostly along the Hellenic Arc, while ratios corresponding to uniaxial extension ($\Phi \to 1$, red colors) or pure-shear ($\Phi \approx 0.5$, white colors) are mainly observed elsewhere. Un-damped SATSI yielded a different pattern in south Aegean related to uniaxial compression

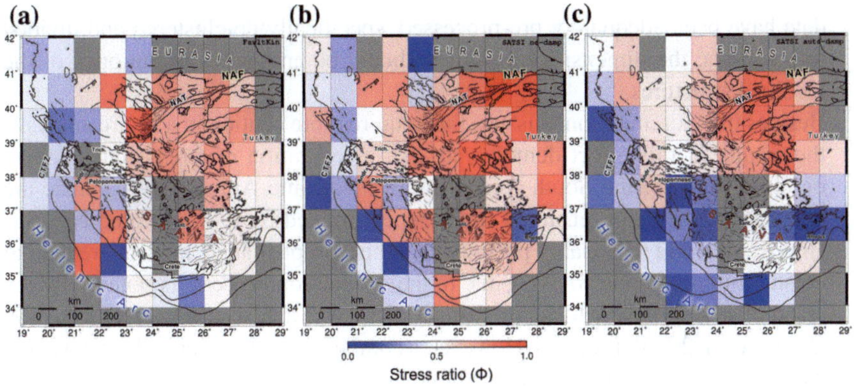

Fig. 20 Map of the resolved stress ratio (Φ) using **a** FaultKin, **b** SATSI without damping and **c** SATSI with damping ($e = 2$), where the color scale blue-white-red indicates uniaxial compression ($\Phi = 0$), pure-shear stress ($\Phi = 0.5$) and uniaxial extension ($\Phi = 1$). Grey colour denotes cells without data or solutions

instead of extension, which is likely the result of over-smoothing of the model. However, it is worth noting that reversal of stress shape occurs along the South Aegean Active Volcanic Arc (SAAVA). Since stress shape controls the orientations and magnitudes of tractions on the rupture surface, and thus it critically affects faults reactivation (Allmendinger et al. 2012), further research is needed on this issue.

Summarizing the abovementioned remarks, the following general conclusions can be drawn:

1. The resolving power of the methods is not directly related to the multitude of observations, but it is mainly connected with the data heterogeneity, basically inherent with complex tectonics.
2. Reversal of the stress field with respect to the strain field is frequently observed (i.e. large Kagan, or misfit angles), which is presumably an artifact of the respective stress model.
3. Given the distribution of the Kagan angles values, it is implied that FaultKin and un-damped SATSI procedure yielded consistent solutions for the largest part of the model.
4. The damped SATSI procedure failed to reproduce a robust regime for the largest part of the model due to over-damping. This does not reject the validity of the method but rather indicates that it is more dependent on proper data arrangement, likely into smaller, partially overlapping spatial cells or grouped into distinct source zones.
5. The methods are considered capable to efficiently resolve stresses when the criteria of the Wallace-Bott hypothesis and assumptions H_2, H_3 and H_4 are fulfilled, i.e. stress homogeneity over space and time and that input focal mechanisms are independent from each other. These issues are efficiently handled only by the MIM procedure. The other methods can perform satisfactorily only when the

data have been adequately pre-processed, specifically de-clustered and divided into subsets, the dimensions of which should be analogous to the ones of the involved structures.

6. Besides the aforementioned issues that require more caution in a future analysis, the results of this effort demonstrate that a separation of stress states occurs on either sides of the South Aegean Active Volcanic Arc (SAAVA), since the prominent $\mathbf{S_3}$ axes of tensional stress appear rotated by ~90°, oriented NW-SE. Hence, we could reasonably presume that this abrupt change, possibly related to crustal thinning along SAAVA due to the subduction process, is potentially the result of, or responsible for slab tearing in NE-SW (Sachpazi et al. 2016), or ~E-W (Jolivet et al. 2015) direction. However, a combination of geological-geophysical data is required to establish such a hypothesis.

Lastly, Table 4 summarizes the pros and cons of the three methods, as derived by their application to the FMS dataset in Greece. Some common issues include that the applied algorithms cannot take into account displacement data or the mechanical properties of the rupture surface. These are handled by the more advanced, third generation methods, also known as Geomechanical models (Sect. 3.3). However, they can provide useful insights into both the regional and local stress field in combination with ground displacement measurements.

Acknowledgements The map figures were drawn using the open-source Generic Mapping Tools (GMT) software (Wessel and Smith 1998; http://www.soest.hawaii.edu/gmt/).

Table 4 Summary of pros and cons of the methods used in this study

FaultKin	Pros	Simple, straightforward algorithm User friendly interface with high quality graphics and plenty of structural analysis capabilities Smaller RMS angular differences against the FMS dataset Easy to handle different input formats The focal mechanisms input follows the Aki-Richards convention Can perform slip tendency analysis given the regional stress field Compatible with other software on structural data analysis
	Cons	Time consuming: requires manual handling per dataset Unable to resolve highly heterogeneous data Uncertainties are large and cannot be quantified by the method Cannot resolve multiple stress states per dataset Output not easy to handle for other applications Unexpected behavior when imported P- and T-axes are not exactly perpendicular Does not ensure independency of focal mechanisms observations Negative eigenvalues (stress magnitudes) are unsigned in the output

(continued)

Table 4 (continued)

SATSI/MSATSI	Pros	Straightforward algorithm, can apply in the four dimensions (spatial and temporal domain) Un-damped solutions are similar to those of FaultKin, but likely more stable due to the resampling potential Smooth variations across the stress model Small uncertainties which are quantified through a resampling procedure Easy to apply (automatic application to grid) Takes into account the resolved stress states in neighboring cells Can resolve highly heterogeneous data, when properly arranged
	Cons	Requires manual handling of data preparation. Requires predefined a grid of cells, the configuration of which is subjective Large Φ uncertainties Cannot resolve multiple stress states per cell Over-smoothing due to damping can create artifacts Does not ensure independency of focal mechanisms observations The input does not follow the Aki-Richards convention, which is not practical when using focal mechanisms
MIM	Pros	Provides more insights into the distribution of the stress tensors Considers the effects of heterogeneity and faults interaction Quantified uncertainties in terms of model-data misfit angles Reduction of redundant and erroneous data Discrimination between different stress states in the same set of data Narrow Φ filtering is possible The focal mechanisms input follows the Aki-Richards convention Nodal planes ambiguity is eliminated since both nodal planes are considered independently by resampling Can perform slip tendency analysis given a regional stress field
	Cons	Complex theoretical background that may be difficult for beginners Time consuming input preparation Requires manual handling per dataset Computational performance requirements increase with the number of observations Only a limited number of observations can be analyzed by the available 32bit version of the code Low graphics resolution, not suitable for publication; complementary graphics software is prerequisite Output not easy to handle for other applications

References

Aki K, Patton H (1978) Determination of seismic moment tensor using surface waves. Tectonophysics 49:213–222

Aki K, Richards PG (2002) Quantitative seismology. University Science Books, USA

Aleksandrowski P (1985) Graphical determination of principal stress directions for slickenside lineation populations: an attempt to modify Arthaud's method. J Struct Geol 7:73–82

Allmendinger RW (1989) Notes on fault slip analysis. In: Short course on quantitative interpretation of joints and faults. Geological Society of America Boulder CO, 59 pp

Allmendinger RW (2016) FaultKin 75 for Mac OS X Windows and Linux. User's Manual, 31 pp

Allmendinger RW, Cardozo N, Fisher D (2012) Structural geology algorithms: vectors and tensors in structural geology. Cambridge University Press. ISBN-13: 978–0748754755, p 290

Allmendinger RW, Gephart JW, Marrett RA (1989) Quantitative interpretation of joints and faults. In: Short course on quantitative interpretation of joints and faults. Geological Society of America Boulder CO, 59 pp

Anderson EM (1905) The dynamics of faulting. Trans Edinb Geol Soc 8:387–402

Anderson EM (1951) The dynamics of faulting and dyke formation with applications to Britain, 2nd edn. Oliver and Boyd Edinburgh

Angelier J (2002) Inversion of earthquake focal mechanisms to obtain the seismotectonic stress IV—A new method free of choice among nodal lines. Geophys J Int 150:588–609

Angelier J, Slunga R, Bergerat F, Stefansson R, Homberg C (2004) Perturbation of stress and oceanic rift extension across transform faults shown by earthquake focal mechanisms in Iceland. Earth Planet Sci Lett 219:271–284

Angelier J (1989) From orientation to magnitudes in paleostress determination using fault slip data. J Struct Geol 11:37–50

Angelier J (1979) Determination of the mean principal directions of stresses for a given fault population. Tectonophysics 56:T17–T26

Angelier J (1984) Tectonic analysis of fault slip data sets. J Geophys Res 89:5835–5848

Angelier J, Tarantola A, Valette B, Manoussis S (1982) Inversion of field data in fault tectonics to obtain the regional stress I Single phase fault populations: a new method of computing the stress tensor. Geophys J R Astron Soc 69:607–621

Angelier J, Mechler P (1977) Sur une methode graphique de recherche des contraintes principales egalment utilisable en tectonique et en seismologie: La methode des diedres droits. Bull Soc Geol Fr 19:1309–1318

Armijo R, Carey E, Cisternas A (1982) The inverse problem in microtectonics and the separation of tectonic phases. Tectonophysics 82:145–160

Arthaud F (1969) Méthode de détermination graphique des directions de raccourcissement d'allongement et intermédiare d'une population de failles. Bull Soc Géol Fr 11:729–737

Barth A, Reinecker J, Heidbach O (2008) World Stress Map project guidelines: stress derivation from earthquake focal mechanisms. GFZ-Potsdam Potsdam Germany

Bernard P, Zollo A (1989) Inversion of near-source S polarization for parameters of double couple point like sources. Bull Seism Soc Am 79:1779–1809

Bingham C (1964) Distributions on the sphere and on the projective plane. PhD Dissertation, Yale University, New Haven, CT

Bingham C (1974) An antipodally symmetric distribution on the sphere. Ann Stat 2:1201–1225

Bock Y, Melgar D (2016) Physical applications of GPS geodesy: a review. Rep Prog Phys 79:119 pp

Bogiatzis P, Ishii M (2014) Constraints on the moment tensor of the 2011 Tohoku-Oki earthquake from earth's free oscillations. Bull Seismol Soc Am 104:875–884

Bott MHP (1959) The mechanics of oblique slip faulting. Geol Mag 96:109–117

Bürgmann R, Pollard D, Martel S (1994) Slip distributions on faults: effects of stress gradients inelastic deformation heterogeneous host-rock stiffness and fault interaction. J Struct Geol 16:1675–1688

Byerlee J (1978) Friction of rocks. Pure Appl Geophys 116:615–626
Carey E, Brunier B (1974) Analyse théorique et numérique d'un modèle mécanique élémentaire appleqúe à l'étude d'une population de failles. C R Hebd Séanc Acad Sci Paris D 279:891–894
Célérier B (1988) Constraint on stress tensor from slip on a single fault plane. University of Texas at Austin Institute for Geophysics, Technical report No 73, 88 pp
Célérier B (2010) Remarks on the relationship between the tectonic regime the rake of the slip vectors the dip of the nodal planes and the plunges of the P, B, T-axes of earthquake focal mechanisms. Tectonophysics 482:1042–1049
Choi P-Y, Angelier J, Souffache´ B (1996) Distribution of angular misfits in fault-slip data. J Struct Geol 18:1353–1367
Chousianitis K, Ganas A, Evangelidis CP (2015) Strain and rotation rate patterns of mainland Greece from continuous GPS data and comparison between seismic and geodetic moment release. J Geophys Res Solid Earth 120(5):3909–3931
Compton RR (1966) Analysis of Pliocene-Pleistocene deformation and stresses in northern Santa Lucia Range California. Geol Soc Am Bull 77:1361–1380
Dahlen FA, Tromp J (1998) Theoretical global seismology. Princeton University Press pp 944
Dahm T, Krüger F (2014) Moment tensor inversion and moment tensor interpretation. Information Sheet IS39
Delvaux D, Sperner B (2003) Stress tensor inversion from fault kinematic indicators and focal mechanism data: the TENSOR program. In: Nieuwland D (ed), New insights into structural interpretation and modelling, vol 212. Special Publications, Geological Society, London, pp 75–100
Dufumier H (1996) On the limits of linear moment tensor inversion of teleseismic body wave spectra. Pageoph 147:467–482
Dufumier H, Cara M (1995) On the limits of linear moment tensor inversion of surface wave spectra. Pageoph 145:235–257
Eberhart-Phillips D (1986) Three-dimensional velocity structure in northern California coast ranges from inversion of local earthquake arrival times. Bull Seismol Soc Am 76:1025–1052
Etchecopar A, Vasseur G, Daignieres M (1981) An inverse problem in microtectonics for the determination of stress tensors from fault striation analysis. J Struct Geol 3:51–65
Fry N (1999) Striated faults: visual appreciation of their constraint on possible paleostress tensors. J Struct Geol 21:7–21
Ganas A, Drakatos G, Pavlides SB, Stavrakakis GN, Ziazia M, Sokos E, Karastathis VK (2005) The 2001 Mw = 64 Skyros earthquake conjugate strike-slip faulting and spatial variation in stress within the central Aegean Sea. J Geodyn 39:61–77
Gapais D, Cobbold PR, Bourgeois O, Rouby D, De Urreiztieta M (2000) Tectonic significance of fault-slip data. J Struct Geol 22:881–888
Gasperini P, Vannucci G (2003) FPSPACK: a package of FORTRAN subroutines to manage earthquake focal mechanism data. Comput Geosci 29:893–901
Gephart JW, Forsyth DW (1984) An improved method for determining the regional stress tensor using earthquake focal mechanism data: application to the San Fernando earthquake sequence. J Geophys Res 89:9305–9320
Gephart JW (1985) Principal stress directions and the ambiguity in fault plane identification from focal mechanisms. Bull Seismol Soc Am 75:621–625
Gephart JW (1988) On the use of stress inversion of fault-slip data to infer the frictional strength of rocks. EOS Am Geophys Union Trans 69:1462
Gephart JW (1990) Stress and the direction of slip on fault planes. Tectonophysics 8:845–858
Godano M, Deschamps A, Lambotte S, Lyon-Caen H, Bernard P, Pacchiani F (2014) Focal mechanisms of earthquake multiplets in the western part of the Corinth rift (Greece): influence of the velocity model and constraints on the geometry of the active faults. Geophys J Int 197:1660–1680
GSRM. http://gsrm2.unavco.org/model/model.html. Accessed April 2017
Hardebeck JL, Hauksson E (2001) Stress orientations obtained from earthquake focal mechanisms: what are appropriate uncertainty estimates. Bull Seism Soc Am 91(2):250–262

Hardebeck JL, Michael AJ (2006) Damped regional-scale stress inversions: methodology and examples for southern California and the Coalinga aftershock sequence. J Geophys Res 111:B11310

Haslinger F, Kissling E, Ansorge J, Hatzfeld D, Papadimitriou E, Karakostas V, Makropoulos K, Kahle H-G, Peter Y (1999) 3D crustal structure from local earthquake tomography around the Gulf of Arta (Ionian region NW Greece). Tectonophysics 304:201–218

Hatzfeld D, Kassaras I, Panagiotopoulos DG, Amorese D, Makropoulos K, Karakaisis GF, Coutant O (1995) Microseismicity and strain pattern in Northwestern Greece. Tectonics 14:773–785

Heidbach OJ, Reinecker M, Tingay B, Müller B, Spencer K, Fuchs F, Wenzel (2007) Plate boundary forces are not enough: Second- and third-order stress patterns highlighted in the World Stress Map database. Tectonics 26(6):TC6014

Heidbach O, Rajabi M, Reiter K, Ziegler M (2016) World Stress Map 2016. GFZ Data Services. https://doi.org/10.5880/WSM.2016.002

Heidbach O, Tingay M, Barth A, Reinecker J, Kurfeß D, Müller B (2010) Global crustal stress pattern on the World Stress Map database release 2008. Tectonophysics 482:3–15

Herrmann RB, Ammon CJ (1997) Faulting parameters of earthquakes in the New Madrid, Missouri, region. Eng Geol 46:299–311

Huang Q (1988) Computer-based method to separate heterogeneous sets of fault-slip data into sub-sets. J Struct Geol 10:297–299

Jackson J, McKenzie D (1988) The relationship between plate motions and seismic moment tensors and the rates of active deformation in the Mediterranean and Middle East. Geophys J 93(88):46–73

Jolivet L, Menant A, Sternai P, Rabillard A, Arbaret L, Augier R, Laurent V, Beaudoin A, Bernhard A, Grasemann B, Huet B, Labrousse L, Le Pourhiet L (2015) The geological signature of a slab tear below the Aegean. Tectonophysics 659:166–182

Julien PH, Cornet FH (1987) Stress determination from aftershocks of the Campania-Lucania earthquake of November 23 1980. Ann Geophys 5B(3):289–300

Kabsch W (1978) A discussion of the solution for the best rotation to relate two sets of vectors. Acta Crystallogr Sect A(34):827–828

Kagan YY (1991) 3-D rotation of double-couple earthquake sources. Geophys J Int 106:709–716

Kamb WB (1959) Ice petrofabric observations from Blue Glacier Washington in relation to theory and experiment. J Geophys Res 64:1891–1909

Kapetanidis V (2017) Spatiotemporal patterns of microseismicity for the identification of active fault structures using seismic waveform cross-correlation and double-difference relocation. PhD Thesis, National and Kapodistrian University of Athens

Kapetanidis V, Deschamps A, Papadimitriou P, Matrullo E, Karakonstantis A, Bozionelos G, Kaviris G, Serpetsidaki A, Lyon-Caen H, Voulgaris N, Bernard P, Sokos E, Makropoulos K (2015) The 2013 earthquake swarm in Helike Greece: seismic activity at the root of old normal faults. Geophys J Int 202:2044–2073

Karakonstantis A, Papadimitriou P (2010) Earthquake relocation in Greece using a unified and homogenized seismological catalogue. Bulletin of the Geological Society of Greece, Proceedings of the 12th International Congress XLIII(4):2043–2052

Kassaras I, Kapetanidis V, Karakonstantis A (2016a) On the spatial distribution of seismicity and the 3D tectonic stress field in western Greece. Phys Chem Earth 95:50–72

Kassaras I, Kapetanidis V, Karakonstantis A, Ganas A (2016b) Regional patterns of earthquake sources in Western Greece: Insights on the 3-D stress tensor and seismic velocity structure arrangement. Paper presented in the General Assembly of the ESC, 4–10 September 2016, Trieste, Italy, paper no 556

Kassaras I, Kapetanidis V, Karakonstantis A, Kaviris G, Papadimitriou P, Voulgaris N, Makropoulos K, Popandopoulos G, Moshou A (2014a) The April–June 2007 Trichonis Lake earthquake swarm (W Greece): New implications toward the causative fault zone. J Geodyn 73:60–80

Kassaras I, Kapetanidis V, Karakonstantis A, Kouskouna V, Ganas A, Chouliaras G, Drakatos G, Moshou A, Mitropoulou V, Argyrakis P, Lekkas E, Makropoulos K (2014b) Constraints on the dynamics and spatio-temporal evolution of the 2011 Oichalia seismic swarm (SW Peloponnesus Greece). Tectonophysics 614:100–127

Kassaras I, Karakonstantis A, Vlachou K, Kapetanidis V, Kaviris G, Papadimitriou P, Voulgaris N, Lagios E, Makropoulos K (2012) Study of the Geodynamics in Aitoloakarnania (W Greece) based on joint Seismological and GPS. In: Paper presented in the 33rd ESC general assembly Moscow Russia, pp 164–165

Kassaras I, Karakonstantis A, Vlachou K, Kapetanidis V, Kaviris G, Vlachou K, Papadimitriou P, Voulgaris N, Makropoulos K (2013) On the geodynamics of western Greece deduced by massive seismological observations. Paper presented in the AHS/IAPSO/IASPEI joint assembly, Gothenburg, Sweden, 22–26 July 2013

Kassaras I, Louis F, Makropoulos K, Magganas A, Hatzfeld D (2009) Elastic-Anelastic properties of the Aegean Lithosphere-Asthenosphere inferred from long period rayleigh waves. In: Anderson JE, Coates RW (eds) The Lithosphere: geochemistry geology and geophysics. Nova Publishers, NY, USA, pp 267–294. ISBN: 978-1-60456-903-2

Kaven JO, Maerten F, Pollard DD (2011) Mechanical analysis of fault slip data: implications for paleostress analysis. J Struct Geol 33:78–91

Konstantinou KI, Mouslopoulou V, Liang W-T, Heidbach O, Oncken O, Suppe J (2017) Present-day crustal stress field in Greece inferred from regional-scale damped inversion of earthquake focal mechanisms. J Geophys Res Solid Earth 122:506–523

Kreemer C, Blewitt G, Klein EC (2014) A geodetic plate motion and global strain rate model. Geochem Geophys Geosyst 15:3849–3889

Laigle M, Hirn A, Sachpazi M, Clément C (2002) Seismic coupling and structure of the Hellenic subduction zone in the Ionian islands region. Earth Planet Sci Lett 200:243–253

Leavers VF (1992) Shape detection in computer vision using the hough transform. Springer, London

Lejri M (2015) Subsurface stress inversion modeling using linear elasticity: sensitivity analysis and applications. PhD Thesis, Geophysics [physicsgeo-ph] Universite de Montpellier pp 317 (https://hal.archives-ouvertes.fr) HAL Id: tel-01324110 version 1

LePichon X, Angelier J (1979) The Hellenic arc and trench system: a key to the neotectonic evolution of the Eastern Mediterranean area. Tectonophysics 60:1–42

Lisle RJ (1987) Principal stress orientations from faults: an additional constraint. Ann Tecton 1:155–158

Lund B, Slunga R (1999) Stress tensor inversion using detailed microearthquake information and stability constraints: application to Olfus in southwest Iceland. J Geophys Res 104:14947–14964

Lyon-Caen H, Armijo R, Drakopoulos J, Baskoutass J, Delibassis N, Gaulon R, Kouskouna V, Latoussakis J, Makropoulos K, Papadimitriou P, Papanastassiou D, Pedotti G (1988) The 1986 Kalamata (South Peloponnesus) earthquake: detailed study of a normal fault evidences for east-west extension in the Hellenic Arc. J Geophys Res Solid Earth 93:14967–15000

Maerten F, Maerten L, Cooke M (2010) Solving 3D boundary element problems using constrained iterative approach. Comput Geosci 14(4):551–564

Maerten F, Resor PG, Pollard DD, Maerten L (2005) Inverting for slip on three-dimensional fault surfaces using angular dislocations. Bull Seismol Soc Am 95:1654–1665

Makropoulos K, Diagourtas D, Kassaras J, Kouskouna V, Papadimitriou P, Ziazia M (1996) The November–December 1994 Lefkas (W Greece) earthquake sequence: results from in situ seismological survey. In: Book of abstracts XXV general assembly of ESC, Reykavik, Iceland, p 108

Mariolakos I, Papanikolaou D (1987) Deformation pattern and relation between deformation and seismicity in the Hellenic arc. Bull Geol Soc Greece (in Greek) XIX:59–76

Marrett RA, Allmendinger RW (1990) Kinematic analysis of fault-slip data. J Struct Geol 12:973–986

Martel S, Shacat C (2006) Mechanics and interpretations of fault slip. In: Abercrombie R, DiToro G, Kanamori H, McGarr A (eds) Radiated energy and the physics of earthquake faulting, vol 170. American Geophysical Union Monograph, pp 207–216

Martínez-Garzón P, Kwiatek G, Ickrath M, Bohnhoff M (2014) MSATSI: A MATLAB package for stress inversion combining solid classic methodology anew simplified user-handling and a visualization tool. Seismol Res Lett 85(4):896–904

Maury JFH, Cornet F, Dorbath L (2013) A review of methods for determining stress fields from earthquake focal mechanisms: application to the Sierentz 1980 seismic crisis (Upper Rhine graben). Bull Soc Geol Fr 184:319–334
McKenzie DP (1969) The relation between fault plane solutions for earthquakes and the directions of the principal stresses. Bull Seism Soc Am 59:591–601
McKenzie DP (1972) Active tectonics of the Mediterranean region. Royal Astron Soc Geophys J 30:109–182
Menke W (1984) Geophysical data analysis: discrete inverse theory. Academic Press, Orlando, Florida
Mercier JL, Carey-Gailhardis E (1989) Regional state of stress and characteristic fault kinematics instabilities shown by aftershock sequences: the aftershock sequences of the 1978 Thessaloniki (Greece) and 1980 Campania-Lucania (Italia) earthquakes as examples. Earth Planet Sci Lett 92(2):247–264
Michael AJ (1984) Determination of stress from slip data: faults and folds. J Geophys Res 89:11517–11526
Michael AJ (1987) Use of focal mechanisms to determine stress: a control study. J Geophys Res 92:357–368
Michael AJ (1991) Spatial variations in stress within the 1987 Whittier Narrows California aftershock sequence: new techniques and results. J Geophys Res 96:6303–6319
Milano G, Di Giovambattista R (2011) Seismicity at the border between Central and Southern Apennines (Italy): re-evaluation of the early 1984 instrumental earthquake. Tectonophysics 499:92–104
Molnar P (1983) Average regional strain due to slip on numerous faults of different orientations. J Geophys Res 88:6430–6432
Muller OH, Pollard DD (1977) The state of stress near Spanish Peaks Colorado determined from a dike pattern. Pure Appl Geophys 115:69–86
Okada Y (1992) Internal deformation due to shear and tensile faults in a half-space. Bull Seismol Soc Am 82(2):1018–1040
Ortner H, Reiter F, Acs P (2002) Easy handling of tectonic data: the programs TectonicVB for Mac and TectonicsFP for Windows. Comput Geosci 28:1193–1200
Otsubo M, Yamaji A (2006) Improved resolution of the multiple inverse method by eliminating erroneous solutions. Comput Geosci 32:1221–1227
Otsubo M, Sato K, Yamaji A (2006) Computerized identification of stress tensors determined from heterogeneous fault-slip data by combining the multiple inverse method and k-means clustering. J Struct Geol 28:991–997
Otsubo M, Yamaji A, Kubo A (2008) Determination of stresses from heterogeneous focal mechanism data: an adaptation of the multiple inverse method. Tectonophysics 457:150–160
Papadimitriou P, Kassaras I, Rigo A, Lyon-Caen H, Hatzfeld D, Makropoulos K, Drakopoulos J (1994) Source parameters of large and small earthquakes in Corinth Gulf (C. Greece) In: Proceedings of the XXIV general assembly of the ESC, pp 848–858
Papanikolaou ID, Papanikolaou DI, Lekkas EL (2009) Advances and limitations of the environmental seismic intensity scale (ESI 2007) regarding near-field and far-field effects from recent earthquakes in Greece: implications for the seismic hazard assessment. In: Reicherter K Michetti AM, Silva PG (eds) Palaeoseismology: historical and prehistorical records of earthquake ground effects for seismic hazard assessment, vol 316. Geological Society of London, Special Publications, pp 11–30
Papazachos B, Papazachou C (2003) The earthquakes of Greece, 3rd edn. Ziti Public Thessaloniki
Pérouse E, Sébrier M, Braucher R et al (2016) Transition from collision to subduction in Western Greece: the Katouna-Stamna active fault system and regional kinematics. Int J Earth Sci (Geol Rundsch) 106(3):967–989
Philippson A (1998) La tectonique de l' Egeide. Ann Geogr 112–141
Plateaux R, Angelier J, Bergerat F, Cappa F, Stefánsson R (2010) Stress changes induced at neighboring faults by the June 2000 earthquakes South Iceland seismic zone. Terra Nova 22:79–86

Pollard DD, Saltzer SD (1993) Stress inversion methods: are they based on faulty assumptions? J Struct Geol 15:1045–1054

Pollard D, Segall P (1987) Theoretical displacements and stresses near fractures in rock: with applications to faults joints veins dikes and solution surfaces. In: Atkinson B (ed) Fracture mechanics of rocks. Academic Press, San Diego

Reches Z (1987) Determination of the tectonic stress tensor from slip along faults that obey the Coulomb yield condition. Tectonics 6:849–861

Reches Z, Baer G, Hatzor Y (1992) Constraints on the strength of the upper crust from stress inversion of fault slip data. J Geophys Res 97:12481–12493

Richardson RM (1992) Ridge forces absolute plate motions and the intraplate stress field. J Geophys Res 97:11739–11748

Rigo A, Lyon-Caen H, Armijo R, Deschamps A, Hatzfeld D, Makropoulos K, Papadimitriou P, Kassaras I (1996) A microseismic study in the western part of the Gulf of Corinth (Greece): implications for large-scale normal faulting mechanisms. Geophys J Int 126:663–688

Royden LH, Papanikolaou DJ (2011) Slab segmentation and late Cenozoic disruption of the Hellenic arc. Geochem Geophys Geosyst 12(3):Q03010

Sachpazi M, Laigle M, Charalampakis M, Diaz J, Kissling E, Gesret A, Becel A, Flueh E, Miles P, Hirn A (2016) Segmented Hellenic slab rollback driving Aegean deformation and seismicity. Geophys Res Lett 43(2). https://doi.org/10.1002/2015gl066818

Sakellariou D, Tsampouraki-Kraounaki K (2016) Offshore faulting in the Aegean Sea: a synthesis based on bathymetric and seismic profiling data. In: Bulletin of the Geological Society of Greece, Proceedings of the 14th International Congress, Thessaloniki, May 2016

Scholz CH (2002) The mechanics of earthquakes and faulting. Cambridge Press, ISBN: 1107079039

Segall P, Pollard D (1980) Mechanics of discontinuous faults. J Geophys Res 85:4337–4350

Serpetsidaki A, Elias P, Ilieva M, Bernard P, Briole P, Deschamps A, Lambotte S, Lyon-Caen H, Sokos E, Tselentis G-A (2014) New constraints from seismology and geodesy on the Mw = 64 2008 Movri (Greece) earthquake: evidence for a growing strike-slip fault system. Geophys J Int 198:1373–1386

Shan Y, Li Z, Lin G (2004) A stress inversion procedure for automatic recognition of polyphase fault/slip data sets. J Struct Geol 26(5):919–925

Shuster MD (1993) A survey of attitude representations. J Astronaut Sci 41(4):439–517

Sperner BB, Müller O, Heidbach D, Delvaux J, Reinecker K, Fuchs K (2003) Tectonic stress in the Earth's crust: advances. In: Nieuwland DA (ed) The World Stress Map project: New insights in structural interpretation and modeling, London Geological Society, pp 101–116

Stein S, Wysession M (2003) An introduction to seismology earthquakes and earth structure. Blackwell, Malden, MA, 510 pp. ISBN: 978-0-86542-078-6

Stucchi M, Rovida A, Gomez Capera AA, Alexandre P, Camelbeeck T, Demircioglu MB, Gasperini P, Kouskouna V, Musson RMW, Radulian M, Sesetyan K, Vilanova S, Baumont D, Bungum H, Fäh D, Lenhardt W, Makropoulos K, Martinez-Solares JM, Scotti O, Živčić M, Albini P, Batllo J, Papaioannou C, Tatevossian R, Locati M, Meletti C, Viganò D, Giardini D (2013) The SHARE European Earthquake catalog (SHEEC) 1000-1899. J Seismol 17(2):523–544

Tape W, Tape C (2012) Angle between principal axis triples. Geophys J Int 108:813–831

Thomas AL (1993) Poly3D: a three-dimensional polygonal-element displacement discontinuity boundary element computer program with applications to fractures faults and cavities in the Earth's crust. MSc thesis, Stanford University, California

Tibi R, Bock G, Xia Y, Baumbach M, Grosser H, Milkereit C, Karakisa S, Zünbül S, Kind R, Zschau J (2001) Rupture processes of the 1999 August 17 Izmit and November 12 Düzce (Turkey) earthquakes. Geophys J Int 144(2):F1–F7

Toda SR, Stein V, Sevilgen, Lin J (2011) Coulomb 33 graphic-rich deformation and stress-change software for earthquake tectonic and volcano research and teaching-user guide. US Geological Survey, Open File Report 2011-1060, 54 pp

Turcotte DL, Schubert G (2002) Geodynamics, 3rd edn. Cambridge University Press

Twiss RJ, Unruh JR (1998) Analysis of fault slip inversions: do they constrain stress or strain rate? J Geophys Res 103(B6):12205–12222
Vavryčuk V (2014) Iterative joint inversion for stress and fault orientations from focal mechanisms. Geophys J Int 199(1):69–77
Vavryčuk V (2015) Earthquake mechanisms and stress field. In: Encyclopedia of earthquake engineering. Springer, Berlin. https://doi.org/10.1007/978-3-642-36197-5_295-1
Vvedenskaya AV (1956) Determination of displacement fields in earthquakes with the aid of dislocation theory. Bull Acad Sci USSR 3:1956
Wallace RE (1951) Geometry of shearing stress and relation to faulting. J Geol 59:118–130
Wessel P, Smith WHF (1998) New improved version of generic mapping tools released. Eos Trans Am Geophys Union 79:579–579
Wiemer S (2001) A software package to analyze seismicity: ZMAP. Seismol Res Lett 72:373–382
Yamaji A (2000) The multiple inverse method: a new technique to separate stresses from heterogeneous fault-slip data. J Struct Geol 22:441–452
Yamaji A (2016) Genetic algorithm for fitting a mixed Bingham distribution to 3D orientations: a tool for the statistical and paleostress analyses of fracture orientations. Island Arc 25:72–83
Yamaji A, Otsubo M, Sato K (2006) Paleostress analysis using the Hough transform for separating stresses from heterogeneous fault-slip data. J Struct Geol 28:980–990
Yamaji A, Sato K (2011) A spherical code and stress tensor inversion. Comput Geosci 38(1):164–167. https://doi.org/10.1016/j.cageo.2011.04.016
Yamaji A, Sato K, Otsubo M (2011) Multiple inverse method user's manual © 2004, pp 42
Zoback ML (1992) First- and second-order patterns of stress in the lithosphere: The World Stress Map project. J Geophys Res 97:11703–11728
Zoback ML, Zoback MD, Adams J, Assumpção M, Bell S, Bergman E, Blümling P, Brereton NR, Denham D, Ding J, Fuchs K, Gay N, Gregersen S, Gupta HK, Gvishiani A, Jacob K, Klein R, Knoll P, Magee M, Mercier JL, Müller BC, Paquin C, Rajendran K, Stephansson O, Suarez G, Suter M, Udias A, Xu Zh, Zhizhin M (1989) Global patterns of tectonic stress. Nature 341:291–298

Relative Locations of Clustered Earthquakes in the Sea of Marmara and States of Local Stresses in the East of the Central Marmara Basin

Yasemin Korkusuz Öztürk and Nurcan Meral Özel

Keywords Stress tensor inversion · Fault plane solution · Seismicity · Tectonics Marmara · Relative location · Hypodd

1 Introduction

The North Anatolian Fault Zone (NAFZ), starting from the Karliova triple junction, is an outcome of a westward migrating series of destructive earthquakes in the northern part of the Anatolian Block during the last century (Kinscher et al. 2013; Barka 1996). The Western section of the NAF bifurcates into two branches in the Western Bolu, so the Northern outgrowth enters the Sea of Marmara at the İzmit Gulf while the Southern one follows the south of İznik Lake and extends towards the Southern Marmara Region (Çemen et al. 2000). A transition zone from the right lateral strike-slip NAFZ to the Aegean extension zone (AEZ) arises in the Eastern Sea of Marmara (Armijo et al. 2005; Flerit et al. 2003).

The 9 August 1912 Mürefte Mw7.4 (Aksoy et al. 2012) and 17 August 1999 İzmit Mw7.4 (Barka et al. 2002) earthquakes are the last devastating occurrences in the western and eastern Marmara Regions, respectively. Even though elongations of these earthquakes are still argued (Korkusuz Öztürk et al. 2015; Bouchon et al. 2002; Özalaybey et al. 2002; Pinar et al. 2001; Wright et al. 2001), there is a 70–150 km long seismic gap between the extensions of these two earthquakes (Geli et al. 2008; Barka et al. 2002; Hubert-Ferari et al. 2000; Ambraseys and Finkel 1987), and no devastating event have been recorded in the Central Marmara Sea, since the 22 May 1766, large İstanbul earthquake. Hence, a major earthquake is expected in the Sea

Y. Korkusuz Öztürk (✉) · N. Meral Özel
Department of Geophysics, Kandilli Observatory & Earthquake Research Institute, Boğaziçi University, Çengelköy, Istanbul 34684, Turkey
e-mail: yasemin.korkusuz@boun.edu.tr; yaseminkrksz@gmail.com

N. Meral Özel
e-mail: ozeln@boun.edu.tr

S. D'Amico (ed.), *Moment Tensor Solutions*, Springer Natural Hazards,
https://doi.org/10.1007/978-3-319-77359-9_20

of Marmara (Erdik 2013; Erdik et al. 2004; King et al. 2001). Therefore, sensitive location and stress tensor inversion analyzes are very important.

The present day states of local stresses in the Sea of Marmara are very complex due to its strain-partitioning, rotated stress axes (Korkusuz Öztürk et al. 2015) and lack of strain accumulation in the Central Marmara Sea (Ergintav et al. 2014; Ergintav et al. 2007). Contrary to previous studies (Le Pichon et al. 2001; Gürbüz et al. 2000), supporting a pure right lateral strike-slip stress regime in the whole Marmara Region, recent seismic studies reveal many extensional fault structures in the Eastern Marmara Sea (Korkusuz Öztürk et al. 2015; Karabulut et al. 2011; Bulut et al. 2009; Pinar et al. 2009; Sato et al. 2004; Tunç et al. 2011; Karabulut et al. 2002; Özalaybey et al. 2002; Örgülü and Aktar 2001). On the other hand, studies within the central Marmara Sea have been limited by virtue of the lack of seismicity and insufficient number of seismic and geodetic stations.

Some recent studies use P-wave first motion polarities technique to figure out the complex stress and strain conditions in the Sea of Marmara and its surroundings (Korkusuz Öztürk et al. 2015; Örgülü 2011; Pinar et al. 2009; Polat et al. 2002; Gürbüz et al. 2000), while others, who are working with relatively large events, use moment tensor inversion techniques (Pınar et al. 2003; Kiratzi 2002) for the same purpose. Moreover, so far a relative location study, done in the Yalova area (Bulut and Aktar 2007), represents a northward dipping fault with 56° from the Hypodd analysis of aftershock activity of the 1999 İzmit earthquake. Therefore, no relative location analysis is done in the Sea of Marmara for the natural seismicity owing to low signal detection capabilities which increase errors in the locations of events and accurate identifications of earthquake clusters. Nevertheless, the signal detection capability in the Marmara Region eventually rises due to the increase in the number of seismic stations on land and sea, deployed by some recent projects (TURDEP, MARsite, Satreps).

In this study, we obtain orientations of maximum, intermediate and minimum principal stress axes (σ_1, σ_2 and σ_3) for a cluster of earthquakes in the Eastern Central Marmara basin through the assumption of a homogenous local stress field which let us linearize the stress-strain relationship and degrade the elasticity tensor to the Young's modulus. We select 27 earthquakes to determine simultaneous focal mechanism solutions within the stress tensor inversion procedure (Horiuchi et al. 1995). Unlike a previous work (Örgülü 2011), presenting a dextral strike-slip stress state in the northwestern part of the Central Marmara basin, we reveal an extensional stress state in the east of it. We also calculate relative locations of hypocenters of earthquakes for not only those in the Central Marmara basin (CMBC), but also in the Eastern Ganos offshore (EGOC), Eastern Tekirdağ basin (ETBC), Çınarcık basin (CBC), Yalova (YC) and Gemlik clusters (GC) which we have recently been analyzed for the individual relocations and orientations of principal stress axes (Korkusuz Öztürk et al. 2015). Even if the relative location procedure may not allow us to define dipping angles of the fault segments, it can be fruitful in terms of the definition of differences in seismic zones in the Sea of Marmara and the opportunity compare them with the findings from both geodetic and other seismic studies.

2 Tectonic Setting and Regime

Locating in the collision zone of the European and African plates in the east, and being pulled by the Hellenic Arc in the west, the Anatolian Block escapes towards the west with a counter-clockwise rotation, mainly via the right lateral strike-slip North Anatolian Fault Zone (NAFZ) (Ketin 1948). The NAFZ mostly consists of Tethyan accretionary complexes extending from east to west (Şengör et al. 2005). The slip-rate from east to west is ~25 mm per year (McClusky 2003; Dewey and Şengör 1979; McKenzie 1978, 1972), and it is almost similar for the Marmara Region (Reilinger et al. 2006). Furthermore, the Marmara Region is situated in the transition zone between the North Anatolian dextral and Aegean extension zones, so it has a complex tectonic setting.

The Marmara Region is identified with a pure dextral strike-slip stress regime from the analysis of M > 5 and micro earthquakes between the periods 1943–1997 and 1995–2002 (Polat et al. 2002; Gürbüz et al. 2000), respectively. On the other hand, it represents transtensional stress states with its nonignorable extensional components for newer data sets (Bulut et al. 2009; Sato et al. 2004), although the R-value, the ratio between stress tensors (Korkusuz Öztürk et al. 2015), for the Marmara Region is an indicator of a compression (Korkusuz 2012; Örgülü 2011; Kiratzi 2002).

On the local scale, the most recent studies compatibly indicate a transpression in the Ganos offshore that is the Western Sea of Marmara (Korkusuz Öztürk et al. 2015; Örgülü 2011). The Eastern Tekirdağ basin cluster, a 4 km-long-area between Tekirdağ and Central basins, is the southern portion of the Western High region with its asymmetric half-graben structure (Okay et al. 2004). The Çınarcık basin is located between two deviant normal faults (Laigle et al. 2008) with its 20 km wide and 1.250 km deep structure (Okay et al. 2000). Especially the Northern segment of it, attributed as the main fault branch, is more complex (Le Pichon et al. 2001). Moreover, the Northern part of the Çınarcık basin is more steep and regular compare with the Southern part of it (Le Pichon et al. 2001). Additionally, an extensional stress regime exits in Yalova region, which is famous for its hot springs (Korkusuz Öztürk et al. 2015; Örgülü 2011), and Gemlik region is on the southern branch of the western portion of the NAF that passes through the Gemlik Gulf.

3 Data

In this study, we identify a new cluster of earthquakes in the Central Marmara Sea (Fig. 1, red rectangle). We individually examine 53 ML $\geq$ 1.5 earthquakes for the time interval between 2006.09.02 and 2014.09.04, using data from 100 land and 5 sea bottom seismometers which are merged from local networks operated by KOERI and TUBITAK. We determine P and S wave arrival times and P wave first motion polarities using even base data. Therefore, we not only determine hypocenters of those events, but also simultaneous focal mechanism solutions and stress tensor

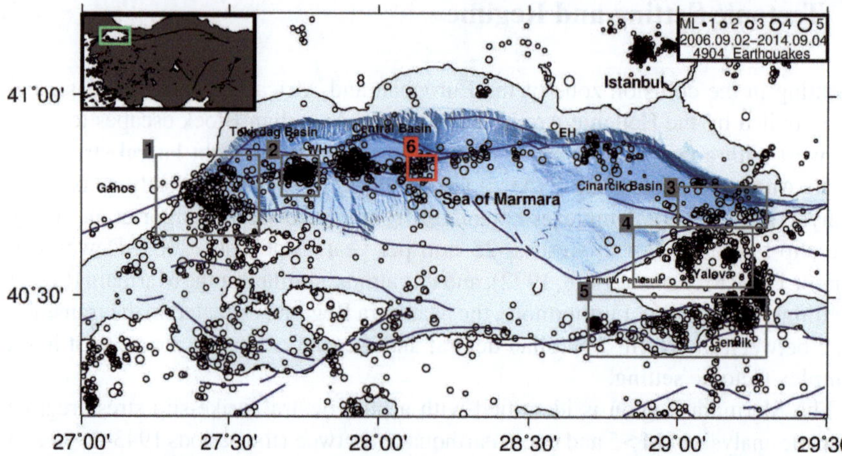

Fig. 1 Seismic activity of the Marmara Region from the earthquake catalogue of KOERI is shown for the time period between 02.09.2006 and 04.09.2014. Black circles are events with ML ≥ 1.0. Black rectangles including are our previously obtained earthquake clusters; Eastern Ganos offshore (1), Eastern Tekirdağ basin (2), Eastern Çınarcık basin (3), Yalova (4) and Gemlik clusters (5) (Korkusuz Öztürk et al. 2015). The red rectangle represents the Eastern Central Marmara basin (6). Fault traces are from Armijo et al. (2002), and bathymetry data are from Rangin et al. (2001)

orientations. We choose earthquakes with at least 10 high quality P wave FMPs and maximum 1 inconsistent station (misfit), so the total number of earthquakes of this cluster decreases from 53 to 27.

For the relative location procedure, we use the earthquake catalogue from Korkusuz Öztürk et al. (2015), which consists of 398 events for five selected clusters, and the list of locations of 27 currently analyzed events as mentioned above. The starting date of the data set is 2006.09.02 for both catalogues, while the ending dates are 2011.03.31 for the former and 2014.09.04 for the latter catalogue. As a result, we obtain relative locations of ML ≥ 1.5 earthquakes for the clusters given in Fig. 1.

4 Methods

4.1 Individual Locations, Focal Mechanism Solutions and Stress Tensor Inversion

First, we determine P and S phase arrival times and observe P-wave FMPs using PQL-II program (Boaz 2009) for 53 events within the selected cluster. No filter is used in order not to affect the polarity determinations by altering the waveforms.

Next, we relocate earthquakes using the crustal model from Karabulut et al. (2003) and select the Vp/Vs ratio as 1.75, since they give the minimum location errors.

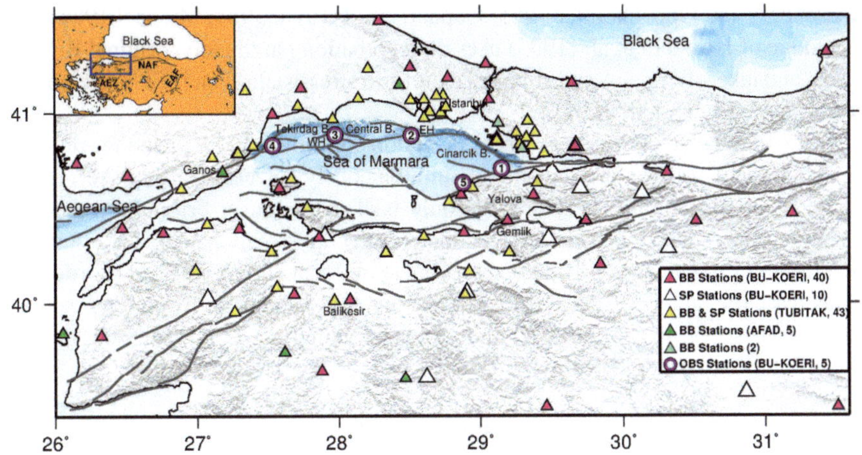

Fig. 2 The map of seismic station distribution. Pink and white triangles are BB (40) and SP (10) stations of KOERI, respectively. Yellow triangles are BB and SP (43 in total) stations of TUBITAK. Dark and light green triangles are BB (5) stations of AFAD and BB (2) stations of other constitutions are demonstrated, respectively. Purple circles with the numbers from 1 to 5 are real-time continuous observation points of KOERI (Korkusuz Öztürk et al. 2015). The fault traces on the main figure and legend are from Armijo et al. (2002) and Şaroğlu et al. (1992), respectively. The topographic data are from Reuter et al. (2007) and bathymetry data are from Le Pichon et al. (2001)

Stations further than 200 km are included neither in the relocation process nor in the focal mechanism solutions (e.g. for P-wave FMPs). The HYPOCENTER approach is used for the determination of earthquake locations (Lienert and Havskov 1995; Lienert 1991; Lienert et al. 1986). The method uses different location approaches depending on the properties of the earthquake (distance, depth, etc…). Now that local earthquakes are used in this study, travel times of Pg, Pn, Sg and Sn are calculated. In addition, using observed arrival times of P and S waves at the seismic stations and a pre-determined crustal model, the location algorithm tries to find the best point and origin time of the source through a complex inverse theory. Even if arrival phases at the stations are perfectly picked, most of the time some small errors are possible. The reason is that velocity models always include some error and may not cover the whole area, or may not be three dimensional. Therefore, nonlinearities of the ray path and the inversion matrix of the location procedure always lead to some errors in any location procedure though they are minimized.

Third, the FOCMEC program is used for 1° grid search to select the best events and test the reliability of simultaneously determined focal mechanisms for 5° grid interval during the stress tensor inversion.

Consequently, the stress tensor inversion method of Horiuchi et al. (1995) is used in order to achieve orientations of principal stress axes and simultaneously determined FPSs. The reason for the use of this method is that even though many stress tensor inversion techniques are developed (e.g. Otsubo et al. 2008; Sato 2006; Shan et al. 2004; Lisle et al. 2001; Horiuchi et al. 1995; Rivera and Cisternas 1990;

Reasenberg and Oppenheimer 1985; Gephart and Forsyth 1984; Michael 1984), the technique of Horiuchi et al. (1995) uses direct location parameters, original P-wave FMPs and take-off angles, and supplies a uniform stress state even if FPSs are poorly constrained (Pınar et al. 2009).

In the initial step of the stress tensor inversion algorithm, all FPSs and the number of inconsistent stations are calculated simultaneously for each earthquake for the 5° grid search interval. Therefore, the only background error originates from the location process. Next, a grid search is done for the 0.2° rotation angle (ω_p), and the best (minimum) values of misfit polarities are stored for every 3° interval in order not to decrease the memory usage. Those stored values are identified by couples each of which consists of an azimuth ($\emptyset_p$) and a plunge angle (θ_p). In this method, the 3D hydrostatic term is neglected by;

$$\sigma_1 + \sigma_2 + \sigma_3 = 0 \tag{1}$$

Here, $\sigma_1, \sigma_2, \sigma_3$ represent the maximum, intermediate and minimum principal compressive stress axis, respectively. Owing to the usage of actual P-wave FMPs, orientations of principal stress axes are found without absolute values of stress axes. The total number of inconsistent polarities (misfits) is derived from the function;

$$N_{tot} = \sum_i M_i \left(\theta_p, \emptyset_p, \omega_p, R\right) \tag{2}$$

M_i denotes the minimum number of misfit polarity of the ith earthquake as a function of the inclination angle (θ_p) and azimuth ($\emptyset_p$) of P axis, respectively. ω_p is the rotation angle and $R = (\sigma_1 - \sigma_2)/(\sigma_1 - \sigma_3)$, R-value, is the ratio between the stress tensors. The method has a benefit of calculation of Eq. (2), which is done for each individual earthquake, being almost independent of the rotation angle around the fault perpendicular vector plane for the ith event. In the third step, alignments of principal stress axes ($\sigma_1, \sigma_2, \sigma_3$) are calculated using the best (minimum) number of misfit polarities, and locations of inclination angle and azimuth couples. Finally, fault planes are discriminated from the auxiliary planes for most of the earthquakes through the stress tensor inversion algorithm, as it uses the information of stress orientations.

4.2 *Relative Locations*

Hypodd program (Waldhauser 2001), works with the double-difference algorithm of Waldhauser and Ellsworth (2000), is separately performed for the relative locations of recently analyzed six earthquake clusters in the Sea of Marmara. The method is much more powerful than regular individual or joint location algorithms, which are based on Geiger's technique (Geiger 1910) that utilizes a first order Taylor Series to linearize the travel time equation. The Hypodd technique derives more clear fault traces from the outputs of dense locations of events (Waldhauser 2001). For the relative hypocentral parameters between events i and j, it uses the Frechet's (1985) equation;

$$\frac{\partial t_k^{ij}}{\partial m}\Delta m^{ij} = dr_k^{ij} \tag{3}$$

where partial derivatives of t_k with respect to m are the components of the slowness vector from receiver to the source, and $\Delta m^{ij} = (\Delta x^{ij}, \Delta y^{ij}, \Delta z^{ij}, \Delta \tau^{ij})$ is the difference between the relative hypocentral parameters of the two events. "k" is the phase, and x, y, z are the coordinates of the location and τ is the origin time for an earthquake (e.g. i or j). Thus, Δx, Δy, Δz, and $\Delta \tau$ are the required changes in the hypocentral parameters of the location to create a better fitted model with the data. $r_k = \left(t^{obs} - t^{cal}\right)_k$ is the travel time difference (residual) between theoretical and observed data. Hence, the double difference algorithm (Waldhauser and Ellsworth 2000), is the residual between theoretical and observed differential travel time between an event pair;

$$dr_k^{ij} = \left(t_k^i - t_k^j\right)^{obs} - \left(t_k^i - t_k^j\right)^{cal} \tag{4}$$

All event pairs are combined through Eq. (3) for each station, and a system of linear equations is formed by the combination of data from all stations. In addition, absolute travel times, t, from our created catalogues are used in Eq. (4). The linear equation system is defined as;

$$\boldsymbol{WGm} = \boldsymbol{Wd} \tag{5}$$

$\boldsymbol{G}$ is a matrix with the dimension of $Mx4N$. Here, M is the number of observations of the double-difference, and N is the number of events. $\boldsymbol{W}$ is the diagonal matrix weighting every station, $\boldsymbol{m}$ is the vector of length 4 N, which include desired changes in hypocentral parameters, and $\boldsymbol{d}$ is the data vector comprising double-differences.

As a result, the Hypodd method defines more accurate cluster locations due to its precision. In addition, since the clusters in this study are well-conditioned and the number of earthquakes is not very large, SVD (Singular Value Decomposition) is used to solve the system Waldhauser and Ellsworth (2000). In other words, LSQR (The Least Squares) technique is used in some studies for large systems via application of some constraints.

Further, the ratio between P and S wave velocities is selected as 1.75 and the crustal model from Karabulut et al. (2003) is used, similar to the individual location procedure. Pick quality factors change between 0, 0.25, 0.5, 0.75 and 1.0. Also, the following criteria are used for each cluster analysis: the maximum epicentral separation is 5 km, maximum number of nearest neighbors per event is 100, and minimum number of links to define a neighbor is 8.

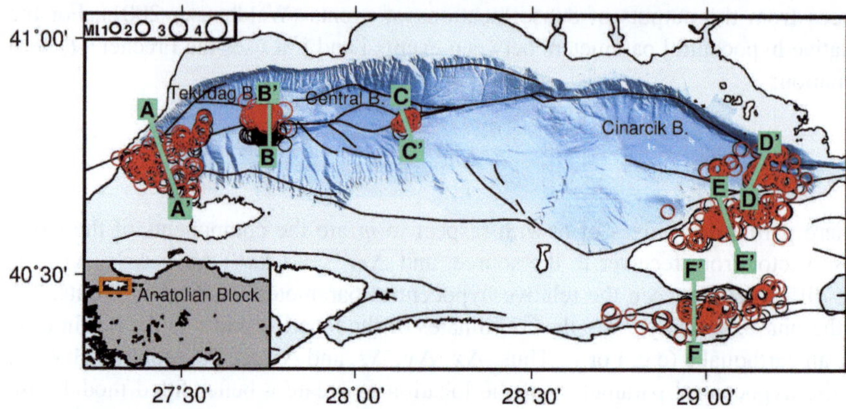

Fig. 3 Epicenters of earthquakes selected within the six visible clusters for individual (black circles) and relative (red circles) locations. Green profiles are for the fault perpendicular cross sections

5 Results and Discussions

5.1 Eastern Ganos Offshore Cluster (1-EGOC)

In the Eastern Ganos Offshore Cluster, we relatively locate 85 earthquakes where previously revised individual locations and derive principal stress axes orientations (Korkusuz Öztürk et al. 2015). The acquired azimuth/plunge values of maximum, intermediate and minimum principal stress axes are; 315°/35°, 141°/50°, and 221°/7°, respectively (Table 2). This dextral strike-slip stress regime evidence coincides with the previous investigations (Örgülü 2011; Polat et al. 2002; Gürbüz et al. 2000) and the 9 August 1912 M7.4 Ganos Earthquake (Aksoy et al. 2010), while not with the one (Pınar et al. 2003) supports a compressional regime. Besides, extensional FPSs in the east of the cluster overlaps with 1935 Erdek-Marmara Island earthquakes (Başarır Baştürk et al. 2016).

After the derivation of relative locations, changes in horizontal (Fig. 3) and vertical (Fig. 4) locations are negligible. Although some small changes are observed in the focal mechanism solutions after the Hypodd locations, still the western part of the cluster exhibits mostly thrust FPSs, the northeast part represents mostly oblique FPSs with some extension, and also a few reverse FPSs, and the south east of the region dominantly represents right lateral strike-slip and extensional motions.

Furthermore, earthquakes shallower than 2 km depth exist in consequence of Hypodd results although they are deeper in individual locations (Fig. 4). A total of 16 out of 85 well-located events are shallower than 3.5 km depth, sedimentary basin (Bayrakçı et al. 2013), and only 4 of them have very small strike-slip components locating at ~2.0 km depth. Hence, the earthquakes in the sedimentary basement mainly have normal or thrust faulting mechanisms as we expect. The depth of the seismogenic zone is still bounded at 17 km depth. Besides, even though some vis-

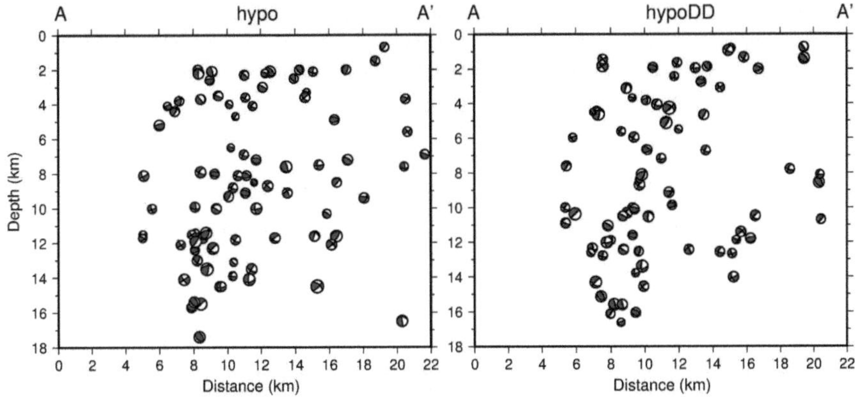

Fig. 4 Fault perpendicular cross sections for AA' profile (Fig. 3) on the Eastern Ganos offshore cluster. The left figure is for individual locations obtained via Hypocenter software, while the right one is derived from the usage of Hypodd relative location technique

ible curvature like earthquake alignments exist in the northwest of the area, there is no clear fault orientation as it is expected concerning the complexity of the fault geometry in the Ganos offshore with its multi segmentation. Moreover, the Ganos onshore is attributed as a locked segment up to 9 ± 2 or 14 km from the geodetic investigations (Ergintav et al. 2014; Schmittbuhl et al. 2015, respectively), as it is a seismically inactive region as well (Fig. 1).

5.2 Eastern Tekirdağ Basin Cluster (2-ETBC)

In this cluster, we derive double-difference locations of 75 earthquakes which we obtain individual hypocentral locations (Table 1) and principal stress axes beforehand (Table 2). Hence, azimuth/plunge values of the orientations of maximum, intermediate and minimum principal compressive stress axes are calculated as; 293°/70°, 143°/30°, and 30°/0°, respectively.

Most of the earthquakes in this region have pure normal faulting mechanisms even though some right lateral strike-slip FPSs with some normal and reverse components also exist (Korkusuz Öztürk et al. 2015). The obtained extensional stress regime overlaps with the recent geological investigations (Ergintav et al. 2014; Armijo et al. 2002), while studies, supporting a pure strike-slip structure, also exist (Sato et al. 2004).

Epicenters of the events in the ETBC move from south to north via application of the relative location process (Fig. 3). After this movement, locations of epicenters appear very compatible with the given fault structure (Armijo et al. 2002). Scrutinizing the hypocentral alignments of the events indicates that many earthquakes are aligned along a line which is around 84° from horizontal (dip angle). But many

Table 1 The summary of average error amounts of azimuthal GAP in station coverage, depth (Erdp), latitude (Erlt) and longitude (Erln). Also, minimum (Min), maximum (Max), average (Av) and total (Tot) number of used polarity, and the total number of inconsistent stations (Misfit Polarity) for a whole earthquake cluster

Cluster	Av-GAP (°)	Av-Erdp (km)	Av-Erlt (km)	Av-Erln (km)	Min/Max/Av/Tot-Polarity	Misfit Polarity
1-EGOC	65 ± 21	2.6 ± 0.93	1.0 ± 0.23	1.2 ± 0.32	10/58/21/1768	22
2-ETBC	58 ± 23	3.1 ± 0.85	1.2 ± 0.21	1.4 ± 0.29	10/65/28/2063	28
3-ECBC	64 ± 22	2.8 ± 0.69	0.8 ± 0.16	1.2 ± 0.22	10/44/20/1494	12
4-YC	55 ± 22	2.5 ± 0.85	0.9 ± 0.23	1.2 ± 0.24	10/63/23/2336	23
5-GC	75 ± 39	3.0 ± 0.80	1.1 ± 0.32	1.4 ± 0.28	11/61/25/1565	19
6-ECMBC	67 ± 29	3.2 ± 1.04	1.3 ± 0.30	1.3 ± 0.36	10/45/21/574	7
Total	63 ± 25	2.8 ± 0.84	1.0 ± 0.23	1.27 ± 0.27	10/65/23/9800	111

Table 2 Stress tensor inversion results via the program of Horiuchi et al. (1995). The second column is the number of used and all events recorded in the related clusters. Third, fourth and fifth columns are azimuth and plunge angles of σ_1, σ_2 and σ_3 axes, respectively. The last column is the best derived R-values

Cluster	Used/All events	σ_1 Az-Pl	σ_2 Az-Pl	σ_3 Az-Pl
1-EGOC	85/144	315-35	141-50	221-7
2-ETBC	75/105	293-70	143-30	30-0
3-ECBC	73/116	123-80	298-10	206-5
4-YC	102/124	285-75	115-7	201-2
5-GC	63/100	102-58	285-30	189-5
6-ECMBC	27/53	90-79	304-3	214-5

events recorded in the right side of this alignment, should be taken into account, as well (Fig. 5, right plot). Besides, hypocenters of the shallowest earthquakes move from 2 to 1 km depth, and the depth interval of the main earthquake activity changes from 10–15.5 to 8.5–15 km. On the other hand, a recent study based on only OBS data locates micro-earthquakes between around 10–26 km depths (Yamamoto et al. 2017). Furthermore, small seismic events are located above around 8.5–10 km depth concerning both our relative and individual locations (Fig. 5) and 4–8 mm/year strain rate is measured by recent extensometers (Yamamoto et al. 2016). As a result, the study area may be creeping above ~10 km depth or less. Additionally, the region is attributed as a deep creep above ~15 km by Schmittbuhl et al. (2015). Regarding the thickness of the sedimentary basin that is ~5.5 km (Bayrakçı et al. 2013) and our cross section analyses for the Hypodd findings, it is most probable that the creeping part might be shallower than 10 km depth. Moreover, it is realized that this segment may not have accumulating stress due to not only the possibility of the suggested aseismic creep but also the occurrences of moderate size earthquakes within the last years. It is recognized that recurrence intervals of ML3.5–4.5 earthquakes are a few

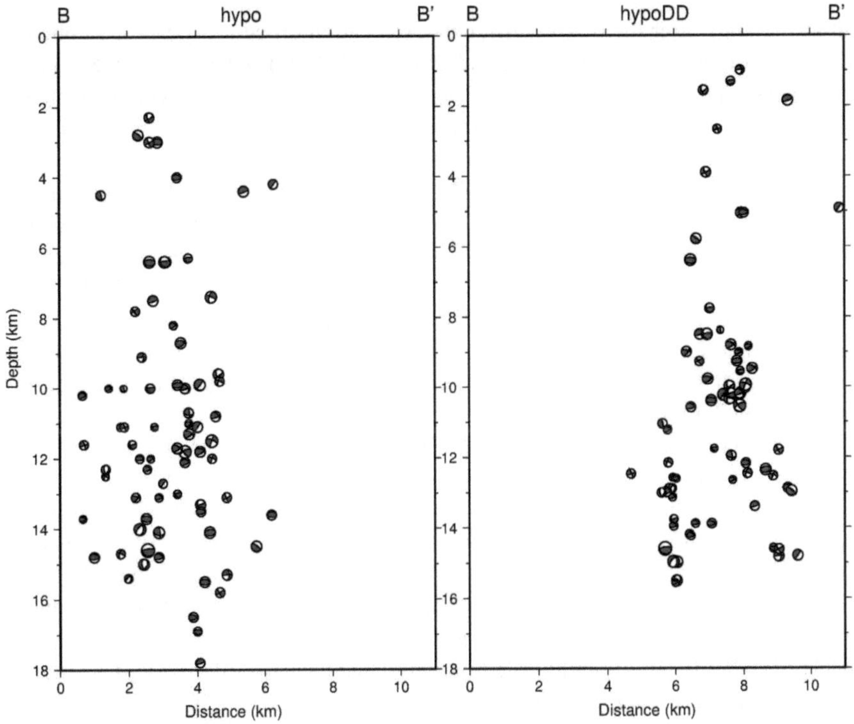

Fig. 5 Fault perpendicular cross sections for BB' profile (Fig. 3) on the Eastern Tekirdağ basin cluster. For more explanation, see figure caption for Fig. 4

months and ML4.5–5.2 events are around one year since 2008, and the depth of some of them is shallower than 10 km (Korkusuz Öztürk et al. 2015; KOERI Catalogue).

5.3 *Eastern Çınarcık Basin Cluster (3-ECBC)*

In the ECBC, we attain relative locations of 73 earthquakes, as we already re-identify their individual locations and the state of stress in the region (Korkusuz Öztürk et al. 2015). Azimuth/plunge values of the orientations of maximum, intermediate and minimum principal stress axes are found out as; 123°/80°, 298°/10°, and 206°/5°, respectively (Table 2).

The Çınarcık basin is mostly attributed to an area with predominant strike-slip movements (Örgülü, 2011; Pınar et al. 2003; Bulut et al. 2009), and even sometimes with both strike-slip and thrust behaviors (Bohnhoff et al. 2012). The bifurcation in the Eastern Çınarcık basin and the strain partitioning may even be rejected concerning the fault structure given by Zachariasen and Sieh (1995) (Pınar et al. 2003). Nevertheless, our focal mechanism solutions still exhibit extensional behaviors without any change after the double-difference relocation process, so they are coherent with the findings from recent individual (Polat et al. 2002) and composite (Sato et al. 2004) FPS studies. Also, recent GPS velocities represent 6 ± 2 mm/yr extension and 15 ± 2 mm/yr right-lateral offset (Ergintav et al. 2014). Besides, 18 September 1963 M6.3 earthquake in the Çınarcık basin, the last destructive earthquake in the Northern Çınarcık basin, possibly has a WNW-ESE trended normal faulting mechanism which matches with the stress tensor orientations of this cluster (Başarır Baştürk et al. 2016; Taymaz et al. 1991). In addition to those mostly strike-slip and extensional movements and 70° dip angle, it should be noted that Coulomb models (e.g. Bohnhoff et al. 2013; Uçarkuş et al. 2010; Parsons 2004; Hubert-Ferrari et al. 2000) introduce that the Princes' Island segment, the northern section of the NAF in the Çınarcık basin, has a high stress field as an indicator of that it did not rupture during the 1999 Mw7.4 İzmit earthquake.

The comparison of hypocenter and Hypodd analyses in our study shows that earthquake locations stay stable horizontally (Fig. 3), but some shifts occur vertically (Fig. 6). Hence, the cross section of the fault perpendicular profile, DD' (Fig. 8), represents a negative flower structure, especially, at the southern part of the basin, as also discovered via other investigations (Beyhan and Selim 2007; Aksu et al. 2000). It is also compatible with the bathymetry of the Çınarcık basin regarding the negative flower structure to the south of the basin and the steep part to the north (Figs. 1, 2 and 6). On the other hand, it should be kept in mind that the Çınarcık segment is not very sensitive at the vertical dip direction, and it has around 70° dip angle (Oglesby and Mai 2012). Therefore, the effect of extensional component to the main strike-slip motion in the Çınarcık basin should not be ignored for future studies.

The sediment layer of the ECBC is changing between 1.5 and 3 km (Bayrakçı et al. 2013). The seismic zone becomes more clear after the application of the Hypodd technique. The seismogenic zone below the southern part of the basin seems to be between ~8 and 13 km depths, while shallower, starting at ~5 km depth, in the north (Fig. 6, right plot). On the other hand, it has been previously attributed as between 10 and 12 km depths (Yaltırak et al. 2003). Also, fault parallel (Korkusuz 2012) and fault perpendicular (Fig. 6) cross sections, which are calculated using our joint data set, for ~30 km eastern half of the Çınarcık basin indicate that the basin is locked up to ~7 km depth notwithstanding it has been given as ~10 km by other seismic (Klein et al. 2017; Schmittbuhl et al. 2015; Bohnhoff et al. 2013) and geodetic observations (Ergintav et al. 2014). This difference might come from the fact that we use a very dense network or we analyze only eastern Çınarcık basin due to the low seismicity in the west.

5.4 Yalova Cluster (4-YC)

In YC, we calculate Hypodd locations of 102 earthquakes, as we already have their separate well-defined locations (Table 1) and stress tensors (Table 2) as an outcome of the use of the same network. As a result, azimuth/plunge values of the orientations of

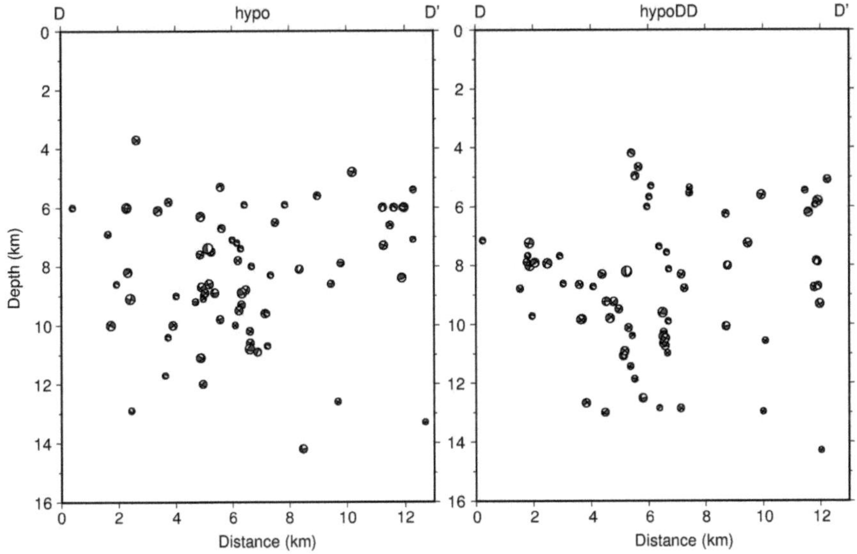

Fig. 6 Fault perpendicular cross sections for DD' profile (Fig. 3) on the Eastern Çınarcık basin cluster. For more explanation, see figure caption for Fig. 4

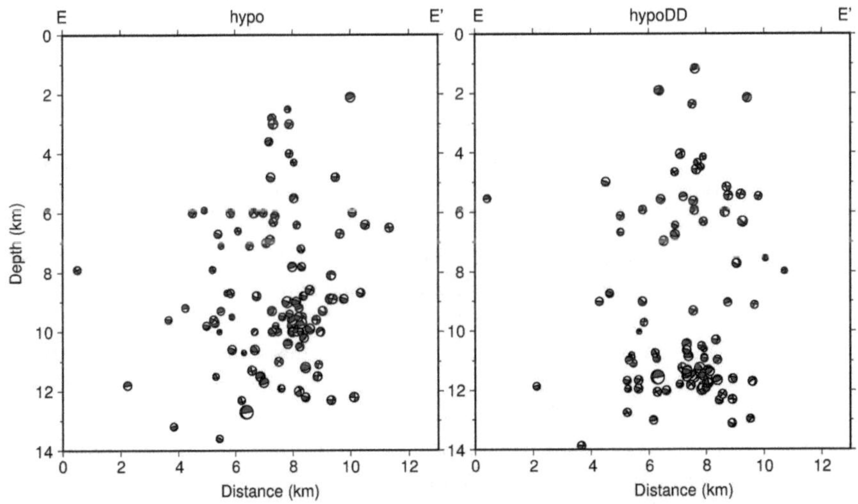

Fig. 7 Fault perpendicular cross sections for EE' profile (Fig. 3) on the Yalova cluster. For more explanation, see figure caption for Fig. 4

maximum, intermediate and minimum principal stress axes are obtained as; 285°/75°, 115°/7°, and 201°/2°, respectively.

Furthermore, an extensional motion subsists on the aftershock activity of the 1999 Izmit Earthquake which has a pure right lateral strike-slip fault plane solution (Pınar

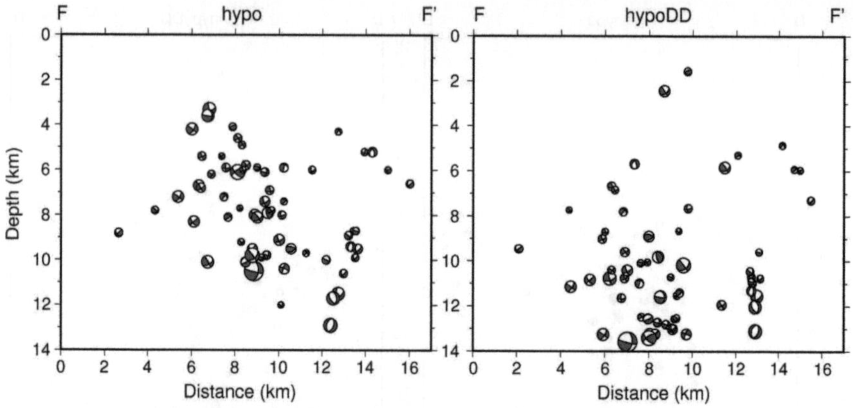

Fig. 8 Fault perpendicular cross sections for FF' profile (Fig. 3) on the Gemlik cluster. For more explanation, see figure caption for Fig. 4

et al. 2009; Bohnhoff et al. 2006; Pinar et al. 2003; Karabulut et al. 2002; Özalaybey et al. 2002). Additionally, as a result of our double-difference locations, our FPS derivations in YC represent normal and some oblique faulting mechanisms, both of which are consistent with other recent individual FPS studies (Korkusuz Öztürk et al. 2015, Karabulut et al. 2011; Örgülü 2011), and composite FPS studies (Bulut et al. 2009; Sato et al. 2004).

Horizontal movements of locations of events in this cluster via implementation of Hypodd are very minor (Fig. 3), while vertical movements from deep to shallow are relatively high (Fig. 7). Moreover, some investigations based on the analysis of aftershock activity of the 17 August 1999 Izmit Earthquake find that earthquakes in Yalova are not deeper than 5 km regarding CMT locations (Pınar et al. 2003, 2001) and some other relocations (Pınar et al. 2009). On the other hand, vertical alignments of earthquakes in our previous (individual locations) and current (relative locations) studies present a seismogenic zone between ~2 and 14 km depth interval (Fig. 7). This seems consistent with results from previous studies indicating zones between 5–12.5 and 3–14 km depth intervals (Ito et al. 2002; Özalaybey et al. 2002, respectively). Also, the squeezed earthquake activity between 10 and 12.5 km depths might originate from the aftershock activity of the 12 March 2008 ML4.6 event (Korkusuz Öztürk et al. 2015; Appendix 1). In addition, we introduce a new relatively silent zone between ~7 and 9 km depths from the results of double-difference locations, as it might be a result of hot springs in the region if this area exits just below the hot springs.

5.5 *Gemlik Cluster (5-GC)*

In the GC, we derive double-difference locations of 63 earthquakes where we have previously obtained individual locations of hypocenters (Table 1) and orientations of the principal stress axes (Table 2). Azimuth/plunge values of the orientations of maximum, intermediate and minimum principal compressive stress axes are calculated as; 102°/58°, 285°/30°, and 189°/5°, respectively. Additionally, the earthquake activity of this cluster starts with the 24 October 2006 ML5.2 event and its aftershock activity.

Focal mechanism solutions and orientations of stress tensors present an extensional motion. In addition, NNS trended normal faulting mechanisms, some of which with little right lateral components, are mainly distributed in the Gemlik Gulf, where right lateral strike-slip and oblique faulting mechanisms are mostly onshore. Well-defined FPSs with low location errors and high number of P-wave first motion polarities of moderate size earthquakes in the Gemlik Gulf also have dominant extension rather than right lateral motion (Korkusuz Öztürk et al. 2015).

Furthermore, only a very small amount of permutation exits in the horizontal positions of earthquakes after the double-difference location technique (Fig. 3), while small changes in places towards deep is visible from the cross sections (Fig. 8). Also, a few strings with 6–10 events may be the representative of that those earthquakes are originated from the branches of the main fault the segment passing through the Gemlik Gulf. In addition, the seismic zone may be situated between ~5 and 13 km depth interval.

5.6 *Eastern Central Marmara Basin Cluster (6-ECMBC)*

In the East of the Central Marmara basin we re-read P and S phases, and determine P-wave first motion polarities for 53 earthquakes. The number of events decreases to 27 by virtue of the criteria that each earthquake must have minimum 10 high quality P-wave first motion polarities and maximum 1 inconsistent station. Next, selected earthquakes are located both individually (hypo) and relatively (Hypodd), and their focal mechanisms are derived. In addition, stress tensor inversion operation is done using individual locations (Figs. 9 and 10).

Owing to the locating at the Northern Marmara Sea, the station coverage of the Central Marmara basin has been an important problem up until now. Nevertheless, we could analyze the ECMBC with an average of 67° azimuthal gap in station coverage by merging data from KEORI and TUBITAK. The average location error in latitude is 1.3 ± 0.3 km, longitude is 1.3 ± 0.4 km and depth is 3.2 ± 1.0 km (Table 1). Most of the FPSs of earthquakes dominate NE-SW trended extension, but some of them also have right lateral components (Fig. 9). Although a previous work (Örgülü 2011) discovers a pure right lateral strike-slip stress structure by the analysis of 13 events within the overall Central Marmara basin, we acquire an extensional state of stress

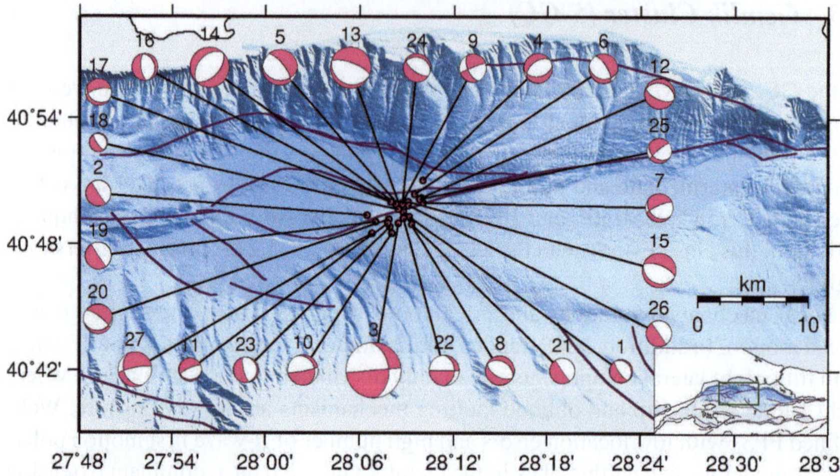

Fig. 9 Fault plane solutions of 27 earthquakes in the Eastern Central Marmara basin cluster, used for the stress tensor calculations

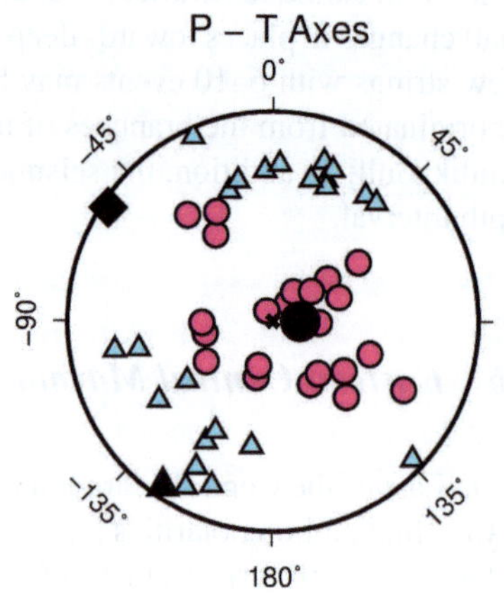

Fig. 10 Lower hemisphere projections of the P (Pink circles) and T (Turquoise triangles) axes of simultaneously identified individual FPSs for six earthquake clusters. Black circles, diamonds and triangles are the maximum, intermediate and minimum principal compressive stress axes, respectively

in the eastern part of the region. The azimuth/plunge values of maximum, intermediate and minimum principal stress axes are; 90°/79°, 304°/3°, 214°/5°, respectively (Fig. 10, Table 2).

Comparison of individual and relative location applications reveals that modest horizontal and vertical shifts exist, making hypocenters of earthquakes closer (Fig. 11). In terms of fault perpendicular cross sections, earthquakes are mostly located between 10.5 and 14 km depths, consistent with the view that the seismic zone is shallower than the Conrad discontinuity (Yamamoto et al. 2017). Also, the Eastern Central Marmara basin cluster harbor a few earthquakes shallower than ~10.5 km; therefore, the upper part may be creeping (Fig. 11) (Klein et al. 2017; Schmittbuhl et al. 2015; Ergintav et al. 2014). Alternatively, no event is located above ~7 km depth, the most upper part may be locked, so only the zone between ~7 and 10.5 km

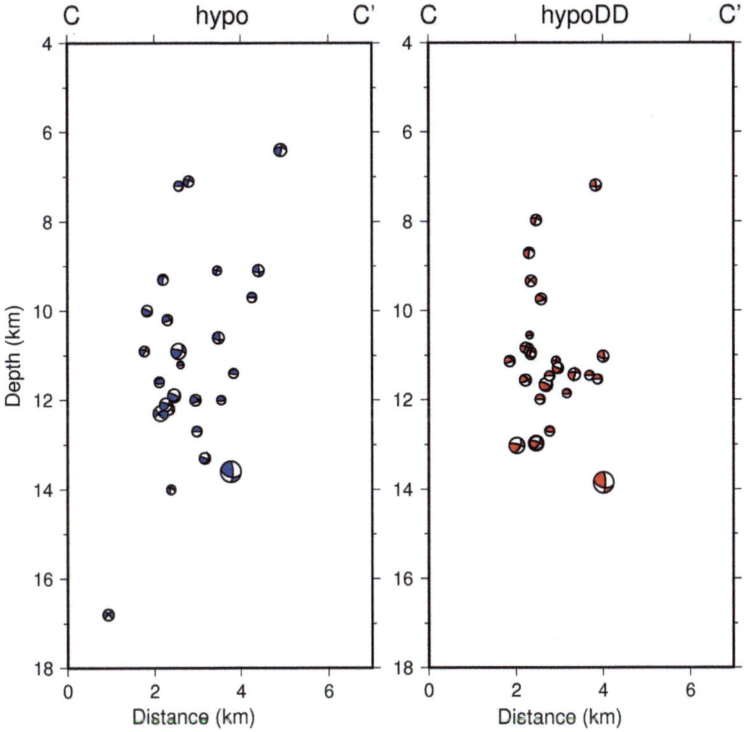

Fig. 11 Fault perpendicular cross sections for CC' profile (Fig. 5) on the Eastern Central Marmara basin cluster. For more explanation, see figure caption for Fig. 6

may be creeping. In addition, the sediment thickness of the region is ~4 km (Bayrakçı et al. 2013), and this might be the reason for the silence of the upper section (0–7 km depth interval). Furthermore, only two M > 7.0 earthquakes, a 989 AD event along the eastern segment of the NAF in the Northern Marmara Sea and a 1343 event on the Central Marmara Fault, have been recorded on the Central Marmara segment during the last 2000 years (Şengör et al. 2005; Ambraseys 2002; Pondard et al. 2007). The calculated strike-slip strain is ~2 mm/yr and the slip deficit is less than 2 m (Ergintav et al. 2014). Hence, the summary of all these current results may be the indicator of possible aseismic creep on the Central Marmara Fault and the segment may not be capable of creating a large earthquake (Ergintav et al. 2014; Meade et al. 2002) even if some energy release might exit or not during the 1766 Earthquake.

6 Conclusion

Since we have previously investigated earthquake clusters in the western and eastern Marmara Sea for the time interval between 2006 and 2011 (Korkusuz Öztürk et al. 2015), seismotectonic results of the Central Marmara Sea complement our research. In this study, we analyze seismotectonics of the Eastern Central Marmara basin via extending the time span of our data set to between 2006 and 2014, due to less seismicity of the region. Hence, for the ECMBC, data from KOERI and TUBITAK-MRC are merged, and P and S phases of 53 $ML \geq 1.5$ earthquakes are re-read. Locations, focal mechanism solutions and stress tensor orientations are obtained for 27 selected events which satisfy the criteria for well-defined locations and FPSs. Alignments of SHmax axes present a NE-SW trended extensional stress structure despite already existing a few oblique FPSs with right-lateral components (Fig. 10).

Hypodd relative locations are derived using our data catalogues from the joint data sets for the Eastern Central Marmara basin cluster and other five already analyzed clusters (Korkusuz Öztürk et al. 2015) in the Sea of Marmara. Through the use of ~150 component data (concerning three component seismometers) for each event and implementation of the hypodd relative location technique for all clusters, some significant information about seismic behavior of the different depth intervals of the fault segments could be analyzed more explicitly.

In the Ganos Offshore cluster, where a dominant dextral stress regime is active (Fig. 12), the seismogenic zone extends over ~17 km depth and the number of the earthquakes is few or almost none within the top 2 km depth, the sedimentary area. Next, Eastern Tekirdağ basin cluster, represented by NE-SW extension, might be creeping above ~10 km depth. Hence, this segment may not be capable of creating large earthquakes, as the loaded stress might be releasing via moderate size earthquakes that occur with a few months or around one year recurrence interval depending on their magnitudes. Furthermore, the upper part of the Eastern Central Marmara basin cluster may be creeping above ~10.5 km or the most upper part may be totally locked above ~7 km depth, the zone between ~7 and 10.5 km is creeping and the zone between ~10.5 and 14 km is seismically active. The Eastern Çınarcık basin cluster that have an extensional stress state might be locked above ~7 or ~10 km depths, and this segment of the main Marmara Fault may be loading stress to create a large earthquake, considering the last largest event occurred in 1766. The Yalova cluster, where a dominant extension exists, hosts a large number of small earthquakes up to ~14 km depth. On the other hand, a ~2 km thick silent zone might exit between ~7 and 9 km depths, and the shallow activity ~6 km depth might come from the existence of the hot springs. Moreover, in the Gemlik cluster, the region where offshore earthquakes are mostly extensional or oblique structures with right-lateral components and onshore ones are right lateral strike-slip patterns, the seismogenic zone is observed between ~5 and 13 km depths.

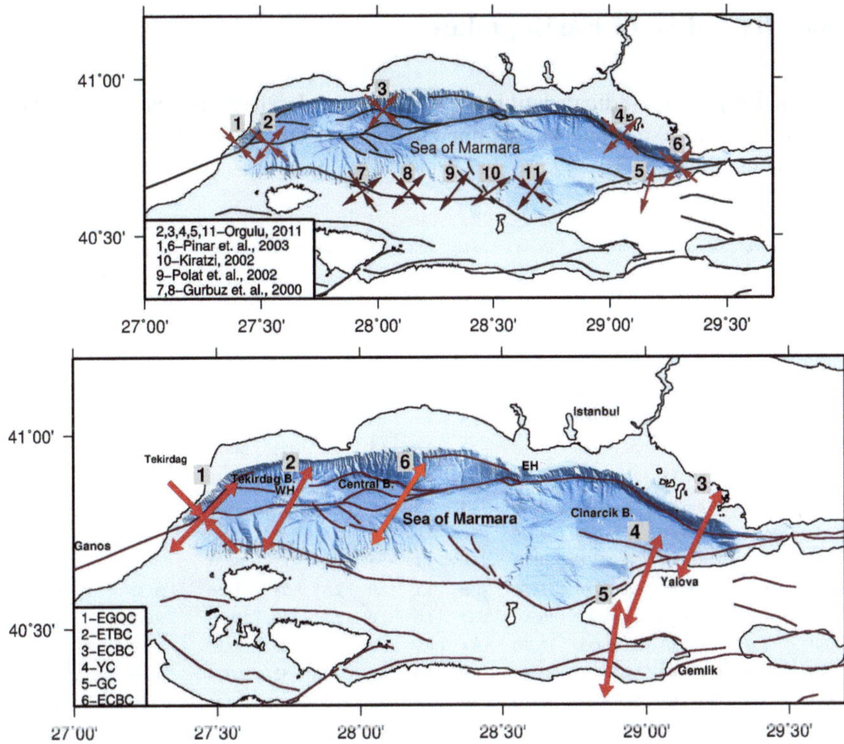

Fig. 12 The map on the top is the representation of orientations of the maximum horizontal stress axes from previous studies; 2, 3, 4, 5 (light brown) for local areas in the Sea of Marmara; 1 and 6 (light brown) for the whole western and the whole eastern Marmara areas; and 7, 8, 9, 10, 11 (dark brown) for the whole region of Marmara (Gürbüz et al. 2000; Kiratzi 2002; Polat et al. 2002; Pınar et al. 2003; Örgülü 2011). The below map is the recent states of stresses in the Sea of Marmara 5 of which are obtained by Korkusuz Öztürk et al. (2015) and one of them (6, light red) is found out during the current study. Horizontal stress alignments for the Eastern Ganos offshore (1), Eastern Tekirdağ basin (2), Eastern Çınarcık basin (3), Yalova (4), Gemlik (5) and Eastern Central Marmara basin (6) clusters

Acknowledgements We thank the National Earthquake Monitoring Center of Kandilli Observatory and Earthquake Research Institute (NEMC-KOERI), and Marmara Research Center of The Scientific and Technological Research Council of Turkey (TUBITAK-MRC) for the data supply through the TURDEP Project. We acknowledge to Semih Ergintav for giving constructive comments, and Onur Tan for his beneficial remarks. Also, we thank Mehmet Yılmazer, Korhan Umut Şemin and Metin Kahraman for software support, Zafer Öğütçü and Zümer Pabuçcu for the supply of earthquake catalogues, and Burak Korkusuz for his contribution to the preparation of this manuscript.

Appendix 1: List of Earthquakes

Erlt, Erln, Erdp values indicate latitude, longitude, and depth errors, respectively. The given Strike/Dip/Rake values are obtained by the stress tensor inversion approach of the Horiuchi et al. (1995).

Date	Time	Latitude	Longitude	Depth	Mag	Erlt	Erln	Erdp	GAP	Strike	Dip	Rake
15.1.2010	23:4	40.821999	28.152000	12.000	2.0	1.0	1.4	2.9	84	144.43	75.09	−91.44
9.5.2010	23:33	40.823002	28.108000	12.300	2.2	1.1	1.1	2.8	57	145.60	90.00	−90.00
3.10.2010	17:49	40.817001	28.141001	13.600	4.7	1.0	1.0	2.2	29	85.00	90.00	−133.05
4.10.2010	16:3	40.851002	28.167000	16.800	2.5	1.8	1.7	5.8	119	233.99	50.00	−107.36
7.10.2010	5:3	40.830002	28.143000	11.900	3.0	1.1	0.9	2.5	37	321.46	70.01	−70.44
7.10.2010	6:20	40.840000	28.158001	10.000	2.5	1.3	1.3	3.3	73	327.85	69.97	−68.20
7.10.2010	23:37	40.833000	28.167000	7.100	2.4	1.6	1.5	5.2	121	246.00	75.02	−117.03
8.10.2010	3:38	40.826000	28.146000	12.000	2.5	1.7	1.3	3.4	102	119.43	55.05	−87.86
8.10.2010	3:41	40.833000	28.145000	9.300	2.4	1.0	1.1	4.2	74	248.80	85.01	−148.25
27.10.2010	15:51	40.808998	28.134001	9.100	2.6	1.0	1.2	2.8	44	79.43	85.00	−113.14
6.4.2011	0:19	40.817001	28.129999	9.100	2.0	1.2	1.5	4.2	61	246.31	76.22	−95.93
21.4.2011	17:49	40.837002	28.165001	12.200	2.6	1.1	1.0	2.6	28	113.66	60.32	−85.68
1.5.2011	8:36	40.828999	28.143000	10.900	3.5	1.0	1.0	2.1	30	289.33	69.99	−91.75
19.5.2011	4:38	40.832001	28.139000	12.300	3.4	0.9	0.8	2.2	30	45.01	50.02	−88.15
22.5.2011	22:39	40.834000	28.150999	12.100	2.9	1.0	0.9	2.0	52	290.75	64.99	−91.52
9.6.2011	20:43	40.832001	28.138000	11.600	2.3	1.6	1.5	2.7	60	173.05	65.00	−85.67
10.6.2011	4:28	40.834000	28.132999	10.900	2.3	1.3	1.2	2.6	45	257.16	70.01	−110.08
14.6.2011	5:37	40.827999	28.141001	11.200	1.6	1.6	1.6	3.1	106	144.43	75.09	−91.44
25.8.2011	7:12	40.818001	28.117001	12.700	2.3	1.0	1.4	2.3	61	145.60	90.00	−90.00
25.8.2011	13:25	40.820000	28.131001	13.300	2.5	1.4	1.7	3.8	79	306.01	74.99	−72.11
20.9.2011	6:24	40.816002	28.155001	9.700	2.2	1.3	1.2	3.9	46	145.60	90.00	−90.00
20.9.2011	6:27	40.820999	28.146000	10.600	2.6	1.0	1.0	2.3	29	90.00	90.00	−121.67
25.3.2012	22:53	40.813999	28.132000	11.400	2.2	1.2	1.2	2.4	76	165.46	75.47	−93.71
10.5.2012	8:34	40.832001	28.145000	10.200	2.4	1.3	1.2	2.6	89	124.59	65.00	−81.06
29.9.2012	17:28	40.835999	28.163000	14.000	2.1	1.8	2.0	3.0	95	230.62	80.00	−122.18
16.9.2013	18:13	40.831001	28.150999	7.200	2.2	0.9	1.0	3.1	55	145.60	90.00	−90.00
7.3.2014	20:51	40.801998	28.125999	6.400	2.8	3.4	3.0	5.4	81	258.79	79.99	−124.08

Appendix 2: List of Stations

See Tables 3, 4, 5 and 6.

Table 3 List of 40 BB stations of Kandilli Observatory & Earthquake Research Institute

No	Station name	Seismometer type	Lat.	Lon.	Altitude	Starting date	Org. name
1	ADVT	CMG-3T	40.4332	29.7383	193	19.05.2006	KOERI
2	ALT	CMG-3T	39.0552	30.1103	1060	31.03.2007	KOERI
3	ARMT	CMG-3ESP	40.5683	28.866	320	17.12.2007	KOERI
4	BALB	CMG-3T	39.64	27.88	120	05.06.2007	KOERI
5	BGKT	CMG-3ESP	41.181	28.773	80	29.05.2007	KOERI
6	CANB	CMG-6TD	40.0167	2703.75	229	12.01.2004	KOERI
7	CAVI	CMG-3ESP	40.2018	29.8377	670	27.09.2007	KOERI
8	CRLT	CMG-3ESP	41.129	27.736	230	16.05.2007	KOERI
9	CTKS	CMG-3ESP	41.2373	28.5072	47	28.09.2007	KOERI
10	CTYL	CMG-3T	41.476	28.2897	77	29.09.2007	KOERI
11	EDC	CMG-3T	40.3468	27.8633	269	27.11.2008	KOERI
12	EDRB	CMG-3T	41.847	26.7437	209	18.04.2007	KOERI
13	ENEZ	CMG-3T	40.7362	26.153	100	21.06.2006	KOERI
14	ERIK	CMG-3ESP	40.6708	26.5132	35	02.07.2008	KOERI
15	EZN	CMG-3ESP	39.8267	26.3258	48	09.11.2007	KOERI
16	GADA	CMG-3T	40.1908	25.8987	130	06.07.2006	KOERI
17	GELI	CMG-3ESP	40.398	26.4742	130	05.07.2008	KOERI
18	GEMT	CMG-3T	40.435	29.189	220	07.07.2006	KOERI
19	GONE	CMG-3ESP	40.0467	27.686	140	27.06.2008	KOERI
20	GULT	CMG-3ESP	40.432	30.515	930	15.09.2007	KOERI
21	HRTX	CMG-3ESP	40.8217	29.668	645	25.06.2008	KOERI
22	ISK	CMG-3T	41.0657	29.0592	132	25.01.2007	KOERI
23	KCTX	CMG-3ESP	40.2625	28.3353	445	11.07.2008	KOERI
24	KDZE	CMG-3T	41.3132	31.443	410	01.07.2006	KOERI
25	KLYT	CMG-3T	41.253	29.042	30	18.05.2006	KOERI
26	KRBG	CMG-3ESP	40.3932	27.2977	75	27.06.2008	KOERI
27	LAP	CMG-3ESP	40.3727	26.7602	200	27.12.2007	KOERI
28	MDNY	CMG-3ESP	40.3708	28.8847	115	09.07.2008	KOERI
29	MDUB	CMG-3T	40.4712	31.1978	1108	22.05.2008	KOERI
30	MFTX	CMG-40T	40.7867	27.2812	924	10.10.1998	KOERI
31	MRMT	CMG-3T	40.6058	27.5837	702	09.10.2008	KOERI
32	PHSR	CMG-40T	41.6308	27.5238	263	09.07.2009	KOERI
33	RKY	CMG-3ESP	40.6875	27.1777	687	16.05.2006	KOERI
34	SILT	CMG-3ESP	41.153	29.643	100	31.05.2007	KOERI
35	SLVT	CMG-3ESP	41.23	28.21	180	01.06.2007	KOERI
36	SPNC	CMG-3ESP	40.686	30.3083	190	27.06.2008	KOERI
37	SVRH	CMG-3T	39.447	31.5232	1000	27.12.2007	KOERI
38	TKR	CMG-3ESP	40.9902	27.5357	140	18.05.2007	KOERI
39	TVSB	CMG-3ESP	39.4497	29.4615	1090	12.02.2009	KOERI
40	YLVX	CMG-40T	40.5667	29.3728	829	21.10.1998	KOERI

Table 4 List of 50 Stations of the TURDEP Project

No	Station Name	Seismometer Type	Lat.	Lon.	Altitude	Starting Date	Org. Name
1	ALET	CMG-LE3D	41.06624	28.60557	91	20.10.2009	TUBITAK
2	ALTM	CMG-3ESPC	41.08800	28.74000	18	25.07.2006	TUBITAK
3	ARCE	CMG-J	40.82616	29.36025	45	09.07.2009	TUBITAK
4	ATIM	CMG-3T	40.08300	27.56333	230	24.01.2008	TUBITAK
5	BAHT	CMG-LE3D	41.08783	28.69179	142	12.10.2009	TUBITAK
6	BEY2	CMG-LE3D	41.00294	28.64531	170	18.03.2010	TUBITAK
7	BEYT	CMG-LE3D	41.00459	28.63510	214	07.10.2009	TUBITAK
8	BOZM	CMG-40T	40.53400	28.78200	119	28.04. 2007	TUBITAK
9	BUYM	CMG-3T	40.85233	29.11800	231	30.11.2006	TUBITAK
10	BZGM	CMG-40T	40.17267	26.9865	180	11.07.2008	TUBITAK
11	CAN	CMG-3T	40.0275	2703.77	200	24.08.2004	TUBITAK
12	CALI	CMG-L4C	40.16519	28.92000	179	10.02.2011	TUBITAK
13	CMHM	CMG-3ESPC	40.01200	27.97000	205	05.09.2006	TUBITAK
14	EREM	CMG-3ESPC	40.04533	28.89117	657	29.03.2007	TUBITAK
15	ESKM	CMG-40T	40.60683	28.94533	20	31.08.2006	TUBITAK
16	GBZM	CMG-3ESPC	40.78600	29.45000	270	08.07.2005	TUBITAK
17	GOZT	CMG-J	40.89078	29.25363	213	13.07.2009	TUBITAK
18	IBBT	CMG-J	40.86608	29.32315	115	09.07.2009	TUBITAK
19	IGDM	CMG-3ESPC	40.26400	29.20133	165	27.03.2007	TUBITAK
20	ISU2	CMG-LE3D	40.99775	28.72363	96	23.03.2010	TUBITAK
21	ISUT	CMG-LE3D	40.98875	28.72382	85	08.10.2009	TUBITAK
22	KLCM	CMG-3ESPC	40.63300	29.39800	138	01.09.2006	TUBITAK
23	KKZM	CMG-40T	41.11600	27.34400	150	28.07.2006	TUBITAK
24	KMRM	CMG-3ESPC	40.41800	27.06900	40	09.09.2006	TUBITAK
25	KNLM	CMG-40T	40.27000	27.52600	178	05.09.2006	TUBITAK
26	KRCM	CMG-3ESPC	40.26517	28.33233	443	30.03.2007	TUBITAK
27	KURN	CMG-J	40.95496	29.33150	210	07.16.2009	TUBITAK
28	KVKM	CMG-3ESPC	40.60400	26.88767	75	03.07.2007	TUBITAK
29	MADM	CMG-40T	40.65367	27.66467	40	06 09.2006	TUBITAK
30	MARM	CMG-40T	40.96700	27.96100	43	27.07 2006	TUBITAK
31	MSDM	CMG-40T	40.34950	28.60033	201	01.04.2007	TUBITAK
32	MYCM	CMG-40T	41.03250	27.71333	99	03.07.2008	TUBITAK
33	NEVM	CMG-3T	39.95400	27.26300	329	17.10.2006	TUBITAK
34	NUKT	CMG-LE3D	41.02651	28.75885	56	07.10.2009	TUBITAK
35	OMRT	CMG-LE3D	40.97066	28.60385	56	09.10.2009	TUBITAK
36	SABA	CMG-J	40.89365	29.38134	161	13.07.2009	TUBITAK

(continued)

Table 4 (continued)

No	Station Name	Seismometer Type	Lat.	Lon.	Altitude	Starting Date	Org. Name
37	SAKI	CMG-J	40.83204	29.27543	50	16.07.2009	TUBITAK
38	SGTM	CMG-40T	40.76683	27.10783	295	11.04.2001	TUBITAK
39	SLVM	CMG-3T	41.07312	28.14034	30	05.05.2008	TUBITAK
40	SNLM	CMG-3ESPC	41.22800	28.20900	173	26.07.2006	TUBITAK
41	TEPT	CMG-LE3D	41.06394	28.50939	91	13.10.2009	TUBITAK
42	TRNM	CMG-40T	40.50500	27.77800	80	00.01.2006	TUBITAK
43	YNKM	CMG-3ESPC	40.82567	27.39667	370	27.03.2007	TUBITAK
44	BALY	CMG-3T	39.74033	27.61933	645	24.05.2007	AFAD
45	BOZC	CMG-3T	39.84190	26.0528	195	18.10.2006	AFAD
46	DURS	CMG-3T	39.60100	28.47000	960	06.06.2007	AFAD
47	ELBA	CMG-3T	41.14667	28.43050	331	29.09.2005	AFAD
48	SART	CMG-3T	40.68883	27.18000	679	11.09.2007	AFAD
49	EADA	CMG-6T	40.82117	29.29050	6	19.02.2009	SENTEZ
50	SYY1	CMG-6T	40.94417	29.12667	30	05.05.2008	SENTEZ

Table 5 List of 10 Short-period stations of Kandilli Observatory & Earthquake Research Institute

No	Station name	Seismometer type	Lat.	Lon.	Altitude	Starting date	Org. name
1	BADT	SS1	4051.14	29.0705	175	28.02.2001	KOERI
2	BNT	SS1	40.3542	27.8950	353	1984	KOERI
3	DST	L4-C	39.6040	28.6192	625	07.1970	KOERI
4	ESKT	CMG-40 V	39.5222	30.8497	1289	03.10.2002	KOERI
5	EYL	Willmore	40.5658	30.1250	1160	11.10.1989	KOERI
6	GPA	CMG-40 V	40.2863	30.3183	560	1970	KOERI
7	HRT	Willmore	4049.30	2940.08	645	1980	KOERI
8	IZI	Willmore	40.3368	29.4728	910	01.08.1989	KOERI
9	ORLT	Mark	40.0462	28.8958	649	04.01.2002	KOERI
10	OSM	L4-C	4036.12	2942.00	820	23.08.2000	KOERI

Table 6 List of 5 SBO stations of Kandilli Observatory & Earthquake Research Institute

No	Station name	Seismometer type	Lat.	Lon.	Altitude	Starting Date	Org. Name
1	OBS1	CMG-3T	40.705638	29.149183	−1260	31.12.2010	KOERI
2	OBS2	CMG-3T	40.878619	28.514247	−810	03.11.2010	KOERI
3	OBS3	CMG-3T	40.884783	27.975100	−1204	21.01.2011	KOERI
4	OBS4	CMG-3T	40.828184	27.535460	−1114	22.12.2010	KOERI
5	OBS5	CMG-3T	40.631132	28.880385	−368	09.06.2010	KOERI

References

Aksoy ME, Meghraoui M, Vallée M, Çakır Z (2010) Rupture characteristics of the AD 1912 Mürefte (Ganos) Earthquake segment of the North Anatolian Fault (Western Turkey). Geol Soc Am 38:37–57

Aksoy ME, Meghraoui M, Çakır Z, Battlo J (2012) The 9 August 1912 Mürefte Earthquake, PANAF, Istanbul Technical University, Turkey

Aksu AE, Calon TJ, Hiscott RN (2000) Anatomy of the North Anatolian Fault Zone in the Marmara Sea, Western Turkey: extensional basins above a continental transform. GSA Today, June 2000

Ambraseys NN, Finkel CF (1987) The Saros-Marmara Earthquake of 9 August 1912. Earthq Eng Struct Dyn 15:189–211

Ambraseys N (2002) The seismic activity of the Marmara Sea region over the last 2000 years. Bull Seismol Soc Am 92:1–18

Armijo R, Meyer B, Navarro S, King G, Barka A (2002) Asymmetric slip partitioning in the Sea of Marmara Pull-Apart: a clue to propagation processes of the North Anatolian Fault? Terra Nova 14:80–86

Armijo R, Pondard N, Meyer B, Uçarkus G, de Lepinay BM, Malavieille J, Dominguez S, Gustcher M, Schmidt S, Beck C, Çagatay N, Çakir Z, Imren C, Eris K, Natalin B, Özalaybey S, Tolun L, Lefevre I, Seeber L, Gasperini L, Rangin C, Emre O, Sarikavak K (2005) Submarine fault scarps in the sea of Marmara Pull-A part (North Anatolian Fault): implications for seismic hazard in Istanbul. AGU Geochem Soc 1525–2027

Barka A (1996) Slip distribution along the North Anatolian Fault associated with the large earthquakes of the period 1939 to 1967. Bull Seismol Soc Am 86:1238–1254

Barka A, Akyüz A, Altunel HS, Sunal G, Çakır Z, Dikbaş A, Yeli B, Armijo R, Meyer B, de Chabalier JB, Rockwel T, Dolan JR, Hartleb R, Dawson T, Christoferson S, Tucker A, Fumal T, Langridge R, Stenner H, Lettis W, Bachhuber J, Page W (2002) The surface rupture and slip distribution of the 17 August, 1999 Izmit Earthquake, M = 7.4, North Anatolian fault. Bull Seismol Soc Am 92:43–60

Başarır Baştürk N, Özel NM, Caciagli M (2016) Seismic parameters re-determined from historical seismograms of 1935—Erdek—Marmara Island and 1963—Çınarcık Earthquakes. Earth Planets Space 68:158. https://doi.org/10.1186/s40623-016-0528-8

Bayrakci G, Laigle M, Becel A, Hirn A, Taymaz T, Yolsal-Çevikbilen S (2013) SEISMARMARA team, 3-D sediment-basement tomography of the Northern Marmara trough by a dense OBS network at the nodes of a grid of controlled source profiles along the North Anatolian Fault. Geophys J Int 194:1335–1357. https://doi.org/10.1093/gji/ggt211

Beyhan G, Selim HH (2007) Tectonics of the North Anatolian Fault located in the sea of Marmara according to seismic reflection data. In: International earthquake symposium, Kocaeli

Boaz R (2009) PQL II. http://www.passcal.nmt.edu/content/pql-ii-program-viewing-data

Bohnhoff M, Grosser H, Dresen G (2006) Strain partitioning and stress rotation at the North Anatolian Fault Zone from aftershock focal mechanisms of the 1999 Izmit Mw = 7.4 earthquake. Geophys J Int 160:373–385

Bohnhoff M, Dresen G, Bulut F, Nurlu M, Akin D, Kılıç T, Ito H, Malin P (2012) GONAF—a deep Geophysical Observatory at the North Anatolian Fault, Kashiwazaki, 8 Nov 2012

Bohnhoff M, Bulut F, Dresen G, Malin PE, Eken T, Aktar M (2013) An earthquake gap south of Istanbul. Nat Commun. https://doi.org/10.1038/ncomms2999

Bouchon M, Toksöz MN, Karabulut H, Bouin MP, Dietrich M, Aktar M, Edie M (2002) Space and time evolution of rupture and faulting during the 1999 Izmit (Turkey) earthquake. Bull Seismol Soc Am 92:256–266. https://doi.org/10.1785/0120000845

Bulut F, Aktar M (2007) Accurate relocation of İzmit Earthquake (Mw = 7.4, 1999) aftershocks in Çınarcık Basin using double difference method. GRL 34:L10307–L10307. https://doi.org/10.1029/2007gl029611

Bulut F, Bohnhoff M, Ellsworth WL, Aktar M, Dresen G (2009) Microseismicity at the North Anatolian Fault in the Sea of Marmara Offshore Istanbul, NW Turkey. J Geophys Res. https://doi.org/10.1029/2008jb006244

Cemen I, Gokten E, Varol B, Kilic R, Ozaksoy V, Erkmen C, Pinar A (2000) Turkish earthquakes reveal dynamics of fracturing along a major strike-slip fault zone, EOS. Trans Am Geophys Union 81:309–320

Dewey JF, Şengör AMC (1979) Aegan and surrounding regions: complex multiplate and continuum mechanics in a convergence zone. Bull Geol Soc Am 84:3137–3180

Erdik M, Demircioğlu MB, Şeşetyan K, Durukal E, Siyahi B (2004) Earthquake hazard in Marmara Region, Turkey. Soil Dyn Earthq Eng 24:605–631

Erdik M (2013) Earthquake risk in Turkey. Science 341:724. https://doi.org/10.1126/science.1238945

Ergintav S, Doğan U, Gerstenecker C, Çakmak R, Belgen A, Demirel H, Aydın C, Reilinger R (2007) A snapshot (2003–2005) of the 3D postseismic deformation for the 1999, Mw = 7.4 Izmit Earthquake in the Marmara Region, Turkey, by first results of joint gravity and GPS monitoring. J Geodyn 44:1–18

Ergintav S, Reilinger RE, Çakmak R, Floyd M, Çakır Z, Doğan U, King RW, McClusky S, Özener H (2014) Istanbul's earthquake hot spots: geodetic constraints on strain accumulation along faults in the Marmara seismic gap. Geophys Res Lett. https://doi.org/10.1002/2014gl060985

Flerit F, Armijo R, King GCP, Meyer B, Barka A (2003) Slip partitioning in the Sea of Marmara pull-apart determined from GPS velocity vectors. Geophys J Int 1–7 (2003)

Frechet J (1985) Sismogenese et doublets sismiques, These d'Etat, Universite Scientifique et Medicale de Grenoble, 206 pp

Geiger L (1910) Herdbestimming bei Erdbeben aus den Ankunftszeiten, K. Ges. Wiss. Gött. 4:331–349

Geli L, Henry P, Zitter T, Dupré S, Tryon M, Çağatay MN, Mercier de Lépinay B, Le Pichon X, Şengör AMC, Görür N, Natalin B, Uçarkuş G, Özeren S, Volker D, Gasperinin L, Burnard P, Bourlange S (2008) Gas emissions and active tectonics within the submerged section of the North Anatolian Fault Zone in the Sea of Marmara. The Marmara Scientific Party

Gephart JW, Forsyth DW (1984) An improved method for determining the regional stress tensor using earthquake focal mechanism data: application to the San Fernando earthquake sequence. J Geophys Res 9305–9320

Gürbüz C, Aktar M, Eyidoğan H, Cisternas A, Haessler H, Barka A, Ergin M, Türkelli N, Polat O, Üçer SB, Kuleli S, Baris S, Kaypak B, Bekler T, Zor E, Bicmen F, Yoruk A (2000) The seismotectonics of the Marmara Region (Turkey): results from a microseismic experiment. Tectonophysics 1–17

Horiuchi S, Rocco G, Hasegawa A (1995) Discrimination of fault planes from auxiliary planes based on simultaneous determination of stress tensor and a large number of fault plane solutions. J Geophys Res 8327–8338

Hubert-Ferrari A, Barka A, Jacques E, Nalbant S, Meyer B, Armijo R, Tapponnier P, King GP (2000) Seismic hazard in the Marmara Sea region following the 17 August 1999 Izmit Earthquake. Nature 404:269–273

Ito A, Ucer SB, Baris S, Honkura Y, Nakamura A, Kono T, Pektas R, Komut T, Hasegawa A, Isikara AM (2002) Aftershock activity of 1999 Izmit Earthquake, Turkey, revealed from microearthquake observations. Bull Seismol Soc Am 418–427

Karabulut H, Bouin M, Bouchon M, Dietrich M, Cornou C, Aktar M (2002) The seismicity in the eastern Marmara Sea after the 17 August 1999 Izmit Earthquake. Bull Seismol Soc Am 387–393

Karabulut H, Özalaybey S, Taymaz T, Aktar M, Selvi O, Kocaoğlu A (2003) A tomographic image of the shallow crustal structure in the eastern Marmara. Geophys Res Lett. https://doi.org/10.1029/2003gl018074

Karabulut H, Schmittbuhl J, Özalaybey S, Lengline O, Kömeç-Mutlu A, Durand V, Bouchon M, Daniel G, Bouin MP (2011) Evolution of the seismicity in the eastern Marmara Sea a decade before and after the 17 August 1999 Izmit Earthquake. Tectonophysics 17–27

Ketin I (1948) Über die tektonisch-mechanischen Folgerungen aus den großen Anatolischen Erdbeben des letzten Dezenniums. Geol Rundschau 36:77–83
King GCP, Hubert-Ferrari A, Nalbant S, Meyer B, Armijo R, Bowman D (2001) Coulomb interactions and the 17 August Izmit, Turkey earthquake. Earth Planet Sci 557–569
Kinscher J, Krüger F, Woith H, Lühr BG, Hintersberger E, Irmak TS, Baris S (2013) Seismotectonics of the Armutlu peninsula (Marmara Sea, NW Turkey) from geological field observation and regional moment tensor inversion. Tectonophysics 608:980–995
Kiratzi AA (2002) Stress tensor inversions along the Westernmost North Anatolian Fault Zone and its continuation into the North Aegean Sea. Geophys J Int 360–376
Klein E, Duputel Z, Masson F, Yavasoglu H, Agram P (2017) Aseismic slip and seismogenic coupling in the Marmara Sea: what can we learn from onland Geodesy? Geophys Res Lett 44:3100–3108. https://doi.org/10.1002/2017GL072777
Korkusuz Y (2012) The present day stress states in the Marmara Region. MSc thesis, Boğaziçi University, Kandilli Observatory and Earthquake Research Institute, Department of Geophysics. https://tez.yok.gov.tr/UlusalTezMerkezi/tarama.jsp
Korkusuz Öztürk Y, Özel NM, Özbakır AD (2015) States of local stresses in the Sea of Marmara through the analysis of large numbers of small earthquakes. Tectonophysics 665:37–57. https://doi.org/10.1016/j.tecto.2015.09.027
Laigle M, Becel A, Voogd B, Hirn A, Taymaz T, Özalaybey S (2008) A first deep seismic survey in the Sea of Marmara: deep basins and whole crust architecture and evolution. Earth Planet Sci Lett 270:168–179
Le Pichon X, Şengör AMC, Demirbağ E, Rangın C, İmren C, Armıjo R, Görür N, Çağatay N, Mercıer De Lepınay B, Meyer B, Saatçiler R, Tok B (2001) The active main Marmara Fault. Earth Planet Sci Lett 192:595–616
Lienert BRE (1991) Report on modifications made to hypocenter. Technical Report, Institute of Solid Earth Physics, University of Bergen, Bergen, Norway
Lienert BRE, Berg E, Frazer LN (1986) Hypocenter: an earthquake location method using centered, scaled, and adaptively least squares. Bull Seismol Soc Am 76:771–783
Lienert BR, Havskov J (1995) A computer program for locating earthquakes both locally and globally. Seismol Res Lett 66(5)
Lisle R, Orife T, Arlegui L (2001) A stress inversion method requiring only fault slip sense. J Geophys Res Solid Earth 106:2281–2289
Meade BJ, Hager BH, McClusky SC, Reilinger RE, Ergintav S, Lenk O, Barka A, Ozener H (2002) Estimates of seismic potential in the Marmara region from block models of secular deformation constrained by GPS measurements. Bull Seismol Soc Am 92:208–215
Michael AJ (1984) Determination of stress from slip data: faults and folds. J Geophys Res 11517–11526
McClusky S, Reilinger R, Mahmoud S, Ben Sari D, Tealeb A (2003) GPS constraints on Africa (Nubia) and Arabia plate motions. Geophys J Int 126–138
McKenzie D (1972) Active tectonics of the Mediterranean region. Geophys J Roy Astron Soc 30:109–185
McKenzie D (1978) Active tectonics of the Alpine-Himalayan belt: the Aegan Sea and surrounding regions. Geophys J Roy Astron Soc 55:217–254
Oglesby DD, Mai PM (2012) Fault geometry rupture dynamics and ground motion from potential earthquakes on the North Anatolian Fault under the Sea of Marmara. Geophys J Int 188:1071–1087. https://doi.org/10.1111/j.1365-246x.2011.05289.x
Okay AI, Kaşlılar-Özcan A, İmren C, Boztepe-Güney A, Demirbağ E, Kuşçu İ (2000) Active faults and evolving strike-slip basins in the Marmara Sea, Northwest Turkey: a multichannel seismic reflection study. Tectonophysics 189–218
Okay AI, Tüysüz O, Kaya Ş (2004) From Transpression to Transtension: changes in morphology and structure around a bend on the North Anatolian Fault in the Marmara Region. Tectonophysics 259–282

Otsubo M, Yamaji A, Kubo A (2008) Determination of stresses from heterogeneous focal mechanism data: an adaptation of the multiple inverse method. Tectonophysics 150–160
Örgülü G (2011) Seismicity and source parameters for small-scale earthquakes along the splays of the North Anatolian Fault (NAF) in the Marmara Sea. Geophys J Int 385–404
Örgülü G, Aktar M (2001) Regional moment tensor inversion for strong aftershocks of the August 17, 1999 Izmit Earthquake (Mw = 7.4). Geophys Res Lett 28(2):371–374
Özalaybey S, Ergin M, Aktar M, Tapırdamaz C, Biçmen F, Yörük A (2002) The 1999 Izmit Earthquake sequence in Turkey: seismological and tectonic aspects. Bull Seismol Soc Am 376–386
Parsons T (2004) Recalculated probability of M ≥ 7 earthquakes beneath the Sea of Marmara, Turkey. J Geophys Res 109. https://doi.org/10.1029/2003jb00266
Pınar A, Honkura Y, Kuge K (2001) Seismic activity triggered by the 1999 Izmit Earthquake and its implications for the assessment of future seismic risk. Geophys J Int 146:F1–F7
Pınar A, Kuge K, Honkura Y (2003) Moment tensor inversion of recent small to moderate sized earthquakes: implications for seismic hazard and active tectonics beneath the Sea of Marmara. Geophys J Int 133–145
Pinar A, Ucer SB, Honkura Y, Sezgin N, Ito A, Baris S, Kalafat D, Matsushima M, Horiuchi S (2009) Spatial variation of stress field along the fault rupture zone of the 1999 Izmit Earthquake. Earth Planets Space 1–14
Polat O, Haessler H, Cisternas A, Philip H, Eyidoğan H, Aktar M, Frogneux M, Comte D, Gürbüz C (2002) The Izmit (Kocaeli), Turkey Earthquake of 17 August 1999: previous seismicity, aftershocks, and seismotectonics. Bull Seismol Soc Am 361–375
Pondard N, Armijo R, King GCP, Meyer B, Flerit F (2007) Fault interactions in the Sea of Marmara pull-apart (North Anatolian Fault): earthquake clustering and propagating earthquake sequences. Geophys J Int 171(3):1185–1197
Rangin C, Demirbag E, Imren C, Crusson A, Normand A, Le Drezen E, Le Bot A (2001) Marine atlas of the Sea of Marmara (Turkey), Ifremer, Brest, France
Reasenberg PA, Oppenheimer D (1985) FPFIT, FPPLOT, and FPPAGE: Fortran computer programs for calculating and displaying earthquake fault plane solutions. Depth of Interior, US Geological Survey, Open-file report, 109, 85–739
Reilinger R, McClusky S, Vernant P, et al (2006) GPS constraints on continental deformation in the Africa-arabia-Eurasia continental collision zone and implications for the dynamic of plate interactions. J Geophys Res. http://doi.org/10.1029/2005JB004051
Reuter HI, Nelson A, Jarvis A (2007) An evaluation of void filling interpolation methods for SRTM data. Int J Geogr Inf Sci 21(9):983–1008
Rivera L, Cisternas A (1990) Stress tensor and fault plane solutions for a population of earthquakes. Bull Seismol Soc Am 80(3):600–614
Sato T, Kasahara J, Taymaz T, Ito M, Kamimura A, Hayakawa T, Tan O (2004) A study of microearthquake seismicity and focal mechanisms within the Sea of Marmara (NW Turkey) using ocean bottom seismometers (OBSs). Tectonophysics 303–314
Sato K (2006) Incorporation of incomplete fault-slip data into stress tensor inversion. Tectonophysics 319–330
Schmittbuhl J, Karabulut H, Lengline O, Bouchon M (2015) Seismicity distribution and locking depth along the main Marmara Fault, Turkey. Geochem Geophys Geosyst 17:954–965. https://doi.org/10.1002/2015GC006120
Shan YH, Lin G, Li Z (2004) An inverse method to determine the optimal stress from imperfect fault data. Tectonophysics 387:205–215
Şaroğlu F, Emre Ö, Kuşçu İ (1992) Active fault map of Turkey, General Directorate of the Mineral Research and Exploration, Ankara, Turkey, 2 sheets, 1.2 000 000 scale
Şengör AMC, Tüysüz O, İmren C, Sakınç M, Eyidoğan H, Görür N, Le Pichon X, Rangin C (2005) The North Anatolian Fault: a new look. Annu Rev Earth Planet Sci 33:37–112
Taymaz T, Jackson J, McKenzie D (1991) Active tectonics of the north and central Aegean Sea. Geophys J Int 433–490

Tunç B, Çaka D, Irmak TS, Woith H, Tunç S, Bariş Ş, Özer MF, Lühr BG, Günther E, Grosser H, Zschau J (2011) The Armutlu network: an investigation into the seismotectonic setting of Armutlu-Yalova-Gemlik and the surrounding regions. Ann Geophys. https://doi.org/10.4401/ag-4877

Uçarkuş G, Çakır Z, Armijo R (2010) Western termination of the Mw 7.4, 1999 Izmit Earthquake rupture: implications for the expected large earthquake in the Sea of Marmara. Turk J Earth Sci 20:383–398. https://doi.org/10.3906/yer-0911-7

Waldhauser F (2001) Hypodd—a program to compute double-difference hypocenter locations

Waldhauser F, Ellsworth WL (2000) A double-difference earthquake location algorithm: method and application to the northern Hayward fault. Bull Seismol Soc Am 90:1353–1368

Wright T, Fielding E, Parsons B (2001) Triggered slip: observations of the 17 August 1999 Izmit (Turkey) Earthquake using radar interferometry. Geophys Res Lett 28:1079–1082. https://doi.org/10.1029/2000GL011776

Yaltırak C, Erturaç MK, Tüysüz O, Saki-Yaltırak K (2003) Marmara Denizi'nde Tarihsel Depremler: Yerleri, Büyüklükleri, Etki Alanları ve Güncel Kırılma Olasılıkları, Kuvaterner Çalıştayı, 174–180

Yamamoto R, Kido M, Ohta Y, Takahashi N, Yamamoto Y, Kalafat D, Pinar A, Ozeren S, Kaneda Y (2016) Creep rate measurement and fault modelling at the North Anatolian Fault, beneath the Sea of Marmara, Turkey, by means of acoustic ranging, paper presented at Japan Geoscience Union meeting 2016, Chiba, Japan, S-CG59-07

Yamamoto Y, Takahashi N, Pinar A, Kalafat D, Citak S, Comoglu M, Polat R, Kaneda Y (2017) Geometry and segmentation of the North Anatolian Fault beneath the Marmara Sea, Turkey, deduced from long-term ocean bottom seismographic observations. J Geophys Res Solid Earth 122:2069–2084. https://doi.org/10.1002/2016JB013608

Zachariasen J, Sieh K (1995) The transfer of slip between, En Echelon Strike-Slip Faults: a case study from the 1992 Landers Earthquake, Southern California. J Geophys Res 100–15:281–301

Focal Mechanisms of Earthquakes and Stress Field of the Earth Crust in Azerbaijan

G. J. Yetirmishli and S. E. Kazimova

1 Introduction

The mechanism of the earthquake focus is one of the most important parameters characterizing the seismic event. In modern seismology, it is associated with the sudden movement of rocks accompanied by the emission of seismic waves along the surface of weakened strength, and reflects simultaneously the spatial orientation of the axes of the principal stresses, possible planes of discontinuities and motions in the earthquake source, and represent almost the bulk of information on the stress state of the earth's interior (Melnikova and Radzyminovich 1999). It is data on the stress and strain fields, together with the geological, structural-tectonic structure, that make it possible to solve the problem of creating models of deformation processes in the tectonic structures of the earth's crust (Sychev 2005).

At the present stage, a dense network of highly sensitive digital seismic stations in the amount of 35, allowing to register all seismic events with magnitude M > 0.1 within Azerbaijan, allowed to obtain new data on the mechanism of earthquake foci, which allowed more detailed study of the regularities of the modern deformation process in the lithosphere of Azerbaijan.

The urgency of studying the stress-strain state of the earth's crust is necessary for solving practical problems such as seismic hazard assessment of industrial and civil structures, development and operation of large deposits of minerals such as oil and gas.

G. J. Yetirmishli · S. E. Kazimova (✉)
Azerbaijan National Academy of Sciences, Republican
Seismic Survey Center, Huseyn Javid 123 Str., 1001, Baku, Azerbaijan
e-mail: sabina.k@mail.ru

S. D'Amico (ed.), *Moment Tensor Solutions*, Springer Natural Hazards,
https://doi.org/10.1007/978-3-319-77359-9_21

The purpose of this study is to calculate the seismic moment tensor of strong earthquakes in Azerbaijan for the period 2012–2015. On the basis of broadband records (BH) of modern digital stations using newly developed algorithms and software products.

2 Geological Structure of the Region

The territory of Azerbaijan is located in the east of the Caucasian segment of the Alpine geosynclinal region. From the east to Azerbaijan, there is a huge meridional depression of the Caspian Sea. The northern part of Azerbaijan encompasses the eastern section of the intricate southern wing of the meganticlinorium of the Greater Caucasus, as well as the area of its southeastern immersion. The central part of Azerbaijan belongs to the region of the Kura intermountain basin, characterized by the accumulation of a powerful sequence of Neogene-Anthropogen Molasses. These deposits in the strip to the south of the longitudinal Alazan-Agrichay depression are crumpled into steep, partly folded to the south, folds, and within the Kura-Araks lowland form flattened brachianticlinal uplifts (Khain 2005).

Globally tectonically, the study area covers three large geoblocks in the central part of the Crimean-Caucasian-Kopetdag Alpine fold system: the Deshte-Lut block located in the east of Iran, the Middle Caspian-Turan plate located on the eastern coast of the Middle Caspian and the Main Zagrosky overthrust located on the south for the fall of Iran. In the environment of these large geoblocks are the main geostructural elements of Azerbaijan (Aslanov 2009).

As is known, the earthquake is a mechanical process of environmental destruction caused by a sharp break and displacement of strained rocks along a fault that appears near the surface of the Earth. In the process of analyzing the mechanism of the source of earthquakes, the choice of the map of tectonic disturbances plays an important role.

Thus, based on the following works (Shikhalibeyli 1996, Kangarli et al. 2017, Rzaev 2013), we compiled a diagram of the main elements of the discontinuous tectonics of the studied region, used in the analysis of this article (Fig. 5).

3 Method for Calculating the Seismic Moment Tensor and Preparing Real Seismograms

For this work, an algorithm was used using the waveform inversion method—Time-Domain Moment Tensor INVerseCode (TDMT INVC) (Dreger 2002). The main source of seismograms is the Republican Seismic Survey Center of ANAS. Seismograms are downloaded in SEED format and converted into SAC format. Broadband seismograms are selected with a distance restriction (70–350 km).

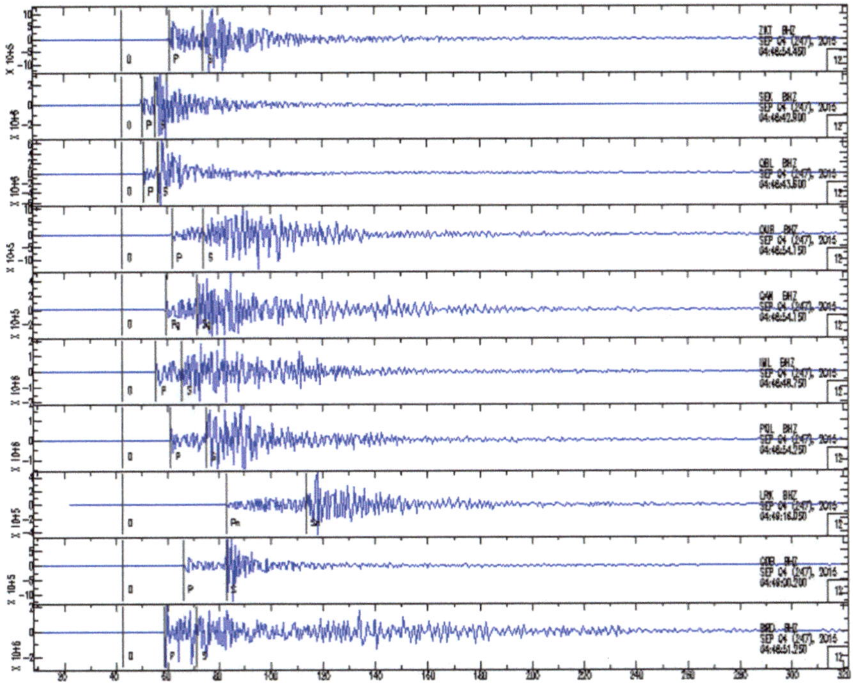

Fig. 1 Broadband wave record of the earthquake occurred on September 4, 2015, with a magnitude of 5.9 in the SAC format

They must have a sufficient duration (includes the interval from P-waves to S-waves) and quality (sufficiently high signal-to-noise ratio, no clipping). Preparation of seismograms for inversion includes: removal of the entry of the P-wave; Deconvolution (restoration of true soil displacements); Definition of epicentral distance, forward and reverse azimuths; Calculation of radial and transverse components; Filtering. Deconvolution is produced in the time domain. For band pass filtering, a Butterworth 4-order filter is used (Fig. 1).

In the course of the script, after reading the parameters files, the coordinates of the stations and the output files of the location program, the following operations are performed: 1. Correction of the earthquake waveforms for the amplitude-frequency response of the recording instruments and the transition from north-south, east-west coordinates, vertical to radial coordinates, Tangential, vertical; 2. Strip filtering by Butterworth filter 4 orders; 3. Calculation of Green's functions used in the inversion of synthetic wave forms for the entire set of depths and distances of the source-receiver (obtained on the basis of the program for locating the real-time subsystem); 4. Transformation of the components of Green's functions in the frequency domain into a representation of the Green's functions in the time domain by means of the inverse Fourier transform; As a result, files with records of Green's function components are created for a given set of distances; 5. Strip filtering of the Green's functions by the

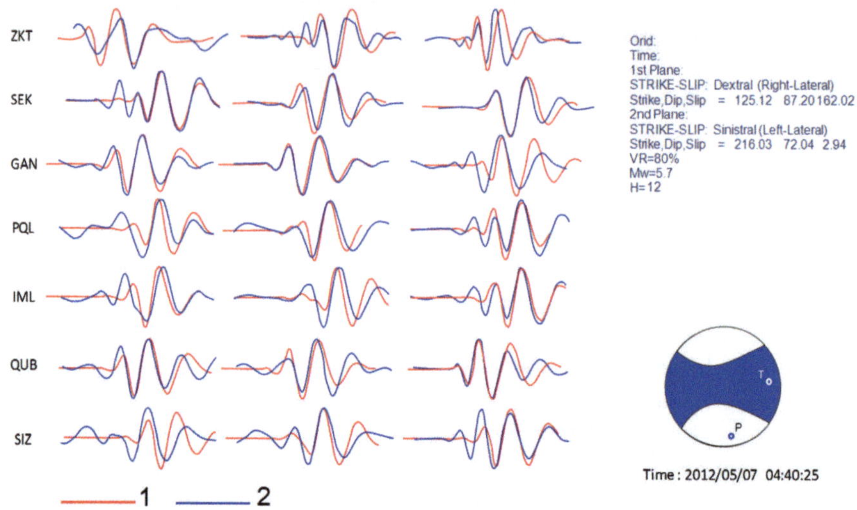

Fig. 2 Wave forms: real (**1**) and synthetic (**2**), and also the solution of the focal mechanism by the example of the earthquake that occurred on May 7, 2012 with a magnitude of 5.6

4th order Butterworth filter; 6. Inversion and calculation of the tensor and Mw using the TDMT program (Fig. 2) (Yetirmishli et al. 2015).

Although the focus of a strong earthquake is an extended object, when studying its characteristics, a point source model is used in the first approximation. This is justified when considering the wavelengths of much larger geometric dimensions of the focus and time periods significantly exceeding the duration of the rupture process in the source. In practice, these provisions are carried out only approximately.

Thus, based on the above, the mechanisms of strong earthquake foci that occurred in 2012–2015 were constructed and analyzed. With a magnitude higher than 4.0, and characteristic features of seismotectonic deformation have been established in individual seismogenic zones of the Republic, namely in the Zakatala, Sheki, Gabala, Oguz, Hajigabul and Ismayilli regions, and also in the Caspian Sea zone (Fig. 3, Table 1). Figure 4 shows the stereogram focus mechanisms in Scheme basic elements discontinuous tectonics studied region. (Shikhalibeyli 1996, Kengerli 2007, Rzayev 2013).

Faults: 1-Dashgil-Mudrese, 2-Vandam, 3-Geokchay, 4-Siazan, 5-Zangi-Kozluchay, 6-Herman, 7-Adzhichay-Alat, 8-West Caspian, 9-Arpa-Samur, Ganjachay-Alazan, 11-Kazakh-Signan, 12-North-Azhinoor, 13-Iori, 14-Kurinsky, 15-Mingachevir-Saatli, 16-Bashlybel, 17-Palmiro Absheron, 18-Ahty-Nyugedi-Kilyazin, Talysh, 20-Yardimli, 21-Predtalyshsky, 22-Central Caspian, 23-Absheron-Pribalkhan, 24-Sangachal-Ogurchi, 25-Milsky-Chikishlyar, 26-Yashmin flexure, 26a-Gizilagach, 27-Shakhov, Azizbayov, 28-Karabogaz-Safidrud.

Table 1 Parameters of the mechanisms of foci for earthquakes 2012–2015 with ml ≥ 4.0

№	Date, *d m y*	t_0, *h m s*	*h*, *км*	*ml*	*Mw*	Coordinates of foci		Nodal planes					
						Latitude	*Longitude*	*NP1*			*NP2*		
								STK	*DP*	*SLIP*	*STK*	*DP*	*SLIP*
1	20120507	04-40-25	9	5.6	5.9	41.50	46.58	125	87	162	216	72	2
2	20120507	14-15-13	12	5.7	5.3	41.56	46.63	130	48	−117	349	48	−62
3	20120518	14-46-33	13	5.0	5.1	41.53	46.62	354	47	−68	144	47	−111
4	20121007	11-42-50	41	5.3	5.1	40.70	48.35	128	45	−81	295	45	−98
5	20121014	10-13-36	8	5.7	5.6	41.66	46.27	116	58	−141	2	58	−39
6	20130406	09-19-36	26	4.0	3.8	40.98	46.67	286	81	108	40	20	24
7	20130418	20-38-51	25	4.5	4.6	41.10	47.28	54	85	6	323	83	175
8	20130430	09-29-34	10	4.2	4.3	40.44	48.02	254	72	−38	357	53	−158
9	20140210	12-06-19	46	5.7	5.5	40.23	48.62	125	59	−57	253	44	−132
10	20140607	06-05-20	61	5.6	5.4	40.13	51.66	119	66	−59	243	38	−139
11	20140629	17-26-07	9	5.2	5.0	41.54	46.54	241	79	−15	334	75	−169
12	20140929	01-38-07	11	5.5	5.1	41.13	47.94	265	64	−43	17	53	−146
13	20141003	09-45-06	8	4.0	4.0	41.24	45.61	354	58	120	127	43	52
14	20141004	04-59-32	6	5.0	4.9	41.11	47.94	268	82	−25	1	65	−171
15	20150904	04-49-36	16	5.9	5.5	40.97	47.43	153	90	−180	63	90	0

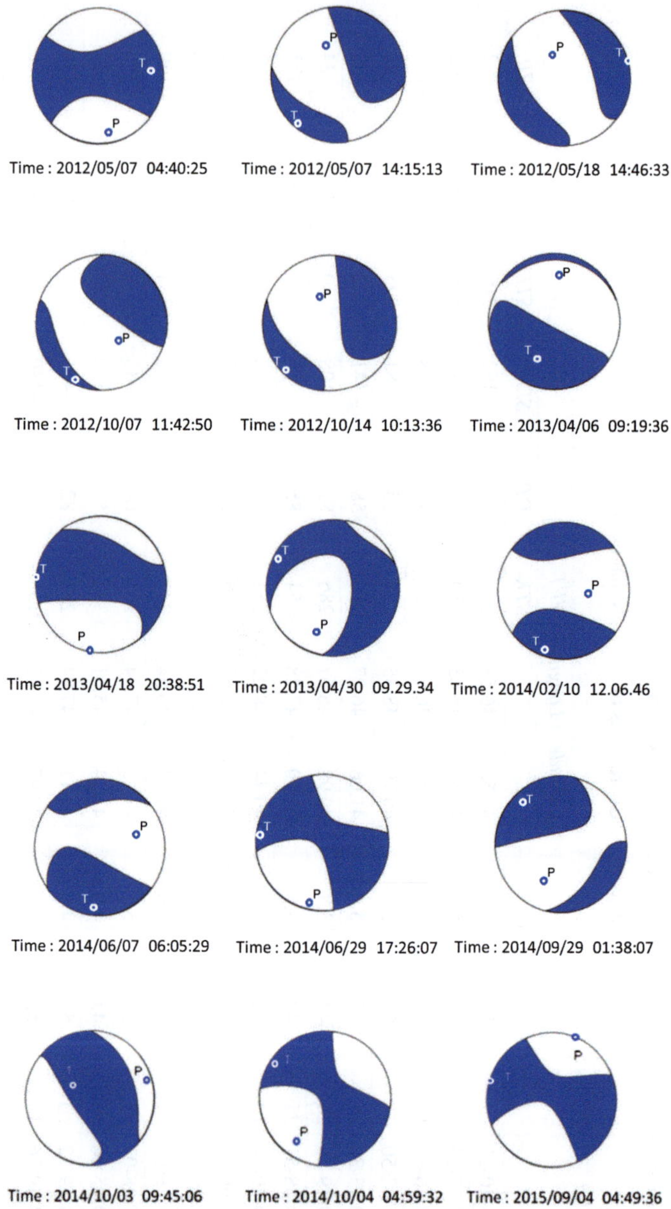

Fig. 3 Stereograms earthquake focal mechanisms occurred in the territory of Azerbaijan in 2012–2015 with a magnitude above 4.0

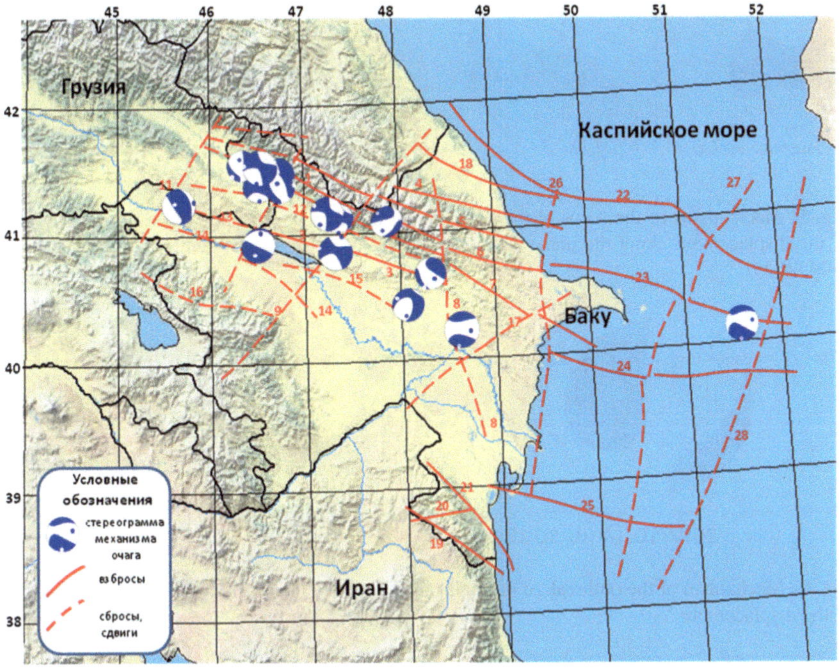

Fig. 4 Circuit structure and fault earthquake focal mechanism 2012–2015 with ml ≥ 4.0

4 Consider Each Earthquake Separately

Thc carthquakc of May 7, 2012 occurred in Zagatala region with ml — 5.6 was characterized by near horizontal (PLP = 10°) compressive and tensile (PLT = 14°) stresses. The type of motion along both steep (DP1 = 87°, DP2 = 72°) planes is a shift. The plane NP1 has a southeast (STK1 = 125°) strike with the type of movement right-sided shift, and NP2—south-west (STK2 = 216°), with the type of movement left-sided shift. A comparison of the strike of nodal planes with fault lines shows the agreement of the first nodal plane NP1 with the right-sided Kazakh-Signak and Ganjachai-Alazan transverse faults, which allows us to consider the plane NP2 as acting. On the same day, there was another significant earthquake with ml = 5.7. The mechanism of the source of this earthquake occurred under the action of near-horizontal tensile stresses (PLT = 1°). Type of movement along the first nodal plane NP1 reset with elements of right-sided shift, on the second—a reset with elements of left-side shift. On May 18, 2012 another earthquake with ml = 5.0 was recorded in this zone. The mechanism of this earthquake is identical with the previous one, the type of displacement movement with shift elements (Fig. 5).

Further in the same year, there are two more notable earthquakes—in the Ismayilli region on October 7 with ml = 5.3 and Balakan district with ml = 5.7. Both these earthquakes occurred under the action of near-horizontal tensile stresses (PLT = 0°).

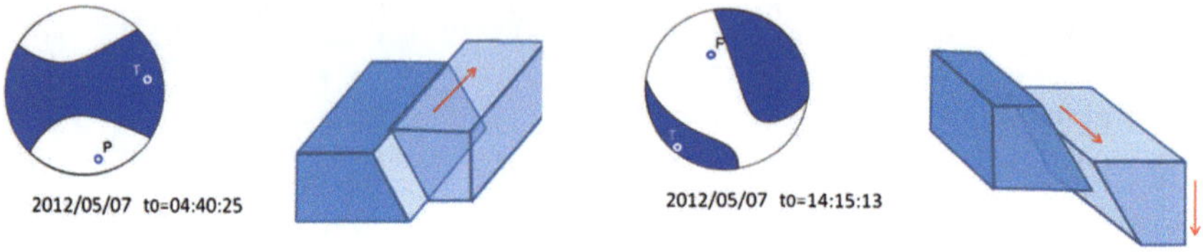

Fig. 5 Mechanisms of the focus of the Zagatala earthquakes in 2012, as well as block diagrams of the displacement along the plane NP1 (2012.05.07 t_0 = 14: 15: 13) and NP2 (2012.05.07 t_0 = 04:40:25)

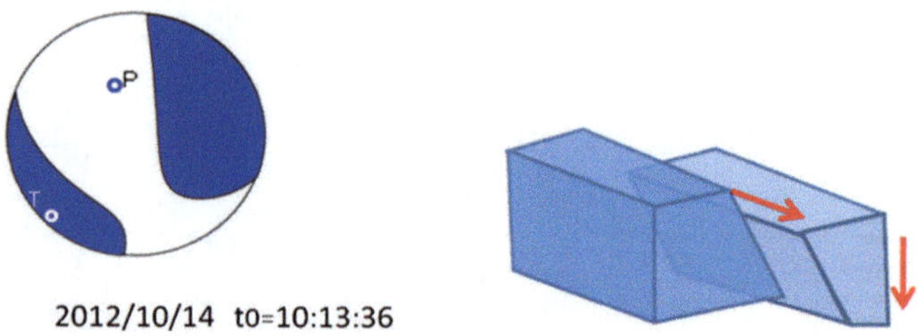

Fig. 6 Mechanism of the outbreak of the Balakan earthquake in 2012, as well as a block diagram of the displacement

The type of earthquake movement is reset with shift elements (Fig. 6). The motivation of the Ismayilli earthquake is related to the activity of the North-Azhinoor fault.

In 2013 there were no earthquakes with a magnitude greater than 5, however, we identified three notable earthquakes that occurred in Mingachevir (2013.04.06), Sheki (2013.04.18) and Kyurdyar (2013.04.30) regions. The type of motion along the first plane of the Mingachevir earthquake is a rebound with elements of right-sided displacement, along the second plane- a left-sided shear-thrust. The angles of the inclination of the planes vary from 81° to 20°. NP1 has a northwest strike, the plane NP2 is northeasterly. Comparison of azimuths of longitudinal axes with the strike of nodal planes of the mechanism of the focus shows the agreement of only one (NP1) of them, corresponding to the Geokchai fault. The type of shift in the Sheki earthquake, as in previous years, along both planes, a shift with elements of the uplift (Vandam fault), and the Kyurdiar earthquake a fault with shift elements corresponding to the Mingachevir-Saatlinsky fault.

For the notable earthquake of February 10, 2014, the compression stresses of the east-north-eastern orientation (AZM = 87°) occurred in Hajigabul region are near-vertical (PLP = 61°), and the tensile stresses of south-south-west orientation (AZM = 192°) are horizontal (PLP = 8°). Type of motion on both planes (DP1 = 59, DP2 = 44°)—reset with shift elements. The plane NP1 has a south-eastern strike (STK1 = 125°), NP2—south-west (STK2 = 253°). A comparison of the strike of the nodal planes with the fault lines in Fig. 6 shows the agreement of the second nodal plane NP2 with two longitudinal Kura and orthogonal Ismail-Gabala faults. Therefore, there is reason to consider it valid.

The Caspian earthquake that took place on June 7, 2014 arose under the action of near-horizontal tensile stresses (PLT = 15°), and therefore are decisive. The type of motion along both planes is a reset with shift elements. The plane NP1 has a south-east strike (STK1 = 119°), and the plane NP2 has a south-west strike (STK2 = 243°). The type of motion along both planes (DP1 = 66°, DP2 = 38°)—a fault with shear elements coordinated with the Absheron-Pribalkhan fault.

The June 29, 2014 earthquake that occurred in the Zagatala-Balaken area with ml = 5.3 was characterized by near-horizontal (PLP = 18°) compressive and tensile (PLT = 3.0°) stresses oriented to the southwest (AZM = 197°) and north (AZM = 288°). The type of motion along both steep (DP1 = 79°, DP2 = 75°) planes is shear. The plane NP1 is near (STK1 = 241°), and NP2 is near-term (STK2 = 334°). Comparison of the stretch of nodal planes with fault lines in Fig. 6 shows the agreement of the first nodal plane NP1 with the right-sided Kazakh-Signak and Ganjachai-Alazan transverse faults, which makes it possible to consider the plane NP1 as acting.

September 29 and October 4, 2014, two earthquakes with ml = 5.5 and ml = 5.0, respectively, occurred to the NE from Gabala. Near-vertical (PLP = 48°) compressive stresses oriented to the south-west (AZM = 265°) prevailed in the focus of the first earthquake. The type of motion along both steep (DP1 = 64°, DP2 = 53°) planes is a reset-shift. The plane NP1 is latitudinal (STK1 = 265°), and NP2 is meridional (STK2 = 17°). A comparison of the strike of nodal planes with fault lines in Fig. 6 shows the agreement of the second nodal plane NP2 with the Ismail-Gabala orthogonal fault.

The advance in the focus of the second earthquake arose under the action of near-horizontal (PLP = 23°) compressive stresses. The type of motion along both planes is the shift with the reset elements. The plane NP1 is latitudinal (STK1 = 268°), and NP2 is meridional (STK2 = 1°). A comparison of the strike of nodal planes with fault lines in Fig. 6 shows the agreement of the first nodal plane NP1 with the Arpa-Samursky transverse fault. Most likely this plane is the one acting.

October 3, 2014 there was an earthquake in the Gazakh region with ml = 4.0. For this earthquake, the compressive stresses were predominant. Type of movement along NP1 oriented to the north—uplift with elements of right-sided shift, in NP2 oriented to the south-east—uplift with elements of left-side shear. A comparison of the strike of the nodal planes with the fault lines shows the agreement of both nodal planes with the Zangi-Kozluchai uplift thrust.

One of the most powerful earthquakes in the last 10 years is the earthquake that occurred in the Oguz region on September 4, 2015. The earthquake occurred under the action of close-in tensile and compressive stresses. It can be seen from Table 1 that the first nodal plane of rupture extends in the SE direction (153°) with a southwestward incline at 90°, the second nodal plane has a NE (63°) with a southeast incline at an angle of 90°. At the same time, the compressive stresses in the earthquake focus were oriented in the northeastern direction (azimuth 18) and acted horizontally (angle with 0° horizon), and the tensile forces were directed in the west-south-west direction (287°) at an angle of 0° to the horizon. The type of movement of this earthquake is a shift with the left-side horizontal component (Fig. 7). The epicenter of the Oguz earthquake is confined to the Arpa-Samur fault and can be interpreted as

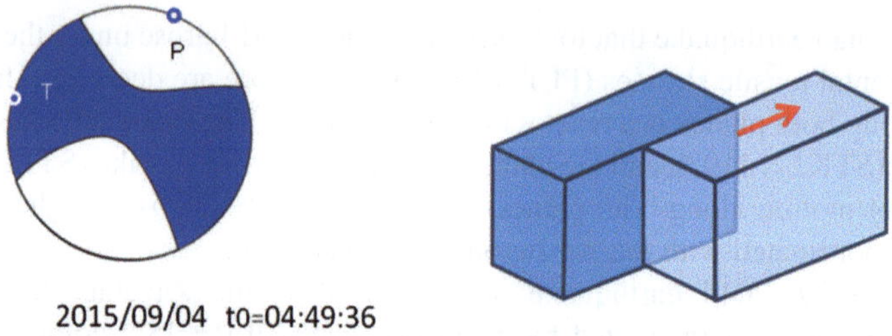

Fig. 7 Mechanism of the outbreak of the Oguz earthquake in 2015, as well as a block diagram of the displacement

a left-sided shear deformation in the zone of geodynamic influence of the left-sided Arpa-Samursky fault.

5 The Discussion of the Results

The division of the sample into faults, shifts and upsets made it possible to identify both significant differences and close coincidences of focal parameters of shocks with different types of motions in the source.

Figures 8 and 9 represent the histograms of the relative differences in the parameters of the axes of the principal stresses and nodal planes, obtained by comparing the solutions of the mechanisms of the 15 earthquakes under consideration. Nearly 80% of the jerks of the tensile axes and 40% of the jogs of the compression axes do not exceed 20° in the PL. Orientation of the tension axes predominantly dominates in the SW, NW direction, however, for the compression axes, the orientation in the NE-SW direction is determined (67%) and in the NW-SE direction (33%). The angles of incidence for motions of different types exceed 45°, which indicates a rather steep dip in the fault zones. This is consistent with the fact that in the investigated zone, the majority of transverse faults have incidence angles of 50–90°, that is, angles sufficiently close to the vertical. Analysis of the SLIP immersion angles showed that for the first nodal plane 27% changes within −15°–(−43°) and 40% within −57°–(−80°). For the second nodal plane, 60% of the foci change within −62°–(−171°). Variations in the angles of immersion of the axes of principal stresses, together with significant standard deviations, bear witness to the significant spatial inhomogeneity of the lithosphere.

On the basis of the results obtained a schematic map of the orientation of the axes of compression and extension of the earthquakes. The analysis of the orientation of the compression axes has the NW-SE orientation in the Zagatala region, N-S in Sheki and further smoothly changing clockwise to the NE-SW direction in the Caspian Sea. The axis of enlargement is mainly oriented to the NE-SW and N-S directions, which is connected with the heat of the Caucasus.

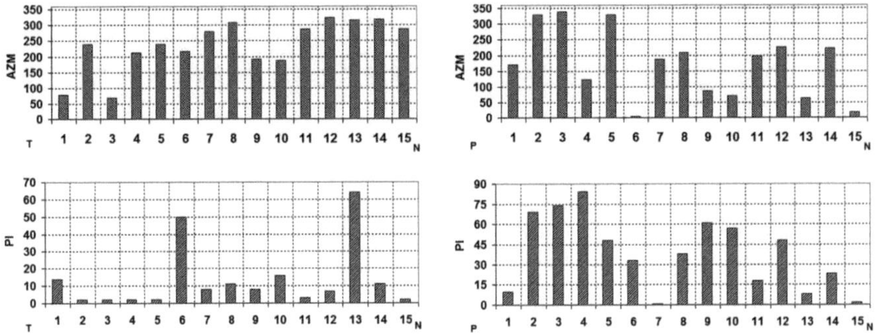

Fig. 8 Histograms of relative differences in the parameters of the axes of principal stresses

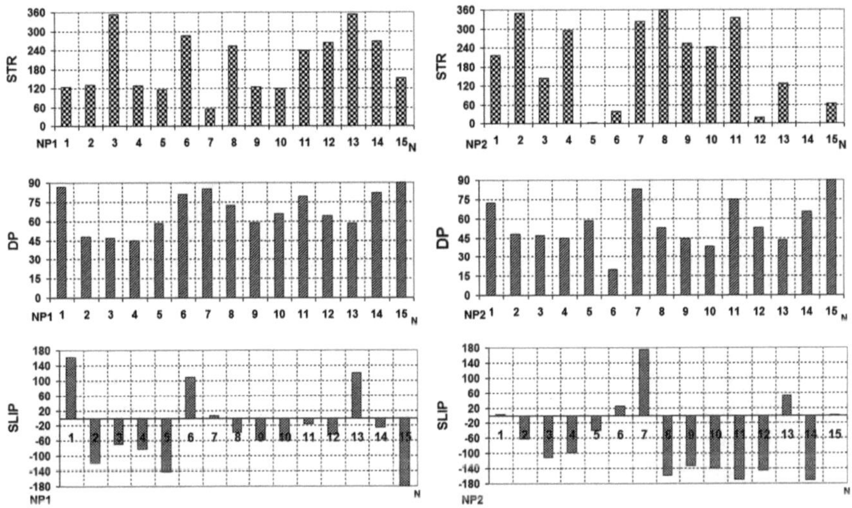

Fig. 9 Histograms of relative differences in parameters of nodal planes

Analyzing the sequence of seismic processes, it can be noted that the examined foci have a definite relationship. It should be noted that the SLIP modules and azimuths of strike STK ups and downs do not exclude the possibility of moving such types along the planes of some faults. Perhaps Zagatala earthquake was the first impetus, which is caused by a series of strong earthquakes in the Balakan, Sheki, Oguz, Gabala and Ismayilli regions. All these zones are in similar seismotectonic conditions. The structural elements of the Tfansk anticlinorium, the Zagatala-Govdag synclinoria, the Vandam anticlinorium and the superimposed Alazan-Agrichay trough take part in the geological structure of these zones. These structures are divided from each other by deep sublatitudinal faults (Shikhalibeyli 1996).

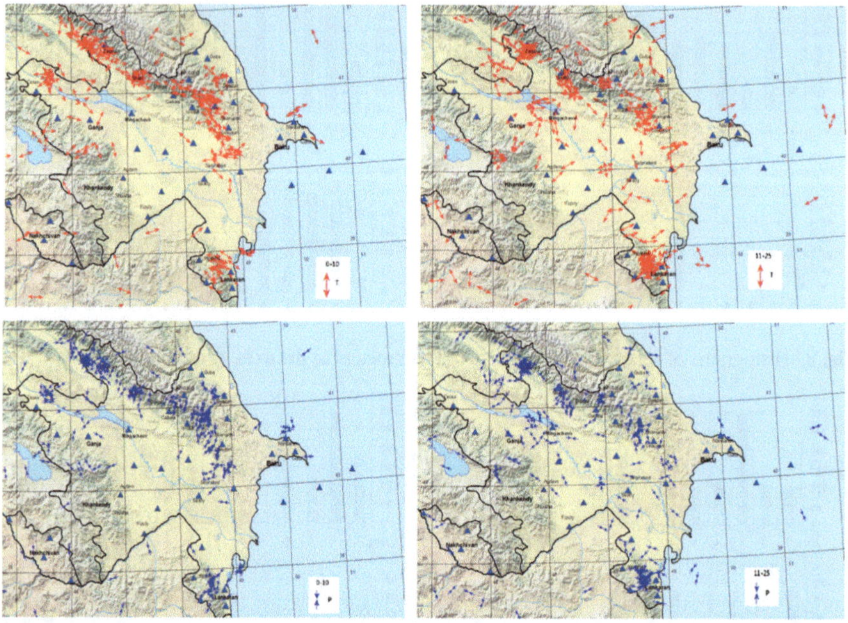

Fig. 10 Scheme of distribution of axes of compression and tension of focal mechanisms in 2012–2015

The results of the analysis of the distribution of the earthquake are made in the territory of Azerbaijan, all strong earthquakes ($M \geq 5.0$) are confined to the surface area of the crystalline basement (Mammadli 2012). Zagatala and Balakan earthquakes also confirmed this conclusion. An exception is the Ismayilli earthquake, whose depth is $H = 41$ km, while the surface of the crystalline basement is at a depth of 10–12 km (Fig. 11). Note that this is not the only earthquake in the Ismayilli region with such a depth of focus. And before that, a number of weak and medium strength earthquakes with a depth of about 40 km were recorded here. We believe that this anomalous seismotectonic phenomenon is for this region and it requires additional studies (Yetirmishli et al. 2013). In Fig. 10 can also be noted of the focal zone of Iranian earthquakes, the depth of which varies within 5–20 km. Note that on August 11, 2012 in the area at a depth of 10 km, there were two earthquakes with a magnitude of 6.4, which resulted in numerous aftershocks, which continue to this day.

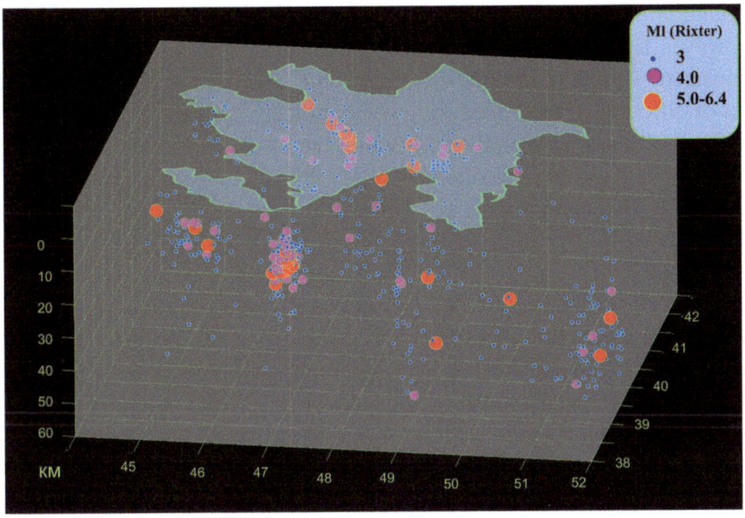

Fig. 11 Three-dimensional scheme of distribution of earthquake hypocenters in Azerbaijan for 2012–2015

6 Conclusion

Thus, based on the algorithm of the wave form inversion—Time-Domain Moment Tensor INVerseCode (TDMT INVC), the seismic moment tensor of strong earthquakes in Azerbaijan for the period 2012–2015 was calculated. On the basis of broadband records (BH) of modern digital stations, which made it possible to establish the characteristic features of seismotectonic deformation in individual seismo genic zones of the Republic, namely in the Zagatala, Sheki, Gabala, Oguz, Hajigabul and Ismayilli regions, as well as in the Caspian Sea zone.

Based on the foregoing, it can be concluded that seismic activity of the territory of the Greater Caucasus and the central part of the Caspian Sea in 2012–2015. Associated with the activation of the Kazakh-Signak with a fall at an angle of 72° to SE, Ganjachay-Alazan with the fall of NW at an angle of 58°, Arpa-Samur with a dip at an angle of 85°–90° (at depths of 16–25 km) to SE, Ismayilli -Gabala with the fall of SE at an angle of 44° (at a depth of 46 km) and 53° (at a depth of 11 km) transverse and Geokchay with a fall in NE at an angle of 81°, Mingechevir-Saatli with a fall in SE at an angle of 53° and Absheron-Pribalhan with a fall to SW at an angle of 66° longitudinal faults. Perhaps Zagatala earthquakes occurred under the influence of right-sided Kazakh-Signak and Ganjachai-Alazan transverse faults prompted subsequent strong earthquakes in this area. Statistical analysis of the mechanisms of the foci of strong earthquakes showed the predominance of tensile stresses associated mainly with the activity of transverse and orthogonal faults throughout the territory of Azerbaijan.

Orientation of the tension axes predominantly dominates in the SW, NW direction, however, for the compression axes, the orientation in the NE-SW direction is determined (67%) and in the NW-SE direction (33%). The angles of incidence for motions of different types exceed 45°, which indicates a rather steep dip in the fault zones. This is consistent with the fact that in the investigated zone, the majority of transverse faults have incidence angles of 50–90°, that is, angles sufficiently close to the vertical. These are mainly the Kazakh-Signak, Arpa-Samur and Ganjachai-Alazans cross faults, the angle of incidence of which at depths of 9–20 km varies between 58 and 87°.

The results of the analysis of the distribution of strong earthquake foci by depth allowed us to conclude that all strong earthquakes ($M \geq 5.0$) in the territory of Azerbaijan are confined to the surface area of the crystalline basement. Zagatala and Balakan earthquakes also confirmed this conclusion. An exception is the Ismayilli earthquake, whose depth is $H = 41$ km., While the surface of the crystalline basement is here at a depth of 10–12 km. The analysis of the orientation of the compression axes has shown the NW-SE orientation in the Zagatala region, NS in Sheki and further smoothly changing clockwise to the NE-SW direction in the Caspian Sea. The axis of stretching is mainly oriented to NE-SW and NS directions, which is associated with the heat of the Kura depression along the Central Caucasus thrust under the zone of the Greater Caucasus.

References

Aslanov BS (2009) Tectonics of the major geostructural elements of Azerbaijan and their expression in the gravitational field // The dissertation of the scientific degree of Doctor of Geological and Mineralogical Sciences, Tashkent, p 227

Dreger DS (2002) Time-Domain Moment Tensor INVerseCode (TDMT_INVC) // University of California, Berkeley Seismological Laboratory, p 18

Kangarli TN, Aliyev FA, Aliyev AM, Vahabov UG (2017) Active tectonics and focal mechanisms of earthquakes in the pseudosubduction active zone of the North and South-Caucasus microplates (within Azerbaijan) // Geophys J Institute of Geophysics of UKrainian NAS, No 4, vol 39, pp 101–104

Kazimova SE, Kazimov IE (2017) The influence of one-dimensional velocity sections on determining the key parameters of the earthquake sources in Azerbaijan. Izv Phy Solid Earth 53(1):69–82. ISSN 1069-3513

Khain VE, Alizade AA (2005) Geology of Azerbaijan // Volume IV Tectonics (ed), The Nafta Press, Baku, pp 214–314

Mammadli TYa (2012) Seismogenerating zones of Azerbaijan and their deep structural features // Seismoprognosis observations in the territory of Azerbaijan, vol 10, Baku, pp 287–295

Rzayev AG, Yetirmishli GJ, Kazimova SE (2013) Reflection of geodynamic regime in variations of the geomagnetic field (for example, the southern slope of the Greater Caucasus) Earth Sciences. News. No 4, Baku, pp 3–15

Shikhalibeyli ESh (1996) Some problematic issues of the geological stucture and tectonics of Azerbaijan, Baku: Elm, p 215

Sycheva NA (2005) Investigation of the features of earthquake focal mechanisms and seimoteconic deformations of the Northern Tien Shan according to the KNET Digital Seismic Network. Cand Sci Phys Math Dissertation, Moscow: IPE RAS, p 176

Yetirmishli GJ, Kazimova SE (2014) Focal parameters of the zagatala earthquake on may 7, 2012. Azerbaycan arazisinde seysmoproqnoz müşahidelerin kataloqu Seismoprogn. Observ Territ Azerb 11(1):14–8

Yetirmishli GJ, Mammadli TY, Kazimova SE (2013) Features of seismicity of Azerbaijan part of the greater Caucasus. J Georgain Geophys Soc Issue (A) Phys of Solid Earth 16a:55–60

Yetirmishli GJ, Abdullayeva RR, Kazimova ES, Ismayilova SS (2015) Strong earthquakes on the territory of Azerbaijan for the period of 2012-2014. Seismoprognosis observations in the territory of Azerbaijan 12(1):19–2

Seismotectonic Crustal Strains of the Mongol-Baikal Seismic Belt from Seismological Data

Alena Seredkina and Valentina Melnikova

1 Introduction

The Mongol-Baikal region is one of the most seismically active intracontinental areas of the Central Asian seismic zone. The seismic belt passes through different ancient and neotectonic structures, traversing units of different ages. Thereby, the major earthquakes in the region are connected with the most powerful, repeatedly renewed fault systems. From 1905 to the present, 9 seismic events with $M \geq 6.9$ have occurred in the considered area (Fig. 1). Some of them, such as the Muya 1957, M_{LH} 7.6 (Kondorskaya and Shebalin 1977), the Tsetserleg 1905, M_w 8.0, the Bolnai 1905, M_w 8.4 (Schlupp and Cisternas 2007), the Fuh-Yun 1931, M_{LH} 7.8 (Kondorskaya and Shebalin 1977) and the Gobi-Altai 1957, M_w 8.0 (Chen and Molnar 1977), are referred to the world's disasters.

A large number of publications have been concerned with fundamental issues of modern geodynamics in the Mongol-Baikal region suggesting different concepts to explain the kinematics of the tectonic movements (Apel et al. 2006; Calais et al. 2003; Emmerson et al. 2006; Lukhnev et al. 2010; Molnar and Tapponier 1975; Petit and Déverchère 2006; Solonenko et al. 1968; Zonenshain and Savostin 1979 etc.). It is presently known that by the nature, orientation and geodynamic type of movements, the complex heterogeneous structure of the crust is generally divided into two essentially different large tectonic provinces: the Baikal rift zone (BRZ) and Mongolia. In the active tectonics of the BRZ a key role is played by rift-forming processes with predominance of the near horizontal extension oriented across the major structural elements, while the formation of new structures in Mongolia occurs under conditions of subhorizontal meridional or NE-SW compression (Heidbach et al. 2010; Logatchev and Zorin 1992; Molnar and Deng 1984).

A. Seredkina (✉) · V. Melnikova
Institute of the Earth's Crust SB RAS, Irkutsk, Russia
e-mail: ale@crust.irk.ru

S. D'Amico (ed.), *Moment Tensor Solutions*, Springer Natural Hazards,
https://doi.org/10.1007/978-3-319-77359-9_22

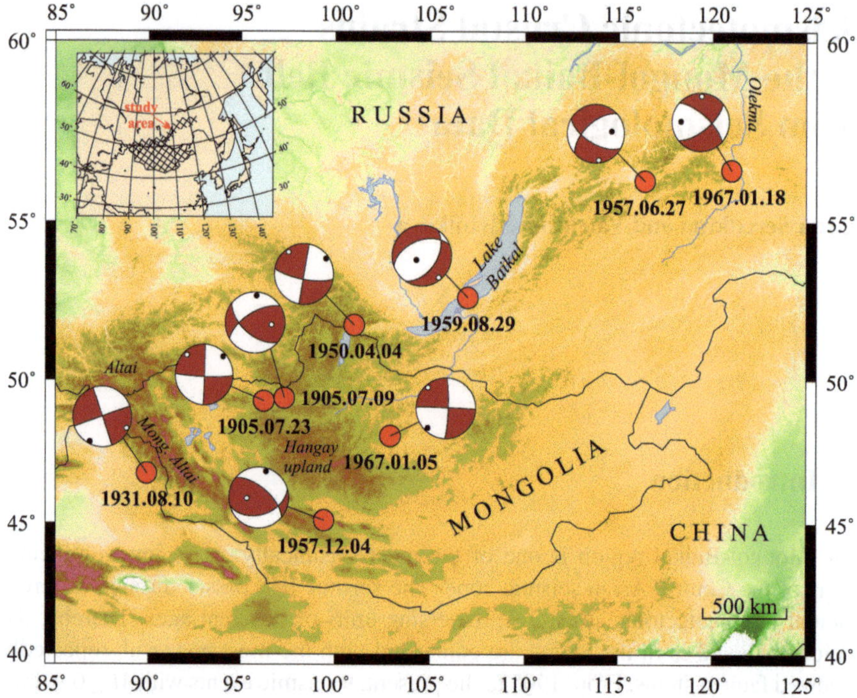

Fig. 1 Scheme of the Mongol-Baikal region and focal mechanisms in a lower hemisphere projection of strong $M > 7.0$ earthquakes (dates in yyyy.mm.dd format). Topography is based on the ETOPO1 topographic database (Amante and Eakins 2009)

One of the main and most effective methods of revealing the pattern of stress-strain state of the crust in the Mongol-Baikal region are the methods based on the usage of data on seismic moment tensor (SMT) of an earthquake. The latter is known to combine the focal mechanism of an earthquake and a value of scalar seismic moment (Kostrov 1975; Randall 1971) and in the double-couple approximation is introduced as

$$M_{0ij} = M_0(n_i b_j + n_j b_i),$$

where M_0 is a scalar seismic moment, determined from seismograms, n_i and b_j—unit vectors in the direction of slip and normal to the rupture site, indices $i, j = 1, 2, 3$ correspond to the axes of the Cartesian coordinates.

The problem of determining the SMT of regional earthquakes and its tectonophysical interpretation has been considered in a vast amount of papers (Chen and Molnar 1977; Delouis et al. 2002; Doser 1991a, b; Okal 1976; Schlupp and Cisternas 2007 etc.). The main goal of such research works was to evaluate a participation degree of major earthquakes in a large-scale regional deformation (Kurushin et al. 1997), define its type and calculate the average velocity of seismotectonic deformation of the crust over a large area (Chen and Molnar 1977). However, in a single SMT of smaller earthquakes some concrete information about tectonic process development can also be found. Therefore, such data need to be statistically processed. The results obtained make it possible to perform seismotectonic analysis at another scale level (Rebetsky 2007; Yunga 1990).

The study region annually experiences hundreds of thousands of seismic events at various magnitudes, while the earthquakes with $M \geq 5.5$ make up less than 1% of the total number of events. As usual, the SMT data for such earthquakes are promptly published by different seismological agencies (Global-CMT, NEIC, and USGS). Some publications provide the SMT data obtained in special studies (Doser 1991a, b; Delouis et al. 2002; Emmerson et al. 2006; Barth and Wenzel 2010). However, this important characteristic of seismic source is only calculated for a few regional earthquakes, since routine processing of most seismic events is restricted to determining their focal mechanisms from the P-wave first-motion polarities recorded by short-period seismic instruments. In this study we shortly review the available information on strong earthquakes, complement the dataset of the source parameters of regional seismic events by the new SMT solutions for medium earthquakes and consider the application of the obtained results to reconstruction of seismotectonic deformation field of the crust.

2 Tectonic Settings and SMT of Strong Earthquakes

The BRZ was originated on the site of a vaulted uplift under the influence of active tectonic movements that began in the Eocene-Oligocene and continue to the present. In the Miocene-Pliocene, at the "rapid rifting" stage, the process of the S-W and N-E oriented rifting began from southern Baikal. It resulted in the formation of the basic structures of the zone and its modern morphological appearance (Logatchev 1999). Development of the rift zone on the flanks encountered limitations in the form of orthogonal structures of the Mongol-Okhotsk fold belt in the S-W and the ancient Aldan shield in the N-E. As a result, the BRZ is presently the major isolated intracontinental structure with specific features of internal structure and peculiarities of developing geodynamic processes.

To the strongest earthquakes of the region refer: the Mondy 1950, M_w 6.9 (Delouis et al. 2002), the Muya 1957, M_{LH} 7.6, the Mid-Baikal (Srednebaikalsk) 1959, M_{LH} 6.8 and the Tas-Yuryaksk 1967 M_{LH} 7.0 (Kondorskaya and Shebalin 1977). Their epicenters are located in the axial part of the rift within the seismically active band, the configuration of which agrees with the orientation of the main rift-forming elements. Some parts of this territory are significantly different in terms of surface relief, neotectonic structure, geological and geophysical characteristics. The various kinematics of displacements in the zones of large faults is reflected in the focal mechanisms of the strongest earthquakes confined to them (Fig. 1). For example, in the focus of the Srednebaikalsk earthquake the normal fault corresponding to fault tectonics nature of central parts of the BRZ is observed. The strike-slips on the rift flanks were fully manifested in the focal mechanism of the strongest in the Eastern Siberia Muya earthquake and in the nature of seismotectonic deformations of the surface (Fig. 2a).

The territory of Mongolia at the neotectonic stage of development became a part of a vast complexly built intracontinental uplift. The uplifting of this region continued throughout the entire Cenozoic. In terms of general geological and geophysical features and the intensity of the latest tectonic movements, it is divided into two different segments—the western and eastern areas (Khilko et al. 1985). Western Mongolia with high mountain ranges and depressions refers to the region of epi-platform mountain building (of the newest activation). Here there is a contrasting, sharply dissected relief, thickened crust (in comparison with the eastern part) and complexly differentiated vertical and horizontal tectonic movements of large amplitude. As a rule, such movements characterize zones of large ancient faults. Eastern Mongolia with flat relief, low-amplitude and poorly differentiated movements refers to the area of a young platform. The processes of the Cenozoic tectonic activation here are significantly weakened (Baljinnyam et al. 1993; Khilko et al. 1985).

In seismotectonics of western Mongolia, the predominant type of active faulting is a strike-slip (Bayasgalan and Jackson 1999). Thereby, long-lived activated faults (suture zones), distinctly expressed in the relief, are spatially and genetically related to paleoseismogenic structures, modern seismic dislocations and the majority of earthquakes of different energy (Baljinnyam et al. 1993; Florensov and Solonenko 1965; Khilko et al. 1985; Molnar and Deng 1984; Okal 1976; Ritz et al. 1995 etc.). The suture zones delineate the newest structures of various ranks differing in type, velocity and direction of movements.

In the territory of Mongolia, several tectonic provinces characterized by increased seismic activity are distinguished. To such refer the Mongolian and Gobi Altai regions, the Ubsu-Nur Lake area, the Bolnai zone, the Mogod and Hovsogol areas. Each of these areas experienced the largest earthquakes ($M \geq 6.9$). Part of them ($M \geq 8$) left cracks distinctly expressed in the relief of the ground surface (Fig. 2b, c).

The SMT of strong earthquakes is widely used for estimation of parameters of large-scale deformations in the Mongol-Baikal region (Delouis et al. 2002; Doser 1991a, b; Huang and Chen 1986; Molnar and Deng 1984; Molnar et al. 1998; Schlupp and Cisternas 2007 etc.). The sum of the SMTs, normalized to the shear modulus (μ) and the volume of the considered area (V), defines seismotectonic deformation (STD) of macroscopic volume of the crust. In this case, the calculation of the average STD speeds is carried out by entering a major earthquake in the denominator of time (Kostrov 1975; Riznichenko 1985). Thus, on the basis of the SMTs of strong earthquakes of the region, high average velocities and main features of the large-scale deformation field were determined. It was found that in western Mongolia strike-slip components of displacements due to SW-NE compression and NW-SE extension prevail (Chen and Molnar 1977; Molnar et al. 1995 etc.). In the Baikal region the subhorizontal NW-SE extension is in connection with rifting (Delvaux et al. 1997; Logatchev 1999; San'kov et al. 2009). Along with geological and geodetic data, the obtained information served as the basis for attempts to explain modern geodynamic processes in the study region (Calais et al. 2003; Emmerson et al. 2006; Lukhnev

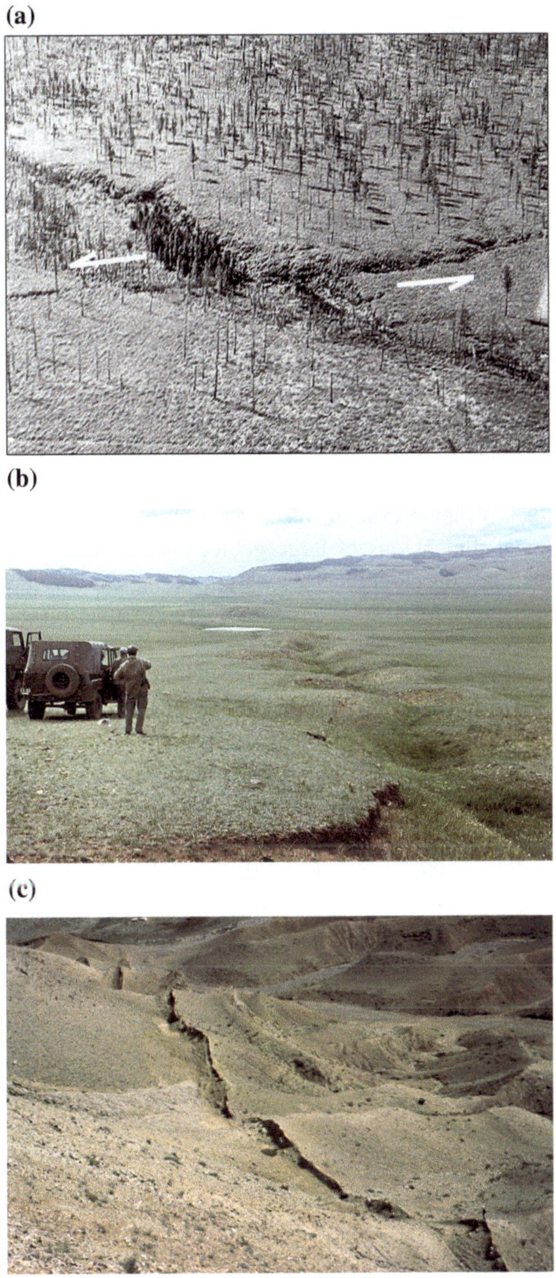

Fig. 2 **a** Seismotectonic deformation of the ground surface caused by the Muya earthquake 1957, M_{LH} 7.6. A view from the S. Strike-slip shown by arrows. Photo by R.A. Kurushin (1962). **b** Main shear rupture of the Bolnai earthquake 1905, M_w 8.4. A view from the W. Photo by R.A. Kurushin (1976). **c** Seismotectonic deformation of the ground surface caused by the Gobi-Altai earthquake 1957, M_w 8.1. A view from the S-W. Photo by P. Molner (1995)

et al. 2010; Molnar and Tapponier 1975; Petit and Déverchère 2006; Tapponnier and Molnar 1979; Zonenshain and Savostin 1979 etc.). The most popular point of view was that the global compression of the crust in the N-E direction here is caused by the convergence of India and Eurasia, and the extension in the BRZ is caused by the action of a local mantle energy source or a combination of global and local forces (Molnar et al. 1995; Zorin 1999).

It is generally accepted, that the continuous deformation process occurring in the depths of the lithosphere is only partially realized in a form of the seismogenic deformations forming seismicity. To evaluate a participation degree of the strongest earthquakes in the regional tectonic deformation, the latter can also be determined via the calculation the SMT with geological data (Baljinnyam et al. 1993; Kurushin et al. 1997). As input parameters, combinations of different fault segments, as well as a scalar seismic moment, calculated from the fault parameters are used. The scalar seismic moment is defined as $M_0 = \mu SD$, where μ is a shear modulus, *S*—a fault area, *D*—an average displacement on a fault.

In the process of modern seismotectonic deformations of the crust strong earthquakes play a crucial role. The rupture areas of such events are comparable with linear dimensions of large geotectonic structures, and a type of displacements in their origins can reflect the nature of the tectonic process development within these structures. We can estimate a degree of manifestation of strong earthquake focal parameters in ruptures on the ground surface by comparing the values of M_0 obtained from seismological and geological data (Table 1). Taking into account uncertainties of determining M_0 with different methods (average uncertainty is about 30% (Molnar et al. 1995)), we note significant disagreements in the considered parameters (Table 1), that can be caused by different reasons. First of them is unequal manifestation of the rupture length at the depth and the ground surface. Also we should take into account inaccuracies of seismic wave attenuation estimations and many other factors. In each case special studies are required.

3 SMT of Medium Earthquakes from Surface Wave Data

In order to complete the dataset of SMTs of medium earthquakes and to perform more accurate seismotectonic analysis, we considered 89 earthquakes with $mb = 4.0 - 5.8$, which occurred in the study area from 2000 to 2016 (Fig. 3). We applied the inversion of surface wave amplitude spectra, based on the method described in Bukchin (1990) and Lasserre et al. (2001). The reliability of this method in focal mechanism determination of seismic events of various energies occurring in regions with different tectonic settings has already been demonstrated (Bukchin et al. 1994; Gomez et al. 1997a, b; Lasserre et al. 2001; Seredkina and Kozmin 2017; Seredkina and Melnikova 2013).

The main assumptions of the method, i.e. an instantaneous pure double-couple point seismic source with known origin time and epicentral location (Bukchin 1990) and medium with weak lateral inhomogeneity (Woodhouse 1974; Babich et al. 1976),

Table 1 Data on the strongest earthquakes

N	Earthquake, date; coordinates	Magnitude	Dislocation parameters		M_0 10^{20}, N m		References
			Length, km	Slip, m	Seismological data	Geological data	
1	Tsetserleg, July 9, 1905; 49.5° N, 97.3° E	M_w 7.5–8.0 M_{LH} 7.6	130	3.2	2.2–55	2.7	Ok, KS, MD, Kh, BB, SC
2	Bolnai, July 23, 1905; 49.3° N, 96.2° E	M_w 7.9–8.4 M_{LH} 8.2	370	7	8.8–72.7	17.1	Ok, KS, MD, Kh, BB, SC
3	Fuh-Yun, Aug. 10, 1931; 46.5° N, 90.5° E	M 7.9–8M_{LH} 7.8	180	15	8.5–9.7	18	KS, CM, MD, BB
4	Mondy, April 4, 1950; 51.8° N, 101.0° E	M_w 6.7–7.0 M_{LH} 7.0	5.4	~0.8	0.09–0.25	0.03	KS, Da, De
5	Muya, June 27,1957; 56.2° N, 116.4° E	M_w 7.1 M_{LH} 7.6	35	4	1.4–9.7	0.9	So, KS, CM, MD, Db
6	Gobi-Altai, Dec. 4, 1957; 45.1° N, 99.4° E	M_w 8.0–8.1 M_{LH} 8.1	270	10	13–32	17.8	FS, O, KS, CM, Kh, R, MK
7	Mogod, Jan. 5, 1967; 48.1° N, 102.9° E	M_w 7.0–7.1 M_{LH} 7.8	40	1.3	0.32–0.38	0.34	KS, CM, TM, MD, HC, BJ
8	Srednebaikalsk, Aug. 29 1959; 52.68° N, 106.98° E	M_w 6.2 M_{LH} 6.8	10	~0.03	0.02	0.02	KS, Da
9	Tas-Yuryaksk, Jan. 18, 1967; 56.59° N, 120.96° E	M_w 5.6 M_{LH} 7.0	5	~0.06	0.002	0.002	KS, Db, IM

Notes Coordinates are taken from Kondorskaya and Shebalin (1977). Depth of the faults penetration is 20 km. References: *BB* Baljinnyam et al. (1993); *BJ* Bayasgalan and Jackson (1999); *CM* Chen and Molnar (1977); *De* Delouis et al. (2002); *Da, Db* Doser (1991a, b); *FS* Florensov and Solonenko (1965); *HC* Huang and Chen (1986); *IM* Imaev et al. (2000); *Kh* Khilko et al. (1985); *MD* Molnar and Deng (1984); *MK* Molnar et al. (1998); *KS* (Kondorskaya and Shebalin 1977); *O* Okal (1976); *Ok* Okal (1977); *R* Ritz et al. (1995); *SC* Schlupp and Cisternas (2007); *So* Solonenko et al. (1968); *TM* Tapponnier and Molnar (1979)

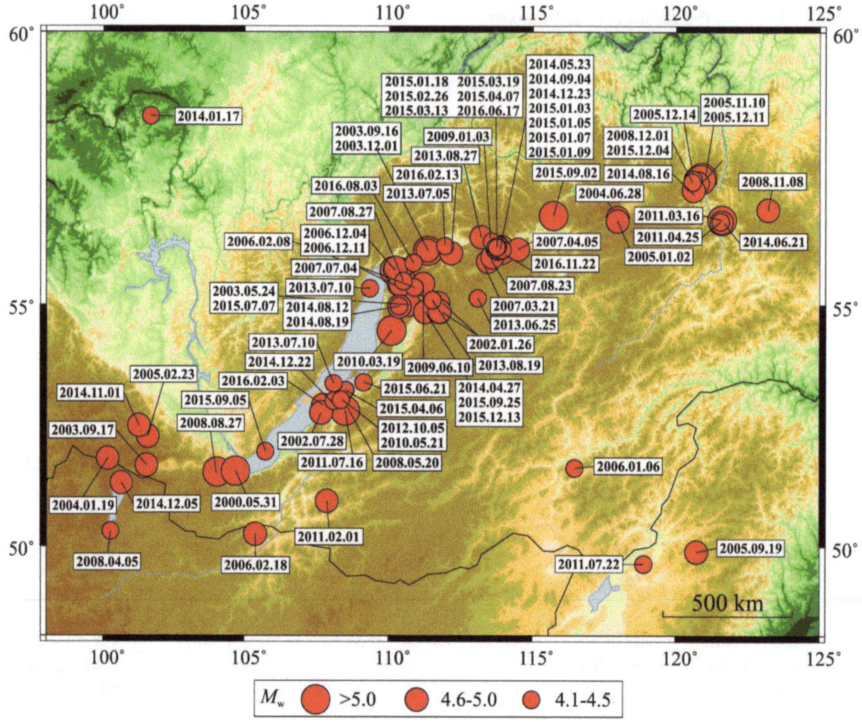

Fig. 3 Epicenters of the studied earthquakes (2000–2016) with their dates in yyyy.mm.dd format. Moment magnitudes were determined in this study

allow us to define a seismic source completely by its double-couple depth, seismic moment M_0 and focal mechanism parameters (i.e. strike, dip and slip angles of the nodal planes or directions of principal compression and tension axes). The focal mechanism and hypocentral depth are determined by a systematic exploration of the 4-D parametric space, and M_0 is estimated by minimizing the residual between the observed and calculated surface wave amplitude spectra for each combination of all the other parameters (Lasserre et al. 2001).

The data used for the calculations were records of the earthquakes at the IRIS broad-band digital stations (Fig. 4). The positions of the stations were selected in such a way that they fell in different azimuths from the epicenter and there were at least four such stations. The records with low signal/noise ratio and polarization anomalies were rejected from the inversion. The minimum and maximum epicentral distances were 2000 and 13,000 km, respectively.

Amplitude spectra of fundamental modes of Rayleigh and Love waves were calculated using a frequency-time analysis procedure (FTAN) (Levshin et al. 1989). The period range for each event was chosen to fit the point source approximation and to avoid uncertainties concerned with surface wave radiation pattern of a shallow focus

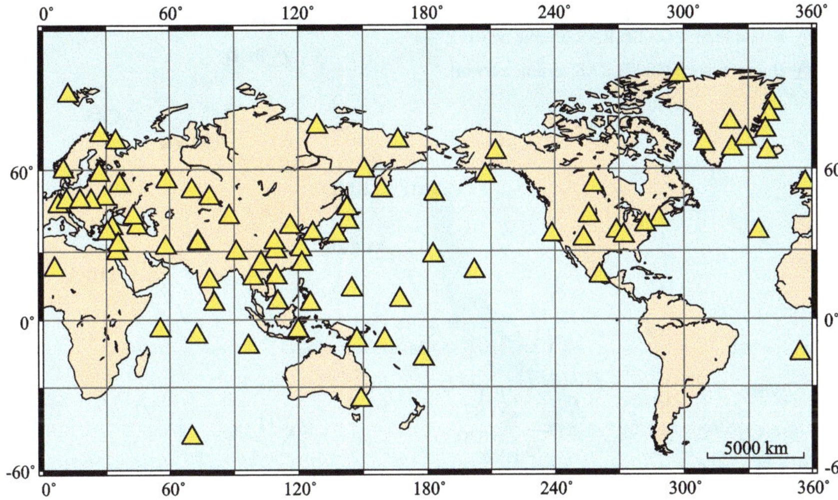

Fig. 4 Seismic stations used for calculation of surface wave amplitude spectra

source (Bukchin 2006; Bukchin et al. 2010). For the entire set of the earthquakes, the period ranged from 25 to 110 s.

Surface wave amplitude spectra only do not provide a unique focal mechanism solution: four equivalent solutions exist due to the sign ambiguity of moment tensor components and to the symmetry of the radiation patterns of surface wave amplitude spectra with respect to the epicenter (Mendiguren 1977). To constrain the uniqueness of the solution, we additionally used P-wave first-motion polarities obtained from the records at short-period seismic stations of the Baikal regional network (Fig. 5). Only reliable polarities of direct (Pg) and head (Pn) waves were selected for the further analysis. According to Lasserre et al. (2001) for each possible focal mechanism solution determined from surface wave data, we calculated the radiation pattern of P-waves and compared predicted and observed polarities.

We estimated the quality of the results obtained by a normalized joint residual function (ε), characterizing the deviation of the synthetic surface wave amplitude spectra from the observed one and the ratio between the P-wave first-motion polarities, which are inconsistent with the calculated radiation pattern and the total number of the polarities. Also, we calculated a partial residual function fixing one of the sought parameters (depth, for instance—ε_h) (Lasserre et al. 2001). For each earthquake the normalized joint residual function was less than 0.35 that evidences good data fitting.

The crustal structure beneath the seismic stations was specified by the 3SMAC model (Nataf and Ricard 1996). We approximated the crust in the vicinity of the source by the 3SMAC and CRUST 2.0 models (Bassin et al. 2000) whichever provided the lower normalized residual. To describe the upper mantle structure and to

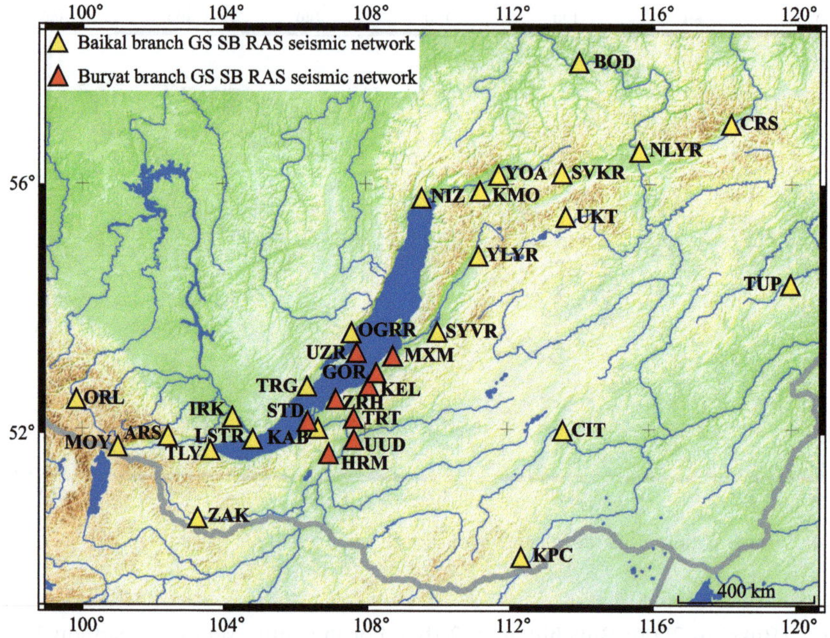

Fig. 5 Regional seismic stations used for determining P-wave first-motion polarities

calculate the attenuation of surface waves, we used the PREM model (Dziewonski and Anderson 1981).

Figure 6 illustrates the method of data processing for the particular case of the March 18, 2016 M_w 4.6 earthquake. For this seismic event we selected 19 seismic stations located at different azimuths from the epicenter. An example of the FTAN analysis is presented for the Enshi seismic station (ENH, China). The initial and filtered vertical—(LHZ) and transverse-component (LHT) seismograms are normalized with respect to the maximum observed amplitude. The FTAN diagrams are shown together with our estimates of the corresponding group velocity dispersion curves of Rayleigh and Love waves, indicated by thick pink lines. We used the dispersion curve to design a time-dependent filter to extract the fundamental mode with minimum noise contamination. Finally, we determined the focal mechanism solution of the considered earthquake with the following parameters of the nodal planes (strike, dip and slip angles): NP1: 266°, 74°, −72°; NP2: 37°, 24°, −136°. We obtained a seismic moment of 0.1×10^{17}Nm and a hypocentral depth of 5–6 km. The moment magnitude M_w was calculated from the seismic moment according to Hanks and Kanamori (1979) and its value is 4.7. Our best double-couple solution is characterized by the normalized joint residual function of 0.28. Moreover, all the available first-motion polarities of P-waves match the determined focal mechanism.

The focal mechanism solutions obtained for the entire set of the earthquakes are presented in Fig. 7 and for the seismic events occurred from 2000 to 2014 are mainly

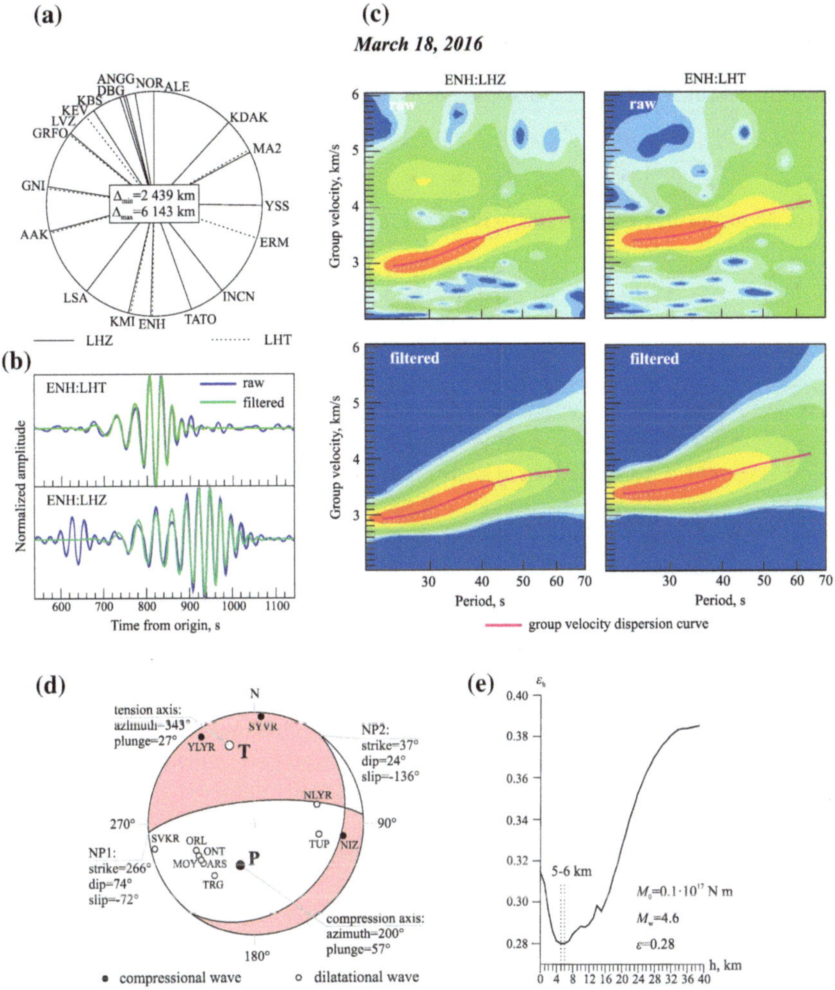

Fig. 6 Example of the data processing for the March 18, 2016 M_{W} 4.6 earthquake: **a**—azimuthal distribution of seismic stations; **b**—raw and filtered records for LHZ- and LHT-components of the ENH station (azm = 182°, Δ = 2827 km); **c**—group velocity versus period diagrams before and after filtration; **d**—focal mechanism solution (lower hemisphere) with P-wave first-motion polarities (abbreviations are in accordance with Fig. 5); **e**—partial residual function as a function of depth

discussed in (Seredkina and Melnikova 2014; Seredkina et al. 2015; Melnikova et al. 2017) including detailed comparison of the results with the Global-CMT solutions and other available information (Emmerson et al. 2006; Barth and Wenzel 2010). Thus, we do not focus on the solutions of individual earthquakes. It should be noted, that SMTs of 56 seismic events were calculated for the first time due to implementation of the surface wave method that significantly complete the existing dataset on

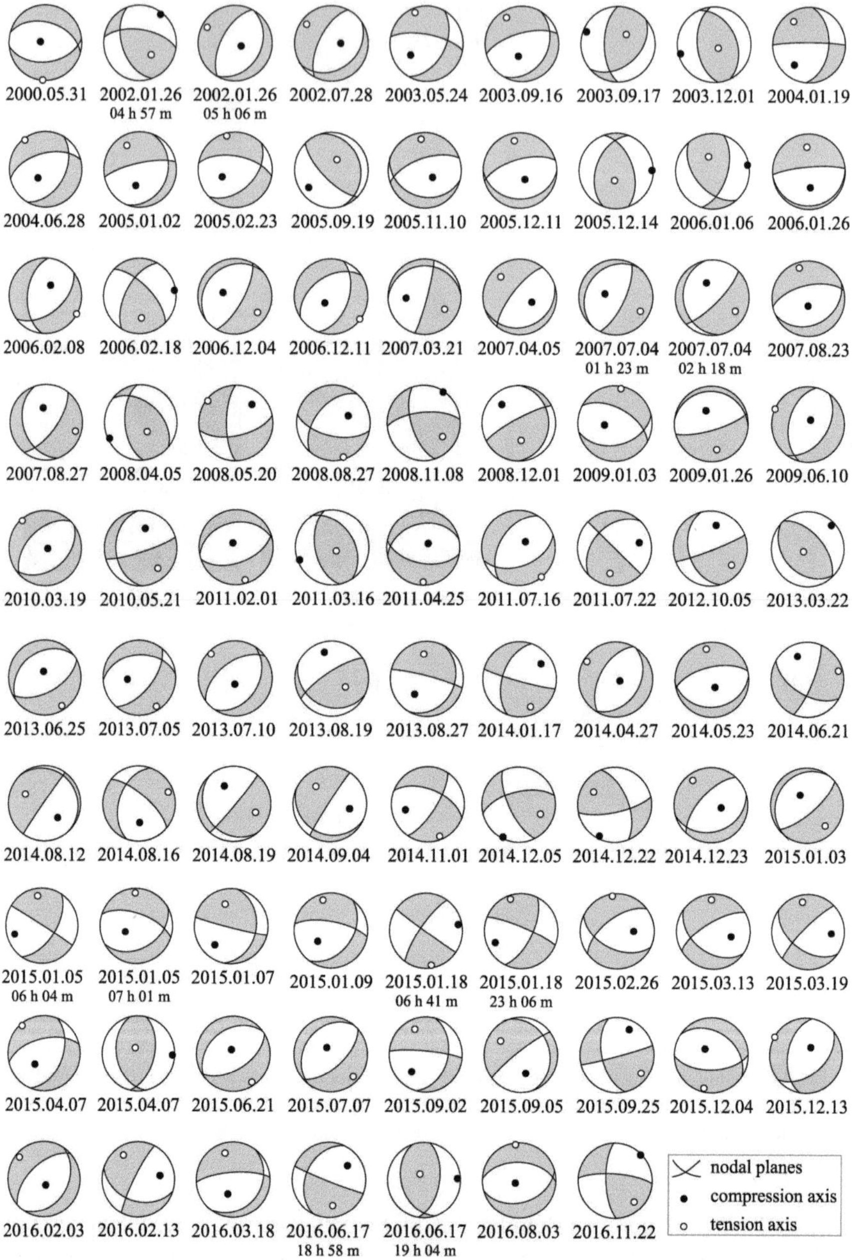

Fig. 7 Focal mechanism solutions in a lower hemisphere projection of the studied earthquakes

focal parameters of regional earthquakes and allow us to carry out seismotectonic reconstructions more precisely.

4 Seismotectonic Deformations from the Data on Focal Mechanisms of Medium Earthquakes

Reconstruction of the stress and strain fields of the crust in active regions is an integral part of studying general trends and peculiarities of the Earth's interior and developing geodynamical models. The strain tensor inversion schemes are widely used for these purposes (Angelier 1979; Barth and Wenzel 2010; Gephart and Forsyth 1984; Gushchenko 1979; Heidbach et al. 2010). Initial data for calculations are results of geological researches, earthquake focal mechanisms and other indicators of the present-day stresses. The main assumption is space-time homogeneity of the stress-strain field and a conjunction of a slip direction with a tangential strain vector.

Another approach of statistical analysis of individual focal mechanism solutions consists in assessment of the average STD tensor (Kostrov 1975; Lukk et al. 1995; Rebetsky 2007; Riznichenko 1985; Yunga 1979, 1990 etc.). According to these methods, in the seismoactive volume of the crust irreversible deformations are considered to exist due to the slips at the earthquake origins. A contribution of the individual seismic event to the entire deformation is determined by a unit SMT.

In this study the STD field of the crust was reconstructed using the method by Yunga (1979, 1990, 1997). The strain tensor is calculated as an arithmetic mean of the set of individual focal mechanism matrices:

$$\langle \varepsilon_{ij} \rangle = \frac{1}{\mu V} \sum_{\alpha=1}^{N} M_0^\alpha m_{ij}^\alpha,$$

where μ is a shear modulus, V—an elementary volume, M_0^α—a scalar seismic moment of the earthquake indexed as α, $\alpha = 1,\ldots, N$, and N is a total number of earthquakes, and m_{ij}^α—a matrix of an individual focal mechanism.

Proceeding from the assumption that seismic strain is self-similar at different scales, the STD tensor is given by

$$\langle \varepsilon_{ij} \rangle = \frac{1}{\mu V} \sum_{\alpha=1}^{N} M_0^\alpha \langle M_{ij} \rangle.$$

The matrix $\langle M_{ij} \rangle$ of an average mechanism is

$$\langle M_{ij} \rangle = \sum_{\alpha=1}^{N} w^\alpha m_{ij}^\alpha / \sum_{\alpha=1}^{N} w^\alpha,$$

where w^{α} is an empirical linear weighting function depending on magnitude of an earthquake M^{α} (Yunga 1997):

$$w^{a} = 0.07M^{\alpha} + 0.45.$$

The resulting matrix of the average mechanism can be described by three eigenvalues ($M_1 \geq M_2 \geq M_3$) and three principal axes of tension (T), compression (P), and zero (B). Inasmuch, as the slip surfaces are planar, a trace of the resulting matrix is zero ($M_1 + M_2 + M_3 = 0$).

The dispersion of the individual mechanism solutions relative to the average mechanism is described by an intensity parameter χ ranging from 0 for sets with totally random fault plane orientations to 1 for identical ones. This parameter identifies a relative measure of the uniformity of individual focal mechanisms. The general deformation pattern within a dataset is defined by a Lodé-Nadaì coefficient (Yunga 1979; Lukk et al. 1995):

$$\mu = 3M_2/\,(M_1 - M_3)$$

Deformation is mostly along the tension axis at $\mu < 0$ and along the compression axis at $\mu > 1$; in the case of $\mu = 0$ it is pure shear, the distribution of the individual P and T axes relative to the respective principal axes of the average mechanism are approximately equal (Yunga 1979).

To represent the results of the seismotectonic reconstruction we use a graphical approach and show the projections of principal compression and tension axes. The STD regimes are classified in accordance with the classification scheme by Yunga (1990, 1997) identifying 11 regimes (Fig. 8). First of all, it includes four main regimes: compression (T, thrust faulting), extension (N, normal faulting), strike-slip faulting S and vertical faulting V. Two regimes can be regarded as ultimate ones. Two-sided compression (TT) and extension (NN) regimes are formed by two thrust or normal faults respectively. The oblique faulting deformation (O) is a transition from vertical to strike-slip regime and equidistant from all the main regimes. Finally, four transitional regimes are defined: the transition from vertical faulting to compression (TV) or extension (NV) and from strike-slip faulting to compression (TS, transpression) or extension (NS, transtension).

To calculate the STD field, we selected more than 300 earthquakes with known focal mechanisms occurred from 1950 to 2016 (Fig. 9). Most of fault plane solutions were taken from the ISC catalog (International… 2014). In Mongolia we gave preference to the Global-CMT solutions for the seismic events with more than one focal mechanism available. To improve the resolution in the Baikal rift the SMT data for medium events described above, as well as earlier fault plane solutions obtained from the data on P-wave first-motion polarities for $M > 3.0$ local earthquakes (Solonenko et al. 1993; Melnikova and Radziminovich 1998; Earthquake Catalogs 2003–2012) were added. Altogether, 18 groups of spatially proximal earthquakes were selected from the initial dataset (Fig. 9), which all satisfied the uniformity condition (the intensity parameter $\chi \geq 0.5$). Individual focal mechanisms were summed using the

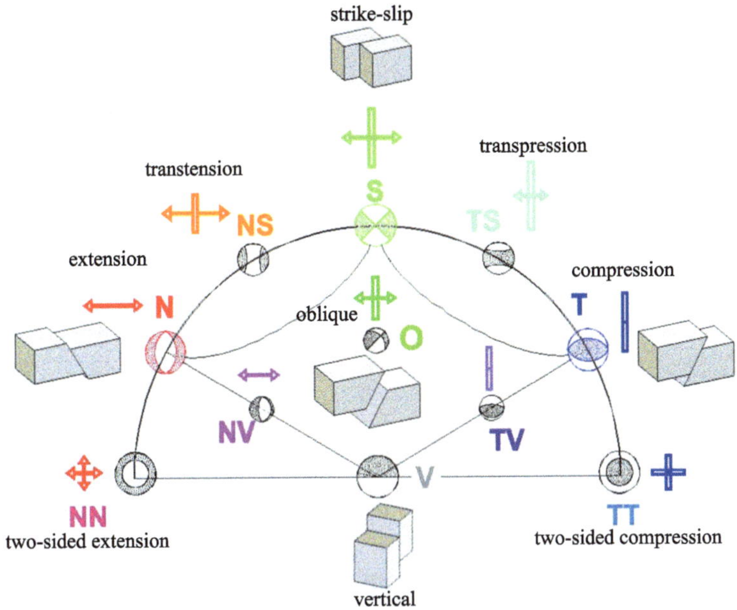

Fig. 8 Classification of the STD regimes by Yunga (1997)

method of "nodal points" (Sycheva et al. 2009) due to different density and irregular distribution of the initial data in the study area. It should be noted that two groups of events in southern Mongolia consist of only three earthquakes, but it was possible to calculate their average mechanism because individual focal mechanisms in these groups were almost identical ($\chi = 0.83$ and 0.93 respectively).

The regions of recent mountain building in southern Siberia and western Mongolia are characterized by transpression (transition from strike-slip to compression) and strike-slip regimes (Fig. 10). The orientation of principal compression varies from nearly N-S in the west to N-E in the east. Though the STD field pattern of the Gobi-Altai was reconstructed on the basis of a few earthquake focal mechanisms the results are in a good agreement with general tendencies of the crust deformations and the data obtained by different methods (Calais et al. 2003; Melnikova 2008; Karagianni et al. 2015). The Sayans and the southwestern flank of the Baikal rift are characterized by transition from transpression to strike-slip regime. Normal-fault deformation is observed in the central segment and major part of the northeastern rift flank. Strike-slip motions prevail in the junction of the Baikal rift and the Olekma-Stanovoi zone. The orientation of principal tension and compression there is the same as in the central part of the BRZ.

The results obtained (Fig. 10) show general tendencies of the crust deformation of the study area. Their reliability is supported by previous studies of the lithospheric strains and deformations, and by the available seismological, structural, and geodetic data (Apel et al. 2006; Calais et al. 2003; Delvaux et al. 1997; Frankel et al. 2010;

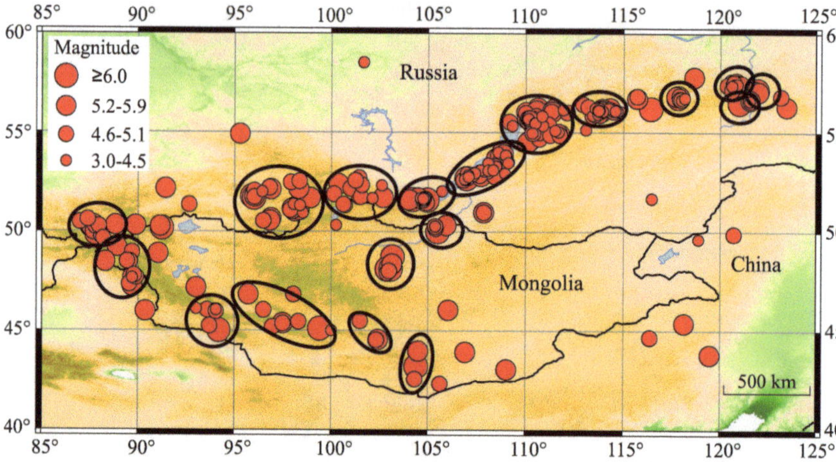

Fig. 9 Map of the earthquake epicenters used for STD field calculations. Black ellipses contour the selected earthquake groups

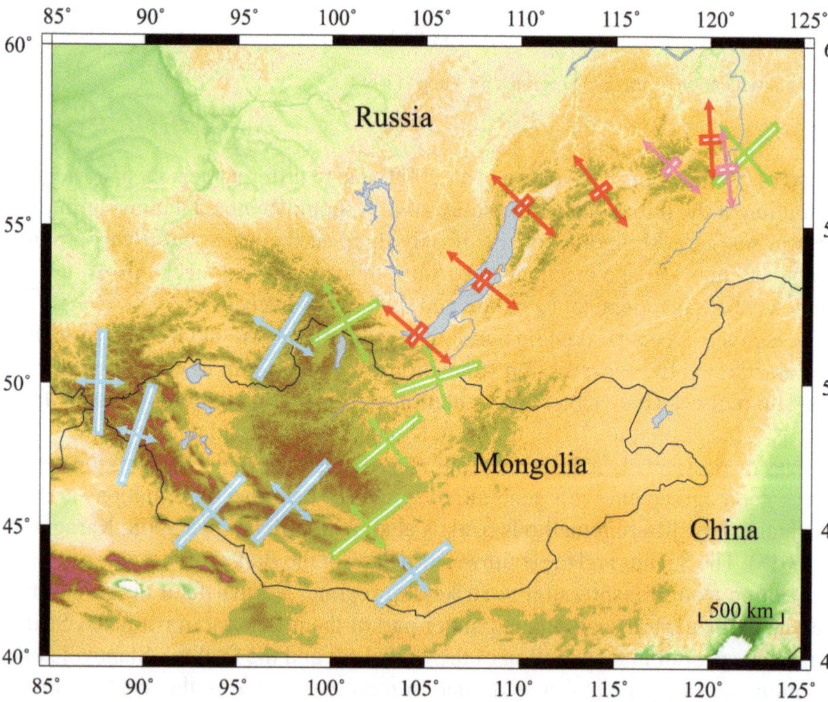

Fig. 10 The calculated STD field in accordance with classification scheme on Fig. 8

Heidbach et al. 2010; Karagianni et al. 2015; Lukhnev et al. 2010; Melnikova 2008; Rebetsky et al. 2012; San'kov et al. 2009).

5 Discussion and Conclusions

Deformation of real geological environment and relations between geological structure and an earthquake are determined by a number of factors associated with processes of different nature. Some of them are regional and long-lasting, while the others are known to be local and short.

The SMTs of medium regional seismic events determined in this study and their statistical analysis make it possible to study the structure of the seismic zone more thoroughly, and regionalize it according to the types of seismotectonic regimes. In our case, these data also reflect the main seismotectonic tendencies in the crust movements of the Mongol-Baikal region and are consistent with the character of active faulting in the area.

Though we considered seismicity and STD of the crust, undoubtedly, they are influenced by mantle processes, which can be indirectly interpreted by analyzing the distribution of the velocity inhomogeneities. In our previous studies (Kozhevnikov et al. 2014; Seredkina et al. 2016) it was found that most of the shallow large earthquakes occurred in the regions with low S-wave velocities in the uppermost mantle (western Mongolia and mountains in southern Siberia) and relatively high lateral S-wave velocity variations observed at the northeastern flank of the BRZ. The dominant STD regimes are compression (a mixture of thrust and strike-slip mechanisms) in the former case and extension in the latter case, as it was indicated in this study.

New combined data about the deep velocity structure and STD patterns can help us to construct the geodynamical models of the study area and, for example, give some evidences of the rift origin. The presence of a low-velocity zone here, which lies immediately below the crust and is separated from the asthenosphere by a thin layer with typically platformic velocities (Seredkina et al. 2016) prompts that hot plastic asthenospheric material may be rising along the zone of weakness at the border of the Siberian platform (Krylov et al. 1981; Artyushkov et al. 1990). This process may be controlled by the general direction of plate motion and by the asthenosphere thickness difference between the Mongol-Okhotsk fold belt and the Siberian platform. The hot material may be spreading at the crust base off the stable platform block. Penetration of the asthenospheric material may have been responsible for a low-angle regional uplift (growth of the Baikal dome), while its southeastern flow produced extension stresses in the crust. The extension stress acts jointly with stronger compression due to viscous friction associated with the NE–SW motion of Asia.

To conclude, on the basis of literature data and our own SMT determinations of regional earthquakes, it has been shown that using this characteristic in seismotectonic analysis of the territory of the Mongol-Baikal region makes it possible to reveal main features of modern stress-strain state of the crust and evaluate seismogenic component of tectonic movements. Increasing the number of the SMT definitions contribute to a better understanding of not only the fine internal structure of earthquake foci, but also more detailed seismotectonic zoning of seismically active territory.

References

Amante C, Eakins BW (2009) ETOPO1. 1 Arc-minute global relief model: procedures, data sources and analysis. NOAA technical memorandum NESDIS NGDC-24. National Geophysical Data Center, NOAA. https://doi.org/10.7289/v5c8276m

Angelier J (1979) Determination of mean principal directions of stresses for a given fault population. Tectonophysics 56(3–4):T17–T26. https://doi.org/10.1016/0040-1951(79)90081-7

Apel EV, Burgmann R, Steblov G, Vasilenko N, King R, Prytkov A (2006) Independent active microplate tectonics of northeast Asia from GPS velocities and block modeling. Geophys Res Lett 33:L11303. https://doi.org/10.1029/2006gl026077

Artyushkov EV, Letnikov FA, Ruzhich VV (1990) A new formation mechanism of the Baikal basin. In: Logachev NA (ed) Geodynamics of intracontinental mountain provinces. Nauka, Novosibirsk, pp 367–376 (in Russian)

Babich V, Chikachev B, Yanovskaya T (1976) Surface waves in a vertically inhomogeneous elastic half-space with weak horizontal inhomogeneity. Izvestiya Akad Nauk SSSR, Fizika Zemli 4:24–31 (in Russian)

Baljinnyam I, Bayasgalan A, Borisov BA, Cisternas A, Dem'yanovic,h MG, Ganbaatar L, Kochetkov VM, Kurushin RA, Molnar P, Philip H, Vashchilov YY (1993) Ruptures of major earthquakes and active deformation in Mongolia and its surroundings. In: Geological society of america memoirs, vol 181, 64 pp, Boulder, Colorado. http://dx.doi.org/101130/MEM181-p 1

Barth A, Wenzel F (2010) New constraints on the intraplate stress field of the Amurian plate deduced from light earthquake focal mechanisms. Tectonophysics 482(1–4):160–169. https://doi.org/10.1016/j.tecto.2009.01.029

Bassin C, Laske G, Masters G (2000) The current limits of resolution for surface wave tomography in North America. EOS Trans AGU 81:F897

Bayasgalan A, Jackson J (1999) Field examples of strike-slip fault terminations in Mongolia and their tectonic significance. Tectonics 18:394–411

Bukchin BG (1990) Determination of source parameters from surface wave recording allowing for uncertainties in the properties of medium. Izvestiya Akad. Nauk SSSR Fizika Zemli 25:723–728

Bukchin BG (2006) Specific features of surface wave radiation pattern by a shallow source. Izvestiya, Phys Solid Earth 42(8):712–717

Bukchin B, Clevede E, Mostinskiy A (2010) Uncertainty of moment tensor determination from surface wave analysis for shallow earthquakes. J Seismol 14(3):601–614. https://doi.org/10.1007/s10950-009-9185-8

Bukchin BG, Levshin AL, Ratnikova LI, Dost B, Nolet G (1994) Estimation of spatio-temporal source parameters for the 1988 Spitak, Armenia, earthquake from broadband surface wave records. Computational seismology and geodynamics, vol 2. AGU, Washington DC, pp 156–161

Calais E, Vergnolle M, San'kov V, Lukhnev A, Miroshnichenko A, Amarjargal S, Déverchère J (2003) GPS measurements of crustal deformation in the Baikal-Mongolia area (1994–2002): implications for current kinematics of Asia. J Geophys Res 108(B 10):2501. https://doi.org/10.1029/2002jb002373

Chen WP, Molnar P (1977) Seismic moments of major earthquakes and the average rate of slip in central Asia. J Geophys Res 82:2945–2969

Delvaux D, Moeys R, Stapel G, Petit C, Levi K, Miroshnichenko A, Ruzhich V, San'kov V (1997) Paleostress reconstructions and geodynamics of the Baikal region, Central Asia, Part 2. Cenozoic rifting. Tectonophysics 282:1–38

Delouis B, Déverchère J, Melnikova V, Radziminovitch N, Loncke L, Larroque C, Ritz JF, San'kov V (2002) A reappraisal of the 1950 (Mw 6.9) Mondy earthquake, Siberia, and its relationship to the strain pattern at the south-western end of the Baikal rift zone. Terra Nova 14:491–500

Doser DI (1991a) Faulting within the western Baikal rift as characterized by earthquake studies. Tectonophysics 196:87–107

Doser DI (1991b) Faulting within the eastern Baikal rift as characterized by earthquake studies. Tectonophysics 196:109–139

Dziewonski AM, Anderson DL (1981) Preliminary reference earth model. Phys Earth Planet Inter 25(4):297–356. https://doi.org/10.1016/0031-9201(81)90046-7
Earthquake Catalogs, source mechanisms, Pribaikalia and Transbaikalia (1997–2006) Eurasian Earthquakes in 2000–2006 (in Russian). GS RAS, Obninsk. 2003–2012
Emmerson B, Jackson J, McKenzie D, Priestley K (2006) Seismicity, structure andrheology of the lithosphere in the lake Baikal region. Geophys J Int 167:1233–1272. https://doi.org/10.1111/j1365-246x.2006.03075.x
Florensov NA, Solonenko VP (eds) (1965) The Gobi-Altai Earthquake. U.S. Department of Commerce, Washington, D.C., p 424
Frankel KL, Wegmann KW, Bayasgalan A, Carson RJ, Bader NE, Adiya T, Bolor E, Durfey CC, Otgonkhuu J, Sprajcar J, Sweeney KE, Walker RT, Marstellar TL, Gregory L (2010) Late Pleistocene slip rate of the Hoh Serh-Tsagaan Salaa fault system, Mongolian Altai and intracontinental deformation in Central Asia. Geophys J Int 183:1134–1150. https://doi.org/10.1111/j.1365-246x.2010.04826.x
Gephart JW, Forsyth DW (1984) An improved method of determining the regional stress tensor using earthquake focal mechanism data: Application to the San Fernando earthquake sequence. J Geophys Res 89:9305–9320
Gomez JM, Bukchin B, Madariaga R, Rogozhin EA, Bogachkin B (1997a) Rupture process of the 19 August 1992 Susamyr, Kyrgyzstan, earthquake. J Seismol 1(3):219–235. https://doi.org/10.1023/a:1009780226399
Gomez JM, Bukchin B, Madariaga R, Rogozhin EA (1997b) A study of the Barisakho, Georgia, earthquake of 1992 October 23 from broad-band surface and body waves. Geophys J Int 129(3):613–623. https://doi.org/10.1111/j.1365-246x.1997.tb04497.x
Gushchenko OI (1979) The method of the kinematic analysis of structures of destruction at reconstruction of fields of tectonic stresses. Fields of stress a lithosphere (in Russian). Nauka, Moscow, pp 7–25
Heidbach O, Tingay M, Barth A, Reinecker J, Kurfeß D, Müller B (2010) Global crustal stress pattern based on the World Stress Map data base release 2008. Tectonophysics 482(1–4):3–15. https://doi.org/10.1016/j.tecto.2009.07.023
Hanks T, Kanamori H (1979) A moment magnitude scale. J Geophys Res 84(B5):2348–2350. https://doi.org/10.1029/jb084ib05p02348
Huang J, Chen W–P (1986) Source mechanisms of the Mogod earthquake sequence of 1967 and the event of 1974 July 4 in Mongolia. Geophys J Rov Astron Soc 84, 361–379
Imaev VS, Imaeva LP, Koz'min BM (2000) Seismotectonics of Yakutia, GEOS, Moscow, 227 pp (in Russian)
International Seismological Centre (2014) On-line Bulletin. Int Seismol Centre. Thatcham, United Kingdom. http://www.isc.ac.uk
Karagianni I, Papazachos SB, Scordilis EM, Karakaisis GF (2015) Reviewing the active stress field in Central Asia by using a modified stress tensor approach. J Seismol 19:541–565. https://doi.org/10.1007/s10950-015-9481-4
Khilko SD, Kurushin RA, Kochetkov VM, Misharina LA, Melnikova VI, Gileva NA, Lastochkin SV, Baljinnyam I, Monkhoo D (1985) Earthquakes and foundations of seismic zoning in Mongolia. Nauka, Moscow, 224 pp (in Russian)
Kondorskaya, N.V. and Shebalin, N.V (1977) New catalog of large earthquakes occurred on the territory of the USSR from the ancient times to 1975. Nauka, Moscow, 535 pp (in Russian)
Kostrov BV (1975) Mechanics of the source of tectonic earthquake. Nauka, Moscow, 176 pp (in Russian)
Kozhevnikov VM, Seredkina AI, Solovei OA (2014) 3D mantle structure of Central Asia from Rayleigh wave group velocity dispersion. Russ Geol Geophys 55(10):1239–1240. https://doi.org/10.1016/j.rgg.2014.09.010
Krylov SV, Mandelbaum MM, Mishenkin BP, Mishenkina ZP, Petrik GV, Seleznev VS, 1981. The Baikal interior from seismic data. In: Puzyrev NN (ed) Nauka, Novosibirsk, 105 pp (in Russian)

Kurushin RA, Bayasgalan A, Olziybat M, Enhtuvshin B, Molnar P, Bayarsayhan C, Hudnut K, Lin J (1997) The surface rupture of the 1957 Gobi-Altay, Mongolia, earthquake. Spec Pap Geol Soc Am 320:143

Lasserre C, Bukchin B, Bernard P, Tapponier P, Gaudemer Y, Mostinsky A, Dailu R (2001) Source parameters and tectonic origin of the 1996 June 1 Tianzhu (Mw = 5.2) and 1995 July 21 Yongen (Mw = 5.6) earthquakes near the Haiyuan fault (Gansu, China). Geophys J Inter 144(1):206–220. https://doi.org/10.1046/j.1365-246x.2001.00313.x

Levshin AL, Yanovskaya TB, Lander AV, Bukchin BG, Barmin MP, Ratnikova LI, Its EN (1989) Surface waves in vertically inhomogeneous media. In: Keilis-Borok VI (ed) Seismic surface waves in a laterally inhomogeneous Earth. Kluwer Academic Publishing, Dordrecht, pp 131–182

Logatchev NA (1999) Main structural features and geodynamics of the Baikal rift zone. Phys Mesomech 2(1–2):163–170 (in Russian)

Logatchev NA, Zorin YuA (1992) Baikal rift zone: structure and geodynamics. Tectonophysics 208:273–286

Lukhnev AV, San'kov VA, Miroshnichenko AI, Ashurkov SV, Calais E (2010) GPS rotation and strain rates in the Baikal-Mongolia region. Russ Geol Geophys 51(7):785–793

Lukk AA, Yunga SL, Shevchenko VI, Hamburger MW (1995) Earthquake focal mechanisms, deformation state, and seismotectonics of the Pamir-Tien Shan region. Central Asia J Geophys Res 100(B10):20321–20343

Melnikova VI (2008) Crust strain parameters in the Baikal Rift Zone, from seismic data. Author's Abstract, Doctor Thesis, IEC SB RAS, Irkutsk, 37 pp (in Russian)

Melnikova VI, Radziminovich NA (1998) Mechanisms of Baikal earthquakes of 1991–1996. Russ Geol Geophys 39(11):1598–1607

Melnikova V, Seredkina A, Radziminovich Y, Melnikov A, Gilyova N (2017) The February 1, 2011 Mw 4.7 earthquake: Evidence of local extension in western Transbaikalia (Eastern Siberia). J Asian Earth Sci 135:110–121. https://doi.org/10.1016/j.jseaes.2016.12.031

Mendiguren JA (1977) Inversion of surface wave data in source mechanism studies. J Geophys Res 82(5):889–894. https://doi.org/10.1029/jb082i005p00889

Molnar P, Deng QP (1984) Faulting associated with large earthquakes and the average rate of deformation in central and eastern Asia. J Geophys Res 89:6203–6227

Molnar P, Kurushin RA, Kochetkov VM, Demjanovich MG, Borisov BA, Vashchilov YY (1995) Deformation and faulting at large earthquakes in Mongolia-Siberian region. Deep-seated structure and geodynamics of Mongolia-Siberian region. Nauka, Novosibirsk, pp 5–55 (in Russian)

Molnar P, Tapponier P (1975) Cenozoic tectonics of Asia: effect of continental collision. Science 89:119–126

Molnar P, Kurushin RA, Bayasgalan A, Khadiat KV (1998) Dislocations of the 1957 Gobi Altai earthquake. Nauka, Novosibirsk, 148 pp (in Russian)

Nataf H-C, Ricard Y (1996) 3SMAC: on a priori tomographic model of the upper mantle based on geophysical modeling. Phys Earth Planet Inter 95(1–2):101–122. https://doi.org/10.1016/0031-9201(95)03105-7

Okal FA (1976) A surface-wave investigation of the rupture mechanism of the Gobi-Altai (December 4, 1957) earthquake. Phys Earth Planet Inter 12:319–328

Okal E (1977) The July 9 and 23, 1905, Mongolian earthquakes: A surface wave investigation. Earth Planet Sci Lett 34:326–331

Petit C, Déverchère J (2006) Structure and evolution of the Baikal rift: a synthesis. Electron J Earth Sci 7(11):Q11016. https://doi.org/10.1029/2006gc001265

Randall MJ (1971) Elastic multipole theory and seismic moment. Bull Seismol Soc Am 61(5):1321–1326

Rebetsky YL (2007) Tectonic stresses and rock strength. Akademkniga, Moscow, 406 pp (in Russian)

Rebetsky YuL, Kuchai OA, Sycheva NA, Tatevossian RE (2012) Development of inversion methods on fault slip data stress state in orogenes of the Central Asia. Tectonophysics 581:114–131. https://doi.org/10.1016/j.tecto.2012.09.027

Ritz JF, Brown ET, Bourle's DL, Philip H, Schlupp A, Raisbeck GM, Yiou F, Enkhtuvshin B (1995) Slip rates along active faults estimated with cosmic-ray-exposure dates: Application to the Bogd fault, Gobi-Altay, Mongolia. Geology 23:1019–1022

Riznichenko, YV (1985) Problems of seismology. Nauka, Moscow, 408 pp. (in Russian)

San'kov VA, Lukhnev AV, Radziminovich NA, Melnikova VI, Miroshnichenko AI, Ashurkov SV, Calais E, Déverchère J (2005) A quantitative estimate of modern deformations of the Earth's crust in the Mongolian block (based on GPS-geodesy and seismotectonic data). Dokl Earth Sci 403(6):946–949

San'kov VA, Lukhnev AV, Miroshnichenko AI, Ashurkov SV, Byzov LM, Dembelov MG, Calais E, Déverchère J (2009) Extension in the Baikal rift: present day kinematics of passive rifting. Dokl Earth Sci 425:205–209. https://doi.org/10.1134/s1028334x09020056

Schlupp A, Cisternas A (2007) Source history of the 1905 great Mongolian earthquakes (Tsetserleg, Bolnay). Geophys J Int 169:1115–1131

Seredkina A, Kozhevnikov V, Melnikova V, Solovey O (2016) Seismicity and S-wave velocity structure of crust and upper mantle in the Baikal rift and adjacent regions. Phys Earth Planet Inter 261:152–160. https://doi.org/10.1016/j.pepi.2016.10.011

Seredkina AI, Melnikova VI (2013) The seismic moment tensor of earthquakes in the Pribaikalye Rerion based on surface waves. Dokl Earth Sci 451(1):746–749. https://doi.org/10.1134/s1028334x13070064

Seredkina AI, Melnikova VI (2014) Seismic moment tensor of Pribaikalye earthquakes from the surface-wave amplitude spectra. Izvestiya Phys Solid Earth 50(3):103–114. https://doi.org/10.1134/s1069351314030094

Seredkina AI, Melnikova VI, Gileva NA, Radziminovich YB (2015) The Mw 4.3 January 17, 2014, earthquake: very rare seismic event the Siberian platform. J Seismol 19(3):685–694. https://doi.org/10.1007/s10950-015-9487-y

Seredkina AI, Kozmin BM (2017) Source parameters of the Taimyr earthquake of June 9, 1990. Dokl Earth Sci 472(2):343–346. https://doi.org/10.1134/s1028334x1702026x

Solonenko VP, Treskov AA, Zhilkin VM, Zorin YA, Korostin PV, Kurushin RA, Pavlov OV, Pshennikov KV, Fomina EV, Khilko SD, Khromovskikh VS, Shmotov AP (1968) Seismotectonics and seismicity of the rift system of Pribaikalye. Nauka, Moscow, 220 pp (in Russian)

Solonenko AV, Solonenko NV, Melnikova VI, Kozmin BM, Kuchay OA, Suhanova SS (1993) Strain and rakes of earthquake origins in Siberia and Mongolia. Seism seism North Eurasia. 1:113–122 (in Russian)

Sycheva NA, Bogomolov LM, Yunga SL (2009) Geoinformatics in the statistical approach to strain modeling. Geoinformatika. Model Geol Objects Process 1:33–43

Tapponnier P, Molnar P (1979) Active faulting and Cenozoic tectonics of the Tien Shan, Mongolia and Baikal regions. J Geophys Res 84:3425–3459

Woodhouse JH (1974) Surface waves in a laterally varying layered structure. Geophys J Roy Astron Soc 37(3):461–490. https://doi.org/10.1111/j.1365-246x.1974.tb04098.x

Yunga SL (1979) On the mechanism of deformation of a seimo-active volume of the Earth's crust. Izv Acad Sci USSR Phys Solid Earth 15:693–699 (in Russian)

Yunga SL (1990) Seismic strain studies: methods and results. Nauka, Moscow, 192 pp (in Russian)

Yunga SL (1997) Classification of seismic moment tensors using their isometric projection onto a sphere. Dokl Earth Sci 352(2):253–255

Zonenshain LP, Savostin LA (1979) Introduction to geodynamics. Nedra, Moscow, 311 pp (in Russian)

Zorin YuA (1999) Geodynamics of the western part of the Mongolia-Okhotsk collisional belt, Trans-Baikal region (Russia) and Mongolia. Tectonophysics 306:33–56. https://doi.org/10.1016/s0040-1951(99)00042-6

The Stress State of Seismic Areas of the Central and East Asia

Yu. L. Rebetsky, A. Yu. Polets, O. A. Kuchay and N. A. Sycheva

This chapter will present the results of inversion of modern state of stress, obtained by the method of cataclastic analysis of discontinuous displacements (MCA), for three seismically active regions of Asia: the Northern Tien Shan, the Altai-Sayan region and the Kuril-Kamchatka region. The stress data on the earthquake focal mechanisms are taken from global and regional seismic catalogues. The Northern Tien Shan and the Altai-Sayan region are the intracontinental orogens. The Kuril Islands and Kamchatka are active continental margins. The similarities and differences in the state of stress between these two geodynamic types of deformation of the lithosphere are shown in this chapter.

The goal of this chapter is to show the capabilities of the MCA. In contrast to other methods of tectonic stress inversion, the MCA allow one to obtain data not only on the parameters of stress ellipsoid, but also on the stress values. Various ways of visualizing parameters reflecting the spatial variations of tensor components will be presented in this chapter.

Y. L. Rebetsky
Laboratory of Tectonophysics, The Schmidt Institute of Physics of the Earth of the Russian Academy of Sciences (IPE RAS), Bolshaya Gruzinskaya str., 10-1, Moscow 123242, Russia
e-mail: reb@ifz.ru

A. Y. Polets (✉)
Laboratory of Seismology, Institute of Marine Geology and Geophysics Far Eastern Branch of the Russian Academy of Sciences (IMGG FEB RAS), Nauki st., 1B, Yuzhno-Sakhalinsk 693022, Russia
e-mail: polec84@mail.ru

O. A. Kuchay
Laboratory of Deep Seismic Studies and Regional Seismicity, Trofimuk Institute of Petroleum Geology and Geophysics of Siberian Branch of Russian Academy of Sciences (IPGG SB RAS), Koptug ave. 3, Novosibirsk 630090, Russia
e-mail: kuchayoa@ipgg.sbras.ru

N. A. Sycheva
GPS laboratory, Research Station Russian Academy of Sciences (RS RAS), Bishkek 720049, Kyrgyzstan
e-mail: nelya@gdirc.ru

S. D'Amico (ed.), *Moment Tensor Solutions*, Springer Natural Hazards,
https://doi.org/10.1007/978-3-319-77359-9_23

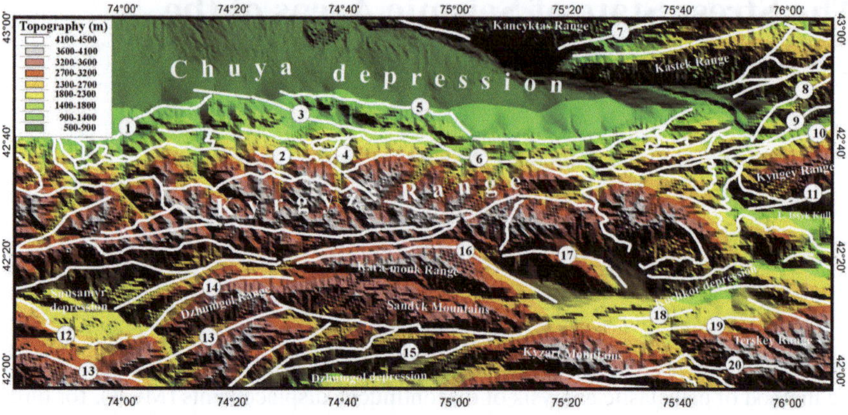

Fig. 1 Topography and major faults (Rebetsky and Kuzikov 2016) of the Northern Tien Shan in the vicinity of the Research Station Russian Academy of Sciences (RS RAS) in Bishkek. Faults: 1. Aksu; 2. Chonkurchak; 3. Baitik; 4. Alamedin; 5. Issyk-Ata; 6. Sahmshi; 7. Karakunuz; 8. Trans-Ili; 9. North Kemin; 10. South Kemin; 11. Toguz–Bulak (North Issyk-Kul); 12. Arami; 13. Kyzyl-Oy; 14. Karakul; 15. Ulun-Bulak (Uzun-Bulak-Oikain zone); 16. Kara-Kol; 17. Ortok; 18. Akchop; 19. South Kochkor; 20. Kara-Kurdzhur

Again, it should be noted that the longitudinal stress is positive under tension and negative under compression. This positive-negative sign convention is accepted in classical mechanics. This is the opposite of the criterion used in mining.

1 The Stress State of the Northern Tien Shan Crust

1.1 Introduction

The study area is located in the Tien Shan region of the Ural-Mongolian orogenic system between two microplates, Tarim in the south and Junggar in the north (Kazakh platform). This mountainous region extends in a ~N–S direction for a distance of more than 2500 km across Kyrgyzstan, Kazakhstan, Uzbekistan, Tajikistan, and China (Fig. 1). The Tien Shan folded area comprises the Northern Tien Shan Caledonides, Southern Tien Shan Hercynides, and Central Tien Shan Caledonides and Hercynides. The Tien Shan folded area is bounded from the south by the Pamir–Himalayan segment of the Mediterranean orogenic belt, which separates the study area from the Hindustan Platform.

The orography of the Northern Tien Shan is dominated by generally E–W-trending mountain ranges and intervening depressions formed as a result of folding and thrusting. The major tectonic features of the study area during the neotectonic phase include the Kyrgyz mega-anticline (Kyrgyz Range) and Chuya depression in the north. The Kyrgyz Range is bounded to the south by the Suusamyr, Kochkor, and Dzhumgol depressions separated by small ranges, to the west by the Talas, Greater and Lesser Karatau Ranges, and to the east by the Kungey-Alatau and Terskey-Alatau Ranges divided by the Issyk-Kul intermontane depression.

The observed seismic structures are interpreted to be marginal faults or their segments. Most of the earthquake epicenters in the Northern Tien Shan lie in a relatively narrow belt along the northern slope of the Kyrgyz Range, which extends eastward to the eastern plunge of the Kungey Range and Trans-Ili Range in the north (Fig. 1). Despite their high density, the earthquake epicenters tend to be clustered in small areas. The seismicity in the Northern Tien Shan is generally restricted to the upper crust or, rather, the pre-Mesozoic base. The analysis of the earthquake focal depth distribution revealed that all foci in the study area lie in depths ranges 0–5 and 5–10 km. Deep-focus earthquakes are more definitely confined to particular zones. Almost all large earthquakes in the Northern Issyk-Kul zone are confined to the Trans-Ili and Kungey Alatau ranges where focal depths are over 30 km. This is also true for the Sarykamysh and Kadji-Say seismic regions of the South Issyk-Kul zone with focal depths as great as 25–30 km.

1.2 Results of Natural Stress Inversion

The reconstruction of the state of stress in the Earth's crust for the North Tien Shan was performed using the MCA and earthquake focal mechanism data obtained by the Kyrgyz Seismic Network (KNET, Research Station of the Russian Academy of Sciences in Bishkek).

The KNET network (Fig. 2), consisting of 10 stations with Streckeisen STS-2 broadband seismometers was deployed in 1991 (Sycheva et al. 2005). The catalogue of focal mechanisms contains over 1056 earthquakes with magnitudes (M_b) between 1.16 and 5.40 for the observation period 1994–2012. The only focal mechanisms available are for 2, 2, and 9 events occurred within the study area in 1994, 1995, and 1996, and no solutions are present for the study area in 1997. The KNET network began operating efficiently in 1998.

The Northern Tien Shan is characterized by events with highly variable focal mechanisms. As seen in Fig. 3, most of the P-axes representing the direction of compressional stress are sub-horizontal and only 10% of the focal mechanisms tend to dip steeply at 55°. The T-axes have a general gentle dip while 33% of them dipping at a higher angle.

Preliminary analysis of the catalogue focal mechanisms showed that the distribution of earthquake density and magnitudes allowed stress inversion with a lateral averaging size of 10–15 km at a 0.05° * 0.05° grid. The calculation was performed for horizontal lines up to 10 km thick, the midpoints of which are located at depths of 5, 10, 15, and 20 km (i.e., these lines have depth intersections). As a result, the stress parameters were determined for 286, 467, 407, and 142 domains, corresponding to crustal depths of 5, 10, 15, and 20 km.

Fig. 2 The distribution and focal mechanisms of 1056 earthquakes are based on KNET data for 1994–2012 (the total number of earthquakes is 7720). KNET stations are depicted by the pentangle (10 stations). Magnitudes of earthquakes are shown in the upper left corner

First stage of MCA, parameters of the stress ellipsoid. Figures 3, 4 and 5 show the orientations of the principal stress axes constructed in the direction of their dipping. It is worth noting that the extension is positive following the classical mechanics and the indexes of the principal stresses are standardly defined by the following relationship $\sigma_1 \geq \sigma_2 \geq \sigma_3$. Therefore the direction of the axes σ_1 determines the minimum compression (deviatoric tension) and the direction of the axes σ_3 determines the maximum compression.

The reconstruction shows that the axes of the algebraically minimum principal stress (maximum compressional stress) σ_3 have a~N–S trend and are oriented at 330–360° N and 150–180° S (Fig. 5a). These axes are always north- and northwest-dipping at a depth of 0–10 km. In deeper layers, the south-dipping orientation of these axes is observed for approximately 30% of the crust domains. The average dip angle is 0–20° for about 50% of determinations. In the uppermost layer, the orientation of the c axes is sub-vertical for 10% of the domains, as can be seen in rose diagrams in the upper left corner of Fig. 5. Similar domains are found to be rare at a depth of 5–15 km and completely absent at depths of 10–20 and 15–25 km (Fig. 5b–d). The upper crust domains with sub-vertical axes largely correspond to the eastern part of the Chuya depression, northern and southern slopes of the Kyrgyz Range to the northeast of the Suusamyr depression.

The axes of the algebraically maximum stresses σ_1 (minimum compressive or tensional stress) vary considerably compared with those of the maximum compressional stresses. In the upper two layers, they have a~E-W trend and are oriented at 240–270° W and 60–90° E (Fig. 4a, b). The west-dipping orientation becomes predominant at depths of 10–20 and 15–25 km (Fig. 4c, d). At depths of 0–10 and 5–15 km, the sub-horizontal dip (approximately 20°) of the σ_1 axes prevails over a steep dip (approximately 50–70°). In deeper layers, both high and low dip angles of the σ_1 axes tend to be dominant. The domains with dominant sub-vertical σ_1 axes

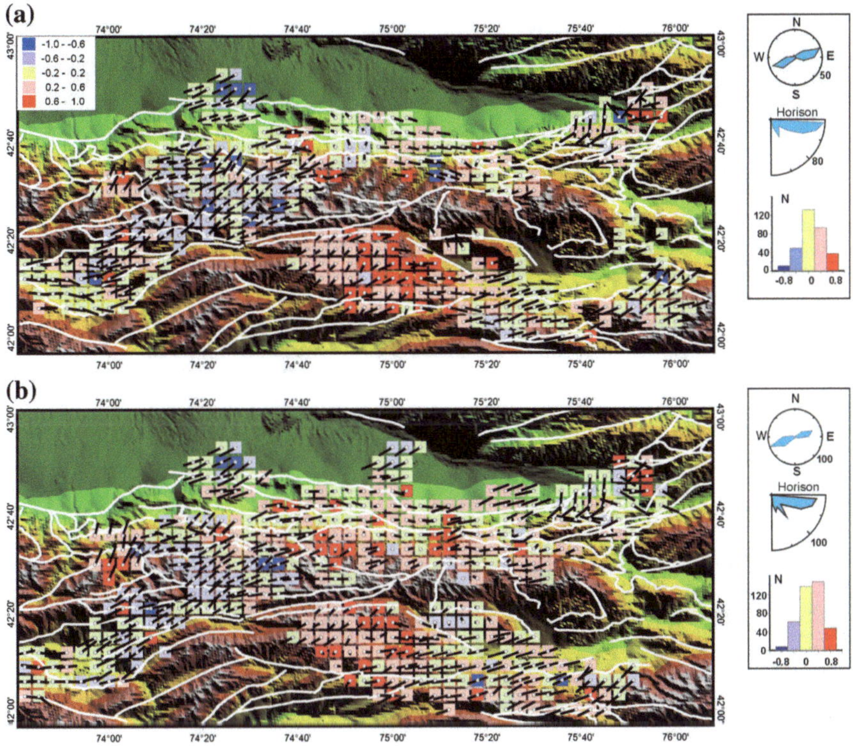

Fig. 3 Horizontal projections of plunges of the principal stress axis, σ_1 (minimum compression) and values of the Lode-Nadai coefficient defining the shape of the stress ellipsoid, μ_σ, (the legend is located in the upper left corner of the plot) for different depth layers: **a** 0–10 km; **b** 5–15 km; **c** 10–20 km; **d** 15–25 km. The open circles in the middle of the axes mark subhorizontal orientations (the angle of plunge is less than 7.5°). The upper and middle rose diagrams (the right side of the plot) show the predominant strike directions (averaged over 20°) and plunges (over 10°) of this stress axis. The bottom bar diagram (the right side of the plot) shows the predominant range of values of the Lode-Nadai coefficient ($-1 \leq \mu_\sigma \leq 1$)

are found at depths of 10–20 and 15–25 km in the central part of the study area. Such changes in the principal stress orientation from sub-horizontal to sub-vertical can be explained by the evolution of the surface topography of the crust rather than differences in the rock properties.

Figure 3 shows the value of the Lode-Nadai coefficient, which characterizes the shape of the stress ellipsoid. The values of the Lode-Nadai coefficient μ_σ ranging from –0.2 to 0.2 calculated for the entire Northern Tien Shan region correspond to the stress tensor of pure shear. At the same time, there are large spatial domains where the Lode-Nadai coefficient is close to +1 and –1, indicating regimes of uniaxial compression and tension.

Figure 4 shows the distribution of the relative value of the vertical component of deviatoric stresses $\sigma^*_{zz} + p^*$ normalized by the value of the maximum shear stress τ

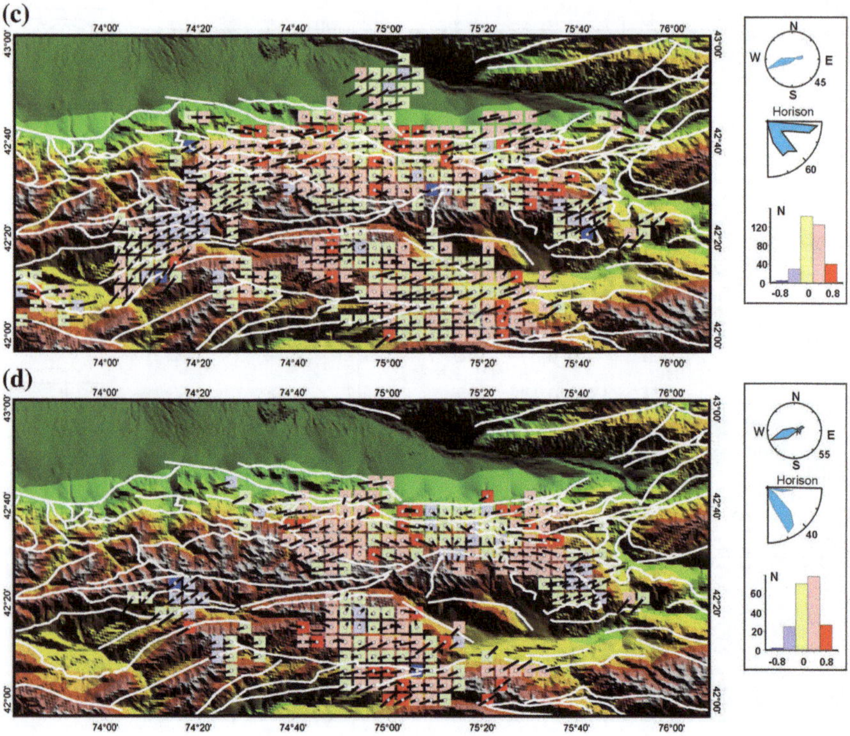

Fig. 3 (continued)

in the Earth's crust of the Northern Tien Shan. As can be seen from this figure the vertical stresses vary considerably. The parameter's values lie in the range −1.25 to 1.15. The highest values are observed in the Earth's crust of the central part of the Kirghiz ridge and to the south of it, and the smallest values are observed in the eastern and especially in the western segment of the investigated region. The greatest difference in values is observed at depths of 5–10 km. At great depths, earthquakes, whose focal mechanisms carry information about stresses, mainly occur in the areas with positive values of the ratio $\left(\sigma_{zz}^{*} + p^{*}\right)/\tau$, i.e. where there are areas of horizontal compression (Fig. 5).

Figure 5 shows the geodynamic zones of the Earth's crust according with the types of stress, i.e., the geodynamic regimes. The inset in the upper left corner in Fig. 5a shows the division of the quadrant of location of the principal stress axes into six zones. The proximity of the axis to zenith in each of six zones defines the type of geodynamic or stress regime: 1, horizontal compression; 2, horizontal compression + shear; 3, horizontal shear; 4, horizontal tension + shear; 5, horizontal tension; 6, vertical shear. It should be noted that the geodynamic regime of horizontal extension determines only the subhorizontal and subvertical orientations of the maximum

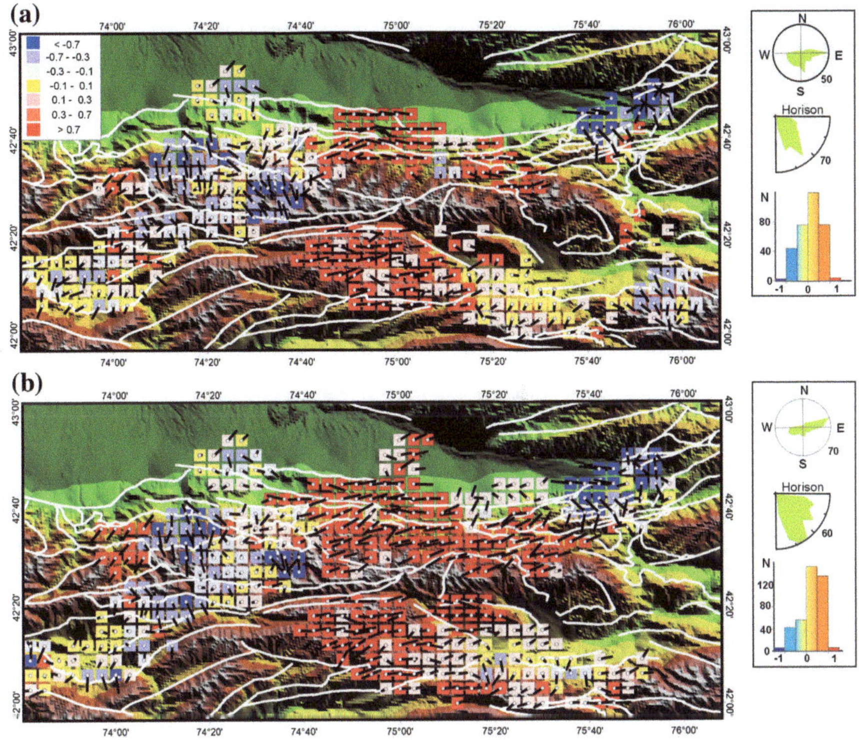

Fig. 4 Horizontal projections of plunges of the principal stress axis, σ_2 and relative value of the vertical component of deviatoric stresses, $\left(\sigma_{zz}^* + p^*\right)/\tau$ (the legend is located in the upper left corner of the plot) for different depth layers: **a** 0–10 km; **b** 5–15 km; **c** 10–20 km; **d** 15–25 km. The rose diagrams show the predominant strike directions and plunges of this stress axis. The bar diagram shows the predominant range of values $\left(\sigma_{zz}^* + p^*\right)/\tau$

deviatoric tension and maximum deviatoric compression. This does not mean that the elongation will be in the horizontal direction.

Based on the types of geodynamic or stress regimes (Fig. 5) the study area is dominated by the combination of horizontal shear and compression. In addition to these two types of stress regimes, there are three areas of localized horizontal extension with sub-vertical maximum and sub-horizontal minimum compressional stress (deviatoric tension) axes, two of which occur in the western segment of the Kyrgyz Range, and the third in the eastern portion of Chuya depression. In the first two areas, the axes of the maximum deviatoric tension are oriented in the ENE direction and in the ~E–W direction in the third area. The central part of the Northern Tien Shan is dominated by horizontal compression. The crust in this region is bounded by areas dominated by the combination of horizontal shear, compression and tension. On the Fig. 5 the diagram shows the prevailing geodynamic regime of stress state.

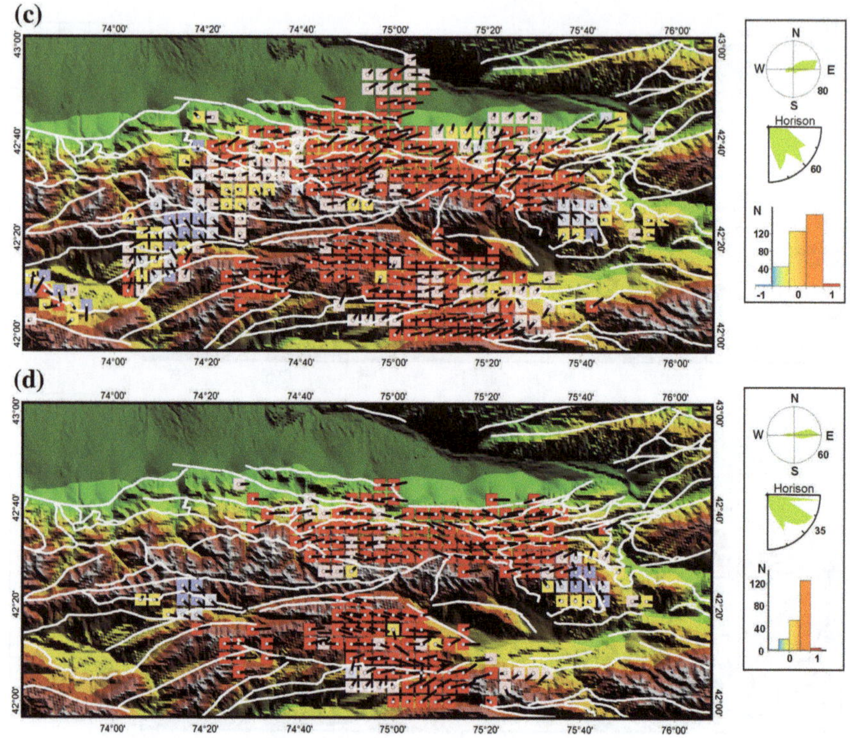

Fig. 4 (continued)

The results of the tectonophysical stress inversion for the Northern Tien Shan show that the crust in the central part of the study area where the topography is represented by high-elevation mountain ranges is dominated by horizontal compression. The crust of the Suusamyr and Kochkor depressions is characterized by a horizontal shear regime. The eastern termination of the Chuya depression and the western segment of the Kyrgyz Range are dominated by horizontal tension.

Characteristic parameters of the stress state in a geographic coordinate system. As noted above, the reconstruction of natural stresses using the algorithm of the first stage of the MCA makes it possible to compute the components of the stress ellipsoid. The major and minor axes on a horizontal plane of the ellipsoid indicate the orientations of maximum and minimum compressive stresses (or tensional stress in some cases) acting in a horizontal direction. Zoback (1992) referred these axes to as S_H and S_h (maximum and minimum horizontal compression).

Figure 6 shows that the maximum horizontal compressive stresses are often in a ~E–W direction and trend 150°–170° S or equivalently 330°–350° N. The most frequent deviations are observed in the upper 0–10 km. In the Chuya depression, ~E-W orientations of the maximum compressive stresses are reported not only for the upper 0–10 km but also for 5–15 km depth. The results in Fig. 6 can be used in

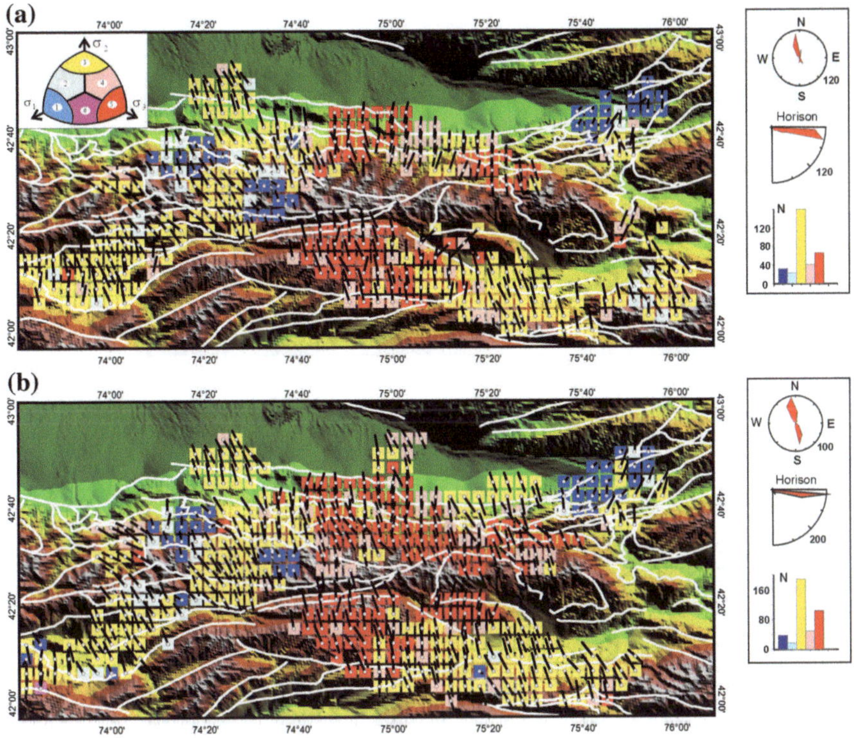

Fig. 5 Horizontal projections of plunges of the principal stress axis, σ_3 (maximum compression) and types of geodynamic regimes of stress state (the legend is located in the upper left corner of plot) for different depth layers: **a** 0–10 km; **b** 5–15 km; **c** 10–20 km; **d** 15–25 km. The diagram in the upper left corner shows the geodynamic shape of the stress state towards the zenith vector: 1. horizontal compression; 2. horizontal compression and shear; 3. horizontal shear; 4. horizontal tension and shear; 5. horizontal tension; 6. vertical shear. The rose and bar diagrams show the predominant trends of strike directions and plunges of this stress axis and the prevalent geodynamic regimes

addition to the data on the types of geodynamic regimes (Fig. 5) to determine the direction of maximum shortening in domains with a horizontal compression regime or the direction of maximum elongation or minimum shortening in domains with a horizontal tension regime (Fig. 7).

Possible stresses exerted by the underlying mantle on the crust can be illustrated in a geographic coordinate system as a depth difference of horizontal motions (Fig. 8). The difference in the rates of horizontal displacement of the crust relative to the mantle produces shear stresses. The maximum shear stresses would act on sub-horizontal planes and their magnitudes would reflect the difference in the rates of horizontal motions of crustal layers relative to the mantle. The directions of shear stresses would reflect the direction of the relative displacement of individual crustal layers or the crust as whole with respect to the mantle. Because such motions could

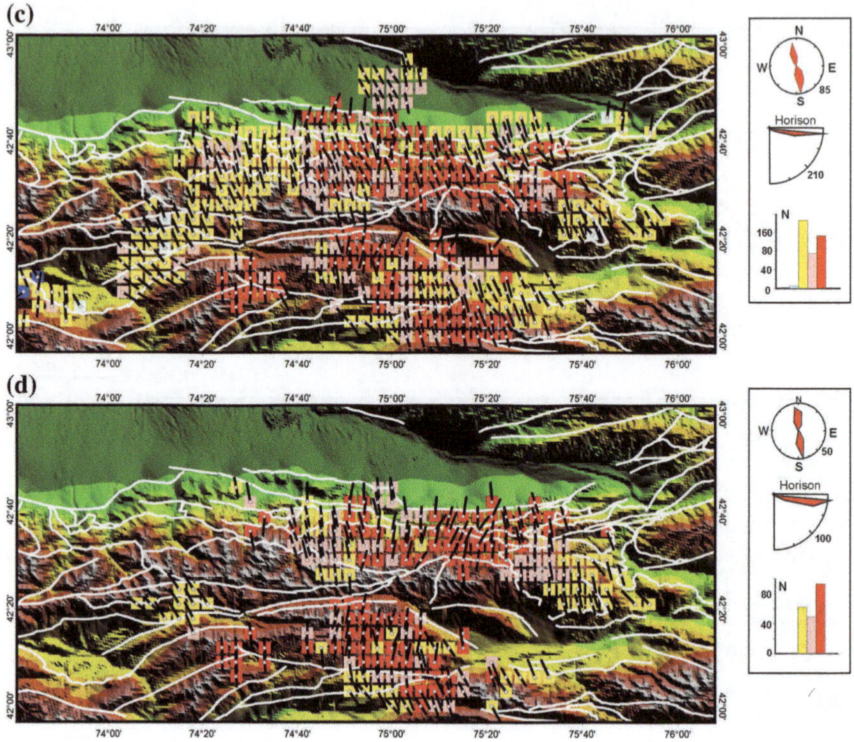

Fig. 5 (continued)

be manifested as underthrusting, the shear stresses acting on horizontal planes should be also referred to as underthrusting shear stresses τ_z. The Fig. 8 shows the directions and relative magnitudes of underthrusting shear stresses acting on horizontal planes oriented normal to the center of the Earth. As can be seen from the figure the stress level of these stresses varies considerably. An additional analysis shows that the values of these shear stresses can be either close to zero or close to the level of the maximum shear stresses.

As can be seen, these stresses have a mosaic orientation and there are only few domains with a general northward orientation of underthrusting shear stresses at all depths layers. The most uniform stress orientations are observed in the upper 0–10 km. The lower crustal layers here exhibit a north-directed displacement with respect to the uppermost crustal layer. An alternative explanation is that the uppermost crustal layer is displaced to the north with respect to the lower layers). The deeper layers are dominated by the SSW direction of shearing and display a larger scatter of orientations. The number of domains with a ~E-W orientation of shear stresses can essentially increase. At a depth of 10–20 km, the domains where shear stresses are oriented in the east, south, and west directions are present in almost the same

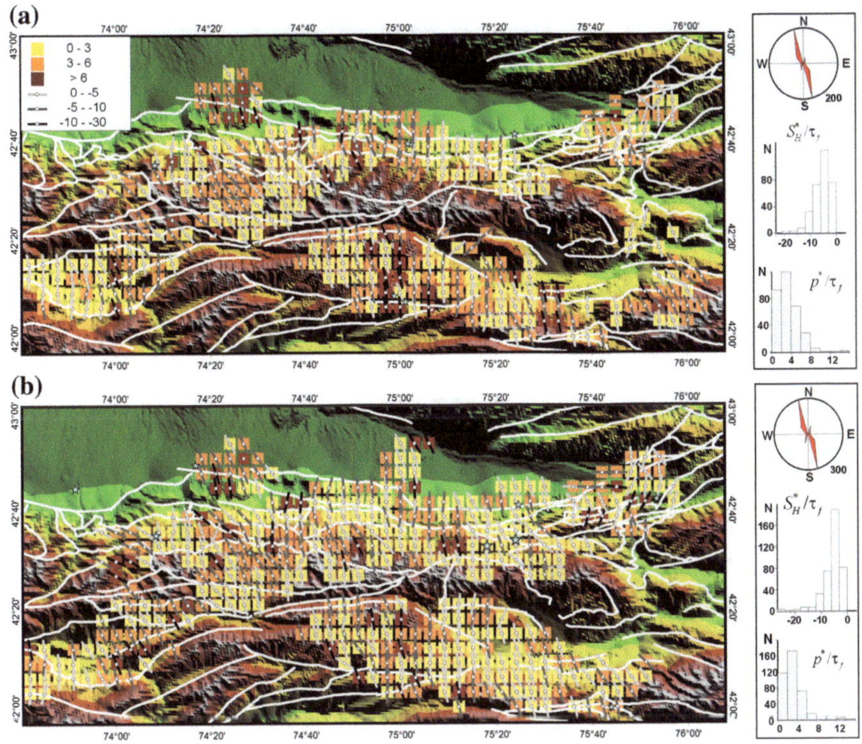

Fig. 6 Directions of the maximum compressive stresses, S_H acting on a horizontal plane and reduced value of the effective isotropic pressure, p^*/τ_f (the legend is located in the upper left corner of the plot), for different depth layers: **a** 0–10 km; **b** 5–15 km; **c** 10–20 km; **d** 15–25 km. Different shades of gray for S_H^* axes characterize the relative value of effective stresses (normalized by the internal cohesive, τ_f). The rose diagram shows the dominant trend of the direction of maximum horizontal compressive stress; the bar diagram shows the reduced values of stresses S_H^* and p^*/τ_f

proportions. At a depth of 15–25 km, the representative stress orientations are to the south and west.

Such a change in the stress orientation in deeper layers suggests that the main concentrator of inhomogeneous horizontal motions occurs in the mid-crust and its impact should diminish with depth. It should be noted that in the case of the predominance of south and SSW directions of underthrusting shear stresses.

It is important to note that there are large domains or crustal blocks, which are characterized by differently oriented stresses (underthrusting shear stresses) exerted by the mantle on the crust. This fact is possibly related to the absence of a uniaxial shearing motion of the crust near the Moho or may result from a rotation of crustal blocks. The rotation is hypothesized to result in shearing stresses of opposite directions and signs acting on horizontal planes at the deep crustal level.

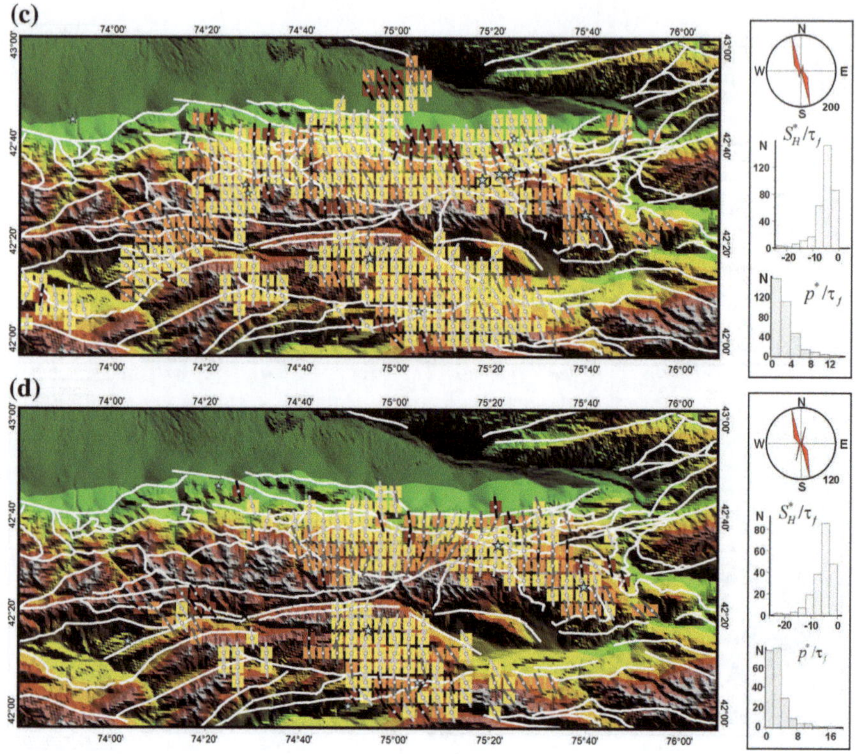

Fig. 6 (continued)

Second stage of stress inversion, reduced stresses magnitudes. In accordance with the MCA algorithm, in the second stage of stress inversion, the spherical and deviatory components of the stress tensor are evaluated up to normalization (reduced stress) by an unknown value of the cohesion strength τ_f, which is the averaged strength parameter of rock masses. The scale of strength averaging corresponds to the scale of stress averaging, which, for this region, is equivalent to a linear scale of 10–15 km. Because of the scaling factor, the averaged strength of rock masses may differ significantly from the values obtained for rock samples 20–100 MPa (Byerlee 1978).

The results of the second stage of the MCA reconstruction for the crust in the Northern Tien Shan showed that the relative magnitudes of maximum shear stresses (τ) and effective confining pressure is non-uniformly distributed in the study area. Effective isotropic pressure ($p^* = p - p_{fl}$, p_{fl}—fluid pressure) is the difference between the rock pressure ($p = -(\sigma_1 + \sigma_2 + \sigma_3)/3$) and the fluid pressure in the pore space. As seen in Figs. 6 and 7, the reduced effective pressure p^*/τ_f and maximum shear stresses τ/τ_f has quite a mosaic distribution. In the upper 0–10 km, the domains with the maximum effective pressure correspond to the Kara Moynok Range and southern slope of the central Kyrgyz Range (Fig. 1). However, in deeper layers, such

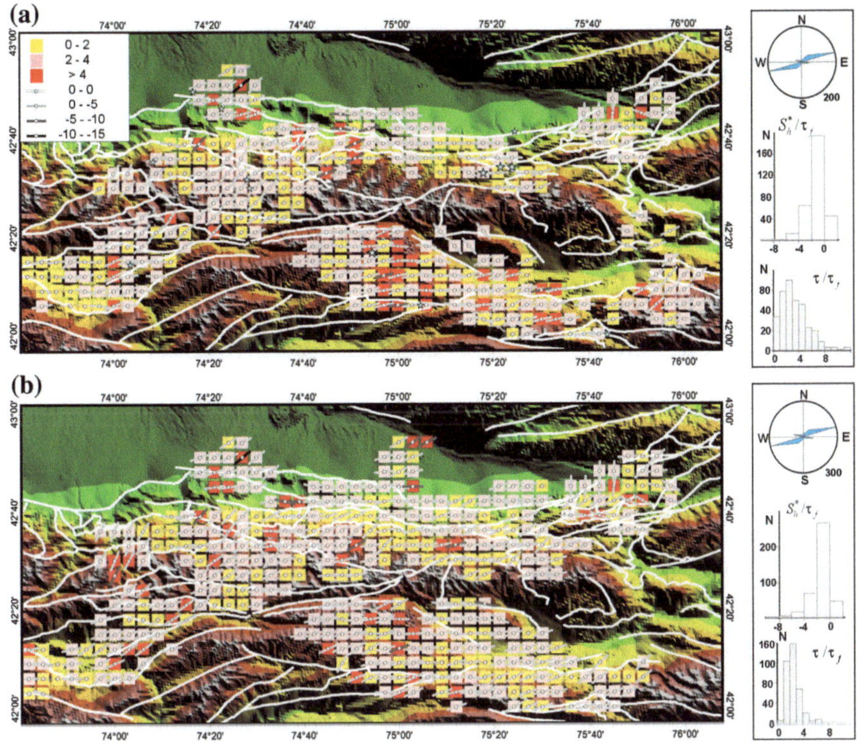

Fig. 7 Directions of minimum compressive stresses, S_h acting on a horizontal plane reduced value of the maximum shear stress, τ/τ_f (the legend is located in the upper left corner of the plot), for different depth layers: **a** 0–10 km; **b** 5–15 km; **c** 10–20 km; **d** 15–25 km. Different shades of gray for S_h^* axes characterize the relative value of effective stresses (normalized by the internal cohesive, τ_f). The rose diagram shows the dominant trend of the direction of minimum horizontal compressive stress; the bar diagrams show the reduced values of stresses S_h^* and τ/τ_f

domains are observed in the north of the central segment of the study area (northern slope in the central part of the Kyrgyz Range and Chuya depression). Large domains (100 * 100 km) with a relatively high effective pressure adjoin domains where these values are half as high.

Figures show that strong earthquakes occur in regions with moderate to low effective pressure. Because yielding state of rocks is defined by the Coulomb-Mohr equation relating effective normal and shear stresses over the brittle fracture plane, the effective isotropic pressure is correlated with maximal shear stresses. Higher values of p^*/τ_f are found in regions characterized by high values of τ/τ_f. The regions with low values of p^*/τ_f are characterized by low values of τ/τ_f. Such stress distribution can be related to the assumption of the MCA method that the stress state is close to the limiting state in regions of high seismicity and that the cohesive strength

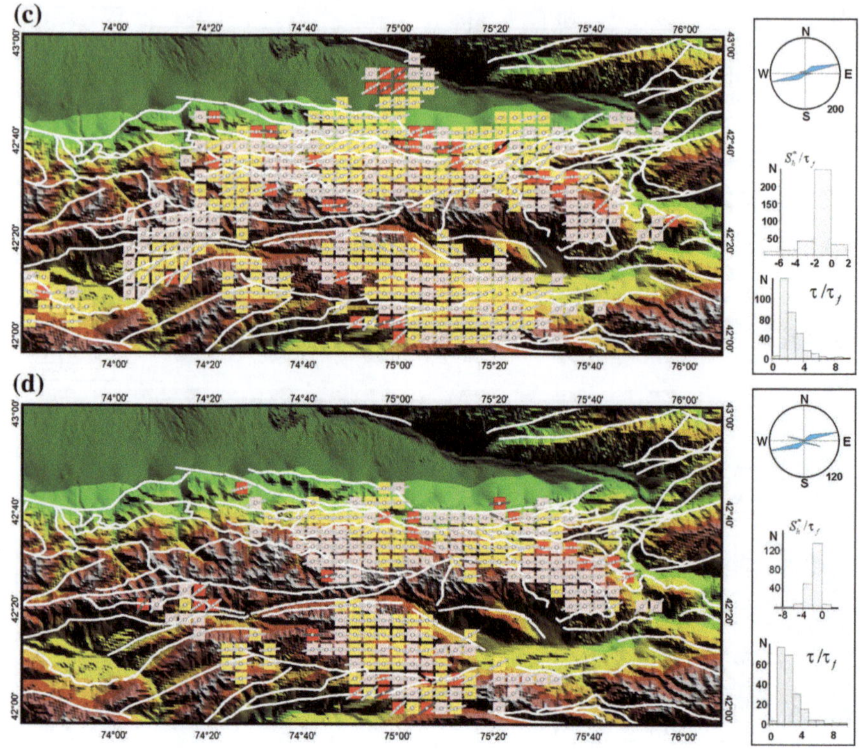

Fig. 7 (continued)

τ_f, averaged over a few tens of kilometers, shows little variability within the same region.

Figure 8 shows the values of effective vertical stress, normalized by τ_f. It should be noticed that this field differs from the data presented in Fig. 4 (the normalized vertical component of deviatoric stress $\left(\sigma_{zz}^* + p^*\right)/\tau$). As can be seen from Fig. 8, the areas of high effective vertical stress are distributed over the whole study area except the central part of the Kirghiz ridge (in Fig. 4 this is the area of the highest values of $\left(\sigma_{zz}^* + p^*\right)/\tau$).

It should be pay attention to the different type of lines for the directions of the axes of maximum horizontal compression S_H and the axis of minimum horizontal compression S_h (Figs. 6 and 7). The type of the lines determines the relative level of effective stress S_H^* and S_h^* ($S_i^* = S_i + p_{fl}$, $i = H, h$). In this case, the normalization was carried out by the value of internal cohesion τ_f.

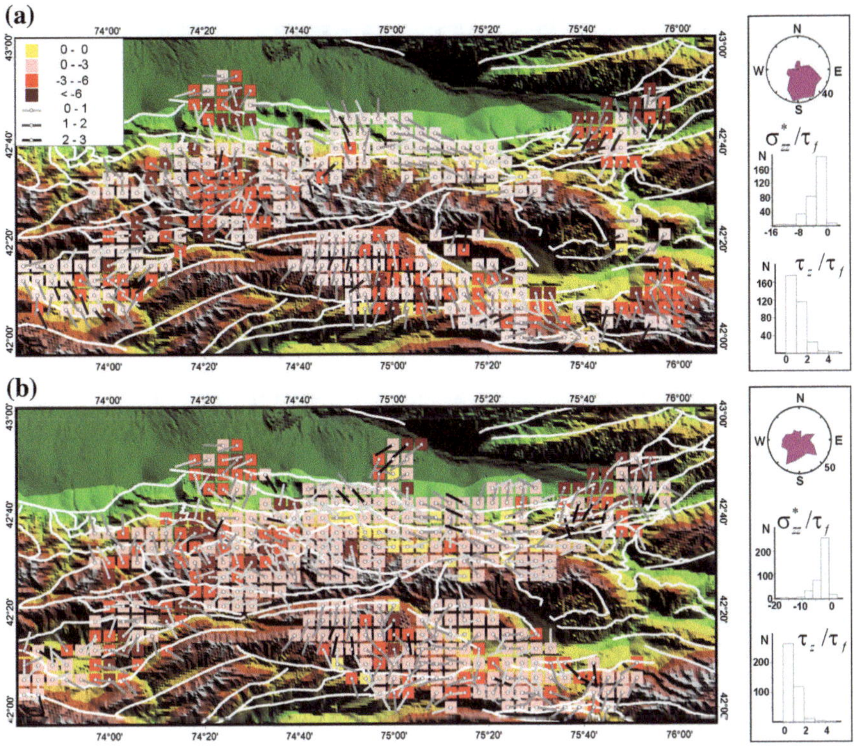

Fig. 8 Orientations of underthrusting shear stresses on the horizontal plane with the normal to the center of the Earth and their relative values (different shades of grey) normalized by the cohesive strength, τ_z/τ_f and reduced value of effective vertical stresses, σ^*_{zz}/τ_f, for different depth layers: **a** 0–10 km; **b** 5–15 km; **c** 10–20 km; **d** 15–25 km. The rose diagram shows the predominant strike directions of the maximum horizontal compressive stress and the bar diagram shows the distribution of values of τ_z/τ and σ^*_{zz}/τ_f

1.3 Conclusions

To summarize, we can conclude that the stress state in the crust of the Northern Tien Shan is not as simple as it may look from a preliminary analysis of GPS observations of surface motions. The complexity of the stress state is not visible only from the orientation of the compressive principal stress axes, which has a stable position in the region) and defines a preferred NNW and SSE dip direction, corresponding to GPS velocity vectors. The complexity of the stress state in the study area is manifested in the areal distribution of the most important parameters characterizing the deformation of a rock massif.

The complexity of the stress state in the study area is manifested in the areal distribution of the most important parameters characterizing the deformation of a rock massif. It was shown that the central part of the study area is dominated by

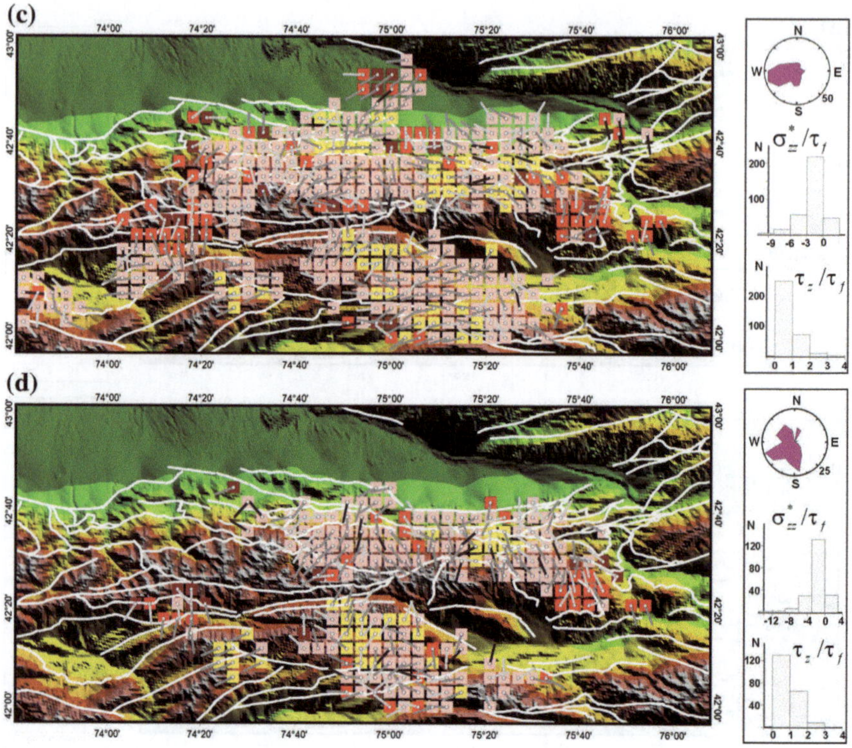

Fig. 8 (continued)

horizontal compression, while multiple domains characterized by horizontal shear and superimposed compression or pure horizontal shear are also present. To the west and east, the stress regimes is dominated by horizontal shear, superimposed shear and tension, or pure tension. while the combinations of horizontal shear with compression or pure horizontal compression are rarely observed. Understanding the mechanisms for generation of tectonic stresses in the Earth's crust of mountainous regions has important implications for geodynamic modeling of the evolution of the lithosphere. Measurements and modeling of in situ rock stresses are critical for studying the effects of loads acting on the lithosphere and lithospheric response to its tectonic loading. A detailed investigation of natural stress fields has be recently made for the crust of the Altai-Sayan (Rebetsky et al. 2012, 2013) and High Asia (Rebetsky and Alekseev 2014). The results revealed a relationship between the type of the crustal stress state and surface topography (uplifts manifested as mountain ranges and plateaus or depressions).

When presenting the parameters of the stress field for the Northern Tien Shan, we limited ourselves to the results of the stress inversion of the first two stages of the MCA algorithm. This was due to the fact that in the study region there were no earthquakes with a magnitude greater than the scale of stress averaging ($M_b > 6.5$) for a period corresponding to the seismological catalog of earthquake focal mechanisms.

2 The Stress State of the Earth's Crust in the Altai-Sayan Mountain Region

2.1 Introduce

The fold-mountain Altai-Sayan region is part of the Ural-Mongolian (Central Asian) fold belt stretching from the Urals to the Pacific Ocean, and is situated between the fold-mountain systems of Central Asia (Tien Shan and Pamir) and near-platform Baikal rift systems. The present Altai-Sayan mountain region began to form in place of denudation plains and low mountains in the Cenozoic. Late Cenozoic activation manifested itself in the accumulation of Neogene–Quaternary continental glacial, lacustrine, and alluvial sediments of the Zaisan, Chuya, Uvs Nuur, Junggar, and other basins.

Data on active faults (Trifonov et al. 2002) show a complex system of reverse faults, thrusts (with a dip angle of the fault plane less than 30°), and strike-slips. Large strike-slip faults are also observed on the northern and southern margins of the Altai and Sayans: the Kuznetsk Altai fault and the Biysk and Katun faults (Fig. 9).

Since 1963, all earthquakes with $M_b > 3.4$ in the Altai-Sayan region are recorded by a network of regional seismic stations. The location accuracy of the epicenters fits into 5–15 km. The focal depth in the Altai-Sayan region cannot be determined reliably because of the large distance between the seismic stations. The epicentral field of the earthquakes in the period of instrumental recording (from 1963 to 2003) indicates the overall stability of the observed pattern of distributions of seismic events (Blagovidova et al. 1986).

The stress inversion was performed on the basis of seismic data on earthquake focal mechanisms obtained from records of the local network of seismic stations. The catalog of focal mechanisms with $M_b > 3$ included 308 events during the period from 1963 to 2003. The depth of earthquake foci is set equal to 15 km since there are no reliable determinations of the source depth. Experimental data for determining the focal mechanisms used in the present study were earthquake records from seismic stations of the Altai-Sayan Seismological Expedition, Expedition at the Institute of the Earth Crust of the SB RAS, Kazakh Seismological Expedition, Kazakh National Nuclear Center, and Mongolia, and data on the sign of compressional displacements published in the Seismological Bulletins.

Preliminary analysis of this catalog showed that the distribution density of the earthquake epicenters and their magnitude range allowed stress inversion with a typical lateral averaging size of 50–70 km and to a depth of 30–40 km (the crust as a whole). In the calculation of the stress parameters, the initial seismic data were

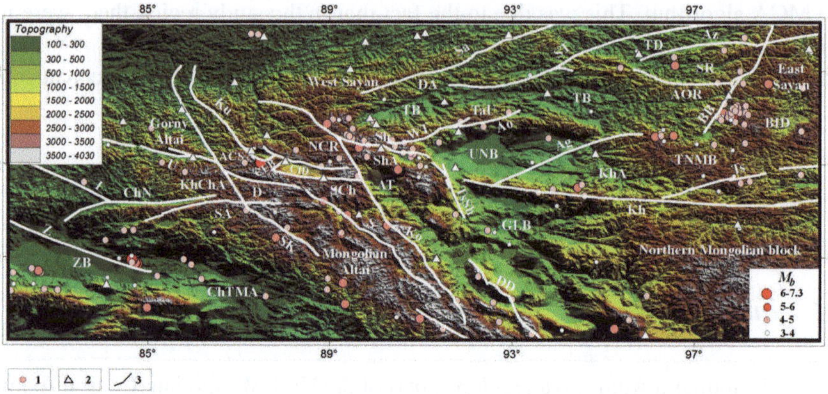

Fig. 9 Topography, tectonic provinces, and major active faults of the Altai-Sayan mountain region. 1. distribution of earthquake epicenters from the catalog of focal mechanisms used for the tectonophysical stress inversion (308 events, the legend is located in the lower-right corner of the plot); 2. location of the regional network of seismic stations; 3. here and below, the major active faults according to Trifonov et al. (2002). Mountain uplifts: DA, Dzhebash anticlinorium, SR, Sangilen Ridge of the Tuva–Northern Mongolia block; TaU, Tannu-Ola uplift; TNMB, Tuva-Northern Mongolia block; KhChA, Kholzun-Chuya anticlinorium; KhA, Khangai anticlinorium; ChTMA, Chingiz—Tarbagatai megaanticlinorium; ShA, Shapshal anticlinorium; NCR, North Chuya Range; CUB, Chagan-Uzun block. Subsidence areas: AT, Achit-Nuur trough; ACS, Anui-Chuya synclinorium; BID, Busingol intermontane depression; GLB, Great Lakes basin; JB, Junggar basin, ZB, Zaisan basin; TD, Todzha depression; TB, Tuva basin; UNB, Uvs Nuur basin; CID, Chuya intermontane depression. Major faults: Ag, Agardag; Az, Azas; Ae, Aelegest; AOR, Obruchev Rift; BB, Belin-Busingol fault system; B, Belin; D, Dzhazator; DD, Dzhun-Dgirgelent; Z, Zaisan; WT, Western Tannu-Ola; I, Irtysh; Ko, Kobdo, Ku, Kurai; TsSh, Tsagan-Shibetin system, Sa, Salen; S, Sagsai; ST, Sayan-Tuva; U, Umsk; Kh, Khangai (Bolnay); Ts, Tsetserleg; ChNR, Chingiz-Narym; Sh, Shapshal; SA, Southern Altai; SK, Southern Katun; SCh, Southern Chuya

processed at the nodes of a 25° × 0.25° grid located at a depth of 15 km. All procedures of generating homogeneous samples of earthquake focal mechanisms were completed for 640 grid nodes with not less than six earthquakes in each of such samples. The grid nodes corresponded to 640 quasi-homogeneous domains, which were assigned the obtained parameters of the natural stress tensor.

2.2 *Results of Natural Stress Inversion*

Parameters of the tectonic stress ellipsoid. Figure 10c shows projections of the dipping axes of the principal stresses of the algebraically minimum stress σ_3 on a horizontal plane. Note that in most of the study region, these axes have a fairly flat south-dipping orientation (see the circle diagram). The average dip angle of these axes is 26°. The orientation of the other principal axis σ_1 has a roughly east–west strike with a broader change in the dip angle (Fig. 10a). Along with this statistically representative distribution of the strike and dip angles of the principal stress axes, there are many local areas where the orientation of the axes undergoes significant deviations.

In the eastern part of the Gornyi Altai, the subhorizontal position of the axis of maximum compression σ_3 is confined mainly to mountain areas and uplifts (Katun and Dzhebash anticlinoria), and its subvertical position to basins and depressions (Tuva and Uvs Nuur basins). In the Western Sayans, the orientation of the axes of this principal stress is extremely unstable in both strike and dip. We suggest that this nature of the stress state is associated with the dominance of intermontane valleys and basins over mountain uplifts in this region of the crust. In the crust of weakly seismic intermontane depressions, earthquakes frequently have fault-type mechanisms, which, combined with the reverse-fault and strike-slip fault mechanisms in the crust of mountain ridges, leads to instability of stress inversion.

In the central zone of the Eastern Sayans, the maximum compression axes σ_3 have a stable subhorizontal position and change strike from NNW in the north to NE in the south. Further south, in the East-Tuva mountain uplift and the Tuva–Northern Mongolia block, they dip more steeply to the northeast. These two regions are separated by a narrow zone stretching along the Sangilen plateau with a roughly east-west orientation of the axis of maximum compression. In mountain areas of the Mongolian Altai, the maximum compression axes again take a subhorizontal position with a NS strike. In the very east of the study region, these axes again have a mosaic orientation, which changes from roughly EW direction to roughly NS direction.

From the circle diagrams presented in Fig. 10c, it follows that the subhorizontal (less than 30°) dip of the maximum compression axes is observed for 60% of the domains, and 12% of the domains has a steep (more than 60°) dip of these axes. Additional analysis showed that in 75% of the domains of stress inversion in the areas of uplifts, the orientation of the maximum compression axis is subhorizontal (dip less than 45°). In 75% of the domains of stress inversion in areas of large intermontane depressions or foredeeps, the orientation of the maximum compression axis is subvertical (dip of more than 45°).

After the first stage, the Lode-Nadai coefficient is calculated (Fig. 10a). In 70% of the Earth's crust domains, its values are close to pure shear $|\mu_\sigma| \leq 0.2$. Among the exceptions with the largest area are the Tuva and Uvs Nuur basins, where there are several crustal regions with the stress tensor near uniaxial compression.

Based on the orientation of the principal stress axes considered above, it is possible to perform geodynamic zoning of the study region. Figure 10c shows a diagram of division of the crust into six geodynamic types (types of stress state) obtained from an analysis of the position of the vector toward the zenith relative to the principal stress axes.

The Gornyi Altai is dominated by domains of horizontal shear and its combination with extension. Local areas of horizontal extension are observed for the crust of the Chuya and Kurai intermontane depressions and near the western and the eastern ends of the South Chuya fault. The crust of the Western Sayans, dominated by basins and large intermontane depressions (Tunka and Uvs Nuur Achit Nuur and Khemchik–Tuva depressions), is mainly subjected to horizontal extension and its

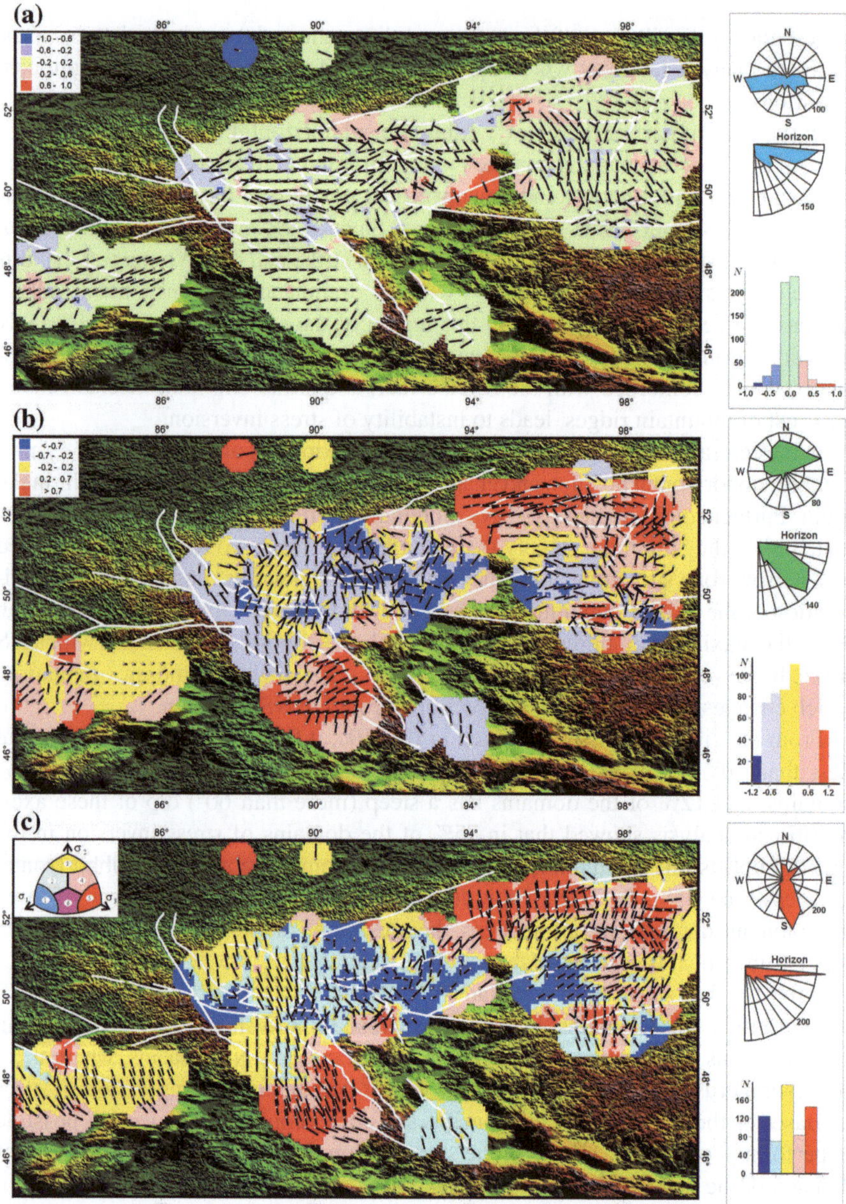

Fig. 10 Horizontal projections of plunges of the principal stress axis σ_1 (**a**), σ_2 (**b**), σ_3 (**c**) and values of the Lode-Nadai coefficient (**a**), reduced deviatoric vertical stresses, $\left(\sigma_{zz}^{*} + p^{*}\right)/\tau$ (**b**), geodynamic type of stress state (**c**) for the crust of Altae-Sayan region. The rose diagrams show the predominant strike directions and plunges of this stress axis. The bar diagram shows the predominant range of the μ_σ, $\left(\sigma_{zz}^{*} + p^{*}\right)/\tau$ values and the types of stress, i.e., the geodynamic regimes

combination with shear. To the south toward the Mongolian Altai and to the east toward the Eastern Sayans, the type of geodynamic regime is replaced by horizontal compression and its combination with shear. A vast region of horizontal extension is observed for the Tuva–Northern Mongolia block. The Zaisan basin is in a horizontal shear setting.

As seen from the diagram in Fig. 10c (the number of domains with different types of geodynamic regime), the state of horizontal shear is the most representative (about 30% of the domains), and the states of horizontal compression and extension are typical of a similar number of domains—about 20 and 25%.

Figure 10b shows the distribution of the relative value of the vertical component of the deviatoric stress in the earth's crust of the study region $\left(\sigma_{zz}^{*} + p^{*}\right)/\tau$. It can be seen that the values of this stress depend little on the surface morphology (mountain uplifts, intermontane depressions). This is probably due to the fact that the maximum shear stresses τ don't depend on the surface morphology.

Stress state in a geographic coordinate system. The thickness of the seismogenic layer—the Earth's crust within which the stress parameters are determined—is much smaller than the lateral dimensions of the study region. Neglecting the surface and bottom topography of the crust, we can speak of stress distribution in a plate with the characteristic lateral dimensions related to thickness as 30–50 to 1 (a thin plate). In this case, it is convenient to represent the stress tensor parameters averaged for the plate crust by the normal stresses of maximum and minimum compression acting in the lateral direction and by the shear stresses applied to the plate bottom. We note that such shear stresses should be referred to as subduction stresses reflecting the interaction between the crust and the mantle.

Figure 11a, b shows the directions of the maximum horizontal compression axes S_H and minimum horizontal compression axes S_h coacting in the lateral direction, which have a more ordered orientation than the projections of the dip axes of the principal stresses (Fig. 10). This figure highlights the north–south and northwest strike of the maximum lateral compression axes in the crust of the Zaisan basin, in the Gornyi and Mongolian Altai, in the Kurtushiba anticlinorium, and in the central zone of the Eastern Sayans. This orientation of the compression axes corresponds to the strike of the fold and anticline hinges within the Katun ridge and in the southern part of the Kurtushiba anticlinorium. Some correspondence exists in the region with the NE trending axis of maximum lateral compression at the junction of the Central Tuva basin with the central zone of the Eastern Sayans.

The orientation of the underthrust shear stresses indicates that the maximum impact of the underlying mantle on the crust is observed in the Gornyi Altai, in the Kurtushiba anticlinorium, in the central part of the Mongolian Altai, and within the Eastern Tuva highland and the Sangilen ridge (Fig. 11c). In these areas, the underthrust shear stresses are mainly oriented from east to west and from northeast to southwest. Increased values of these stresses are more frequently found in the crust of uplifts (Gornyi and Mongolian Altai, Tuva–Northern Mongolia blocks), mountain ranges, and reduced values occur in basins (Zaisan, Tuva).

From the diagram of the distribution of the number of domains with different values of the azimuth, which determines the direction of subduction shear stresses τ_z, it is

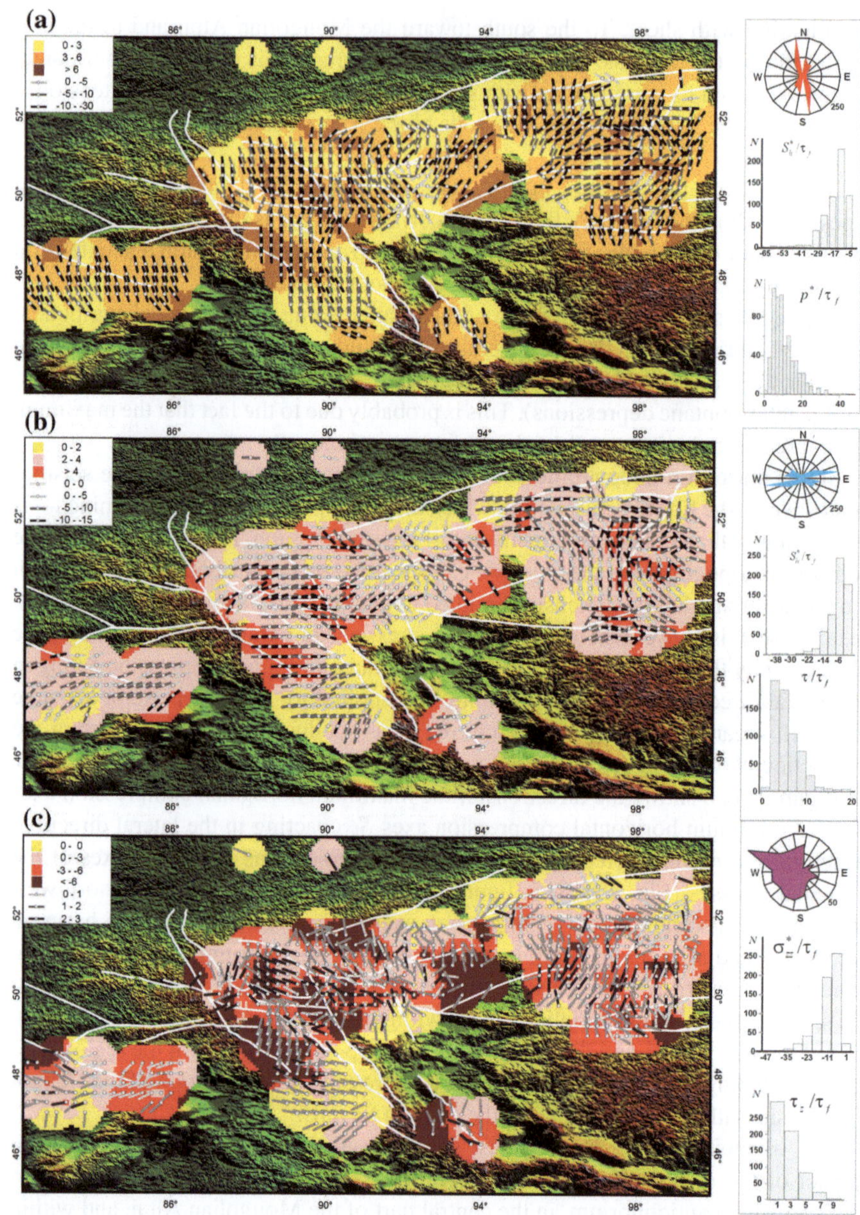

Fig. 11 Directions of the axes of maximum and minimum horizontal compression, S_H (**a**), S_h (**b**), the underthrust shear stresses, τ_z (**c**) and value of effective pressure, p^* (**a**), maximum shear stress, τ (**b**), effective vertical stresses, σ_{zz}^* (**c**). Different shades of gray for the stress axes characterize the value of effective stresses S_H^*, S_h^* and τ_z normalized by τ_f. The rose diagram shows the predominant strike directions of this stress axis. The bar diagrams show the predominant range of the p^*, τ and σ_{zz}^* values normalized by τ_f

evident that the predominant direction of these shear stresses is the range of azimuths of 180°–360° with the maximum near 270°–290°. This orientation determines the predominant westeast direction of sliding of the crust relative to the mantle.

Reduced stresses magnitudes. The results of the second stage of reconstruction showed that the effective confining pressure p^*/τ_f and reduced maximum shear stresses τ/τ_f is non uniformly distributed in the study region (Fig. 11a, b). Extensive areas of increased effective pressure extend along the Kobdo fault within the Gornyi Altai and the Alash fault in the Western Sayans. In the Eastern Sayans, they are associated with the Belin-Busingol fault system trending roughly northsouth and the east–west Obruchev fault. In the Mongolian Altai, there is a change from increased effective pressure to low pressure from north to south.

It should be noted that increased values of p^*/τ_f are observed for the crust regions subjected to both horizontal extension (areas along the Kobdo fault for the Gornyi and Mongolian Altai) and horizontal compression (the central part of the Eastern Sayans). Moreover, in a number of areas subjected to a more stable (in the area of occurrence) horizontal compression regime with a northwest trending σ_3 axis (the Zaisan basin, the southern part of the Mongolian Altai, and the northern margin of the Eastern Sayans), decreased effective pressure is observed.

In general, the reduced maximum shear stress is distributed similarly to the effective pressure. In areas where increased values of p^*/τ_f (Fig. 11a) are observed, there are also increased values of τ/τ_f (Fig. 11b). In areas of reduced values of p^*/τ_f the values of τ/τ_f are also reduced. This distribution is related to the assumption adopted in the MCA that the stress state in highly seismic areas is nearly limiting, which is expressed as tangency of the large Mohr circle to the outer envelope of (brittle fracture point) in the Mohr diagram. Since the limiting state for rocks is determined by the Coulomb-Mohr relation between the spherical stress tensor and the stress deviator (normal and shear stresses in the brittle fracture plane), a consequence of this is the relationship between the effective pressure and the maximum shear stress (Fig. 11).

Figure 11c shows the distribution of the effective vertical stress σ_{zz}^* normalized by the internal cohesion τ_f in the Earth's crust of the study region. As well as in Fig. 10c, this component, reflecting the influence of the weight of overlying rocks, does not depend on the surface morphology. In this case, this is due to the peculiarities of the spatial distribution of the fluid pressure in the fractured-pore space of the massif.

It should be noted that like in the previous section, the different shades of gray for the axes of maximum S_H and minimum S_h horizontal compression, as well as for the directions of the underthrusting shear stresses τ_z, determine the intensity of their reduced values with normalization by the value of internal cohesion τ_f.

It should be noted that the strongest earthquakes ($M_b \geq 5$) of the study region occur in areas of lower pressure and are not observed in areas of increased effective pressure (see Fig. 11a). This is consistent with the idea of the greater efficiency of brittle fracture in areas of smaller confining compression (Rebetsky 2007; Rice 1980).

In Fig. 11c shows the distribution in the region cortex of the effective vertical stress normalized to the adhesion strength. As well as in Fig. 10b, this component, reflecting the influence of the weight of overlying rocks, does not depend on the morphology of the surface. In this case, this is due to the peculiarities of the spatial distribution of the fluid pressure in the fractured-pore space of the massif.

Note that, just like in the previous section, the shades of gray in the color of the lines for the axes of maximum and minimum horizontal compression, as well as for the directions of the underthrust shear stresses determine the intensity of their reduced values with normalization to the adhesion strength.

The third stage of the stress inversion. Determination of the stress values. In the third stage of the MCA, the internal cohesion averaged for the corresponding scale level is evaluated to transform from reduced stresses to their absolute values.

In order to determine the values of the stresses, it is necessary to estimate the value of the internal cohesion τ_f according to the MCA algorithm (see expressions (36–38), Chapter "The Method of Cataclastic Analysis of Discontinuous Displacements"). On the one hand, for this it is necessary to know the dynamic parameters of the source area of a large earthquake (E_s—energy radiated in elastic waves, M_o—seismic moment (the data obtained by seismologists after the processing of seismic records)) whose size exceeds the averaging scale of stresses. On the other hand, it is necessary to calculate the relative value of the shear stress drop $\left\{\Delta\tau_n/\tau_f\right\}$ in the source area of this earthquake.

In the Altai-Sayan region, such an earthquake was the 2003 Chuya earthquake ($M_w = 7.3$) (Gol'din 2004). The scale of strength averaging for Altay-Sayan crust corresponds to the scale of stress averaging, which, for this region, is equivalent to a linear scale of 50–70 km. Therefore, within the MCA, an algorithm for estimating τ_f was developed which is implemented in the third stage of reconstruction. For this earthquake, the Harvard University Web site gives the seismic moment and seismic wave energy $M_o = 10^{20}$ Nm and $E_s = 4.8 \times 10^{15}$ Nm, respectively, which allows the released stress to be evaluated from the formulas of (Kostrov 1975; Kostrov and Das 1988). According to our calculations, the released stress $\Delta\tau_n$ for the Chuya earthquake was about 4 MPa (in the calculations, the shear modulus of the crust was taken as $\mu \approx 3 \times 10^4$ MPa).

On the other hand, from the results of the first two stages of reconstruction using the MCA for the Chuya earthquake, it is possible to calculate the released stress up to the unknown normalizing coefficient $(1/\tau_f)$, and thus, data on the released stress $\Delta\tau_n$ of the Chuya earthquake make it possible to evaluate τ_f.

The fault parameters of the Chuya earthquake, according to the seismological data, were the following: Strike1 = 131°, Dip1 = 71°, Slip1 = 158°; Strike2 = 228°, Dip2 = 70°, Slip2 = 20°. Using the MCA algorithm, three states of stress were selected from the results of the stress inversion for stress drop estimation. They were obtained in the areas of the earth's crust (domains) which were the closest to the source of the Chuya earthquake (see Table 1).

The states of stress for these three domains and the positions of the points that correspond to the normal and shear stresses for two nodal planes of the Chuya earthquake are presented in Fig. 12. It should be noted that one of the nodal planes of the 2003 Chuya earthquake (for each analyzed state of stress) falls into the brittle

Table 1 Stress parameters for domains which were the closest to the source of the Chuya earthquake

№	Long.	Lat.	σ_1		σ_2		σ_3	
			Azimut	Plung	Azimut	Plung	Azimut	Plung
1	87.50°	50.50°	264	0	174	86	354	4
2	87.75°	50.00°	221	27	324	23	90	54
3	88.00°	50.25°	90	18	188	23	325	60

(continued)

Table 1 (continued)

μ_σ	$\frac{p^*}{\tau_f}$	$\frac{\tau}{\tau_f}$	$\frac{\Delta\tau_n}{\tau_f}$
−0.27	7.13	3.93	0.96
−0.04	6.24	3.66	0.67
0.39	7.43	4.47	0.40

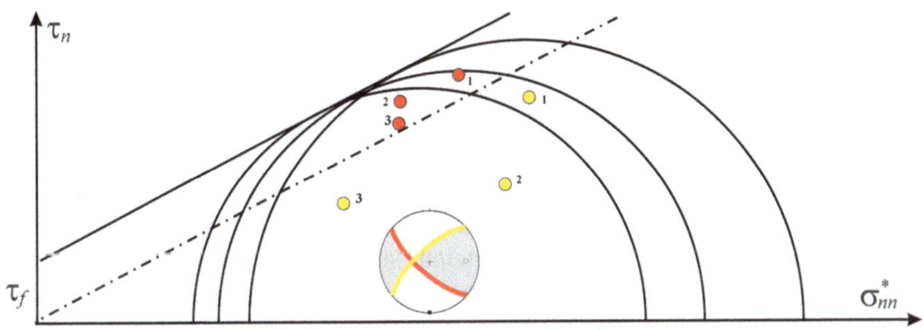

Fig. 12 The Mohr diagram for three states of stress (see Table 1) with the position of the points on it. The points correspond to the stresses for two nodal planes of the 2003 Chuya earthquake. Red color—the points that correspond to the actual fault plane, and yellow color—the points that correspond to the second nodal plane. The solid line is the breaking strength, the dashed line is the line of minimum frictional resistance. The actual fault plane of 2003 Chuya earthquake is marked with a red line

failure zone, and the second nodal plane with the same states of stress falls into the area below the line of dry friction.

For these domains, the average normalized value of the released shear stress is 0.68. The above data on the released stress $\Delta\tau_n = 4$ MPa make it possible to determine the effective (average for an averaging scale of 50–70 km) internal cohesion τ_f of rock massive, which for the investigated region of the crust was about 6 MPa ($\Delta\tau_n/\tau_f \approx 0.667$).

Note the following important fact: The released stress obtained in our calculations is lower than the internal cohesion. This is not a contradiction because the presence of surface anisotropy leads to the fact that the direction of displacement along the fault often does not coincide with the shear stress direction on the fault before its activation (the coincidence of these directions corresponds to the well-known postulate introduced in (Bott 1959; Wallace 1951). A consequence of this discrepancy should be a decrease in the stress release.

Conversion of the stress in the crust of the study region from relative to absolute values gives a range of 4.5–240 MPa for the effective pressure p^* and a range of 7–120 MPa for the maximum shear stress τ (Fig. 11a, b). It is seen that the range of the effective isotropic pressure of 20–60 MPa and the range of the maximum shear stress of 18–30 MPa are the most representative.

2.3 Conclusions

The tectonophysical calculations of the present state of stress of the crust of the Altai-Sayan region show that the interpretation of the present state of orogens must take into account the residual stresses of horizontal compression occurring in rocks due to their elastic unloading in uplift areas of rapid erosion and denudation (Rebetsky 2008). The role of this stress-generating mechanism is particularly important for intraplate orogens, which are characterized by high differentiation of vertical movements (uplift in ridges and subsidence in intermontane depressions), pronounced block divisibility (Bogachkin 1981), and significant surface erosion. In blocks in which additional stresses of lateral compression were accumulated at relatively great depths only under gravitational stress and which are moved closer to the surface, erosional unloading can be followed by a change in the principal stress index. This gives rise to the geodynamic regime of horizontal compression. In this case, intermontane valleys and large basins, which accumulate a considerable part of eroded material, undergo continuous subsidence of the paleosurface and are therefore subjected to horizontal extension.

These phenomena determine the coexistence of the horizontal compression regime in the crust of uplift areas and the horizontal extension regime in the crust of large intermontane valleys, basins, and depressions. This combination of geodynamic regimes is observed not only for the Altai-Sayan orogen, but also for the North Tien Shan (Rebetsky et al. 2012, 2013, 2016a). In our calculations, the alternation of areas with different geodynamic regimes and the block nature of the distribution of the orientation of subduction shear stresses correspond to a linear scale of 500–700 km and 50–100 km, respectively, which reflects the dependence of the present state of stress of the Earth's crust in the study region on the deep crustal and mantle heterogeneities.

3 The Stress State of the Kuril-Kamchatka Suduction Zone

3.1 Introduce

The seismic focal area of the Kuril Islands and Kamchatka is the most seismically active in Russia. Here, large earthquakes occur most often: one of them is November 15, 2006 Simushir earthquake, $M_w = 8.3$ (Fig. 13). Another one is January 13, 2007 large Simushir earthquake that occurred approximately in the same area with the magnitude $M_w = 8.1$ (Zlobin and Polets 2009; Polets and Zlobin 2013). In the southern segment of the island arc, October 4, 1994, Shikotan earthquake, $M_w = 8.3$ (Polets and Zlobin 2014). It occurred prior to Simushir large earthquakes.

The goal of this study was to investigate the state of stress of the Earth's crust of the Kuril Islands and Kamchatka before the 2006 Simushir earthquake (Rebetsky 2009; Rebetsky and Tatevossian 2013).

In order to evaluate the characteristic features of the stress state in the preparation areas of strong earthquakes, a regional catalogue of earthquake focal mechanisms was compiled on the basis of the global catalogue of seismological data available on the web site of the Global CMT Project. It includes 1220 earthquakes with magnitudes higher than 4.7 that occurred from 1976 to June 2006 at depths of up to 100 km (Fig. 13). The magnitude range of the regional catalogue and the density of the earthquake epicenter distribution allowed us to reconstruct the characteristics of the stress state averaging over 30–50 km. Since more than 600 events occurred at depths of more than 30 km and 250 events occurred deeper than 50 km, the stress state parameters were reconstructed in the nodes of a $0.2° \times 0.2°$ grid at depths of 20, 40, and 60 km. Such a step size was selected in order to smooth the spatial variations of the stresses already at this stage of the calculations and to draw the domains analyzed closer to the source of the Simushir earthquake. The data on the smothed stresses for this earthquake will be used during the further calculations.

It should be noted that 99% of the data on the earthquake focal mechanisms refers to the Earth's crust west of the oceanic trench. There was about 12–15 events to the east of the trench, within the Kuril-Kamchatka seismic focal zone according to Fig. 13. At the same time, most of these earthquakes correspond to normal faults. The same mechanism had the January 13, 2007 earthquake, $M_w = 8.3$.

All the procedures for the formation of homogeneous samples of earthquake focal mechanisms (first stage of the stress inversion) were accomplished for 778, 835, and 663 quasi-homogeneous domains corresponding to depths of 20, 40, and 60 km; each of the samples contained at least six earthquakes. It should be noted that the grid step was smaller than the characteristic size defining the scale of the stress averaging (50–100 km).

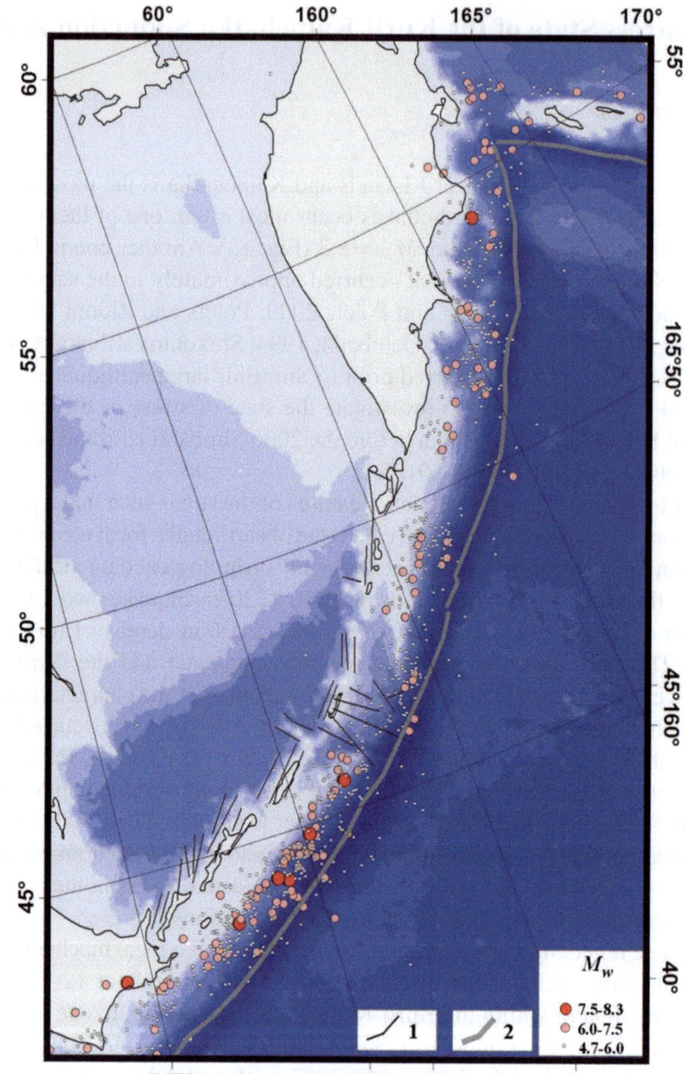

Fig. 13 The distribution of epicenters of the earthquake focal mechanisms used in the tectonophysical inversion of natural stress of the Earth's crust in the Kuril-Kamchatka region 1. the faults; 2. the oceanic Trench

3.2 Results of Natural Stress Inversion

Parameter of the stress ellipsoid. This paper presents only the results of the reconstruction for the 20 km depth level. The reconstruction of the parameters of the stress ellipsoid and the seismotectonic strain increments performed during the first stage of MCA allowed us to construct maps for the orientation of the principal axes of the stress tensor, the type of the stress tensor, and the characteristics of the stress ellipsoid.

Figures 14c and 15c shows the axes of the algebraically minimum principal stresses σ_3 constructed in the direction of their plunging and projected onto a horizontal plane. Note that the results of the reconstruction were thinned for the more convenient graphical presentation. The axes of maximum compression σ_3 plunge gently to the southeast (the average strike azimuth is 135°, and the average dip is approximately 19°). The dip of these axes decreases toward the island arc and reverses at some points. A rather abrupt change in the σ_3 axis strike was observed at the junction of the Kamchatka segment of the subduction zone with the Aleutian segment. The revealed regularity of the state of stress does not change with depth.

There is only one area where σ_3 dipping more than 60° in the Earth's crust of the central part of the island arc. Thus, there was not enough data on the normal-fault type focal mechanisms east of the trench to determine the parameters of the state of stress.

The axes of the algebraically maximum stress, σ_1 (Figs. 14a and 15a), plunge steeply toward the back-arc basin. Their average dip is 62°. These axes show more considerable deviations from the average strike (290°) compared with σ_3. Two areas deserve special attention: the junction of the Kamchatka and Aleutian segments of the subduction zones and the southwestern segment near Hokkaido Island, where the orientation of the axes of these stresses changes by more than 45°.

The reconstructed direction of the principal stress axes indicates a gentle dip beneath the oceanic plate for the axis of the maximum compression, a steep dip beneath the subcontinental plate for the axis of the minimum compression, and the orientation of the intermediate principal stress along the oceanic trench lineament (Figs. 14b and 15b). This is typical of the principal stresses in subduction zones (Rebetsky et al. 2016b)

The distribution of the stress tensor characteristic defined by the Lode–Nadai coefficient appears to be very homogeneous (Figs. 14a and 15a). There is almost everywhere a pure shear tensor, when the algebraically maximum and minimum principal deviatoric stresses are similar in absolute value but opposite in sign (the intermediate principal deviatoric stress is zero). The tensor type (stress ellipsoid) is different in the aforementioned southwestern and northeastern end segments of the region.

Figures 14b and 15b shows the changes in the value of the lithostatic pressure (the weight of ocean water, the topography of the sea floor and islands were taken into account) in the calculation points. The lithostatic pressure increases with distance from the oceanic trench.

It should be noted that, for all subduction zones studied by us, the type of stress tensor is almost everywhere a pure shear, the prevalent geodynamic regime is horizontal compression. The deviations from this rule may be observed only in the areas where the subduction zones change its strike and within the nearby areas (on the other side of the trench or the island arc).

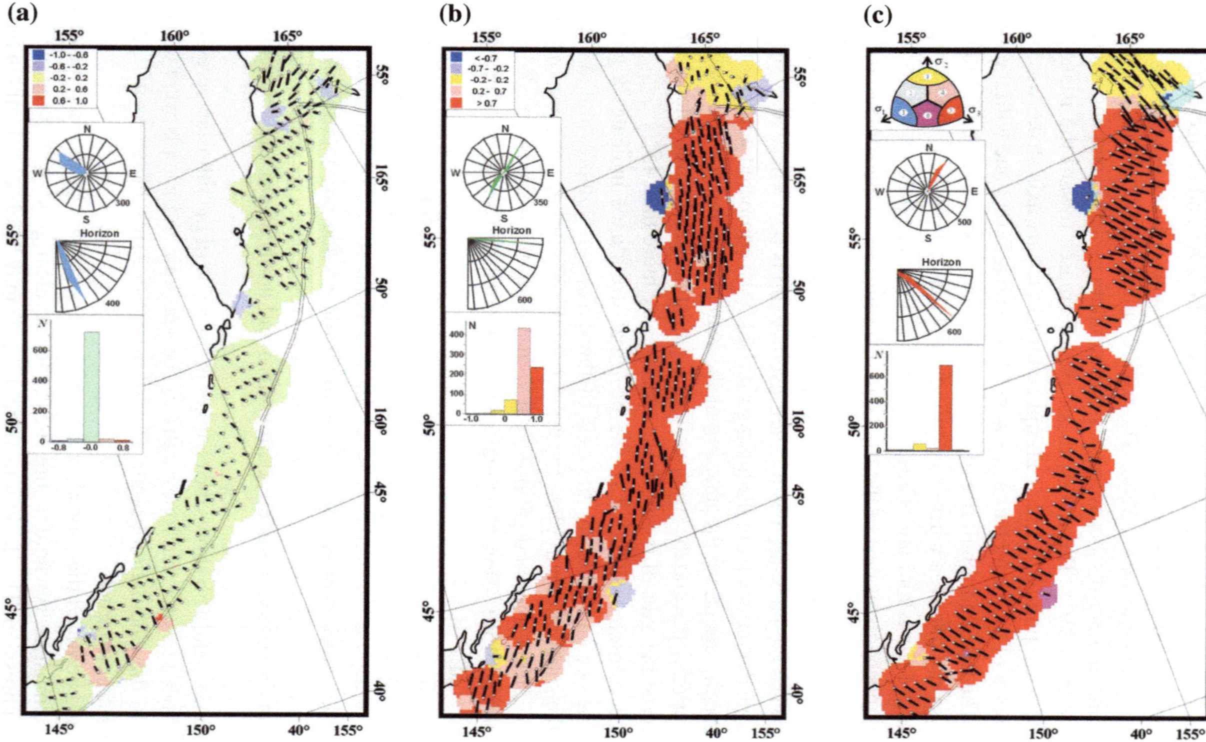

Fig. 14 Horizontal projections of plunges of the principal stress axis σ_1 (**a**), σ_2 (**b**), σ_3 (**c**) and values of the Lode-Nadai coefficient (**a**), reduced deviatoric vertical stresses, $\left(\sigma_{zz}^* + p^*\right)/\tau$ (**b**), geodynamic type of stress state for the crust of Kuril-Kamchatka seismofocal area, depth 20 km. The rose diagrams show the predominant strike directions and plunges of this stress axis. The bar diagrams show the predominant range of the μ_σ, $(\sigma_{zz} + p)/\tau$ values and the types of stress, i.e., the geodynamic regimes

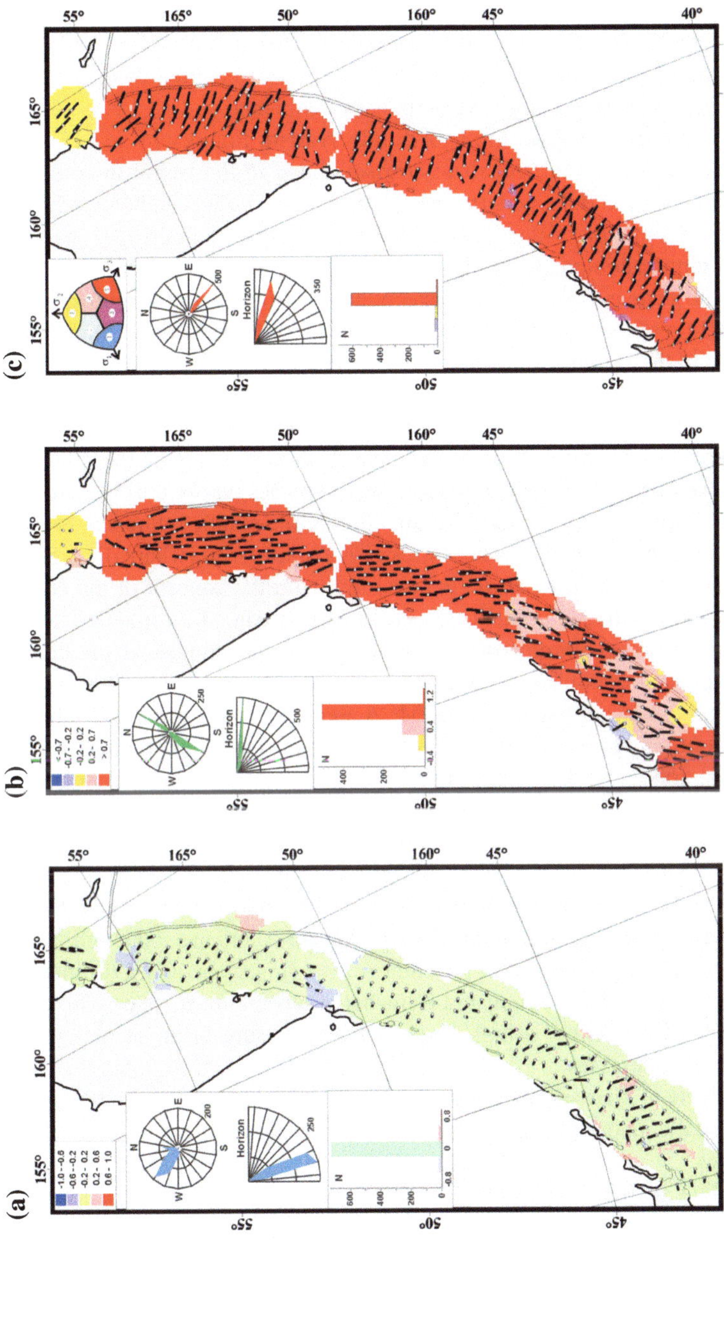

Fig. 15 Horizontal projections of plunges of the principal stress axis σ_1 (**a**), σ_2 (**b**), σ_3 (**c**) and values of the Lode-Nadai coefficient (**a**), reduced deviatoric vertical stresses, $(\sigma_{zz} + p)/\tau$ (**b**), geodynamic type of stress state for the crust of the Kuril-Kamchatka seismofocal zone, depth 60 km. The rose diagram shows the predominant trends and plunges of this stress axis. The bar diagrams show the predominant range of the μ_σ, $(\sigma^*_{zz} + p^*)/\tau$ values and the types of stress, i.e., the geodynamic regimes

Such a distribution of the principal stress axes corresponds to the prevalent regime of horizontal compression (Figs. 14c and 15c). Some deviations from this regime were again observed in the northeastern and southwestern segments of the arc. The former segment contains extended domains with the regime of horizontal slipping under the conditions of additional compression, and the latter segment shows evidence for the slip regime. The slip regime was also detected in the crust along the Kamchatka shore starting from the Shipunsky Peninsula.

Stress state in a geographic coordinate system. Figures 16 and 17 show the orientations of the normal and shear stresses acting in the lateral direction. These are the stresses S_H and S_h, underthrusting shear stress τ_z. As can be seen from these figures, the orientation of the normal stresses of the maximum (Figs. 16a and 17a) and the minimum compression (Figs. 16b and 17b) acting in the horizontal direction is almost the same everywhere. The exception, as in the previous figures, is the area of the junction of the Kamchatka and Aleutian seismic focal zones. Such a uniform orientation of these stresses was not observed in the orogenic crust (see the previous sections).

The stable orientation of the principal stress axes in the subduction zones is reflected in the stable orientation of the shear stresses τ_z acting on the subhorizontal planes (Figs. 16c and 17c). Such stresses on the planes whose normals are directed toward the Earth's center indicate the direction of the mantle influence on the crust. In the region considered, these shear stresses are almost everywhere directed normal to the trench. Such an orientation is consistent with the concept of the convection mechanism of the mantle influence on the moving oceanic lithospheric plate.

Reduced stresses magnitudes. The homogeneous samples of the earthquake focal mechanisms compiled for each of the domains (depth of 20 and 60 km) allowed us to calculate, in accordance with the MCA algorithm, the normalized values of the effective confining pressure, p^*/τ_f (Figs. 16a and 17a) and maximum shear stresses, τ/τ_f (Figs. 16b and 17b). The region under investigation showed a rather mosaic distribution of the effective confining pressure (along Iturup and Urup islands and the Kamchatka shore). On the other hand, there are several large areas (150–250 km) where this parameter is relatively stable, and the values from 4 to 12 correspond to its medium level. Such areas were detected near Shikotan, Simushir, and Paramushir islands and near the junction of the Kamchatka and Aleutian subduction zones.

It should be emphasized that the ratio σ^*_{zz}/τ_f (Figs. 16c and 17c) increases in the junction area of the Aleutian and Kamchatka seismic focal zones. This probably means that there is either a high level of fluid pressure or an increased level of maximum shear stresses.

Magnitude of stresses. At this stage the reconstruction was based on the use of seismological data on the energy parameters of the first of the two strong Simushir earthquakes.

According to the data available on the web site of Harvard University, the energy released in the seismic waves, E_s, and the seismic moment of the earthquake of November 15, 2006, M_o, were 7.4×10^{16} J and 3.4×10^{21} J, respectively. The magnitude of the stress drop for the large Simushir earthquake can be calculated using the following Eq. (38), from the Chapter "The Method of Cata-

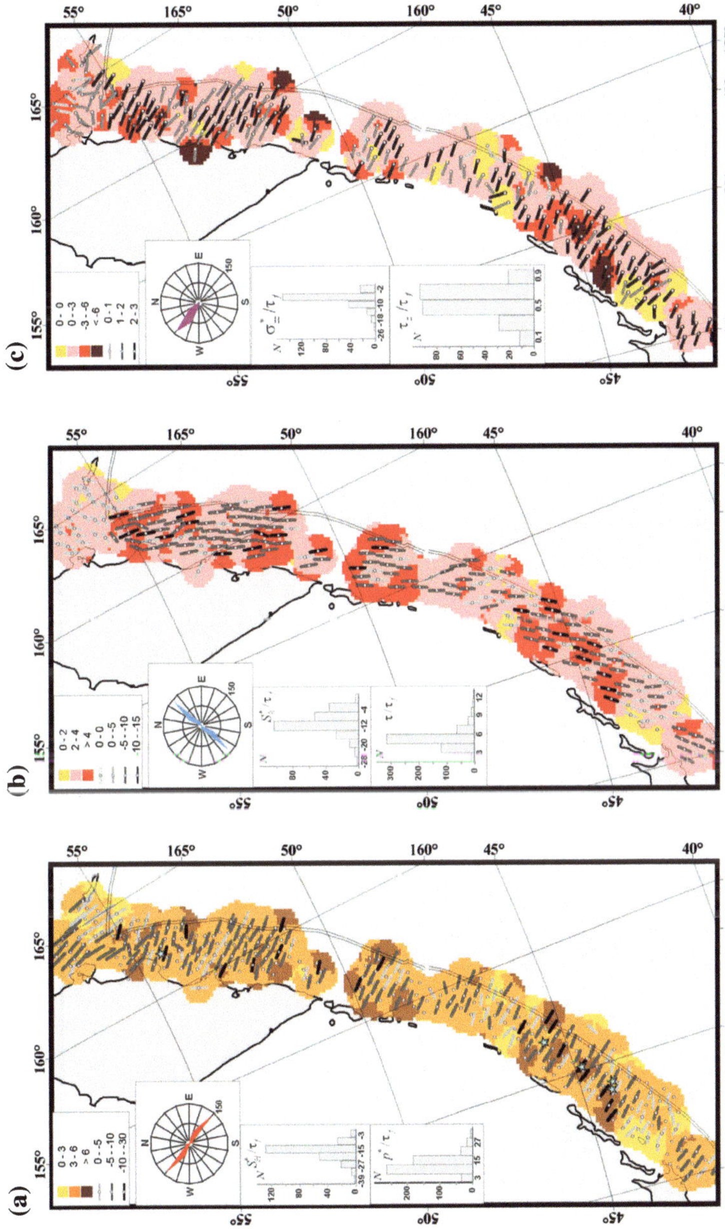

Fig. 16 Directions stress axis S_H (**a**), S_h (**b**), τ_z (**c**) and reduced values of the maximum shear stresses τ (**a**), effective pressure p^* (**b**), effective vertical stresses σ^*_{zz} (**c**) (stress values, normalized by the τ_f) for the crust of the Kuril-Kamchatka seismofocal zone, depth 20 km. Different shades of gray for the stress axes characterize the value of effective stresses S^*_H, S^*_h and τ_z normalized by τ_f. The rose diagram shows the predominant strike directions of this stress axis. The bar diagrams show the predominant range of the p^*, τ and σ^*_{zz} values normalized by τ_f

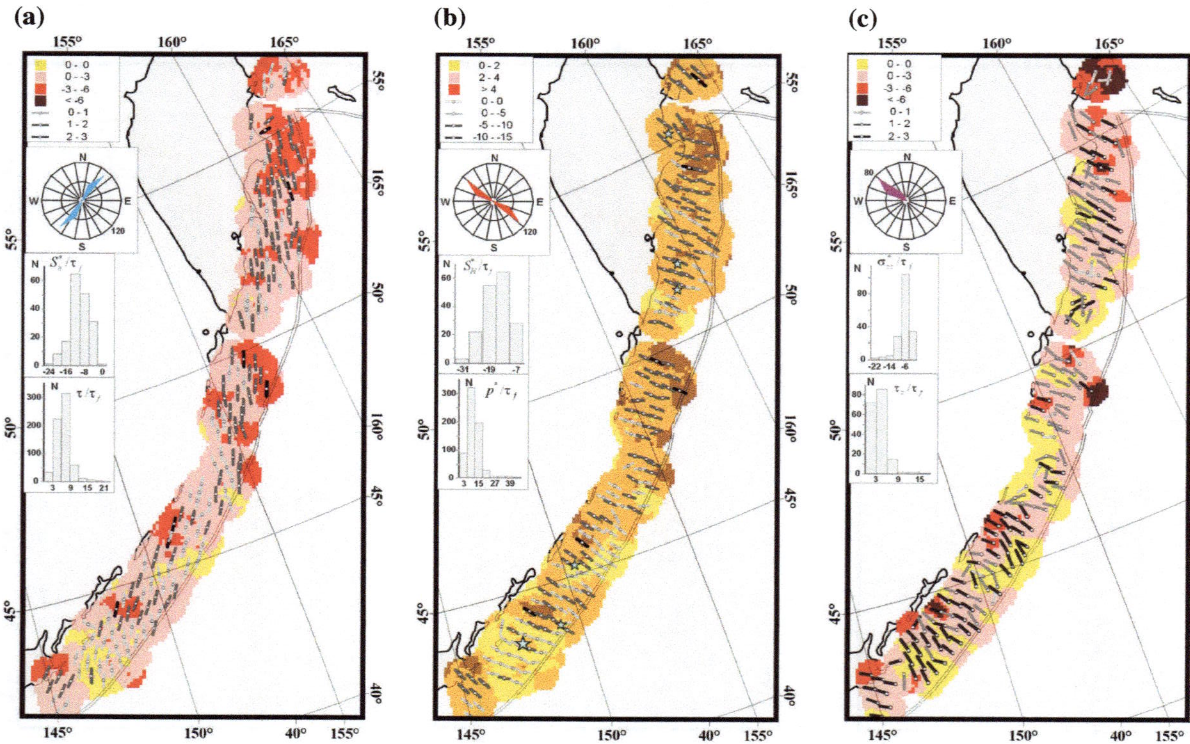

Fig. 17 Directions stress axis S_H (**a**), S_h (**b**), τ_z (**c**) and reduced values of the effective pressure p^* (**a**), maximum shear stresses τ (**b**), vertical effective stresses σ_{zz}^* (**c**) (stress values, normalized by the τ_f) for the crust of the Kuril-Kamchatka seismofocal zone, depth 60 km. Different shades of gray for the stress axes characterize the value of effective stresses S_H^*, S_h^* and τ_z normalized by τ_f. The rose diagram shows the predominant strike directions of this stress axis. The bar diagrams show the predominant range of the p^*, τ and σ_{zz}^* values normalized by τ_f

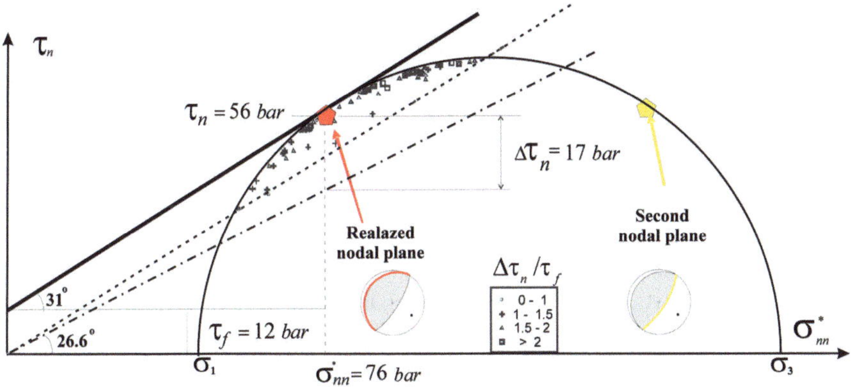

Fig. 18 Selecting one of the nodal planes as the actual fault plane on the Mohr diagram. Different marks show calculated values of the stress drop of earthquakes in the source area of the strong Simushir earthquake. The points that correspond to normal and shear stresses are also shown. The actual fault plane is close to the point of maximum strength of the rock massif

clastic Analysis of Discontinuous Displacements". Assuming that the shear modulus is $\mu = 4 \times 10^{10}$ Pa, we can find the stress drop value $\Delta\tau = 17 \times 10^5$ Pa. The analysis of the aftershock zone accounting for the possible geometry of the earthquake source (http://earthquake.usgs.gov/eqcenter/eqachives/poster/2006/20061115_image.php) showed that the rupture occurred on one of the nodal planes corresponding to the shallow subduction of the oceanic plate beneath the continental plate.

On the other hand, the magnitude of the relaxed stresses can be estimated up to the normalization to the unknown value of the effective cohesion of the rock mass, τ_f, using the results of the first two stages of the MCA reconstruction and Eq. (36) from the Chapter "The Method of Cataclastic Analysis of Discontinuous Displacements". Fig. 16 shows the distribution of the stress drop, calculated by the algorithm of the third stage of the MCA, in the source area of this earthquake. Figure 18 shows the reduced Mohr diagram for the stresses corresponding to the onset of the source ripping. It shows the parameters of the stress vectors acting on each of the nodal planes. It can be seen that the shallow nodal plane is very close to the plane of the effective internal friction, whereas the second near-vertical nodal plane lies outside the field of the brittle failure (the field between the outer envelope and the line of the static friction).

The source zone of the analyzed earthquake extends for approximately 240 km along the trench and includes 145 domains for which the components of the stress tensor were obtained (Figs. 14, 15, 16 and 17). These domains cover ~85% of the source zone. Since the absence of stress data is usually related to the absence of earthquake focal mechanism data, it can be supposed that not only the limiting state (the contact of the failure envelope with the large Mohr circle; Fig. 18) but also the state of activation of preexisting fractures (the large Mohr circle does not intersect the line of the minimum dry friction) were not attained in these areas. In accordance with Eq. (36), Chapter "The Method of Cataclastic Analysis of Discontinuous Displacements", the normalized values of the relaxed stresses, $\Delta\tau/\tau_f$, can be calculated for these domains.

Using the parameters of the stress state of these domains and summing the contribution to the relaxed stresses for the source area within each of the domains, we obtain from Eq. (36), Chapter "The Method of Cataclastic Analysis of Discontinuous Displacements" $\Delta\tau/\tau_f = 1.42$. Then, using the value of the relaxed stresses previously obtained from the seismological data, the effective cohesion can be obtained as $\tau_f = 1.2$ MPa As can be seen, the internal cohesion is in 5 times less in the subduction zone than in the Altai and Sayan orogens (see the previous section).

Given the τ_f value, the absolute values of the effective stresses can be determined. If we multiply the stress components, shown in Figs. 16 and 17, by the value of τ_f we will obtain the absolute values of the maximum shear stresses τ, the underthrusting shear stress τ_z, effective isotropic pressure p^* and effective stresses S_H^* and S_h^*. It should be noticed that the level of effective isotropic pressure and maximum shear stresses does not exceed 60 and 30 MPa, respectively. This is almost in 3–4 times less than the level of stress in the orogenic crust of the Altai-Sayan.

3.3 Conclusions

The analysis of the character of the stress state in the preparation zone of the Simushir earthquake of November 15, 2006, did not reveal areas with high underthrust shear stresses. Such areas reflect the degree of the influence of the mantle on the crust (Figs. 16c and 17c). They could have corresponded to the zones where the crust exerts an especially strong resistance to displacements induced by mantle flows. The absence of such areas indicates the relatively uniform distribution of the friction forces in the lower crust. This region is almost free of local areas (mosaics) with elevated levels of effective isotropic pressure (Figs. 16b and 17b). Such areas are confined to the boundaries of this zone. The same is true for the distribution of the maximum shear stresses (Figs. 16a and 17a), because these two characteristics are interrelated. Note also that the onset of ripping was confined to the boundary of the segment with the minimum values of the effective confining pressure, where the maximum gradient of this parameter was observed.

These observations are in good agreement with the concept developed by us (Rebetsky and Marinin 2006; Rebetsky 2007) concerning distinguishing the areas with high seismic potentials. According to this concept, the high efficiency of brittle failure is observed in the areas with medium levels of deviatoric and isotropic stresses. The maximum of the relaxed stresses is observed there. In the areas of elevated stress levels, the plastic or cataclastic (numerous weak events) flows are more efficient. The size of the area with medium stress levels defines the magnitude of the expected event. The realization of destruction in such an area requires that the incipient rupture should overcome the resistance of those zones where brittle failure is less efficient (local islands with elevated levels of isotropic stresses and elevated friction forces). Because of this, on the one hand, it is important that this area is sufficiently homogenous with respect to the stress level, and, on the other hand, a high initial energy pulse is needed at the site of the rupture generation. Such a situation appears in areas with an elevated stress gradient. The propagation of destruction in such areas from the zone of high stresses to the zone of low stresses is accompanied by an increase in the release of elastic strain energy.

4 Discussion

There is a certain similarity in the state of the Earth's crust of the Tien Shan and the Altai-Sayan orogens according to the results of the tectonic stresses inversion: (1) the submeridional orientation of the axes of maximum compression in the regions of mountain uplifts with their subhorizontal position; (2) the prevalent regime of horizontal compression and shear in the Earth's crust of mountain rises; (3) the geodynamic regime of horizontal tension and shear in the Earth's crust of large intermontane depressions and in the fore troughs; (4) the changes of the Lode-Nadai values in the whole range from -1 to $+1$; (5) the mosaic distribution of the orientation of the underthrusting shear stress at the predominantly latitudinal (Altai-Sayan) and meridional (Tien-Shan) orientations.

There are large differences between the stress state for the orogenic crust and for the active plate boundaries.

The distribution of the stress axes in the Earth's crust of the Kuril and Kamchatka is much more ordered than in the orogenic crust. The unidirectional orientation of the underthrusting shear stress from the ocean to the continent and a relatively shallow 25–35° dipping of the compression axes toward the oceanic lithosphere are important facts of the reconstruction. The main value of the Lode-Nadai coefficient corresponds to a pure shear.

The different level of internal cohesion is the most important difference in stress state of these regions: 6 MPa is in the Earth's crust of orogens and 1.2 MPa is in the Earth's crust of the active margins. Such a different level of internal cohesion determines a higher (4–5 times) level of deviatoric stresses in the orogenic crust.

Data on the values of stress in the earth's crust of seismically active regions provide additional important information not only for clarifying the mechanism of formation of deformation structures, but also for revealing different types of seismic regime.

References

Blagovidova TYa, Zhalkovskii ND, Muchnaya VI, Filina AG, Tsibul'chik ID (1986) Seismicity of the Altai-Sayan region according to instrument data. Soviet Geol Geophys 27(1):120–126

Bogachkin BM (1981) History of the Tectonic evolution of Gorny Altai in the Cenozoic. Nauka, Moscow, p 131 (in Russian)

Bott MHP (1959) The mechanics of oblique slip faulting. Geol Mag 96:109–117

Byerlee JD (1978) Friction of rocks. Pure Appl Geophys 116:615–626

Gol'din SV (2004) Dilatancy, repacking, and earthquakes. Izv Phys Solid Earth 40(10):817–832

Kostrov BV (1975) Mechanics of Tectonic earthquake source. Nauka, Moscow, p 174 (in Russian)

Kostrov BV, Das Sh (1988) Principles of Earthquake source mechanics. Cambridge University Press, Cambridge, p 256

Mises R, von Z (1928) Angew Math Mech 8:161−185

Polets AYu, Zlobin TK (2013) Properties of spatial and temporal distribution of the focuses of strong earthquakes and types of seismic dislocations before the catastrophic Simushir earthquakes of 2006–2007. Dokl Earth Sci 450(2):633–637

Polets AYu, Zlobin TK (2014) Estimation of the stress state of the Earth's crust and the Upper Mantle in the area of the Southen Kuril Islands. Russ J Pac Geol 8(2):126–137

Rebetsky YuL (2007) Tectonic stress and strength of mountain ranges. Nauka, Moscow, p 406 [in Russian]

Rebetsky YuL (2008) Possible mechanism of horizontal compression stress generation in the Earth's crust. Dokl Earth Sci 423A(9):1448–1451

Rebetsky YuL (2009) State of stress of the Earth's crust of the Kuril Islands and Kamchatka before the Simushir earthquakes. Russ J Pac Geol 3(5):477–490. https://doi.org/10.1134/S1819714009050108

Rebetsky YuL, Alekseev RS (2014) The field of recent tectonic stresses in Central and Southeastern Asia. Geodyn Tectonophys 5(1):257–290. https://doi.org/10.5800/GT-2014-5-1-0127

Rebetsky YuL, Kuzikov SI (2016) Active faults of the northern Tien Shan: tectonophysical zoning of seismic risk. Russ Geol Geophys 57:967–983. http://dx.doi.org/10.1016/j.rgg.2016.05.004

Rebetsky YuL, Marinin AV (2006) Preseismic stress field before Sumatra-Andaman earthquake of 26.12.2004. A model of metastable state of rocks. Geol i Geofiz 47(11):1173–1185

Rebetsky YuL, Tatevossian RE (2013) Rupture propagation in strong earthquake sources and tectonic stress field. Bull Soc Geol Fr 184(4–5):335–346

Rebetsky YuL, Sycheva NA, Kuchay OA, Tatevossian RE (2012) Development of inversion methods on fault slip data. Stress state in orogenes of the central Asia. Tectonophysics 581:114–131. https://doi.org/10.1016/j.tecto.2012.09.027

Rebetsky YuL, Kuchai OA, Marinin AV (2013) Stress state and deformation of the Earth's crust in the Altai-Sayan mountain region. Russ Geol Geophys 54(2):206–222. http://dx.doi.org/10.1016/j.rgg.2013.01.011

Rebetsky YuL, Sycheva NA, Sychev VN, Kuzikov SI, Marinin AV (2016a) The stress state of the northern Tien Shan crust based on the KNET seismic network data. Russ Geol Geophys 57:387–408. http://dx.doi.org/10.1016/j.rgg.2016.03.003

Rebetsky YuL, Polets AYu, Zlobin TK (2016b) The state of stress in the Earth's crust along the northwestern flank of the Pacific seismic focal zone before the Tohoku earthquake of 11 March 201. Tectonophysics 685:60–76. http://dx.doi.org/10.1016/j.tecto.2016.07.016

Rice J (1980) The mechanics of earthquake rupture. Elsevier, Amsterdam, pp 555–649

Sycheva NA, Yunga SL, Bogomolov LM, Mukhamadieva VA (2005) Seismotectonic deformations and recent tectonics of the Tien Shan (based on the KNET digital seismic network). Phys Solid Earth 11:62–78 [in Russian]

Trifonov VG, Soboleva OV, Trifonov RV, Vostrikov GA (2002) Recent geodynamics of the Alpine-Himalayan collision belt. Geos, Moscow, p 225 [in Russian]

Wallace RE (1951) Geometry of shearing stress and relation to faulting. J Geol 59:18–130

Zlobin TK, Polets AYu (2009) Source zones of the catastrophic Simushir earthquakes on November 15, 2006 (Mw = 8.3) and January 13, 2007 (Mw = 8.1) and the deep crust structure beneath the Middle Kuril Segment. Russ J Pac Geol 3(5):460–469

Zobak ML (1992) First- and second modern pattern of stress in lithosphere: the world stress map project. J Geopys Res 97(B8):11703–11728

The Significance of Crustal Velocity Model in Moment Tensor Solutions: A Case Study of Yedisu Earthquakes

Fatih Turhan, Musavver Didem Cambaz and Jiří Zahradník

Keywords North anatolia fault zone · Yedisu segment · Moment tensor solution
Waveform inversion · Crustal velocity model

1 Introduction

Eastern Turkey is situated at the collision zone of Arabian and Euroasian plates. As a result of continuous intense deformation the region contains diverse structures, such as suture zones, metamorphic complexes and young orogenesis (Fig. 1). NAFZ is a right-lateral strike-slip fault, running for about 1500 km from eastern Anatolia to Aegean Sea generating destructive earthquakes. It has mostly a single-fault trace character from Karlıova to Mudurnu, and it splits into several branches in the Marmara Sea (Barka and Kadinsky-Cade 1988; Şengör et al. 2005). The study area has a long history of devastating earthquakes (Barka 1984; Barka et al. 1987; Akyüz et al. 2009; Sançar and Akyüz 2014). Earthquakes in the last century produced more than 1000 km surface rupture on the North Anatolian Fault Zone (Sançar and Akyüz 2014). Historical and instrumental earthquakes indicate that the Yedisu and the surroundings is one of the most active regions in Turkey. Two most destructive earthquakes have occurred causing significant loss of life and property in the town of Erzincan on 26 December 1939 (Ms = 8.2) and 13 March 1992 (Ms = 6.8) (Aktar et al. 2004). The area have attracted considerable attention especially after these

F. Turhan · M. D. Cambaz (✉)
Kandilli Observatory and Earthquake Research Institute, Boğaziçi University, Istanbul, Turkey
e-mail: didem.samut@boun.edu.tr

J. Zahradník
Faculty of Mathematics and Physics, Charles University, Prague, Czech Republic

S. D'Amico (ed.), *Moment Tensor Solutions*, Springer Natural Hazards,
https://doi.org/10.1007/978-3-319-77359-9_24

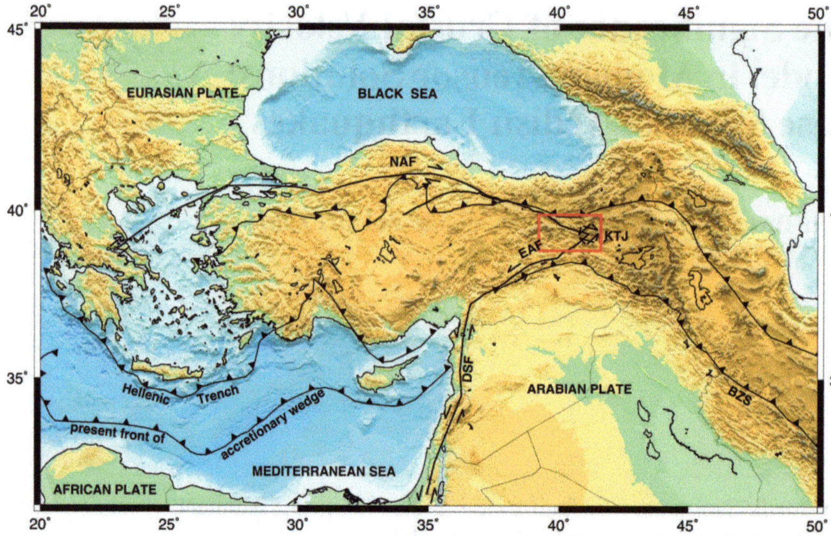

Fig. 1 Topography of Turkey and surrounding regions. Black lines with triangles indicate the suture zones and convergent boundaries. The inset marks the study region. EAF: East Anatolian Fault, NAF: North Anatolian Fault, BZS: Bitlis-Zagros suture zone, DSF: Dead Sea Fault, KTJ: Karlıova Triple Junction. Red rectangle shows the study area

devastating earthquakes. A number of different studies were performed in the area (Kaypak and Eyidoğan 2002; Örgülü et al. 2003; Aktar et al. 2004; Gökalp 2007) in order to understand the seismic and seismotectonic structure in detail. Limited geodetic information exists on the present day tectonics of the area. A relatively dense Global Positioning system (GPS) monitoring network was established in 2003 in order to investigate contemporary neotectonic deformation on the eastern NAFZ and its neighbourhood (Özener et al. 2010).

Moment tensor solutions of earthquakes provide valuable information in understanding the active tectonics and stress variations (Gephart and Forsyth 1984; Zoback 1992; Montone et al. 1999; Pondrelli et al. 2002). They have also a great importance in generating shake-maps and planning emergency responses, hazard assessment and seismotectonic studies (Presti et al. 2013; Herman et al. 2014; Wald et al. 1999; Convertito et al. 2009; Gallovic et al. 2009; Convertito and Herrero 2004; Morrato et al. 2007). The success of moment tensor solution relies mainly on the accuracy of the velocity model and hypocenter location (Robertson 2008). The precision of the hypocenter locations depends on the quality of data recordings, spatial distribution of stations, methodologies used in earthquake location and the P and S wave velocity models (Chiu et al. 1997). It is obvious that the determination and use of a specific crustal velocity model not only in the location of hypocenters, but in the estimation of moment tensor solutions, is a critical point of obtaining reliable mechanisms. The selection of the crustal velocity model still plays the key point for a healthy solution of moment tensor and earthquake parameters.

In this study, moment tensor solution of earthquakes in Yedisu, the easternmost part of the NAFZ near the vicinity of the Karlıova Triple Junction (KTJ), at the intersection of NAFZ and EAFZ, are examined in relation to crustal velocity model. Four different velocity models that represent the study area are selected. In order to estimate the optimum velocity model, earthquakes from the Kandilli Observatory and Earthquake Research Institute (KOERI) catalogue, occurring in the vicinity of Yedisu, were re-located by using these four different velocity models. Earthquake relocation results were investigated in terms of the depth and residual (RMS, ERH, ERZ) variations. Focal mechanism solution is a convenient device to determine tectonic factors affecting the basin morphology, such as crustal extension and shortening direction. In this study, we investigated moment tensor solutions of earthquakes with the selected velocity model. Moment tensor solutions of 21 earthquakes from 2009 to 2016 with M > 4.0 were computed by using this optimum velocity model. This study will provide beneficial input for further research on seismotectonics and seismic hazard analysis on one of the most striking deformation belts of the Earth.

2 Tectonic Setting

North Anatolian Fault Zone presented a considerable amount of seismicity during the westward migration of earthquake sequence (Zabcı et al. 2017). Akyüz et al. (2009) defines the NAFZ between Karlıova and Erzincan basin in three segments; Yedisu, Elmalı and Ilıpınar Segments. Yedisu segment is the one that did not rupture during the migration of earthquake sequence from east to west and thus described as the Yedisu Seismic Gap (Fig. 2) in various studies (Barka et al. 1987; Barka and Kadinsky-Cade 1988; Akyüz et al. 2009; Sançar and Akyüz 2014). Erzincan region is the most active seismic region not only in historical but also in instrumental time-period in Turkey (Table 1). December 26, 1939 Erzincan (M = 8) earthquake was the largest earthquake that occurred along the NAFZ (Barka et al. 1987). Following 1939 Erzincan earthquake, earthquakes migrated to west along the NAFZ. Several seismic, paleoseismologic morphologic studies were performed in the study area mainly after the 1992 Erzincan earthquake (Kaypak and Eyidoğan 2002; Aktar et al. 2004; Gökalp 2007; Akyüz et al. 2009). Barka et al. (1988) reviewed and described the geometry of these faults. It concludes that the seismic behavior of the segments at the easternmost part of the NAFZ is not in accordance with the general characteristic of the NAFZ (Sançar and Akyüz 2014). Each segment exhibits its own characteristic (Barka 1984; Barka et al. 1987; Barka and Kadinsky-Cade 1988; Akyüz et al. 2009; Sançar and Akyüz 2014). Earthquakes at the eastern part of Erzincan exhibit a more complicated distribution in time. Most of the earthquakes occurring near the vicinity of Erzincan is related with the basin that is opening between the Erzincan basin and the Ovacık fault (Barka et al. 1987).

Table 1 Earthquakes (M > 6) occured around Karlıova triple junction since 1866 (Akoğlu 2008)

Date	Location		Mag (M)	Depth (km)	Fault length (km)	Strike	Dip	Rake	Slip (cm)	Location
12.05.1866[b]	41	39	6.8	0–18	31.98	52	90	0	−100	Bingöl
26.12.1939[a]	39.51	39.80	7.9	0–18	310.4	−67/−86	90	−180	150−650	Erzincan
17.08.1949[e,k]	40.62	39.57	6.8	0–18	35.4	96/113	90	−180	50–150	Karlıova
19.08.1966a[c]	41.56	39.17	6.8	0–18	34.98	−53	70	150	60	Varto
20.08.1966b[c]	40.98	39.42	6.1	0–18	11.98	106	90	−180	25	Varto
26.07.1967[d, f, n]	40.38	39.54	6.0	0–18	–	111	90	−170	–	Pülümür
22.05.1971[f]	40.52	38.83	6.7	0–18	44.98	38	86	−8	45	Bingöl
13.03.1992[l, m]	39.63	39.72	6.8	1.5–18	30.16	115/119	63	−164	90–143.3	Erzincan
15.03.1992[h, j]	39.93	39.53	6.0	1–18	8	−125	61	57	20	Pülümür
27.01.2003[g, i]	39.66	39.58	6.1	1–18	14.62	−118	90	0	20	Pülümür
01.05.2003[k]	40.53	39.04	6.4	1–18	21.98	137	90	−180	40	Bingöl

[a]Ambraseys (1997)
[b]Ambraseys and Finkel (1995)
[c]Ambraseys and Jackson (1998)
[d]Ambraseys and Zatopek (1968)
[e]Barka (1996)
[f]Barka and Reilinger (1997)
[g]Bekler et al. (2003)
[h]Fuenzalida et al. (1997)
[i]Milkereit et al. (2004)
[j]Nalbant (1996)
[k]Pamir and Ketin (1940)
[l]Seymen and Aydin (1976)
[m]Taymaz et al. (1991)
[n]Wallace (1968)

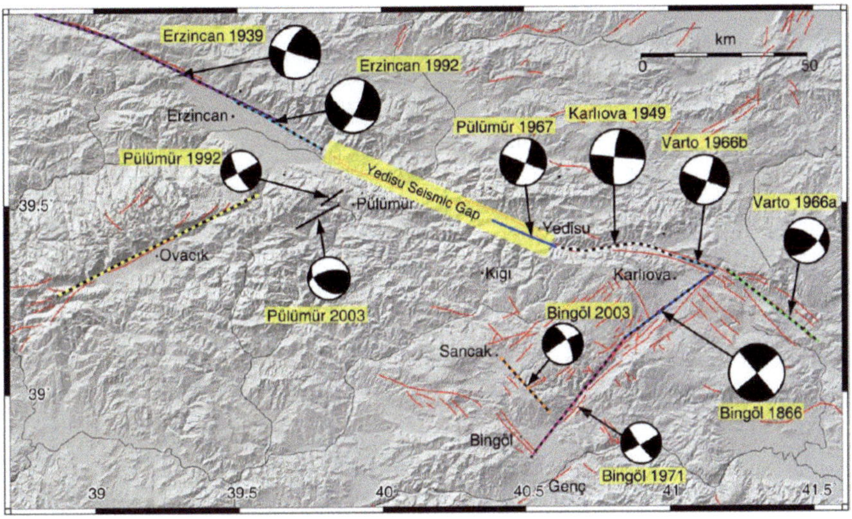

Fig. 2 Focal mechanism of the earthquakes (M > 6) that occured in the study area since 1866. Various colours indicate estimated rupture surfaces of these earthquakes (modified from Akoğlu 2008; Sançar and Akyüz 2014)

3 Data and Method

In Turkey, mainly after the August 1999, Mw7.4 İzmit earthquake, public interest in earthquake safety spawned the creation of state funding targeting to improve seismic network data quality. Seismic network quality and station reliability improved across all major network operators in Turkey, especially after 2004. This high-quality data made the creation of more comprehensive and higher quality solutions of hypocenter location parameters and hence moment tensor solutions. However, the use of most appropriate crustal velocity model is still one of the most important aspects in the determination of hypocenter locations and MT solution calculations.

In order to investigate the effect of crustal velocity model, we examined different velocity models from previous studies (Table 2). Following the 1992 Erzincan earthquake a number of different studies were performed in the study area in order to obtain detailed seismic velocity information. Kaypak and Eyidoğan (2002) used a temporal network of 10 stations installed just after the 1992 Erzincan earthquake. They have used VELEST code (Kissling et al. 1995) in order to calculate minimum 1-D velocity structure of the Erzincan basin for both P and S wave. Türkelli et al. (2003) have used broadband waveform data collected by the Eastern Turkey Seismic Experiment (ETSE). A grid search approach and phase data from 66 very well located events were used in order to determine the optimal 1-D model for Eastern Turkey. Gökalp (2007) collected the aftershock data of the 13 March 1992 earthquake by using the 15 temporarily installed stations in the study area. They inverted the travel time data in order to compute the 3D Vp and Vs structure and obtained a detailed

Table 2 Crustal velocity models (Türkelli et al. 2003; Kalafat et al. 1987; Kaypak and Eyidogan 2002; Gökalp et al. 2007)

Türkelli et al. (2003)		Kalafat et al. (1987)		Kaypak and Eyidogan (2002)		Gökalp (2007)	
0	4.93	0	4.5	0	1.95	0	4.14
2	6.3	5.4	5.91	1	2.48	2	4.27
42	7.69	31.6	7.8	2	4.79	3	6.01
		89.2	8.3	3	5.44	6	6.17
				6	6.01	8	6
				7	6.26	12	5.36
				20	6.94	15	6.15
				25	7.07	19	6.23
				30	8	21	6
						35	7

velocity structure which is sensitive to the low velocity layer in the region. According to the tomography study of Gökalp (2007) a wide range of velocity variation exists in the area. They indicate that 3D velocity structure beneath the region is heterogeneous with a significant low velocity layer throughout the basin and high velocities occur at south and east of Erzincan basin. They concluded that the large-scale fractures can cause the low velocities and the ophiolitic units can be the possible reason for the high velocities. Kalafat et al. (1987) is not a site specific model for the study area, but it is a more general crustal velocity model that is used in the routine hypocenter location procedure of Regional Earthquake-Tsunami Monitoring Center (RETMC) in the Kandilli Observatory and Earthquake Research Institute (KOERI).

Moment tensor solutions in Yedisu region were calculated using the iterative deconvolution of the point sources (ISOLA package; Sokos and Zahradnik 2008). ISOLA is based on an extension of the method proposed by Kikuchi and Kanamori (1991) to regional and local distances and provides precise information on the source parameters. The computation options include inversion to retrieve the full MT, deviatoric MT, and pure double-couple MT. Position and origin time of the point sources are obtained by grid search. Green's functions were calculated by using the discrete wavenumber method (Bouchon 1981; Coutant 1989). The crustal structure and velocity model play an important role in estimating Green's function. The moment tensor inversion is retrieved by the least square inversion using waveforms from both local and regional distances. In order to examine the solution quality, ISOLA provides a number of parameters such as variance reduction (VR), condition number (CN), focal mechanism variability index (FMVAR), and space time variability (STVAR) (Sokos and Zahradnik 2013). Variance reduction defines the residual between the observed and synthetic waveform, and is one of the most important parameters for checking the quality of waveform fit. A detailed description of the package is presented in ISOLA User's Guide (Sokos and Zahradnik 2006).

Waveform database from 2009 to 2016 for this study is provided from the Kandilli Observatory and Earthquake Research Institute (KOERI) and Disaster and Emergency Management Presidency Earthquake Department (AFAD) broadband networks. As the station coverage upgraded, both of the networks have more than ten stations in the region and the combination of the dataset provides a good coverage and wealthy solution for moment tensor inversion. According to the results of earthquake re-location with respect to different crustal models, the crustal model Gökalp (2007) was used in the computation of moment tensor solutions.

4 Variations of Earthquake Re-location Parameters

It is one of the major foci of seismology to determine reliable earthquake locations but the earthquake depths are determined less reliably than the epicenters when the shear wave velocity model is not appropriately determined (Chiu et al. 1997). We made a test in order to estimate the effect of the crustal velocity models and re-located the earthquakes by using four different velocity models. Kalafat et al. (1987), Kaypak et al. (2002), Türkelli et al. (2003) and Gökalp (2007) were used in the study. Earthquake location software; zSacWin (Yılmazer 2012), that is routinely used for obtaining the earthquake location parameters in RETMC- KOERI, was used for the re-location test.

Table 3 shows the results of hypocenter re-location parameters according to four different velocity models. Latitude and longitude values do not differ significantly, in general. Residuals (RMS, ERH, ERZ) and earthquake solution quality (Q) parameter presented slight variations according to the different velocity models. We observed higher residual values (RMS, ERH, ERZ) with the Kaypak et al. (2002) model, generally. The hypocenter depth values presented significant variations according to the change of velocity model. Türkelli et al. (2003) gives the highest depth values, (Kaypak and Eyidogan 2002) gives the lowest depth values, in general. Aktar et al. (2004) studied the 3-D seismic velocity model for the upper crust beneath the Erzincan basin using the tomographic inversion technique. They have determined the depth distribution of the hypocenters within a seismogenic zone that lies in between 4 and 11 km. In our re-location test it was the Gökalp (2007) velocity model providing the hypocenter depth within this interval so, we continued with this model.

5 Variations of MT Solutions

Four different crustal models for the easternmost part of the NAFZ were tested in order to investigate the variations of MT solutions. Results of an example earthquake were presented in Table 4. Variance reduction (VR), one of the most important parameters for checking the quality of waveform fit, was obtained as 0.87, 0.79, 0.80 and 0.86 for the velocity models of Gökalp, Kalafat, Kaypak and Türkelli,

Table 3 Results of earthquake re-locations for the events 07.11.2010 (Ml = 4.1), 06.03.2011 (Ml = 4.2), 04.04.2012 (Ml = 4.2) and 07.09.2015 (Ml = 4.5) with respect to the selected velocity models in Table 2

Velocity models	Date	OT	Lat	Lon	Depth	RMS	ERH	ERZ	Q
Türkelli et al. (2003)	07.11.2010	05:51:31.41	39.5040	40.2110	16.9	0.41	0.7	0.3	C1
Gökalp (2007)	07.11.2010	05:51:29.53	39.5010	40.2193	8.0	0.44	0.7	0.4	C1
Kalafat et al. (1987)	07.11.2010	05:51:29.77	39.4992	40.2108	3.5	0.43	0.7	0.3	C1
Kaypak and Eyidogan (2002)	07.11.2010	05:51:30.12	39.5025	40.2100	2.2	0.53	0.9	0.4	D1
Türkelli et al. (2003)	06.03.2011	07:58:39.55	39.2258	40.3777	16.3	0.46	0.9	0.4	C1
Gokalp (2007)	06.03.2011	07:58:37.75	39.2178	40.3753	8.1	0.45	0.7	0.4	C1
Kalafat et al. (1987)	06.03.2011	07:58:38.00	39.2167	40.3667	3.4	0.4	0.7	0.2	C1
Kaypak and Eyidogan (2002)	06.03.2011	07:58:38.40	39.2037	40.3725	2.0	0.6	1.1	0.5	D1
Türkelli et al. (2003)	04.04.2012	14:18:38.22	39.3307	41.0403	17.0	0.5	0.8	0.3	C1
Gokalp (2007)	04.04.2012	14:18:36.46	39.3240	41.0398	10.8	0.45	0.7	0.4	C1
Kalafat et al. (1987)	04.04.2012	14:18:36.16	39.3233	41.0445	5.0	0.54	1.0	0.4	D1
Kaypak and Eyidogan (2002)	04.04.2012	14:18:37.09	39.3238	40.9998	1.3	0.68	1.4	0.4	D1
Türkelli et al. (2003)	07.09.2015	00:37:36.86	39.1727	40.2410	20.3	0.43	0.7	0.3	C1
Gokalp (2007)	07.09.2015	00:37:35.15	39.1752	40.2278	10.5	0.45	0.7	0.4	C1
Kalafat et al. (1987)	07.09.2015	00:37:35.77	39.1753	40.2028	7.2	0.46	0.7	0.2	C1
Kaypak and Eyidogan (2002)	07.09.2015	00:37:35.49	39.1728	40.1983	5.0	0.58	1.2	0.2	D1

respectively. Other parameters such as FMVAR, STVAR, CN for checking the quality of the MT solution were also satisfactory with almost similar values for the four different crustal models. Significant variations were observed in the magnitude and depth values. Depth values of the MT solutions vary in between 3 and 6 km and moment magnitude values vary in the range of 4.7–4.9 for four different velocity models. For MT solutions, it is important to use a proper velocity model in order to obtain an accurate fault mechanism and best waveform fit. Unless a specific tuned velocity model is used for each source-station path, the waveform modelling will be feasible up to epicentral distances equivalent to a few minimum shear wavelengths (Fojtikova and Zahradnik 2014). Also a single station waveform inversion might not fit well unless using specifically calibrated velocity models (Dias et al. 2016). We selected the Gökalp 2007 as the optimum model according to the results of MT solutions. Figure 3 shows the results of MT solution together with the observed and synthetic waveform fits, correlation vs. depth plot and the time space correlation plot for the 30.07.2009 Erzincan earthquake (Mw = 4.9).

Table 4 Results of MT solutions for the event 30.07.2009 with respect to the selected velocity models in Table 2

Velocity models	Date	Time	ISOLA											
			Centroid Depth (km)	Mw	Strike (°)	Dip (°)	Rake (°)	M_0 (Nm)	DC (%)	VR	CN	FMVAR	STVAR	Frequency Band
Kaypak and Eyidogan (2002)	30.07.2009	07:37:50	3	4.7	320	33	−101	1.40E + 16	96.8	0.8	3.2	7 ± 3	0.05	0.05–0.08
Türkelli et al. (2003)	30.07.2009	07:37:50	4	4.9	325	31	−96	3.01E + 16	86.5	0.86	2.1	6 ± 3	0.06	0.05–0.08
Gökalp (2007)	30.07.2009	07:37:50	5	4.9	328	35	−89	2.59E + 16	93.5	0.87	1.9	5 ± 3	0.05	0.05–0.08
Kalafat et al. (1987)	30.07.2009	07:37:50	6	4.8	324	38	−94	1.89E + 16	77.1	0.79	2.2	10 ± 8	0.06	0.05–0.08

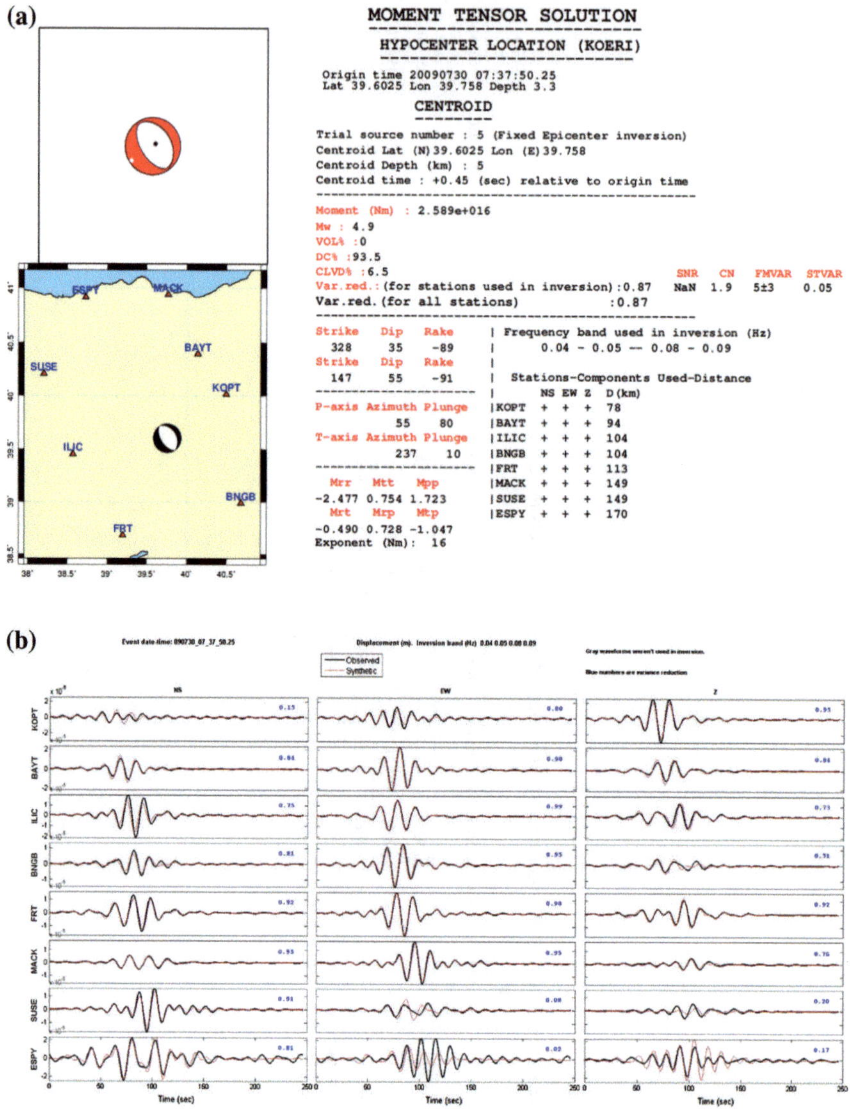

Fig. 3 **a** MT solution parameters **b** observed and synthetic waveform fits **c** correlation versus depth plot **d** time space correlation plot

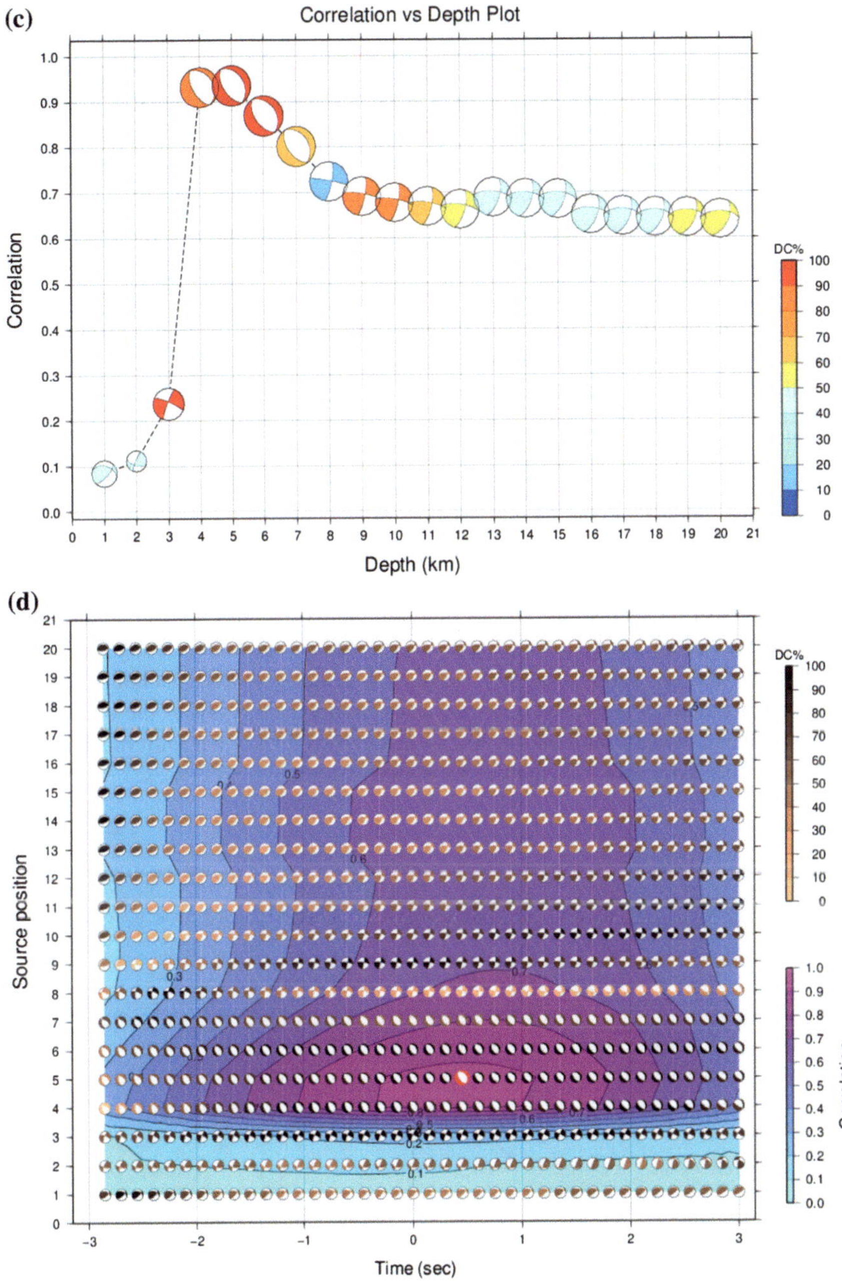

Fig. 3 (continued)

6 Conclusion and Discussions

Moment tensor solutions of earthquakes in the easternmost part of the NAFZ, around the Yedisu segment were investigated in this study. Results of the moment tensor solutions are presented in Fig. 4 together with the hypocenter location and the quality control parameters in Table 5. Four different velocity models were tested in order to see the effect of crustal velocity models on the re-locations and moment tensor solutions. The results of the epicenter location parameters do not vary significantly. But the depth of the earthquakes varies explicitly. We interpreted the increase of hypocenter location depth as a good indicator of the depth constraint due to the very detailed structure of the velocity model. Thus, we used the Gökalp (2007) model for the further MT solutions of earthquakes.

MT solutions of earthquakes do not vary significantly with local crustal models. When we compared the quality control parameters of the MT solutions such as the variance reduction, condition number FMVAR, and STVAR, we generally observed similar values. However, the Moment Magnitude (Mw) and depth values vary, in each solution. The variations of depth also influence the strike, dip and rake values of the mechanism.

The results of the MT solutions are in agreement with fault zone characteristics and the active tectonics of Turkey. Right-lateral strike slip motion along the NAFZ is consistent with the fault motions. The normal faulting source mechanism of the historical earthquakes in the study area was explained as a result of the opening as a basin between Ovacık fault and Erzincan basin. Similar normal fault mechanisms obtained in the same area indicate the existence of the same processes.

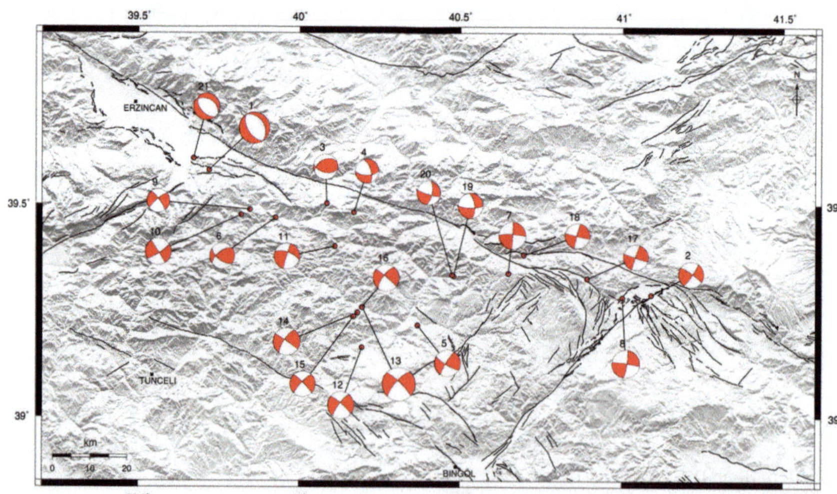

Fig. 4 Moment tensor solutions of earthquakes in the study area

Table 5 Hypocenter location and the quality control parameters of the earthquakes from MT solutions

No.	Date	Time	KOERI							ISOLA																
			Lat (°N)	Lon (°E)	Ml	Mb	Mw	Depth (km)		Mw	Nodal Plane 1			Nodal Plane 2			P Axis		T Axis		M_0 (Nm)	DC (%)	VR	CN	FMVAR	STVAR
								Hypocenter	Centroid		Strike (°)	Dip (°)	Rake (°)	Strike (°)	Dip (°)	Rake (°)	Azimuth	Plunge	Azimuth	Plunge						
1	30.07.2009	07:37:50	39.6025	39.7580	5.0	4.8	-	3.3	5.0	4.9	328	35	-89	147	55	-91	55	80	237	10	2.59E+16	93.5	0.87	1.9	5±3	0.05
2	22.05.2010	02:30:32	39.3107	41.1265	4.1	4.1	-	5.0	5.0	4.1	304	82	173	35	83	8	139	1	259	11	2.03E+15	99.3	0.84	1.5	3±1	0.14
3	07.09.2010	02:48:51	39.5252	40.1242	3.8	4.1	-	3.8	7.0	4.0	102	52	114	247	44	62	176	4	73	71	1.18E+15	87.3	0.79	2.4	18±13	0.20
4	07.11.2010	05:51:29	39.5032	40.2093	4.1	-	-	5.0	8.0	3.8	89	61	-149	342	63	-33	304	41	36	2	7.34E+14	76.2	0.74	1.6	16±6	0.23
5	06.03.2011	07:58:38	39.2383	40.4063	4.2	4.3	-	5.0	10.0	4.2	120	79	171	212	81	11	346	1	76	14	2.77E+15	89.8	0.70	1.9	5±5	0.23
6	24.08.2011	15:47:24	39.4908	39.9650	4.3	4.4	-	5.0	8.0	4.1	126	61	146	234	60	33	180	1	89	43	2.10E+15	14.4	0.87	2.1	33±11	0.44
7	08.10.2011	03:25:19	39.3603	40.6873	4.0	-	-	2.2	7.0	4.3	272	81	178	2	88	9	137	5	228	7	3.90E+15	98.9	0.90	2.2	3±1	0.26
8	04.04.2012	14:18:36	39.3045	41.0388	4.2	-	-	5.0	7.0	4.4	278	87	173	8	83	3	323	3	233	8	4.23E+15	72.1	0.62	1.9	5±5	0.12
9	18.11.2012	15:44:20	39.5100	39.8872	4.0	-	-	5.0	5.0	3.7	323	80	-167	230	77	-10	187	16	96	2	4.44E+14	48.3	0.84	3.1	10±6	0.23
10	20.11.2012	17:56:28	39.4970	39.8600	4.2	-	-	5.0	4.0	4.2	327	89	170	57	80	1	13	6	282	7	2.72E+15	69.1	0.81	2.7	14±10	0.10
11	09.02.2015	22:52:49	39.4237	40.1490	4.1	-	4.1	4.7	4.0	4.1	197	87	-26	288	64	-177	149	20	245	16	1.82E+15	74.8	0.80	3.2	15±7	0.06
12	07.09.2015	00:37:35	39.1822	40.2135	4.5	-	4.6	8.4	11.0	4.3	310	80	-167	218	77	-10	174	16	84	2	3.49E+15	46.4	0.80	1.7	11±7	0.38
13	02.12.2015	23:27:06	39.2815	40.2353	5.5	-	5.3	5.0	8.0	5.3	314	80	-164	222	75	-10	179	18	87	4	1.07E+17	77.0	0.67	1.4	8±4	0.18
14	04.12.2015	15:37:40	39.2683	40.2203	4.3	-	4.3	5.0	7.0	4.3	304	88	-162	214	72	-2	171	14	78	11	3.33E+15	37.5	0.72	1.8	7±5	0.11
15	04.12.2015	16:00:14	39.2595	40.2080	4.1	-	4.2	5.0	7.0	4.1	315	83	-168	224	78	-7	180	13	89	4	1.67E+15	39.5	0.70	1.9	4±6	0.12
16	07.12.2015	23:27:16	39.2810	40.2337	4.3	-	4.3	5.0	6.0	4.2	309	84	-160	217	70	-6	175	18	82	10	2.84E+15	82.9	0.72	2.1	8±4	0.12
17	25.05.2016	11:40:57	39.3482	40.9295	4.0	-	-	5.0	7.0	4.0	108	87	176	198	86	3	153	1	63	5	1.11E+15	73.0	0.79	2.3	5±4	0.26
18	21.06.2016	06:25:21	39.4037	40.7345	4.1	-	-	5.0	7.0	4.0	104	87	175	194	85	3	149	1	59	6	1.46E+15	58.0	0.76	2.2	4±3	0.25
19	19.07.2016	09:39:45	39.3543	40.5177	4.0	-	-	1.3	7.0	4.0	353	63	-17	91	74	-152	315	31	220	7	1.17E+15	75.3	0.83	2.1	7±3	0.36
20	20.07.2016	01:56:43	39.3573	40.5123	3.8	-	4.1	2.5	7.0	3.8	11	65	-12	106	79	-154	331	26	237	10	5.69E+14	92.2	0.79	2.0	7±4	0.26
21	16.12.2016	21:21:48	39.6328	39.7052	4.4	-	-	1.3	7.0	4.3	299	42	-106	140	50	-76	111	78	220	4	4.10E+15	91.8	0.71	2.0	7±8	0.12

Acknowledgements Waveform data used in this study were obtained from the Kandilli Observatory and Earthquake Research Institute (KOERI) and Prime Ministry Disaster and Emergency Management Presidency (AFAD). We thank KOERI and AFAD for providing data for this study. Some of figures were made by using the Generic Mapping Tools (GMT) software (Wessel et al. 2013).

References

Akoğlu A (2008) Analysis and Modelling of Earthquake Surface Deformation With SAR Interferometry. Case studies from Turkey and the World. PhD Thesis, İstanbul Technical University, 148p

Aktar M, Dorbath C, Arpat E (2004) The seismic velocity and fault structure of the Erzincan basin, Turkey, using local earthquake tomography. Geophys J Int 156:497–505

Akyüz S, Altunel E, Zabcı C, Sançar T, Çakır Z, Yalçıner Ç, Gutsuz P (2009) Kuzey Anadolu Fay Zonu Üzerindeki Yedisu Sismik Boşluğu ve Karlıova Fayı (Bingöl) Üzerinde Paleosismolojik ve Morfotektonik Araştırmalar. TUBITAK Project No: 106Y174. Final Report

Ambraseys NN (1997) The little-known earthquakes of 1866 and 1916 in Anatolia (Turkey). J Seismolog 1(3):289–299

Ambraseys NN, Finkel C (1995) The seismicity of Turkey and adjacent areas: a historical review, 1500-1800. Eren, Beyoğlu, İstanbul

Ambraseys NN, Jackson JA (1998) Faulting associated with historical and recent earthquakes in the Eastern Mediterranean region. Geophys J Int 133(2):390–406

Ambraseys NN, Zatopek A (1968) The Varto Ustukran (Anatolia) earthquake of 19 August 1966 summary of a field report. Bull Seismol Soc Am 58(1):47–102

Barka AA (1984) Some neotectonic features of the Erzincan Basin, Eastern Turkey. (In Turkish with English abstract). In: Symposium on earthquake hazards in NE Turkey. Spe. Publication, Ataturk University, Erzurum

Barka AA, Toksöz MN, Gülen L, Kadinsky Cade K (1987) segmentation, seismicity and earthquake potential of the eastern part of the north Anatolian fault zone. Yerbilimleri 14:337–352

Barka A, Kadisnky Cade K (1988) Strike slip fault geometry in Turkey and its influence on earthquake activity. Tectonics

Barka A (1996) Slip distribution along the North Anatolian fault associated with the large earthquakes of the period 1939 to 1967. Bull Seismol Soc Am 86(5):1238–1254

Barka AA, Reilinger R (1997) Active tectonics of the Eastern Mediterranean region: deduced from GPS, neotectonic and seismicity data. Ann Geophys 40(3):587–610

Bekler FN, Kekovalı K, Kalafat D, Pınar A (2003) Pülümür (Turkey) Earthquake Kandilli Observatory and Earthquake Research Institute, Seismology Division, İstanbul 27 Jan 2003

Bouchon M (1981) A simple method to calculate Green's functions for elastic layered media. Bull Seismol Soc Am 71:959–971

Chiu J, Chiu S, Kim SG (1997) The Significance of the crustal velocity model in local earthquake locations from a case example of a PANDA experiment in the Central United States. BSSA. 87:1537–1552

Convertito V, Herrero A (2004) Influence of focal mechanism in probabilistic seismic hazard analysis. Bull Seism Soc Am 94:2124–2136

Convertito V, De Matteis R, Cantore L, Zollo A, Iannaccone G, Caccavale M (2009) Rapid estimation of ground-shaking maps for seismic emergency management in the Campania region of southern Italy. Nat Hazards 52:97–115

Coutant O (1989) Program of numerical simulation AXITRA. Tech. rep, LGIT, Grenoble, France

Dias F, Assumpção M, Facincani EM, França GS, Assine ML, Filho ACP, Gamarra RM (2016) The 2009 earthquake, magnitude 4.8 mb, in the Pantanal Wetlands, west-central Brazil, Anais da Academia Brasileira de Ciências

Fojtíková L, Zahradník J (2014) A new strategy for weak events in sparse networks: the first-motion polarity solutions constrained by single-station waveform inversion. Seism Res Lett 85:1265–1274. https://doi.org/10.1785/0220140072

Fuenzalida H, Dorbath L, Cisternas A, Eyidoğan H, Barka A, Rivera L, Haessler H, Philip H, Lyberis N (1997) Mechanism of the 1992 Erzincan earthquake and its aftershocks, tectonics of the Erzincan Basin and decoupling on the North Anatolian Fault. Geophys J Int 129:1–28

Gallovic F, Zahradnik J, Krizova D, Plicka V, Sokos E, Serpetsidaki A, Tselentis GA (2009) From earthquake centroid to spatialtemporal rupture evolution: Mw 6.3 Movri Mountain earthquake, 8 June 2008, Greece. Geophys Res Lett 36:L21310. https://doi.org/10.1029/2009GL040283

Gephart J, Forsyth W (1984) An improved method for determining the regional stress tensor using earthquake focal mechanism data: application to the San Fernando earthquake sequence. J Geophys Res 89:9305–9320

Gökalp H (2007) Local earthquake tomography of the Erzincan Basin and the surrounding area in Turkey. Ann Geophys 50(6):707–724

Herman MW, Herrmann RB, Furlong KP, Benz HM (2014) Using regional moment tensors to constrain the kinematics and stress evolution of the 2010–2013 Canterbury earthquake sequence, South Island, New Zealand. Tectonophysics 633:1–15

Kalafat D, Gürbüz C, Üçer B (1987) Batı Türkiye'de Kabuk ve Üst Manto Yapısının Araştırılması. Deprem Araştırma Bülteni 59:43–64 (in Turkish)

Kaypak B, Eyidoğan H (2002) Erzincan havzası ve dolayının üst-kabuk hız yapısının belirlenmesi. İTÜ Dergisi/d 1:107–122

Kikuchi M, Kanamori H (1991) Inversion of complex body waves, III. Bull Seismol Soc Am 81:2335–2350

Kissling E, Kradolfer U, Maurer H (1995) VELEST User's Guide

Milkereit C, Grosser H, Wang R, Wetzel H-U, Woith H, Karakisa S, Zunbul S, Zschau J (2004) Implications of the 2003 Bingol Earthquake for the Interaction between the North and East Anatolian Faults. Bull Seismol Soc Am 94(6):2400–2406

Montone P, Amato A, Pondrelli S (1999) Stress map of Italy. J Geophys Res 104:25595–25610

Moratto L, Orlecka-Sikora B, Costa G, Suhadolc P, Papaioannou C, Papazachos CB (2007) A deterministic seismic hazard analysis for shallow earthquakes in Greece. Tectonophysics 442:66–82

Nalbant SS (1996) Depremlerin oluşturduğu deformasyon ve gerilme alanlarının modellenmesi PhD Thesis. İstanbul University

Örgülü G, Aktar M, Türkelli N, Sandvol E, Barazangi M (2003) Contribution to the seismotectonics of the Eastern Anatolian Plateau from moderate and small size events. Geophys Res Lett 30:24. https://doi.org/10.1029/2003GL018258

Özener H, Arpat E, Ergintav S, Dogru A, Cakmak R, Turgut B, Doğan U (2010) Kinematics of the eastern part of the North Anatolian Fault Zone. J Geodyn 49:141–150

Pamir HN, Ketin İ (1940) Das Erdbeben in der Türkei vom 27./28. Dezember 1939. Geol. Rund. 31:77–78

Pondrelli S, Morelli A, Ekstrom G, Mazzaa S, Boschi E, Dziewonski AM (2002) European-Mediterranean regional centroidmoment tensors: 1997–2000. Phys. Earth Planet. In. 130:71–101

Presti D, Billi A, Orecchio B, Totaro C, Faccenna C, Neri G (2013) Earthquake focal mechanisms, seismogenic stress, and seismotectonics of the Calabrian Arc. Italy. Tectonophysics 602:153–175. https://doi.org/10.1016/j.tecto.2013.01.030

Robertson E (2008) Seismic moment tensor solutions from GeoNet data to provide a moment magnitude scale for New Zealand. GNS Science. EQC. http://www.eqc.govt.nz/research/research-papers/seismic-moment-tensor-solutions-geonet-magnitude-scale-for-nz. Accessed Nov 2017

Sançar T, Akyüz HS (2014) Paleoseismology of the Ilıpınar Segment (Karlıova, Bingöl), The North Anatolian Fault Zone, Türkiye Jeoloji Bülteni 2(57) (in Turkish)

Seymen I, Aydin A (1976) Bingöl Deprem Fayı ve Bunun Kuzey Anadolu Fay Zonu ile İlişkisi (The Bingöl Earthquake fault and its relation to the North Anatolian Fault): MTA Dergisi 79:1–8

Sokos E, Zahradnik J (2006) A MATLAB GUI for Use with ISOLA fortran codes user's guide. http://seismo.geology.upatras.gr/isola/. Accessed Nov 2017

Sokos E, Zahradník J (2008) ISOLA a Fortran code and a Matlab GUI to perform multiple-point source inversion of seismic data. Comput Geosci 34:967–977. https://doi.org/10.1016/j.cageo.2007.07.005

Sokos E, Zahradník J (2013) Evaluating centroid-moment tensor uncertainty in the new version of ISOLA software. Seis Res Lett 84:656–665. http://dx.doi.org/ https://doi.org/10.1785/0220130002

Şengör AMC, Tüysüz O, İmren C, Sakınç M, Eyidoğan H, Görür N, Le Pichon X, Rangin C (2005) The North Anatolian fault: a new look. Annu Rev Earth Planet Sci 33:37–112. https://doi.org/10.1146/annurev.earth.32.101802.120415

Taymaz T, Eyidogan H, Jackson J (1991) Source parameters of large earthquakes in the East Anatolian fault zone (Turkey). Geophys J Int 106(3):537–550

Türkelli N, Sandvol E, Zor E, Gök R, Bekler T, Al-Lazki A, Karabulut H, Kuleli S, Eken T, Gürbüz C, Bayraktutan S, Seber D, Barazangi M (2003) Seismogenic Zones in Eastern Turkey. Geophys Res Lett 30(24):8039. https://doi.org/10.1029/2003gl18023

Wald DJ, Quitoriano V, Heaton TH, Kanamori H, Scrivner CW, Worden CB (1999) TriNet ShakeMaps: rapid generation of instrumental ground motion and intensity maps for earthquakes in southern California. Earthq. Spectra 15:537–555

Wallace RE (1968) Earthquake of August 19, 1966, Varto Area, eastern Turkey. Bull Seismol Soc Am 58(1):11–45

Wessel P, Smith WHF, Scharroo R, Luis J, Wobbe F (2013) Generic mapping tools improved version released. EOS Trans Am Geophys Union 94:409–410. https://doi.org/10.1002/2013EO450001

Yılmazer M (2012) zSacWin: a rapid earthquake processing and archiving system, user guide v1.0. December 2012. Supported by the research fund of the Boğaziçi University Project Number 5725 P

Zabcı C, Akyüz HS, Sançar T (2017) Paleoseismic history of the eastern part of the North Anatolian Fault (Erzincan. Implications for the seismicity of the Yedisu Seismic Gap. J Seismol, Turkey). https://doi.org/10.1007/s10950-017-9673-1

Zoback ML (1992) First- and second-order patterns of stress in the lithosphere: the world stress map project. J Geophys Res 97:11703–11728

An Overview of the Seismicity and Tectonics of the Black Sea

Doğan Kalafat

Keywords Black sea · Oceanic basins · Seismicity · Focal mechanisms · Stress analysis

1 Introduction

In the past few decades, the Black Sea has been the subject of intense geological and geophysical studies, including deep seismic sounding, reflection profiling, gravity and magnetic surveys for scientific and petroleum exploration purposes (Nikishin et al. 2010; Soson et al. 2010; Stephenson and Schellart 2010).

The region is controlled by the collision of the Arabian and Eurasian plates of mid-Miocen age. This type of thrust generation as a result of compressional regime extends east-west.

Avagyan and Sosson 2010 deal with the recent tectonic stress evolution in the area, especially the Lesser Caucasus and adjacent regions. The present stress field, derived from the kinematics of active faults, corresponds to a strike–slip regime with both transtensional and transpressional characteristics. However, both the micro fault kinematic data and the distribution of Neogene to Quaternary volcanic clusters demonstrate changes of the stress field orientation through time since the beginning of the Arabia–Eurasian collision. A NW–SE orientation of compression was dominant between the Palaeogene and the late Early Miocene, a NE–SW orientation prevailed between the Late Miocene and the Quaternary.

According to previous GPS studies, the shortening rate axes are almost perfectly oriented at 45° to the fault plane for right lateral slip on the North Anatolian Fault (NAF). Shortening axes are normal to the thrust structures in the Caucasus (Allmendinger et al. 2007; McClusky et al. 2000; Reilinger and McClusky 2011), as shown in Fig. 1.

D. Kalafat (✉)
Boğaziçi University Kandilli Observatory and ERI, 34684 Çengelköy-Istanbul, Turkey
e-mail: kalafato@boun.edu.tr

S. D'Amico (ed.), *Moment Tensor Solutions*, Springer Natural Hazards,
https://doi.org/10.1007/978-3-319-77359-9_25

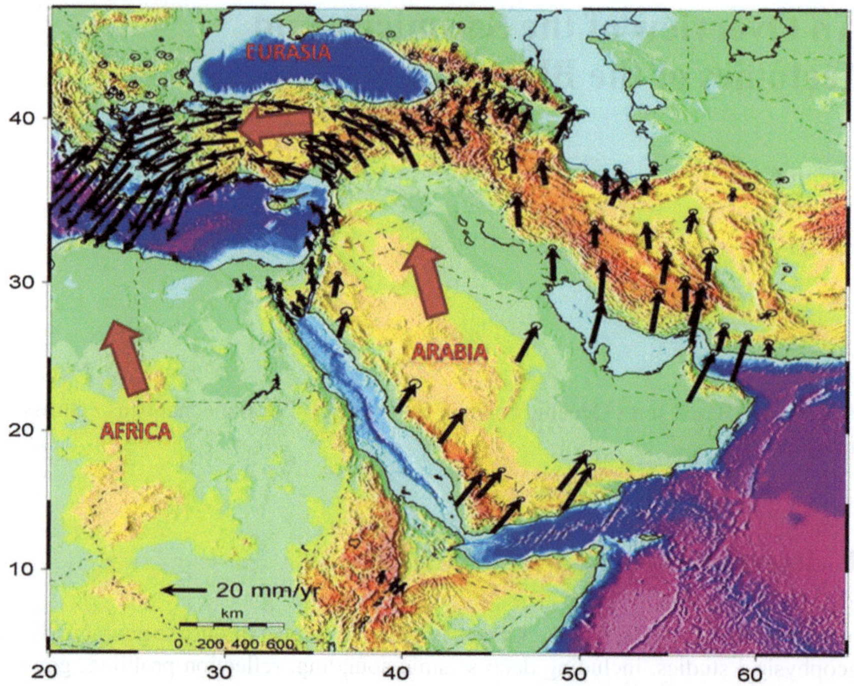

Fig. 1 Main tectonic structures, GPS Velocity distribution and movement directions (Reilinger and McClusky 2011)

The Black Sea, consists of the two oceanic basins separated by mid sea ridges (Fig. 2). There are obvious implications for the geometry of extension and rifting within the western and eastern Black Sea basins and the role of broader plate configuration and kinematics in controlling controlling the structures (Okay et al. 1994; Sosson et al. 2010).

The deepest part of the basins have oceanic crust below a thick cover of sediments, at a depth of about 10 km. The margins and the ridges have continental crusts (Nikishin et al. 2015b; Sosson et al. 2010; Yegorova and Gobarenko 2010; Rangin et al. 2002).

According to the previous studies in the region using multi-channel deep seismic reflection-refraction, gravity and magnetic data, the eastern and western basins with oceanic crust, show the different structural features (Nikishin et al. 2015a; Tüysüz et al. 2012; Munteanu et al. 2011; Dinu et al. 2005; Rangin et al. 2002) (Fig. 3).

In this paper, we present the seismicity and fault mechanisms of earthquakes in and around the margins of the Black Sea.

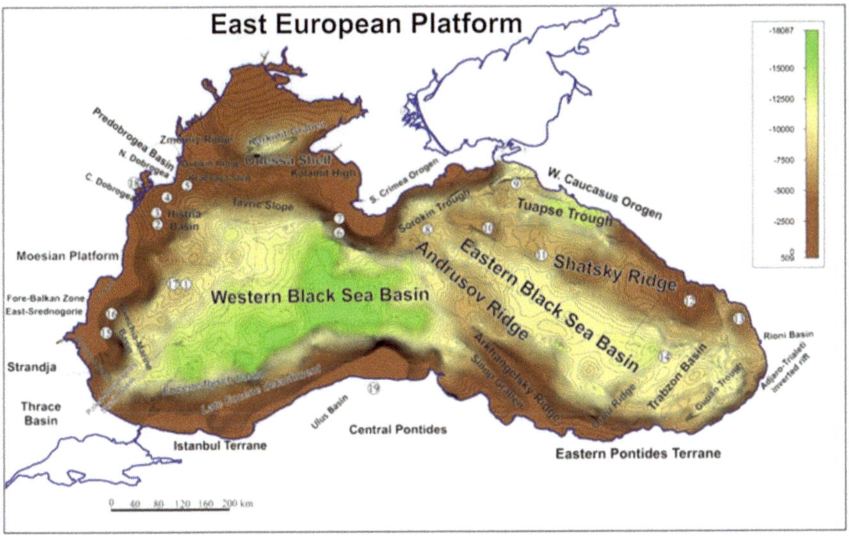

Fig. 2 Basement topography of the Black Sea Basin, and the prominent features (Nikishin et al. 2015a)

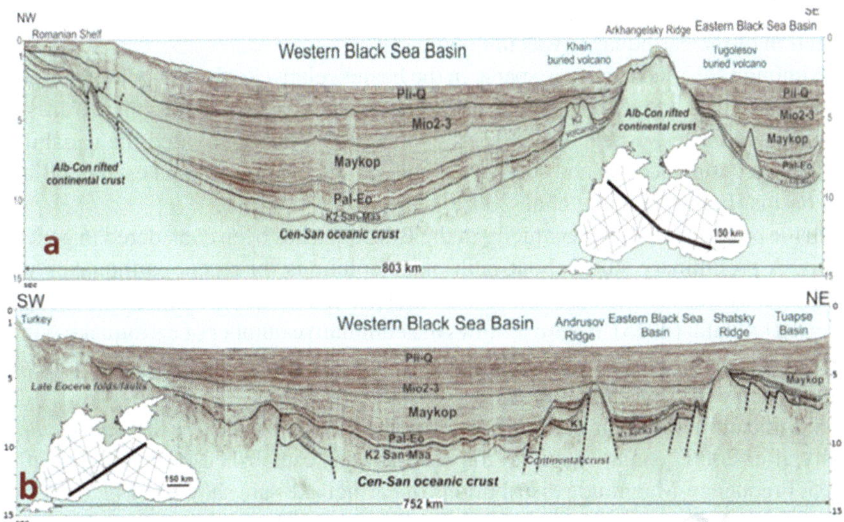

Fig. 3 Seismic section along two seismic lines in the NW (**a**) and NE **b** directions. The sections show the sedimentary layers and basement topography (Nikishin et al. 2015a)

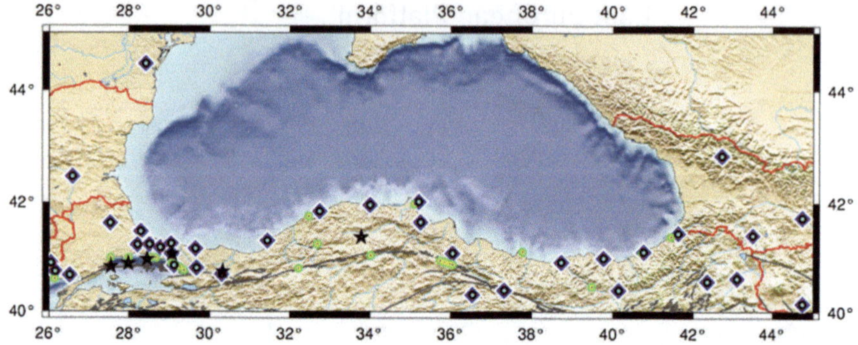

Fig. 4 Seismic stations in the region (diamonds show broad-band BB, Green circles the strong-motion SM), Stars show short period SP-with strong motion SM or BB stations)

2 Seismicity

Previously, the former Soviet Union, Bulgaria, Romania, Turkey have published various earthquake catalogs about seismicity. But because of the lack of sufficient seismic stations in the region, catalogs were not homogeneous and complete. Therefore, the detection of the earthquakes was not complete.

Although seismic activity is sparse in the basins relative to the surrounding region of the Caucasus and Turkey, the broad-band seismic networks established in surrounding countries, especially in Turkey since 2005, have provided the capability for the detection, location and source mechanism studies of earthquakes in the Black Sea basins (Fig. 4) (Kalafat et al. 2005).

In the previous studies, seismicity in the Black Sea has been considered in general terms. A preliminary study about fault plane solutions for recent earthquakes was published by Bayrak et al. (2013). In this study, all the available data in the catalogue are used (Kalafat (2015). Figure 5 shows the cumulative number of earthquakes since 1976. Changes of slope in the curve in Fig. 5 are due to the increased number of stations and detection capabilities.

Significant increase in the number of earthquakes was seen in the region especially after 2006, primarily due to the new stations established in this region after 2005. Frequency-Magnitude distributions of earthquakes are shown in Fig. 6 for two different time intervals: 1900–1976 and 1976–2015. The recent period shows the great improvement in detection, with the magnitude of completeness $M_L = 2.7$ in the Black Sea region. For our study we used earthquakes $M \geq 3.0$.

Frequency-Magnitude distribution was analyzed by two different methods. Maximum Likelihood (MLM; Wiemer 2001; Wiemer and Schorlemmer 2007) and Entire Magnitude Range (EMR; Woessner and Wiemer 2005), and are shown in Fig. 6. The results were similar.

There are a significant number of earthquakes in the Black Sea, mostly of magnitude $M_w = 4.0$ or smaller (Fig. 7).

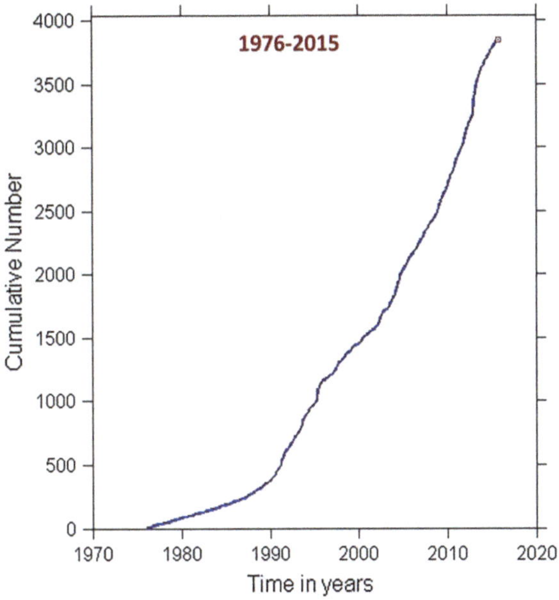

Fig. 5 Number of earthquake occurrence—time correlation (distribution of the earthquakes by years between 1976 and 2015)

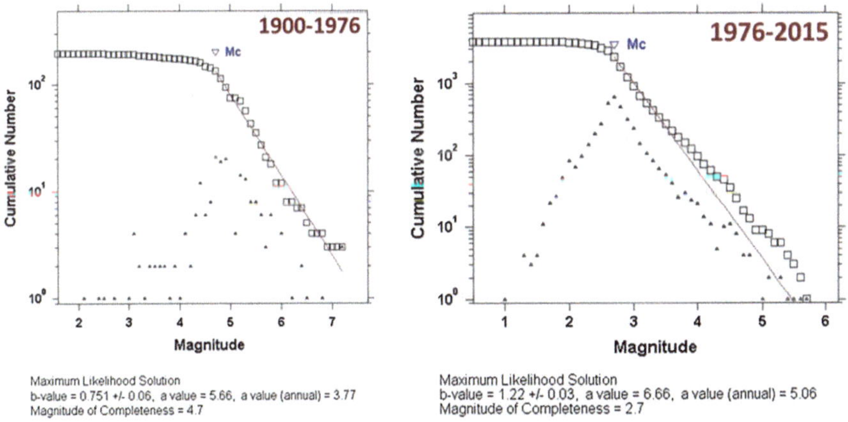

Fig. 6 Frequency-Magnitude relation for two different time periods and detection threshold (Mc)

The seismicity increases toward the margins, with the largest events at the margins. The seismic activity in the region is greater offshore of Georgia, along the basin edges between Crimea and Russia, Bulgaria-Romania, the Bulgaria-Turkey-Trachea borderline and the middle part of Southern Black Sea—offshore Turkey. The Western and Eastern basin of the Black Sea—inner basin have very few earthquakes. The

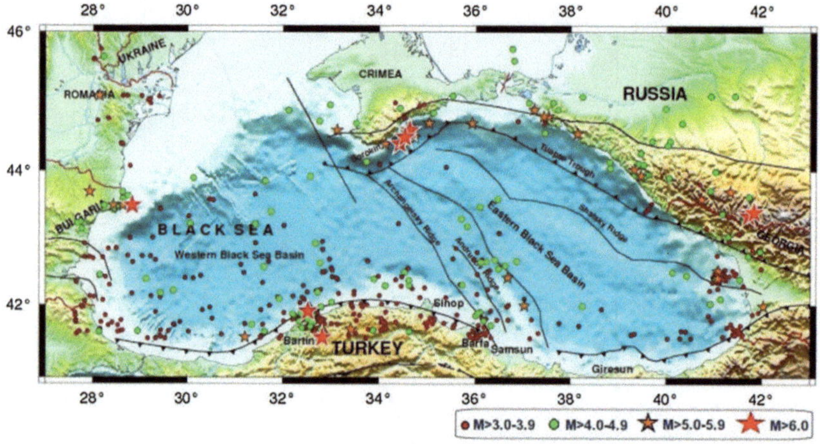

Fig. 7 Recent seismic activity of the Black Sea region (red stars show M $\geq$ 6.0; orange stars show M $\geq$ 5.0; green circles show M $\geq$ 4.0 and red circles show M $\geq$ 3.0 earthquakes in the period of 1900–2015)

seismicity pattern in the northeastern Turkey and the Caucasus corresponds to the plate interactions.

3 Source Mechanism

Using the data from broad-band stations, we calculated CMT for earthquakes with magnitudes higher than M = 3.7, which occured between 2006 and 2015 (Fig. 8 and Table 1). We used regionally recorded broad-band velocity waveforms for CMT solutions. Usually we calculated moment tensors with Moment Tensor Inversion Techniques (Dreger 2002; Sokos and Zaharadnik 2013). We used the earthquakes recorded by at least 4 digital broad-band seismic stations. The earthquakes between 1968 and 2006 we used global data or published results.

The focal mechanisms indicate primarily N-S compression with some E-W component. The mechanisms are consistent with plate motions, where westward motion of the Anatolian Plate and N-S deformation of the Caucasus take up most of the motion of the Arabian Plate. Only small motions (about 1 mm per year) are transmitted through the Pontides, and the Black Sea is being compressed in an N-S direction.

The fault plane solutions of the earthquakes occurred in the Black Sea especially, in recent 6 years, generally have oblique strike-slip faulting in northeastern part (Bulgaria-Romania borderline offshore waters). After 2008 a series of important earthquakes occurred especially in the southern coasts of the Black Sea and offshore Georgia.

Table 1 The fault plane solutions in the Black Sea region

No.	Date			O. Time		Coordinates		Depth	Mag.	Fault parameters			P axis		T axis		Region	Sources
	Day	Mon.	Year	Hour	Min.	Lat.	lon.	h-km	Mw	Strike	Dip	Rake	Azimuth	Plunge	Azimuth	Plunge		
1	3	9	1968	8	19	41.78	32.43	5	6.0	28.0	38.0	80.0	349.4	27.8	232.8	40.4	Offshore Amasra-Bartın Black Sea (southern part)	25
2	3	9	1978	0	21	43.65	38.04	15	5.7	117.0	78.0	89.0	207.8	33.0	25.7	57.0	Offshore Russia (northern part)	HRV
3	9	11	2002	2	18	44.77	37.21	15	5.5	136.0	90.0	−87.0	52.0	44.7	−140.0	44.7	Offshore Russia (northern part)	HRV
4	1	11	1986	3	18	41.24	40.21	29	4.8	301.0	87.0	171.0	346.5	4.2	255.9	8.5	Offshore Rize -Black Sea (southern part)	36
5	19	5	2006	23	1	42.63	35.98	15	4.5	145.0	71.0	56.0	259.7	18.9	15.4	51.7	Black Sea (central part)	KT

(continued)

Table 1 (continued)

No.	Date			O. Time		Coordinates		Depth	Mag.	Fault parameters			P axis		T axis		Region	Sources
	Day	Mon.	Year	Hour	Min.	Lat.	lon.	h-km	Mw	Strike	Dip	Rake	Azimuth	Plunge	Azimuth	Plunge		
6	20	6	2008	8	23	41.09	38.91	5	4.3	52.0	84.0	42.0	176.9	23.4	283.2	33.0	Offshore Tirebolu-Giresun (southern part)	KT
7	5	7	2008	8	0	45.51	30.96	17	4.9	45.0	42.0	59.0	336.5	6.9	228.6	68.6	Offshore Odeea-Ukraine-Crimea (southern part)	HRV
8	9	8	2008	1	53	42.71	32.50	29	3.7	61.0	51.0	−88.0	344.2	83.8	149.6	6.0	Black Sea (central part)	KT
9	16	3	2009	16	48	41.62	37.64	20	3.7	5.0	84.0	−136.0	237.0	34.2	129.0	24.5	Black Sea (southeastern part)	KT
10	5	8	2009	7	49	43.42	28.60	20	5.0	211.0	65.0	171.0	75.6	11.4	170.6	23.5	Offshore Bulgaria-Black Sea (western part)	HRV

(continued)

Table 1 (continued)

No.	Date			O. Time		Coordinates		Depth	Mag.	Fault parameters			P axis		T axis		Region	Sources
	Day	Mon.	Year	Hour	Min.	Lat.	lon.	h-km	Mw	Strike	Dip	Rake	Azimuth	Plunge	Azimuth	Plunge		
11	17	3	2011	2	13	43.42	36.33	56	3.9	71.1	72.5	157.6	120.0	2.3	28.8	27.7	Black Sea (central part)	KT
12	29	11	2011	2	17	44.53	37.07	34	4.1	350.1	77.3	128.9	51.3	22.1	298.2	44.0	Offshore Russia-Black Sea (northern part)	KT
13	23	12	2012	13	31	42.46	41.02	16	5.8	214.0	76.0	177.0	78.5	7.8	170.2	11.9	Offshore Georgia-Black Sea (eastern part)	HRV
14	25	12	2012	22	44	42.51	40.95	16	5.4	302.0	85.0	4.0	256.8	0.7	166.8	6.4	Offshore Georgia-Black Sea (eastern part)	HRV
15	25	12	2012	22	55	42.39	40.93	14	4.7	353.0	82.0	168.0	39.2	2.7	308.5	14.1	Offshore Georgia-Black Sea (eastern part)	KT

(continued)

Table 1 (continued)

No.	Date			O. Time		Coordinates		Depth	Mag.	Fault parameters			P axis		T axis		Region	Sources
	Day	Mon.	Year	Hour	Min.	Lat.	lon.	h-km	Mw	Strike	Dip	Rake	Azimuth	Plunge	Azimuth	Plunge		
16	7	3	2013	21	50	40.83	36.24	10	3.9	98.7	88.7	131.5	155.5	30.9	43.1	32.5	Taşova-Sepetli-Amasya	KT
17	19	3	2013	12	44	42.00	29.49	19	4.3	196.0	59.0	117.0	267.0	10.1	154.7	64.7	Black Sea (southwestern part)	KT
18	26	3	2013	23	35	43.25	41.57	18	4.9	285.0	32.0	72.0	208.0	14.1	63.2	72.9	Offshore Georgia (eastern part)	HRV
19	28	5	2013	0	9	43.28	41.65	12	5.2	294.0	38.0	116.0	185.7	9.6	305.9	71.5	Offshore Georgia (eastern part)	HRV
20	21	7	2013	21	43	41.13	33.49	6	4.0	317.6	86.0	−25.4	271.1	20.3	6.6	14.4	İhsangazi-Kastamonu	KT
21	24	11	2013	20	49	40.78	31.88	8	4.9	51.0	82.0	143.0	104.4	18.8	2.4	31.3	Ulumescit-Bolu	KT
22	2	3	2014	3	34	44.33	34.36	57	4.2	185.0	75.0	33.0	312.1	10.9	49.5	33.7	Offshore Crimea-Black Sea (northern part)	KT

(continued)

Table 1 (continued)

No.	Date			O. Time		Coordinates		Depth	Mag.	Fault parameters			P axis		T axis		Region	Sources
	Day	Mon.	Year	Hour	Min.	Lat.	lon.	h-km	Mw	Strike	Dip	Rake	Azimuth	Plunge	Azimuth	Plunge		
23	4	9	2014	3	53	41.66	32.49	4	3.9	80.0	74.3	98.7	162.8	28.4	2.8	60.0	Kayadibi-Bartın	KT
24	11	7	2014	22	14	41.77	35.92	13	3.9	62.0	72.0	73.0	165.2	25.1	308.0	59.5	Offshore Fener-Bafra-Samsun (southern part)	KT
25	20	7	2014	14	7	41.76	35.92	18	3.7	215.0	50.0	91.0	304.3	5.0	133.0	84.9	Offshore Fener-Bafra-Samsun (southern part)	KT
26	20	9	2014	23	24	41.77	35.91	24	3.7	98.0	69.0	161.0	326.0	2.2	57.2	27.9	Offshore Fener-Bafra-Samsun (southern part)	KT

(continued)

Table 1 (continued)

No.	Date			O. Time		Coordinates		Depth	Mag.	Fault parameters			P axis		T axis		Region	Sources
	Day	Mon.	Year	Hour	Min.	Lat.	lon.	h-km	Mw	Strike	Dip	Rake	Azimuth	Plunge	Azimuth	Plunge		
27	18	10	2014	15	34	42.53	35.40	16	3.7	176.0	39.0	83.0	9.0	6.2	306.6	82.4	Offshore Sinop-Black Sea (southern part)	KT
28	28	1	2015	3	39	41.36	32.46	8	3.6	33.7	56.5	44.8	154.0	2.0	246.7	53.6	Özbaşı-Bartın	KT
29	8	5	2015	19	14	41.72	33.35	9	4.0	184.0	41.0	36.0	130.3	151.1	18.7	53.7	Başakçay-Azdavay-Kastamonu	KT
30	15	7	2015	8	30	43.33	28.37	30	4.1	197.0	69.0	−37.0	155.0	40.6	−107.8	8.3	Offshore Bulgaria-Balgarevo-Black Sea (eastern part)	KT
31	22	8	2015	5	27	41.30	40.65	17	3.8	275.0	51.0	179.0	133.2	25.8	237.6	27.0	Offshore Pazar-Rize -Black Sea (southern part)	KT

Sources (**25**. Alptekin, Ö., J.N. Nabalek ve M.N. Toksöz (1985). 3 Eylül 1968 Bartın Depreminin Kaynak Mekanizması ve Karadeniz'in Aktif Tektoniği Hakkında Düşünceler, DAB, Sayı 60, 5–38 (in Turkish); **HRV**. Harvard Centroid-Moment Tensor Project CMT, Harvard University, MA, USA (1977–2013); **36**. Kalafat D., 1998. Anadolu'nun Tektonik Yapılarının Deprem Mekanizmaları Açısından irdelenmesi, Deprem Araştırma Bülteni, Sayı 77, 1–217, Temmuz 1998 (in Turkish); **KT**—Kalafat 2015)

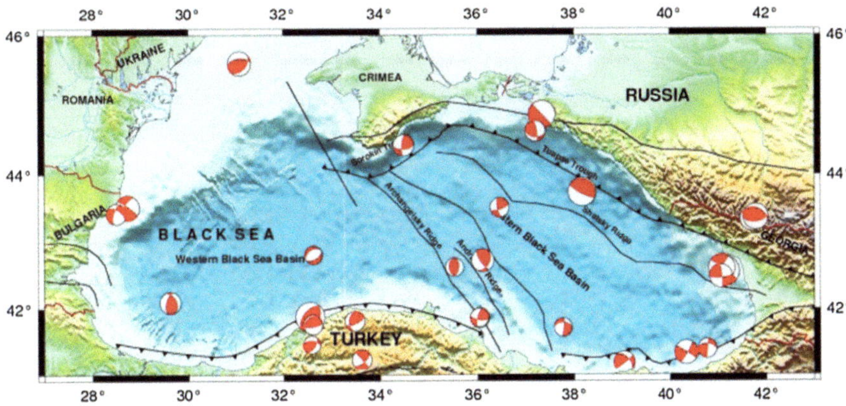

Fig. 8 Fault mechanisms in the region (time period: 1968–2015)

In the southern Black Sea and in the middle part, especially offshore of Bartın, around Kastamonu, offshore of Samsun, the reverse faulting component dominated solutions expressing compressional tectonic regime. The strikes of earthquakes generally lie parallel to the basin. This fault geometry shows that a compressional stress is effective in the region.

The earthquakes in central part of the Black Sea located on faults associated with the ridges.

These earthquakes show that the region is not aseismic and produces earthquakes from time to time, though not very frequently. Offshore of Bafra, earthquake activity started in July 2014 and continued until September 2014 at times causing anxiety in the region.

4 Conclusions

The important earthquakes in the region (Mw > 6.0) generally occurred in southeastern Crimea, the Georgia mainland, offshore of Bulgaria, offshore of Bartın (coast of the Black Sea). Since 2005 no other important earthquakes have occurred apart from those. The depths of the earthquakes were generally between 10 and 35 km.

Largest earthquakes occurred at the boundaries of Black Sea. The mechanism of the large events at the margins indicates and oblique deformation of the region combined with North-South compression and East-West shear.

Stress pattern based on focal mechanism solutions of the earthquakes in the region show that largest compressional axis (P_{max}) is oriented in **NNW-SSE** direction. In the eastern part of Black Sea, especially in offshore Georgia, the largest compression axis (P_{max}) is found to be **NNE-SSW** (Fig. 9).

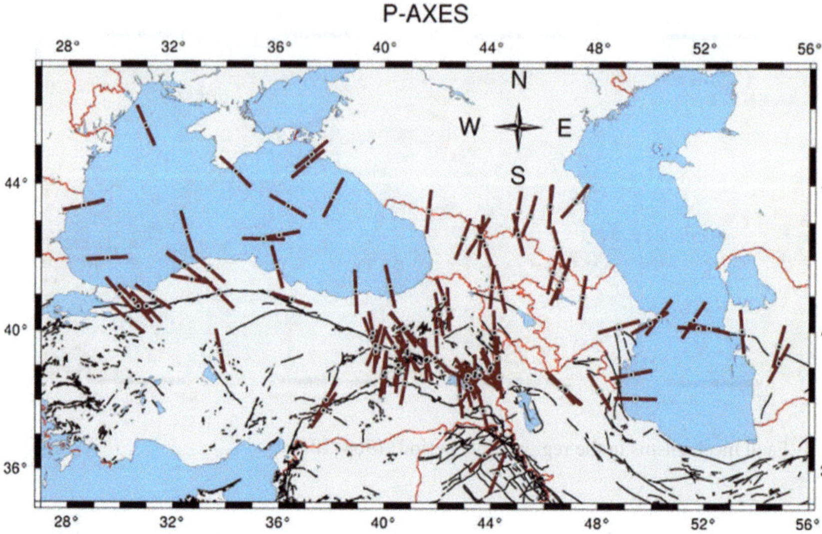

Fig. 9 Stress analysis of the region P_{max} compressional axes

With improved instrumentation many small events have been detected at the shelf margins of the Black Sea. It is most likely that this events are located on the faults formed during the opening of the Black Sea.

The deep basin of the Black sea is relatively aseismic. The earthquakes in the central part of the Black Sea are associated with the ridges oriented in SE-NW directions.

Acknowledgements This study was supported by the Department of Science Fellowship and Grant programs (2014–2219) of TUBITAK (The Scientific and Technological Research Council of Turkey) and by Massachusetts Institute of Technology (MIT) The Earth Resources Laboratory (ERL). I express my thanks and appreciation to Prof.M.Nafi Toksöz for his encouragement and assistance.

References

Allmendinger R, Reilinger R, Loveless J (2007) Strain and rotation rate from GPS in Tibet. Anatolia Altiplano Tectonics 26(TC3013):1–18

Alptekin Ö, Nabalek ve JN, Toksöz MN (1985) 3 Eylül 1968 Bartın DeprememininKaynak Mekanizması ve Karadeniz'in Aktif Tektoniği Hakkında Düşünceler. DAB, Sayı 60:5–38 (in Turkish)

Avagyan A, Sosson M et al (2010) Recent tectonic stress evolution in the Lesser Caucasus and adjacent regions. In: Sosson M, Kaymakci N, Stephenson RA, Bergerat F, Starostenko V (eds) Sedimentary Basin Tectonics from the Black Sea and Caucasus to the Arabian Platform, vol 340. Geological Society, London, Special Publications, pp 393–408

Bayrak Y, Irmak TS, Doğan B, Çınar H, Yılmaz Ş, Kalafat D, Karakaş A, Bayrak E (2013) A Preliminary results of fault plane solutions of the recent earthquakes in and around Eastern Black Sea region. In: 7th Congress of Balkan Geophysical Society-Tirana, Albania, vol 18669, pp 7–10 October 2013

Dinu C, Wong HK, Tambrea D, Matenco L (2005) Stratigraphic and structural characteristics of the Romanian Black Sea shelf. Tectonophysics 410:417–435

Dreger D (2002) Manual of the time- domain moment tensor inverse code (TDMT_INVC), Release 1.1. Berkeley Seismology Laboratory, p 18

HRV (1977–2013) Harvard Centroid—Moment tensor project CMT. Harvard University, MA, USA

Kalafat D (1998) Anadolu'nun Tektonik Yapılarının Deprem Mekanizmaları Açısından irdelenmesi. Deprem Araştırma Bülteni Sayı 77:1–217, Temmuz 1998 (in Turkish).

Kalafat D (2015) Statistical Evaluation of Turkey Earthquake Data (1900-2015): a case study. East Anatolian J Sci II(I):14–36.

Kalafat D, Öğütçü Z, Yılmazer M, Suvarıklı M, Güneş Y, Kekovalı K, Geçgel V (2005) Türkiye'de ilk defa gerçekleştirilen uydu bağlantılı genişbantlı alt ağlar: Mavi Ağ (Blue Net), Turuncu Ağ (Orange Net), Beyaz Ağ (White Net) Projeleri, Aktif Araştırma Grubu 9.Toplantısı *ATAG9 Bildiri Özetleri Kitabı*, pp 34–36., Cumhuriyet Üniversitesi MF 22-24 Eylül 2005 Sivas

McClusky S, Balassanian S, Barka A, Demir C, Ergintav S, Georgiev I, Gurkan O, Hamburger M, Hurst K, Kahle H, Kastens K, Kekelidze G, King R, Kotzev V, Lenk O, Mahmoud S, Mishin A, Nadariya M, Ouzounis A, Paradissis D, Peter Y, Prilepin M, Reilinger R, Sanli I, Seeger H, Tealeb A, Toksöz MN, Veiss G (2000) Global positioning system constraints on plate kinematics and dynamics in the eastern Mediterranean and Caucasus. J Geophys Res 105(B3):5695–5719

Munteanu I, Matenco L, Dinu C, Cloetingh S (2011) Kinematics of back-arc inversion of the Western Black Sea Basin. Tectonics 30: 21

Nikishin AM, Ershov AV, Nikishin VA (2010) Geological history of western Caucasus and adjacent foredeeps based on analysis of the regional balanced section. Dokl Earth Sci 430(2):155–157

Nikishin AM, Okay AI, Tüysüz O, Demirer A, Amelin N, Petrov E (2015a) The Black Sea basins structure and history: new model based on new deep penetration regional seismic data. Part 1: basins structure and fill. Mar Pet Geol 59:638–655

Nikishin AM, Okay A, Tüysüz O, Demirer A, Wannier M, Amelin N, Petrov E (2015b) The Black Sea basins structure and history: new model based on new deep penetration regional seismic data. Part 2: Tectonic history and paleogeography. Mar Pet Geol 59:656–670

Okay AI, Şengör AMC, Görür N (1994) Kinematic history of the opening of the Black Sea and its effect on thesurrounding regions. Geology 22:267–270

Rangin C, Bader AG, Pascal G, Ecevitoğlu B, Görür N (2002) Deep structure of the Mid Black Sea High (offshore Turkey) imaged by multi-channel seismic survey (BLACKSIS cruise). Mar Geol 182(2002):265–278

Reilinger R, McClusky S (2011) Nubia-Arabia-Eurasia plate motions and the dynamics of Mediterranean and Middle East tectonics. Geophys J Int 186:971–979

Sokos E, Zahradník J (2013) Evaluating centroid-moment-tensor uncertainty in the new version of ISOLA software. Seismol Res Lett 84:656–665

Sosson M, Kaymakci N, Stephenson R, Bergerat F, Starostenko V (eds) (2010) Sedimentary Basin Tectonics from the Black Sea and Caucasus to the Arabian Platform, vol 340. Geological Society, London, Special Publications

Stephenson R, Schellart WP (2010) The Black Sea back-arc basin: insights to itsorigin from geodynamic models of modern analogues. In: Sosson M, Kaymakci N, Stephenson RA, Bergerat F, Starostenko V (eds) Sedimentary Basin Tectonics from the Black Sea and Caucasus to the Arabian Platform, vol 340. Geological Society, London, Special Publications, pp 11–21

Wiemer S (2001) A software package to analyze seismicity: ZMAP. Seismol Res Lett 72:373–382

Wiemer S, Schorlemmer D (2007) ALM: An asperity—based likelihood model for California. Seismol Res Lett 78:134–140

Woessner J, Wiemer S (2005) Assessing the quality of earthquake catalogues: estimating the magnitude of completeness and its uncertainty. Bull Seism Soc Am 95(2):684–698

Yegorova T, Gobarenko V (2010) The crustal structure of the Black Sea from the reinterpretation of deep seismic sounding data acquired in the 1960s, vol 340. Geological Society, London, Special Publications, pp 43–56

Coulomb Stress Changes in the Area of December 2013–January 2014 Sannio-Matese Seismic Sequence (Southern Italy)

Santanu Baruah and Sebastiano D'Amico

Keywords Southern Italy · Coulomb stress · Focal mechanism

1 Introduction

The Italian Apennines are seat of extensional deformation, concentrated along the inner part of the mountain belt due to the opening of the Tyrrhenian back-arc basin during the late Miocene and the following rolling back subduction of Adriatic plate with an extension velocity of about 3 mm year^{-1} (D'Agostino et al. 2008; Faccenna et al. 2004). Apennines chain is a zone of high seismic hazard (http://zonesismiche.mi.ingv.it; "Mappa di pericolositàsismica del territorionazionale"; D'Amico et al. 2013a) and it has been affected by a number of earthquakes in the past century suffering intensity X or higher several times in the past centuries (Boschi et al. 2000; CPTI Working Group 2004). The most recent examples are the 1980, $M = 6.9$, Irpinia events (Pino et al. 2008; Secomandi et al. 2013); the 1997–1998 Umbria-Marche (Caccamo et al. 2007) and the 2009 L'Aquila (D'Amico et al. 2010a, 2013b) sequences.

On December 29, 2013 (17:08 UTC) a moderate magnitude earthquake (ML 4.9) struck the Sannio-Matese area in the Central-Southern Apennines. It produced damage in the epicentral area corresponding to a maximum intensity of V–VI on

S. Baruah
CSIR-North East Institute of Science and Technology (Govt. of India), Jorhat 785006, Assam, India

S. D'Amico (✉)
Department of Geosciences, University of Malta, Msida Campus, Msida 2080, MSD, Malta
e-mail: sebastiano.damico@um.edu.mt

S. D'Amico (ed.), *Moment Tensor Solutions*, Springer Natural Hazards,
https://doi.org/10.1007/978-3-319-77359-9_26

the European Macroseismic Scale; a lot of people were evacuated from their homes and emergency actions have been carried out by the Civil Protection authorities. This event can be considered as the mainshock of a seismic sequence with several smaller events that took place in the area for about one month. In the following weeks 31 aftershocks, with local magnitude greater than or equal to 2.5, and more than 250 aftershocks with smaller magnitude were recorded by INGV permanent seismic network.

The main goal of this paper is to investigate Coulomb stress changes in the area of December 2013–January 2014 Sannio-Matese seismic sequence (Fig. 1). For the seismic sequence a detailed moment tensor catalogue has been computed by performing broadband waveform inversion using the Cut And Paste (CAP) method (D'Amico et al. 2014). The main advantage of the CAP method (Zhao and Helmberger 1994; Zhu and Helmberger 1996) is that each waveform is broken up into Pnl and surface wave segments, which are weighted differently during the inversion procedure (see D'Amico et al. 2010a, 2011, 2012, 2013b; D'Amico 2014 for details) (Fig. 2).

2 Moment Tensor Solutions in the Local Seismotectonic Framework and Coulomb Stress Changes

The tectonic framework of the Apennines is the result of geodynamic processes related to the Neogene–Quaternary convergence between Africa and Eurasia (e.g. Dewey et al. 1989), coupled to a rapid roll-back of the Adriatic–Ionian slab (Malinverno and Ryan 1986; Royden et al. 1987; Doglioni 1991; Gueguen et al. 1998; Faccenna et al. 2001). According to Patacca et al. (1990) the thrusting is presently active on the Adriatic side of the Northern Apennines, whereas in the Central-Southern Apennines held during Late Pliocene–Early Pleistocene. Events of magnitude up to 7 characterize the entire Apennine chain, mainly occurring along NW–SE striking normal faults, that move in response to a prevailing NE–SW extension of the minimum horizontal stress (Montone et al. 2004; Pondrelli et al. 2006). In the northern most portion of the Apennines the most recent events occurred in 2012 and 2009; on May 2012, the Emilia Romagna region was hit by an earthquake of Mw 6.1 (Caporali and Ostini 2012), and in April 2009 a Mw of 6.3 occurred in Central Apennines destroying the city of L'Aquila (D'Amico et al. 2010b).

The sectors of the Southern Apennines chain are characterized by an important uplift (7 mm/year) coupled to a predominant NE–SW extension (Westaway 1993; Cinque et al. 1993), thus NW–SE striking normal faults developed in response to this extension oriented approximately parallel to the chain axis. In general, the Southern Apennines area is characterized by strong and frequent earthquakes as remarked by the historical events of Mw $\approx$ 7 occurred in Irpinia (1694), Sannio (1688), Basilicata (1857), and by the Medieval seismic sequences of 1349 and 1456. The largest instrumental event in the area (Irpinia, 1980 Ms = 6.9) showed a fracture process linked

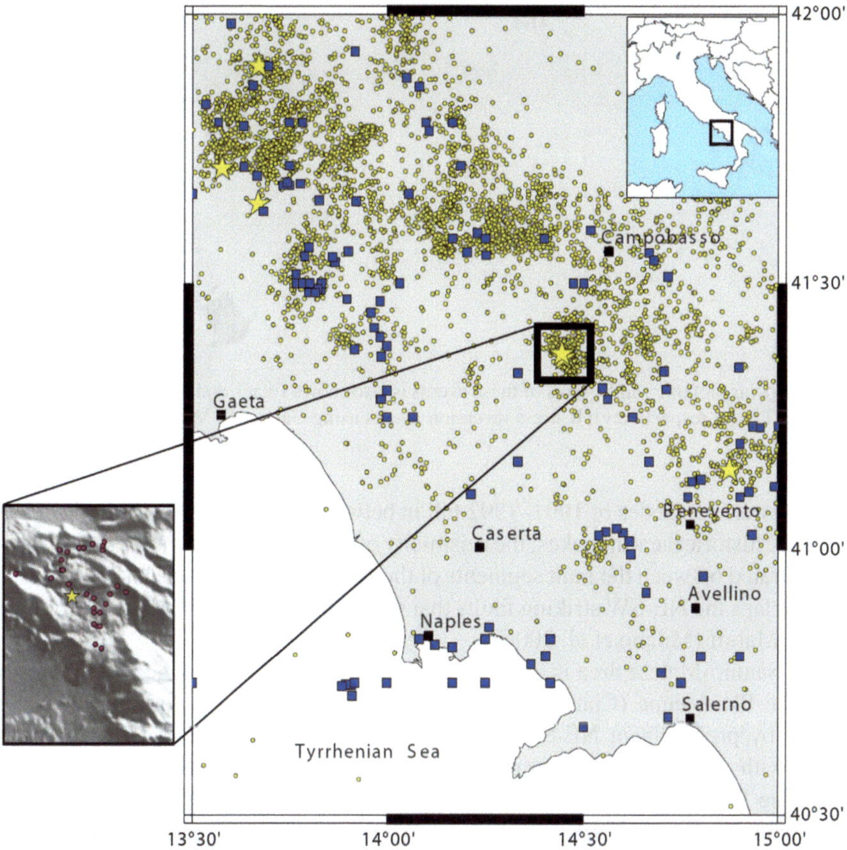

Fig. 1 Map of the study area showing the background seismicity. The distribution of epicenters of the seismic events (2005–2011) is reported. Yellow stars and blue squares represent events having magnitude greater than four and historical earthquakes (intensity at the epicentre greater that VIII), respectively; the yellow circles are the epicentres of minor earthquakes. Inset in lower left shows the location of the events related to December 2013–January 2014, Matese earthquake sequence for which focal solutions are available (D'Amico et al. 2014)

to a NW–SE normal faulting with a complex aftershocks distribution (Milano et al. 2006 and references therein).

In the Sannio-Matese area, five large events with $I_{MCS} > X$ occurred in 1456, 1688, 1702, 1735 and 1805, causing extensive damage. During the historical time, strong earthquakes, seismic swarms and sequences of tectonic origin (lasting from a few days up to several months) struck the area; seismic sequences of low magnitude also occurred in 1885, 1903, 1905 and, recently, in 1990 and in 1997–1998 (Milano et al. 2006). The epicentral distribution of the 1990 and 1997–1998 seismic sequences, the swarm episodes of the 1991–1992 and the location of the isolated events, disclose NE-SW striking active faults (Milano et al. 2006). In particular, the 1990 sequence

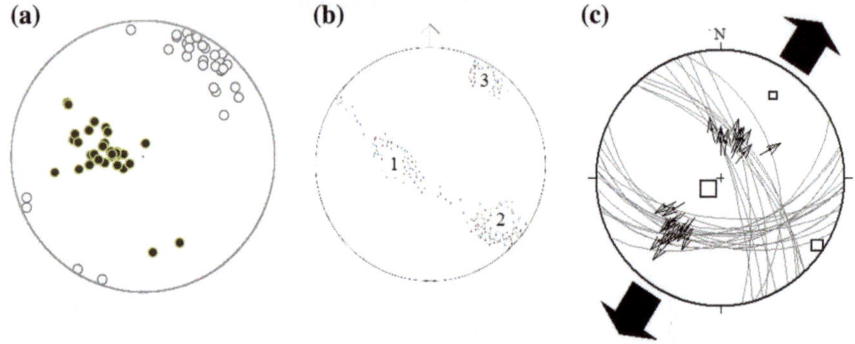

Fig. 2 **a** polar plot of P- and T-axes of these events is also shown (black and white dots correspond to P- and T-axes, respectively); **b** stress inversion results using Gauss and Micheal method (**c**)

and the swarm episodes of 1991–1992 fall in between the fault segments of the 1688 and 1732 historical earthquakes; the seismicity of the 1997–1998 seismic sequence is concentrated between the fault segments of the 1688 and 1805 historical earthquakes and overlaps the NE-SW striking faults that separate the Sannio Mountains from the Matese Massif (Milano et al. 2006).

The Sannio-Matese area is a part of the Apenninic Chain characterized by uplift since the Pleistocene (Ciaranfi et al. 1983) with a prevalent deformation regime created by predominant NE-SW and subordinate NW-SE extensional long lineaments with Apenninic/anti-Apenninic directions (NW-SE/NE-SW) (Patacca and Scandone Patacca 1989). The Matese Massif is a carbonate structure intersected by a main system of WNW-ESE trending tectonic lineaments and subordinate N-S and E-W trending structures. The Sannio Mountains, located to the east of the Matese Massif, represent the sector of the Apenninic Chain that slopes down to the Bradanic Foredeep. These mountains are mainly formed by terrigenous deposits cut by minor normal faults. The Sannio Mountains and the Matese Massif are divided by a N-S to NW-SE elongated morphological depression. Figure 3 reports all the moment tensor solutions derived for the Sannio-Matese seismic sequence recently compiled (D'Amico et al. 2014). On the whole the solutions feature mainly a normal fault mechanism well matching with the local seismotectonic setting and the regional stress tensor indicates good levels of stress homogeneity and solution constraint.

Individual fault-plane solutions may not represent precise regional tectonic stress pattern. In such situations, stress inversion techniques using a number of fault-plane solutions estimate the regional stress system much more precisely, especially in seismically active zones (e.g. Angelier 2002; Angelier and Baruah 2009; D'Amico et al. 2010b, 2012, 2013b; Baruah et al. 2013). In this work, we used two methods, the Michael method (1984, 1987) and the Gauss method (Zalohar and Vrabec 2007), for the inversion analysis. For stress tensor inversion, we used the inversion algorithm of Michael (1984, 1987) (hereafter Michael method) and Zalohar and

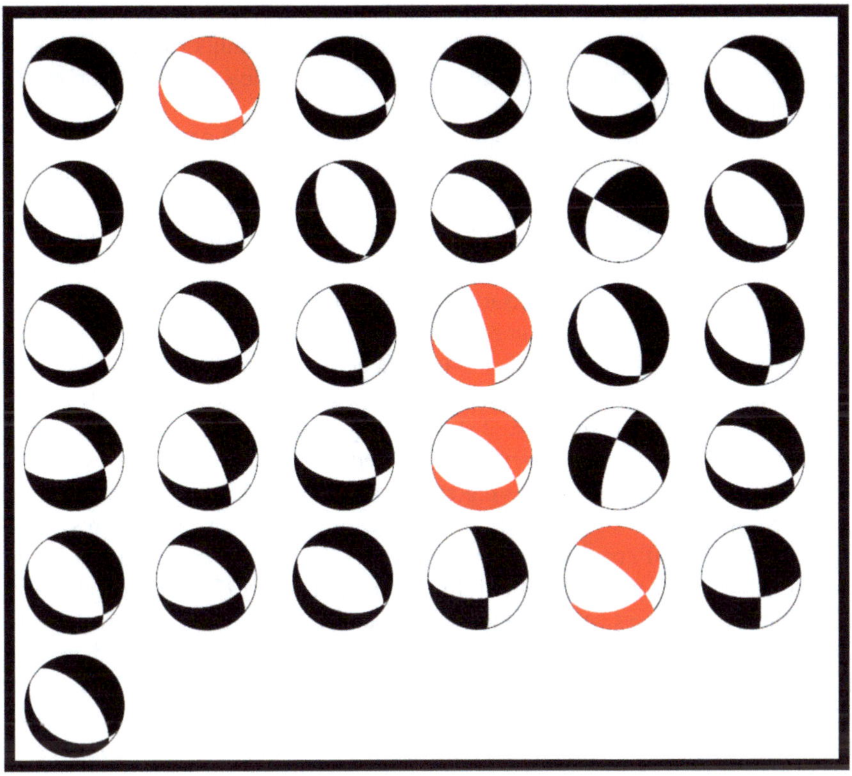

Fig. 3 Focal mechanisms estimated by using CAP waveform inversion method (refdrawn from D'Amico et al. 2014)

Vrabec (2007) (hereafter Gauss method). The Michael method uses the statistical technique of bootstrap resampling and allows determination of the three principal stresses (s1 = maximum, s2 = intermediate and s3 = minimum) as well as a relative stress magnitude R = (s2 − s3)/(s1 − s3), 0 < R < 1. These parameters are determined by finding the best-fitting stress tensor to the observed focal mechanisms under the assumption that:

- stress is uniform in each zone of investigation during the observed time interval;
- the earthquakes are shear-dislocations on pre-existing faults;
- slip occurs in the direction of the resolved shear stress on the fault plane.

On the other hand, the Gauss method involves the concept of the best-fit stress tensor. Here in this algorithm, a new function called "compatibility function" is defined as a Gaussian function that depends on the compatibility measure taking into account both angular misfit between the resolved shear stress and actual direction of movement on the fault plane, and the ratio between the normal and shear stress on the fault plane. The detailed description can be found in Zalohar and Vrabec

(2007). As in such studies confidence level is intended to give the assurance whether the statistical model is correct or not, the stress inversion results which satisfy 95% confidence level have been considered. Figure 3 shows the stress inversion results obtained in this study.

The Coulomb static stress change ΔS_f using Coulomb 3.3 software (http://earthquake.usgs.gov/research/modeling/coulomb/) has been used for the computation of the coulomb stress pattern. We use the moment tensor solution of the main shock computed by D'Amico et al. (2014) as the causative fault plane for the earthquake having strike = 317°, dip = 64°, rake = −7 ° and Mw = 4.9. We present here the map of Coulomb stress change at a depth of 12 km and the regional stress pattern of as shown in Fig. 4. The Coulomb stress change is defined as $\Delta S_f = \Delta\tau - \mu'\Delta\sigma$ (e.g. King et al. 1994; Stein et al. 1994; Toda et al. 1998), where $\Delta\tau$ is the shear stress change resolved on a given failure plane (assumed to be positive in the direction of fault slip), $\Delta\sigma$ is the normal stress change, positive if compressive, and μ' is the effective coefficient of friction, calculated from $\mu' = \mu(1 - B)$, where, B is the Skempton's coefficient varying between 0 and 1. Skempton's coefficient B is defined as the ratio of the induced pore pressure to the change of stress loading under undrained condition (Skempton 1954). It can also be explained by the compressibility of rock and pore fluid (Brown and Korringa 1975; Zhang et al. 2009). We computed the stress change in an elastic half space due to slip on a rectangular fault at the respective depth assuming a shear modulus of 3.32×10^{10} dyne-cm^2, a Possion's ratio of 0.25 and an effective coefficient of friction of 0.4 (Stein et al. 1994). Failure is encouraged if ΔS_f is positive, and discouraged if it is negative. Thus, aftershocks are expected to occur in the zone of increased ΔS_f, which is the case for all four felt aftershocks. We expect most of the smaller and felt stronger aftershocks within this increased ΔS_f zone. Figure 4 shows the coulomb stress change in the area. Mapped lobes of increased and decreased Coulomb stress around a fault exhibit increased and decreased rates of seismicity for all these earthquakes. It is found that the aftershocks were occurred in the region of maximum stress released. The correlation between the increase in static Coulomb stress and aftershocks activity has thus provided the strongest evidence that stress changes promote seismicity.

3 Concluding Remarks

Earthquake source parameters play a key role in several seismological observations and represent a key information in understanding the geodynamic processes both a local and regional scale. Moment tensor solutions provide the necessary source parameters to constrain regional seismo-tectonic deformations and the stress field. In this paper, we used 31 moment tensor solutions (D'Amico et al. 2014) in order to obtain the regional stress tensor and the coulomb stress change in the study area.

The better understanding of earthquake source properties and the local stress variations can contribute to an improved seismic hazard evaluation. Indeed, as is well known, high population density and building inadequacy can make moderate

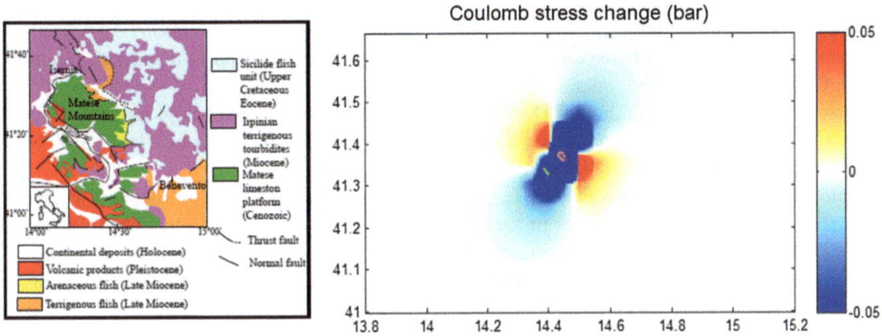

Fig. 4 C stress change (right panel) computed for the study area. The left panel Geological sketch of the Maltese–Benevento area redrawn from D'Amico et al. (2014)

and intermediate magnitude earthquakes the cause of death, injury, and significant property damages.

Acknowledgements D'Amico S. was partially supported by University of Malta through the research project "*Earthquake source mechanisms in the Central Mediterranean in relation to active tectonics*". This research was carried out using computational facilities procured through the European Regional Development Fund, Project ERDF-080 "A supercomputing laboratory for the University of Malta" (http://www.um.edu.mt/research/scienceeng/erdf_080). The authors thank the Istituto Nazionale di Geofisica e Vulcanologia (Italy) for providing high quality data. Some Figures were created using the Generic Mapping Tools (GMT) by Wessel and Smith (1991).

References

Angelier J (2002) Inversion of earthquake focal mechanisms to obtain the seismotectonic stress, IV: a new method free of choice among nodal planes. Geophys J Int 150:588–609

Angelier J, Baruah S (2009) Seismotectonics in northeast India: a stress analysis of focal mechanism solutions of earthquakes and its kinematic implications. Geophys J Int 178:303–326

Baruah S, Baruah S, Kayal JR (2013) State of tectonic stress in Northeast India and adjoining South Asia region. Bull Seismol Soci Am 103:894–910

Boschi E, Guidoboni E, Ferrari G, Mariotti D, Valensise G, Gasperini P (eds) (2000) Catalogue of strong Italian earthquakes from 461 B.C. to 1997. Ann. Geophys 43:609–868

Brown RJS, Korringa J (1975) On the dependence of the elastic properties of a porous rock on the compressibility of the pore fluid. Geophysics 40:608–616

Caccamo D, D'Amico S, Parrillo F, Barbieri FM, Laganà C (2007) Umbria-Marche sequence (central-Italy): a study about its aftershock sequence. Bollettino di Geofisica Teorica ed Applicata 48:385–398

Caporali A, Ostini L (2012) Analysis of the displacement of geodetic stations during the Emilia seismic sequence of May 2012. Ann Geophys 55:767–772. https://doi.org/10.4401/ag-6115

Ciaranfi N, Guida M, Iaccarino G, Pescatore T, Pieri P, Rapisardi L, Ricchetti G, Sgrosso I, Torre M, Toriorici L, Turco E, Scarpa R, Cuscito M, Guerra I, Iannaccone G, Panza GF, Scandone P (1983) Seismotectonic elements of the Southern Apennines. Boll Soc Geol It 102(2–3):201–222

Cinque A, Patacca E, Scandone P, Tozzi M (1993) Quaternary kinematic evolution of the Southern Apennines. Relationship between surface geological features and deep lithospheric structures. Ann Geofis 36(2):249–260

D'Agostino N, Avallone A, Cheloni D, D'Anastasio E, Mantenuto S, Selvaggi G (2008) Active tectonics of the Adriatic region from GPS and earthquake slip vectors. J Geophys Res 113:B12413. https://doi.org/10.1029/2008JB005860

D'Amico S, Orecchio B, Presti D, Zhu L, Herrmann RB, Neri G (2010a) Broadband waveform inversion of moderate earthquakes in the Messina Straits, southern Italy. Phys Earth Planet Int 179:97–106. https://doi.org/10.1016/j.pepi.2010.01.012

D'Amico S, Koper KD, Herrmann RB, Akinci A, Malagnini L (2010b) Imaging the rupture of the Mw 6.3 April 6, 2009 L'Aquila, Italy earthquake using backprojection of teleseismic P-waves. Geophys Res Lett 37:L03301. http://dx.doi.org/10.1029/2009GL042156

D'Amico S, Orecchio B, Presti D, Gervasi A, Guerra I, Neri G, Zhu L, Herrmann RB (2011) Testing the stability of moment tensor solutions for small moderate earthquakes in the Calabrian-Peloritan arc region. Boll Geo Teor Appl 52:283–298. https://doi.org/10.4430/bgta0009

D'Amico S, Gervasi A, Guerra I, Neri G, Orecchio B, Presti D, Totaro C (2012) Seismicity and focal mechanisms at the Calabro-Lucanian boundary along the Apennine chain (southern Italy). Rend Online Soc Geol It 21:778–779

D'Amico S, Lombardo G, Panzera F (2013a) Seismicity of the Mediterranean Region and mitigation of earthquake losses. Phys Chem Earth 63:1–2. https://doi.org/10.1016/j.pce.2013.07.001

D'Amico S, Orecchio B, Presti D, Neri G, Wu W-N, Sandu I, Zhu L, Herrmann RB (2013b) Source parameters of small and moderate earthquakes in the area of the 2009 L'Aquila seismic sequence (central Italy). Phys Chem Earth 63:77–91. https://doi.org/10.1016/j.pce.2013.02.005

D'Amico S, Cammarata L, Cangemi M, Cavallaro D, Di Martino R, Firetto Carlino M (2014) Seismic moment tensors and regional stress in the area of the December 2013–January 2014, Matese earthquake sequence (Central Italy). J Geodyn. https://doi.org/10.1016/j.jog.2014.09.008

D'Amico S (2014) Source parameters related to a small earthquake swarm off-shore of Malta (central Mediterranean). Dev Earth Sci 2:8–13

Dewey JF, Helman ML, Turco E, Hutton DHW, Knott SD (1989) Kinematics of the western Mediterranean Alpine Tectonics. Geol Soc Spec Publ 45:265–283. https://doi.org/10.1144/GSL.SP.1989.045.01.15

Doglioni C (1991) A proposal for the kinematic modelling of W-dipping subductions—possible applications to the Tyrrenian-Apennines system. Terra Nova 3:423–434. https://doi.org/10.1111/j.1365-3121

Faccenna C, Piromallo C, Crespo-Blanc A, Jolivet L, Rossetti F (2004) Lateral slab deformation and the origin of the western Mediterranean arcs. Tectonics 23(TC1012):1012–1033

Faccenna C, Becker TW, Lucente FP, Jolivet L, Rossetti F (2001) History of subduction and back-arc extension in the central Mediterranean. Geophys J Int 145:809–820

Gueguen E, Doglioni C, Fernandez M (1998) On the post-25 Ma geodynamic evolution of the western Mediterranean. Tectonophysics 298:259–269

King GCP, Stein RS, Lin J (1994) Static stress changes and the triggering of earthquakes. Bull Seismol Soc Am 84:935–993

Malinverno A, Ryan WBF (1986) Extension in the Tyrrhenian Sea and shortening in the Apennines as result of arc migration driven by sinking of the lithosphere. Tectonics 5(2):227–245

Michael AJ (1984) Determination of stress from slip data: faults and folds. J Geophys Res 89:11517–11526

Michael AJ (1987) Stress rotation during the Coalinga aftershock sequence. J Geophys Res 92:7963–7979

Milano G, Di Giovambattista R, Ventura G (2006) Seismicity and stress field in the Sannio-Matese area. Ann Geophys 49(Supplemento 1):347–356

Montone P, Mariucci MT, Pondrelli S, Amato A (2004) An improved stress map for Italy and surrounding regions (central Mediterranean). J Geophys Res 109(B10410):1–22. https://doi.org/10.1029/1003JB002703

Patacca E, Scandone P (1989) Post-Tortonian mountain building in the Apennines. The role of the passive sinking of a relic lithospheric slab. In: Boriani A, Bonafede M, Piccardo GB, Vai GB (eds) The Lithosphere in Italy (Acc. Naz. Lincei), vol 80, pp 157–176

Patacca E, Sartori R, Scandone P (1990) Tyrrhenianbasin and Apenninic arcs. Kinematic relations since late Tortonian times. Mem Soc Geol Ital 45:425–451

Pino NA, Palombo B, Ventura G, Perniola B, Ferrari G (2008) Waveform modeling of historical seismograms of the 1930 Irpinia earthquake provides insight on "blind" faulting in Southern Apennines (Italy). J Geophys Res 113:B05303. https://doi.org/10.1029/2007JB005211

Pondrelli S, Salimbeni S, Ekström G, Morelli A, Gasperini P, Vannucci G (2006) The Italian CMT dataset from 1977 to the present. Phys Earth Planet Int 159(3–4):286–303. https://doi.org/10.1016/j.pepi.2006.07.008

Royden L, Patacca E, Scandone P (1987) Segmentation and configuration of subducted lithosphere in Italy: an important control on thrust-belt and foredeep-basin evolution. Geology 5:714–717

Secomandi M, D'Amico S, Paoletti V, Rapolla A (2013) Macroseismic attenuation in the Campanian Area (Southern Italy). Izvestiya - Phys Solid Earth 49:416–425. https://doi.org/10.1134/S1069351313020079

Skempton AW (1954) The pore water coefficient A and B. Geotechnique 4:143–147

Stein RS, King GCP, Lin J (1994) Stress triggering of the 1994 M = 6.7 Northridge, California earthquake by its predecessors. Science 265:1432–1435

Toda S, Stein RS, Reasenberg PA, Dieterich JH (1998) Stress transferred by the Mw = 6.5 Kobe, Japan, shock: effect on aftershocks and future earthquake probabilities. J Geophys Res 103:24543–24565

Westaway R (1993) Quaternary uplift of Southern Italy. J Geophys Res 98(B12):21741–21772

Wessel P, Smith W (1991) Free software helps map and display data. Eos Trans AGU 72:441

Zalohar J, Vrabec M (2007) Paleostress analysis of heterogeneous fault-slip data: the Gauss method. J Struct Geol 29:1798–1810

Zhao LS, Helmberger D (1994) Source estimation from broad-band regional seismograms. Bull Seismol Soc Am 85:590–605

Zhang Y, Huang F, Lai G (2009) Research on Skempton's coefficient B based on the observation of groundwater of Changping station. Earthq Sci 22:631–638

Zhu L, Helmberger D (1996) Advancement in source estimation technique using broadband regional seismograms. Bull Seismol Soc Am 86:1634–1641

Active Faulting in the Earth's Crust of the Baikal Rift System Based on the Earthquake Focal Mechanisms

Vladimir A. Sankov and Anna A. Dobrynina

1 Introduction

The destruction of the lithosphere with the formation of fault zones is one of the leading geological processes determining the structure of the continents, both in the past and at the present stage. Seismicity providing information on the structure and dynamics of formation of large fault zones in real time reflects the modern fault formation in the crust. For its study, both the epicentral field of earthquakes (see, for example, Sherman 2009) and the data on the position of their hypocenters are actively used (see, for example, Kaven and Polland 2013). To determine the orientation of modern faults of various orders, one can also use data on the earthquake focal mechanism solutions preliminarily distinguishing the true fault planes in the source. In the case of strong earthquakes, the geological data (the outcrop of the fault on the surface, the existing of faults with similar geometry, etc.), the data on the orientation of the aftershock field, the shape of the first isoseits, and other data are indirect features that help to choose one plane or another as the true fault plane. These approaches are inapplicable in study of weak earthquakes (magnitude $M \leq 4.0$) and the only information available on them is concerned with their waveforms.

At present, a number of methods for determining the parameters of motions in the earthquake sources have been developed. Conditionally, all these methods can be divided into two groups according to the initial and output data. The first group includes the methods based on analysis of waveforms that allow the determination of the geometrical size of the rupture, its velocity and direction (Gorbunova and Kalmet'eva 1988; Moskvina 1971, 1990; Solonenko and Solonenko

V. A. Sankov · A. A. Dobrynina (✉)
Institute of the Earth's Crust of Siberian Branch of Russian Academy of Sciences,
Lermontov st., 128, 664033 Irkutsk, Russia
e-mail: dobrynina@crust.irk.ru

V. A. Sankov
Irkutsk State University, Karl Marks st., 2, Irkutsk, Russia

A. A. Dobrynina
Geological Institute of Siberian Branch of Russian Academy of Sciences,
Sakh'yanova st., 6a, Ulan-Ude, Russia

S. D'Amico (ed.), *Moment Tensor Solutions*, Springer Natural Hazards,
https://doi.org/10.1007/978-3-319-77359-9_27

1987; Ben-Menahem 1961; Benioff et al. 1961; Bouchon et al. 2002; Isaks et al. 1967; Kikuchi and Kanamori 2003; Khattri 1969; Fukao 1972; Warren 2014, etc.). These methods are based mainly either on the Doppler effect, or on the inversion of the seismic moment tensor and the modeling of waveforms. The second group includes methods based on the reconstruction of the regional stress field using focal mechanism solutions and allowing the selection of the fault plane in the source (Nikitin and Yunga 1977; Gushchenko and Kuznetsov 1979; Gephart and Forsyth 1984; Angelier 1989; Rebetsky 2007; etc.).

Recently, absolute preference has been given to inversion methods that make it possible to determine comprehensively the focal mechanism, the rupture velocity, the fault plane (or the direction of rupture propagation), and the source time function (Bouchon et al. 2002; Kikuchi and Kanamori 2003; etc.). At the same time, these methods have a number of significant practical limitations, such as the availability of the sufficient quantity of broadband waveforms, good azimuthal distribution of seismic stations around the source, the knowledge of the velocity structure of the crust, etc. The same requirements are preserved even when the methods based on the Doppler effect are used (Khattri 1969; Fukao 1972; Moskvina 1971, 1990; Solonenko and Solonenko 1987; Gorbunova and Kalmet'eva 1988; Warren 2014).

The above limitations currently do not allow use to use inversion methods in the study of the earthquakes in the Baikal rift system. Therefore, in this work, the method of cataclastic analysis of shear fractures (MCA) was used to select the fault planes (Rebetsky et al. 2012). The criterion of identification of the plane in the earthquake source suggested in MCA is correlated with the consequence of Coulomb's law, that determines the implemented nodal planes with a greater value of the stress drop (Rebetsky et al. 2012). Dobrynina and Sankov (2010) compare the true fault planes in the earthquake sources of the South Baikal and Kichera seismic sequences of 1999 obtained by the method of azimuthal hodographs (Gorbunova and Kalmet'eva 1988) and MCA. Consistent (in most cases, ~80%) determinations are the evidences of their reliability (Dobrynina and Sankov 2010).

The main objectives of this work are to characterize the modern deformation field in the Baikal rift system and to identify active structures in the seismically active layer of the crust based on the comparison of the true fault planes in earthquake foci and fault tectonics.

2 Tectonic Settings and Seismicity

The Baikal rift system (North Eurasia) (Fig. 1) stretches along the edge of the Siberian platform, 1600 km from North Western Mongolia, through the mountain structures of East Siberia to South Yakutia, and consists of a linear system of uplifts and basins, bordered predominantly by the normal faults (Logatchev and Florensov 1978). In neotectonic sense, the Baikal rift system is confined to the boundary of the North Eurasian and Amurian lithospheric plates (Fig. 1). The high contemporary mobility in between the plates determines the seismic process taking place there.

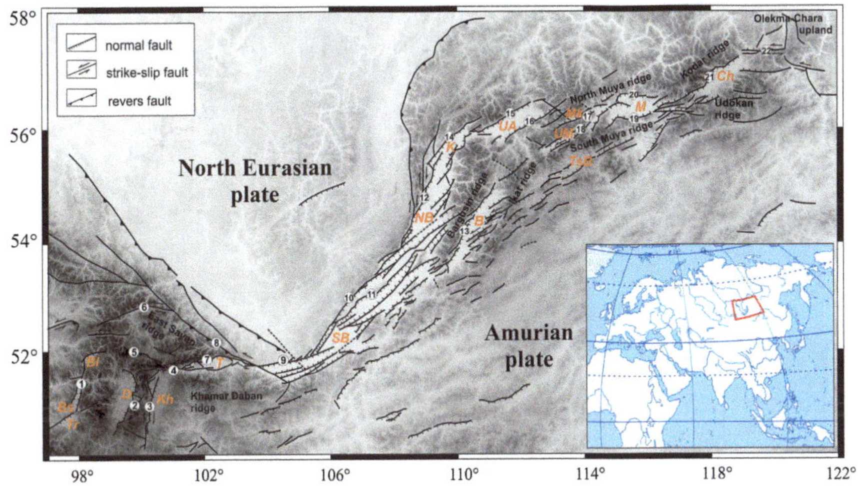

Fig. 1 The scheme of neotectonic structures of the Baikal rift system and its surroundings. *Notation* the letters denote rift basins: *Ch* Chara, *M* Muya, *Mk* Muyakan, *UM* Upper Muya, *UA* Upper Angara, *TsB* Tsipa-Baunt, *B* Barguzin, *NB* North Baikal, *SB* South Baikal, *T* Tunka, *Kh* Khubsugul, *D* Darkhat, *Bs* Busingol, *Bl* Belin, *Tr* Terekhol. 1–3—faults: 1—normal, 2—thrust and reverse, 3—assumed. The faults are denoted by numbers in the circles: 1—Busingol, 2—Darkhat, 3—Khubsugul, 4—Baikal-Mondy, 5—Yamat, 6—Okino-Zhombolok, 7—Tunka, 8—Main Sayan Fault, 9—Obruchev,10—Primorskiy, 11—Morskoy, 12—North Baikal, 13—Barguzin, 14—Kichera, 15—Akulinskiy, 16—Upper Angara, 17—Muyakan, 18—Upper Muya, 19—South Muya, 20—North Muya, 21—Kodar, 22—Khani

The South Baikal basin is the "historical" core of the Baikal rift system (Logatchev 1993). It is confined to the boundary of the Siberian platform and mobile area and it is the longest and the deepest basin of the Baikal rift system. The extension of the lithosphere of the South Baikal basin started in the border of the Cretaceous and Paleogene with the formation of a large sedimentary basin. During the Oligocene and Miocene the rifting processes progressed in both south-west and north-east directions from the South Baikal basin by the formation of a series of basins—the North Baikal, Upper Angara, Barguzin, Muya and Chara in the north-eastern part of the rift and the Tunka, Khubsugul, Darkhat, Busingol basins—on the south-western part of the Baikal rift system. The rift system is contiguous with the Siberian platform in its middle part only. In the south-western and the north-eastern parts it is separated from the platform by the transition structures—the Sayan and Baikal-Patom uplifts.

The rift grabens and half-grabens are bordered by active normal faults. The displacement amplitudes of the major normal faults reach 1500–2000 m. There are data on the exiting of the strike-slip displacements on the wings of the Baikal rift system (San'kov et al. 2000; Jolivet et al. 2013 etc.). The faults with the sub-latitudinal orientation have a left-lateral strike-slip component and the faults with the sub-meridional orientation have a right-lateral strike-slip component.

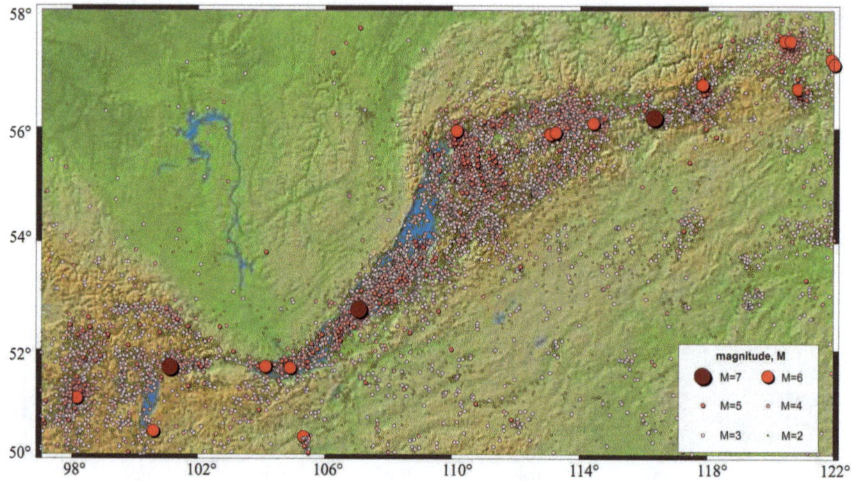

Fig. 2 The seismicity of the Baikal rift system in 1950–2012 (data base from Baikal Regional Seismological Center, Geophysical Survey of the Russian Academy of Sciences). The digital relief is given according to the data published in (Ryan et al. 2009)

The results of the microstructural studies (Sherman and Dneprovsky 1989; Delvaux et al. 1997; San'kov et al. 1997 etc.), the GPS measurements (Calais et al. 2003; San'kov et al. 2009), the investigations of earthquake focal mechanisms and the calculations of the seismotectonic deformations (Petit et al. 1996; Solonenko et al. 1997; Melnikova and Radziminovich 2007) indicate that in the center of the rift system the extension regime dominates. It is characteristic of the rift wings to increase the role of the strike-slip type of tectonic stresses. An oblique extension relative to the axis of the rift structures for the northeastern part of the rift system was established and a strike-slip with compression for its south-western part was revealed.

The Baikal rift system is characterized by a high level of seismic activity (Fig. 2). Since 1950, thirteen earthquakes with magnitude $Ms \geq 6$ occurred here, while magnitudes up to 8.2 is assumed according to historical data (Kondorskaya and Shebalin 1982). The number of weak and moderate events (with magnitude $M_L \geq 1.4$) is fairly large—about 3–4 thousands earthquakes per year (Radziminovich et al. 2013). The epicenters form extensive belts along strike of the rift structures (Fig. 2). Most earthquakes in the southern part of the Baikal rift system (South Baikal and Tunka basins) occur at depths of 10–25 km (Radziminovich 2010, and references therein). According to data from local seismic station networks, the hypocenters of the South and Central Baikal Lake are located at depths of 9–21 km with maximum peak in the range 14–19 km (Suvorov and Tubanov 2008). The deepest earth- quakes are located in the lower crust and in the upper mantle at depths of 32–46 km (Déverchère et al. 1991).

Palaeoseismicity. Numerous palaeoearthquakes were found within the Baikal rift system by the seismologists of the V.P. Solonenko school, their age is estimated as

Holocene (Solonenko et al. 1966; Solonenko 1977; McCalpin and Khromovskikh 1995; Chipizubov et al. 2010). These dislocations are morphologically expressed, as a rule, in the form of escarps, open fractures and more rarely—linear strike-slip systems, including pressure ridges and extension fractures. They are located in the zones of known active faults. According to the kinematics of displacements, palaeo seismic dislocations are divided into normal faults, open fractures, less often—strike-slips, and sometimes—reverse faults.

3 The Cataclastic Analysis Method

The MCA uses of a sequence of four reconstruction stages. As a result, the tensors of stresses and increments of seismotectonic deformations (STD), the values of the stress drop and lithostatic and fluid pressures are calculated. The selection of the fault plane is one of the intermediate stages of calculations.

At the initial stage, the homogeneous sampling of slip faults has to be compiled from the catalog of focal mechanism solutions. The following requirements must be met:

(1) the areas of elastic discharge of events should overlap;
(2) for obtaining the sought tensor of STD increments in the direction of the axis of maximum elongation, the contribution from each slip gives only an elongation deformation and in the direction of maximum shortening only the deformation of the shortening;
(3) in the direction of the axis of the algebraically maximal principal stresses the deformation of the elongation is only formed and in the direction of the axis of the algebraically minimal principal stresses the deformation of the shortening is formed.

For obtaining homogeneous sampling, the parameters of stress tensors and increments of seismotectonic deformations are coordinated. Herewith, among all possible variants of the stress tensors, one is chosen for which the maximum of dissipation of the internal energy accumulated by elastic deformations is reached. Otherwise, the stress tensor is selected in such a way that for each earthquake there are a decrease in energy. Thus, as a result of the first stage of reconstruction, there is some uniformly deforming domains with known orientations of three main axes of the stress tensors and the increment of the STD (Euler angles), with the coefficients of the form of both tensors known (the Lode-Nadai coefficients μ_σ).

From these data, the reduced normal ($\overline{\sigma}_{nn}$) and tangential ($\overline{\tau}_n$) stresses acting on the nodal planes are calculated:

$$\overline{\sigma}_{nn} = (1 - \mu_\sigma) \cdot l_{n1}^2 - (1 + \mu_\sigma) \cdot l_{n3}^2 + \mu_\sigma, \quad (1)$$

$$\overline{\tau}_n = (1 - \mu_\sigma) \cdot l_{n1} l_{t1} - (1 + \mu_\sigma) \cdot l_{n3} l_{t3}, \quad (2)$$

where l_{nk} and l_{tk}—directing the cosines of the normal vector n to the fault plane and the vector t that coincides with the direction of the shearing stresses acting on this plane in the coordinate system associated with the axes of principal stresses ($k = 1$, 2, 3).

The criterion for selecting the fault plane is as follows:

$$(\overline{\tau}_n + k_S\overline{\sigma}_{nn}) - (\overline{\tau}_s + k_S\overline{\sigma}_{ss}) > 0, \quad (3)$$

where k_S—coefficient of dry friction (assumed to be equal to the coefficient of internal friction of rocks), τ_i, σ_{ii}—tangential and normal stresses acting on the nodal planes n and s, respectively. This criterion is based on the principle of maximum stresses discharge, that allow us choosing one of the nodal planes (n or s), for which the greatest value of the stress drop is reached. This plane is considered as energetically more profitable (Rebetsky et al. 2012).

4 Seismological Data

As the initial data, the focal mechanism solutions of the earthquakes occurred in the Baikal rift system for the period from 1950 to 2010 were used. The main part of the focal mechanism solutions for earthquakes of the Baikal rift system is obtained from the polarities of the first arrivals of P-waves. Most definitions of focal mechanisms are available for moderate and strong earthquakes (Fig. 3). In the present work, 733 focal mechanism solutions for individual earthquakes and 202 composite (group) determinations were used (Pavlov 1987; Solonenko and Florensov 1985; Melnikova and Radziminovich 1998, 2003, 2004; Melnikova et al. 2006, 2007, 2008, 2012, 2013; Radziminovich et al. 2005, 2009a, b; Misharina 1961, 1969; Misharina et al. 1975, 1977, 1985, Misharina and Solonenko 1981; Solonenko et al. 1993; Solonenko and Solonenko 1987; Dobrynina 2015; Barth and Wenzel 2009; Delouis et al. 2002; Doser 1991; Petit et al. 1996), including the solutions obtained by international seismological agencies by inversion methods (International Seismological Center, Harvard University, US Geological Survey).

Figure 4 shows the focal mechanisms and classification diagrams for different regions of Baikal rift system that clearly characterize the quantitative relationships between events with different types of movement in the source—strike-slips, dip-slips and thrusts. In the central part of the Baikal rift system (regions II and III), "rift" events predominate—the compression stress axis is near vertical, the tension stress axis is horizontal, and the nodal planes are oriented according to the orientation of the main rift structures (Fig. 4). The proportion of strike-slip and thrust events is negligible. A different picture is observed for the flanks of the Baikal rift system. On the southwestern flank (region I), strong earthquakes (with $M \geq 6$) are strike-slips. There has also been an increase in the proportion of thrust events compared with the central part of the Baikal rift system. For the northeastern flank of the rift (regions IV and V), the almost complete absence of thrust events is typical, the earthquakes

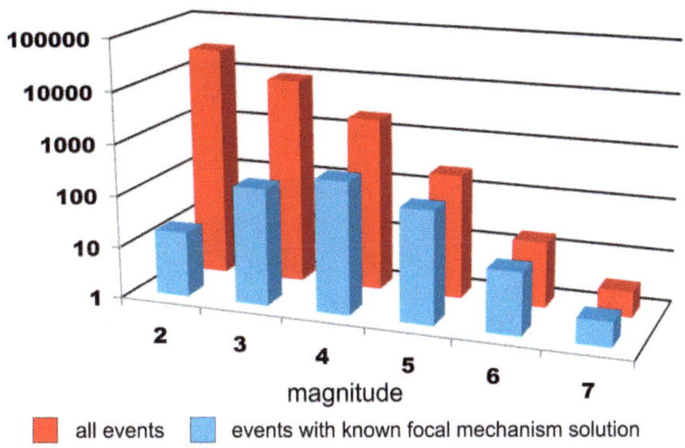

Fig. 3 The ratio of the total number of earthquakes (red) and earthquakes with a known solution of the focal mechanism (blue)

of the normal fault type predominate, while for the strong earthquakes the strike-slip component prevails (Fig. 4).

5 Results and Discussion

In this paper, the territory of the Baikal rift system and North Western Mongolia (97°–122°N and 50°–58°E) is considered. For the analysis, earthquakes that occurred during the period from 1959 to 2010, with a range of magnitudes from 3.5 to 5.5, were chosen since in this magnitude range there is the greatest number of focal mechanism solutions (Fig. 3). In addition to the individual of focal mechanism solutions, we also used the composite ones obtained for the groups of weak earthquakes. When calculating, the whole group was taken for one earthquake with the magnitude of 3.5. Calculations of stress tensors and increments of STD were performed for the whole territory in steps of 0.15° × 0.15°. The results are obtained for 970 domains (Fig. 5). Due to the peculiarities of the epicentral field in the territory considered (earthquakes are confined to the rift basins, spurs and, to a lesser extent, to the mountain framing of the basins), there are gaps in the distribution of domains: the Darkhat basin, the south of the Khubsugul basin, the eastern termination of the Tunka basin system, the North Baikal basin, the center of the North Muya range, the eastern part of the South Muya ridge, the northeastern closure of the Chara basin and the eastern part of the Udokan range (Fig. 5). There is also a reverse phenomenon: the dissemination of the results of calculations to territories without initial data, such as the southern part of the Siberian platform (Fig. 5). This is explained by the grid spacing (0.15°) and a sufficiently large averaging radius—60 km.

As a result of the first stage of reconstruction, the type of stress state, the Lode-Nadai coefficients and the orientation of the principal stress axes were obtained. Analysis of the type of stress state shows that the regime of horizontal extension predominates on the territory of the rift system—883 domains, 91% (Fig. 6). The regime of horizontal shear is observed in the area of the Busingol basin. In the

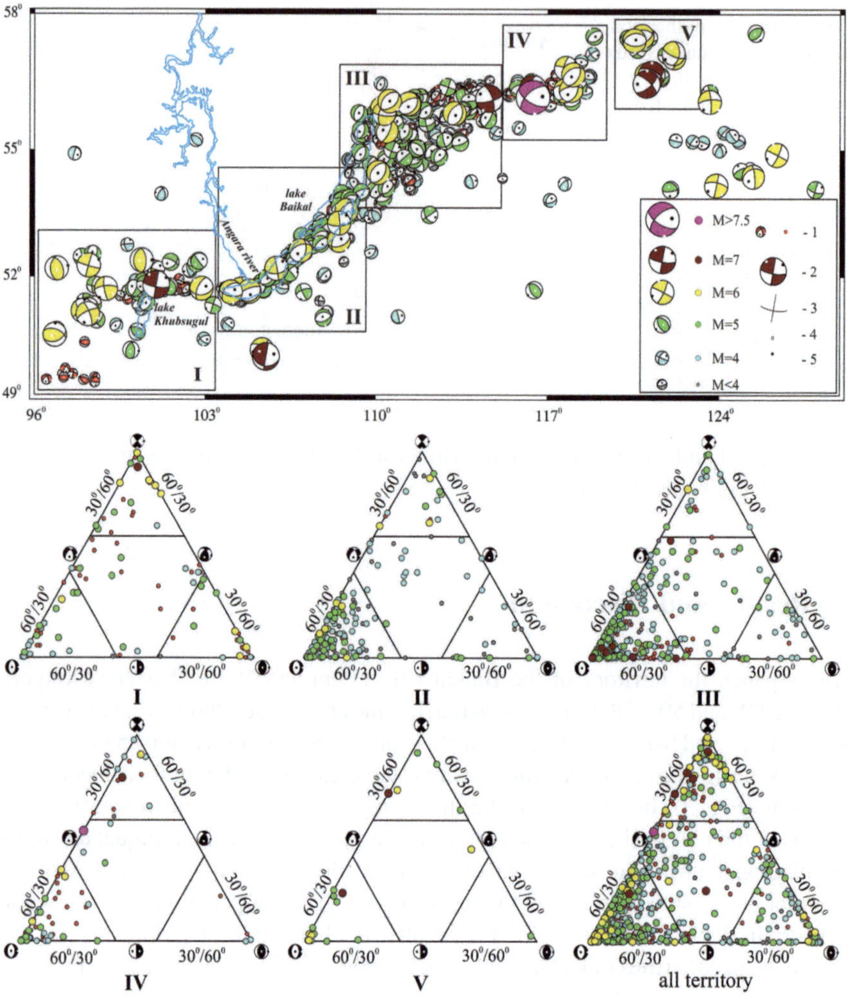

Fig. 4 The focal mechanisms of earthquakes of the Baikal rift (lower hemisphere) for the period 1950-2010 (top) and classification diagrams (bottom). Different colors and sizes show earthquakes of different magnitudes. Rectangles indicate the areas for which the diagrams are plotted, the area numbers are shown in Latin numerals. Legend: 1—composite focal mechanism solutions, 2—the focal mechanism, 3—nodal planes, 4, 5—axes of the main stresses of tension and compression, respectively

region of the Eastern Sayan, a variety in the types of stress states are noted—from horizontal extension to horizontal compression (Fig. 6). This is explained by the change in the orientation of the rift structures from the meridional (the Busigol, Darkhat and Khubsugul basins) to the latitudinal (the system of the Tunka basins, the South Baikal basin) and, as a consequence, a wide variety of focal mechanisms in this area. The regime of horizontal extension with a strike-slip is also observed in the

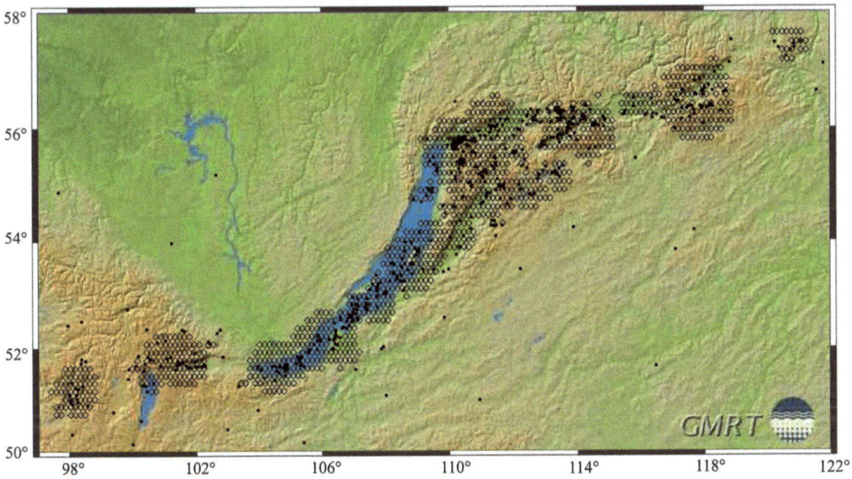

Fig. 5 The domains in which tensors of stress and increment STD (rhombic) are obtained. The circles show the earthquakes used for calculations. The digital relief is given according to the data published in (Ryan et al. 2009)

Kodar and Udokan uplifts (framing of the Chara basin) and within the Olekma-Chara highlands (Fig. 6). Also, one domain with a vertical shear stress regime is observed for the Kodar uplift (Fig. 6). The most interesting is the presence of several domains with horizontal shear regimes localized in the mountain frame of the South Baikal basin. They are obtained as a result of the formation of homogeneous samples of earthquakes with a significant strike-slip component, confined mainly to the boards of the South Baikal basin. In the central part of the basin, their influence on the resultant geodynamic type of stress state is insignificant due to a large number of typical "rift" events (Fig. 4). The obtained results reflect the character of deformation of the crust of the territory investigated (Fig. 4, insert).

Orientation of principal stress axes. The results of the determination of the orientation of the axes of compression and extension (shortening and elongation) do not contradict the results of generalization of data on focal mechanisms (Solonenko et al. 1993; Petit et al. 1996; Parfeevets et al. 2002) and STD calculations by different methods (Kuchai 1990; Melnikova and Radziminovich 2007) (Fig. 7). The axis of maximum compression lies subvertically almost on the entire territory of the Baikal rift system (Fig. 7a). Only on the south-western flank it has gentle angles and extends in the direction of SW-NE.

The axis of minimal compression (extension) is almost everywhere subhorizontal, in the central part of the Baikal rift system and on the part of its NE flank it is located across the rift structures (Fig. 7b). On the territory of the SW flank and in the eastern part of the NE flank of the Baikal rift system, it is obliquely directed toward the general strike of these segments of the rift zone (Fig. 7b).

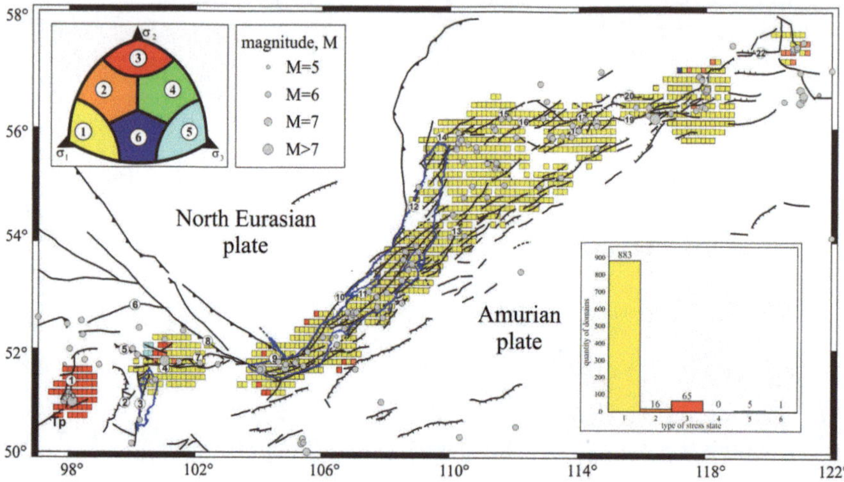

Fig. 6 The types of geodynamic regimes: 1—the horizontal extension regime (axis σ_1 is sub-vertical); 2—the horizontal extension with horizontal shear; 3—the horizontal shear regime (axis σ_2 is sub-vertical); 4—the horizontal compression with horizontal shear; 5—the horizontal compression regime (axis σ_3 is sub-vertical); and 6—the vertical shear stress state (axis σ_2 is sub-horizontal, and σ_1 and σ_3 have angles of axes plunges near 45°). Insert—histogram distribution of domains with different geodynamic regimes. The notation of tectonic structures as is in the caption to Fig. 1

The intermediate axis also has, for the most part, a gentle dip angle to the horizon, that increases in the area of the Busingol basin (SW flank of the Baikal rift system) and in the region of the Olekma river, beyond the rift structures of the NE flank of the Baikal rift system (Fig. 7c). The direction of the extension axis (SW-NE) is fairly stable in the rift zone. Significant variations in the strike of the intermediate axis are observed within the SW flank of the Baikal rift system.

In a generalized form, the spatial characteristics of the current stress state of the crust of the Baikal rift system can be represented using an approach similar to that chosen by the authors of the World Stress Map project (Heidbach et al. 2010), that uses the axes of maximum horizontal compression. On Fig. 8 the trajectories of the axes of the maximum Sh_{max} and the minimum Sh_{min} of the horizontal compression of the crust of the Baikal rift system are shown. The Sh_{max} trajectories in the central part of the Baikal rift system coincide with the general orientation of the rift structures, and the Sh_{min} trajectories are roughly perpendicular to them. In accordance with this, on this segment of the rift modern fault movements along the faults are manifested. Within the meridional section of the southwestern flank, the trajectories of both horizontal axes are located obliquely to the orientation of the rift basins, which determines the tendency to manifest modern strike-slip movements along the faults. A similar trend is also observed in the eastern segment of the sublatitudinal section of this flank, in the area of the junction of the Tunka and South Baikal basins. The same can be noted for the eastern termination of the NE flank of the Baikal rift system.

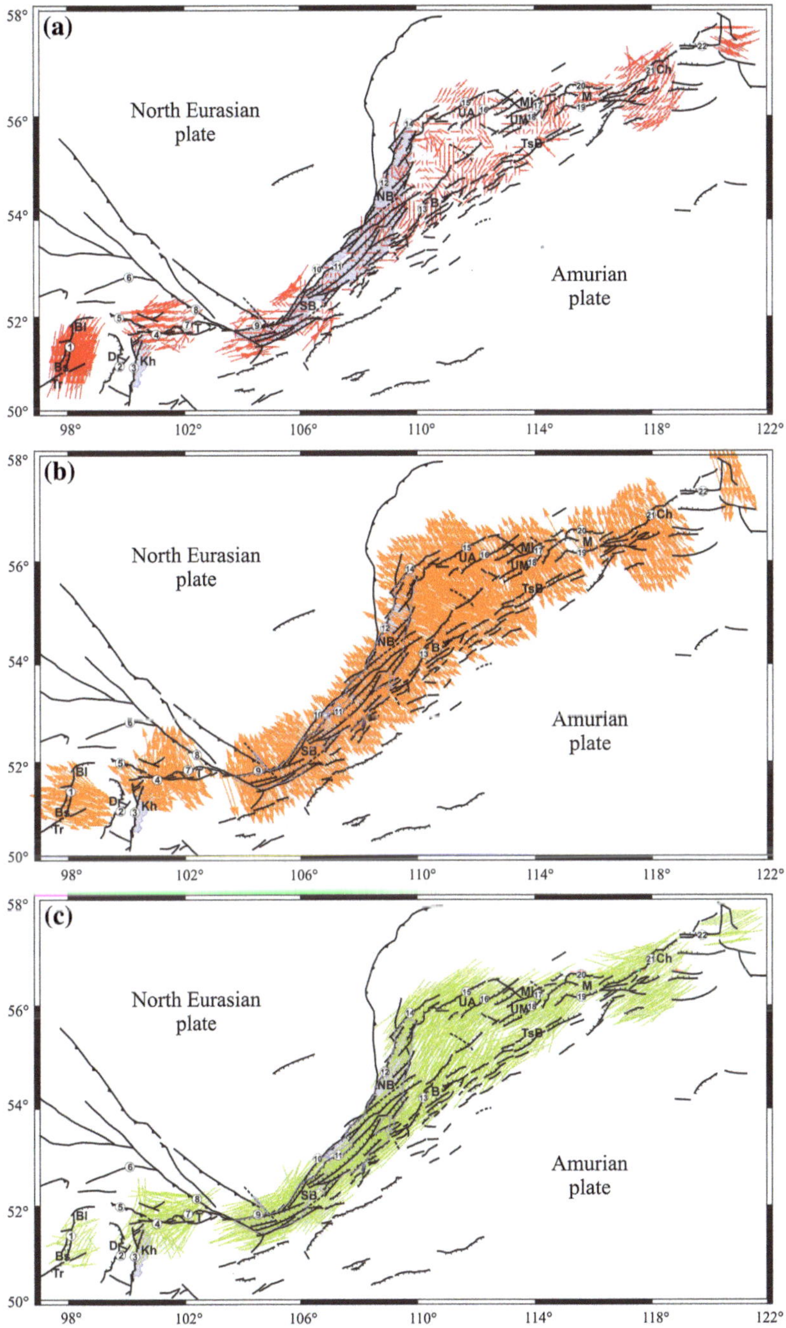

Fig. 7 The orientation of the main normal stress axes of maximum compression (**a**), minimum compression (**b**) and intermediate axes (**c**). The length of the arrow depends on the axis dip angle to the horizon. The dots represent the vertically located stress axes. The notation of tectonic structures as is in the caption to Fig. 1

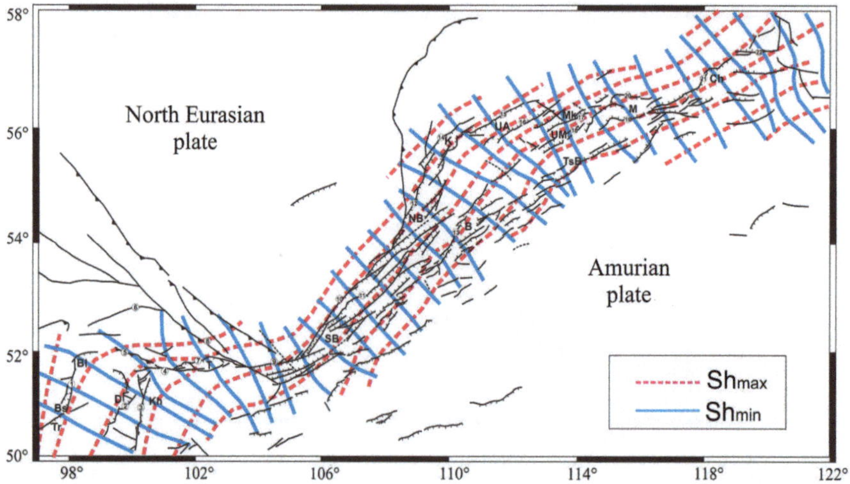

Fig. 8 The trajectories of the axes of the maximum Sh_{max} and the minimum Sh_{min} of the horizontal compression of the crust. The notation of tectonic structures as is in the caption to Fig. 1

Previously, the trajectories of the main normal compressive and extension stresses for the study area were based on seismological (Gushchenko 1979; San'kov et al. 1997) and geological and structural data (Sherman and Dneprovsky 1989). In general, the obtained picture of stress trajectories of maximum and minimum horizontal compression does not contradict the previously obtained data. At the same time, as a result of this study, a new level of detalization of the spatial characteristics of the stress-strain state was achieved. Especially it concerns the south-western part of the Baikal rift system where the stress field is characterized by the greatest difficulty due to a fundamental turn of orientation of the rift structures.

The Lode-Nadai coefficient. Additional data on the nature of crust deformations were obtained from the results of calculations of the Lode-Nadai coefficient (Fig. 9). Practically the entire territory of the Baikal rift system is deformed under the conditions of shear (in the understanding of the mechanics of continuous media). There are two mode regions with an uniaxial compression and close to it deformation regime—in the area of the junction zone of the Khubsugul and Tunka segment of the rift system and in the area of the northern end of the Ikat range. And, if in the first case the compression axis is located obliquely or subhorizontal, then in the second case it is vertical. Areas with a transitional regime (from shear to uniaxial extension) are found in the north-eastern part of the Baikal rift system—in the vicinity of the Muya, Chara basins, in the Olekma-Chara highland. Also, they are manifested locally in the central part of the Baikal basin, near its southern termination. According to these data, in this zone there is a region deformed under conditions of uniaxial extension.

The manifestations of compression areas within the rift system are interesting from the point of view of interpretation of geological and geophysical data. Thus, the presence of conditions of modern horizontal compression in the junction of

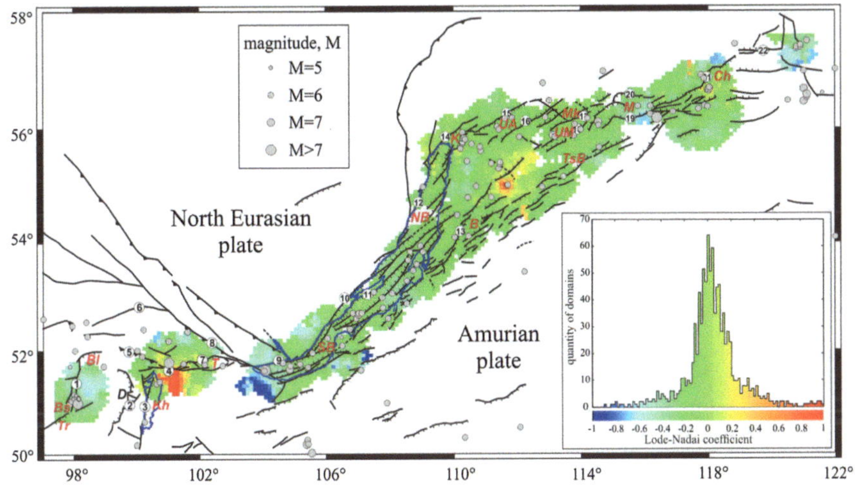

Fig. 9 The types of stress state of the crust of the Baikal rift system, expressed in terms of the Lode-Nadai coefficient according to the data of the MCA. The averaging radius is 60 km. The notation of tectonic structures as is in the caption to Fig. 1

latitudinal and meridional structures on the SW flank of the Baikal rift system helps explain the abnormally high uplift of the Munku-Sardyk massif and the western part of the Tunka Alps.

According to the leveling data (Pogorazdova 1989), the area of junction of the northern part of the Ikat range with the North Muya and South Muya ranges is characterized by high velocities of modern vertical movements. Uplift velocities over a large area exceed 8 mm/year and reach 14.9 mm/year. The detected vertical compression in the crust at the level of the earthquake focal layer correlates with the mechanism of fast glaciostatic uplift of ranges in this region in Holocene from the beginning of the melting of the glaciers of the last glaciation proposed by Levi et al. (1997). The process of uplift continues, apparently, and at the present stage.

The earthquake fault planes. For 500 earthquakes (including 131 groups of events with composite focal mechanism solutions), the nodal planes were divided into fault and auxiliary planes (Sankov, Dobrynina, San'kov and Dobrynina 2009) (Fig. 10). Outwardly, the orientation of the earthquake fault planes in most cases, sometimes—up to the details, inherits existing neotectonic structures. This follows both from analysis of the orientations of the fault planes for apart sections of the rift system (Fig. 11), and from the comparison of orientations of the mapped active faults, paleo seismic dislocations and earthquake fault planes for large segments of the Baikal rift system (Fig. 12). Two orientations of the fault planes (sublatitudinal and northeastern) prevail in the southern part of the Baikal basin (up to the delta of the Selenga river). It is precisely these orientations of active faults that are observed in the boards of the basin and are identified by seismic profiling data (Levi et al. 1997). Minimal variations in the orientation of the earthquake fault planes are characteristic of the

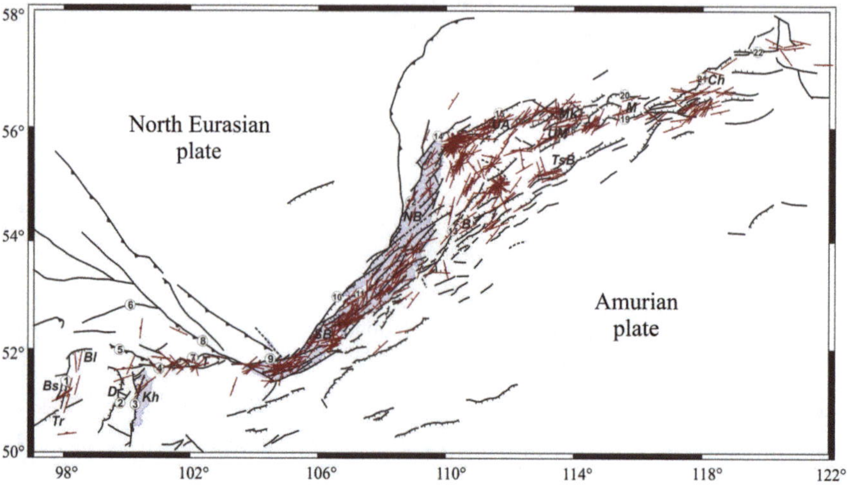

Fig. 10 The fault planes of earthquakes of the Baikal rift system. The marks on the planes indicate the dipping direction of the fault plane. The length of the lines is the same for all earthquakes without taking into account their magnitude and orientation. The notation of tectonic structures as is in the caption to Fig. 1

central part of the Baikal basin, where the extension is strictly perpendicular to the neotectonic structure. The characteristic dip of the earthquake fault planes for the southern and central parts of the Baikal basin is 40–60°. At the same time, the fault planes dipping to the north (the secondary faults) are steeper. Steep fault planes predominate in the North Baikal basin. Apparently, in conditions of the predominance of vertical block movements, along the surfaces of steeply dipping faults, prerequisites are created for the realization of both normal and reverse movements, which are observed according to data on the focal mechanisms of earthquakes (Dobrynina 2015).

Two orientations of active faults prevail in the Barguzin basin. From the west, the basin is bordered by faults of the NNE orientation, and its eastern board is cut by the faults of the NE orientation. According to diagram (Fig. 11) the earthquake fault planes in zones of NNE faults have the predominant dip angles 65–70°, and in NE fault zones—55–60°.

On the NE flank of the Baikal rift system, the spur areas have the most complex structure in the field of seismogenic faults, and in the basins the earthquake fault planes are parallel to the known neotectonic faults.

The spur between the Upper-Angara and Muya basins is characterized by the development of NE and ENE active faults. The largest of faults have a dip in the northern rumba according to geological and structural data. In accordance with the diagram (Fig. 11), main faults have a less steep dipping (angle 55–65°) than the secondary ones accompanying them (the angle is up to 80–85°).

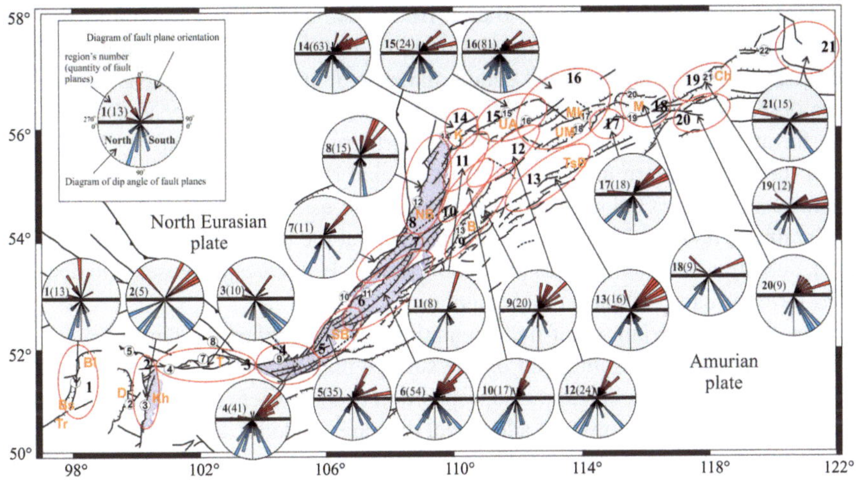

Fig. 11 The diagrams of orientation (top) and dip angles (bottom, dipping in the northern rumba—on the left, in the south ones—on the right) of the fault planes of the earthquakes of the Baikal rift system. The regions for which the diagrams were plotted were singled out and denoted by figures. The notation of tectonic structures as is in the caption to Fig. 1

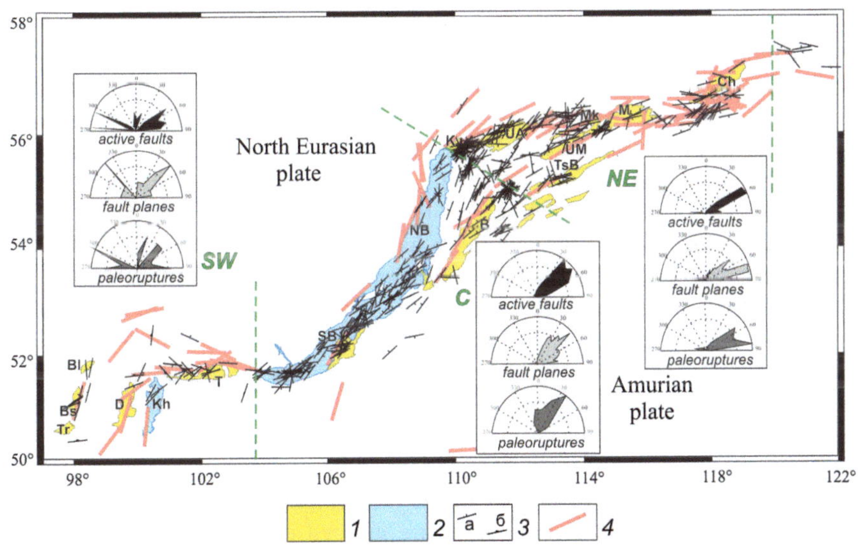

Fig. 12 The proportion of the orientation of active faults, paleo seismic dislocations and earthquake fault planes. *Legend* 1—Cenozoic basins; 2—the lakes; 3—the earthquake fault planes: a—dip-slips, b—thrusts; 4—palaeo seismic dislocations (shown by the same lines, regardless of the length of the seismogenic rupture and amplitude). On insert—diagrams of orientation of active faults, paleo seismic dislocations and earthquake fault planes for different parts of the Baikal rift system

The south-western part of the Baikal rift system and the termination of the NE flank of the Baikal rift system are an exception to the general trend. In the area of the Busingol basin, the earthquake fault planes with uncharacteristic NE orientation were observed. A wide variety of the earthquake fault planes appeared within the East Sayan range, the Tunka basin and the northern Khubsugul area.

At the north-eastern termination of the Baikal rift system, in the vicinity of the Chara basin and the Olekma-Chara highland, north-eastern and submeridional faults predominate; however, in the present stress field at the level of the earthquake focal layer, sublatitudinal fault planes are mainly active. Holocene palaeoseismic dislocations have the same sublatitudinal orientations.

The obtained deviations of the orientations of the earthquake fault planes from the neotectonic structure mapped on the surface may indicate a change in the time of the tectonic stress field and the superposition of the present structure with respect to the Late Cenozoic. However, additional data are needed to solve this problem.

6 Conclusions

The application of the method of cataclastic analysis of shear fracture (MCA) (Rebetsky et al. 2012) allow us to determine the spatial characteristics of seismotectonic deformations and the fault planes of the earthquakes for the Baikal rift system as a whole and for its individual structures. Within the central part and on most of the northeastern flank of the Baikal rift system, extension conditions prevail with the orientation of earthquake fault planes parallel to the main faults. The average dip angles of fault planes in the zones of these faults for some rift basins are 45–60°, which allows us to assume the decreasing of the dip angle of the faults in the middle part of the crust. Antithetic faults have a more steep dip angle. Strike-slip deformations predominate at the distal terminations of the flanks of the Baikal rift system, and the fault planes are more varied in morphogenetic terms—along with the dip-slip faults the strike-slips and thrusts are observed. Orientation of the fault planes in different parts of the Baikal rift system coincides with the orientation of the Holocene palaeoruptures, as well as with the orientation of earthquake dislocations, which manifested itself at the historical stage. At the same time, on the flanks of the rift system, consistency the fault plane and the neotectonic structure is less evident. The data obtained on the deviations of the orientations of the modern faults (the fault planes) from the neotectonic structure mapped on the Earth's surface may indicate a change in the time of the tectonic stress field and the partial superposition of the present structure with respect to the Late Cenozoic. These deviations are most clearly visible within the southwestern flank of the rift system.

Acknowledgements The reported study was funded by RFBR according to the research project № 17-05-00826.

References

Angelier J (1989) Inversion of filed data in fault tectonics to obtain the regional stress II. Using conjugate fault sets within heterogeneous families for computing palaeostress axes. Geophys J 96:139–149

Barth A, Wenzel F (2009) New constraints on the intraplate stress field of the Amurian plate deduced from light earthquake focal mechanisms. Tectonophysics. https://doi.org/10.1016/j.tecto.2009.01.029

Benioff H, Press F, Smith S (1961) Excitation of the free oscillations of the earth by earthquakes. J. Geophys Res 66:605–618

Ben-Menahem A (1961) Radiation of seismic surface-waves from finite moving sources. Bull Seismol Soc Am 51:401–435

Bouchon M, Toksoz MN, Karabulut H et al (2002) Space and time evolution of rupture and faulting during the 1999 Izmit (Turkey) earthquake. Bull Seismol Soc Am 92:256–266

Calais E, Vergnolle M, San'kov V, Lukhnev A, Miroshnitchenko A, Amarjargal S, Déverchère J (2003) GPS measurements of crustal deformation in the Baikal-Mongolia area (1994–2002: Implications for current kinematics of Asia). J Geophys Res 108(B10). https://doi.org/10.1029/2002jb002373

Chipizubov AV, Smekalin OP, Semenov RM, Imaev VS (2010) Paleoseismicity of the Pribaikalie. Seism Instr 46:136. https://doi.org/10.3103/S0747923910020040

Delouis B, Deverchere J, Melnikova V et al (2002) A reappraisal of the 1950 (Mw 6. 9) Mondy earthquake, Siberia, and its relationship to the strain pattern at the south–western end of the Baikal rift zone. Terra Nova 6:491–500

Delvaux D, Moyes R, Stapel G, Petit C, Levi K, Miroshnitchenko A, Ruzhich V, Sankov V (1997) Paleostress reconstruction and geodynamics of the Baikal region, Central Asia. Part II: Cenozoic rifting. Tectonophysics 282:1–38

Déverchère J, Houdry F, Diament M, Solonenko NV, Solonenko AV (1991) Evidence for a seismogenic upper mantle and lower crust in the Baikal rift. Geophys Res Lett 18(6):1099–1102

Dobrynina AA (2015) Focal mechanisms of the North Baikal earthquakes. In: XXI All-Russia youth conference "Lithosphere structure and Geodynamics", Irkutsk, Russia, pp 245–246 (in Russian)

Dobrynina AA, Sankov VA (2010) Rupture velocities and propagations in sources of Baikal rift system's earthquakes. Geophys Invest 11(2):52–61 (in Russian with English abstract)

Doser DI (1991) Faulting within the western Baikal rift as characterized by earthquake studies. Tectonophysics 196:87–107

Fukao Y (1972) Source process of a large deep—focus earthquakes and its tectonic implications. Earth Planet Int 5:61–76

Gephart JW, Forsyth DW (1984) An improved method for determining the regional stress tensor using earthquake focal mechanism data: application to the San Fernando earthquake sequence. J Geophys Res 89(B11):9305–9320

Gorbunova IV, Kalmet'eva Z (1988) Experimental characteristics of weak earthquake radiation. Ilim, Frunze, 129 pp (in Russian)

Gushchenko OI (1979) The method of the kinematic analysis of structures of destruction at reconstruction of fields of tectonic stresses. In: Fields of stress a lithosphere. Nauka, Moscow, pp 7–25 (in Russian)

Gushchenko OI, Kuznetsov VA (1979) Determination of the orientations and the ratio of principal stresses on the basin of tectonic fault slip data. In: Stress fields in the lithosphere. Nauka, Moscow, pp 60–66 (in Russian)

Heidbach O, Tingay M, Barth A, Reinecker J, Kurfeß D, Muller B (2010) Global crustal stress pattern based on the World Stress Map database release. Tectonophysics 482(1–4):3–15. https://doi.org/10.1016/j.tecto.2009.07.023

International Seismological Centre (2014) On-line Bulletin. International Seismological Centre, Thatcham, United Kingdom. http://www.isc.ac.uk

Isaks BL, Sykes LR, Oliver J (1967) Spatial and temporal clustering of deep and shallow earthquakes in the Fiji–Tonga–Kermadec region. Bull Seismol Soc Am 57:935–958

Jolivet M, Arzhannikov S, Chauvet A, Arzhannikova A, Vassallo R, Kulagina N, Akulova V (2013) Accomodating largescale intracontinental extension and compression in a single stress-field: a key example from the Baikal Rift System. Gondwana Res 24:918–925

Kaven JO, Polland DD (2013) Geometry of crustal faults: identification from seismicity and implications for slip and stress transfer models. J Geophys Res Solid Earth 118. https://doi.org/10.1002/jgrb.50356

Khattri KN (1969) Determination of the earthquake fault plane, fault area and rupture velocity from spectra of long period P waves and the amplitude of SH waves. Bull Seismol Soc Am 59:615–630

Kikuchi M, Kanamori H (2003) Note on teleseismic body—wave inversion program. http://www.eri.u-tokyo.ac.Jp/ETAL/KIKUCHI

Kondorskaya N, Shebalin N (eds) (1982) New catalogue of strong earthquakes in the USSR. World Data Centre A, US Department of Commerce, Washington, DC

Kuchai OA (1990) Features of seismotectonic deformation of Lake. Baikal and its mountain frame. Research on the creation of scientific foundations for the forecast of earthquakes in Siberia. Operative information. IK SB RAS, Irkutsk, pp 17–22 (in Russian)

Levi KG, Miroshnitchenko AI, San'kov VA, Babushkin SM, Larkin GV, Badardinov AA, Wong HK, Colman S, Delvaux D (1997) Active faults of the Baikal depression. Bull Centre Rech Elf Explor Prod 21(2):399–434

Logatchev NA (1993) History and geodynamics of the Lake Baikal rift in the context of the Eastern Siberia rift system: a review. Bull Cent Rech Explor Prod Elf Aquitaine 17(2):353–370

Logatchev NA, Florensov NA (1978) The Baikal system of rift valleys. Tectonophysics 45:1–13

McCalpin JP, Khromovskikh VS (1995) Holocene paleoseismicity of the Tunka fault, Baikal rift Russia. Tectonics 14(3):594–605

Melnikova VI, Radziminovich NA (1998) Focal mechanisms of the earthquakes of the Baikal region for 1991–1996. Russ Geol Geophys 39(11):1598–1607

Melnikova VI, Radziminovich NA (2003) Baikal and Transbaikalie (IV. Earthquake focal mechanisms catalog). Earthquakes of the Northern Eurasia in 1997. FOP, Obninsk, 280 pp (in Russian)

Melnikova VI, Radziminovich NA (2004) Baikal and Transbaikalie (IV. Earthquakes focal mechanism catalog). Earthquakes of the Northern Eurasia in 1998.FOP, Obninsk, 268 pp (in Russian)

Melnikova VI, Radziminovich NA (2007) Parameters of seismotectonic deformations of the Earth's crust in the Baikal rift zone based on seismological data. Dokl Earth Sci 416:1137–1139

Melnikova VI, Radziminovich NA, Tatomir NV, Dobrynina AA (2006) Baikal and Transbaikalie (IV. Earthquakes focal mechanism catalog). Earthquakes of the Northern Eurasia in 2000. FOP, Obninsk, 376 pp (in Russian)

Melnikova VI, Radziminovich NA, Gileva NA, Chipizubov AV, Dobrynina AA (2007) Activation of rifting processes in the Northern Cis-Baikal region: a case study of the Kichera earthquake sequence of 1999, Izvestiya. Phys Solid Earth 43(11):905–921

Melnikova VI, Radziminovich NA, Tatomir NV, Dobrynina AA (2008) Baikal and Transbaikalie (IV. Earthquake focal mechanisms catalog). Earthquakes of the Northern Eurasia in 2002. FOP, Obninsk, 428 pp (in Russian)

Melnikova VI, Gileva NA, Arefiev SS, Bykova VV, Masalskii OK (2012) The 2008 Kultuk earthquake with $Mw = 6.3$ in the south of Baikal: spatial-temporal analysis of seismic activation. Izvestiya, Phys Solid Earth 48(7–8):594–614. http://dx.doi.org/10.1134/s1069351312060031

Melnikova VI, Gilyova NA, Masalsky OK (2013) Baikal and Transbaikalia. Earthquakes of the Northern Eurasia, 2007. GS of the RAS, Obninsk, pp 160–169 (in Russian)

Misharina LA (1961) Aftershocks of the middle Baikal earthquake of August 29, 1959. Russ Geol Geophys 2:105–110 (in Russian)

Misharina LA (1969) Investigation of the focal mechanism of weak earthquakes in the northeastern sector of the Baikal rift. In: Proceedings of the third all-union symposium on the seismic regime, Part I, pp 147–166 (in Russian)

Misharina LA, Solonenko NV (1981) The focal mechanism of earthquakes in the southwestern flank of the Baikal rift zone. In: Seismic studies in Eastern Siberia, Nauka, pp 3–11 (in Russian)
Misharina LA, Solonenko NV, Leontyeva LR (1975) Local tectonic stresses in the Baikal rift zone from observations of groups of weak earthquakes. Baikal Rift. Nauka, Novosibirsk, pp 9–21 (in Russian)
Misharina LA, Solonenko NV, Vertlib MB (1977) Some features of the epicentral field of the Baikal rift zone in comparison with the focal mechanism of earthquakes. In: Seismicity and seismogeology of Eastern Siberia. Nauka, Moscow, pp 43–61 (in Russian)
Misharina LA, Solonenko AV, Melnikova VI, Solonenko NV (1985) Stresses and faults in earthquake foci. In: Solov'ev SL (ed) Geology and seismicity of the BAM zone. Nauka, Siberian Branch, Novosibirsk, 192 pp (in Russian)
Moskvina AG (1971) On the possibility of determining certain characteristics of the earthquake focus from the spectra of body waves. Phys Earth 111(11) (in Russian)
Moskvina AG (1990) Use of the Doppler effect to determine the parameters of the earthquake source. Phys Earth 1:35–45 (in Russian)
Nikitin LV, Yunga SL (1977) Methods of theoretical determination of tectonic deformation and stress in seismically active areas. Izvestia USSR. Phys Earth 11:54–67 (in Russian)
Parfeevets AV, Sankov VA, Miroshnichenko AI, Luhnev AV (2002) Evolution of the stress state of the Earth's crust of the Mongolian-Baikal mobile belt. Pac Geol 21:14–28 (in Russian)
Petit C, Déverchère J, Houdry F, Sankov VA, Melnikova VI, Delvaux D (1996) Present-day stress field changes along the Baikal rift and tectonic implications. Tectonics 15:1171–1191
Pavlov OV (ed) (1987) The Angarakansky seismic swarm in the Baikal rift zone. Nauka, Novosibirsk, 81 pp (in Russian)
Pogorazdova SV (ed) (1989) Map of modern vertical movements of the Earth's crust according to geodetic data on the territory of the USSR, scale 1: 5000000. Factory No. 11 PA "Azerbaijanerogeodesiya" GUSK USSR, Baku (in Russian)
Radziminovich NA (2010) Focal depths of earthquakes in the Baikal region: a review. Izvestiya Phys Solid Earth 46(3):216–229. https://doi.org/10.1134/S1069351310030043
Radziminovich N, Deverchere J, Melnikova V et al (2005) The 1999 Mw 6. 0 earthquake sequence in the Southern Baikal rift, Asia, and its seismotectonic implications. Geophys J Int 161:387–400
Radziminovich NA, Gileva NA, Radziminovich YaB, Kustova MG, Chechelnitskii VV, Melnikova VI (2009) The 2003 September 16 Kumora Earthquake with Mw = 5.6, KR = 14.3 and I0 = 7. Earthquakes of the Northern Eurasia in 2003, Obninsk, Geophysical Survey of RAS, pp 293–309 (in Russian)
Radziminovich NA, Melnikova VI, Tatomir NV, Dobrynina AA (2009) VII. Catalogs of focal mechanism solutions. Earthquakes of the Northern Eurasia in 2003, Obninsk, Geophysical Survey of RAS, pp 293–309 (in Russian)
Radziminovich NA, Gileva NA, Melnikova VI, Ochkovskaya MG (2013) Seismicity of the Baikal rift system from regional network observations. J Asian Earth Sci 62:146–161
Rebetsky YL (2007) Tectonic stresses and strength of natural mountain massifs, p 406. Moscow: Akademkniga
Rebetsky YuL, Kuchai OA, Sycheva NA, Tatevossian RE (2012) Development of inversion methods on fault slip data. Stress state in orogenes of the Central Asia. Tectonophysics 581:114–131
Ryan WBF, Carbotte SM, Coplan JO, O'Hara S, Melkonian A, Arko R, Weissel RA, Ferrini V, Goodwillie A, Nitsche F, Bonczkowski J, Zemsky R (2009) Global multi-resolution topography synthesis. Geochem Geophys Geosyst 10:Q03014. https://doi.org/10.1029/2008gc002332
San'kov VA, Dobrynina AA (2009) Active ruptures in the earth's crust of the Baikal rift system at the level of the earthquake focal layer. In: Sklyarova EV (ed) Geodynamic evolution of the lithosphere of the Central Asian mobile belt (from the ocean to the continent): proceedings of the meeting, T 2, no 7. IEC SB RAS, Irkutsk, pp 67–70 (in Russian)
San'kov VA, Miroshnitchenko AI, Levi KG, Lukhnev AV, Melnikova AI, Delvaux D (1997) Cenozoic stress field evolution in the Baikal rift zone. Bull Centre Rech Elf Explor Prod Elf Aquitaine 21(2):435–455

San'kov V, Déverchère J, Gaudemer Y, Houdry F, Filippov A (2000) Geometry and rate of faulting in the North Baikal Rift, Siberia. Tectonics 19:707–722

San'kov VA, Lukhnev AV, Miroshnichenko AI, Ashurkov SV, Byzov LM, Calais E, Déverchère J (2009) Extension in the Baikal rift: present-day kinematics of passive rifting. Doklady Earth Sci 425:205–209

Sherman SI (2009) Tectonophysical model of the seismic zone: experience of development on the example of the Baikal rift system. Phys Earth 11:8–21 (in Russian)

Sherman SI, Dneprovsky YuI (1989) The field stresses of the earth's crust and the geological-structural methods of their study. Nauka Publ House, Siberian Branch, Novosibirsk (in Russian)

Solonenko VP (ed) (1977) Seismic zoning of the East Siberia and its geology and geophysics base. Nauka, Novosibirsk (in Russian)

Solonenko VP (ed) (1981) Seismogeology and detailed seismic zoning of the Baikal region. Nauka, Novosibirsk, 169 pp (in Russian)

Solonenko VP, Florensov NA (eds) (1985) Earthquakes and seismic hazard mapping basements of Mongolia. Nauka, Moscow, 224 pp (in Russian)

Solonenko NV, Solonenko AV (1987) Aftershock sequences and earthquake swarms in the Baikal rift zone. Nauka, Novosibirsk (in Russian)

Solonenko VP, Treskov AA, Kurushin RA et al (1966) Living tectonics, volcanoes and seismicity of the Stanovoi Upland. Nauka, Moscow, 231 pp (in Russian)

Solonenko AV, Solonenko NV, Melnikova VI, Kozmin BM, Kuchai OA, Sukhanova SS (1993) Strains and displacements in earthquake foci of Siberia and Mongolia. In: Seismicity and seismic zoning of Northern Eurasia, Moscow, vol 1, pp 113–122 (in Russian)

Solonenko A, Solonenko N, Melnikova V, Shteiman E (1997) The seismicity and earthquake focal mechanisms of the Baikal rift zone. Bull Centre Rech Elf Explor Prod 21:207–231

Suvorov DB, Tubanov TsA (2008) Distribution of sources of close earthquakes in the crust beneath the Central Baikal. Russ Geol Geophys 49(8):611–620

Warren LM (2014) Dominant fault plane orientations of intermediate-depth earthquakes beneath South America. J Geophys Res Solid Earth 119:5762–5785. https://doi.org/10.1002/2013JB010856

Quaternary Stress Field and Faulting in the Western Part of the Catanzaro Trough (Calabria, Southern Italy)

F. Brutto, F. Muto, M. F. Loreto, Sebastiano D'Amico, N. De Paola, V. Tripodi and S. Critelli

Keywords Calabrian arc · Extensional tectonics · Seismotectonics

1 Introduction

The Calabria region is characterized by a complex geo-structural setting being located within the western Mediterranean subduction system (Fig. 1; Gueguen et al. 1998; Faccenna et al. 2001; Carminati et al. 2005). According to GPS (D'Agostino and Selvaggi 2004; Serpelloni et al. 2007, 2010) and seismic tomography (Selvaggi and Chiarabba 1995; Neri et al. 2009) studies, the subduction and roll-back of the Ionian oceanic slab is still ongoing beneath the Calabrian region, which is part of the southern Apennine system (e.g. D'Amico et al. 2014, and references therein). Moreover, the gravitational roll-back of the Ionian slab could be responsible for strong fault kinematics variation at shallow depth, along the convergent margin (Serpelloni et al. 2010). Thus, the area is also characterized by an intermediate and deep seismicity clustered and aligned along a narrow (less than 200 km) and steep (about 70°) Wadati-Benioff zone striking NE-SW and dipping towards the NW down to 500 km of depth (Neri et al. 2009).

F. Brutto (✉) · F. Muto · V. Tripodi · S. Critelli
Department Biology, Ecology and Earth Sciences, University of Calabria, 87036 Arcavacata di Rende (CS), Italy
e-mail: fabrizio.brutto@gmail.com

M. F. Loreto
Institute of Marine Sciences – National Research Council, U.O.S. Bologna, Via Gobetti 101, 40129 Bologna, Italy

S. D'Amico
Department of Geosciences, University of Malta, Msida Campus, Msida, MSD 2080, Malta

N. De Paola
Department of Earth Sciences, University of Durham, Durham DH1 3LE, UK

S. D'Amico (ed.), *Moment Tensor Solutions*, Springer Natural Hazards,
https://doi.org/10.1007/978-3-319-77359-9_28

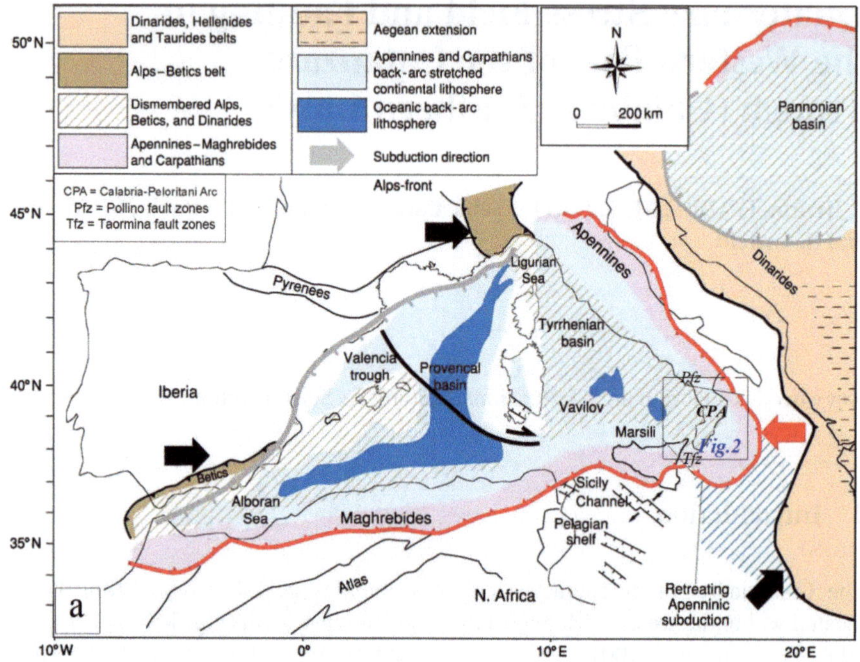

Fig. 1 Present geodynamic setting of the western Mediterranean region (modified from Carminati et al. 2005)

The Calabrian Arc represents the area with the highest seismogenetic potential of the whole Italy (Gasparini et al. 1982; Westaway 1993; Castello et al. 2005; Basili et al. 2008; Rotondi 2010; Presti et al. 2013). Both historical and instrumental seismological data show that the main catastrophic earthquakes may have ruptured along NE-SW oriented normal fault systems (Fig. 2): the Serre and Scilla Fault systems, responsible for the 1783 seismic sequence with Mw ranging between 6.5 and 7 (Galli and Bosi 2002; Galli et al. 2007; Ferranti et al. 2008); the Crati Graben faults system, associated with the 1184, 1870, and 1980 events with Mw = 6.7, 6.2 and 4.4, respectively (Valensise and Pantosti 1992; Tortorici et al. 1995); the Messina Fault responsible for the large Mw > 7.1 1908 earthquake (Tinti et al. 1999; Billi et al. 2008; Polonia et al. 2012). Finally, the Sant'Eufemia Fault has been mapped by Loreto et al. (2013), and its recent activity has been associated with the 1905 earthquake, Mw = 7, whose epicentre was located within the Gulf of Sant'Eufemia, SE Tyrrhenian Sea (Rovida et al. 2011).

In this seismotectonic frame, the Catanzaro Trough (Fig. 2; Brutto 2016; Brutto et al. 2016), is located between the northern and southern Calabrian Arc, extending from the Ionian to the Tyrrhenian Sea. It has a strong seismic activity, as shown by events with epicentral intensity of XI MCS (Mercalli–Cancani–Sieberg) scale and magnitude M > 6.5 (Pondrelli et al. 2006).

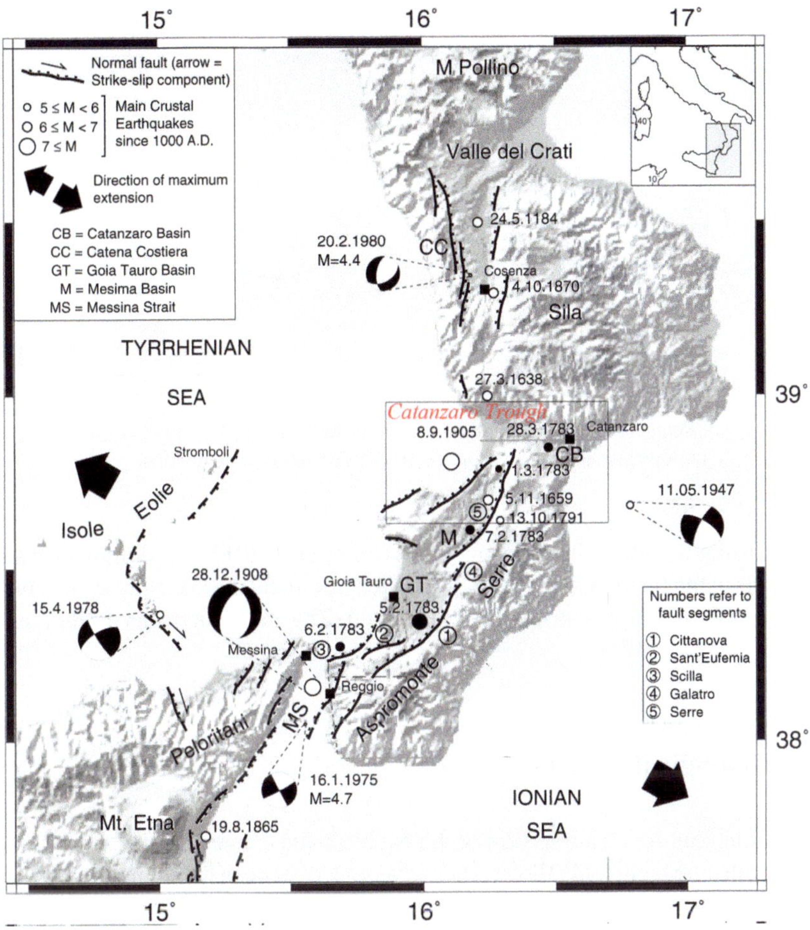

Fig. 2 Seismotectonic map of the Calabrian Arc. The 1783 earthquake sequence is marked with black dots (modified from Jacques et al. 2001)

Using focal mechanisms coming recent activity and instrumental earthquakes, it has been possible to retrieve important faults information related to the orientation of motion along transform faults (e.g. Wilson 1965; Sykes 1967; Ekström and Engdahl 1989) or understanding the main kinematic features of subducting slabs in different areas (Isacks and Molnar 1971); as well as to characterize seismic sequences (e.g. D'Amico et al. 2011).

In order to address the seismotectonic setting of the Catanzaro Trough, an integrated analysis of onshore geological and offshore geophysical data with an updated database of crustal focal mechanisms has been performed. The latter was compiled using several sources, relying on the inversion of historical earthquakes (Sirovich and Pettenati 2007; Loreto et al. 2012) and on focal mechanisms solutions, com-

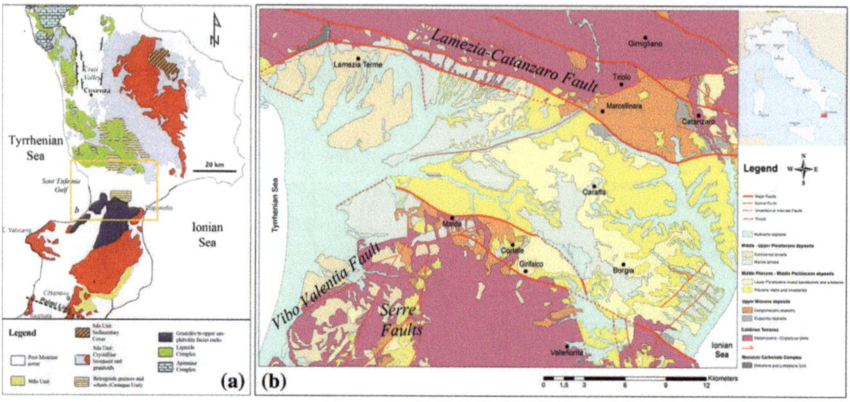

Fig. 3 **a** Simplified geological map of the Calabrian Arc (modified after Cirrincione et al. 2015); **b** Sketch of geological map of the Catanzaro Trough (Longhitano et al. 2014)

puted using waveform inversion methods (Presti et al. 2013). In addition to available datasets in the literature, in this study, about 8 new focal mechanisms were computed by using the CAP waveform inversion method (Zhu and Helmberger 1996; Zhu et al. 2006).

2 Geological Setting

The Calabrian Arc is a fragment of the Alpine chain connecting the southern Apennine with the Maghrebide Block. The convergence between the Nubian and Euroasian plates controls the NW-subduction and the SE-ward roll-back of the Ionian oceanic crust that caused the migration of the Calabrian block (Malinverno and Ryan 1986; Faccenna et al. 2005). The rapid trench migration also controlled the opening of the Tyrrhenian back-arc basin during the Upper Miocene-Pleistocene (Finetti et al. 1996; Neri et al. 2009), and the fragmentation of the arc into structural highs and sedimentary basins (Ghisetti 1979; Tansi et al. 2007; Critelli et al. 2017, 2013; Tripodi et al. 2013).

The Calabrian Arc hosts numerous longitudinal N-S to NE-SW trending and transversal E-W to WNW-ESE trending basins, which are bounded by normal and transcurrent main fault systems, respectively (Ghisetti 1979; Monaco and Tortorici 2000). Amongst the transversal basins, the *Catanzaro Trough* (*Study area* in Fig. 3a) is an atypical one, acting as a connection zone both between the northern and southern Calabrian Arc and between the western Tyrrhenian back-arc and eastern Ionian fore-arc basins.

The Catanzaro Trough is filled by Neogene-Quaternary sedimentary succession (Fig. 3b; Chiarella et al. 2012; Longhitano et al. 2014), unconformably overlie igneous-metamorphic units (Cavazza and DeCelles 1998).

The basal portion of the sedimentary succession is depicted by the Middle-Upper Miocene transgressive sequence consisting of continental conglomerates, fan-delta sediments and marine clays at the top and passing upward by means of an abrupt erosional surface to the Messinian sequence. The Messinian deposits are made up of thin and discontinuous limestone–gypsum strata (Barone et al. 2008; Govers et al. 2009; Muto et al. 2014; Brutto et al. 2016).

Pliocene conglomerate and marly deposits overlie the Messinian deposits (Zecchin et al. 2012, 2015). The conglomerate deposits contain evaporite derived clasts that gradually evolve to a fining and thinning upward sequence, whereas marls units show an affinity with the Trubi formation cropping out along the peri-Ionian Neogene basins (Cavazza and DeCelles 1998). These deposits are overlain by sands and sandstones consisting of tidalite-bearing, cross-stratified mixed siliciclastic–bioclastic deposits up to 60 m thick (Chiarella et al. 2012; Chiarella et al. 2016; Longhitano et al. 2014).

Finally, Middle–Upper Pleistocene marine terraces represent the top of the Catanzaro Trough sedimentary succession, which extensively outcrop along the western coast of the Catanzaro Trough. These Quaternary terraced deposits are generally made of siliciclastic sands and coarse sandstones with poor fossiliferous content.

The presence of various kinematics and orientations referred to trending fault systems suggests that the Catanzaro Trough underwent a more complex tectonic history compared to that of other transversal basins. Indeed, the complex interaction between these fault systems suggests an alternation of compressive, related to the transcurrent kinematics, and extensional events from the Miocene up to the Quaternary (Brutto et al. 2016).

The northern margin of the Catanzaro Trough shows the most prominent *transversal* faults system in the area, represented by a regional WNW-ESE-trending left-lateral shear zone (Van Dijk et al. 2000). This structural lineament consist of three S-dipping, right-stepping *en' echelon* major fault segments, where the southernmost segment is represented by the Lamezia-Catanzaro Fault (Fig. 3b; sensu Tansi et al. 2007). Moving southward, the NE-trending Serre and Vibo Valentia normal faults represent the most important onshore *longitudinal* fault system of the study area (Fig. 3b; Monaco and Tortorici 2000). These faults belong to the longitudinal fault belt that deform, together with the offshore NE-SW oriented Sant'Eufemia Fault (Loreto et al. 2013), the Pleistocene–Holocene sediments. This fault system forms in response to the intense regional uplift of the arc (Bianca et al. 2011; Monaco et al. 1997; Monaco and Tortorici 2000), driven by the opening of the southern Tyrrhenian back-arc basin.

3 Data and Methods

A multidisciplinary and multi-scale approach has been performed to analyze the active structures in the onshore and offshore areas of the Catanzaro Trough. New geological and structural data have been collected, and integrated with information

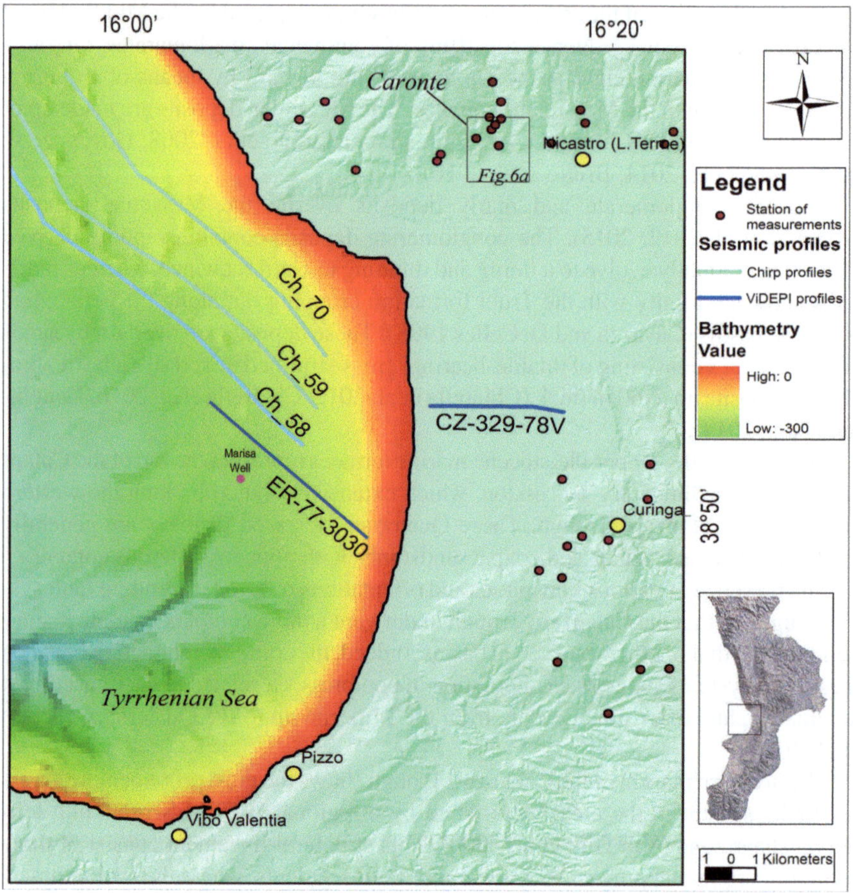

Fig. 4 Location map of the chirp seismic (high resolution) profiles (light blue line; part of the ISTEGE project), ViDEPI multichannel profiles (blue thick lines, part of ViDEPI project;), Eni wells (pink dots) and station of structural measurements (violet dots)

derived from onshore and offshore geophysical data (multichannel seismic reflection and sparker profiles, sub-bottom chirp profiles, and morpho-bathymetry) (Fig. 4). To constrain the active tectonics and stress field, an updated dataset of crustal focal mechanisms was compiled, including data available from the literature and 8 new fault plane solutions computed through the CAP waveform inversion method (D'Amico et al. 2011).

3.1 On-land Structural Data

A structural dataset was collected in the target area of the western Catanzaro Trough (Fig. 4), comprising more than 300 fault measurements (Brutto 2016) in the Quaternary deposits. Fault orientation and associated kinematic indicator data have been used to perform a stress inversion analysis using the DAISY software (Salvini 2002). The results constrain the stress field active in the study area from the Pleistocene to the Holocene time. The distribution of structural measurements has been analyzed and processed by using the DAISY software (Salvini 2002).

3.2 Geophysical Data Coming from ViDEPI and ISTEGE Projects

Mesostructural analysis previously described has objective limits as regards investigation of offshore and onshore area, characterized by recent and incoherent deposits. In order to improve the tectono-stratigraphic knowledge of western part of Catanzaro Trough, two crossing multichannel seismic (mcs) profiles have been analyzed. These data (see Fig. 4, dark blue lines) have been acquired in the Sant'Eufemia Gulf by Eni S.p.A. (Italian oil company) at the beginning of the 80' and made available in the frame of the VIDEPI project (unmig.sviluppoeconomico.gov.it/videpi/).

The sub-bottom *Chirp* profiles (Fig. 4, light blue lines), acquired aboard the R/V OGS-Explora in the frame of ISTEGE project in the summer 2010, allowed to define the shallow structural and stratigraphic features of the northeastern part of the Sant'Eufemia Gulf. The boundary and depth of shallow deposits were defined assuming a constant sound velocity of about 1550 m/s within shallowest sediments, see Loreto et al. (2012) for more details.

Management and interpretation of seismic profiles were performed using the Kingdom software (IHS Global Inc.), made available by ISMAR—CNR D.B. Bologna.

3.3 Earthquake Focal Mechanisms

In order to analyze the seismotectonic setting of the Catanzaro Trough, historical and instrumental seismic data and information were collected and analyzed, including:

1. historical earthquakes inversion performed in Sirovich and Pettenati (2007) and Loreto et al. (2012), which is based on expressing the radiation from an earthquake in terms of dimensionless values of a KF kinematic function (full details in Levine 1996; Sirovich and Pettenati 2004; Gentile et al. 2004);
2. all the available crustal focal mechanisms with a calculated epicentre depth, less deep than 30 km, performed by Presti et al. (2013) for the study region,

using waveform inversion methods. The selected *moment tensor solutions* used to reconstruct the database is based on (i) Italian CMT catalogue (http://www.bo.ingv.it/RCMT/Italydataset; Pondrelli et al. 2006 and European-Mediterranean RCMT Catalogue (http://www.bo.ingv.it/RCMT/; Pondrelli et al. 2004) covering the time period from 1977 to 2011, (ii) time-domain moment Tensor covering the time period from 2004 to 2011 (Scognamiglio et al. 2009), and (iii) Cut and Paste technique (CAP).
3. In addition to the mentioned above datasets, 8 new focal mechanisms with shallow epicentre depth (<30 km) computed through the CAP waveform inversion method have been included.

4 Results

4.1 Structural Data

The structural analysis was carried out at first on major structures occurring in the area (Brutto et al. 2016), although a large number of measurements derive from minor faults collected to the main structural system. Field scale structures have been acquired in the multilayered Pleistocene, more rarely Holocene successions and in the crystalline units of Calabride Complex, when the faults of basement units are directly juxtaposed with Quaternary deposits. Moreover, the faults displacement is estimated on order of approximately a few centimeters, although locally it reaches also meters. Structural measurements of faults and associated kinematic indicators, collected in the Quaternary deposits of the northern and southern sectors of the western Catanzaro Trough, have been plotted and analysed to reconstruct the chronological evolution of faulting and stress field during the Quaternary in the area (Fig. 5; DAISY software; Salvini 2002). Overall, the structural data display a scattered distribution (Fig. 5). Hence, we have divided the whole dataset into sub-datasets on the basis of fault orientation and kinematics (Fig. 5).

The NE-SW faults show a kinematics ranging from normal to strike-slip (Fig. 5). These data are collected mostly in well exposed Terme di Caronte site, located to the west of Lamezia Terme City and along the southern portion of the study area, near Curinga and Maida Villages. They represent the most widespread structural features in the area, although both N-S trending normal and NW-SE trending right-lateral faults are also numerically abundant within the dataset. Indeed, this faults system is collected diffusely along the north-eastern margin of the area, along the tectonic contact between cristalline basement and basin infill.

The NE-SW trending faults offset the Middle- Upper Pleistocene deposits (Fig. 6a and b). Within the dataset, reverse faults display nearly dip-slip striaes, whilst left-lateral faults display a local transpressional and transtensional kinematics with pitch ranging from 20° to 40° (Fig. 5). Cross-cutting relationships observed along the strike-slip faults suggest their later reactivation as normal faults. NE-SW normal

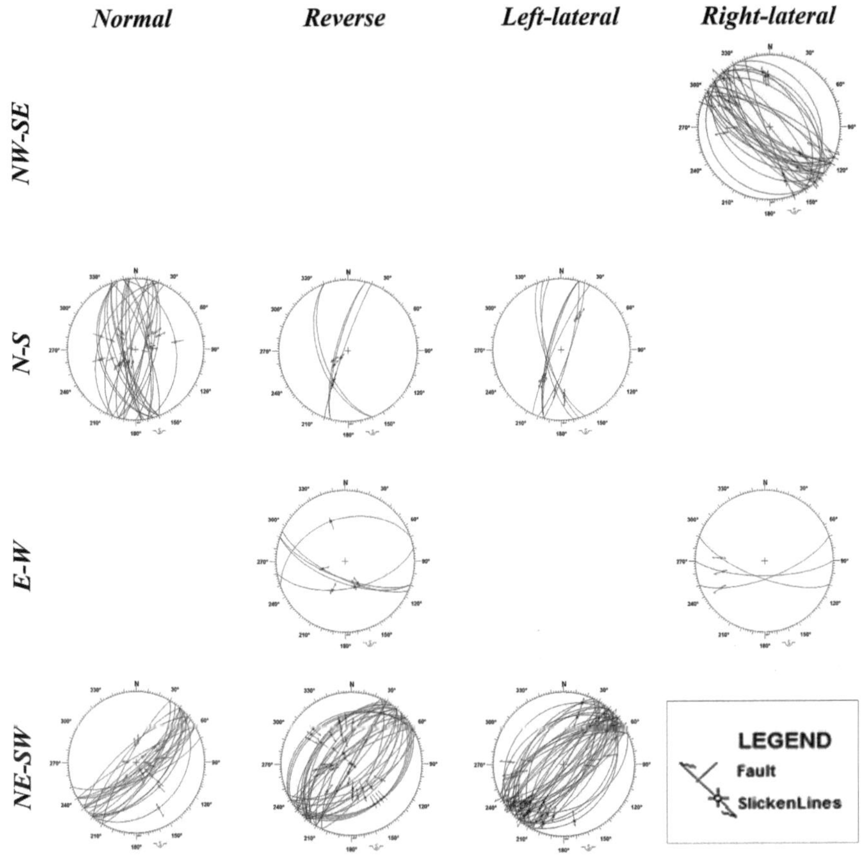

Fig. 5 Structural data, offsetting the Quaternary deposits, acquired within the study area based on kinematics and fault direction

faults show oblique-normal slickensides along sub-vertical fault planes (Fig. 5), with pitches ranging from 60° to 90°. These dip-slip normal kinematic indicators often overprint the pre-existing strike-slip-oblique kinematic indicators (Fig. 5).

The N-S trending faults show sub-vertical planes, equally dipping toward the West and East direction, with a predominance of normal dip- and oblique-slip kinematics (Fig. 5).

The presence of minor E-W oriented reverse and right lateral faults, and NW-striking right lateral faults has also been revealed in the Quaternary sedimentary units of the basin (Fig. 5). The E-W and NW-SE structural lineaments display sub-vertical fault planes. The slickensides clearly revealed oblique kinematics, and we rarely observed pure strike-slip faults.

At the mesoscale, cross-cutting relationships suggest a reactivation of the NW-SE left lateral strike-slip faults as right lateral faults, thus documenting complex and varying tectonic stages within the area. Compared to the NE-SW and N-S trending

Fig. 6 **a** NE-SW oriented normal to oblique faults offsetting the Triassic dolostone and the Middle Pleistocene terraces, **b** Continuation of the same NE-SW fault plane; (inset) stereographic projection of NE-SW fault plane showed in the figure **a** and **b**. See Fig. 4 for location

fault systems, the E-W and NW-SE trending fault systems only affect the lower portion of the Quaternary sedimentary units in the basin.

Stress inversion analysis was applied to the fault populations measured in the Quaternary units, to estimate a regional stress tensor. The faults were grouped and analysed for each sub-datasets based on their relative age obtained from cross-cutting relationships observed in the field (Brutto 2016; Brutto et al. 2016). The results of stress inversion analyses can be summarized as follows:

1. The NE-SW and N-S trending normal faults are observed in the field to offset the Middle Pleistocene deposits. Stress inversion analyses for this set of faults show an extensional stress field, characterised by a vertical maximum principal stress, σ_1, and an horizontal, N270° E-trending, minimum principal stress σ_3. The orientation of the σ_3 defines an E-W oriented extensional regime active during the formation of such faults (Fig. 7).
2. Two distinct fault populations, cutting through the Lower Pleistocene deposits, can be identified on the basis of their orientation and kinematics (Fig. 8). The first population comprises one set of conjugate, NE-SW trending left-lateral and NW-SE trending right-lateral faults. Stress inversion analyses of this fault population returns a stress field consistent with strike slip faulting, characterized by a sub-horizontal, ca. N-S oriented, maximum principal stress axis (σ_1), and a sub-horizontal, E-W oriented minimum principal stress axis σ_3 (Fig. 8a). The second fault population is represented by NE-SW trending reverse faults, consistent with a sub-orizzontal σ_1 axis, and a sub-vertical σ3 axis in the NW-SE orientation (Fig. 8b).

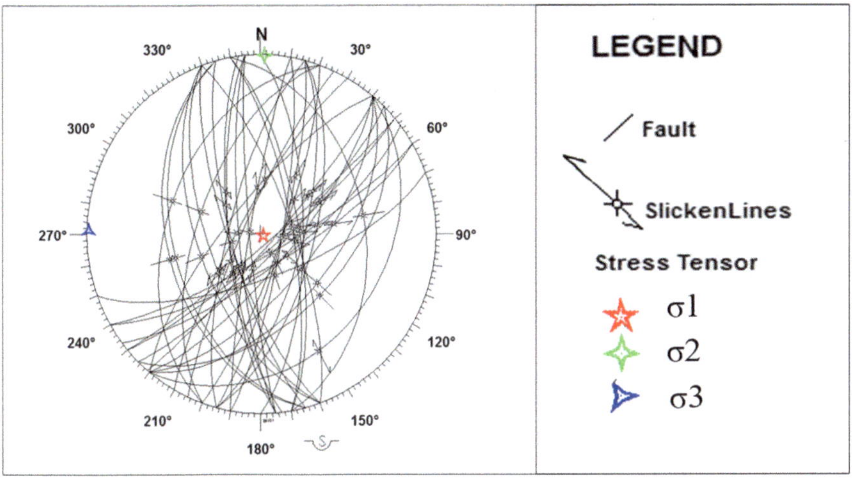

Fig. 7 Stereographic projection of the post Lower Pleistocene associated faults with stress field (Schmidt's net, lower hemisphere)

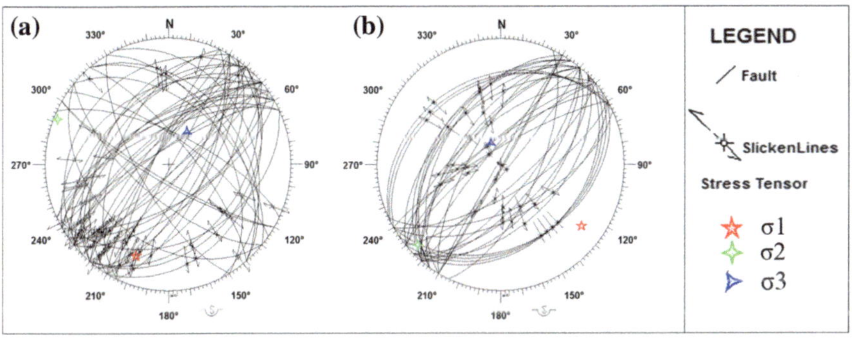

Fig. 8 Stereographic projection of the Piacenzian—Lower Pleistocene associated faults with stress field of faults (Schmidt's net, lower hemisphere): **a** NE-SW left lateral and NW-SE right lateral associated faults, **b** NE-SW reverse faults

4.2 Geophysical Data

Structural data have been integrated with further onshore and offshore geophysical data (morpho-bathymetric data, Chirp and ViDEPI seismic profiles).

Despite the low quality of ViDEPI profiles, the E-W oriented CZ-329-78 mcs profile (Fig. 9b) shows thickness increase and small dislocations of sedimentary layers amongst shot points 160 and 190. These features are associated to the activity of a recently identified W-dipping normal fault, named *San Pietro Lametino Fault*, affecting the units from Upper Miocene to Pleistocene (Fig. 9a, b). This study, combined with an unpublished VIPEDI profile and fieldwork survey carried out in the

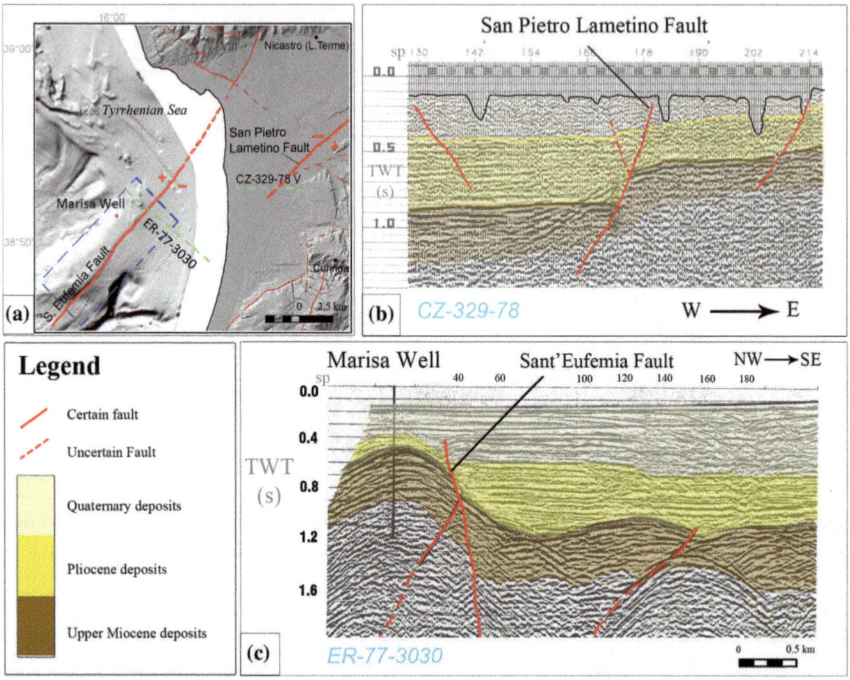

Fig. 9 Geophysical data: **a** high resolution morpho-bathymetry and digital elevation model (DEM, cell size: 10 m) of the western Catanzaro Trough. Blue dashed box represent elongated high, whereas dashed green lines represent two ViDEPI mcs profiles; **b** multi-channel seismic CZ-329-77 (from ViDEPI project), **c** multi-channel seismic ER-77-3030 (from ViDEPI project)

western portion of the Catanzaro Trough, allowed us to extent northward this fault deducing a NE-SW fault direction.

Similarly, the ER-77-3030 ViDEPI mcs profile (Fig. 9c) images a gentle fold, which axis is NE-SW oriented as also noticeable as an elongated high on morpho-bathymetric map (Fig. 9a), likely controlled by a deeper E- or SE-ward verging thrust fault. The eastern side of the anticline (Fig. 9c) is bordered by a normal fault, namely Sant'Eufemia Fault, that crosses the entire Upper Miocene—Quaternary sedimentary succession, contributing thus to the partial E-ward deepening of the Sant'Eufemia Gulf.

Figure 10 includes three segments of Chirp profiles, acquired orthogonally and further north of the above described NE- elongated high (Fig. 9a), imaging the recent and well stratified Quaternary deposits dislocated by a normal fault that deepen sediments SE-ward. This normal fault, characterized by mainly NE-SW direction, represent the upward termination of a much deeper structure corresponding to the wide NE-trending Sant'Eufemia Fault recognized on ViDEPI (Fig. 9c) and GSE10_07 mcs profile (Loreto et al. 2013). The analysis of chirps has revealed a little displacement in the shallowest sediments (ranging between TWT 0.075 s and TWT 0.2 s of

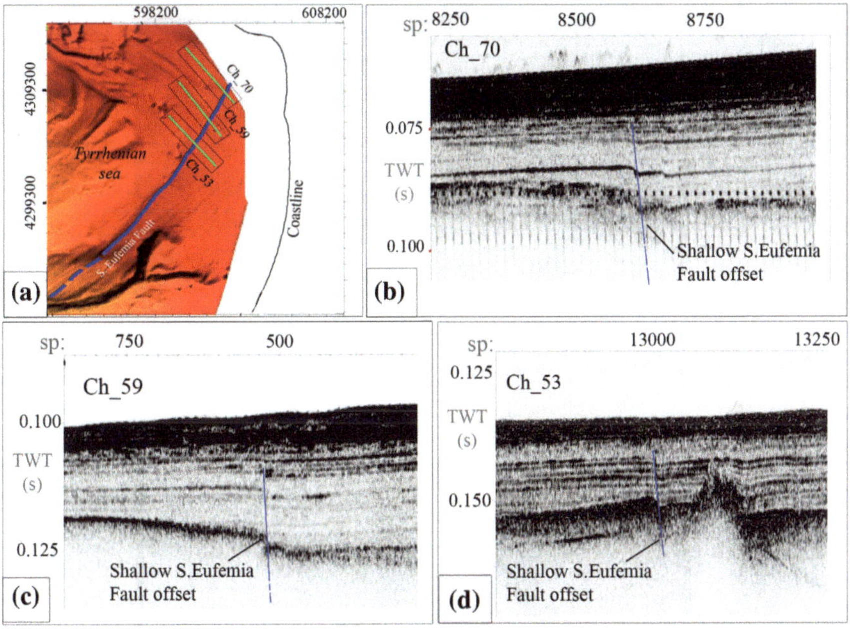

Fig. 10 **a** Location of three chirp profiles (green lines), whereas blue line represent the interpretation of the Sant'Eufemia Fault, **b** Ch_70; **c** Ch_59 and **d** Ch_53 Chirp profile

depth). The dislocations of recent deposits and, locally, of the seafloor (see Loreto et al. 2013) witness the recent activity of this structure considered one of the best candidate as seismogenic source of the 1905 Calabrian earthquake (Loreto et al. 2013; Sandron et al. 2015).

4.3 *Earthquakes Focal Mechanisms Solution*

For this study the obtained database includes a number of focal mechanisms, displayed in Fig. 11, coming from published and unpublished database sources:

- Two historical earthquakes inversions represented by the 28 March 1783 and the 8 September 1905 events (blue in Fig. 11), modelled in Sirovich and Pettenati (2007) and in Loreto et al. (2012), respectively.
- Five crustal focal mechanisms computed for the study region by using waveform inversion methods performed by Presti et al. (2013) (red in Fig. 11; Li et al. 2007; D'Amico et al. 2010, 2011).
- New inverted focal mechanisms, related to 8 seismic events, computed by using the CAP waveform inversion method (green in Fig. 11; D'Amico et al. 2010, 2011).

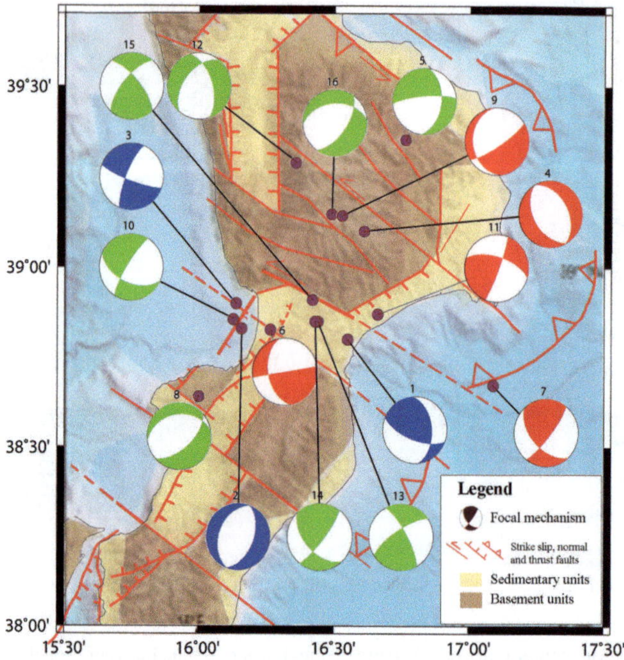

ID	Date	O.T.	Lon	Lat	Depth	Strike	Dip	Rake	Mag	M-Type	FT	Reference
1	28/03/1783		16.55	38.80	20.2	111	65	229	7.2	M	SS	Sirovich and Pettenati 2007
2	08/09/1905		16.16	38.83	22.9	21	50	270	7	Mo	NF&NS	Loreto et al., 2012
3	08/09/1905		16.14	38.90	17.0	110	80	-14	5.8	Mo	SS	Loreto et al., 2012
4	18/10/2001	11.02.44	16.61	39.10	10.0	332	44	-88	4.01	Mb	NF&NS	Presti et al. 2013
5	23/04/2005	19.11.42	16.83	39.46	19.0	265	78	-45	4.3	Mw	SS	This study
6	07/09/2005	12.40.33	16.32	38.71	16.0	80	90	-42	5.06	Mw	U	Presti et al. 2013
7	27/09/2005	22.33.09	17.10	38.62	29.0	38	79	141	3.09	Mw	SS	Presti et al. 2013
8	30/10/2005	19.09.47	15.93	38.53	24.0	48	59	-108	3.7	Mw	NF&NS	This study
9	18/01/2008	13.01.00	16.55	39.15	15.5	57	78	-67	3.09	Mw	NF&NS	Presti et al. 2013
10	16/06/2010	22.39.41	16.13	38.86	14.0	212	80	-140	3.8	Mw	SS	This study
11	15/10/2010	5.21.19	16.66	38.87	15.0	287	62	173	4.04	Mw	SS	Presti et al. 2013
12	28/12/2014	21.43.38	16.36	39.29	16.0	218	50	-47	4.3	Mw	NF&NS	This study
13	06/02/2015	19.09.00	16.44	38.85	11.0	331	76	18	3.22	Mw	SS	This study
14	06/02/2015	19.13.00	16.43	38.85	29.0	38	79	141	3.09	Mw	TF&TS	This study
15	30/05/2015	12.18.28	16.42	38.91	11.0	320	70	25	3.1	Mw	SS	This study
16	03/08/2015	7.27.49	16.49	39.15	19.0	35	60	-121	4.1	Mw	NF&NS	This study

Fig. 11 (above) Selected crustal earthquake focal mechanisms for the large Calabrian Arc area including Calabria, eastern and western Tyrrhenian Sea, coming from Presti et al. (2013), in red; from Sirovich and Pettenati (2007), in blue, and elaborated in *this study* in green. Table (below), *FT* is faulting type definition: *NF* and *NS* normal faulting and normal faulting with a strike-slip component, *SS* strike-slip faulting, *TF* and *TS* thrust faulting and thrust faulting with a strike-slip component, and *U* unknown stress regime (Zoback 1992)

4.4 Historical Earthquakes Inversion

Inversion of macroseismic field of 1905 and 1783 earthquakes uses as input to the KF kinematic function (blue focal mechanisms in Fig. 11). The 1905 seismic event, represented by 447 intensity data points, provided two probable fault sources (see Fig. 9 in Loreto et al. 2012): (1) a 40 km-long dip-slip fault, N21°-oriented; and (2) a more than 40 km-long strike-slip fault, N110°-striking, both fault sources are located within the Sant'Eufemia Gulf and affect the western Calabria region. The ambiguity of the inversion has been later solved comparing models with geophysical data, which revealed the presence of a normal fault N31° oriented, SE-dipping (Loreto et al. 2013), named Sant'Eufemia Fault; allowing of associating the source to the fault.

Likewise, the fault source of the 28 March 1783 earthquake was defined by using the regional macroseismic intensity field produced by the event (Sirovich and Pettenati 2004, 2007). The obtained fault source model is represented by a strike-slip fault ca. 45 km-long, with N111° direction, displaying a right lateral kinematics with reverse component of motion.

4.5 Moment Tensor Solution

Five crustal focal mechanisms computed for the study region were selected from Presti et al. (2013) (red focal mechanisms in Fig. 11): Italian Centroid Moment Tensor (ItCMT); Time-Domain Moment Tensor (TDMT); D'Amico et al. (2011) (CAP-BGTA); D'Amico et al. (2010) (CAP-BSSA).

These five crustal focal mechanisms show sub-vertical fault planes with direction ranging from NE-SW to NW-SE, equally distributed between normal and strike slip kinematics. The normal slip focal mechanisms 4 and 9 (see ID in the table of Fig. 11) show a main fault planes strike N332 and N57, respectively. The strike slip focal mechanisms 7 and 11 show a right-lateral kinematics, even though the orientation of two computed faults is different (N38 and N287, respectively). Amongst these, we have computed a focal mechanism related to a fault plane (6 event ID) with an undetermined kinematics.

4.6 Earthquakes Focal Mechanisms of Recent Earthquakes

In this study a number of new focal solutions have been computed by using the CAP Method (Zhu and Helmberger 1996). In the CAP method (Zhu and Helmberger 1996) each waveform is broken up into Pnl (i.e., the first arrivals from a seismic source located in the crust eventually including waves reflected and multi-reflected from the top of the sharpest discontinuity) and into surface wave, which are weighted differently during the inversion procedure. The surface waves, although large in amplitudes, are easily influenced by heterogeneities of the surface crust, whereas Pnl waves are controlled by the average crustal velocity and are therefore more stable. Ground velocity data are preferred to ground displacement mainly because of the use of weak-motion data and because, for earthquakes of magnitude less than 4, there is a high signal-to-noise ratio only at higher frequencies (Presti et al. 2013).

All these features make the CAP method effective for earthquakes over a wide range of magnitudes as also proven by several tests and comparisons (down to a minimum of 2.6; D'Amico et al. 2010, 2011, 2014). In the table in Fig. 11 are listed the focal solutions computed in this study. For more details about the CAP method refer to Zhu and Helmberger (1996), Tan et al. (2006) and D'Amico et al. (2010, 2011),

The computed focal mechanisms show similarly to the previous sections almost equally distribution between normal/oblique and strike slip kinematics.

The fault planes obtained by using waveform inversion belong to two fault systems, the first one shows orientation ranging from N32° to N48° (8, 10, 12, 14, 16 events ID), with dip-slip kinematics except the 10 and 14 events ID (Fig. 10). The second one is oriented between N265° and N331° (5, 13, 15 events ID), where all three focal mechanisms display left lateral kinematics (Fig. 11).

5 Discussion

On-land and off-shore structural and geophysical data were analysed and integrated with historical and instrumental seismicity, to identify the active Quaternary faults and define a sismotectonic model of the Central Calabrian Arc (Catanzaro Trough).

Analysis of faults offsetting the Lower-Middle Pleistocene deposits and interpretation of the ViDEPI Multichannel and ISTEGE high-resolution seismic (Chirp) profiles allowed us to reconstruct the Quaternary tectonics of central Calabrian Arc. In particular, NE–SW and N-S oriented normal faults have been identified as the main fault systems in the study area (Figs. 6 and 7), which controlled its tectonic evolution. They diffusely dislocate the late Quaternary deposits, favoring the development of NE-SW elongated sedimentary basins, as the *Lamezia Basin* (Fig. 12) which considered an emerging onshore to offshore graben-like system (Brutto et al. 2016; Brutto 2016). Conjugate systems with NW-SE right-lateral and NE-SW left lateral faults (Fig. 8), were also observed to displace the Lower Pleistocene deposits. However, due to the lack of outcrop exposure of younger deposits displaced by these structures, neither their very recent activity or their regional relevance can be unambiguously proved and assessed.

Onshore geophysical data allow the definition of two NE-trending and NW-dipping fault segments, the San Pietro Lametino Fault (Fig. 9b; Brutto et al. 2016) and the Vibo Valentia Fault (Tortorici et al. 2002). These faults are arranged in a left-stepping, *en'echelon* geometry (*overstepping lineaments*), generating the prominent morphological escarpments along the southwestern side of the Catanzaro Trough (Fig. 12), These findings carry some relevant implications in terms of tectono-stratigraphic reconstruction of the area and seismic hazard.

Offshore geophysical data show that the main fault trend is represented by the Sant'Eufemia Fault. This fault locally deforms recent sediments (Figs. 9c and 10), up to the seafloor (Loreto et al. 2013), suggesting that it is currently active. Our novel data show that this fault can be prolonged onshore, up to the Terme di Caronte

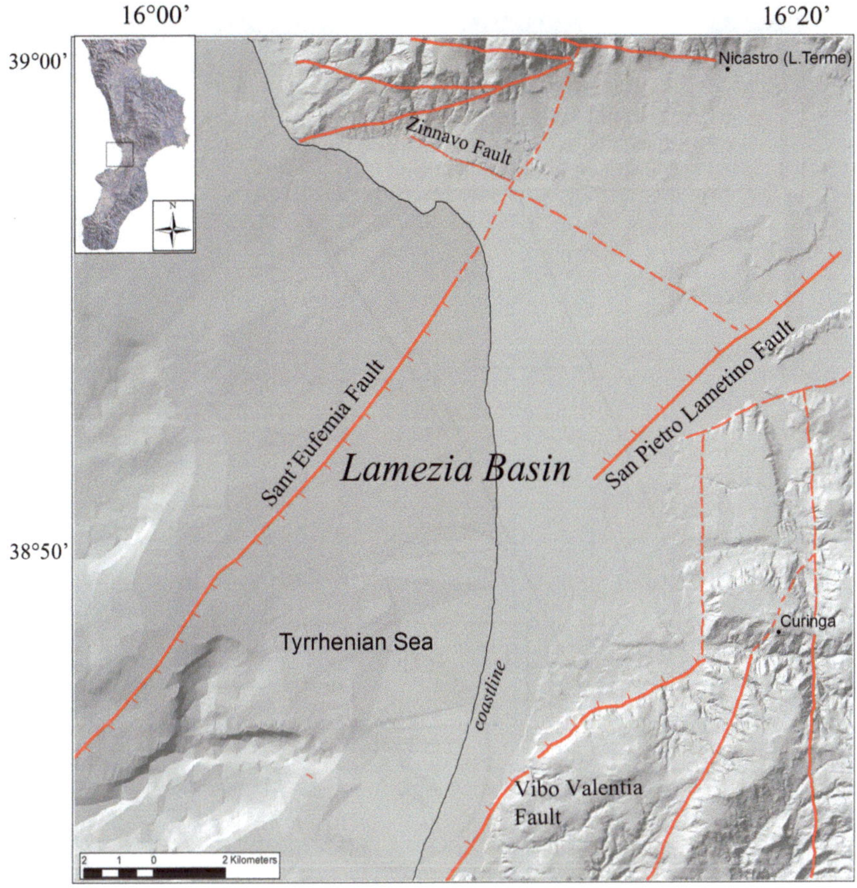

Fig. 12 Structural map with major faults of the western Catanzaro Trough

locality, where there is field evidence well supported by structural data (see Fig. 6). Based on the integration of off-shore interpretations and new field observations and data, we propose that the Sant'Eufemia Fault may extend up to 30 km-length. These findings carry some relevant implications in terms of seismic hazard, as they suggest that a longer fault segment than previously hypothesized may rupture, producing a more energetic seismic event.

Crustal earthquake focal mechanisms, computed for the study area by using the waveform inversion methods (Li et al. 2007; D'Amico et al. 2010, 2011; Presti et al. 2013), and the inversion of the historical 28th March 1783 and 8th September 1905 earthquakes (Sirovich and Pettenati 2007; Loreto et al. 2012) show that an extensional regime is affecting the entire Calabria region. Field observations and structural data collected in the Quaternary deposits of the study area are in strong agreement with both instrumental and historical seismological data (Fig. 7). New focal mecha-

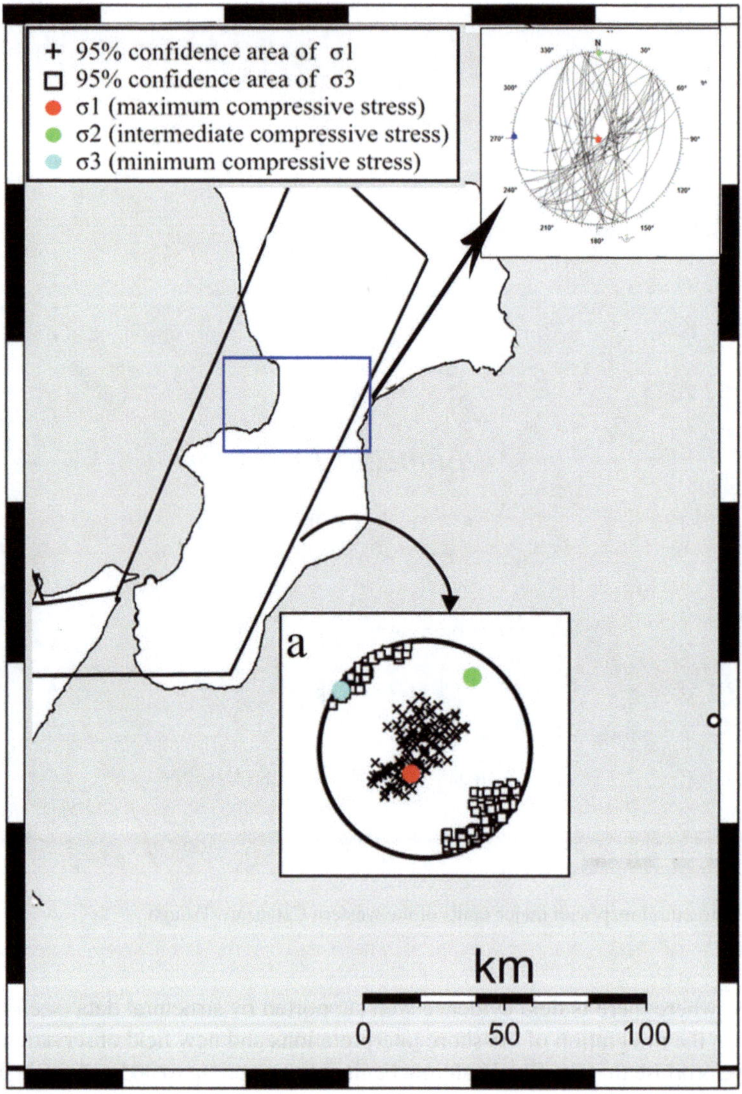

Fig. 13 Orientations of the principal stress axes obtained by inversion of **a** the earthquake focal mechanisms from Presti et al. 2013 and **b** NE-SW and N-S inverted structural data and shown in the present work

nisms of selected extensional earthquakes obtained in this paper show two main fault trends with NW-SE and NE-SW orientation (table in Fig. 11), consistent with field observations and structural data.

The NW-trending faults are characterized by both normal and transcurrent kinematics likely related to the structural elements bordering the Catanzaro Trough to the north and to the south (Lamezia-Catanzaro Fault, sensu Tansi et al. 2007; Soverato-Lamezia Fault Zones, sensu Van Dijk et al. 2000; Maida- Punta Stalettì Faults System, sensu Ghisetti 1979). Whereas, the NE-SW lineaments show dominant normal kinematics, with minor strike-slip component of motion. The focal mechanisms showing normal kinematics well agree with the onshore structural and geophysical (seismic profiles) data, defined in the previous sections (Figs. 7, 8 and 9).

Finally, the extensional regime obtained from the stress inversion applied by Presti et al. (2013), computed through waveform analyses by using the Gephart and Forsyth (1984) and Gephart (1990) methods, have been combined with the stress field derived by structural data selected in this work. The combined results, by bibliographic and new data, show a sub-horizontal σ_3, with N304°E trend (Fig. 13a, Presti et al. 2013), defining a WNW-ESE oriented extensional stress regime, that closely match the stress field newly elaborated in this work (Figs. 7 and 13b). The observed extensional stress regime fits perfectly with the Quaternary WNW-ESE oriented extension affecting the entire Apennine chain. This is also in agreement with the graben and half-graben systems, which are widespread in the Calabrian Arc, e.g. the Crati and the Mesima Grabens, the Gioia Tauro plain and the recently recognized Lamezia Graben.

Furthermore, our geophysical and field/mesostructural data indicate that two fault populations were active during the Early to Middle (?) Pleistocene. These faulting events may be associated with transcurrent and compressive tectonic episodes, characterized by sub-horizontal, ca. N-S oriented and NW-SE oriented maximum principal stress axis (σ_1), respectively.

6 Conclusion

In the frame of the Calabrian Arc evolution, we analyzed the stress field controlling the evolution of the Catanzaro Trough.

The combination of geological, geophysical and seismotectonic data shows that, during the Early Pleistocene, a contraction episode occurred within the basin, possibly related to the regional N- to NE-oriented regional compression. This gradual increase of compressive inter-plate stress is linked with transpressional tectonics and associated block rotations. Later, since the Middle Pleistocene to Holocene age, the study area experienced a WNW-ESE oriented extensional phase, in response to one of the last extensional stages controlled by the Tyrrhenian back-arc basin opening. NE-SW and N-S oriented normal faults control this tectonic phase, which partitioning obliquely the western Catanzaro Trough producing up-faulted and down-faulted blocks that arranged as graben-type systems. Amongst these systems we identified a new one named Lamezia Basin. Based on these considerations and on structural data, the extensional boundary defined for the central Calabrian Arc can be widen further offshore.

In agreement with Jacques et al. (2001), who state that the main destructive earthquake sources bound the NNE-SSW trending steep range fronts (Coastal and Serre Range, Aspromonte and Messina Strait), the NE-SW trend normal faults play a relevant role as part of recent seismotectonic processes controlling the Late Quaternary geodynamics of the central Calabrian Arc. Thus representing the source of the main destructive historical and very recent earthquakes occurred in the area. Although the role of NW-SE strike-slip faults have to be still completely understood.

Considering these features, the improvement of geological and structural knowledge of the western Catanzaro Trough and its offshore could provide new insights about the evaluation of the seismic and tsunami hazard in the frame of a new urban development planning and the existing infrastructures safeguarding.

Acknowledgements This work is part of Ph.D. work Thesis (Brutto 2016). We are grateful to the OGS (Trieste) for user licence of geophysical dataset coming from the ISTEGE project. We wish to thank CNR-ISMAR of Bologna for permission to consult seismic reflection profiles carried out in the S. Eufemia Plain and to publish its line drawings.

References

Barone M, Dominici R, Muto F, Critelli S (2008) Detrital mode in late Miocene wedge-top basin northeasten Calabria, Italy: compositional record of wedge—top partitioning. J Sediment Res 78:693–711

Basili R, Valensise G, Vannoli P, Burrato P, Fracassi U, Mariano S, Tiberti MM, Boschi E (2008) The database of individual seismogenic sources (DISS), version 3: summarizing 20 years of research on Italy's earthquake geology. Tectonophysics 453(1–4):20–43. https://doi.org/10.1016/j.tecto.2007.04.014

Bianca M, Catalano S, De Guidi G, Gueli AM, Monaco C, Ristuccia GM, Stella G, Tortorici G, Tortorici L, Troja SO (2011) Luminescence chronology of Pleistocene marine terraces of Capo Vaticano Peninsula (Calabria, Southern Italy). Quatern Int 232:114–121

Billi A, Funiciello R, Minelli L, Faccenna C, Neri G, Orecchio B, Presti D (2008) On the cause of the 1908 Messina tsunami, southern Italy. Geophys Res Lett 35:L06301. https://doi.org/10.1029/2008GL033251

Brutto F (2016) Geological and structural evolution of tectonically active areas of Central Calabria Arc. Unpublished Ph.D. thesis, University of Calabria

Brutto F, Muto F, Loreto MF, Tripodi V, Critelli S (2016) Transition from strike-slip to extensional tectonics in the Plio-Pleistocene Catanzaro Trough (Calabria, south Italy). Rend Online Soc Geol It 38:9–12

Carminati E, Negredo AM, Valera JL, Doglioni C (2005) A Subduction-related intermediate-depth and deep seismicity in Italy: insights from thermal and rheological modeling. Phys Earth Planet Inter 149:65–79

Castello B, Selvaggi G, Chiarabba C, Amato A (2005) CSI Catalogo della sismicità italiana 1981–2002. Versione 1.0 (INGV-CNT, Roma)

Cavazza W, Decelles PG (1998) Upper Messinian siliciclastic rocks in southeastern Calabria (S Italy): paleotectonic and eustatic implications for the evolution of the central Mediterranean region. Tectonophysics 298:223–241

Chiarella D, Longhitano SG, Muto F (2012) Sedimentary features of the Lower Pleistocene mixed siliciclastic-bioclastic tidal deposits of the Catanzaro Strait (Calabrian Arc, south Italy). Rendiconti online Società Geologica Italiana 21:919–920

Chiarella D, Moretti M, Longhitano SG, Muto F (2016) Deformed cross-stratified deposits in the Early Pleistocene tidally-dominated Catanzaro strait-fill succession, Calabrian Arc (Southern Italy): Triggering mechanisms and environmental significance. Sediment Geol 344:277–289

Cirrincione R, Fazio E, Fiannacca P, Ortolano G, Pezzino A, Punturo R (2015) The Calabria-Peloritani Orogen, a composite terrane in Central Mediterranean; its overall architecture and geodynamic significance for a pre-Alpine scenario around the Tethyan basin Periodico di Mineralogia 84:I 3B

Critelli S, Muto F, Tripodi V, Perri F (2013) Link between thrust tectonics and sedimentation processes of stratigraphic sequences from the southern Apennines foreland basin system, Italy. Rend Soc Geol Ital 25:21–42

Critelli S, Muto F, Perri F, Tripodi V (2017) Interpreting provenance relations from sandstone detrital modes, southern Italy foreland region: Stratigraphic record of the Miocene tectonic evolution. Mar Pet Geol 87:47–59

D'Agostino N, Selvaggi G (2004) Crustal motion along the Eurasia-Nubia plate boundary in the Calabrian Arc and Sicily and active extension in the Messina Straits from GPS measurements. J Geophys Res: Solid Earth (1978–2012), 109(B11)

D'Amico S, Orecchio B, Presti D, Zhu L, Herrmann RB, Neri G (2010) Broadband waveform inversion of moderate earthquakes in the Messina Straits, southern Italy. Phys Earth Planet Inter 179:97–106

D'Amico S, Orecchio B, Presti D, Gervasi A, Zhu L, Guerra I, Neri G, Herrmann RB (2011) Testing the stability of moment tensor solutions for small earthquakes in the Calabro-Peloritan Arc region (southern Italy). Bollettino di Geofisica Teorica ed Applicata 52(2):283–298

D'Amico S, Cammarata L, Cangemi M, Cavallaro D, Di Martino R, Firetto Carlino M (2014) Seismic moment tensors and regional stress in the area of the December 2013-January 2014, Matese earthquake sequence (Central Italy). J Geodyn 82:118–124. https://doi.org/10.1016/j.jog.2014.09.008

Ekström G, Engdahl ER (1989) Earthquake source parameters and stress distribution in the Adak Island region of the central Aleutian Islands, Alaska. J Geophys Res 94:15499–15519

Faccenna C, Funiciello F, Giardini D, Lucente P (2001) Episodic back-arc extension during restricted mantle convection in the Central Mediterranean. Earth Planetary Science Letters 187:105–116

Faccenna C, Civetta L, D'Antonio M, Funiciello F, Margheriti L, Piromallo C (2005) Constraints on mantle circulation around the deforming Calabrian slab. Geophys Res Lett 32:L06311. https://doi.org/10.1029/2004GL021874

Ferranti L, Monaco C, Morelli D, Antonioli F, Maschio L (2008) Holocene activity of the Scilla Fault, Southern Calabria: insights from coastal morphological and structural investigations. Tectonophysics 453(1):74–93

Finetti I, Lentini F, Carbone S, Catalano S, Del Ben A (1996) Il sistema Appenninico Meridionale-Arco Calabro–Sicilia nel Mediterraneo Centrale: studio geofisico–geologico. Bollettino della Società Geologica Italiana 115:529–559

Galli P, Bosi V (2002) Paleoseismology along the Cittanova fault: implications for seismotectonics and earthquake recurrence in Calabria (southern Italy). J Geophys Res 107:B3. https://doi.org/10.1029/2001JB000234

Galli P, Scionti V, Spina V (2007) New paleoseismic data from the Lakes and Serre faults: seismotectonic implications for Calabria (Southern Italy). Ital J Geosci 126:347–364

Gasparini C, Iannacone G, Scandone P, Scarpa R (1982) Seismotectonics of the Calabrian Arc. Tectonophysics 82:267–286

Gentile F, Pettenati F, Sirovich L (2004) Validation of the automatic nonlinear source inversion of the U.S. Geological Survey intensities of the Whittier Narrows, 1987 earthquake. Bull Seismol Soc Am 94(5):1,737–1,747

Gephart JW, Forsyth DW (1984) An improved method for determining the regional stress tensor using earthquakes focal mechanisms data: application to the San Ferdinando earthquake sequence. J. Geophysical Research 89:9305–9320

Gephart JW (1990) Stress and the direction of slip on the fault planes. Tectonics 9:845–858

Ghisetti F (1979) Evoluzione neotettonica dei principali sistemi di faglie della Calabria centrale. Bolletino Società Geologica Italiana 98:387–430

Govers R, Meijer P, Rijgsman W (2009) Regional isostatic response to Messinian salinity crisis events. Tectonophysics 463:109–129

Gueguen E, Doglioni C, Fernandez M (1998) On the post-25 Ma geodynamic evolution of the western Mediterranean. Tectonophysics 298:259–269

Isacks B, Molnar P (1971) Distribution of stresses in the descending lithosphere from a global survey of focal-mechanism solutions of mantle earthquakes. Rev Geophys Space Phys 9:103–174

Jacques E, Monaco C, Tapponier P, Tortorici L, Winter T (2001) Faulting and earthquake triggering during the 1783 Calabria seismic sequence. Geophys J Int 147:499–516

Levine D (1996) Users guide to the PGAPack parallel genetic algorithm library. Rep. Argonne National Laboratory, ANL-95/18, Argonne, IL, p 73

Li H, Michelini A, Zhu L, Bernardi F, Spada M (2007) Crustal velocity structure in Italy from analysis of regional seismic waveforms. Bull Seismol Soc Am 97:2024–2039. https://doi.org/10.1785/0120070071

Longhitano SG, Chiarella D, Muto F (2014) Three-dimensional to two-dimensional cross-strata transition in the lower Pleistocene Catanzaro tidal strait transgressive succession (southern Italy). Sedimentology 61:2136–2171. https://doi.org/10.1111/sed.12138

Loreto MF, Zgur F, Facchin L, Fracassi U, Pettenati F, Tomini I, Burca M, Diviacco P, Sauli C, Cossarini G, De Vittor C, Sandron D, The Explora Technicians Team (2012) In search of new imaging for historical earthquakes: a new geophysical survey offshore Western Calabria (Southern Tyrrhenian Sea, Italy). Bollettino di Geofisica Teorica e Applicata 53:385–401. https://doi.org/10.4430/bgta0046

Loreto MF, Fracassi U, Franzo A, Del Negro P, Zgur F, Facchin L (2013) Approaching the potential seismogenic source of the 8 September 1905 earthquake: new geophysical, geological and biochemical data from the S. Eufemia Gulf (S Italy). Mar Geol 343:62–75

Malinverno A, Ryan WBF (1986) Extension in the Tyrrhenian sea and shortening in the apennines as result of arc migration driven by sinking of the lithosphere. Tectonics 5:227–245

Monaco C, Tapponnier P, Tortorici L, Gillot PY (1997) Late Quaternary slip rates on the Acireale-Piedimonte normal faults and tectonic origin of Mt. Etna (Sicily). Earth Planetary Science Letters 147:125–139

Monaco C, Tortorici L (2000) Active faulting in the Calabrian Arc and eastern Sicily. J Geodyn 29:407–424

Muto F, Spina V, Tripodi V, Critelli S, Roda C (2014) Neogene tectonostratigraphic evolution of allochthonous terranes in the eastern Calabrian foreland (southern Italy). Ital J Geosci 133:455–473

Neri G, Orecchio B, Totaro C, Falcone G, Presti D (2009) Subduction beneath southern Italy close the ending: results from seismic tomography. Seismol Res Lett 80(1):63–70

Polonia A, Torelli L, Gasperini L, Mussoni P (2012) Active faults and historical earthquakes in the Messina Straits area (Ionian Sea). Nat Hazards Earth Syst Sci 12:2311–2328. https://doi.org/10.5194/nhess-12-2311-2012

Pondrelli S, Morelli A, Ekström G (2004) European-Mediterranean regional centroid-moment tensor catalog: solutions for years 2001 and 2002. Phys Earth Planet Inter 145(1–4):127–147

Pondrelli S, Salimbeni S, Ekström G, Morelli A, Gasperini P, Vannucci G (2006) The Italian CMT dataset from 1977 to the present. Phys Earth Planet Inter 159(3–4):286–303

Presti D, Billi A, Orecchio B, Totaro C, Faccenna C, Neri G (2013) Earthquake focal mechanisms, seismogenic stress, and seismotectonics of the Calabrian Arc, Italy. Tectonophysics 602:153–175. https://doi.org/10.1016/j.tecto.2013.01.030

Rotondi R (2010) Bayesian nonparametric inference for earthquake recurrence time distributions in different tectonic regimes. J Geophys Res: Solid Earth 115:B01302

Rovida A, Camassi R, Gasperini P, Stucchi (a cura di) M (2011) CPTI11, la versione 2011 del Catalogo Parametrico dei Terremoti Italiani. Milano, Bologna. http://emidius.mi.ingv.it/CPTI

Sandron D, Gentile GF, Gentili S, Saraò A, Rebez A, Slejko D (2015) The Wood-Anderson of Trieste (Northeast Italy); One of the last operating torsion seismometers. Seismol Res Lett 86

Salvini F (2002) The structural data integrated system analyzer. Version 3.43b (11/12/2002). Dept. of Earth Sciences, University "Roma Tre", Rome, Italy, Free software available upon request at: http://www.dea.uniroma3.it/begin.htm
Scognamiglio L, Tinti E, Michelini A (2009) Real-time determination of seismic moment tensor for the Italian Region. Bull Seismol Soc Am 99(4):2223–2242. https://doi.org/10.1785/0120080104
Selvaggi G, Chiarabba C (1995) Seismicity and P-wave velocity image of the southern Tyrrhenian subduction zone. Geophys J Int 121(3):818–826
Serpelloni E, Vannucci G, Pondrelli S, Argnani A, Casula G, Anzidei M, Baldi P, Gasperini P (2007) Kinematics of the Western Africa-Eurasia plate boundary from focal mechanisms and GPS data. Geophys J Int 169:1180–1200
Serpelloni E, Bürgmann R, Anzidei M, Baldi P, Ventura BM, Boschi E (2010) Strain accumulation across the Messina Straits and kinematics of Sicily and Calabria from GPS data and dislocation modeling. Earth Planet Sci Lett 298(3):347–360
Sirovich L, Pettenati F (2004) Source inversion of intensity patterns of earthquakes: a destructive shock in 1936 in northern Italy. J Geophys Res 109:B10309. https://doi.org/10.1029/2003JB002919
Sirovich L, Pettenati F (2007) A new kinematic approach to calculate seismic hazard scenarios (intensity and peak ground displacement); an example in SE Sicily. In: Proceedings joint assembly AGU 2007, 22–25 May, Acapulco 2007; ref. no. S51A-01
Sykes LR (1967) Mechanism of earthquakes and nature of faulting on the mid-oceanic ridges. J Geophys Res 72:5–27
Tan Y, Zhu L, Helmberger D, Saikia C (2006) Locating and modeling regional earthquakes with two stations. J Geophys Res 111:B01306
Tansi C, Muto F, Critelli S, Iovine G (2007) Neogene-Quaternary strike-slip tectonics in the central Calabria Arc (southern Italy). J Geodyn 43:397–414
Tinti S, Armigliato A, Bortolucci E, Piatanesi A (1999) Identification of the source fault of the 1908 Messina earthquake through tsunami modelling. Is it a possible task? Phys Chem Earth 24(5):417–421
Tortorici L, Monaco C, Tansi C, Cocina O (1995) Recent and active tectonics in the Calabrian arc (Southern Italy). Tectonophysics 243:37–55
Tortorici G, Bianca M, Monaco M, Tortorici L, Tansi C, De Guidi G, Catalano S (2002) Quaternary normal faulting and marine terracing in the area of Capo Vaticano and S. Eufemia Plain (Southern Calabria, Italy). Studi Geologici Camerti 2:1–16
Tripodi V, Muto F, Critelli S (2013) Structural style and tectono-stratigraphic evolution of the Neogene-Quaternary Siderno Basin, southern Calabrian Arc, Italy. Int Geol Rev 4:468–481
Valensise G, Pantosti D (1992) A 125 Kyr-long geological record of seismic source repeatability: the Messina Straits (southern Italy) and the 1908 earthquake (MS 7 1/2). Terra Nova 4:472–483
Van Dijk JP, Bello M, Brancaleoni GP, Cantarella G, Costa V, Frixa A, Golfetto F, Merlini S, Riva M, Torricelli S, Toscano C, Zerilli A (2000) A regional structural model for the northern sector of the Calabrian Arc (southern Italy). Tectonophysics 324:267–320
Westaway R (1993) Seismic moment summation for historical earthquakes in Italy: tectonic implications. J Geophys Res 97:15.437–15.464
Wilson JT (1965) A new class of faults and their bearing on continental drift. Nature 207:343–347
Zecchin M, Caffau M, Civile D, Critelli S, Di Stefano A, Maniscalco R, Muto F, Sturiale G, Roda C (2012) The Plio-Pleistocene evolution of the Crotone Basin (southern Italy): interplay between sedimentation, tectonics and eustasy in the frame of Calabrian Arc migration. Earth-Sci Rev 115:273–303
Zecchin M, Praeg D, Ceramicola S, Muto F (2015) Onshore to offshore correlation of regional unconformities in the Plio-Pleistocene sedimentary successions of the Calabrian Arc (central Mediterranean). Earth-Science Reviews 142:60–78
Zhu L, Helmberger D (1996) Advancement in source estimation technique using broadband regional seismograms. Bull Seismol Soc Am 86:1634–1641

Zhu L, Mitchell BJ, Akyol N, Cemen I, Kekovali K (2006) Crustal thickness variations in the Aegean region and its implications for the extension of continental crust. J Geophys Res 111:B01301. https://doi.org/10.1029/2005JB003770

Zoback ML (1992) First- and second-order patterns of stress in the lithosphere: the World Stress Map project. J Geophys Res 97(11):703–711

A Seismogenic Zone Model for Seismic Hazard Studies in Northwestern Africa

J. A. Peláez, J. Henares, M. Hamdache and C. Sanz de Galdeano

1 Introduction

Known seismic activity in the northern Algeria-Morocco region (Fig. 1), especially during the last 50 years, includes several damaging earthquakes. In particular, in the El Asnam region (nowadays Cheliff) have been located the most destructive and damaging earthquakes recorded in northern Algeria. They were the 1954 September 9, M_W 6.8, and the 1980 October 10, M_S 7.3 (v.g., EERI 1983), earthquakes. The most significant recent event was the Zemmouri/Algiers 2003 May 21, M_W 6.9, earthquake, located approximately 50 km offshore, to the northeast of Algiers (v.g., Hamdache et al. 2004; ASCE 2004). The northern Morocco area has also experienced some damaging earthquakes occurred near Al Hoceima on 1994 May 26, with M_W 6.0 (v.g., Calvert et al. 1997), on 2004 February 24, with M_W 6.4 (v.g., Stich et al. 2005), and the most recent one, on 2016 January 25, with M_W 6.3. In Morocco, the most recent destructive event was the M_W 5.7 Agadir earthquake, which occurred on 1960 February 29 (v.g., AISI 1962). This event caused extensive damage and claimed about 12,000 casualties, attributed primarily to the weakness of the particular brittle masonry constructions of both old and new buildings.

It is generally accepted that seismic hazard values, by analyzing the most energetic seismicity and its spatial distribution characteristics, are an important tool to provide to the design engineers and planners with critical information on earthquake-prone

J. A. Peláez (✉) · J. Henares
Department of Physics, University of Jaén, Jaén, Spain
e-mail: japelaez@ujaen.es

M. Hamdache
Département Études et Surveillance Sismique, CRAAG, Algiers, Algeria

C. Sanz de Galdeano
Instituto Andaluz de Ciencias de la Tierra, CSIC, Granada, Spain

S. D'Amico (ed.), *Moment Tensor Solutions*, Springer Natural Hazards,
https://doi.org/10.1007/978-3-319-77359-9_29

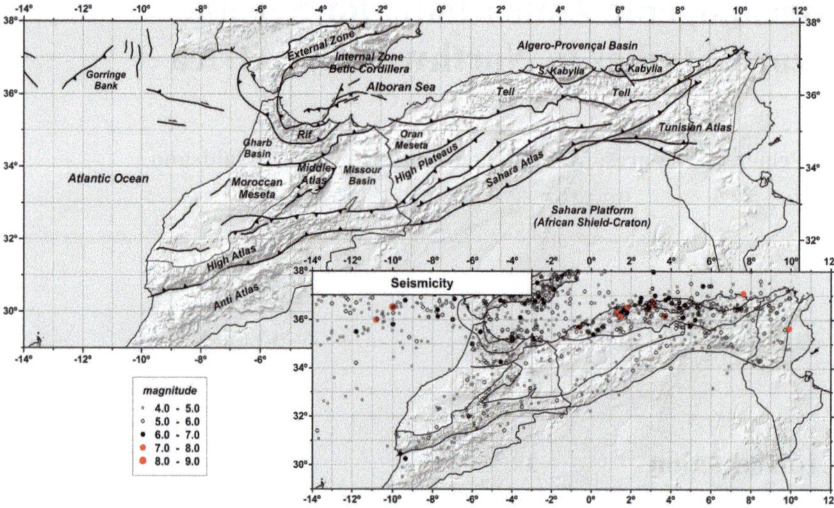

Fig. 1 Global tectonic sketch for the studied region including seismicity. Only shallow (h < 30 km) earthquakes above magnitude M_W 4.0 are depicted

areas and their damaging potential. Therefore, the interest of the scientific community in seismology and seismotectonics of Algeria and Morocco is steadily increasing, especially with regard to the mitigation of destructive events and the seismic risk assessment of urban areas.

The aim of this study is to propose and characterize a new seismogenic source zone model for the Algeria-Morocco region. This is considered as a fundamental key to perform a probabilistic seismic hazard in a certain region.

Different authors proposed previously seismogenic zone models for to carry out different researches in the studied region. For example, Hamdache (1998), Aoudia et al. (2000), Hamdache and Retief (2001), and Peláez et al. (2003), developed zone models aimed to seismic hazard studies in northern Algeria. The zone models by Hamdache (1998) and Hamdache and Retief (2001) cover only the Tell region, while the Aoudia et al. (2000) zone model cover strictly the Algerian territory, not including the nearby Moroccan, Tunisian and offshore seismicity. The proposed model by Peláez et al. (2003) was used later for additional seismic hazard studies (Peláez et al. 2005, 2006; Hamdache et al. 2012). In addition to these zonations, mention must be made to that included in the different Algerian building codes (RPA 1999, 2003), which consider only the seismicity level.

Zone models developed specifically to seismic hazard studies in Morocco are those proposed by Tadili (1991), Vaccari et al. (2001), Aït Brahim et al. (2007), Carbon et al. (2010) and Chourak et al. (2013), being the last one an update of the Aït Brahim et al. (2007) zone model. Moreover, here we must also mention the zonation included in the different Moroccan building codes (RPS 2000, 2011), which is also based in the seismicity level, as that used in the Algerian regulations.

Zone models proposed specifically for seismic hazard studies in the Iberian Peninsula but covering some of the northernmost part of the studied region are, for example, those proposed by Peláez and López Casado (2002), Vilanova and Fonseca (2007), Campos Costa et al. (2008), Silva et al. (2014) and García Mayordomo (2015), among others. Some of them have been considered to some extent in the delimitation of the proposed zone model in the shared region.

Finally, we must quote two models developed for the entire region. They are those proposed in the works by Jiménez et al. (1999) and Mourabit et al. (2014), both of them aimed to seismic hazard studies. The model proposed by Mourabit et al. (2014) is a genuinely developed model for the entire region aimed to a deterministic seismic hazard assessment, while the Jiménez et al. (1999) model is the junction of different original models developed for different regions, namely, Portugal, Spain, Morocco, Algeria and Tunisia.

Taking into account the development of the concept of probabilistic seismic hazard (v.g., Cornell 1968; McGuire 1976; Abrahamson 2006), for the proposed model, the available geological and tectonic data related to the region have been used. A wide-ranging literature assisted us in deriving a broad overview of the most important tectonic domains, which are discussed below.

The zone model, including several seismogenic zones, was delineated, taking care to adjust them to the boundaries of the domains, and including both the well-known tectonic structures and the known seismicity. The delineation of the proposed model for the northern Algeria-Morocco region is then based on the seismic activity distribution, the geology and the stress pattern. To do this, and using the stress field inversion approach, the predominant focal mechanism solution was computed for each zone in order to characterize them. Previously to the definition of the model, the most recent seismic catalog was considered, and a focal mechanisms catalog was specifically compiled for this work.

Separate sub-catalogs for each zone were created, and the seismicity parameters (mean annual activity rate, b-value of the frequency-magnitude Gutenberg-Richter relationship, and the regional maximum possible earthquake) were computed and analyzed in detail. In addition, as previously quoted, the predominant focal mechanism solution was also computed for each seismic source using the method by Delvaux and Sperner (2003) and Delvaux and Barth (2010).

2 Geological and Tectonic Setting of the Studied Region

The studied region (Fig. 1) covers the northwestern part of the Africa plate, within the Nubia plate, and its contact with the Eurasian plate, the western Mediterranean area, and also the different domains present in Morocco, Algeria and Tunisia. This contact and the main features of the different concerned domains are described briefly.

2.1 The Contact Between the Nubia and Eurasia Plates

The contact between the Nubia and the Eurasia plates (specifically with the Iberian subplate) can be divided in three sectors. From west to east they are the following: (a) the sector situated from the Middle Atlantic Ridge (from the Azores Islands) to the Strait of Gibraltar, (b) the Alboran Sea, and (c) the marine northern border of Algeria and Tunisia. They are discussed below.

(a) *Middle Atlantic Ridge to the Strait of Gibraltar*

In the western and central parts of this sector the contact between the plates is well defined, being more or less rectilinear. It begins in the Azores triple junction forming a ridge to the east, then passing to the E-W striking Gloria Fault, practically not affected by transversal faults. This simplicity finishes more to the east, in the proximities of the Gorringe Ridge, an area striking in the NE-SW direction. From this point to the east, that includes our study area, the limit between the plates is not well defined. The E-W striking faults are less marked, and they appear divided in several WNW-ESE faults whose continuities are not well known. Moreover, several NE-SW important transversal reverse faults affect this area, contributing to the blurring of the limits of both plates. Two of these faults are those named Gorringe and Horseshoe Faults. These transversal faults generally, but not always, verge to the NW (Alonso Chaves et al. 2011; Rosas et al. 2012). Finally, in the proximities of the Strait of Gibraltar, in the Gulf of Cadiz domain, the limit between the Nubian and Eurasian plates cannot be well situated.

Only the eastern part of this sector is considered in this study.

(b) *Alboran Sea*

The Alboran Sea occupies the western end of the Mediterranean Sea, situated between the Rif, in Morocco, and the Betic Cordillera, in southern Spain. This marine area is the western prolongation of the Algero-Provençal Basin, opened during the early Miocen. While the Algero-Provençal Basin is formed by oceanic crust, the Alboran Sea presents thinned continental crust because in it, the process of opening and extension had lesser importance. Owing to this process, the Betic-Rif Internal Zones were westwards expulsed, and during their drift, they underwent a noticeable compression. At the same time, the Strait of Gibraltar was formed, defining an arc which corresponds to an incipient subduction (López Casado et al. 2001) or to a clearly subduction (Pedrera et al. 2011), depending on the authors.

The Alboran Sea domain is not homogeneous, being the more conspicuous structures affecting this domain the sinistral faults that, from the SE of Spain, cross it in a NE-SW approximate direction (Grevemeyer et al. 2015). These faults form the Alboran Ridge, where it is situated the tiny islet of Alborán, and continue to the Rif coast, although the precise lines of several of these faults are not well drawn till this moment. Some of them (v.g., the Al Idrisi Fault) continue to the proximities of the town of Al Hoceima (Martínez García et al. 2013), in northern Morocco. On the whole, these faults form part of a long system of faults that, from the Rhin Basin, in Germany (even further north), continues in the Rhone Basin, passes to the east of Spain, and cut the eastern part of the Betic Cordillera. To the south, these faults are prolonged to the Atlas, in Morocco, and from Agadir passes to the Canary Islands (Sanz de Galdeano 1990).

In addition to these faults, other WNW-ESE striking faults can be observed in the Alboran Sea, as that of the Yusuf Ridge.

(c) *Marine northern border of Algeria and Tunisia*

The northern marine border of Algeria and Tunisia is the southern limit of the Algero-Provençal Basin. Here is the transition from the oceanic crust of the basin to the continental crust that, inland, forms the Tell Mountains, a chain that can be considered the eastern prolongation of the Rif.

It is generally considered that the limit between Africa and Eurasia originally corresponded, during the Mesozoic and part of the Tertiary, to an important sinistral strike-slip fault, although nowadays, at least part of this contact presents dextral strike-slip displacements, as it happens in the Gloria Fault (DeMets et al. 2010).

In the Alboran Sea, the limit was completely obliterated during the Miocene, and its former position is approximately marked by the contact between the Internal and External Rif Zones, with rests of the Flysch Basin situated in between. Nowadays, the central thinned crust of the Alboran Basin could be considered as the new area in which the contact of the plates is susceptible to be found.

To the east, the situation is similar, and the former limit, prolongation of the cited limit in the Rif, is to be found in the contact between the Internal (the Kabylias) and External Zones of the Tell domain. The dorsal of the Algero-Provençal Basin could be considered this limit at the present times.

2.2 Main Characteristics of the Cordilleras in the Northwesternmost Part of the Nubia Plate

These cordilleras are the Rif and the Tell, the Atlas and the Moroccan Meseta, and the High Plateaus, this last one forming part of the Variscan belts.

(a) *The Rif and the Tell*

Both cordilleras were formed by the collision triggered by the opening, during the early Miocene, of the Algero-Provençal Basin with the Nubia Plate, likely also combined with a northwards dipping subduction process of this plate.

In these cordilleras exists tectonic nappes (Wildi 1983; Chalouan and Michard 2004; Chalouan et al. 2008), separated by thrusts generally verging southwards, including: (i) the Rif Internal Zone (similar to the Betic Internal Zone) and the equivalent Kabylias (the Great Kabylias, to the east, and the Small Kabylias, to the west) in Algeria, (ii) the Flyschs units, and (iii) the External Zones, in which can be distinguished several domains, progressively more external to the south. The External Zones corresponded to the Mesozoic and Tertiary sedimentary cover of the

Paleozoic African basement and its prolongation to the north, nowadays folded and fractured, forming tectonic nappes.

The main tectonic trend of both Rif and Tell cordilleras is nearly striking E-W, although in the Rif, in the proximities of the Strait of Gibraltar is nearly striking N-S, by accommodation to the arc shape. Conversely, to the east, particularly in the area of Tunisia, the striking is ENE-WSW. In addition, both chains are affected by transversal faults, many of them striking NE-SW with sinistral displacement. Two of them are the Jebha and Nekor Faults, in the Rif, and there exist many others in Algeria and Tunisia, being the more important ones those situated in the Babors area.

(b) *The Atlas*

This name is applied to the important mountain ranges going from the northeast to the southwest of Morocco (Middle and High Atlas), but also corresponds to the long lines of mountains that from the Moroccan Atlas are prolonged till Tunisia, striking WSW-ENE (the Saharian and Tunisian Atlas, respectively).

The Moroccan Atlas is a cordillera formed over a zone of crustal weakness accompanied by fractures (Jacobshagen 1992), existing from the end of the Paleozoic. The process of convergence of Africa and Eurasia, with the subsequent compression, determined the formation of thrusts, and reverse and strike slip faults -moreover folds- that cause the uplift of the region. Some faults cut the entire crust and permitted the extrusion of Quaternary volcanic rocks, a feature that express the continuity of the crustal frailty in this area.

Both the Saharian and Tunisian Atlas can be considered the southernmost expression of the deformation induced by the Africa and Eurasia convergence. This long chain presents well developed folded areas and also south verging thrusts. Moreover, it is affected by transversal NNE-SSW striking sinistral faults.

(c) *The Moroccan Meseta and the High Plateaus* (*High Mesetas*)

Between the Atlas and the Rif and Tell there are two intermediate domains, the Meseta and the High-Plateaus.

The Moroccan Meseta is situated between the Moroccan Atlas (High and Middle Atlas) and the Atlantic Ocean, limiting to the north with the External Zone of the Rif. It is formed by a Paleozoic basement, and to the southwest, by a Mesozoic and Tertiary cover. It is also affected by faults, many of them striking NNE-SSW. On the whole, it is a Paleozoic massif surrounded by Alpine structures, with the exception of its western oceanic border.

The High Plateaus are situated more to the east, between the Tell, to the north, and the High and Sahara Atlas, to the south. In the eastern Algeria and Tunisia areas, the Atlas and the Tell do not leave space to the High Plateaus. As the Moroccan Meseta, the High Plateaus have also a Paleozoic basement deformed during the Hercynian orogeny, but in this region, this basement is much more covered by Mesozoic and Miocene sediments. This area is relatively little deformed during the Alpine times, although in some sectors there are abundant folds and sinistral NE-SW striking faults.

Finally, southwards of the Atlas (High, Sahara and Tunisian Atlas) is the immense African craton (in Morocco, the Antiatlas forms its northern border), a stable shield mainly formed by a basement of Precambrian rocks. It is beyond the scope of the present description.

In summary, the studied region comprise very different geologic domains, from a wide oceanic areas (the Atlantic area) and a young oceanic crust basin (the Algero-Provençal Basin), passing to an area of continental crustal thinning (Alboran Sea), and to thickened crustal areas as part of the Rif and the Atlas. Particularly, these last cordilleras, together with the Tell, correspond to complex structured Alpine sectors, while the Moroccan Meseta and the High Plateaus were mainly deformed during the Hercynian orogeny.

Nevertheless, at present, most of these structures are not tectonically active, and hence, not seismically active. In the sector situated between the Middle Atlantic Ridge and the Gibraltar Strait, the contact between the two plates and the transversal faults can be considered as very active structures. In the Alboran Sea, there are also two active areas: (i) the transversal faults that from the southeastern Spain form the Alboran Ridge, and are prolonged to the southwest, and (ii) the Gibraltar Arc, where some intermediate depth seismicity, although with low magnitudes, is located. In the Rif and the Tell there are important transversal reverse and strike slip faults, responsible of most of the earthquakes of the region. The same occurs in the Atlas, a chain that in Morocco is oblique to the E-W general structures of the region.

Below, when describing the proposed seismic and seismotectonic zone model, these domains will be described from a seismic point of view.

3 Data Sources

In this section, both seismic catalog and focal mechanism data used for this zoning are presented. The seismic catalog is one of the keystones in seismic hazard and forecasting studies, not only for to establish a seismic zonation but also for zoneless studies (Peláez et al. 2016). In addition, focal mechanism data are essential in order to study the seismotectonic of a certain region, and accordingly, to characterize the delimited seismic and seismotectonic zoning model (Sawires et al. 2016).

3.1 Seismic Catalog

To carry out this study, two unified catalogs for northern Morocco (Peláez et al. 2007) and northern Algeria (Hamdache et al. 2010) have been used. The first one includes earthquakes in the area between 27° to 37° N and 15° W to 1° E, initially spanning the years 1045 to 2005. The second catalog covers the area between 32° to 38° N and 3° W to 10° E, initially spanning the years 856 to 2008. These catalogs were aggregated, duplicated earthquakes in the overlapped areas as well as non-crustal events were erased, and finally updated. The final catalog is a unified catalog in terms of the moment magnitude including only Poissonian events, compiled specifically for seismic hazard and forecasting studies in the region (Peláez et al. 2016).

The main data source was the Spanish Instituto Geográfico Nacional (IGN). Other regional sources also contributed to the initial compiled catalogs as the Moroccan Service de Physique du Globe (SPG), and the Algerian Centre de Recherche en Astronomie, Astrophysique et Géophysique (CRAAG). Unlike Spanish IGN, Moroccan and Algerian seismic nets have not covered efficiently the seismicity in this region, especially in past decades. In addition, some data from international agencies were also used, as the U.S. Geological Survey/National Earthquake Information Center—Preliminary Determination of Epicenters (USGS/NEIC-PDE) and the International Seismological Centre (ISC) catalogs. Data concerning moment tensor catalogs and specific published studies were also took into account (Peláez et al. 2007; Hamdache et al. 2010). Among them, the Harvard Centroid Moment Tensor (Harvard CMT), the European-Mediterranean Regional Centroid Moment Tensor (EM RCMT), and the Spanish IGN and the Spanish Instituto Andaluz de Geofísica (IAG) moment tensor catalogs.

In order to carry out the magnitude unification, several empirical relationships between reported magnitudes, maximum intensities and moment magnitude were employed, some of them specifically developed for this catalog. Overall, the global catalog can be considered complete for magnitudes above M_W 6.0 since 1885, and for magnitudes above M_W 5.0 since 1900, approximately, with mean rates of 0.21 and 2.15 events by year, respectively. Magnitudes above M_W 4.0 are complete only approximately in the last 15 years, with a mean rate of about 30 events by year (Peláez et al. 2016). More characteristics on the compiled catalog can be seen in previous works (Peláez et al. 2007, 2016; Hamdache et al. 2010). The characterization of this seismicity from both spatial and energetic points of view is described in the recent work by Hamdache et al. (2017).

Earthquakes included in the catalog from magnitude M_W 4.0 are depicted in Fig. 1. The largest density of earthquakes is located offshore along all the contact between the Nubia and Eurasia plates, and onshore, in the Rif and Tell, i.e., in the northernmost part of Morocco, Algeria and Tunisia. From available data, largest events with magnitude above M_W 7.0 are confined mainly to the Tell and the Gorringe Ridge region (Fig. 1).

3.2 Focal Mechanism Data

Initially, in order to establish the focal mechanisms database, the catalogs by Henares et al. (2000), Henares and López Casado (2001) and Custódio et al. (2016) have been used. The catalog by Henares et al. (2000) and Henares and López Casado (2001) for the Ibero-Moghrebian region, updated to December 2000, includes 486 solutions (453 correspond to individual and 33 to joint solutions). The catalog by Custódio et al. (2016) for the Azores-western Mediterranean region includes 3058 solutions.

On the whole, the area covered by these two catalogs extends from 32° W to 13° E in longitude, and from 30° to 45° N in latitude.

In addition to the previous catalogs, all the published specifics works including focal solutions for both specific earthquakes and specific areas were incorporated. Alphabetically, they are the works by Ayadi et al. (2008), Beldjoudi et al. (2012), Bahrouni et al. (2014), Beldjoudi et al. (2016), Ben Ayed and Zargouni (1990), Benito et al. (2007), Bensaid et al. (2011), Bensaid et al. (2012), Bezzeghoud et al. (1994), Braunmiller and Bernardi (2005), El Alami et al. (1989, 1992, 1998), Geissler et al. (2010), Gueddiche et al. (1992), Hatzfeld (1978), Hfaiedh et al. (1985), Martin et al. (2015), Medina (2008), Mejri (2012), Ousadou et al. (2013, 2014), Ouyed et al. (1983), Rodríguez Pascua and De Vicente (2001), Sánchez Gómez and Torcal (2002), Stich et al. (2001, 2003a, 2005), Van Der Woerd et al. (2014), Vannucci and Gasperini (2004), Yelles Chaouche et al. (2014) and Yielding et al. (1989). Furthermore, data from the international European Mediterranean Regional Centroid Moment Tensor-RCMT catalog (Pondrelli et al. 2001, 2002, 2004, 2006, 2007, 2011; Pondrelli and Salimbeni 2015; www.bo.ingv.it/RCMT) has been also included.

Finally, in a third stage, the final database has been updated until December 2016 including data computed or compiled from the following national and international agencies:

- Global Centroid Moment Tensor-GCMT (Dziewonski et al. 1981; Ekström et al. 2012; http://www.globalcmt.org).
- Euro-Mediterranean Seismological Center Quick Moment Tensor-EMSC-QMT (http://www.emsc-csem.org/Earthquake/tensors.php).
- International Seismological Center focal mechanism compilation-ISC (http://www.isc.ac.uk/iscbulletin/search/fmechanisms/).
- Instituto Geográfico Nacional focal mechanism database-IGN (Rueda and Mezcua 2005; http://www.ign.es/web/ign/portal/tensor-momento-sismico/-/tensor-momento-sismico/getExplotacion).

The compiled focal mechanism database includes information on the following earthquake parameters: date, time, latitude, longitude, depth, magnitude (M_W), and strike, dip, and rake for the two fault planes. Whenever possible, the hypocentral information was standardized, based on the earthquake catalog of the Spanish IGN. In order to obtain a unified moment magnitude, when necessary, the relationships by CNIG (2013) and Heaton et al. (1986) were used. Finally, when merging focal mechanism data, the information provided by different sources was prioritized using the criterion by Custódio et al. (2016).

Next, the focal mechanisms catalog initially used for this study is depicted. Figure 2 shows the shallow ($h \leq 30$ km) focal data and Fig. 2e the intermediate depth ($30 < h \leq 60$ km) focal mechanisms. To classify the faulting style, the software Win-Tensor™ (Delvaux and Sperner 2003; Delvaux and Barth 2010) was used, assigning to each style a color based on the so-called Frohlich triangle diagram (Frohlich 1992). As can be seen, the Atlantic region (Fig. 2a) presents a predominance of thrust focal (TF) focal mechanisms with some presence of strike-slip (SS) mechanisms. The central part of the region (Fig. 2c) presents a predominance of strike-slip (SS) focal

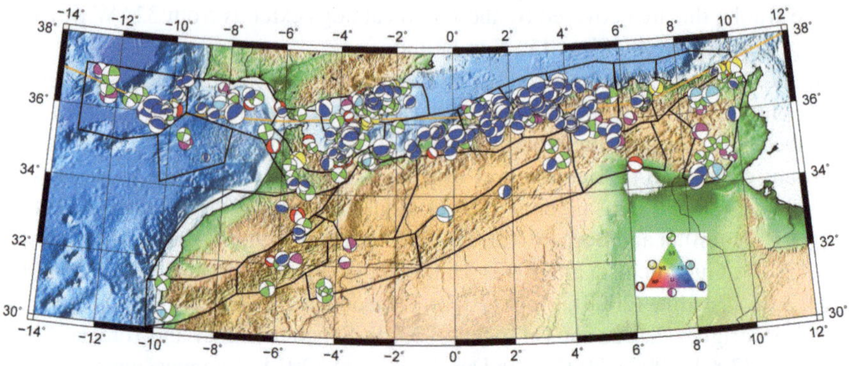

(a)

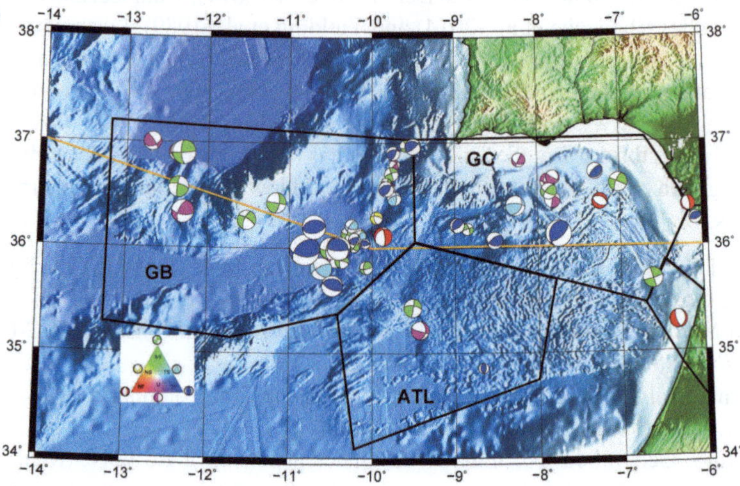

Fig. 2 Shallow depth ($h \leq 30$ km) focal mechanism data compiled in the region. The Frohlich (1992) diagram is showed, and the Zoback (1992) classification is used (*TF* thrust faulting; *TS* thrust to strike slip; *SS* strike slip; *NS* normal to strike slip; *NF* normal faulting). Sphere size is proportional to magnitude. Black: Shallow-depth zones. White: Intermediate-depth zone. Orange: NUVEL-1A plates model boundary (DeMets et al. 1994). **a** Enlargement of Fig. 2 for the Atlantic region. **b** Enlargement of Fig. 2 for the Moroccan Meseta and High Atlas region. **c** Enlargement of Fig. 2 for the Rif, Alboran and western Tell region. **d** Enlargement of Fig. 2 for the middle and eastern Tell, and eastern Sahara and Tell Atlas. **e** Intermediate depth (30<h<60 km) focal mechanism data compiled in the Atlantic and western Alboran region

mechanisms. The eastern part (Fig. 2d) presents a predominance of thrust focal (TF) mechanisms with some presence of strike-slip (SS) mechanisms. Finally, the southern region (Fig. 2b and bottom of Fig. 2d) presents few focal mechanism data and it is difficult to observe any predominance.

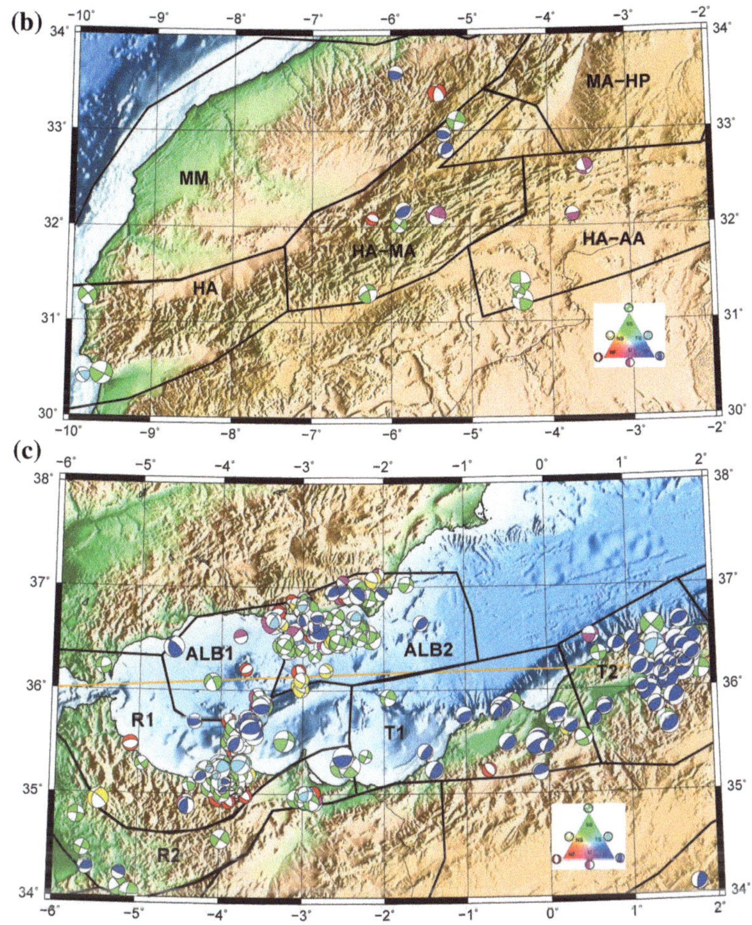

Fig. 2 (continued)

4 Description of the Model

Although already showed in Figs. 2a, b, c, d and e, the proposed zoning model is introduced in Fig. 3. Below, used criteria and their description are showed. The main used criterion in this delimitation has been to define zones that exhibit similar seismotectonic characteristics and a rather homogeneous seismicity release. Initially, the tectonic domains have been divided into sections with a rather homogeneous stress pattern and seismicity recurrence relationship. Then, initial boundaries have been more or less modified, attending to the obtained stress patterns and seismicity parameters, in order to improve the initial zoning. In the proposed zoning, individual faults have not been specifically considered due to nowadays there is not enough information in this region to include them in a probabilistic seismic hazard analysis.

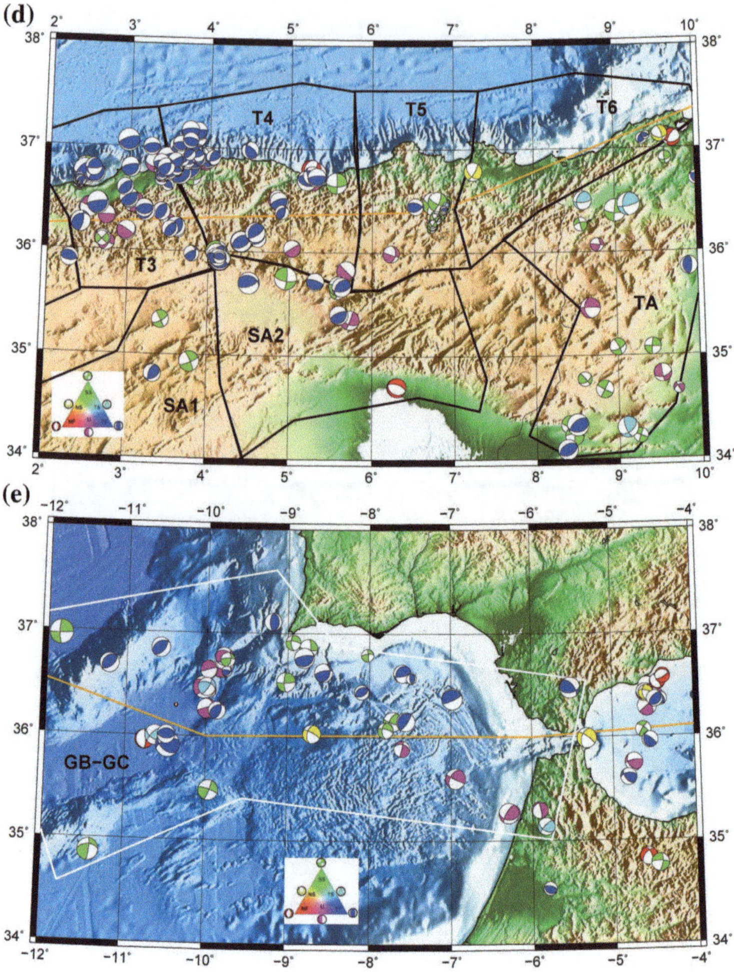

Fig. 2 (continued)

However, in the delimitation of the area zones, known faults were included in an only source, i.e., every care has been taken to ensure that known faults, or at least the same fault segments, were not partially included in two different sources.

Following the Araya and Der Kiureghian (1988) criterion, delimited zones can be considered as seismogenic source zones, the most usual type of sources zones used in probabilistic seismic hazard analyses. They are somehow related to geological structures or domains, but they lack the development of a clear history relating the contemporary seismic activity to a geological structure.

In the following, the proposed seismic sources (Fig. 3) are described to some extent.

• *Tell seismogenic sources (T1, T2, T3, T4, T5 and T6)*

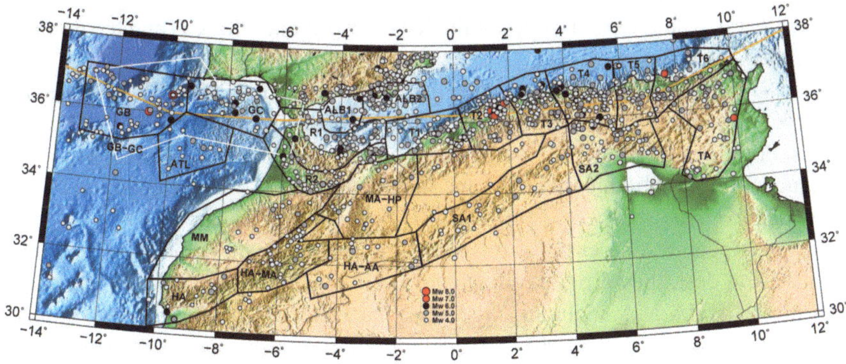

Fig. 3 Proposed zone model. Black: Shallow-depth zones. White: Intermediate-depth zone. Orange: NUVEL-1A plates model boundary (DeMets et al. 1994). Only shallow earthquakes above magnitude M_W 4.0 are depicted

The Tell region has been divided in six different source zones, named from T1 to T6. This is the region with the largest density of earthquakes, as it was quoted before. This source zoning delimitation is mainly related to the different observed earthquake activity, clearly revealed when computing seismicity parameters.

These source zones are based in the works by Aoudia et al. (2000) and Peláez et al. (2003). In fact, considering the work by Aoudia et al. (2000), T1 is based on Z3, T2 on Z1, and T5 and T6 on Z9 source. Moreover, there is a big agreement between this delimitation and that proposed by Peláez et al. (2003) for this region. Thereby, T1 approximately coincides with Sa1, T2 with Sa2, T3 with Sa3 and Sa4, and T4, T5 and T6 with Sa5 to Sa8 seismogenic sources.

The biggest recorded earthquake in the Tell region was the 1980 Chlef, Algeria, M_W 7.3 earthquake, located in the T2 source zone. Another significant earthquake was the 2003 Boumerdes, Algeria, M_W 6.9 earthquake (Hamdache et al. 2004), this one included in the T4 source zone. In other source zones stand out big historical earthquakes as the 1790 Oran, Algeria, in the T1 source, the 1867 Blida, Algeria, in the T3 source, or the 1722 NW Annaba, Algeria, earthquakes, felt with intensities equal to IX-X, X-XI and X, respectively.

• *Rif seismogenic sources (R1, R2)*

The Rif region has been divided in two source zones, named R1, in the northern part, and R2, to the south. They have been developed specifically for this work and are not based in previous studies. Their shapes are based in the tendency of the Rif domain, and on some observed and inferred faults in the southwestern part of the Alboran Sea.

R1 source embraces the Internal Zones and the northern part of the External Zones of the Rif (Intrarif, including the Ketama and Tanger-Loukkos units), and the southwesternmost part of the Alboran Sea domain. Active structures, as the Alboran Ridge and the Jebah, Nekor, and southern segment of the Al Idrisi faults (Galindo Zaldívar et al. 2015), are included in this source zone. R2 source mainly includes the southern part of the External Zones of the Rif (Mesorif and Pre-Rif) and the northern part of the Gharb Basin.

Some significant earthquakes located in this region are the 2004 SW Al Hoceima, Morocco, M_W 6.4, or the most recent 2016 Al Hoceima, Morocco, M_W 6.3 earthquakes, both located in the R1 source zone, and the historical 1624, Fès, Morocco, earthquake, felt with intensity IX-X, and included in the R2 source zone.

- *Atlas seismogenic sources (HA, HA-MA, HA-AA, SA1, SA2 and TA)*

Six seismogenic sources cover this structural large domain (near 2000 km long). From west to east, they are the following: (a) HA, including the eastern part of the High Atlas, (b) HA-MA, including the middle part of the High Atlas and the southern part of the Middle Atlas, (c) HA-AA, including the western part of the High Atlas and the northeastern part of the Anti Atlas, where have been located some small to moderate earthquakes, (d) SA1, mainly including the eastern an middle part of the Sahara Atlas, (e) SA2, mainly including the western part of the Sahara Atlas, and (f) TA, in the easternmost part of the region, including the Tunisian Atlas.

The HA and HA-MA source zones are based, and roughly coincides, on AO and MA-HA sources, respectively, delimited by Aït Brahim et al. (2007). However, HA-AA was developed specifically for this work, not based in previous works. In addition, SA1 and SA2 are based, and roughly coincides, on Z12 and Z11 sources, respectively, delimited by Aoudia et al. (2000), and on Sa10 and Sa9 sources, respectively, delimited by Peláez et al. (2003). Finally, TA source zone is based on St2, delimited by Peláez et al. (2003).

The biggest earthquakes in the Atlas were recorded in the limit between the High Atlas and the Antiatlas, as the historical 1731, and the instrumental 1960 Agadir, Morocco, M_W 5.7 earthquakes, felt with intensity IX and X, respectively. These two earthquakes are included in the HA source zone. In addition to these two earthquakes, it was also remarkable the historical 856 Tunisia earthquake, felt with intensity X. This last earthquake was located near the eastern limit of the Tunisian Atlas, in the so-called North-South Axis (Soumaya et al. 2015). This event is included in the delimited TA source zone. Excluding these three earthquakes, the seismicity recorded in this region could be considered as small to moderate. To the south of this domain, in the African shield-craton, there is not remarkable seismicity.

- *Moroccan Meseta—High Plateaus seismogenic sources (MM and MA-HP)*

This region has been covered by two source zones: (a) MM, delimiting the Moroccan Meseta, and (b) MA-HP, including the northern part of the Middle Atlas and the western part of the High Plateaus (the eastern part could be considered as aseismic), including the Missour Basin.

MM seismogenic source is based, and roughly coincides, on MO source delimited by Aït Brahim et al. (2007). MA-HP is based, and it is included, on Sm2 source zone in Peláez et al. (2003).

Seismicity is scarce in these two sources. In the MM source zone, the biggest recorded event was the 1979 N Khenifra, Morocco, M_W 5.1 (M_S 4.6) earthquake. In the MA-HP zone, only the 1980 E Jerada, Morocco, M_W 5.3 (M_S 4.9) earthquake stands out.

- *North Alboran seismogenic sources (ALB1 and ALB2)*

This region is covered by two seismic sources: (a) ALB1, including the seismicity in the northwestern part of the Alboran Sea, and some known fault, as the northern segment of the Al Idrisi fault, and (b) ALB2, including the seismicity in the north-eastern part of the Alboran Sea, and some onshore and offshore structures, as the Balanegra and Carboneras fault zones (Pedrera et al. 2012).

ALB1 and ALB2 source zones are based in two sources, numbered 40 and 38, respectively, delimited in the preliminary zonation model for Spain by García Mayordomo et al. (2010). In a later zoning model (CNIG 2013; García Mayordomo 2015) used for the update of the Spanish seismic hazard maps (CNIG 2013), the corresponding source to ALB2 was completely restructured.

Both source zones have hosted moderated earthquakes, as the historic 1804 Alboran Sea earthquake, felt in the Spanish coast with intensity VII-VIII, and included in the ALB1 zone, and also the historic 1522 W Alhama de Almería, Spain, earthquake, felt with intensity VIII-IX, included in the ALB2 source zone.

- *Atlantic region seismogenic sources (GB, GC, ATL and GB-GC)*

The Atlantic region (SW Iberian margin) included in this zonation has been divided in three shallow sources and one additional intermediate-depth source zone. Concerning the shallow zones, they are the following: (a) GB, including the seismicity located in the Gorringe Bank, being the northwesternmost delimited seismogenic zone, (b) GC, including the seismicity located in the Gulf of Cádiz, and (c) ATL, covering the scarce seismicity to the south of the previous two source zones. GB zone includes known tectonic structures as the Gorringe, Marquês de Pombal and Horseshoe faults (v.g., Ferranti et al. 2014). In addition, some known tectonic structures (alignments), approximately E-W striking, have been included in the ATL source.

GB and GC source zones are based on AIb1 and AIb2 zones, respectively, previously delineated by Peláez and López Casado (2002), on BG and ATL-RO sources, respectively, previously delineated by Aït Brahim et al. (2007), and on sources 50 and 51, respectively, previously delineated by García Mayordomo et al. (2010).

In addition to the shallow-depth zones, an intermediate-depth source zone named GB-GC has been included. It includes intermediate seismicity (30–60 km depth) located in the region covered by shallow source zones GB and GC. This intermediate zone roughly coincides with the intermediate-depth source previously delineated by Peláez and López Casado (2002).

The biggest events recorded in this region are the following. The GB source zone hosted the historic 1755, SW Cape St. Vincent, earthquake, also known as the 1755 Lisbon earthquake, felt with intensity X, as well as the instrumental 1969, W Cape St. Vincent, M_W 7.8 (M_{bLg} 7.3) earthquake, also known as the 1969 Horseshoe earthquake, located in the Horseshoe Abyssal Plain. The GC source zone hosted the historic 1722 Gulf of Cádiz event, felt in the coast with intensity VIII. In addition, the ATL source zone hosted the 1937, Cape St. Vincent, M_W 5.6 (M_S 5.4) earthquake. Finally, the GB-GC intermediate zone hosted the most recent 2009, SW Cape St. Vincent, M_W 5.5 earthquake.

Clearly, the source zone with a biggest seismic potential is, at the same time, the most distant to the African coast, the GB zone.

One comment must be made at this point regarding the intermediate seismicity in the Alboran Sea. Although it is well known and characterized (v.g., Buforn et al. 1991; Grevemeyer et al. 2015; among many other authors), given their magnitudes below m_b 4.5, and their location and depth range (60–100 km depth), it has not been considered for this study. The influence of this seismicity to the seismic hazard both in the Spanish and the African coasts has been proved in previous studies negligible (Peláez and López Casado 2002; Peláez et al. 2003). In fact, it was not even considered in the more recent seismic hazard studies in this region (CNIG 2013).

Two interesting final remarks concern the seismogenic depth and the most likely seismicity depth of each seismogenic source. Given the region, the distribution and quality of local seismic networks, and also the quality of earthquake records, it is not possible to know in a reliable way these two parameters, very important in seismic hazard studies, only from seismicity data. Depth errors are significant, causing that a great number of earthquakes, even nowadays, have not assigned depth.

5 Seismicity Parameters

The next step in the definition and characterization of a zone model is to characterize each seismogenic source through different parameters in order to know its earthquake recurrence. Usually these parameters are the mean annual activity rate above the magnitude 4.0, a typical threshold magnitude value in seismic hazard computations, denoted λ ($M \geq 4.0$), the *b*-value of the Gutenberg-Richter recurrence relationship, and the regional maximum possible earthquake magnitude m_{max}.

The classical assumption in any time independent probabilistic seismic hazard assessment (PSHA) about the temporal distribution of the number of earthquakes within a specified region is that it can be described by a stochastic Poisson process (v.g., Cornell 1968; Lomnitz 1973; Gardner and Knopoff 1974). Therefore, the probability that *n* earthquakes could be observed during a specified time interval *t* within a specified area is given by

$$P\{N_t = n\} = \frac{(\lambda t)^n}{n!} e^{-\lambda t}$$

being N_t the number of occurrences during the specified time interval, and λ (or $\lambda_{m_{min}}$) a parameter describing the annual activity rate (area dependent) of events with magnitude above a certain threshold value (m_{min}).

Based on observation, it is commonly assumed that the distribution of the number of earthquakes with respect to their magnitudes follows the classical Gutenberg-Richter relationship (Gutenberg and Richter 1942, 1956), defined by the well-known equation

$$\log N(\geq m) = a - bm; \qquad m \geq m_c$$

The statistic $N(\geq m)$ is the cumulative number of events with magnitude larger than or equal to m, whereas a and b are positive constants. The a-parameter, which exhibits significant variations in space, measures the seismicity activity level. It is sensitive to the period of observation and the area of investigation. The b-parameter describes the size distribution of events, and it is related to the tectonic characteristics of the region. Usually, this relationship is used to perform studies of the size distribution of the seismicity, by quantifying the Gutenberg-Richter b-value as an indicator of strength and heterogeneity of the materials, and also reflecting the stress distribution in space and time. The b-value is a statistic measuring the proportion of large to small earthquakes in the considered region; if b is comparatively large, small earthquakes are relatively common, whereas that, lower the value of this parameter is, greater is the probability that a large earthquake happens. Depending on the tectonic setting and the stress pattern in the region, b-values usually range from 0.7 to 1.5 for area sources. For seismically active regions the value is close to the unity (Frohlich and Davis 1993). In contrast, volcanic areas are characterized by b-values greater or less than 1.0, with values as high as 3.0 (McNutt 2005).

In the Gutenberg-Richter frequency-magnitude distribution, m_c is the so called threshold magnitude, for which all events are reported in the considered database. It is widely recommended to use a complete dataset for all magnitude ranges to estimate both the a and b parameters of the frequency-magnitude relationship. Completeness in magnitude varies systematically as a function of space and time, and particularly, temporal changes in completeness magnitude can potentially lead to erroneous b-values estimates (Wiemer and Katsumata 1999). There are other factors producing a bias on the b-value estimate, for example, it is established that the least square procedure introduces systematic errors in its computation (Marzocchi and Sandri 2003).

Several approaches have been developed to estimate m_c (v.g., Wiemer and Wyss 2000; Albarello et al. 2001; Schorlemmer and Woessmer 2008). Among the most widely procedures used nowadays are the so called Maximum Curvature, Goodness of Fit, Entire Magnitude Range, and Mean Base Slope approaches.

Utsu's method (Utsu 1965), for instance, widely used in order to estimate the maximum-likelihood b-value, is based on the equation

$$b = \frac{\log e}{\langle m \rangle - \left(m_c - \frac{\Delta m}{2}\right)}$$

being $\langle m \rangle$ the mean value of the magnitudes, and Δm the interval of the binned magnitudes, usually equal to 0.1 for instrumental events.

In our study, taking into account the quality of the used earthquake catalog, including historical and the more recent instrumental events, the assessment of the b-value was carried out by application of the extension of the so called Aki-Utsu b-value estimator by Kijko and Smit (2012) and Kijko et al. (2016). It is the simplest procedure of all alternative approaches that utilize incomplete earthquake catalogs. To implement this procedure, the earthquake catalog was divided in periods with different level of completeness. The estimator of β, being $\beta = b \cdot \ln 10$, is obtained in the form (Kijko and Smit 2012)

$$\hat{\beta} = \left(\frac{r_1}{\hat{\beta}^{(1)}} + \frac{r_2}{\hat{\beta}^{(2)}} + \cdots + \frac{r_s}{\hat{\beta}^{(s)}} \right)^{-1}$$

being s the total number of subcatalogs, $r_i = n_i/n$, $n = \sum_{i=1}^{s} n_i$ the total number of seismic events in a complete catalog with magnitudes equal to or exceeding the relevant level of completeness $m_c^{(i)}$, and $\hat{\beta}^{(i)}$ is defined by

$$\hat{\beta}^{(i)} = \frac{1}{\overline{\boldsymbol{m}_i} - m_{min}^{(i)}}$$

In this equation, $\overline{\boldsymbol{m}_i} = \sum_{k=1}^{n_i} m_{ik}/n_i$ denotes the mean of the earthquake magnitudes observed within the complete part of the subcatalog i. The equation given $\hat{\beta}^{(i)}$ is the b-value Aki-Utsu estimator (Aki 1965; Utsu 1965) for each one of the individual complete sub-catalogs.

According to Kijko and Smit (2012) and Kijko et al. (2016), from the estimated β-values, the mean activity rate above m_{min} is computed, taking into account that the number of events per unit of time is Poissonian, using the following equation

$$\hat{\lambda}\,(m_{min}) = \frac{n}{\sum_{i=1}^{s} t_i\, exp\left[-\hat{\beta}\left(m_{min}^{(i)} - m_{min}\right)\right]}$$

This is the usual parameter employed for seismic source characterization instead of a-value. It is obvious to obtain one parameter from the other.

The maximum possible earthquake magnitude, m_{max}, is another important parameter with different definitions and references. According to Kijko and Graham (1998), Kijko (2004), Kijko and Singh (2011) and Holschneider et al. (2011), it is defined as the upper magnitude limit for a given region. In addition to these quoted works, other authors have also published studies on the subject, as Kijko and Sellevoll (1989, 1992), Pisarenko (1991), and Pisarenko et al. (1996), among others, using different methods for its computation. Generally, for the evaluation of this parameter two different approaches can be found in the literature: methods based on deterministic and probabilistic formalisms.

The deterministic methods are generally based on empirical relationships between magnitude and diverse tectonic and fault parameters. Wells and Coppersmith (1994), Wheeler (2009) and Mueller (2010), among others, provide a clear discussion of such techniques and relationships.

Statistical procedures for its estimation can be found in Kijko (2004), Kijko and Singh (2011), and Kijko et al. (2016). All procedures are based on the assumption that earthquake magnitudes are independent and identically distributed random variables, with a probability density function (PDF) $f_M\,(m)$, and a cumulative distribution function (CDF) $F_M\,(m)$. Then, the unknown maximum regional earthquake magnitude m_{max} is considered as the upper limit of the magnitude distribution. Some of the estimation procedures, as described by Kijko (2004), Kijko and Singh (2011) and Holschneider et al. (2011), do not make use of any subjective assumptions. The procedures are generic, and provide solutions in different forms, depending on the assumptions of the statistical model and the knowledge of the past seismic activity.

The procedures developed by Kijko (2004) and Kijko and Singh (2011) are grouped into three distinct categories: (a) parametric, (b) non-parametric, and (c) the fit of the cumulative distribution function of earthquake magnitudes. The parametric procedures can be used when the functional model of the frequency-magnitude distribution is known. All the procedures of this category are based on the estimated value is equal to $m_{max} = m_{max}^{obs} + \Delta$, being m_{max}^{obs} the largest observed earthquake magnitude in the area, and Δ a correction factor that depends on seismicity parameters, supporting the intuitive expectations that it is always positive and that its value decreases as the time span of observation increases. In this work, we have focused on four different types of procedures to assess the maximum regional magnitude for each one of the source zones included in the proposed seismogenic model.

First, we consider what is the simplest form of the correction factor, $\Delta = \left[n\, f_M\left(m_{max}^{obs}\right)\right]^{-1}$, where n denotes the number of earthquakes. This notation was derived for the first time by Tate (1959), and it was used for the first time for the assessment of m_{max} by Pisarenko et al. (1996). Applied to the frequency-magnitude Gutenberg-Richter relationship, it takes the form (v.g., Kijko and Graham 1998; Kijko et al. 2016)

$$\hat{m}_{max} = m_{max}^{obs} + \frac{1 - exp[-\beta\left(m_{max}^{obs} - m_{min}\right)]}{n\, exp[-\beta\left(m_{max}^{obs} - m_{min}\right)]}$$

The value obtained from the iterative solution of the previous equation is referred to as the Tate-Pisarenko estimator (in short, the T-P estimator).

The second parametric procedure considered in our study is based on the equation

$$\hat{m}_{max} = m_{max}^{obs} + \frac{E_1(n_2) - E_1(n_1)}{\beta \exp(-n_2)} + m_{min} \exp(-n)$$

where

$$n_1 = \frac{1}{1 - exp[-\beta(m_{max} - m_{min})]}; \quad n_2 = n_1 exp[-\beta(m_{max} - m_{min})]$$

and $E_1(\cdot)$ denotes the exponential integral function (Abramowitz and Stegun 1970). This procedure was introduced by Kijko and Sellevoll (1989), hereafter referred to as the K-S estimator. The m_{max} value is also obtained through an iteration process. It is important to note that in terms of the mean-squared error, this procedure is substantially better than the respective T-P estimator (Kijko and Graham 1998).

The next two considered procedures are extensions of the T-P and K-S estimators, to be used when the uncertainty of the earthquake magnitude distribution is taken into account. When the variation of the seismic activity is considered as a random process, the formalism of compound distributions (for example, when the model parameters are treated as random variables) provides an efficient account for the fluctuating seismicity (Campbell 1982). In this case, and according to Kijko (2004) and Kijko and Singh (2011), if the temporal variation of the b-value is represented by a Gamma distribution with parameters p and q, the T-P and K-S procedures can respectively be modified as follow

$$\hat{m}_{max} = m_{max}^{obs} + \frac{1}{n\,\bar{\beta}\,C_\beta}\left(\frac{p}{p + m_{max}^{obs} - m_{min}}\right)^{-(q+1)}$$

and

$$\hat{m}_{max} = m_{max}^{obs} + \frac{\delta^{1/q} exp\left(\frac{nr^q}{1-r^q}\right)}{\beta}\left(\Gamma\left(-\frac{1}{q}, \delta r^q\right) - \Gamma\left(-\frac{1}{q}, \delta\right)\right)$$

where

$$C_\beta = \frac{1}{1 - r^q};\quad r = \frac{p}{p + m_{max} + m_{min}};\quad p = \frac{\bar{\beta}}{\sigma_\beta^2};\quad q = \left(\frac{\bar{\beta}}{\sigma_\beta}\right)^2$$

being $\bar{\beta}$ the mean β value, σ_β the standard deviation of β-values fluctuation, $\delta = nC_\beta$ and $\Gamma(\cdot)$ the complementary incomplete Gamma function. These two solutions are called the Tate-Pisarenko-Bayes (T-P-B) and the Kijko-Sellevoll-Bayes (K-S-B) estimators, respectively.

All the procedures presented above have been used to estimate the area-characteristic seismic parameters, that is, the mean annual activity rate above magnitude 4.0, λ ($m \geq 4.0$), the b-value of the Gutenberg-Richter relationship, and the maximum possible magnitude in each one of the seismogenic zones defined in the Algeria-Morocco region. These seismic parameters were computed and analyzed for the main tectonic domains in this region by Hamdache et al. (2017). Thus, the results obtained in this work can be considered as an update of the previously obtained for larger areas.

The used final earthquake database was described above. The biggest recorded event in each sub-catalog is showed in Table 1 and, as it was also described previously, each sub-catalog has been divided in a historical and an instrumental part, as shown on Table 2. The time periods, the maximum observed magnitude and the threshold magnitude of completeness for each sub-file are also given.

When data do not fit well the Gutenberg-Richter frequency-magnitude recurrence relationship, Kijko (2004), Kijko and Singh (2011) and Kijko et al. (2016) proposed different procedures to derive an estimation of the maximum regional earthquake magnitude, among of them, some procedures using non-parametric technics and ordered statistics.

Table 1 Number of events (n) above magnitude M_W 4.0 and maximum observed magnitude, including σ, in the defined sources

Shallow sources	n	$M_{W\,max}^{obs}$	σ	Date	Recorded size	Location
ALB1	13	6.7[a]	0.3	13/01/1804, 17:53	VII-VIII	Alboran Sea
ALB2	64	6.5[a]	0.3	22/09/1522, 10:—	VIII-IX	W Alhama de Almería, Spain
ATL	10	5.6[b]	0.2	21/08/1937, 23:55	M_S 5.4	Cape St. Vincent
GB	60	8.7[a]	0.3	01/11/1755, 10:16	X	SW Cape St. Vincent
GC	38	6.9[a]	0.3	27/12/1722, 17:30	VIII	Gulf of Cádiz
HA	15	6.4[c]	0.5	—/—/1731, —:—	IX	Agadir, Morocco
HA-AA	14	5.8[c]	0.5	22/06/1941, 19:—	VIII	Talsinnt, Morocco
HA-MA	36	5.3[d]	0.2	11/08/2007, 20:46	m_b 5.3	S Azrou, Morocco
MA-HP	20	5.3[b]	0.2	29/11/1980, 12:47	M_S 4.9	E Jerada, Morocco
MM	16	5.1[b]	0.2	17/01/1979, 17:43	M_S 4.6, VIII	N Khenifra, Morocco
R1	66	6.4	0.1	24/02/2004, 02:28	M_W 6.4, VIII	SW Al Hoceima, Morocco
R2	59	6.7[c]	0.5	11/05/1624, —:—	IX-X	Fès, Morocco
SA1	26	5.8[c]	0.5	04/04/1924, —:—	VIII	E Djelfa, Algeria
SA2	43	6.0	0.1	12/02/1946, 02:43	M_W 6.0, VIII-IX	M'Sila, Algeria
T1	61	7.0[c]	0.5	09/10/1790, 01:15	IX-X	Oran, Algeria
T2	70	7.3	0.1	10/10/1980, 12:25	M_W 7.3, IX	Chlef, Algeria
T3	90	7.3[c]	0.5	02/01/1867, 07:14	X-XI	Blida, Algeria
T4	89	6.9	0.1	21/05/2003, 18:44	M_W 6.9, IX-X	Boumerdes, Algeria

(continued)

Table 1 (continued)

Shallow sources	n	$M_{W\,max}^{obs}$	σ	Date	Recorded size	Location
T5	34	5.8	0.1	27/10/1985, 19:35	M_W 5.8, VIII	Constantine, Algeria
T6	37	7.0[c]	0.5	27/11/1722, —:—	X	NW Annaba, Algeria
TA	40	7.0[c]	0.5	03/12/856, —:—	X	Tunis, Tunisia
Intermediate source	n	$M_{W\,max}^{obs}$	σ	Date	Recorded size	Location
GB-GC	76	5.5	0.1	17/12/2009; 01:38	M_W 5.5, IV	SW Cape St. Vincent

[a]From intensity data (Bakun and Wentworth 1997, 1998); [b]from M_S magnitude; [c]from I_{max}; [d]from m_b

Table 2 Used sub-catalogs for the computation of the seismicity parameters. Time interval, and m_c and σ are showed

Source	$\Delta t; m_c \pm \sigma$		
	Incomplete sub-catalog	Complete subcatalog (1)	Complete subcatalog (2)
ALB1	1450–1910; 4.0±0.3	1925–2010; 4.2±0.1	
ALB2	1400–1910; 4.0±0.3	1910–1960; 4.0±0.2	1960–2010; 4.0±0.1
ATL		1935–2010; 4.0±0.2	
GB	1350–1900; 4.0±0.3	1915–1990; 4.0±0.2	1990–2010; 4.0±0.1
GC	1700–1910; 4.0±0.2	1910–2010; 4.0±0.2	
HA	1700–1800; 5.2±0.3	1930–2010; 4.0±0.2	
HA-AA		1925–2010; 4.0±0.2	
HA-MA		1935–2010; 4.0±0.2	
MA-HP		1920–2010; 4.0±0.2	
MM		1935–2010; 4.0±0.2	
R1	1650–1910; 4.0±0.3	1910–1960; 4.0±0.2	1960–2010; 4.0±0.2
R2	1000–1910; 4.0±0.3	1910–1950; 4.0±0.2	1960–2010; 4.0±0.1
T1	1750–1910; 4.0±0.5	1910–1980; 4.0±0.3	1980–2010; 4.0±0.1
T2	1850–1910; 4.0±0.3	1910–1975; 4.0±0.2	1980–2010; 4.0±0.1
T3	1150–1895; 4.0±0.5	1900–1960; 4.0±0.2	1960–2010; 4.0±0.1
T4	1350–1905; 4.0±0.3	1910–1960; 4.0±0.2	1960–2010; 4.0±0.1
T5	1600–1910; 4.0±0.3	1910–1960; 4.0±0.2	1960–2010; 4.0±0.1
T6	1550–1910; 4.0±0.2	1920–1960; 4.0±0.2	1960–2010; 4.0±0.1
TA		1915–1960; 4.0±0.3	1960–2010; 4.0±0.2

SA1 and SA2 source zones do not include enough earthquake data. Then, the Robson-Whitlock-Cooke procedure (Robson and Whitlock 1964; Cooke 1979; Kijko and Singh 2011) has been used to compute an estimation of the maximum possible earthquake magnitude. The estimator of the Robson-Whitlock-Cooke procedure, namely R-W-C, is given as follow. Giving an ordered set of magnitudes

$$m_1 \leq m_2 \leq m_3 \leq \cdots \leq m_n \leq m_{max}^{obs}$$

the maximum magnitude and its standard deviation are given by the equations

$$m_{max} = m_{max}^{obs} + 0.5\left(m_{max}^{obs} - m_n\right)$$

$$\sigma_{m_{max}} = 0.5\left[3\sigma_{m_{max}^{obs}}^2 + 0.5\left(m_{max}^{obs} - m_n\right)^2\right]$$

where $\sigma_{m_{max}^{obs}}$ denotes the standard deviation of the maximum observed magnitude.

The obtained results using the different procedures are shown in Table 3. The implementation assumes that the data fit a linear Gutenberg-Richter relationship. As can be seen, the obtained maximum regional magnitude estimates using different parametric procedures are comparable. By the way, the Bayesian version of the T-P and K-S procedures (T-P-B and K-S-B procedures) include the facts that the data are not well fitted by the Gutenberg-Richter relationship, as well as the uncertainty affecting the *b*-value.

6 Inversion of Focal Mechanism Data and Inferred Stress Pattern

The stress pattern for the delineated shallow and intermediate seismic sources model has been computed from the available focal mechanism data and by applying the inversion method developed by Delvaux and Sperner (2003) and Delvaux and Barth (2010) using the Win-Tensor™ software. In the final assessment have been used only earthquakes above magnitude 3.5. Earthquakes below this magnitude usually correspond to aftershocks that can display local instead of regional stresses. In Table 4 can be seen the final number of focal mechanisms used in each seismogenic source facing the available total number of focal mechanism solutions. In some sources, it was not possible to obtain a stress solution. Thus, in the MA-HP source there are not focal mechanism data, the ATL, HA, MM and T6 sources includes very few data, and, in the HA-AA source, also including few data, it was not possible to obtain a solution from Win-Tensor™.

It is important to highlight that it is not necessary to choose between the two nodal planes available for each focal mechanism solution for the inversion process. The improved Right Dihedron method (Delvaux and Sperner 2003), proposed initially

Table 3 Seismicity parameters b, λ ($m \geq 4.0$) and m_{max}, including σ, for the defined sources. In the case of m_{max}, the K-S, K-S-B, T-P, T-P-B and R-W-C procedures (see text) have been used

Source	$b \pm \sigma$	$\lambda \pm \sigma$	$m_{max} \pm \sigma$				
			K-S	K-S-B	T-P	T-P-B	R-W-C
ALB1	1.06 ± 0.29	0.189 ± 0.071	6.80 ± 0.30	6.80 ± 0.45	6.80 ± 0.30	6.85 ± 0.30	
ALB2	0.71 ± 0.09	0.376 ± 0.056	6.65 ± 0.34	6.62 ± 0.32	6.70 ± 0.36	6.64 ± 0.30	
ATL	0.74 ± 0.29	0.137 ± 0.055	6.54 ± 0.96	6.37 ± 0.80	6.10 ± 0.20	6.10 ± 0.20	
GB	0.68 ± 0.09	0.520 ± 0.098	9.20 ± —[a]	9.20 ± —[a]	9.20 ± —[a]	9.20 ± —[a]	
GC	0.68 ± 0.12	0.281 ± 0.066	7.99 ± 1.14	7.49 ± 0.66	7.40 ± —[a]	7.40 ± —[a]	
HA	0.50 ± 0.10	0.187 ± 0.064	6.64 ± 0.55	6.62 ± 0.98	6.78 ± 0.63	6.72 ± 0.59	
HA-AA	0.65 ± 0.27	0.170 ± 0.062	7.05 ± 1.35	6.64 ± 0.98	6.30 ± —[a]	6.30 ± —[a]	
HA-MA	1.08 ± 0.25	0.473 ± 0.142	5.54 ± 0.31	5.50 ± 0.28	5.80 ± —[a]	5.80 ± —[a]	
MA-HP	1.00 ± 0.28	0.222 ± 0.075	5.59 ± 0.35	5.57 ± 0.34	5.80 ± 0.20	5.80 ± 0.20	
MM	1.45 ± 0.43	0.216 ± 0.077	6.47 ± 0.38	5.79 ± 0.72	5.60 ± 0.20	5.60 ± 0.20	
R1	0.89 ± 0.13	0.332 ± 0.058	7.12 ± 0.73	6.80 ± 0.41	6.90 ± 0.10	6.90 ± 0.10	
R2	0.94 ± 0.11	0.245 ± 0.041	7.68 ± 0.36	7.10 ± 0.64	7.20 ± 0.10	7.20 ± 0.10	
SA1							5.90 ± 0.02
SA2							6.10 ± 0.02
T1	0.71 ± 0.11	0.418 ± 0.076	8.90 ± 0.10	7.68 ± 0.84	7.50 ± 0.20	7.50 ± 0.20	
T2	0.66 ± 0.09	0.586 ± 0.105	7.60 ± 0.68	8.20 ± 0.10	7.80 ± 0.10	7.80 ± 0.10	
T3	0.70 ± 0.08	0.426 ± 0.069	7.67 ± 0.62	7.51 ± 0.54	7.80 ± 0.10	7.63 ± 0.60	
T4	0.81 ± 0.08	0.511 ± 0.084	7.27 ± 0.38	7.11 ± 0.23	7.40 ± 0.10	7.26 ± 0.37	
T5	0.60 ± 0.16	0.234 ± 0.051	5.88 ± 0.13	5.88 ± 0.13	5.90 ± 0.14	5.89 ± 0.13	
T6	0.88 ± 0.14	0.193 ± 0.040	7.50 ± —[a]	9.00 ± 0.60	7.50 ± —[a]	7.50 ± —[a]	
TA	0.50 ± —[a]	0.447 ± 0.107	5.94 ± 0.24	5.94 ± 0.24	5.97 ± 0.26	5.97 ± 0.26	

[a]There is not enough data to compute σ

by Angelier and Mechler (1977), was applied in our study jointly with the iterative Rotational Dihedron method, for estimating the different parameters of the reduced stress tensor.

Initially, the improved Right Dihedron method has been used to compute the four parameters of the reduced stress tensor: (a) the principal stress axis σ_1 (maximum compression), (b) σ_2 (intermediate compression), (c) σ_3 (minimum compression), and (d) the stress ratio $R = (\sigma_2 - \sigma_3) / (\sigma_1 - \sigma_3)$. This method allows both to provide the first estimations of the previous four reduced stress tensor parameters and to obtain the initial filtered focal mechanism data. Using this methodology, the nodal planes which are incompatible with the average stress regime are removed. The obtained results in this stage, the compatible focal mechanisms and the computed stress tensor, are then used directly as the starting point for the tensor Rotational Optimization approach.

The iterative grid-search Rotational Optimization procedure is based on a controlled grid search of stress tensors with the aim of minimizing the so-called misfit function ($F5$) in the Win-Tensor^{TM} code (Delvaux and Barth 2010). This function minimizes the angle deviation between the observed and modelled slip lines on the

Table 4 Number of focal mechanism solutions (n), solutions with Mw ≥ 3.5 (n') and maximum magnitude ($M_{W_{max}}$)

Source	n	n'	$M_{w_{max}}$
ALB1	13	9	6.0
ALB2	96	62	5.2
ATL	3	2	4.4
GB	35	18	7.4
GC	18	15	6.3
HA	4	3	5.9
HA-AA	5	4	5.4
HA-MA	9	7	5.1
MA-HP	0	0	–
MM	2	2	4.7
R1	312	96	6.3
R2	24	22	5.5
SA1	5	5	5.1
SA2	10	10	5.5
T1	20	18	5.8
T2	166	84	6.6
T3	117	33	5.8
T4	85	63	6.6
T5	113	5	5.6
T6	4	2	4.5
TA	24	18	5.6
GB-GC	45	37	6.1

planes (respectively the slip vector and the calculated shear) using the stress tensor that is being tested, but it also favours a higher shear stress magnitude and a lower normal stress magnitude (Delvaux and Barth 2010). In this procedure, both nodal planes for each focal mechanism solution are inverted to a stress tensor at the beginning, and the plane that is best explained by the stress tensor is selected among them, and considered as the actual fault plane (Hussein et al. 2013). It considers the movement plane best fit to the general tensor (it has the lowest misfit value), and the auxiliary plane has the highest value of the misfit function. Therefore, the final inversion includes the focal planes that are best fitted by a uniform stress field (Gephart and Forsyth 1984). This category of focal planes that fit a single stress tensor contains the correct choices, and hence, the inferred stress tensor represents the final one. Finally, it is necessary to re-consider the excluded focal planes without changing the stress tensor for testing, after the last optimization, when some of the focal planes can become compatible with the computed stress tensor. If so, it is necessary to re-incorporate them in the database, re-optimize the stress tensor and check again the rejected data. This process is repeated as often as necessary.

This methodology was applied to the focal mechanism data included in the defined seismic sources. The stress inversion results can be seen in Table 5 and Fig. 4, including, among other data, principal stress axes, stress ratio, misfit function, stress regime index-R' ($R' = R$ for a pure extensional regime, $R' = 2 - R$ for a pure strike-slip regime, and $R' = 2 + R$ for a pure compressional regime; Delvaux et al. 1997), minimum and maximum horizontal stress axes (after Lund and Townend 2007), and the stress regime according to the World Stress Map project. These results, clearly characterize the stress regime of each seismogenic source from the available dataset.

The R' index ranges from 0 (radial extension) to 3 (radial compression). Intermediate values are 0.5 for a pure extension, 1.0 for an extensional strike-slip, 1.5 for a pure strike-slip, 2.0 for a strike-slip compressional, and 2.5 for a pure compression regime (Delvaux and Sperner 2003).

In general, the obtained results agree very well with the tectonic framework of the Ibero-Moghrebian region (Fig. 5). Results show different tectonic stress regimes, including extensional, compressional, transpression and strike-slip (Table 5; Fig. 6), although the directions of the horizontal maximum principal stress are similar throughout the region (Figs. 5 and 6) nearly coinciding with a horizontal P axis in the NW-SE direction, as defined by Henares et al. (2003).

For those seismic sources delineated in the Atlantic area (ATL, GB and GC sources), the stress field is pure compressional with a dominant compression trending NW-SE to NNW-SSE (Henares et al. 2003; Custódio et al. 2016). For the intermediate depth seismic source also delineated in the Atlantic area (GB-GC), the result gives an oblique faulting regime (UF), including a horizontal compression in a NW-SE direction. In the central part of the study region, including the Rif and Alboran area (R1, R2 and ALB2 sources), the stress field is in agreement with the preference strike-slip (Stich et al. 2003b; Custódio et al. 2016) incorporating a horizontal compression in the NW-SE to NNW-SSE direction and a perpendicular horizontal extension. Conversely, the seismic source ALB1, in the same previous region, the stress field is pure extensional including a WSW-ENE horizontal extension (Henares et al. 2003).

For those sources delineated along the Algero-Provençal Basin, with the exception of the T5 source, the stress field is pure compressional with a dominant compression trending NNW-SSE (Henares et al. 2003; Stich et al. 2003b; Custódio et al. 2016). For the source T5, the stress field is a compressional strike-slip ($R' = 1.87$) regime striking N-S. The stress pattern for the HA-MA source is a NNW-SSE horizontal compression. The seismic sources located along the Sahara Atlas are predominantly influenced by a compressional stress field with a horizontal compression trending from NNW-SSE to N-S. Finally, in the TA source, in the Tunisian Atlas, the stress field appears as a pure compressional regime trending NNW-SSE (Soumaya et al. 2015).

Table 5 Stress inversion results for the defined sources

Source	eq	n/n_t	σ_1		σ_2		σ_3		R	$\bar{\alpha}$	αmax	$F5$	R'	SH_{min}	SH_{max}	Q	Stress regime
			Az	Pl	Az	Pl	Az	Pl									
ALB1	9	5/18	357	66	153	22	247	9	0.67	2.7	6.4	1.1	0.67	68	158	C	NF-Pure extensional
ALB2	62	34/124	354	29	144	57	256	13	0.32	11.2	44.2	6.6	1.68	81	171	A	SS-Pure strike-slip
GB	18	10/36	122	10	28	21	236	67	0.46	12.6	20.6	7.1	2.46	35	125	B	TF-Pure compressional
GC	15	10/30	321	0	231	9	51	81	0.33	8.5	20.6	4.2	2.33	51	141	B	TF-Pure compressional
HA-MA	7	7/14	159	2	249	12	60	78	0.29	12.4	22.9	7.7	2.29	69	159	C	TF-Pure compressional
R1	96	78/192	338	10	144	80	248	2	0.52	11.8	36.0	7.5	1.48	68	158	A	SS-Pure strike-slip
R2	22	16/44	134	20	320	70	225	2	0.42	17.2	168.0	24.5	1.58	44	134	B	SS-Pure strike-slip
SA1	5	4/10	304	18	36	5	140	71	0.31	2.6	3.6	2.0	2.31	33	123	C	TF-Pure Compressional
SA2	10	6/20	2	21	106	33	245	49	0.12	9.1	22.9	3.7	2.12	90	0	C	UF-Oblique compressive

(continued)

Table 5 (continued)

Source	eq	n/n_t	σ_1		σ_2		σ_3		R	$\bar{\alpha}$	α_{max}	$F5$	R'	SH_{min}	SH_{max}	Q	Stress regime
			Az	Pl	Az	Pl	Az	Pl									
T1	18	16/36	320	14	222	30	72	56	0.25	9.1	31.4	5.0	2.25	52	142	A	TF-Pure compressional
T2	84	76/168	325	5	55	2	168	85	0.33	9.1	33.5	4.9	2.33	55	145	A	TF-Pure compressional
T3	33	28/66	329	21	61	4	161	69	0.62	7.1	26.5	4.9	2.62	55	145	A	TF-Pure compressional
T4	63	53/126	325	20	59	10	174	67	0.54	10.4	25.5	6.3	24	50	140	A	TF-Pure compressional
T5	5	5/10	356	16	191	73	87	4	0.13	2.9	4.7	0.4	1.87	86	176	C	SS-Compressional strike-slip
TA	18	10/36	138	12	235	31	29	56	0.53	8.0	17.5	6.1	2.53	43	133	B	TF-Pure compressional
GB-GC	37	25/74	155	22	273	49	50	32	0.32	10.3	21.0	5.4	1.68	61	151	A	UF-Oblique compressive

eq number of focal mechanism data; *n* number of data explained by the stress tensor; n_t total population of fault plane solutions; σ_i principal stress axes; *Az* azimuth; *Pl* plunge; *R* stress ratio; $\bar{\alpha}$ average misfit angle; α_{max} maximum misfit angle; $F5$ misfit function of Delvaux and Barth (2010); R' stress regime index; SH_{min} and SH_{max} minimum and maximum horizontal stress axes, respectively; Q quality rank

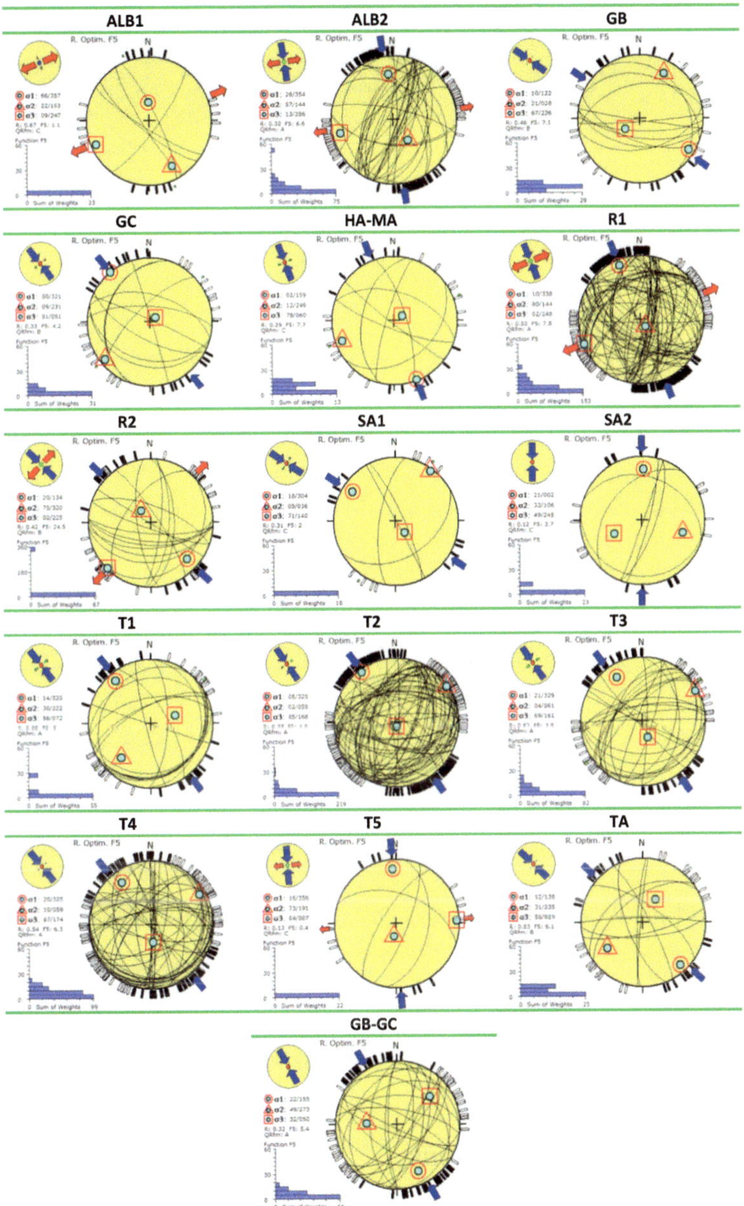

Fig. 4 Lower-hemisphere equal-area stereoplots of the selected focal planes for the defined sources, and associated slip lines. Results are represented by the orientation of the three principal stress axes (a turquoise dot surrounded by a circle for σ_1, a triangle for σ_2, and a square for σ_3). The related SH_{min} and SH_{max} orientations are represented by large arrows outside the stereogram (see Fig. 5). The histogram represents the distribution of the misfit angle $F5$ weighted arithmetically according to the magnitude

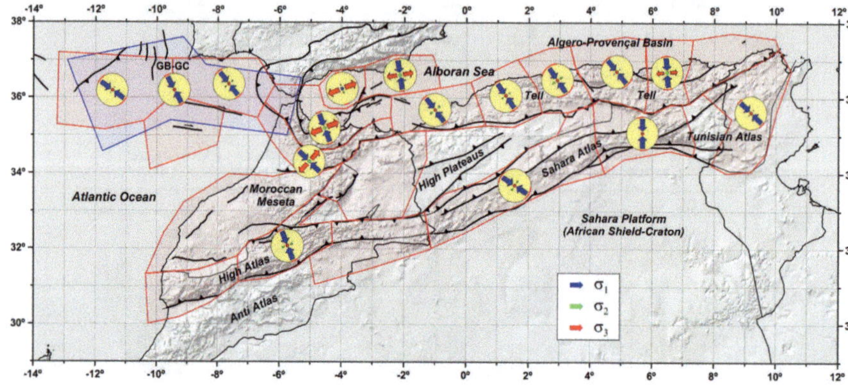

Fig. 5 Plot of the SH_{min} and SH_{max} results for the shallow and intermediate seismogenic zones

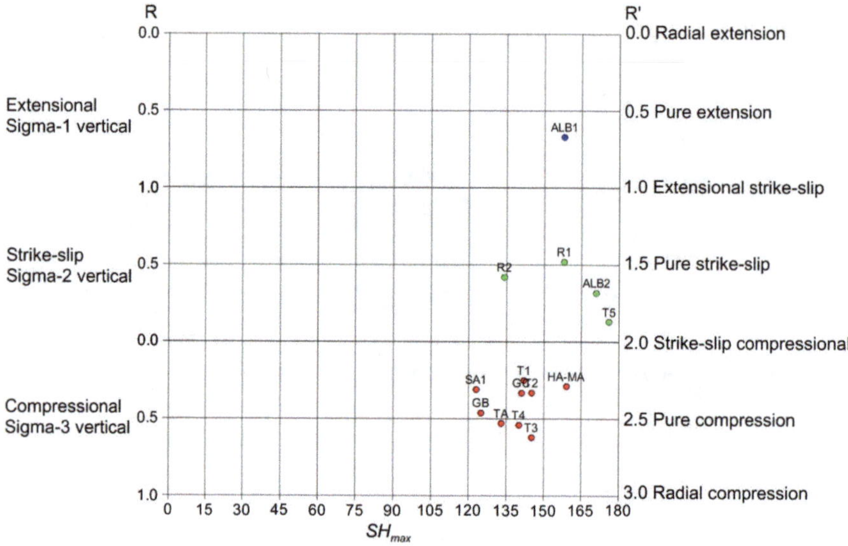

Fig. 6 SH_{max} versus R and R' indexes (Delvaux et al. 1997) for the defined sources

7 Summary and Conclusions

A preliminary seismogenic zone model is proposed for northwestern Africa aimed to seismic hazard studies. In order to establish and characterize this model, an update seismic catalog and a focal mechanisms compilation have been used. The model attempts to consider and connect both geological, tectonic and seismicity data. Delimited area sources, from a tectonic point of view and from the seismicity distribution, are as homogeneous as possible, taking into account the deficiencies of the used databases and the lack and low quality of data in some areas. Without a doubt, the quality of a zone model is influenced by the quality of the used data for its definition.

Finally, 21 shallow (crustal) zones and one additional intermediate depth zone have been considered in order to integrate the seismicity of the region influencing the seismic hazard in the northern Morocco-Algeria region.

After establish this model, and in a feed-back process, it has been characterized from selected seismicity parameters and the inversion of focal mechanism data for each delimited source. Seismicity parameters are the usual ones for seismic hazard studies, i.e., the b-value of the Gutenberg-Richter recurrence relationship, the annual rate of earthquakes above M_W 4.0, λ ($m \geq 4.0$), and the maximum magnitude that the area source can generate (m_{max}). The inversion of the available focal mechanisms in each source leads to know its stress pattern.

Taking into account the scarce available seismicity data in some areas of this region and the low quality of these data, the most robust approaches from a statistical point of view have been used. Although results fluctuate depending on the used methodology, several sources are available to generate earthquakes above magnitude M_W 6.0 and 7.0, as it can be observed in the recent past, when some damaging earthquakes happened in this region. Actually, even sources as T1 and GB could host earthquakes above M_W 8.0.

The obtained stress pattern, already commented in the previous section, can be summarized as follows. The directions of the obtained compression (σ_1) in the different seismic sources in which it was possible to compute the stress pattern, are congruent with the regional directions of compression indicated by many authors (v.g., DeMets et al. 2010). In general, the main directions for σ_1 strike NNW-SSE, although ranging between NW-SE and N-S directions. In mostly cases, σ_1 is almost horizontal, being in agreement with SH_{max}. These directions show the current NNW-SSE convergence between the Eurasia Plate, particularly between the Iberia sub-plate, and the northwest part of the Nubia Plate.

There is a great homogeneity in these directions, with the local exception of that obtained in the Sahara Atlas, likely related with a deficiency of data instead of with a real variation of the stress ellipsoid. This deficiency (low number of focal mechanisms depicting local stresses instead of regional ones) is clear in some seismic sources, where the stress ellipsoid was not computed.

It must be also highlighted that the stress pattern obtained in the seismogenic sources of the Gorringe Bank (GB) and the Gulf of Cádiz (GC) is in agreement with that computed for the source GB-GC, which includes the intermediate seismicity in this area. This can be interpreted as the continuity of the stresses with depth in this region.

An extension (σ_3) concur with the NNW-SSE to N-S SH_{max} in some seismogenic sources. Both features, particularly well identified in the areas of the Alboran Sea and in the Rif, are compatible, expressing a perpendicular escape to the main compression. In the Strait of Gibraltar area, this aspect has been interpreted as linked to a westwards subduction (Pedrera et al. 2011).

The quoted extension is particularly remarkable in the central-northern part of the Alboran Sea, an area with a nearly vertical σ_1, occupying σ_2 the NNW-SSE direction, then, changing their usual respective positions. In this area, the current stress ellipsoid nearly coincides with a radial extension, linked to horizontal and vertical readjustments.

Acknowledgements This research was supported by the Spanish Seismic Hazard and Active Tectonics research group and the Spanish MINECO CGL2015-65602-R and CGL2016-80687-R (AEI/FEDER, UE) projects. Figures 2 and 3 were created using the GMT software package (Wessel et al. 2013).

References

Abrahamson N (2006) Seismic hazard assessment: problems with current practice and future developments. In: First European conference on earthquake engineering and seismology, Geneva, Switzerland

Abramowitz M, Stegun I (1970) Handbook of mathematical functions. Dover Publications

AISI—American Iron and Steel Institute (1962) The Agadir, Morocco earthquake, 29 Feb 1960

Aït Brahim LA, Tadili BA, Peláez JA, El Hajji S, Chourak M, Hamdache M, López Casado C, Martínez Solares JM, Abdelouafi A, Laassiri J (2007) A new seismotectonic zoning of northern Morocco. In: Third National Congress on Seismic Engineering, Girona, Spain

Aki K (1965) Maximum likelihood estimate of *b* in the formula logN = a − bM and its confidence limits. Bull Earthq Res Inst Tokyo Univ 43:237–239

Alonso Chaves FM, García Navarro E, Camacho Cerro MA, Fernández Rodríguez C (2011) Propuesta sismotectónica para la terminación oriental de la Zona de Fractura de Azores - Gibraltar entre el Banco de Gorringe y el Banco del Guadalquivir. Geogaceta 50:11–14

Albarello D, Camassi R, Rebez A (2001) Detection of space and time heterogeneity in the completeness of a seismic catalog by a statistical approach: an application to the Italian area. Bull Seismol Soc Am 91:1694–1703

Angelier J, Mechler P (1977) Sur une méthode graphique de recherche des contraintes principales également ultilisable en tectonique et en séismologie: la méthode des dièdres droits. Bull. Soc. Géol. Fr. XIX:1309–1318

Aoudia A, Vaccari F, Suhadolc P, Meghraoui M (2000) Seismogenic potential and earthquake hazard assessment in the Tell Atlas of Algeria. J Seismol 4:79–98

Araya R, Der Kiureghian A (1988) Seismic hazard analysis: improved models, uncertainties and sensitivities. Report to the National Science Foundation, Earthquake Engineering Research Center, report no. UCB/EERC-90/11

ASCE—American Society of Civil Engineers (2004) Zemmouri, Algeria, M_W 6.8 earthquake of May 21, 2003. Report to the Technical Council on Lifeline Earthquake Engineering, monograph no. 27

Ayadi A, Dorbath C, Ousadou F, Maouche S, Chikh M, Bounif MA, Meghraoui M (2008) Zemmouri earthquake rupture zone (Mw 6.8, Algeria): aftershocks sequence relocation and 3D velocity model. J Geophys Res 113:B09301

Bahrouni N, Bouaziz S, Soumaya A, Ben Ayed N, Attafi K, Houla Y, El Ghali A, Rebai N (2014) Neotectonic and seismotectonic investigation of seismically active regions in Tunisia: a multidisciplinary approach. J Seismol 18:235–256

Bakun WH, Wentworth CM (1997) Estimating earthquake location and magnitude from seismic intensity data. Bull Seismol Soc Am 87:1502–1521

Bakun WH, Wentworth CM (1998) Corrections to "Estimating earthquake location and magnitude from seismic intensity data". U.S. Geol. Surv. Open-File Rept. 98-614

Beldjoudi H, Delouis B, Heddar A, Nouar OB, Yelles Chaouche A (2012) The Tadjena earthquake (Mw = 5.0) of December 16, 2006 in the Cheliff region (northern Algeria): waveform modelling, regional stresses, and relation with the Boukadir Fault. Pure Appl Geophys 169:677–691
Beldjoudi H, Delouis B, Djellit H, Yelles Chaouche A, Gharbi S, Abacha I (2016) The Beni-Ilmane (Algeria) seismic sequence of May 2010: seismic sources and stress tensor calculations. Tectonophysics 670:101–114
Ben Ayed N, Zargouni F (1990) Carte sismotectonique de la Tunisie à l'échelle 1/1000000. Fondation Nationale de la Recherche Scientifique, Tunisia
Benito B, Capote R, Murphy P, Gaspar Escribano JM, Martínez Díaz JJ, Tsige M, Stich D, García Mayordomo J, García Rodríguez MJ, Jiménez ME, Insua Arévalo JM, Álvarez Gómez JA, Canora C (2007) An overview of the damaging and low magnitude Mw 4.8 La Paca earthquake on 29 January 2005: context, seismotectonics, and seismic risk implications for southeast Spain. Bull Seismol Soc Am 97:671–690
Bensaid I, Buforn E, Cherkaoui TE, Medina F, Hahou Y (2011) New fault-plane solutions of Moroccan earthquakes for the 2005–2008 period. Bulletin de l'Institut Scientifique, Rabat, section Sciences de la Terre 33:47–52
Bensaid I, Cherkaoui TE, Medina F, Caldeira B, Buforn E, Emran A, Hahou Y (2012) The 1992 Tafilalt seismic crisis (Anti-Atlas, Morocco). J Seismol 16:35–53
Bezzeghoud M, Ayadi A, Sebaï A, Benhallou H (1994) Seismogenic zone survey by Algerian telemetered seismological network: case-study of Rouina earthquake, 19 January 1992. Phys Earth Planet Inter 84:235–246
Braunmiller J, Bernardi F (2005) The 2003 Boumerdes, Algeria earthquake: regional moment tensor analysis. Geophys Res Lett 32:L06305
Buforn E, Udías A, Madariaga R (1991) Intermediate and deep earthquakes in Spain. Pure Appl Geophys 136:375–393
Calvert A, Gomez F, Seber D, Barazangi M, Jabour N, Ibenbrahim A, Demnati A (1997) An integrated geophysical investigation of recent seismicity in the Al Hoceima region of north Morocco. Bull Seismol Soc Am 87:637–651
Campbell KW (1982) Bayesian analysis of extreme earthquake occurrences. Part I. Probabilistic hazard model. Bull Seismol Soc Am 72:1689–1705
Campos Costa A, Sousa ML, Carvalho A (2008) Seismic zonation for Portuguese national annex of Eurocode 8. In: The 14th world conference on earthquake engineering, Beijing, China
Carbon D, Martin C, Clement C, Bouceta M, Aït Brahim L, Tadili BA, Combes P, Secanell R, Abdellaoui K (2010) A new seismic zonation for Morocco based on probabilistic seismic hazard assessment. In: European Seismological Commission 32nd general assembly, Montpellier, France
Chalouan A, Michard A (2004) The Alpine Rif Belt (Morocco): a case of mountain building in a subdution-subduction-transform fault triple junction. Pure Appl Geophys 161:489–519
Chalouan A, Michard A, El Kadiri K, Negro F, Frizon de Lamotte D, Soto JI, Saddiqi O (2008) The Rif Belt. In: Michard A, Saddiqi O, Chalouan A, Frizon de Lamotte D (eds) Continental evolution: the geology of Morocco. Lect Notes Earth Sci (Springer) 116, 203–302
Chourak M, Aït Brahim L, Bezzeghoud M, Tadili BA, Peláez JA, Hamdache M, López Casado C, Martínez Solares JM (2013) Seismotectonics zoning of Morocco and adjacent region. Geophys Res Abstr 15:EGU 2013-11957
CNIG—Centro Nacional de Información Geográfica (2013) Updating seismic hazard maps for Spain 2012, Madrid, Spain (in Spanish)
Cooke P (1979) Statistical inference for bounds of random variables. Biometrika 66:367–374
Cornell CA (1968) Engineering seismic risk analysis. Bull Seismol Soc Am 58:1583–1606
Custódio S, Lima V, Vales D, Cesca S, Carrilho F (2016) Imaging active faulting in a region of distributed deformation from the joint clustering of focal mechanisms and hypocentres: application to the Azores-western Mediterranean region. Tectonophysics 676:70–89
Delvaux D, Sperner B (2003) New aspects of tectonic stress inversion with reference to the TENSOR program. Geol Soc, London, Spec Publ 212:75–100

Delvaux D, Barth A (2010) African stress pattern from formal inversion of focal mechanism data. Tectonophysics 482:105–128

Delvaux D, Moeys R, Stapel G, Petit C, Levi K, Miroshnichenko A, Ruzhich V, Sankov V (1997) Paleostress reconstructions and geodynamics of the Baikal region, Central Asia. Part II: Cenozoic rifting. Tectonophysics 282:1–38

DeMets C, Gordon R, Argus DF (2010) Geologically current plate motions. Geophys J Int 181:1–80

DeMets Ch, Gordon R, Argus DF, Stein S (1994) Effect of recent revisions to the geomagnetic reversal time scale on estimates of current plate motions. Geophys Res Lett 21:2191–2194

Dziewonski AM, Chou TA, Woodhouse JH (1981) Determination of earthquake source parameters from waveform data for studies of global and regional seismicity. J Geophys Res 86:2825–2852

EERI—Earthquake Engineering Research Institute (1983) El-Asnam, Algeria earthquake of October 10, 1980. A reconnaissance and engineering report

Ekström G, Nettles M, Dziewonski AM (2012) The global CMT project 2004–2010: centroid-moment tensors for 13,017 earthquakes. Phys Earth Planet Inter 200:1–9

El Alami SO, Medina F, Jebli H (1992) Etude du séisme d'Agadir du 5 avril 1992. Gaia. Revista de Geociências 5:1–5

El Alami SO, Ramdani F, Medina F, Jebli H (1989) Le séisme d'Essaouira du 21 novembre 1988: carte des isoséistes, étude des répliques et mécanisme au foyer. Bull Inst Sci 13:43–49

El Alami SO, Tadili BA, Cherkaoui TE, Medina F, Ramdani M, Aït Brahim L, Harnafi M (1998) The Al Hoceima earthquake of May, 26, 1994 and its aftershocks: a seismotectonic study. Ann Geofis 41:519–537

Ferranti L, Passaro S, de Alteriis G (2014) Morphotectonics of the Gorringe Bank summit, eastern Atlantic Ocean, based on high-resolution multibeam bathymetry. Quat Int 332:99–114

Frohlich C (1992) Triangle diagrams: ternary graphs to display similarity and diversity of earthquake focal mechanisms. Phys Earth Planet Inter 75:193–198

Frohlich C, Davis SD (1993) Teleseismic *b* values; Or much ado about 1.0. J Geophys Res 98:631–644

Galindo Zaldívar J, Azzouz O, Chalouan A, Pedrera A, Ruano P, Ruiz Constán A, Sanz de Galdeano C, Marín Lechado C, López Garrido AC, Anahnah F, Benmakhlouf M (2015) Extensional tectonics, graben development and fault terminations in the eastern Rif (Bokoya-Ras Afrou area). Tectonophysics 663:140–149

García Mayordomo J (2015) Creation of a seismogenic zone model for the computation of the seismic hazard map of Spain. Instituto Geológico y Minero de España, Madrid (in Spanish)

García Mayordomo J, Insua Arévalo JM, Martínez Díaz JJ, Perea H, Álvarez Gómez JA, Martín González F, González A, Lafuente P, Pérez López R, Rodríguez Pascua MA, Giner Robles J, Azañón J, Masana E, Moreno X (2010) Integrated seismogenic source-zones model for Spain. First Iberian Meeting on Active Faults and Paleoseismology. Sigüenza, Guadalajara, Spain (in Spanish)

Gardner JK, Knopoff L (1974) Is the sequence of earthquakes in Southern California, with aftershocks removed, Poissonian? Bull Seismol Soc Am 64, 1363–1367

Geissler WH, Matias L, Stich D, Carrilho F, Jokat W, Monna S, IbenBrahim A, Mancilla F, Gutscher MA, Sallarès V, Zitellini N (2010) Focal mechanisms for sub-crustal earthquakes in the Gulf of Cadiz from a dense OBS deployment. Geophys Res Lett 37:L18309

Gephart JW, Forsyth DW (1984) An improved method for determining the regional stress tensor using earthquake focal mechanism data: application to the San Fernando earthquake sequence. J Geophys Res 89:9305–9320

Grevemeyer I, Gràcia E, Villaseñor A, Leuchters W, Watts AB (2015) Seismicity and active tectonics in the Alboran Sea, Western Mediterranean: constraints from an offshore-onshore seismological network and swath bathymetry data. J Geophys Res Solid Earth 120:8348–8365

Gueddiche M, Harjono H, Ben Ayed N, Hfaiedh M, Diament M, Dubois J (1992) Analyse de la séismicité et mise en évidence d'accidents actifs dans le nord de la Tunisie. Bull Soc Geol Fr 163:415–425

Gutenberg B, Richter CF (1942) Earthquake magnitude, intensity, energy, and acceleration. Bull Seismol Soc Am 32:163–191
Gutenberg B, Richter CF (1956) Earthquake magnitude, intensity, energy, and acceleration (Second paper). Bull Seismol Soc Am 46:105–145
Hamdache M (1998) Seismic hazard assessment for the main seismogenic zones in north Algeria. Pure Appl Geophys 152:281–314
Hamdache M, Peláez JA, Kijko A, Smit A (2017) Energetic and spatial characterization of seismicity in the Algeria-Morocco region. Nat Hazards 86:S237–S293
Hamdache M, Peláez JA, Talbi A, López Casado C (2010) A unified catalog of main earthquakes for Northern Algeria from A.D. 856 to 2008. Seismol Res Lett 81:732–739
Hamdache M, Peláez JA, Talbi A, Mobarki M, López Casado C (2012) Ground-motion hazard values for northern Algeria. Pure Appl Geophys 169:711–723
Hamdache M, Peláez JA, Yelles Chauche AK (2004) The Algiers, Algeria earthquake (M_W 6.8) of 21 May 2003: Preliminary report. Seismol Res Lett 75:360–367
Hamdache M, Retief JP (2001) Site-especific seismic hazard estimation in the main seismogenic zones of northern Algeria. Pure Appl Geophys 158:1677–1690
Hatzfeld D (1978) Etude sismotectonique de la zone de collision Ibéro-Maghrébine. Thèse de doctorat. Université Scientifique et Médicale de Grenoble, France
Heaton TH, Tajima F, Mori AW (1986) Estimating ground motions using recorded accelerograms. Surv Geophys 8:25–83
Henares J, López Casado C (2001) Catálogo de mecanismos focales del área Ibero-Mogrebí. Granada, Spain. ISBN 84-699-5695-7
Henares J, López Casado C, Delgado J (2000) Catalogue of focal mechanisms for the Iberian-Maghrebian region. *2ª Asamblea Hispano Portuguesa de Geodesia y Geofísica*, Lagos, Portugal
Henares J, López Casado C, Sanz de Galdeano C, Delgado J, Peláez JA (2003) Stress fields in the Ibero-Maghrebian region. J Seismol 7:65–78
Hfaiedh M, Ben Ayed N, Dorel J (1985) Etude néotectonique et sismotectonique de la Tunisie nord-orientale. Note Serv Géol Tunisie 16:41–56
Holschneider M, Zoller G, Hainzl S (2011) Estimation of the maximum possible magnitude in the framework of a doubly-truncated Gutenberg-Richter model: limits of statistical inference from earthquake catalogs. Bull Seismol Soc Am 101:1649–1659
Hussein HM, Abou Elenean KM, Marzouk IA, Korrat IM, Abu El-Nader IF, Ghazala H, ElGabry MN (2013) Present-day tectonic stress regime in Egypt and surrounding area based on inversion of earthquake focal mechanisms. J Afr Earth Sci 81:1–15
Jacobshagen V (1992) Major fracture zones of Morocco: the South Atlas and the Transalboran fault systems. Geol Rundsch 81:185–197
Jiménez MJ, García Fernández M, The GSHAP Ibero-Maghreb Working Group (1999) Seismic hazard in the Ibero-Maghreb region. Ann Geofis 42, 1057–1065
Kijko A (2004) Estimation of the maximum earthquake magnitude, m_{max}. Pure Appl Geophys 161:1655–1681
Kijko A, Graham G (1998) Parametric-historic procedure for probabilistic seismic hazard analysis. Part I. Estimation of maximum regional magnitude m_{max}. Pure Appl Geophys 152:413–442
Kijko A, Sellevoll MA (1989) Estimation of earthquake hazard parameters from incomplete data files. Part I. Utilization of extreme and complete catalogues. Bull Seismol Soc Am 79:645–654
Kijko A, Sellevoll MA (1992) Estimation of earthquake hazard parameters from incomplete data files. Part II. Incorporation of magnitude heterogeneity. Bull Seismol Soc Am 82:120–134
Kijko A, Singh M (2011) Statistical tools for maximum possible earthquake magnitude estimation. Acta Geophys 59:674–700
Kijko A, Smit A (2012) Extension of the Aki-Utsu *b*-value estimator for incomplete catalogs. Bull Seismol Soc Am 102:1283–1287
Kijko A, Smit A, Sellevoll MA (2016) Estimation of earthquake hazard parameters from incomplete data files. Part III. Incorporation of uncertainty of earthquake-occurrence model. Bull Seismol Soc Am 106:1210–1222

Lomnitz C (1973) Poisson processes in earthquake studies. Bull Seismol Soc Am 63:735–758
López Casado C, Sanz de Galdeano C, Molina Palacios S, Henares J (2001) The structure of the Alboran Sea: an interpretation from seismological and geological data. Tectonophysics 338:79–95
Lund B, Townend J (2007) Calculating horizontal stress orientations with full or partial knowledge of the tectonic stress tensor. Geophys J Int 170:1328–1335
Martin R, Stich D, Morales J, Mancilla F (2015) Moment tensor solutions for the Iberian-Maghreb region during the IberArray deployment (2009–2013). Tectonophysics 663:261–274
Martínez García P, Comas M, Soto JI, Lonergan L, Watts AB (2013) Strike-slip tectonics and basin inversion in the western Mediterranean: the Post-Messinian evolution of the Alboran Sea. Basin Res 25:361–387
Marzocchi W, Sandri L (2003) A review and new insights on the estimation of the *b*-value and its uncertainty. Ann Geophys 46:1271–1282
McGuire RK (1976) Fortran computer program for seismic risk analysis. U.S. Geol. Surv. Open-File Rept. 76-67
McNutt SR (2005) Volcanic seismology. Annu Rev Earth Planet Sci 33:461–491
Medina F (2008) Catalogue of focal mechanisms of Moroccan earthquakes for the period 1959–2007. Documents de l'Institut Scientifique 23, Rabat, Morocco
Mejri L (2012) Tectonique quaternaire, paléosismicité et sources sismogéniques en Tunisie nord-orientale: étude de la faille d'Utique. Thèse de doctorat. Univ. Toulouse, France
Mourabit T, Abou Elenean KM, Ayadi A, Benouar D, Ben Suleman A, Bezzeghoud M, Cheddadi A, Chourak M, ElGabry MN, Harbi A, Hfaiedh M, Hussein HM, Kacem J, Ksentini A, Jabour N, Magrin A, Maouche S, Meghraoui M, Ousadou F, Panza GF, Peresan A, Romdhane N, Vaccari F, Zuccolo E (2014) Neo-deterministic seismic hazard assessment in North Africa. J Seismol 18:301–318
Mueller CS (2010) The influence of maximum magnitude on seismic-hazard estimates in the central and eastern United States. Bull Seismol Soc Am 100:699–711
Ousadou F, Dorbath L, Ayadi A, Dorbath C, Gharbi S (2014) Stress field variations along the Maghreb region derived from inversion of major seismic crisis fault plane solutions. Tectonophysics 632:261–280
Ousadou F, Dorbath L, Dorbath C, Bounif MA, Benhallou H (2013) The Constantine (Algeria) seismic sequence of 27 October 1985: a new rupture model from aftershock relocation, focal mechanisms, and stress tensors. J Seismol 17:207–222
Ouyed M, Yielding G, Hatzfeld D, King GCP (1983) An aftershock study of the El Asnam (Algeria) earthquake of 1980 October 10. Geophys J Int 73:605–639
Pedrera A, Galindo Zaldívar J, Marín Lechado C, García Tortosa FJ, Ruano P, López Garrido AC, Azañón JM, Peláez JA, Giaconia F (2012) Recent and active faults and folds in the central-eastern Internal Zones of the Betic Cordillera. J Iber Geol 38:191–208
Pedrera A, Ruiz Constán A, Galindo Zaldívar J, Chalouan A, Sanz de Galdeano C, Marín Lechado C, Ruano P, Benmakhlouf M, Akil M, López Garrido AC, Chabli A, Ahmamou M, González Castillo L (2011) Is there an active subduction beneath the Gibraltar orogenic arc? Constraints from Pliocene to present-day stress field. J Geodyn 52:83–96
Peláez JA, Chourak M, Tadili BA, Brahim LA, Hamdache M, López Casado C, Martínez Solares JM (2007) A catalog of main Moroccan earthquakes from 1045 to 2005. Seismol Res Lett 78:614–621
Peláez JA, Hamdache M, López Casado C (2003) Seismic hazard in northern Algeria using spatially smoothed seismicity. Results for peak ground acceleration. Tectonophysics 372:105–119
Peláez JA, Hamdache M, López Casado C (2005) Updating probabilistic seismic hazard values of northern Algeria with the 21 May 2003 M 6.8 Algiers earthquake included. Pure Appl Geophys 162:2163–2177
Peláez JA, Hamdache M, López Casado C (2006) Seismic hazard in terms of spectral accelerations and uniform hazard spectra in northern Algeria. Pure Appl Geophys 163:119–135
Peláez JA, Hamdache M, Sanz de Galdeano C, Sawires R, García Hernández MT (2016) Forecasting moderate earthquakes in northern Algeria and Morocco. In: D'Amico S (ed) Earthquakes and their impact on society. Springer, Springer Natural Hazards, pp. 81–95

Peláez JA, López Casado C (2002) Seismic hazard estimate at the Iberian Peninsula. Pure Appl Geophys 159:2699–2713
Pisarenko VF (1991) Statistical evaluation of maximum possible magnitude. Izv.- Earth Phys 27, 757–763
Pisarenko VF, Lyubushin AA, Lysenko VB, Golubieva TV (1996) Statistical estimation of seismic hazard parameters: maximum possible magnitude and related parameters. Bull Seismol Soc Am 86:691–700
Pondrelli S, Ekström G, Morelli A (2001) Seismotectonic re-evaluation of the 1976 Friuli, Italy, seismic sequence. J Seismol 5:73–83
Pondrelli S, Morelli A, Ekström G (2004) European-Mediterranean regional centroid-moment tensor catalog: solutions for years 2001 and 2002. Phys Earth Planet Int 145:127–147
Pondrelli S, Morelli A, Ekström G, Mazza S, Boschi E, Dziewonski AM (2002) European-Mediterranean regional centroid-moment tensors: 1997–2000. Phys Earth Planet Int 130:71–101
Pondrelli S, Salimbeni S (2015) Regional moment tensor review: an example from the European Mediterranean region. In: Beer M, Kougioumtzoglou IA, Patelli E, Siu-Kui Au I (eds) Encyclopedia of earthquake engineering. Springer, pp 1–15
Pondrelli S, Salimbeni S, Ekström G, Morelli A, Gasperini P, Vannucci G (2006) The Italian CMT dataset from 1977 to the present. Phys Earth Planet Int 159:286–303
Pondrelli S, Salimbeni S, Morelli A, Ekström G, Boschi E (2007) European-Mediterranean regional centroid moment tensor catalog: solutions for years 2003 and 2004. Phys Earth Planet Int 164:90–112
Pondrelli S, Salimbeni S, Morelli A, Ekström G, Postpischl L, Vannucci G, Boschi E (2011) European-Mediterranean regional centroid moment tensor catalog: solutions for 2005–2008. Phys Earth Planet Inter 185:74–81
Robson DS, Whitlock JH (1964) Estimation of a truncation point. Biometrika 51:33–39
Rodríguez Pascua MA, De Vicente G (2001) Estado de esfuerzos actual en la Cordillera Bética Oriental. Boletín Geológico y Minero 112:79 95
Rosas FM, Duarte JC, Neves MC, Terrinha P, Silva S, Matias L, Gràcia E, Bartolome R (2012) Thrust-wrench interference between major active faults in the Gulf of Cadiz (Africa-Eurasia plate boundary, offshore SW Iberia): Tectonic implications from coupled analog and numerical modeling. Tectonophysics 548–549:1–21
RPA—Règles Parasismiques Algériennes (1999) Centre National de Recherche Apliquée en Génie Parasismique, Algiers, Algeria
RPA—Règles Parasismiques Algériennes (2003) RPA 99—version 2003. Centre National de Recherche Apliquée en Génie Parasismique, Algiers, Algeria
RPS—Reglement de Construction Parasismique (2000) Ministère de l'Aménagement du Territoire, de l'Urbanisme, de l'Habitat et de l'Environnement, Rabat, Morocco
RPS—Reglement de Construction Parasismique (2011) RPS 2000—version 2011. Ministère de l'Habitat et de la Politique de la Ville. Rabat, Morocco
Rueda J, Mezcua J (2005) Near-real-time seismic moment-tensor determination in Spain. Seismol Res Lett 76:455–465
Sánchez Gómez M, Torcal F (2002) Recent tectonic activity on the south margin of the Guadalquivir basin, between Cabra and Quesada towns (provinces of Jaén and Córdoba, Spain). I Centenario del Observatorio de Cartuja, Granada, Spain
Sanz de Galdeano C (1990) La prolongación hacia el sur de las fosas y desgarres del Norte y Centro de Europa: Una propuesta de interpretación. Rev Soc Geol España 3:231–241
Sawires R, Peláez JA, Ibrahim HA, Fat-Helbary RE, Henares J, Hamdache M (2016) Delineation and characterization of a new source model for seismic hazard studies in Egypt. Nat Hazards 80:1823–1864
Schorlemmer D, Woessner J (2008) Probability of detecting an earthquake. Bull Seismol Soc Am 98:2103–2117
Silva V, Crowley H, Varum H, Pinho R (2014) Seismic hazard and risk assessment of Portugal. In: Second European conference on earthquake engineering and seismology, Istambul, Turkey

Soumaya A, Ben Ayed N, Delvaux D, Ghanmi M (2015) Spatial variation of present-day stress field and tectonic regime in Tunisia and surroundings from formal inversion of focal mechanisms: geodynamics implications for central Mediterranean. Tectonics 34:1154–1180

Stich D, Alguacil G, Morales J (2001) The relative location of multiplets in the vicinity of the Western Almería (southern Spain) earthquake series of 1993–1994. Geophys J Int 146:801–812

Stich D, Ammon CJ, Morales J (2003a) Moment tensor solutions for small and moderate earthquakes in the Ibero-Maghreb region. J Geophys Res 108:2148

Stich D, Batlló J, Morales J, Macià R, Dineva S (2003b) Source parameters of the Mw = 6.1 1910 Adra earthquake (southern Spain). Geophys J Int 155:539–546

Stich D, Mancilla F, Baumont D, Morales J (2005) Source analysis of the Mw 6.3 2004 Al Hoceima earthquake (Morocco) using regional apparent source time functions. J Geophys Res 110:B06306

Tadili BA (1991) Etude du risque sismique au Nord du Maroc. Thèse de doctorat. Université Mohammed I, Morocco

Tate RF (1959) Unbiased estimation: functions of location and scale parameters. Ann Math Stat 30:341–366

Utsu T (1965) A method for determining the value of *b* in a formula log n = a − bM showing the magnitude-frequency relation for earthquakes (with English summary). Geophys Bull Hokkaido Univ 13:99–103

Vaccari F, Tadili BA, El Qadi A, Ramdani M, Aït Brahim L, Limouri M (2001) Deterministic seismic hazard assessment for North Morocco. J Seismol Earthq Eng 3:1–12

Van Der Woerd J, Dorbath C, Ousadou F, Dorbath L, Delouis B, Jacques E, Tapponnier P, Hahou Y, Menzhi M, Frogneux M, Haessler H (2014) The Al Hoceima Mw 6.4 earthquake of 24 February 2004 and its aftershocks sequence. J Geodyn 77:89–109

Vannucci G, Gasperini P (2004) The new release of the database of earthquake mechanisms of the Mediterranean area (EMMA version 2). Ann Geophys 47:307–334

Vilanova SP, Fonseca JFBD (2007) Probabilistic seismic-hazard assessment for Portugal. Bull Seismol Soc Am 97:1702–1717

Wells DL, Coppersmith KJ (1994) New empirical relationships among magnitude, rupture length, rupture width, rupture area and surface displacement. Bull Seismol Soc Am 84:974–1002

Wessel P, Smith WHF, Scharroo R, Luis J, Wobbe F (2013) Generic mapping tools: improved version released. EOS Trans AGU 94(45):409–410

Wheeler RL (2009) Methods of M_{max} estimation east of the Rocky Mountains. U.S. Geol. Surv. Open-File Rept. 2009-1018

Wiemer S, Katsumata K (1999) Spatial variability of seismicity parameters in aftershock zones. J Geophys Res 104:135–151

Wiemer S, Wyss M (2000) Minimum magnitude of completeness in earthquake catalogs: examples from Alaska, the western United States, and Japan. Bull Seismol Soc Am 90:859–869

Wildi W (1983) La chaîne tello-rifaine (Algérie, Maroc, Tunisie): structure, stratigraphie et évolution du Trias au Miocène. Rev Géol Dynam Géog Phys 24:201–297

Yelles Chaouche A, Abacha I, Semmane F, Beldjoudi H, Djellit H (2014) The Béni-Ilmène (north-central Algeria) earthquake sequence of May 2010. Pure Appl Geophys 171:1283–1298

Yielding G, Ouyed M, King GCP, Hatzfeld D (1989) Active tectonics of the Algerian Atlas Mountains-evidence from aftershocks of the 1980 El Asnam earthquake. Geophys J Int 99:761–788

Zoback ML (1992) First and second order patterns of stress in the lithosphere: the world stress map project. J Geophys Res 97:11703–11728

A Trial Modeling of Perturbed Repeating Earthquakes Combined by Mathematical Statics, Numerical Modeling and Seismological Observations

Keisuke Ariyoshi, Shunichi Nomura, Naoki Uchida and Toshihiro Igarashi

1 Introduction: A Review of Previous Studies on Renewal Point Processes

Earthquake recurrence cycle has been thought as repetition of stress accumulation for long term as interseismic stage and stress release for short period as coseismic stage. Without any stress perturbation under constant stress loading rate, earthquakes are expected to become perfectly characteristic in that the recurrence intervals and magnitudes (or moment release amounts) are always constant.

Focusing on the previous record of repeating earthquake activity off Kamaishi, Japan, Matsuzawa et al. (2002) succeeded in forecasting the occurrence time of next repeating earthquake on the ground that recurrence interval was so stable (Igarashi et al. 2003) as to be treated as the normal distribution. This assumption is a good approximation in case of weak stress perturbation condition due to isolated asperities far from large earthquake source regions (Uchida et al. 2005). However, the assumption of the normal distribution is just for simplification, which is lack of corroborative scientific evidence.

From these background, the Headquarters for Earthquake Research Promotion (2013) belonging to the Ministry of Education, Culture, Sports, Science and Technology (MEXT) of the Japanese government proposed a physical model for the earthquake recurrence by combining the Brownian vibrational relaxation process

K. Ariyoshi (✉)
Earthquake and Tsunami Research Project for Disaster Prevention, Japan Agency for Marine-Earth Science and Technology, Yokohama 236-0001, Japan
e-mail: ariyoshi@jamstec.go.jp

S. Nomura
The Institute of Statistical Mathematics, Tachikawa, Japan

N. Uchida
Graduate School of Science, Research Center for Prediction of Earthquakes and Volcanic Eruptions, Tohoku University, Sendai, Japan

T. Igarashi
Earthquake Research Institute, The University of Tokyo, Bunkyō, Japan

S. D'Amico (ed.), *Moment Tensor Solutions*, Springer Natural Hazards,
https://doi.org/10.1007/978-3-319-77359-9_30

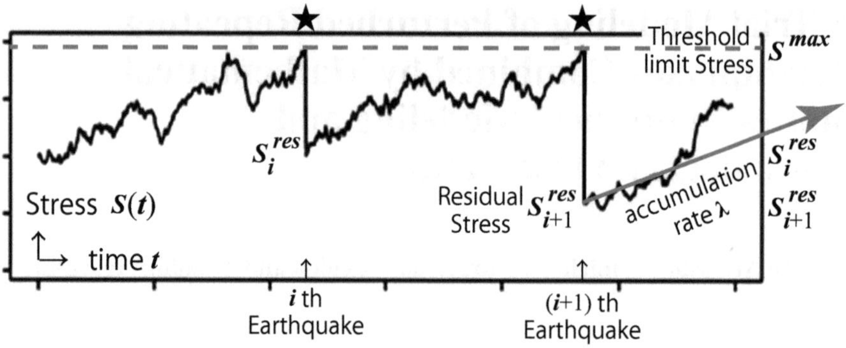

Fig. 1 Schematic illustration of the Brownian Passage Time (BPT) renewal point process, where $S^{\max}$ and S^{res} is the threshold limit of loading shear stress and residual shear stress after the occurrence of earthquakes, respectively (modified from Nomura 2015)

and constant threshold limit of stress $S^{\max}$ (Fig. 1). This model is described as a Brownian Passage Time (BPT) renewal point process, and stress $S(t)$ at the time t is described as

$$dS(t) = \lambda dt + \sigma dW(t), \tag{1}$$

where the stress accumulation rate (λ) is assumed to be constant value derived from the average of stress release amount per recurrence interval for each event, σ is a perturbation rate to be justified as a dispersion, and $W(t)$ is a standard Brownian motion described as stress perturbation mainly derived from nearby earthquakes.

Equation (1) is developed from the Time Predictable Model (Shimazaki and Nakata 1980), which enables us to evaluate the probability of recurrence time of the next earthquake (x) as follows

$$f(x|\mu,\alpha) = \sqrt{\frac{\mu}{2\pi\alpha^2 x^3}} \exp\left\{-\frac{(x-\mu)^2}{2\mu\alpha^2 x}\right\} (x > 0), \tag{2}$$

where

$$\mu = \left(S^{\max} - S^{\text{res}}_{\text{ave}}\right)/\lambda, \tag{3}$$

$$\alpha^2 = \sigma^2/\left\{\lambda\left(S^{\max} - S^{\text{res}}_{\text{ave}}\right)\right\}. \tag{4}$$

$$S^{res}_{ave} = \sum_i^n S^{res}_i \Big/ n. \tag{5}$$

2 Application to Perturbed Repeating Earthquakes on the Basis of Trend Renewal Model

The physical model in Fig. 1 means that actual stress rate is described as the combination of the accumulation rate (λ) and the Brownian relaxation process, which is true for the case of small stress perturbation. With significantly great stress perturbation, however, this assumption is not good approximation. For example, repeating earthquake recurrence interval off Kamaishi was temporarily shorter just after the 2011 Mw 9.0 Tohoku earthquake (Fig. 2). This is because the accumulation rate became significantly higher due to large postseismic slip (Ariyoshi et al. 2015).

To consider the acceleration effect of loading rate due to large postseismic slip, Nomura et al. (2014) has improved the Brownian vibrational relaxation process by describing the accumulation rate λ as a function of time $\lambda(t)$ as follows:

$$dS(t) = \lambda_0 \lambda(t)\, dt + \sigma_0 \sqrt{\lambda(t)} dW(t), \tag{6}$$

where the non-stationary accumulation rate and dispersion is converted to $\lambda_0 \lambda(t)$ and $\sigma_0 \sqrt{\lambda(t)}$, respectively. Equation (6) does not allow Brownian Passage Time (BPT) renewal point process, but it obey BPT distribution by converting time t' as

$$t' = \int_0^t \lambda(s)\, ds. \tag{7}$$

The time-transformed renewal process defined by Eqs. (6) and (7) is called as the Trend renewal process (Lindqvist et al. 2003). Nomura et al. (2014) applied the law of Omori-Utsu formula $\lambda(t) = K(t + c)^{-p}$ to Eq. (7) by fitting the constant parameters p and c on the basis of aftershock seismicity as shown in Fig. 3.

3 Observation and Modeling of Perturbed Seismic Moment Release for the Source Region of Repeaters

Figure 2 suggests that magnitude of repeating earthquakes off Kamaishi is largely characteristic before the occurrence of the 2011 M9 Tohoku earthquake. Just after the occurrence, however, the magnitude is significantly changed to be not only greater (Fig. 4) but also smaller (as shown in the right side of Fig. 2). This phenomenon means that we should treat the moment release amount as non-stationary. In other words, not only the stress accumulation rate described in the Sect. 2 but also the difference of $S^{max} - S^{res}$ should be not constant.

From recent observation results, Uchida et al. (2015) reported that the repeating earthquakes (M2.5–6.1) show postseismic magnitude increases for most sequences located in the area of large postseismic slip at the downdip extension of the M9

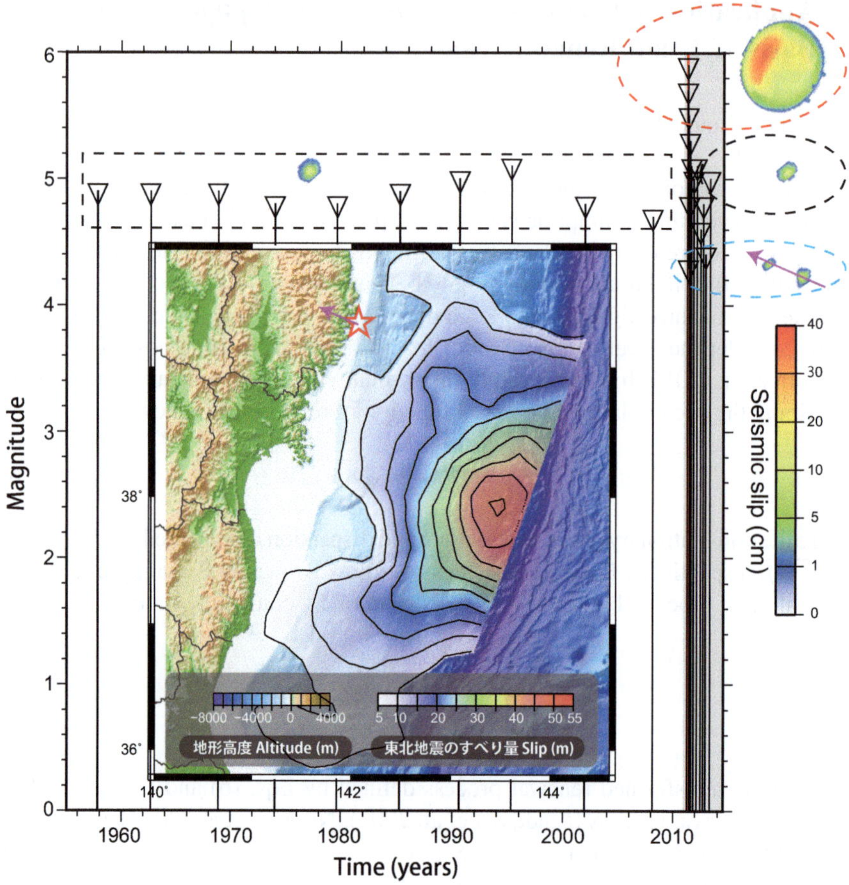

Fig. 2 Precursory Hypocenter (☆) and activity history of a repeating earthquake off Kamaishi. Borders between white and gray background colors areas represent the time at which the 2011 M9 Tohoku Earthquake occurred. Coseismic slip distributions off Kamaishi are shown with color scale on the right side. Arrows indicate the direction of postseismic slip estimated by numerical simulations on the basis of a rate- and state friction law (Ariyoshi et al. 2015)

source region. From numerical simulation as shown in Fig. 2, Ariyoshi et al. (2015) suggested that repeating earthquake occurs partial rupturing of the asperity without significant stress perturbation while complete rupturing and smaller doublet rupturing with significant stress perturbation due to the passage of large postseismic slip. In addition, Fig. 2 suggests that the stress release amount in the asperity is not homogeneous because of complex distribution of coseismic slip.

From recently statistical modeling, Nomura et al. (2017) applied the relation written as

$$\log T = -\log v + (1/6) \log M_o - 2.36 \tag{8}$$

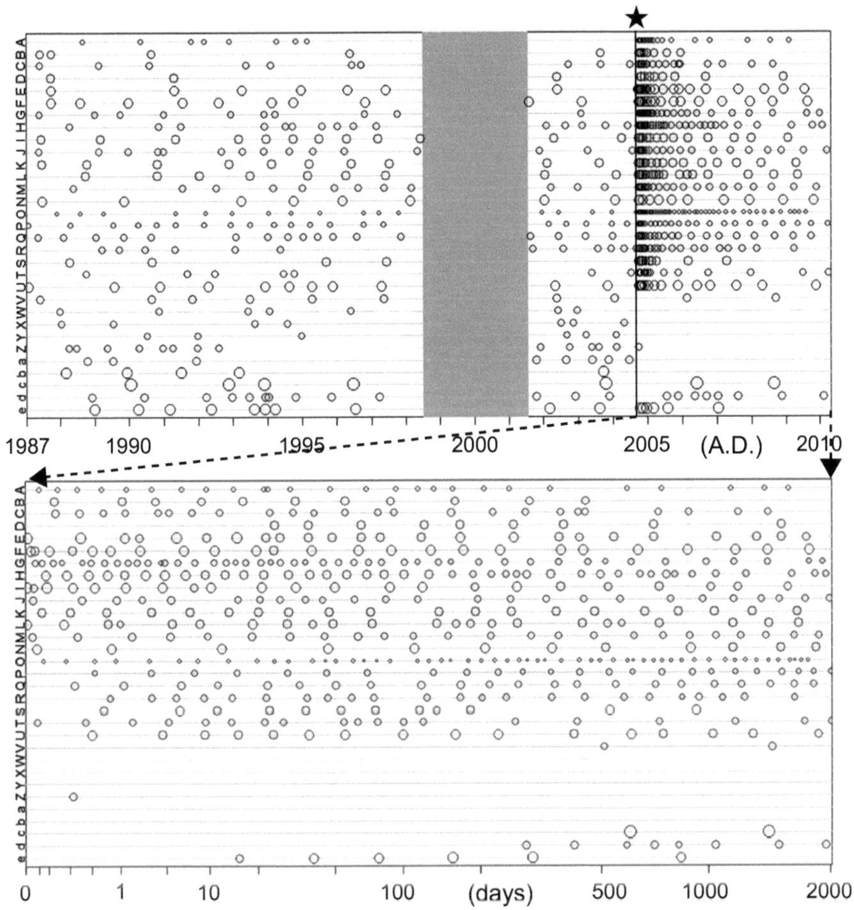

Fig. 3 The occurrence data of repeating earthquake sequences in Parkfield for each source region represented by indices (A–Z, a–e). The black star represents the occurrence time of mainshock (M6) on 28th September in 2004. The relative magnitude of each earthquake is reflected in the relative size of the symbols. The gray area represents unrecorded period. This figure is modified from Nomura et al. (2014)

to the non-stationary renewal model, in order to estimate slip velocity (v) by using the observed recurrence interval (T) and the seismic moment (M_0) as input parameters. This method has succeeded in estimating the spatiotemporal variations of interplate slip rate in northeast Japan. In order to forecast the occurrence time and the magnitude of the forthcoming earthquakes in advance, however, we have to take non-similar earthquakes such as doublet as shown in Fig. 2 and temporarily slow slip events (Ariyoshi et al. 2007) due to spatially inhomogeneous distribution of coseismic slip into account for moment release.

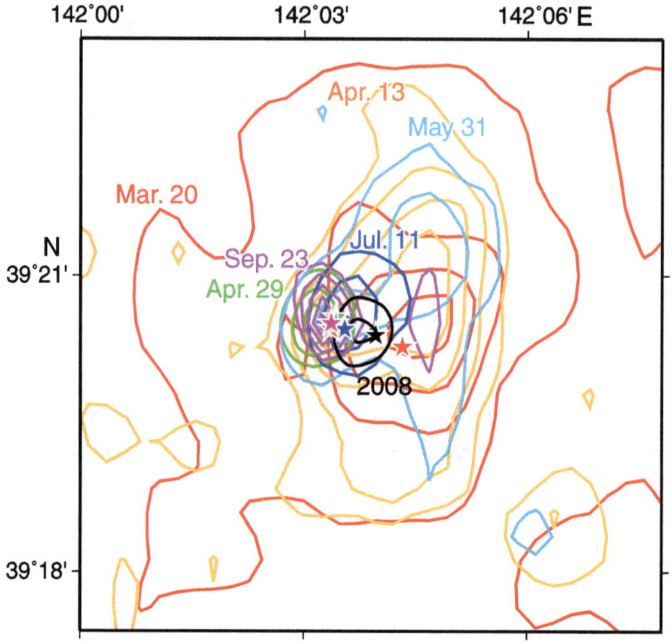

Fig. 4 Coseismic slip distribution (contour) and epicenter (star) just after the 2011 M9 Tohoku Earthquake. Black: 11 January 2008 earthquake, Red: 20 March 2011 for reference, Orange: 13 April 2011, Green: 29 April 2011, Light blue: 31 May 2011, Blue: 11 July 2011, Pink: 23 September 2011. The contours start at 5 cm and intervals are 20 cm for the 20 March 2011 event (red) and 5 cm for the other events. This figure is after Uchida et al. (2015)

From recent earthquake cycle modeling, Ariyoshi et al. (2007) showed that the simulated slip history of isolated small asperity with stress perturbation due to the passage of large postseismic slip far from the source region obeys the slip/time predictable model (Shimazaki and Nakata 1980) as shown in Fig. 5. Focusing on the stage just after the passage of the postseismic slip, Ariyoshi et al. (2015) pointed out that seismic moment release amount becomes temporally smaller due to triggered slow earthquakes and doublet rupturing along the propagation direction of postseismic slip.

4 Summaries and Future Plan

Our concept is summarized in Fig. 6. As our final goal, perfectly forecasting the next earthquake, is to identify its location, occurrence time and magnitude in advance. The location has been precisely determined on the basis of "asperity model" especially for repeating earthquakes as observed off Kamaishi (e.g., Igarashi et al. 2003). Without any stress perturbation, repeating earthquake is expected to be characteristic under the

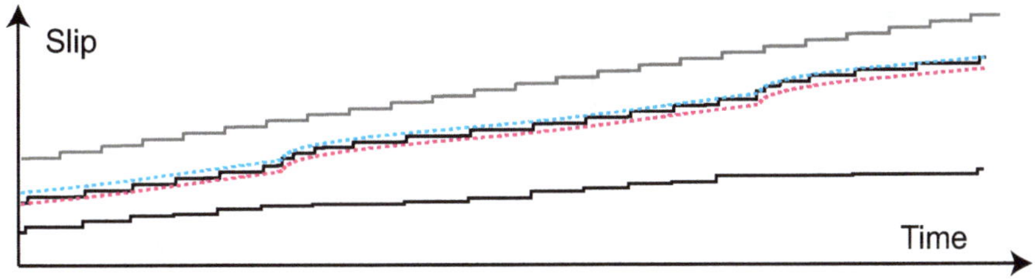

Fig. 5 Simulated time histories of total slip (black) sandwiched between the slip predictable model (cyan) and time predictable model (magenta) in case of the loading rate changed by postseismic slip. The top (gray) and bottom (black) curved represents coseismic slip of the small asperity in case of isolated asperity and two (small and large) asperities, respectively

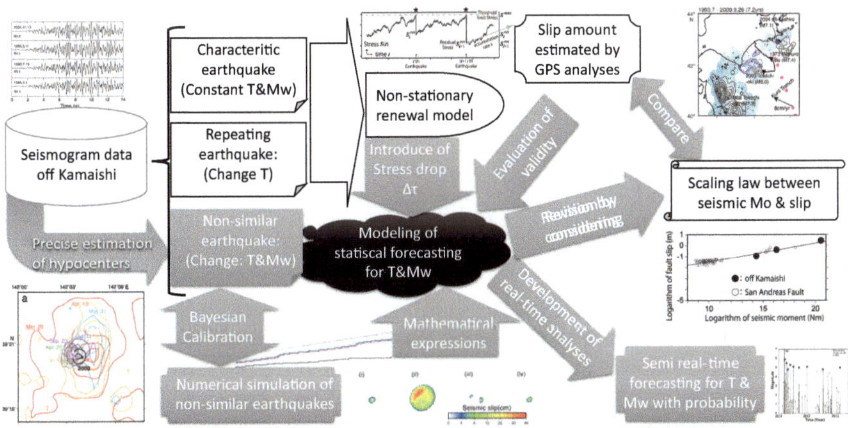

Fig. 6 Overview of seismological analysis for modeling of statistical forecasting occurrence time (T) and seismic moment magnitude (M_w). Gray background represents target study. Some parts of figure are after Igarashi et al. (2003), Uchida et al. (2009, 2015), Nomura et al. (2014) and Ariyoshi et al. (2007, 2015)

constant value of recurrence interval (T) and magnitude (M_w). With non-negligible stress perturbation, the recurrence time is also fluctuated under the condition that cross correlation is still high and approximately characteristic magnitude (± 0.1). The estimation of the recurrence time has been improved by applying non-stationary renewal model (Nomura et al. 2015, 2017).

Just after the 2011 M9 Tohoku earthquake, however, magnitude of aftershocks occurred in the source area of repeating earthquakes tends to be fluctuated not only greater but also smaller with non-similarity in the observed seismograms due to great postseismic slip of megathrust earthquakes, which suggests that we should consider the effect of perturbed magnitude in order to enhance the forecast precision.

To overcome the difficult application to non-similar earthquake in case of the great postseismic slip, we have to understand the mechanism of the non-similar earthquake triggered by the passage of large postseismic slip from numerical simulations

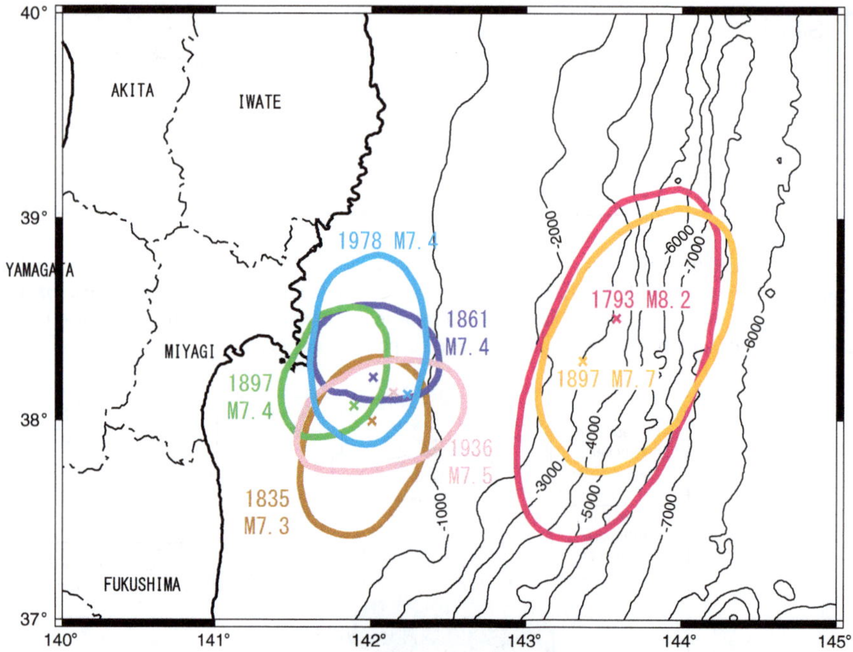

Fig. 7 Location of the source areas of Miyagi-oki (off Miyagi prefecture) earthquakes estimated by Hatori (1999). Colored stars represents the estimated epicenters

(Ariyoshi et al. 2015) based on rate- and state-dependent friction laws (Dieterich 1979; Ruina 1983) and dense network of seismological observations (Igarashi et al. 2010). Considering those simulated and observed results, we describe the temporal change of moment release amount as mathematical expressions toward the non-stationary renewal model.

Applying the scaling law between seismic moment and fault slip amount (Igarashi et al. 2003), we compare the estimated slip history between GPS data analysis and repeating earthquake (Uchida et al. 2009), which enables us to perform semi real-time analysis for evaluating the probability of occurrence time and magnitude of the next repeating earthquake as well as plate convergence rate in the future.

So far, most of repeating earthquakes have been observed as small to middle class ($Mw < 6$), while great interplate earthquakes such as off Miyagi prefecture (Fig. 7) and Nankai Trough (Fig. 8) have appeared to be approximately characteristic magnitude until the occurrence of M9 megathrust earthquakes.

Assuming that the difference of slip distribution for M7 earthquakes in Fig. 7 is caused by stress perturbation during longer period as interseismic stage, we can apply the generation mechanism of non-similar earthquake in the source region of repeating earthquake to the great earthquakes (M7 class) with perturbed source regions and recurrence intervals from a macroscopic perspective.

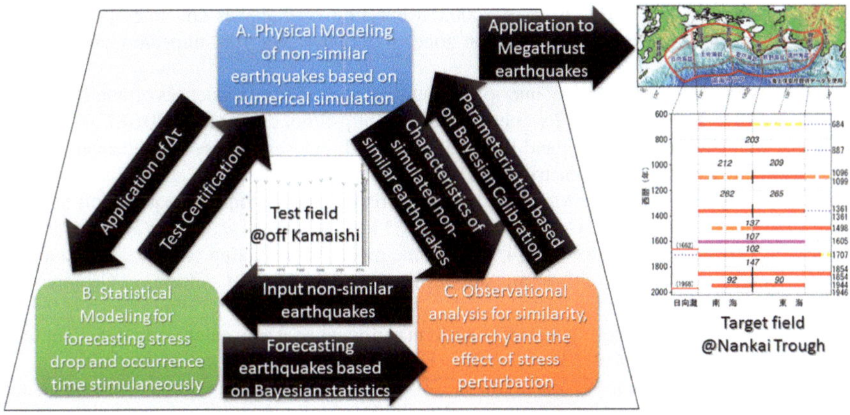

Fig. 8 Schematic diagram of our future study. Map view of Nankai Trough is from the Headquarters for Earthquake Research Promotion (2013)

Figure 8 shows schematic diagram of our future study. First, we adopt repeating earthquake off Kamaishi as test field, in order to combine (A) physical modeling of non-similar earthquakes by numerical simulation on the basis of rate- and state-dependent friction law (Ariyoshi et al. 2015), (B) statistical modeling for forecasting stress drop and occurrence time simultaneously on the basis of non-stationary renewal model, and (C) observational analysis for similarity and hierarchy (Uchida et al. 2015) with stress perturbation due to great postseismic slip. After confirmed in the test field, our modeling and understanding of non-similar earthquake would be applied to Nankai Trough as target field.

Acknowledgements Some figures were drawn using GMT software (Wessel and Smith 1998). Publication of this chapter was partly supported by JSPS KAKENHI Grant Number JP16H06477 in Scientific Research on Innovative Areas "Science of Slow Earthquakes".

References

Ariyoshi K, Matsuzawa T, Hino R, Hasegawa A (2007) Triggered non-similar slip events on repeating earthquake asperities: results from 3D numerical simulations based on a friction law. Geophys Res Lett 34. https://doi.org/10.1029/2006gl028323

Ariyoshi K, Uchida N, Matsuzawa T, Hino R, Hasegawa A, Hori T, Kaneda Y (2015) A trial estimation of frictional properties, focusing on aperiodicity off Kamaishi just after the 2011 Tohoku earthquake. Geophys Res Lett 41:8325–8334. https://doi.org/10.1002/2014GL061872

Dieterich JH (1979) Modeling of rock friction: 1. Experimental results and constitutive equations. J Geophys Res 84:2161–2168

Hatori T (1999) Reexamination of magnitude scale of the historical Sanrikuoki tsunamis. Programme and Abstracts the Seismological Society of Japan 1999 Fall Meeting, C76 (written in Japanese)

Igarashi T, Matsuzawa T, Hasegawa A (2003) Repeating earthquakes and interplate aseismic slip in the northeastern Japan subduction zone. J Geophys Res 108. https://doi.org/10.1029/2002jb001920

Igarashi T (2010) Spatial changes of inter-plate coupling inferred from sequences of small repeating earthquakes in Japan. Geophys Res Lett 37:L20304. https://doi.org/10.1029/2010GL044609

Lindqvist BH, Elvebakk G, Heggland K (2003) The trend-renewal process for statistical analysis of repairable systems. Technometrics 45:31–44

Matsuzawa T, Igarashi T, Hasegawa A (2002) Characteristic small-earthquake sequence off Sanriku, northeastern Honshu, Japan. Geophys Res Lett 29(11). https://doi.org/10.1029/2001gl014632

Nomura S, Ogata Y, Nadeau RM (2014) Space-time model for repeating earthquakes and analysis of recurrence intervals on the San Andreas Fault near Parkfield, California. J Geophys Res 119:7092–7122

Nomura S (2015) Point process models for recurrent earthquakes at active faults and their long-term forecast. Proc Inst Stat Math 63(1):83–104 (written in Japanese with English Abstract)

Nomura S, Ogata Y, Uchida N, Matsu'ura M (2017) Spatiotemporal variations of interplate slip rates in northeast Japan inverted from recurrence intervals of repeating earthquakes. Geophys J Int 208(1):468–481. https://doi.org/10.1093/gji/ggw395

Ruina A (1983) Slip instability and state variable friction laws. J Geophys Res 88:10,359–10,370

Shimazaki K, Nakata T (1980) Time-predictable recurrence model for earthquakes. Geophys Res Lett 7:279–282

The Headquarters for Earthquake Research Promotion (2013) Evaluations of occurrence potentials or subduction-zone earthquakes to date (written in Japanese). http://www.jishin.go.jp/evaluation/long_term_evaluation/subduction_fault/. Accessed 26 July 2017

Uchida N, Matsuzawa T, Hasegawa A, Igarashi T (2005) Recurrence intervals of characteristic M4.8 ± 0.1 earthquakes off Kamaishi, NE Japan—comparison with creep rate estimated from small repeating earthquake data. Earth Planet Sci Lett 233:155–165

Uchida N, Yui S, Miura S, Matsuzawa T, Hasegawa A, Motoya Y, Kasahara M (2009) Quasi-static slip on the plate boundary associated with the 2003 M8.0 Tokachi-oki and 2004 M7.1 off-Kushiro earthquakes, Japan. Gondwana Res 16:527–533

Uchida N, Shimamura K, Matsuzawa T, Okada T (2015) Postseismic response of repeating earthquakes around the 2011 Tohoku-oki earthquake: moment increases due to the fast loading rate. J Geophys Res 120(1):259–274. https://doi.org/10.1002/2013JB010933

Wessel P, Smith WHF (1998) New, improved version of the generic mapping tools released. EOS Trans AGU 79:579

Getting Started with GMT: An Introduction for Seismologists

Matthew R. Agius

1 Introduction

The Generic Mapping Tools (GMT) is a free, open source, computer software package developed in the late nineteen eighties to help scientists, in particular geoscientists, visualise their data (Wessel and Smith 1991). The toolkit has a collection of modules that enable the manipulation of geographic and Cartesian data sets, including two and three-dimensional projecting, filtering, trend fitting, binning, etc. GMT is capable of generating print quality, vector format figures, necessary for publication in journals. The main advantage of GMT is its very powerful scriptable command line use, which makes it possible to automate the commands and yet generate high-quality images. Over the years GMT has improved and has had new commands introduced with each major version: 3.0 (Wessel and Smith 1995), 3.1 (Wessel and Smith 1998), and 5.0 (Wessel et al. 2013), however, its user interaction has not changed.

Unlike many graphic software packages, GMT has no Graphical User Interface (GUI), which in turn makes it a bit tedious for new users to get familiar with its cryptic commands. In this Chapter, a few simple examples of GMT syntax are presented and explained. There are far more commands available, and each with their respective flags and attributes. The aim here is to introduce the software, its use and concept, with a focus for seismologists by showing basic examples of x-y graphs, maps, seismograms, and earthquake focal mechanisms. In the first 6 examples, the GMT syntax is introduced using one-line commands, and then, in the following examples, multi-line commands (scripts) are used to make detailed maps. The reader is expected to have some basic knowledge of Bash (Unix shell) scripting—popular with Linux users—with some familiarity to standard built-in UNIX utilities such as `ls`, `echo`, `awk`, `cat` and `grep`. All the examples are available as Supporting Material and can be used to regenerate the shown plots.

M. R. Agius (✉)
University of Southampton, Waterfront Campus,
European Way, Southampton SO14 3ZH, United Kingdom
e-mail: matthew.agius@soton.ac.uk

S. D'Amico (ed.), *Moment Tensor Solutions*, Springer Natural Hazards,
https://doi.org/10.1007/978-3-319-77359-9_31

1.1 Installing

GMT is available for various computer platforms: Linux, MacOS X and Microsoft Windows, and is available for download from http://gmt.soest.hawaii.edu. Please follow the installation instructions provided on the website. Its source code can also be downloaded, if necessary. Package management software makes installation on Linux and MacOS X easier, for example using the Advanced Package Tool (APT) for the former and MacPorts or Homebrew for the latter. Alternatively for Apple, an application bundle standalone installer (.dmg file) is available for MacOS. A standalone installer is also available for Windows in 32- and 64-bit versions.

To test if GMT has installed and runs, for Linux open a new shell window and type `gmt` followed by a carriage return/enter key. On MacOS X and Windows run the application directly and type `gmt` on the opened terminal. A welcome note should show up on the terminal. To check the version installed type `gmt—version`. The code snippets shown here were tested on GMT version 5.4.2.

2 Syntax

The user interaction with GMT is limited to command lines on the shell terminal, or in a more practical way as scriptable commands stored in a text file and run as a shell script. The advantage of a command-line interface is that it runs faster, and can be compatible with other programs. The combination of GMT and shell-scripting allows for very powerful programming and automation of repetitive tasks.

Many of the GMT modules start with letters 'ps', for example, **psbasemap**, **psxy** or **pscoast**. These modules can be launched from a single executable call to **gmt**, for example, `gmt psbasemap`, followed by certain *arguments* (options) specific to the module operation. The options are a sequence of *flags* identified by a preceding hyphen, for example, **-V** sets the verbosity level output from the module. Most flags have associated *attributes* made up of a combination of letters, numbers and symbols that help specify more precisely the parameters required. For example, **-R**0/10/0/10 sets the range flag with attributes defining the extent of the map/plot region.

The output from most modules is PostScript—a standard page description language used for plotting. Basically, the output is plain ASCII text and can be stored in a file. The output from a GMT module can be conveniently redirected to a file using right angle brackets > or >> at the end of the GMT command followed by a file name. It is recommended that the file name has no blank spaces and ends with a '.ps' extension to automatically be identified as a PostScript file by the operating system.

A call for a GMT module has the following format:

`gmt module -Flag`*`attributes`* `-Flag`*`attributes`* .... `> PostScript.ps`

Importantly note:

- No blank space between the flag and the associated attributes,
- No blank space between attributes of the same flag,
- Flags are separated by a blank space, and
- The order of the flags is not important.

The full list of flags and attributes associated with every module are explained in the manual pages and in the usage messages. In this Chapter, only those flags and attributes used in the examples provided are explained.

2.1 Getting Help

The best source of help is directly from the GMT installation. When a GMT module is run without any flags the manual (man) pages are displayed on the screen. This documentation is the most accurate for the GMT version installed on the machine used. Another good source of help is from the official website http://gmt.soest.hawaii.edu, however, the information may be of a more recent GMT version, or if from another website, of an older version.

The help provided through the console is very technical and concise. To find out which flags can be used with, for example, module **psbasemap**, simply run `gmt psbasemap`. The first line output is a short description: 'Plot base maps and frames', followed by the usage:

```
usage: psbasemap -J<args> -R<west>/<east>/ <south>/ <north>
[/<zmin>/<zmax>][+r] [-B<args>] [-A[<file>]] [-D[g|j|J|
n|x] <refpoint>+w<width>[<unit>] [/<height>[<unit>]] [+j
<justify>] [+o<dx> [/<dy>]] [+s<file>]] |-D[g|j|J|n|x]
<refpoint> +w<width>[<unit>] [/<height>[<unit>]] [+j
<justify>] [+o<dx>[/<dy>]] [+s<file>]] [-Jz|Z<args>] [-F[+c
<clearance(s)>] [+g<fill>] [+i[[<gap>/]<pen>]] [+p[<pen>
]][+r[<radius>]] [+s[<dx>/<dy>/]<fill>]]] [-K] [-L[g|j|J|
n|x]<refpoint> +c[<slon>/]<slat>+w<length>[e|f|M|n|k|u] [+a
<align>] [+f] [+j<justify>] [+l[<label>]] [+o<dx>[/<dy>
]][+u]] [-O] [-P] [-Td[g|j|J|n|x]<refpoint>+w<width>[+f[
<level>]] [+j<justify>] [+l<w,e,s,n>] [+o<dx>[/<dy>]]]
[-Tm[g|j|J|n|x] <refpoint>+w<width> [+d[<dec>[/<dlabel>]]]
[+i<pen>] [+j<justify>] [+l<w,e,s,n>] [+p<pen>] [+t<ints>]
[+o <dx>[/<dy>]]] [-U[[<just>]/<dx>/<dy>/] [c|<label>]] [-V[
<level>]] [-X[a|c|r]<xshift>[<unit>]] [-Y[a|c|r]<yshift>[
<unit>]] [-f[i|o]<info>] [-p[x|y|z]<azim>[/<elev>[/
<zlevel>]][+w<lon0>/<lat0>[/<z0>][+v<x0>/<y0>]] [-t<transp>]
```

and followed by a detailed explanation on the use of each flag (the hyphen followed by a letter: **-J**, **-R**, **-B**, **-A**, etc.). Despite the overwhelming information, many of the flags and attributes are the same for the different modules, and use letters of the alphabet which stand for the argument so that they will be easy to remember. For example, **-R** specifies the minimum and maximum coordinates of the data region.

Attributes accompanying flags are indicated by angle brackets (< >). For example, **-R** requires at least four attributes separated by a forward slash (/). Square brackets ([ ]) indicate that the values are optional. Thus, in the case of **-R** west, east, south, and north are required attributes, whereas zmin and zmax and **+r** are optional. A pipe (|) indicates an 'or' selection from a list of arguments such as the case for flag **-D**[g|j|J|n|x]. Although the attributes are optional (because of the square bracket), only one of the g, j, J, n or x can be used. Flags with attributes indicated as <args> have more complex arguments and are detailed in the documentation.

Not all of the flags listed in the usage are necessary, in fact for **psbasemap** only **-J** and **-R** flags are required, while all the other flags are enclosed in square brackets ([ ]). Note, that plus (+), minus/hyphen (-) and forward slash (/) signs are part of the arguments as shown in the usage. Remember not to put blank spaces within the list of attributes of the same flag. In the case of missing required, or invalid, or ambiguous command arguments, GMT will print an alert message on to the screen with a synopsis of the valid arguments.

The examples shown here are indeed few and basic, but once it becomes clear how GMT scripts work and are written, it will become easier to try any other command available and generate the desired figures. GMT is thoroughly documented, especially online (http://gmt.soest.hawaii.edu), with a technical reference and pages that explain the purpose of each module and the many features. The website provides numerous examples to help new users to quickly become familiar with the operation and philosophy of the system.

2.2 *psbasemap*

psbasemap is considered the basic command for GMT. Essentially, it creates the basic layout of a map or graph, with more overlaying plots to follow. The command takes three essential flags: **-J** sets the projection and size of the base map; **-R** sets the range of the *x*- and *y*-axes; and **-B** sets the boundary frame with attributes such as the title, axes labels, annotations, tick-mark intervals, and grid lines. Other flags such as **-P** sets the output plot to portrait mode and **-V** sets the level of information output from GMT during execution. A sample code for **psbasemap** is shown in Example 1 and illustrated in Fig. 1. The figure shows a grid with *x*- and *y*-axes ranging from 0 to 10. The south and west border frames are set to show labels (-BWeSn), and the background is set to white '+g255'. Each axis has their respective labels, annotations, grids and ticks set.

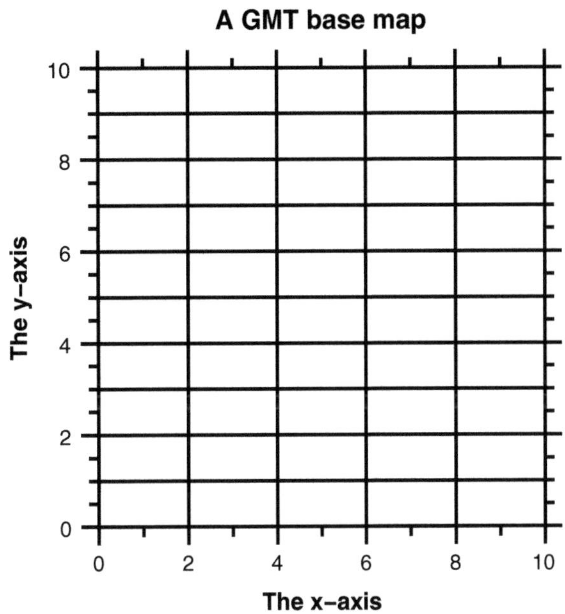

Fig. 1 An example of a base map with *x*- and *y*-axes in the range from 0 to 10. The annotation of major ticks have an interval of 2, minor ticks have an interval of 1 and 0.5, respectively, and the grid spacing interval is 2 and 1 on the horizontal and vertical axes, respectively

The right angle bracket (>) at the end of the **psbasemap** command redirects the PostScript output to a file named Example1.ps. This file contains standard text, and the file extension '.ps' indicates to the operating system that it is a PostScript file viewable by a PostScript reader. Try run Example 1 by typing the commands in your console/command prompt and look out for the information output by GMT. **Note: the entire command should be typed in a single line not as multiple lines—the shell terminal only executes one line at a time.** Check if the file Example1.ps has been created using the list command (`ls`) and view the plot by opening the file Example1.ps by a PostScript viewer (probably easier to open via the GUI). You can experiment with changing the numbers associated with the flags and see how the base map changes.

Example 1 An example of the **psbasemap** command.

```
gmt psbasemap -JX5c/5c -R0/10/0/10 -BWeSn+t"A GMT base
    map"+g255 -Bpya2f0.5g1+l"The y-axis" -Bpxa2f1g2+l"The
    x-axis" -P -V > Example1.ps
```

Command	Flag	Description
psbasemap	-JX	Projection. 'X' sets the projection to linear, with at a specific size of 5 cm along the x axis by 5 cm along the y axis (5c/5c)
	-R	Coordinate ranges (*xmin/xmax/ymin/ymax*). The limits for the x and y axes set from 0 to 10 (0/10/0/10)
	-B	Boundary. Attributes 'WeSn' represent west, east, south and north, respectively (order not important). Each letter sets the border plotted (letters in lower case) and with labels (upper case). **+t** adds a title to the base map. **+g** adds a white background colour (255, see Sect. 4). 'pya2f0.5g1' sets the **p**rimary **a**nnotation of the **y** axis at an interval of 2, minor tick spacing interval of 0.5 (f) and a **g**rid with a spacing of 1. Similarly for the x axis. **+l** adds a label to the respective axis
	-P	Portrait. Set plot into portrait mode, the default is landscape
	-V	Verbose. Set the level of information output from GMT during execution
	>	Redirect output. The generated PostScript will be written to the file `Example1.ps`

2.3 *psxy: Lines*

A popular GMT module is **psxy** for plotting of lines, polygons and scattered symbols. Here are various examples of **psxy**, either using the command by itself (Example 2) or combined with other commands (Examples 7–9). Like many other GMT commands (e.g., **psbasemap**, Example 1), **psxy** requires flags **-J** for projection and **-R** for axes ranges.

psxy requires input data. In general, the processing and display routines within GMT handle any x, y or x, y, z data as input. In many cases, the x and y coordinates are longitude and latitude, respectively, but in other cases, they may be any other variable such as depth and compressional velocity. The data, made up of arbitrary sequences of x, y or x, y, z values, are read from multi-column ASCII tables, that is, each file consists of several records in which each coordinate is confined to a separate column.

Example 2 has **psxy** read a file `prem_vp.dat`, which is a text file with values of depth and compressional velocity (Dziewonski and Anderson 1981), first and second column, respectively. Like Example 1 the projection (**-J**) is 5 by 5 cm. The x axis range is from 0 to 6371—the radius of the Earth, and from 0 to 16 in the y axis in kms^{-1} (**-R**). The PostScript output Example2.ps is shown in Fig. 2.

Example 2 An example of the **psxy** command. This code requires file `prem_vp.dat` to execute.

```
gmt psxy prem_vp.dat -JX5c/5c -R0/6371/0/16
    -BWeSn+t"Preliminary Reference Earth Model"+g255
    -Bpya2f1+l"V@-P@- (kms@+-1@+)" -Bpxa2000f1000+l"Depth (km)"
    -W1p,black -P -V > Example2.ps
```

Module	File/flag	Description
psxy	`prem_vp.dat`	A two-column text file with *x* and *y* values, separated with a blank space
	-JX	Projection. 'X' sets the projection to linear, with at a specific size of 5 cm along the *x* axis by 5 cm along the *y* axis (5c/5c)
	-R	Coordinate ranges (*xmin/xmax/ymin/ymax*). The limits for the *x* and *y* axes are set from 0 to 6371 and 0 to 16, respectively (0/6371/0/16)
	-B	Boundary. Attributes 'WeSn' represent west, east, south, and north, respectively (order not important). Each letter sets the border plotted (letters in lower case) and with labels (upper case). **+t** adds a title to the plot. **+g** adds a white background colour (255). 'pya2f1' sets the **p**rimary **a**nnotation of the **y** axis at an interval of 2 and minor tick spacing interval of 1 (f). The *x* axis has annotation interval of 2000 and ticks at every 1000 (pxa2000f1000). **+l** adds a label to the respective axis. '@-' and '@+' toggle on and off sub- and superscripts
	-W	Pen attributes. Set line **p**oint thickness and **black** colour (1p,black)
	-P	Portrait. Set plot into portrait mode, the default is landscape
	-V	Verbose. Set the level of information output from GMT during execution
	>	Redirect output. The generated PostScript will be written to the file `Example2.ps`

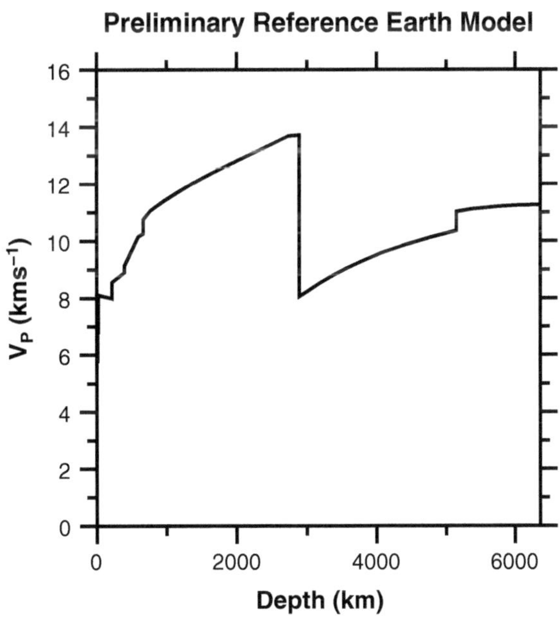

Fig. 2 Graph of compressional velocity (V_P) with respect to depth (Dziewonski and Anderson 1981)

2.4 *pscoast*

To plot maps use the module **pscoast**. GMT provides a variety of map projections such as cylindrical, conic and azimuthal. These projections provide ways to show maps either in equidistant or equal area. Some map projections are shown in Fig. 3. The key flag for setting up the different map projections is **-J** appended by a letter indicating the type of projection, and followed by the size or the scale to be used for the map. The upper or lower case of the letter representing the type of projection indicates if

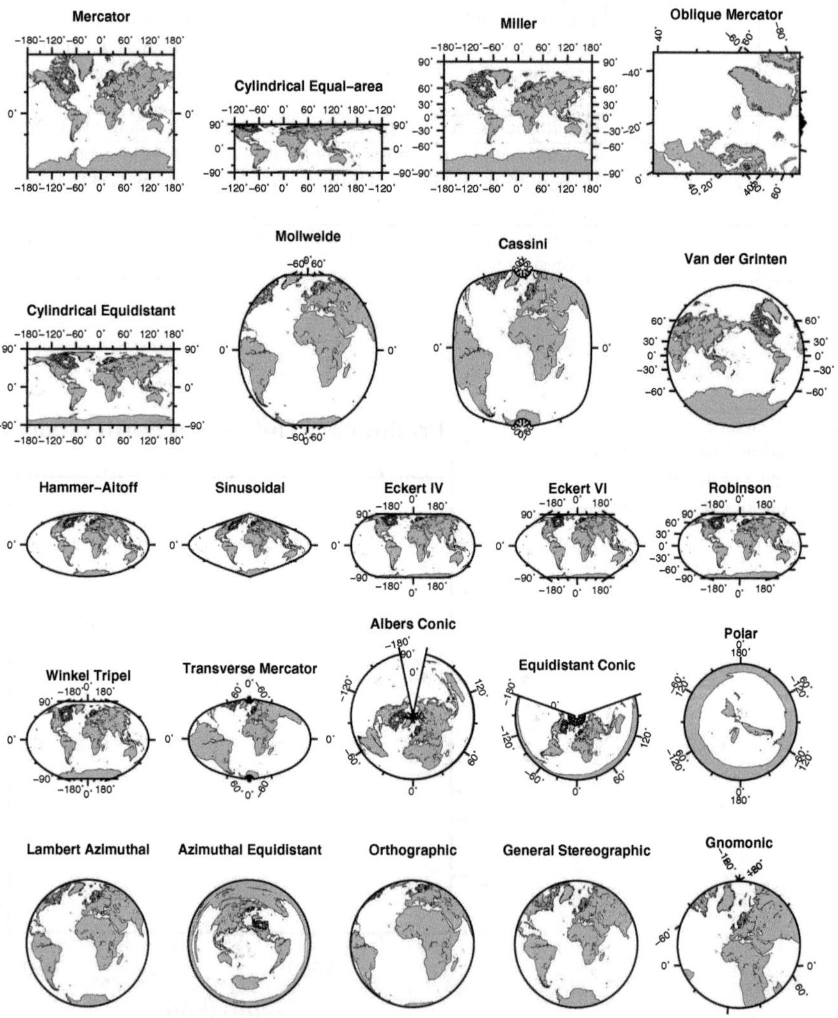

Fig. 3 Examples of various geographical projections such as Mercator, cylindrical, conic, azimuthal and other miscellaneous map projections. See text for details

the following attributes are for a fixed size or a scale value. For example, '-JM3c' produces a Mercator map of width 3 cm, whereas '-Jm3c' produces a Mercator map with a scale of 3 cm per degree. Example 3 shows a series of **pscoast** commands, each with a different map projection and generating its own PostScript file. Try changing the values of attributes of flags **-J**, **-R**, **-B**, **-W**, **-G** and **-S**.

Example 3 Examples of **pscoast** command. Each command is run independently, and creates a corresponding PostScript file.

```
gmt pscoast -JM3c -R-180/180/-80/80 -BWESN+t"Mercator"
    -Bpxa60f30 -Bpya90f45 -Dc -Wthinnest,black -Ggrey -Swhite
    -P -V > Example3-JM.ps

gmt pscoast -JY60/3c -Rg -BWESN+t"Cylindrical Equal-area"
    -Bpxa60f30 -Bpya90f45 -Dc -Wthinnest,black -Ggrey -Swhite
    -P -V > Example3-JY.ps

gmt pscoast -JJ0/3c -Rg -BWESN+t"Miller" -Bpxa60f30 -Bpxa60f30
    -Bpya30f15 -Dc -Wthinnest,black -Ggrey -Swhite -P -V >
    Example3-JJ.ps

gmt pscoast -JOA5/15/30/3c -R10/70/10/50 -BWeSN+t"Oblique
    Mercator" -Bpxa20f10 -Bpya20f10 -Dc -Wthinnest,black -Ggrey
    -Swhite -P -V > Example3-JO.ps

gmt pscoast -JQ0/0/3c -Rg -BWESN+t"Cylindrical equidistant"
    -Bpxa60f30 -Bpya90f45 -Dc -Wthinnest,black -Ggrey -Swhite
    -P -V > Example3-JQ.ps

gmt pscoast -JW0/2.8c -R-80/80/-80/80 -BWESN+t"Mollweide"
    -Bpxa60f30 -Bpya90f45 -Dc -Wthinnest,black -Ggrey -Swhite
    -P -V > Example3-JW.ps

gmt pscoast -JC0/0/2.8c -R-80/80/-80/80 -BWESN+t"Cassini"
    -Bpxa60f30 -Bpya90f45 -Dc -Wthinnest,black -Ggrey -Swhite
    -P -V > Example3-JC.ps

gmt pscoast -JV2.8c -Rg -BWESN+t"Van der Grinten" -Bpxa60f30
    -Bpya30f15 -Dc -Wthinnest,black -Ggrey -Swhite -P -V >
    Example3-JV.ps

gmt pscoast -JH0/2.5c -Rg -BWeSN+t"Hammer-Aitoff" -Bpxa60f30
    -Bpya90f45 -Dc -Wthinnest,black -Ggrey -Swhite -P -V >
    Example3-JH.ps

gmt pscoast -JI0/2.5c -Rg -BWeSN+t"Sinusoidal" -Bpxa60f30
    -Bpya90f45 -Bpya90f45 -Dc -Wthinnest,black -Ggrey -Swhite
    -P -V > Example3-JI.ps

gmt pscoast -JKf0/2.5c -Rg -BWeSN+t"Eckert IV" -Bpxa180f60
    -Bpya90f45 -Dc -Wthinnest,black -Ggrey -Swhite -P -V >
    Example3-JKf.ps

gmt pscoast -JKs0/2.5c -Rg -BWeSN+t"Eckert VI" -Bpxa180f60
    -Bpya90f45 -Dc -Wthinnest,black -Ggrey -Swhite -P -V >
    Example3-JKs.ps

gmt pscoast -JN0/2.5c -Rg -BWeSN+t"Robinson" -Bpxa180 -Bpya30
    -Dc -Wthinnest,black -Ggrey -Swhite -P -V > Example3-JN.ps
```

```
gmt pscoast -JR0/2.5c -Rg -BWeSN+t"Winkel Tripel" -Bpxa180f60
    -Bpya90f45 -Ggrey -Swhite -Dc -Wthinnest,black -P -V >
    Example3-JR.ps

gmt pscoast -JT0/0/2.5c -R-80/80/-80/80 -BWESN+t"Transverse
    Mercator" -Bpxa60f30 -Bpya90f45 -Dc -Wthinnest,black -Ggrey
    -Swhite -P -V > Example3-JT.ps

gmt pscoast -JB0/30/60/90/2.5c -R-180/180/-90/90
    -BWeSn+t"Albers Conic" -Bpxa60f30 -Bpya90f45 -Dc
    -Wthinnest,black -Ggrey -Swhite -P -V > Example3-JB.ps

gmt pscoast -JD0/15/30/45/2.5c -R-180/180/-85/85
    -BWeSn+t"Equidistant Conic" -Bpxa60f30 -Bpya90f45 -Dc
    -Wthinnest,black -Ggrey -Swhite -P -V > Example3-JD.ps

gmt pscoast -JPa2.3c -Rg -BWeSN+t"Polar" -Bpxa60f30 -Bpya90f45
    -Dc -Wthinnest,black -Ggrey -Swhite -P -V > Example3-JP.ps

gmt pscoast -JA0/0/2.5c -Rg -BWESN+t"Lambert Azimuthal"
    -Bpxa60f30 -Bpya90f45 -Dc -Wthinnest,black -Ggrey -Swhite
    -P -V > Example3-JA.ps

gmt pscoast -JE-156/20/2.5c -Rg -BWESN+t"Azimuthal equidistant"
    -Bpxa60f30 -Bpya90f45 -Dc -Wthinnest,black -Ggrey -Swhite
    -P -V > Example3-JE.ps

gmt pscoast -JG0/0/2.5c -R-180/180/-85/85
    -BWESN+t"Orthographic" -Bpxa60f30 -Bpya90f45 -Dc
    -Wthinnest,black -Ggrey -Swhite -P -V > Example3-JG.ps

gmt pscoast -JS0/0/2.5c -R-80/80/-80/80 -BWESN+t"General
    Stereographic" -Bpxa60f30 -Bpya90f45 -Dc -Wthinnest,black
    -Ggrey -Swhite -P -V > Example3-JS.ps

gmt pscoast -JF10/30/2.5c -Rg -BWESN+t"Gnomonic" -Bpxa60f30
    -Bpya90f45 -Dc -Wthinnest,black -Ggrey -Swhite -P -V >
    Example3-JF.ps
```

Module	Flag/attributes	Description
pscoast	-JM3c	Mercator cylindrical projected map of width 3 cm
	-JY60/3c	Cylindrical Equal-Area projected map centred at 60° longitude and width of 3 cm
	-JJ0/3c	Miller cylindrical projected map centred at 0° longitude and width of 3 cm
	-JOA5/15/30/3c	Oblique Mercator cylindrical projected map with a longitude and latitude centred at 5° and 5°, respectively and azimuth of oblique equator at 30° and width of 3 cm. The range flag **-R** for this projection '-R10/70/10/50' specifies the low left longitude, low left latitude, upper right longitude and upper right latitude
	-JQ0/0/3c	Cylindrical equidistant projected map with the central meridian longitude and standard parallel latitude set at 0° and width of 3 cm
	-JW0/2.8c	Mollweide projected map centred at 0° longitude and width of 2.8 cm
	-JC0/0/2.8c	Cassini projected map with a longitude and latitude centre at 0° and width of 2.8 cm

Module	Flag/attributes	Description
	-JV2.8c	Van der Grinten projected map with width of 2.8 cm
	-JH0/2.5c	Hammer projected map centred at 0° longitude and width of 2.5 cm
	-JI0/2.5c	Sinusoidal projected map centred at 0° longitude and width of 2.5 cm
	-JKf0/2.5c	Eckert IV projected map centred at 0° longitude and width of 2.5 cm
	-JKs0/2.5c	Eckert VI projected map centred at 0° longitude and width of 2.5 cm
	-JN0/2.5c	Robinson projected map centred at 0° longitude and width of 2.5 cm
	-JR0/2.5c	Winkel Tripel projected map centred at 0° longitude and width of 2.5 cm
	-JT0/0/2.5c	Transverse Mercator cylindrical projected map centred at 0° longitude and latitude, and width of 2.5 cm
	-JB0/30/60/90/2.5c	Albers conic projected map centred at 0° longitude and 30° latitude, and two standard parallels 60° and 90° latitudes, and width 2.5 cm
	-JD0/15/30/45/2.5c	Conic equidistant projected map centred at 0° longitude and 15° latitude, and two standard parallels 30° and 45° latitudes, and width 2.5 cm
	-JPa2.3c	Polar non-geographical projected map with azimuths clockwise from North and width 2.3 cm
	-JA0/0/2.5c	Lambert azimuthally projected map with 0° longitude and latitude specified projection center and width 2.5 cm
	-JE-156/20/2.5c	Azimuthal equidistant map centred at -156° longitude and 20° latitude and width 2.5 cm
	-JG0/0/2.5c	Orthographic azimuthally projected map centred at 0° longitude and latitude, and width 2.5 cm
	-JS0/0/2.5c	General Stereographic azimuthally projected map centred at 0° longitude and latitude, and width 2.5 cm
	-JF10/30/2.5c	Gnomonic azimuthally projected map centred at 10° longitude and 30° latitude, and width 2.5 cm
	-R	Longitude and latitude ranges (*lon min*/*lon max*/*lat min*/*lat max*). For global map use the letter 'g' (global) instead of limits
pscoast	-B	Boundary. Attributes 'WESN' represent west, east, south, and north, respectively (order not important). Each letter sets the border plotted (letters in lower case) and with labels (upper case). **+t** adds a title to the map. 'pya90f45' sets the **p**rimary **a**nnotation of the **y** axis at an interval of 90 and minor tick spacing interval of 45 (f). The *x* axis have annotation at an interval of 60 and ticks at every 30 (pxa60f30)
	-D	Resolution. Set the resolution of the coastline data to **c**rude (c)
	-W	Pen attributes for the coastal line. Set the line thickness to **thinnest** and to colour **black** (thinnest,black)
	-G	Set the land fill colour (grey)
	-S	Set the sea colour (white)
	-P	Portrait. Set plot into portrait mode, the default is landscape
	-V	Verbose. Set the level of information output from GMT during execution
	>	Redirect output. The generated PostScript will be written to the file `Example-J*.ps`

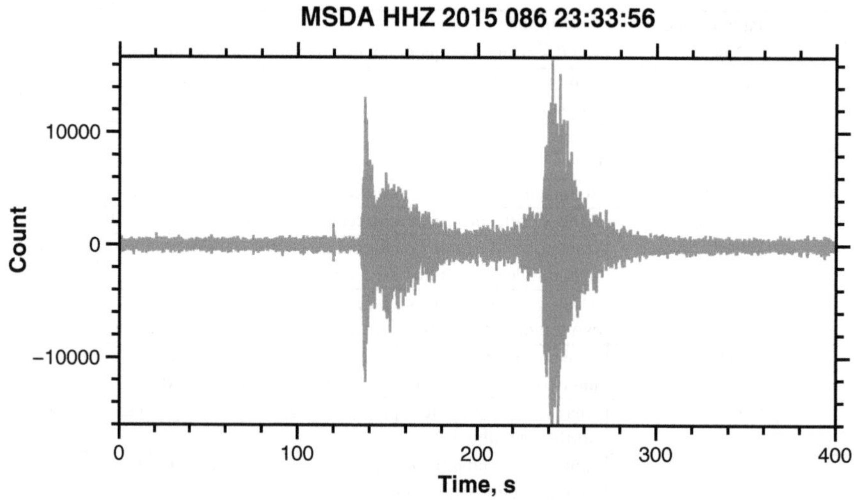

Fig. 4 Seismogram of an earthquake of magnitude 5.2 that struck Crete (Greece) on March 27th, 2015 (Julian day 086), recorded on station MSDA (Malta) (Agius et al. 2015). The waveform shows the vertical component HHZ

2.5 *pssac*

One of the newest module made available in GMT 5 is **pssac**. This is useful for seismologists who would like to plot a seismogram stored in SAC format (Seismic Analysis Code, Goldstein et al. 1998). The command is easy to use and follows the style of the other commands i.e., it requires flags such as **-J** and **-R**, for projection and axes limits, and other optional flags such as **-B** to set boundary frames, titles and labels. Example 4 shows how to plot a single SAC file. **pssac** can be used to plot multiple SAC files by, for example, using the wildcard (*) such as 'MSDA.*.SAC' (Fig. 4).

Example 4 An example of the **pssac** command. This code requires SAC file `MSDA.HHZ.SAC` to execute.

```
gmt pssac MSDA.HHZ.SAC -JX10c/5c -R0/400/-16000/16710
    -BWeSn+t"MSDA HHZ 2015 086 23:33:56"+g255 -Bxaf+l"Time, s"
    -Byaf+l"Count" -W1p,darkgrey -Fr -P -V > Example4.ps
```

Module	Flag	Description
pssac	-JX	Projection. 'X' sets the projection to linear, with at a specific size of 10 cm along the *x* axis by 5 cm along the *y* axis (10c/5c)
	-R	Coordinate ranges. The limits for the *x* and *y* axes defined in *xmin/xmax/ymin/ymax* here set from 0 to 400 s and −16000 to 16710 amplitude counts (0/400/-16000/16710)
	-B	Boundary. Attributes 'WeSn' represent west, east, south, and north, respectively (order not important). Each letter sets the border plotted (letters in lower case) and with labels (upper case). **+t** adds a title to the plot. **+g** adds a white background colour (255). Attributes 'pyaf' without any numbers sets the **p**rimary **a**nnotation and minor ticks (f) of the **y** axis automatically. Similarly for the *x* axis. **+l** adds a label to the respective axis
	-W	Pen attributes. Set line **p**oint thickness and **dark grey** colour (1p,darkgrey)
	-Fr	Performs preprocessing on the data before plotting. '**r**' removes the mean of the seismogram
	-P	Portrait. Set plot into portrait mode, the default is landscape
	-V	Verbose. Set the level of information output from GMT during execution
	>	Redirect output. The generated PostScript will be written to the file `Example4.ps`

2.6 *pswiggle*

Seismic receiver functions are often plotted as wiggles that show the positive and negative amplitudes of a deconvolution of a phase signal from a seismogram. The wiggles can be plotted using **pswiggle**. The command requires a text file that has three fields on every line such as the longitude, time and the amplitude of the deconvolution. In Example 5 a dataset from KIP station (Kipapa, Hawaii, USA) is illustrated. The file `KIP.dat` has 40 waveforms each separated by a right angle bracket (>). This type of file is called a multisegment file. The first field (column) of data in `KIP.dat` is used as an index (1–40), the second field is the time (1–204 s), and the third field is the amplitude of the wiggle. The positive parts of the wiggles can be colour filled to better illustrate coherent amplitudes. In Fig. 5 positive amplitudes at about 45 and 70 seconds are the P410s and P660s seismic phase conversions, respectively.

Example 5 An example of the **pswiggle** command. The code requires file `KIP.dat`.

```
gmt pswiggle KIP.dat -JX5c/-5c -R-5/45/30/80 -BWSn+t"Receiver
    functions (KIP)"+g255 -Bxa40 -Bya10f5+l"Time, s" -W0.5p,0
    -G150 -Z1c -P -V > Example5.ps
```

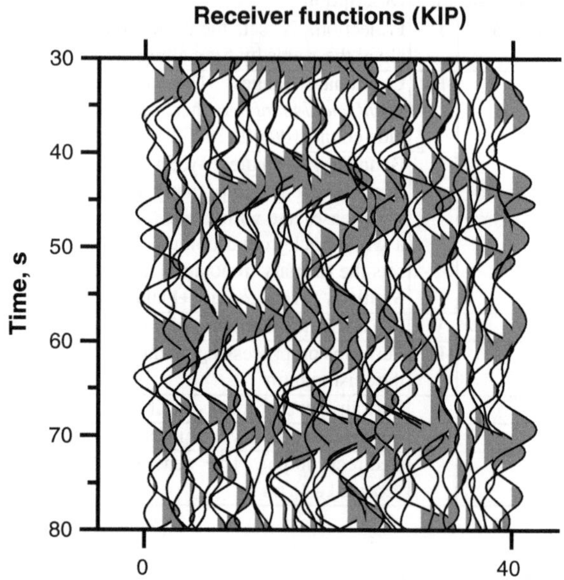

Fig. 5 Receiver functions for station KIP. Each waveform is from a different earthquake. Grey shade represents the positive amplitude of the receiver function. The positive amplitude at about 45 and 70 seconds are the P410s and P660s seismic phase conversions, respectively. Data provided from Agius et al. (2017)

Module	File/flag	Description
pswiggle	`KIP.dat`	A text file containing *x*, *y* and *z* values
	-JX	Projection. 'X' sets the projection to linear, with at a specific size of 5 cm along the *x* axis by 5 cm along the *y* axis (5c/5c)
	-R	Coordinate ranges (*xmin*/*xmax*/*ymin*/*ymax*). The limits for the *x* and *y* axes set from −5 to 45 and 30 to 80 s, respectively (−5/45/30/80)
	-B	Boundary. Attributes 'WSn' represent west, south, and north, respectively (order not important). Each letter sets the border plotted (letters in lower case) and with labels (upper case). **NOTE: Letter 'e' is missing thus no border is plotted on the east side**. **+t** adds a title to the plot. **+g** adds a white background colour (255). 'ya10f5' sets the **p**rimary **a**nnotation every interval of 10 and minor ticks (f) every 5 on the ***y*** axis. 'xa40' sets the ***x*** axis **a**nnotation with an interval of 40. **+l** adds a label to the respective axis
	-W	Pen attributes. Set line half **p**oint thickness and black colour (0.5p, 0)
	-Z	Sets the scale for the amplitude of the wiggle (1c)
	-P	Portrait. Set plot into portrait mode, the default is landscape
	-V	Verbose. Set the level of information output from GMT during execution
	>	Redirect output. The generated PostScript will be written to the file `Example5.ps`

2.7 *pshistogram*

Binning of data and plotting of histograms is facilitated using module **pshistogram**. In Example 6 the submissions of felt earthquake reports are binned in 30-min bins and plotted against time (x axis). This module requires only two flags **-J** for a linear scale on the axes and **-W** to set the bin width used for histogram calculations. In this case, the x axis is projected in time (-JX12cT/5c). This type of projection can be applied to other modules such as **psxy**. The corresponding x values of the data set have to be in the time format like '2011-04-24T00:07'. The file *submissions.dat* has the time of when each report was submitted. Additional settings might be necessary to set the time format correctly on your system, see Sect. 5. Optional flags such as **-B** generate a complete plot with axes, labels and titles as shown in Fig. 6.

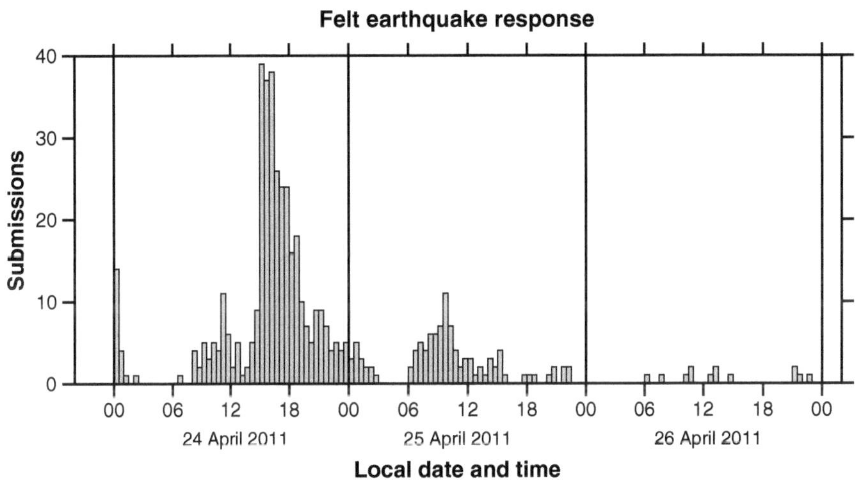

Fig. 6 A histogram showing felt earthquake reports from Malta following a series of tremors that took place between the 24th and 26th of April, 2011. Grey bars are 30-min bins. Figure modified from Agius et al. (2016)

Example 6 An example of the **pshistogram** command. The file `submissions.dat` is required for execution.

```
gmt pshistogram submissions.dat -JX12cT/5c
    -R2011-04-23T20:00/2011-04-27T02:00/0/40 -BWeSn+t"Felt
    earthquake response"+g255 -Bpxa6h+l"Local date and time"
    -BsxalDg1D -Bpya10+l"Submissions" -W1800 -Glightgrey -Z0
    -L0.5p,black -P -V > Example6.ps
```

Module	Flag	Description
pshistogram	-JX	Projection. 'X' sets the projection to linear, with at a specific size of 12 cm along the *x* axis by 5 cm along the *y* axis. Appending 'T' to the *x* axis part sets the projection in time (12cT/5c). The input coordinates have to be absolute time
	-R	Coordinate ranges (*xmin*/*xmax*/*ymin*/*ymax*). The limits for the *x* and *y* axes, however, now the *xmin* and *xmax* are time coordinates where as *ymin* and *ymax* are the range of the bins counts (2011-04-23T20:00/2011-04-27T02:00/0/40)
	-B	Boundary. Attributes 'WeSn' represent west, east, south, and north, respectively (order not important). Each letter sets the border plotted (letters in lower case) and with labels (upper case). **+t** adds a title to the plot. **+g** adds a white background colour (255). 'pya10' sets the **p**rimary **a**nnotation of the ***y*** axis intervals of 10. 'pxa6h' sets the **p**rimary **a**nnotation of the ***x*** axis at every 6 h. A second **-B** flag sets a **s**econdary **a**nnotation of the ***x*** axis every 1 **D**ay and a *g*rid for every day (sxa1Dg1D)
	-W	Bin width. Sets the bin width used for histogram calculations. In this case the width is 1800 seconds equivalent to 30 minutes
	-G	Colour fill for bars (lightgrey)
	-Z	Sets the type of histogram. '0' is for counts. See help pages for other options
	-L	Pen attributes. Draw bar outline with half **p**oint thickness and colour **black** (0.5p,black)
	-P	Portrait. Set plot into portrait mode, the default is landscape
	-V	Verbose. Set the level of information output from GMT during execution
	>	Redirect output. The generated PostScript will be written to the file `Example6.ps`

3 PostScript Layers

The examples illustrated so far are all executed using a single command. This is not always the case. Elaborative figures such as a seismicity map will require plotting the topography, the earthquake and stations locations, put labels on top of the map, add a legend, and so on. It can be achieved by running a series of commands. The first GMT command will have to indicate that more PostScript code will be appended later (**-K**), whereas the subsequent code has to indicate that the executing GMT command is in overlay plot mode (**-O**). Also, the first GMT command makes use of a single right angle bracket (>) forcing the output to create/replace the PostScript file. Subsequent commands make use of a double right angle bracket (≫) to append the output to the existing PostScript file.

Example 7 combines Example 1 and 2 into one (Figs. 1, 2 and 7). First, a **psbasemap** is called to set the projection, coordinates range, bordering frames, title, annotations, ticks and labels. This is followed by calling **psxy** three times, one for each parameter of the Preliminary Reference Earth Model: compressional velocity (V_P), shear velocity (V_S) and density (ρ), which are available in the text file `prem.dat` (Dziewonski and Anderson 1981). **psxy**, however, can only read a text file with lines made up of two fields of data separated by a blank space. **awk** can assist with parsing the text file accordingly.

awk

awk is a processing language designed for text processing and is typically used as a data extraction tool (Aho et al. 1979). It is a standard feature of most Unix-like operating systems, including MacOS. **awk** scans an input file for lines that match any of a set of patterns. An input line is normally made up of fields separated by a blank space, like the file `prem.dat`. Fields are denoted **$1, $2,** , etc., while `$0` refers to the entire line. For instance in Example 7, when **`awk '{print $1,$2}' prem.dat`** is run by itself, it will print out the first two fields of the entire file `prem.dat`—the depth and compressional velocity (V_P). If **$3** is used instead of **$2** then the shear velocity (V_S) is output. The pipe (|) that connects the **awk** with the **gmt psxy** command sends the output of the former program to the latter program for further processing.

An important consideration is the use of the same flags across multiple command lines. The projection flag **-J**, for example, need only have the attributes declared once, with the first command **psbasemap**. Subsequent commands will then only need to put **-J** flag without any attributes, inheriting the previously set values. Similarly, flags **-R** and **-B** (see Examples 7–9).

Example 7 A simple example of multiple GMT commands using **psbasemap** and **psxy**. The file `prem.dat` is required for execution.

```
gmt psbasemap -Jx0.001c/0.3c -R0/6371/0/16 -BWeSn+t"Preliminary
    Reference Earth Model"+g255 -Bpya2f1+l"@;black;V@-P@-
    (kms@+-1@+)@;;, @;red;V@-S@- (kms@+-1@+)@;;,
    @;blue;@~\162@~ (gcm@+-3@+)@;;" -Bpxa2000f1000+l"Depth
    (km)" -P -V -K > Example7.ps
awk '{print $1,$2}' prem.dat | gmt psxy -J -R -W1p,black -V -K
    -O >> Example7.ps
awk '{print $1,$3}' prem.dat | gmt psxy -J -R -W1p,red -V -K -O
    >> Example7.ps
awk '{print $1,$4}' prem.dat | gmt psxy -J -R -W1p,blue -V -O
    >> Example7.ps
```

Fig. 7 Graph of compressional velocity (V_P, black), shear velocity (V_S, red) and density (ρ, blue) with respect to depth (Dziewonski and Anderson 1981)

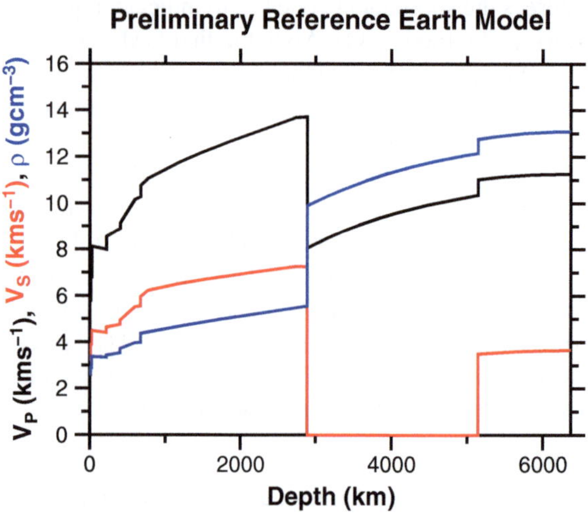

Module	Flag/param.	Description
psbasemap	-Jx	Projection. 'x' sets the projection to linear, with at scale of 0.001 cm per unit along the *x* axis and 0.3 cm along the *y* axis (0.001c/0.3c)
	-R	Coordinate ranges (*xmin*/*xmax*/*ymin*/*ymax*). The limits for the *x* and *y* axes. The *x* axis represents the radius of the Earth whereas the *y* axis represents the velocity or density (0/6371/0/16)
	-B	Boundary. Attributes 'WeSn' represent west, east, south, and north, respectively (order not important). Each letter sets the border plotted (letters in lower case) and with labels (upper case). **+t** adds a title to the plot. **+g** adds a white background colour (255). 'pya2f1' sets the **p**rimary **a**nnotation of the ***y*** axis at an interval of 2 and minor ticks at every 1. 'pxa2000f1000' sets the **p**rimary **a**nnotation of the ***x*** axis at an interval of 2000 and minor ticks at 1000. '@-' and '@+' toggle on and off sub- and super-scripts. '@;black;' and '@;;' toggle on and off the text colour to **black**, or other. '@~' toggles between the selected font and Greek (Symbol). '\162' plots Greek rho symbol. Check the official GMT website for a detailed list of Greek symbols
	-P	Portrait. Set plot into portrait mode, the default is landscape
	-V	Verbose. Set the level of information output from GMT during execution
	-K	More PostScript code will follow. Required for all but the last plot command when writing multi-layer plots
	>	Redirect output. The generated PostScript will be written to the file `Example7.ps`
awk	`prem.dat`	Text file containing four fields: depth in km, compressional velocity (V_P, kms^{-1}), shear velocity (V_S, kms^{-1}) and density (ρ, gcm^{-1}) from Dziewonski and Anderson (1981). The first line has a '#' indicating that this line is a comment and is ignored by **psxy**
	print	Output. Print first and second field (`$1` and `$2`) or others

Modules	Flag/param.	Description
psxy	-J	Projection. The attributes of the flag are inherited from the previous command
	-R	Range. The attributes of the flag are inherited from the previous command
	-W	Pen attributes. Set line **p**oint thickness and **black** colour (1p,black)
	-V	Verbose. Set the level of information output from GMT during execution
	-K	More PostScript code will be appended later. Required for all but the last plot command when writing multi-layer plots
	-O	Selects overlay plot mode. Required for all but the first plot command when writing multi-layer plots
	>>	Redirect output. The generated PostScript will be appended to the file `Example7.ps`

Shell scripts

It is recommended that the GMT commands are put together into a single text file as a script. The file can then be run from the shell, hence the name ***shell script***. The advantages of shell scripts are that the code snippets are easily editable in an editor, run in sequence, and can include additional comments for future reference. Importantly, the text file needs to be set as executable. This can be done by using the command `chmod u+x filename`, and then executed as `./filename`. All the commands shown in this chapter are available as shell scripts in the Supporting Material.

3.1 *Seismicity Map*

In the following example, two maps are shown side by side, one showing the earthquake focal mechanisms and the other showing the seismicity. A variety of GMT commands is used to make Fig. 8, including GMT modules not discussed so far: **psmeca**, **makecpt** and **psscale**. These plot 'beach balls', generate a colour palette, and plot the colour scale, respectively. The flag **-X**, available for most of the GMT modules, is used to shift the second base map to the right by 9 cm (-X9c). Unlike the previous use of **psxy**, here it is used to plot scattered *x*, *y* colour-filled circles representing the earthquakes.

Example 8 An intermediate level example of multiple GMT commands: **psbasemap**, **pscoast**, **psmeca**, **makecpt**, **psxy** and **psscale**. The file `globalcmt.dat` and `earthquake.dat` are required for execution.

```
gmt psbasemap -JM7c -R10/18/30/42.5 -BWESN+t"Focal mechanisms
    across the Central Mediterranean" -Bpxa5f2.5 -Bpya5f2.5 -P
    -V -K > Example8.ps
gmt pscoast -J -R -Df -G200/150/100 -S200/200/255 -V -K -O >>
    Example8.ps
gmt psmeca globalcmt.dat -J -R -Sd0.5c/-1/0 -D0/600 -h13
    -L1p,black -G0 -V -K -O >> Example8.ps
gmt psbasemap -J -R -BWESN+t"Seismicity across the Central
    Mediterranean" -Bxa5f2.5 -Bya5f2.5 -V -K -O -X9c >>
    Example8.ps
```

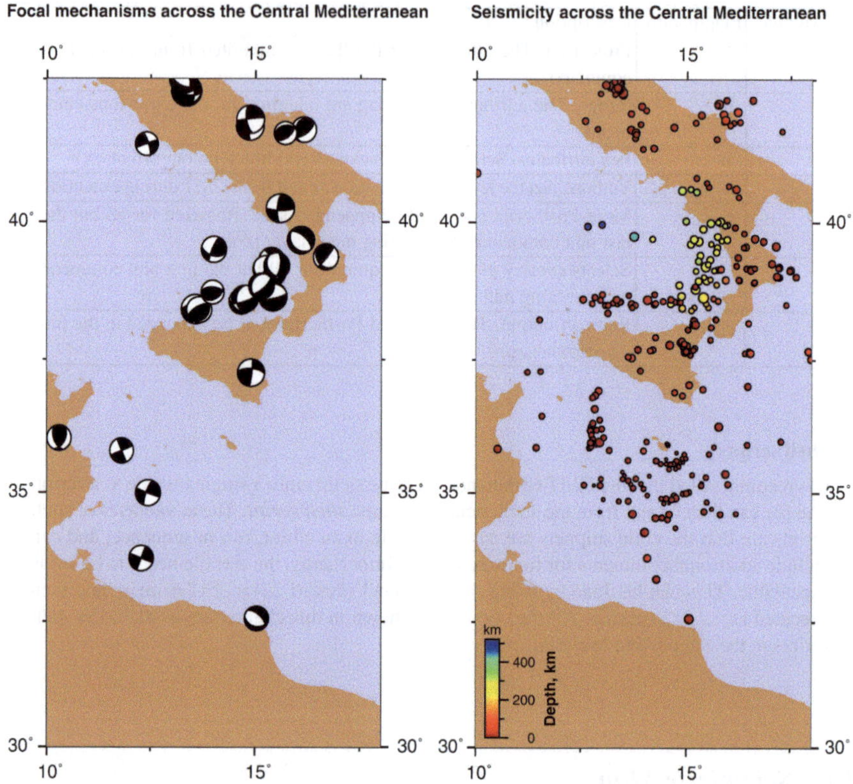

Fig. 8 Seismicity across the Central Mediterranean. Left frame shows the focal mechanisms of earthquakes between 1989 and 2009 retrieved from http://www.globalcmt.org. Right frame shows the regional seismicity for the year 2006–2009 retrieved from http://earthquake.usgs.gov (M > 3) and Agius and Galea (2011)

```
gmt pscoast -J -R -Df -G200/150/100 -S200/200/255 -V -K -O >>
    Example8.ps
gmt makecpt -Cseis -T0/522/1 > depth.cpt
gmt psxy earthquake.dat -J -R -Cdepth.cpt -Scp -W1p,black -V -K
    -O >> Example8.ps
gmt psscale -Cdepth.cpt -J -R -Bxaf+l"Depth, km" -By+l"km"
    -DjBL+w2c/0.3c+o0.2c/0.2c -V -O >> Example8.ps
```

Module	Flag	Description
psbasemap	-JM	Projection. 'M' sets the projection to **M**ercator cylindrical map, with at a specific width of 7 cm
	-R	Longitude and latitude ranges (*lon min*/*lon max*/*lat min*/*lat max*). The limits for the *x* and *y* axes set 10–18 and 30–42.5, respectively
	-B	Attributes 'WESN' represent west, east, south, and north, respectively (order not important). Each letter sets the border plotted (letters in lower case) and with labels (upper case). **+t** adds a title to the map. 'pya5f2.5' sets the **p**rimary **a**nnotation of the **y** axis at an interval of 5 and a minor tick spacing interval of 2.5 (f). Likewise for the *x* axis

Module	Flag	Description
	-P	Portrait. Set plot into portrait mode, the default is landscape
	-V	Verbose. Set the level of information output from GMT during execution
	-K	More PostScript code will follow. Required for all but the last plot command when writing multi-layer plots
	-O	Selects overlay plot mode. Required for all but the first plot command when writing multi-layer plots
	-X	Offset the plot along the x axis. The plot will be offset to the right by 9 cm relative to the previous command
	>	Redirect output. The generated PostScript will be written to the file `Example8.ps`
	>>	Redirect output. The generated PostScript will be appended to the file `Example8.ps`
pscoast	-J	Projection. The attributes of the flag are inherited from the previous command
	-R	Range. The attributes of the flag are inherited from the previous command
	-D	Resolution. Set the resolution of the coastline data to **f**ull
	-G	Set the fill colour for land (RGB: 200/150/100. See Sect. 4)
	-S	Set the fill colour of the sea (RGB: 200/200/255. See Sect. 4)
	-V	Verbose. Set the level of information output from GMT during execution
	-K	More PostScript code will be appended later. Required for all but the last plot command when writing multi-layer plots
	-O	Selects overlay plot mode. Required for all but the first plot command when writing multi-layer plots
	>>	Redirect output. The generated PostScript will be appended to the file `Example8.ps`

3.1.1 psmeca

Earthquake focal mechanisms are easily plotted using module **psmeca**. The module reads data values from a file and generates the PostScript code of the 'beach balls'. It requires at least four arguments: a text file containing the data organised in columns, **-J** for projection, **-R** for axes ranges, and **-S** to select the format of the columns in the data file. The data file formats may be in Harvard CMT convention, in Aki and Richards convention, as a seismic moment tensor, or other focal mechanism formats such as double couple, anisotropic and principal axis. Data suitable for **psmeca** can be downloaded from http://www.globalcmt.org.

Module	File/flag	Description
psmeca	globalcmt.dat	An input file containing earthquake focal mechanism information: X, Y, depth, mrr, mtt, mff, mrt, mrf, mtf, exp, newX, newY, and event title. Data is available from http://www.globalcmt.org
	-J	Projection. The attributes of the flag are inherited from the previous command
	-R	Range. The attributes of the flag are inherited from the previous command
	-S	Select file format, font size and offset. Attribute 'd' plots only the double couple part defined from the seismic moment tensor (Harvard CMT), scaled at 0.5 cm, without labels and no offset (d0.5c/-1/0)
	-D	Plot events between 0 and 600 km depth
	-h	Skip 13 header lines from the input file
	-L	Draws the 'beach ball' outline with pen attributes of one **p**oint thickness and **black** colour (1p,black)
	-G	Set the fill colour of focal mechanism. Compressional quadrants of the focal mechanism beach balls are shaded black (0)
	-V	Verbose. Set the level of information output from GMT during execution
	-K	More PostScript code will be appended later. Required for all but the last plot command when writing multi-layer plots
	-O	Selects overlay plot mode. Required for all but the first plot command when writing multi-layer plots
	>>	Redirect output. The generated PostScript will be appended to the file `Example8.ps`

3.1.2 makecpt

Colours can help visualise data better. GMT utilise the full range of colours acceptable by the PostScript description language and comes a long with a long list of predefined colour tables. Some of these colour tables are useful when for example plotting bathymetry (ocean), or elevation (topo), or seismic velocity (seis), etc. Module **makecpt** generates a colour palette by scaling the colours of a selected table (**-C**) across the set range of values at a specified increment (**-T**). For the full list of colour tables available see the manual pages. The output is a GMT colour palette table in ASCII text. To see the output colour palette execute the **makecpt** command without the redirection at the end.

Module	Flag	Description
makecpt	-C	Select colour table. Palette 'seis' has a broad range of colours starting from red to orange, yellow, green and blue. The colours will represent the depth of the earthquakes
	-T	Set the range of the colour palette (*min/max/inc*). Define the range of the palette from 0 to 522 km and with an interval of 1 km (0/522/1)
	>	Redirect output. The colour palette will be written to file `depth.cpt`

3.1.3 psxy: Symbols

Scattered points on an *x*, *y* graph (or map) are plotted using module **psxy**. Unlike the previous Examples 2 and 7, a special flag **-S** is called to plot each record in the input file as symbols. This flag requires arguments that describe the symbol type and size, which alternatively, can be read from added columns in the input file. In Example 8, the fill colour and the size of the circles is determined by the depth (and its corresponding colour palette) and the magnitude of the earthquakes, respectively.

Module	File/flag	Description
psxy	`earthquake.dat`	A text file with longitude, latitude, depth and magnitude field values
	-J	Projection. The attributes of the flag are inherited from the previous command
	-R	Range. The attributes of the flag are inherited from the previous command
	-C	Select colour palette. Reads the colour palette from file `depth.cpt`, which will define the fill colour of the symbol as determined from the third field in `earthquake.dat`, the depth
	-S	Symbol and size. A **c**ircle is plotted at each longitude and latitude, the first two fields in the text file, with a diameter of **p**oint size scaled by the fourth field, the magnitude (cp)
	-W	Pen attributes. Set line **p**oint thickness and **black** colour (1p,black)
	-V	Verbose. Set the level of information output from GMT during execution
	-K	More PostScript code will be appended later. Required for all but the last plot command when writing multi-layer plots
	-O	Selects overlay plot mode. Required for all but the first plot command when writing multi-layer plots
	>>	Redirect output. The generated PostScript will be appended to the file `Example8.ps`

3.1.4 psscale

Module **psscale** plots a colour scale on maps. As input, it reads a GMT colour palette (**-C**)—that file produced from **makecpt**. The position, dimension and orientation of the scale are determined by the **-D** flag and labels are controlled by the **-B** flag.

Module	Flag	Description
psscale	-C	Select colour palette. Read the colour palette (`depth.cpt`)
	-J	Projection. The attributes of the flag are inherited from the previous command
	-R	Range. The attributes of the flag are inherited from the previous command
	-B	Boundary. Attributes 'xaf' sets automatic **a**nnotation and tick of the ***x*** axis. **+l** adds a label to the respective axis
	-D	Specify position and dimensions of the scale bar. Plot the colour scale in the **B**ottom **L**eft corner (jBL) with a length of 2 cm and width of 0.3 cm (+w2c/0.3c) and **o**ffset by 0.2 cm along the *x* and *y* axis (+o0.2c/0.2c)
	-V	Verbose. Set the level of information output from GMT during execution
	-O	Selects overlay plot mode. Required for all but the first plot command when writing multi-layer plots
	>>	Redirect output. The generated PostScript will be appended to the file `Example8.ps`

3.2 *Topography and Seismic Networks Map*

In this final example, a detailed map of seismic networks deployed across the Tibetan plateau is compiled (Fig. 9). The map shows a high-resolution topography of the region, rivers, national boundaries, labels identifying the areas, an inset map, legend and a scale for the topography. Generating the PostScript of this figure requires numerous sequential calls of different GMT modules. First, a **psbasemap** is called to set the background of the map. A colour palette for a topography ranging from −5 to 5 km is created using **makecpt**. The topography is then plotted using the module **grdimage**. National boundaries and rivers are plotted using **pscoast**. **pstext** is then called to write labels on the map, followed by **psxy** to plot the stations. Next, an inset regional map is created by calling **psbasemap** and **pscoast** again, followed by a call to **psxy** to draw a red box highlighting the area of the larger map. A legend is added by calling module **pslegend**, and finally, **psscale** adds the scale for the topography.

Example 9 introduces two other methods of input to a GMT module. One is using **echo** and piping '|' its output to the GMT module and the other is by redirecting the text between end-of-file (`EOF`) tags using left angle brackets (<<), see Example 9.

Example 9 An advanced level example of multiple GMT commands: **psbasemap**, **makecpt**, **grdimage**, **pscoast**, **pstext**, **psxy**, **pslegend** and **psscale**. Several files are required to plot the stations location (.txt files) and the topography grid file `etopo1_ice_c_i2.bin`, which is available from https://www.ngdc.noaa.gov.

```
gmt psbasemap -Jm0.6c -R82/106/22/38 -BWESN -Bya5f2.5 -Bxa5f2.5
    -P -V -K -X2c -Y5c > Example9.ps
gmt makecpt -Cglobe -T-5000/5000/1 > elevation.cpt
gmt grdimage etopo1_ice_c_i2.bin -J -R -Celevation.cpt -E300 -V
    -K -O >> Example9.ps
gmt pscoast -J -R -Na/1p,black -Df -V -K -O >> Example9.ps #
    National Boundaries
gmt pscoast -J -R -B -Ia/0.5p,turquoise -Df -V -K -O >>
    Example9.ps # Rivers

# Labels
```

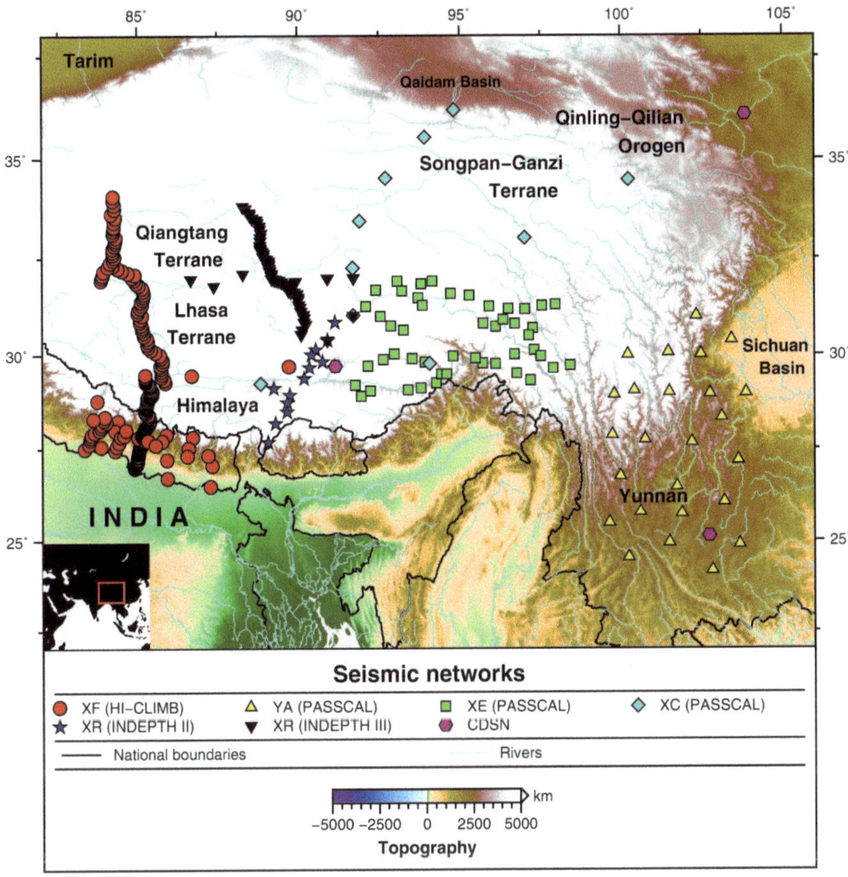

Fig. 9 Seismic networks deployed across the Tibetan plateau between 1991–2005. The topography is a 1 arc-minute global relief model (Amante and Eakins 2009). Black contours are national boundaries. Blue lines indicate rivers. Coloured symbols represent different seismic networks. Figure modified from Agius (2013)

```
echo "104.8 30.2 Sichuan" | gmt pstext -J -R
   -F+f9,Helvetica-Bold -V -K -O >> Example9.ps
echo "105.0 29.6 Basin" | gmt pstext -J -R -F+f9,Helvetica-Bold
   -V -K -O >> Example9.ps
echo "94.75 36.9 Qaidam Basin" | gmt pstext -J -R
   -F+f8,Helvetica-Bold -V -K -O >> Example9.ps
echo "83.5 37.45 Tarim" | gmt pstext -J -R
   -F+f10,Helvetica-Bold -V -K -O >> Example9.ps
echo "87 31.2 Lhasa" | gmt pstext -J -R -F+f10,Helvetica-Bold
   -V -K -O >> Example9.ps
echo "87 30.5 Terrane" | gmt pstext -J -R -F+f10,Helvetica-Bold
   -V -K -O >> Example9.ps
echo "87.5 28.7 Himalaya" | gmt pstext -J -R
   -F+f10,Helvetica-Bold -V -K -O >> Example9.ps
echo "86.4 33.2 Qiangtang" | gmt pstext -J -R
   -F+f10,Helvetica-Bold -V -K -O >> Example9.ps
```

```
echo "86.6 32.5 Terrane" | gmt pstext -J -R
    -F+f10,Helvetica-Bold -V -K -O >> Example9.ps
echo "96 34.9 Songpan-Ganzi" | gmt pstext -J -R
    -F+f10,Helvetica-Bold -V -K -O >> Example9.ps
echo "97 34.2 Terrane" | gmt pstext -J -R -F+f10,Helvetica-Bold
    -V -K -O >> Example9.ps
echo "100 36 Qinling-Qilian" | gmt pstext -J -R
    -F+f10,Helvetica-Bold -V -K -O >> Example9.ps
echo "101 35.3 Orogen" | gmt pstext -J -R -F+f10,Helvetica-Bold
    -V -K -O >> Example9.ps
echo "101 26.2 Yunnan" | gmt pstext -J -R -F+f10,Helvetica-Bold
    -V -K -O >> Example9.ps
echo "85 25.7 I N D I A" | gmt pstext -J -R
    -F+f14,Helvetica-Bold -V -K -O >> Example9.ps

# Stations
gmt psxy XF_2004-2005.txt -J -R -Sc0.25c -G200/0/0 -W0.75p,0 -V
    -K -O >> Example9.ps
gmt psxy YA_2003-2004.txt -J -R -St0.25c -G230/230/0 -W0.75p,0
    -V -K -O >> Example9.ps
gmt psxy XE_2003-2004.txt -J -R -Ss0.25c -G0/200/0 -W0.75p,0 -V
    -K -O >> Example9.ps
gmt psxy XC_1991-1992.txt -J -R -Sd0.25c -G0/200/200 -W0.75p,0
    -V -K -O >> Example9.ps
gmt psxy XR_1994.txt -J -R -Sa0.25c -G0/0/200 -W0.75p -V -K -O
    >> Example9.ps
gmt psxy XR_1997-1999.txt -J -R -Si0.25c -G0/0/0 -W0.75p,0 -V
    -K -O >> Example9.ps
gmt psxy CD.txt -J -R -Sh0.25c -G200/0/200 -W0.75p,0 -V -K -O
    >> Example9.ps

# Inset map.
gmt psbasemap -Jm0.02c -R30/130/-20/60 -Bsn -Bxa1000 -Bya1000
    -V -K -O >> Example9.ps
gmt pscoast -J -R -B -G50 -Swhite -A10000 -V -K -O >>
    Example9.ps

# Inner inset box.
gmt psxy -J -R -O -W1p,red -A -L -V -K -O >> Example9.ps << EOF
82 22
82 38
106 38
106 22
EOF

# Legend
gmt pslegend -F+gwhite+p1p,0 -Dx0c/-4c+w14.4c/4c+l1.2
    -C0.2c/0.2c -V -K -O << EOF >> Example9.ps
H 12 Helvetica-Bold Seismic networks
D 0.2c 1p
N 4
S 0.1c c 0.25c 200/0/0 0.75p,0 0.5c XF (HI-CLIMB)
S 0.1c t 0.25c 230/230/0 0.75p,0 0.5c YA (PASSCAL)
S 0.1c s 0.25c 0/200/0 0.75p,0 0.5c XE (PASSCAL)
S 0.1c d 0.25c 0/200/200 0.75p,0 0.5c XC (PASSCAL)
S 0.1c a 0.25c 0/0/200 0.75p,0 0.5c XR (INDEPTH II)
S 0.1c i 0.25c 0/0/0 0.75p,0 0.5c XR (INDEPTH III)
S 0.1c h 0.25c 200/0/200 0.75p,0 0.5c CDSN
D 0.2c 1p
N 2
S 0.5c - 0.75c 0 1p,0 1.1c National boundaries
S 0.5c - 0.75c turqoise 0.5p,turquoise 1.1c Rivers
D 0.2c 1p
```

```
EOF

gmt psscale -Celevation.cpt -Dx5.4c/-2.8c+w3.5c/0.25c+h+ef
    -F+gwhite -Bxa2500f500+l"Topography" -By+l"km" -V -O >>
    Example9.ps
```

Module	Flag/file	Description
psbasemap	-Jm	Projection. '**m**' sets the projection to **M**ercator cylindrical map, with at a scaling factor of 0.6 cm per unit (0.6c) for the large map and 0.2 cm (0.2c) for the inset map
	-R	Longitude and latitude ranges (*lon min*/*lon max*/*lat min*/*lat max*). The limits for the *x* and *y* axes of the large map are 82–106 and 22–38, respectively
	-B	Attributes 'WESN' represent west, east, south, and north, respectively (order not important). Each letter sets the border plotted with labels (upper case). 'pya5f2.5' sets the **p**rimary **a**nnotation of the ***y*** axis at an interval of 5 and a minor tick spacing interval of 2.5 (f). Likewise for the *x* axis
	-P	Portrait. Set plot into portrait mode, the default is landscape
	-V	Verbose. Set the level of information output from GMT during execution
	-K	More PostScript code will follow. Required for all but the last plot command when writing multi-layer plots
	-O	Selects overlay plot mode. Required for all but the first plot command when writing multi-layer plots
	-X	Offset the plot along the *x* axis. The plot will be offset to the right by 2 cm (2c)
	Y	Offset the plot along the *y* axis. The plot will be offset up by 5 cm (5c)
	>	Redirect output. The generated PostScript will be written to the file `Example9.ps`
pscoast	-J	Projection. The attributes of the flag are inherited from the previous command
	-R	Range. The attributes of the flag are inherited from the previous command
	-B	Boundary. The attributes of the flag are inherited from the previous command
	-N	Political borders. Plot national, Americas states, marine boundaries, or **a**ll in **black** with **p**oint thickness (a/1p,black)
	-D	Resolution. Set the resolution of the coastline data to **f**ull
	-I	Draw rivers. Plot rivers or canals, or **a**ll in **turquoise** with half **p**oint thickness (a/0.5p,turquoise)
	-G	Set the fill colour for land
	-S	Set the colour of the sea
	-A	Minimum area. Feature areas that are greater than 10000 only (minimum area)
	-V	Verbose. Set the level of information output from GMT during execution
	-K	More PostScript code will be appended later. Required for all but the last plot command when writing multi-layer plots
	-O	Selects overlay plot mode. Required for all but the first plot command when writing multi-layer plots
	>>	Redirect output. The generated PostScript will be appended to the file `Example9.ps`
psxy	`CD.txt`	Input text file. File `CD.txt` and others contain the longitude and latitude of each station
	-J	Projection. The attributes of the flag are inherited from the previous command

Module	Flag	Description
	-R	Range. The attributes of the flag are inherited from the previous command
	-S	Symbol and size. Set symbol type: **c**ircle, **t**riangle, **s**quare, **d**iamond, st**a**r, **i**nverted triangle and **h**exagon. A symbol is plotted at each longitude and latitude, the first two fields in the text file, with fixed size of 0.25 cm (e.g., t0.25c)
	-G	Set the fill colour symbol (RGB: 0–255/0–255/0–255. See Sect. 4)
	-W	Pen attributes. Set symbol **p**oint thickness and colour (0.75p,0)
	-V	Verbose. Set the level of information output from GMT during execution
	-K	More PostScript code will be appended later. Required for all but the last plot command when writing multi-layer plots
	-O	Selects overlay plot mode. Required for all but the first plot command when writing multi-layer plots
	>>	Redirect output. The generated PostScript will be appended to the file `Example9.ps`
makecpt	-C	Select colour table. Palette 'globe' has a broad range of colours suitable for the Earth's topography, good for both deep oceans and high elevation. Colour palette starts from dark blue changing to light blue, light brown, green
	-T	Set the range of the colour palette (*min*/*max*/*inc*). Define the range of the palette from -5000 to 5000 m and with an interval of 1 m
	>	Redirect output. The colour palette will be written to the file `elevation.cpt`
psscale	-C	Select colour palette. Read the colour palette `elevation.cpt`
	-B	Boundary. Attributes 'xa2500f500' sets the **a**nnotation of the ***x*** axis at an interval of 2500 and minor ticks spacing interval of 500 (f). **+l** adds a label to the respective axis
	-Dx	Specify position and dimensions of the scale bar. Plot the colour scale 5.4 cm to the east and 2.8 cm down (5.4c/−2.8c). **+w** sets the bar to be 3.5 cm long and 0.25 cm wide (+w3.5c/0.25c). **+h** sets the scale to a horizontal orientation. A sidebar triangle is added to show **f**oreground colour of the palette, indicating that the elevation could be higher (+ef)
	-F	Set background behind scale. Fill the background white (+gwhite)
	-V	Verbose. Set the level of information output from GMT during execution
	-O	Selects overlay plot mode. Required for all but the first plot command when writing multi-layer plots
	>>	Redirect output. The generated PostScript will be appended to the file `Example9.ps`
#		Comment. Any text following the hash (#) sign is ignored by the shell and GMT. This is useful to leave comments for future reference

3.2.1 grdimage

Two-dimensional (2-D) binary grid files such as the one used to visualise the topography in Example 9 are composed of x, y, z data, in this case, the longitude, latitude and depth. Grid files are plotted using the GMT module **grdimage**, which reads a grid file and produces a colour-shaded map. The map is projected according to the selected projection (**-J**) and range (**-R**), using the desired resolution (**-E**) and colour palette (**-C**). High-quality global topographic data is publicly available from https://www.ngdc.noaa.gov (Amante and Eakins 2009).

Module	File/Flag	Description
grdimage	`etopo1_ice_c_i2.bin`	Input grid file. The topography file is available from https://www.ngdc.noaa.gov
	-J	Projection. The attributes of the flag are inherited from the previous command
	-R	Range. The attributes of the flag are inherited from the previous command
	-C	Select colour palette. Read the colour palette (`elevation.cpt`)
	-E	Grid resolution. Set the dot-per-inch (dpi) for the projected grid. 300 dpi is the standard for high-quality printing
	-V	Verbose. Set the level of information output from GMT during execution
	-K	More PostScript code will be appended later. Required for all but the last plot command when writing multi-layer plots
	-O	Selects overlay plot mode. Required for all but the first plot command when writing multi-layer plots
	>>	Redirect output. The generated PostScript will be appended to the file `Example9.ps`

3.2.2 pstext

Text strings of variable size, font type and orientation can be plotted on a map using the GMT module **pstext**. The input to the module requires the coordinates (*longitude* and *latitude*) and a text string, which can either be read from a file, piped through, or redirected. In Example 9 the **pstext** input is piped through (|) from **echo**. The fonts size and type are set using the **-F** flag.

echo

echo is a standard built-in command in Unix and Unix-like operating systems that outputs its arguments back onto the screen. An example syntax is: `echo "Hello World"`, which output 'Hello World' on the following line. When **echo** is used in conjunction with another command through piping (|), the output from **echo** is used as input to the other.

Module	Flag	Description
pstext	-J	Projection. The attributes of the flag are inherited from the previous command
	-R	Range. The attributes of the flag are inherited from the previous command
	-F	Fonts attributes. Specifies the **f**ont size and font type e.g., '+f8,Helvetica-Bold'
	-V	Verbose. Set the level of information output from GMT during execution
	-K	More PostScript code will be appended later. Required for all but the last plot command when writing multi-layer plots
	-O	Selects overlay plot mode. Required for all but the first plot command when writing multi-layer plots
	>>	Redirect output. The generated PostScript will be appended to the file `Example9.ps`

3.2.3 pslegend

To plot a legend call GMT module **pslegend**. Flags **-D** and **-C** set the location, width and height, and the clearance between the legend frame and the internal items, respectively. **pslegend** reads specific legend-related information from an input file or redirected input. Legend instructions allow for the addition of a vertical gap (**G**), vertical line (**V**), set the number of columns (**N**), draw a horizontal line (**D**), add a header (**H**), label (**L**) or symbol (**S**), etc. In Example 9, the legend information is redirected using left angle brackets (**<<**) and end-of-file (**`EOF`**) tags.

Module	Flag	Description
pslegend	-F	Attributes to legend box. Set a **white** back**g**round (+gwhite), with pen attributes (+p) of 1 **p**oint thick black border (1p,0)
	-D	Sets the reference point of the legend. The legend is set 4 cm below the map, with no offset along the *x* axis (x0c/−4c), with a **w**idth of 14.4 cm and height of 4 cm (+w14.4c/4c). The **l**ine-spacing factor of the legend is set to 1.2 (+l1.2)
	-C	Sets the clearance between the legend frame and the internal items to 0.2 cm for both *x* and *y* coordinates (0.2c/0.2c)
	-V	Verbose. Set the level of information output from GMT during execution
	-K	More PostScript code will be appended later. Required for all but the last plot command when writing multi-layer plots
	-O	Selects overlay plot mode. Required for all but the first plot command when writing multi-layer plots
	<< `EOF`	Redirect input. The content to be plotted by **pslegend** is written in the subsequent lines. << `EOF` instructs the shell to redirect the content of the following lines to the **pslegend** command until the next `EOF` tag is read
	`EOF`	End of file. Begin and end tag for the text to be directed as input into the preceding command
	>>	Redirect output. The generated PostScript will be appended to the file `Example9.ps`

Module	Code	Description
pslegend	H	Plots a centred header with a font size 12 and type Helvetica-Bold
	D	Draws a horizontal line with an offset of 0.2 cm and pen thickness of one point (1p)
	N	Set the number of columns
	S	Legend symbol. Set the offset (0.1 cm), the symbol type (**c**ircle, **t**riangle, **s**quare, **d**iamond, st**a**r, **i**nverted triangle, **h**exagon) and horizontal line dash (-)), the size (0.25 cm), colour (R/G/B, see Sect. 4), pen thickness and colour (0.75p,0), text offset to 0.5 cm, and the label text

4 Pen Attributes and Colours

Line thickness and colours, often defined by flags like **-W** and **-G**, can be set in two ways, either by using plain English or by using specific numeric values. For example, flag '-Wthick,black' is equally represented numerically '-W1p,0'. Thickness can be set as 'faint', 'thinnest', 'thinner', 'thin', 'thick', 'thicker', 'thickest', 'fat', 'fatter', 'fattest' and 'obese', or by setting the point thickness: '0', '0.25p', '0.50p', '0.75p', '1.0p', '1.5p', '2p', '3p', '6p', '12p' and '18p', respectively, or any other value. Note: A too thin line can be difficult for printers to render. Similarly, colour can be specified in five different ways:

Name Specify a colour name such as white, black, red, orange, yellow, green, cyan, bluc, magcnta, grcy, brown, with variants of light and dark such as darkgreen, etc.

Grey Specify a grey shade in the range of 0–255, black to white, linearly.

RGB Specify red, green and blue (r/g/b), each ranging from 0–255. 0/0/0 is black, 255/255/255 is white, 255/0/0 is red, etc.

HSV Specify hue-saturation-value (h-s-v), with the former in the 0–360° range while the latter two take on the range 0–1.

CMYK Specify cyan, magenta, yellow and black (c/m/y/k), each ranging from 0–100%.

5 Default Parameters

GMT runs on a set of internal default parameters, which define background colour, frame border properties (e.g., thickness and colour), annotation, label and title fonts etc., (e.g., COLOR_BACKGROUND, COLOR_FOREGROUND, MAP_FRAME _PEN, MAP_FRAME_WIDTH, FONT_ANNOT_PRIMARY, FONT_LABEL, FONT_TITLE). The internal settings used by GMT during the last execution are stored in a file `gmt.conf` located in the current working directory. To view the contents of this file run `cat gmt.conf`. The values of these parameters can be overwritten using the module **gmtset**. For example, to change the font of the title (flag **-B**) to size 10 points, type Helvetica-Bold and black colour use the module **gmtset**:

```
gmt gmtset FONT_TITLE = 10p,Helvetica-Bold,black
```

A change to the default GMT settings may be necessary for Example 6 (see Supporting Material). The *x* axis is projected in time with input coordinates in absolute time and a specific time format. The following **gmtset** commands are run before **pshistogram** to define the time format:

```
gmt gmtset FORMAT_DATE_IN = yyyy-mm-dd
gmt gmtset FORMAT_DATE_OUT = yyyy-mm-dd
gmt gmtset FORMAT_CLOCK_MAP = hh:mm
gmt gmtset FORMAT_DATE_MAP = "dd o y"
gmt gmtset TIME_EPOCH = 2014-01-26T12:00:00
```

To view the full list of GMT default parameters call `gmt gmtdefaults`. All the listed parameters in upper case can be modified.

6 Conclusion

GMT scripts may become lengthy, however, each GMT call can be considered as an independent command that overlays on top of the previous command. Watch out for flags **-K** and **-O** as well as the redirection arrows **>** and **>>**. **-K** is required for all but the last plot command, and **-O** is required for all but the first plot command. Similarly, **>** is used with the first GMT command that creates a PostScript file, whereas **>>** appends to an existing file. Lastly, in this Chapter, only those flags and attributes used in the examples provided are explained. The full list of flags and attributes associated with every module are explained in the manual pages and in the usage messages (Sect. 2.1). There are far more GMT modules and flags that provide the tools necessary to create the desired figure, including 3-D projection, semi-transparent layers and also animations. The use of other modules is very similar to these described here and hold the same style of execution.

References

Agius MR (2013) The structure and dynamics of the lithosphere beneath Tibet from seismic surface-wave analysis. PhD thesis, Trinity College Dublin

Agius MR, D'Amico S, Galea P (2016) The Easter Sunday 2011 earthquake swarm offshore Malta: Analysis on felt reports. Springer International Publishing, pp 631–645

Agius MR, Galea P (2011) A single-station automated earthquake location system at Wied Dalam Station, Malta. Seismol Res Lett 82(4):545–559

Agius MR, Galea P, D'Amico S (2015) Setting up the Malta seismic network: instrumentation, site selection and real-time monitoring. In: Panzera F, Lombardo G (eds) Establishment of an inte-

grated Italy-Malta cross-border system of civil protection. Geophysical aspects. Aracne Editrice int.le S.r.l., pp 11–30
Agius MR, Rychert C, Harmon N, Gabi L (2017) Mapping the mantle transition zone beneath Hawaii from Ps receiver functions: Evidence for a hot plume and cold mantle downwellings. Earth and Planetary Science Letters (In reivew)
Aho AV, Kernighan BW, Weinberger PJ (1979) Awk—a pattern scanning and processing language. Softw Prac Exp 9(4):267–279
Amante C, Eakins BW (2009) ETOPO1 1 arc-minute global relief model: procedures, data sources and analysis. US Department of Commerce, National Oceanic and Atmospheric Administration, National Environmental Satellite, Data, and Information Service, National Geophysical Data Center, Marine Geology and Geophysics Division Colorado
Dziewonski AM, Anderson DL (1981) Preliminary reference earth model. Phys Earth Planet Inter 25(4):297–356
Goldstein P, Dodge D, Firpo M, Ruppert S (1998) What's new in SAC2000? Enhanced processing and database access. Seismol Res Lett 69(3):202–205
Wessel P, Smith WHF (1991) Free software helps map and display data. Eos Trans Am Geophys Union 72(41):441–446
Wessel P, Smith WHF (1995) New version of the generic mapping tools. Eos Trans Am Geophys Union 76(33):329–329
Wessel P, Smith WHF (1998) New, improved version of generic mapping tools released. Eos Trans Am Geophys Union 79(47):579–579
Wessel P, Smith WHF, Scharroo R, Luis J, Wobbe F (2013) Generic mapping tools: improved version released. Eos Trans Am Geophys Union 94(45):409–410

Devices for a Rotational Ground Motion Measurement

Leszek R. Jaroszewicz and Anna Kurzych

1 Introduction

Generally, a rotational ground motion can be induced by earthquakes, explosions, and ambient vibrations. From the above point of view, it is interesting for a study concerning the area of rotational seismology (Lee et al. 2009a). This new scientific field is very attractive for two kinds of research. The first group of interest are scientists practicing geophysical disciplines such as earthquake physics (Teisseyre et al. 2006, 2008), strong-motion seismology (Anderson 2003), seismic hazards (McGuire 2008), seismotectonics (University of Munich 2016), and also geodesy (Carey 1983). The second group of interest are scientists dealing with engineering aspects of rotational events including seismic behaviour of irregular and complex civil structures (Trifunac 2009; Mustafa 2015), as well as rotation motions in mine regions induced by seismic events (Zembaty et al. 2016). Recently, we can observe an increasing interest in these effects also from physicists using Earth-based observatories for detecting gravitational waves where instrument corrections on very slow ground rotations are significant (Ju et al. 2000; Lantz et al. 2009).

Historically, rotational effects connected with the earthquake wave were observed as a rotated obelisk after the 1783 Calabria earthquake which is cited in literature as the first illustration of these phenomena (Kozák 2009). Such effects are physically described on the basis of two different models (Kozák 2006; Igel et al. 2012). The first model includes a historical approach (Mallet 1862) and is based on the rotation of bodies with regard to their underlying structures. The second one should be treated as a new approach (Teisseyre et al. 2006, 2008) and it is derived from the studies on asymmetric and micromorphic theories of continuum mechanics, as well as in nonlinear physics. In this field the interesting but also controversial investigation

L. R. Jaroszewicz (✉) · A. Kurzych
Faculty of Advanced Technologies and Chemistry, Military University of Technology, 2 gen. Witolda Urbanowicza St., Warsaw 00-908, Poland
e-mail: jarosz@wat.edu.pl; leszek@jaroszewicz.com

A. Kurzych
e-mail: anna.kurzych@wat.edu.pl

S. D'Amico (ed.), *Moment Tensor Solutions*, Springer Natural Hazards,
https://doi.org/10.1007/978-3-319-77359-9_32

regarding the existence of rotational seismic waves should be considered theoretically (Teisseyre et al. 2003a; Majewski 2009) as well experimentally (Merkel et al. 2008).

Regarding engineering aspects of rotational events, the investigations are quite new and can be identified with recent progress in the structural heath monitoring of civil engineering structures. The primary interests of researchers are associated mainly with formulating additional seismic loads on structures in terms of the rotational seismic excitations. While building response to translational motion has been thoroughly investigated and implemented into design codes of practice, the study of building response to rotational motions is a relatively new field (Wang et al. 2009). This is because the engineering importance of the rotational effects in strong seismic ground motion was noted much later than the translational seismic effects (Newmark and Hall 1969; Newmark and Rosenblueth 1971). However, this rotation may be responsible for damage in high-rise buildings (Zembaty 2009) as well as in structures where the soil-structure interaction effects are expected to be significant (Trifunac 2006; Kalakan and Graizer 2007).

Generally, the above theoretical investigations or practical observations have not been well correlated with measuring rotational ground motions. However, the early attempts regarding measurements of rotational ground motions can be found more than a century ago. Already in 1912 Galitzin (Galitzin 1912) suggested using two identical pendulums installed on different sides of the same rotational axis for separate measurements of a rotational and translational motion. This idea was used in an instrument designed for the registration of a strong ground motion (Kharin and Simonov 1969), as well as in a set of azimuthal arrays of rotational seismographs for rock bursts in a nearby mine (Droste and Teisseyre 1976). Another example of an early instrument is the construction of a gyroscopic seismometer used to measure a static displacement of $<10^{-3}$ m and a tilt of $<0.5 \times 10^{-6}$ rad at La Jolla, California, during the Borrego Mountain earthquake on April 9th, 1968 (Farrell 1969). Early efforts also included studies on induced explosions using seismological sensors to directly measure point rotations after nuclear explosions (Graizer 1991), as well as commercial rotational sensors based on microelectro-mechanical systems for identifying significant near-field rotational motions from a one-kiloton explosion (Nigbor 1994). Finally, it should be noted that rotations in the ground in the responses of structures have been indirectly deduced from accelerometer arrays using methods valid for seismic waves with wavelengths longer than the distances between sensors (Trifunac 1979, 1982; Olivera and Bolt 1989; Spudich et al. 1995; Bodin et al. 1997; Huang 2003; Suryanto et al. 2006; Wassermann et al. 2009).

2 Fundamental Requirements for Rotational Sensors

Now, practical studies of rotational motions are still at an early stage. There are the two reasons for such situation. First, taking into consideration the application area, rotational sensors should be easy implemented in the measurement system existing in seismology. Such systems are being developed for more than two hundred years and are known as seismographs. Since the moment of an earthquake occurrence is never known, rotational sensors' construction should protect suitable detection and recording motions along with precise timing information. For the above reason a seismograph (Lee 2009) consists of seismometer, precise time source, complicated recording device, independent power supply, their remote control possibility, as well as network action readiness according to special defined procedures. Here, we will concentrate only on the heart of such system—the seismometer, sending readers interested in other seismograph components to Springer's book (see Havskov and Alguacil 2016). Unfortunately, a classical seismometer which is a device sensitive only to linear displacement, velocity or acceleration, cannot detect rotational motions (Havskov and Alguacil 2016), and, thus, a new class of devices is required. Moreover, if such devices are based on a classical seismometer, special attention should be paid to separate linear and rotational motions. Second, a wide range of potential scientific interest regarding rotational seismology gives different expectations regarding the requirements. For the above reason the fundamental requirements for rotational seismometers in seismological and engineering applications have not been formulated or agreed for a long time. Fortunately, recently the situation has changed and the main requirements for mentioned above two fundamental areas of the rotational seismology applications are shown in Table 1. They are based on propositions regarding seismological and engineering applications formulated by Bernauer et al. (2012) and Jaroszewicz et al. (2016), respectively. It should be underlined that the assumed amplitude ranges for two above areas of applications fulfil the conditions for a weak motion and, even better than that, for a strong motion seismology, respectively (Evans et al. 2007).

Final remarks are connected with the frequency range assumed for a new class of rotational seismometers, especially when their construction based on sensors uses inertia as a physical principle of operation. Since such system's frequency response has nonlinear characteristics with a specific resonance frequency, this phenomenon

Table 1 The main requirements for a rotational seismometer

Rotational seismology area of interest	
Seismological applications (Bernauer et al. 2012)	Engineering applications (Jaroszewicz et al. 2016)
(S1): effectively insensitive to linear motion, or at any time, independent measurement of linear and rotational motions must be possible	(E1): effectively insensitive to linear motion, or at any time, independent measurement of linear and rotational motions must be possible
(S2): small and stable with respect to ambient conditions, including changes of temperature	(E2): small and stable with respect to ambient conditions, including changes of temperature
(S3): electrical power supply should be easily managed using batteries, at least in combination with solar panels or fuel cells	(E3): electrical power supply should be easily managed using batteries, at least in combination with solar panels or fuel cells
(S4): be able to measure amplitudes of the order of 10^{-7} rad/s at a frequency range between 0.01 and 0.1 Hz	(E4): be able to measure amplitudes of the order of a few rad/s at a frequency range of between 0.01 and 100 Hz

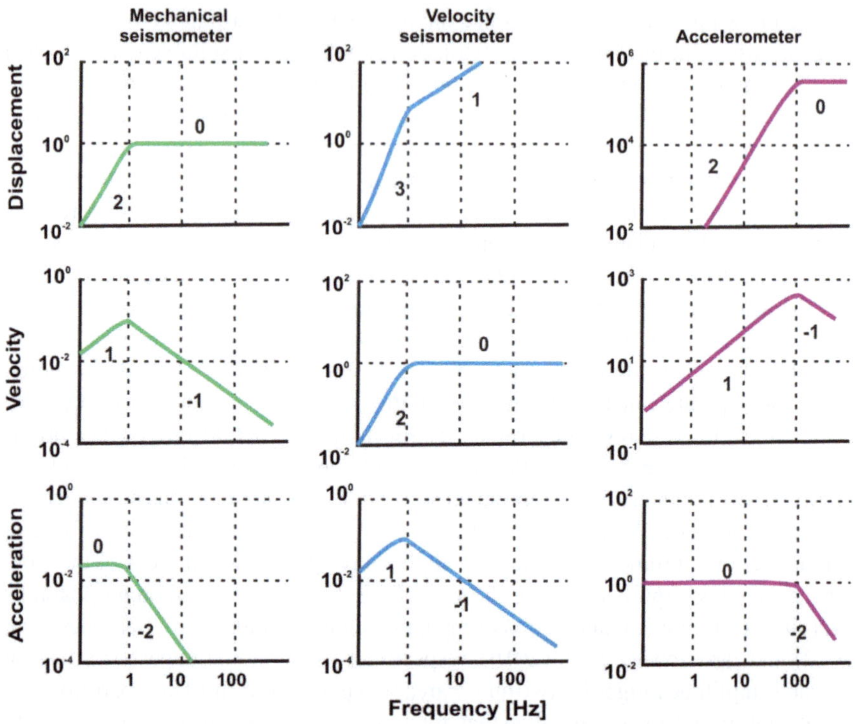

Fig. 1 Amplitude frequency response of a displacement, a velocity and an accelerometer. From top to bottom, the figures show the sensor response to ground displacement, velocity and acceleration, respectively. The asymptotic slope for each segment is indicated (Havskov and Alguacil 2016)

limits the proper system's work and always should be taken under consideration. Generally, the velocity can be easily obtained either from displacement (by time differential operation) or from acceleration (by time integral operation). However, Havskov and Alguacil (2016) proved that it can be a bit confusing if such different sensors are used to measure the above different quantities as shown schematically in Fig. 1.

After Havskov and Alguacil (2016) we concluded that only a direct measurement of given quantity (for example velocity by velocity seismometer) gives possibility to obtain a flat response, but none of the sensors is linear, in the sense that an arbitrary waveform of a ground motion can be exactly reproduced at scale, for any kind of response. The mechanical sensor has a flat displacement response above resonance frequency. The velocity sensor has the same characteristics but for velocity. The most broadband available instrument is the accelerometer since it has a flat response for acceleration in the whole seismic frequency band. But the accelerometer is insensitive to displacement at low frequencies and too sensitive at high frequencies. So for a very sensitive accelerometer, it is difficult to cover the whole amplitude displacement range. In this way, considering application of the classical seismic sensor as the base for the rotational seismometer construction we should consider it as a frequency filter and we must investigate its frequency response for a proper interpretation of the obtained results of the rotational velocity measurement.

Basing on the above general consideration the subsequent two sections contain descriptions of all main types of rotational seismometers with a discussion of their operation principles and a comparison of their fundamental parameters. These devices can be divided into two groups of sensors: rotational seismometers measure rotation in an indirect or direct way. Rotational sensors based on classical seismometers, such as mechanical or acoustical types, belonging to the first group are described in Sect. 3. The rotational sensors which belong to the second group such as electrochemical, mechanical and optical types are described in Sect. 4. These descriptions include a discussion of how far the requirements formulated in Table 1 are fulfilled, and show that the optical type of a rotational seismometer and particularly the type based on a fibre-optic gyroscope (FOG), is the most promising device for future research and application in rotational seismology. On the basis of this conclusion, Sect. 5 presents our experiences with such devices.

3 Rotational Seismometers for an Indirect Measurement of Rotation

It is a practical realisation of the idea of Galitzin (1912) where constructed devices issue a pair of standard seismic sensors (pendulums or geophones) oriented parallel to the chosen axes (one, two or three) and rigidly mounted at the distance l along the perpendicular axis, for instance, the y axis as shown in Fig. 2. The rotation rate Ω_z in radians per second around the vertical axis z is (Havskov and Alguacil 2016):

$$\Omega_z = \frac{\partial v_x}{\partial y} \approx \frac{v_2 - v_1}{l}, \tag{1}$$

where v_1 and v_2 are the velocities measured by sensors 1 and 2 along the x axis.

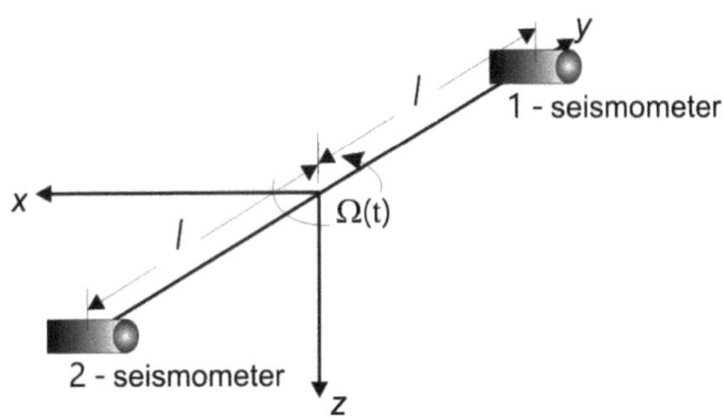

Fig. 2 The operation principle scheme of a rotational seismometer for an indirect measurement; sensors 1 and 2 are velocity seismometers or accelerometers

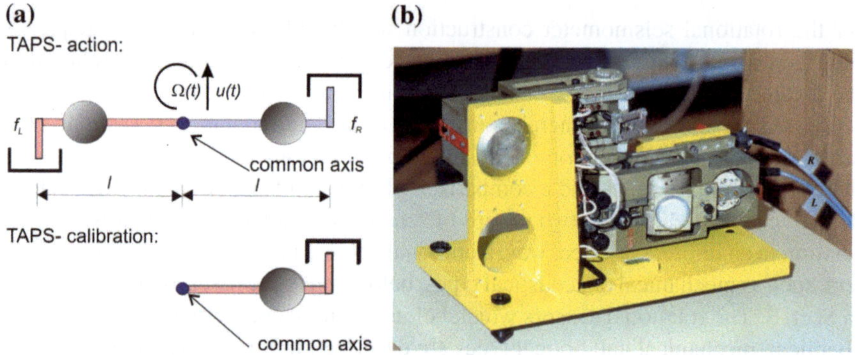

Fig. 3 The TAPS rotational seismometer: **a** scheme during action and calibration; **b** general view

3.1 Rotational Seismometer Using a Pair of Classical Pendulum Seismometers

Teisseyre and Nagahama (1999) constructed a practical example of this type of the system using two antiparallel pendulum seismometers (SM-3 type, ROTOR International Ltd., Kursk, Russia), known as TAPS (Twin-Antiparallel Pendulum Seismometers). These are positioned on a common axis and connected in parallel, but with opposite orientations, as shown in Fig. 3. In the case of ground motion involving the displacement $u(t)$ and the vertical rotation $\Omega_z = \Omega(t)$, the electromotive force $f(t)$ recorded by each SM-3 contains the component of displacement $\pm u(t)$ and the rotational motion $\Omega(t)$ multiplied by the proper length of the pendulum l (Teisseyre and Nagahama 1999):

$$f_{L,R}(t) = \pm u(t) + 1 \cdot \Omega(t), \tag{2}$$

where the signs "+" and "−" represent right (R) and left (L) seismometers, respectively.

From the above relation the rotational and translational components can be easily obtained from the sum and difference of two recorded signals:

$$\Omega(t) = \frac{1}{2l}[f_R(t) + f_L(t)] \text{ and } u(t) = \frac{1}{2l}[f_R(t) - f_L(t)]. \tag{3}$$

However, the fact that the rotational velocity is calculated as a sum of measured signals is the main difficulty in this measurement. For identical seismometers this sum is precisely proportional to the rotation velocity. Unfortunately, in practice the pendulums are never identical, and this fact implies a difference in the measured signals of an order of magnitude lower than the magnitude of the component signal when the signals are intense; for weak signals this difference may be of another order of magnitude lower, and may be nearly of the same size as the noise in the

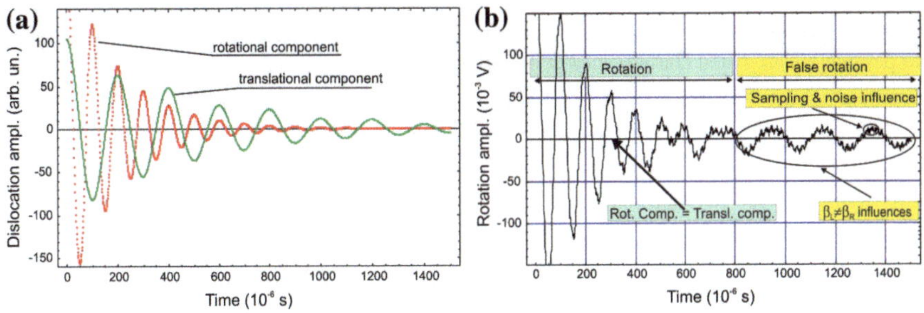

Fig. 4 The simulation influence of difference for pendulums, sampling procedure and noises for TAPS work: **a** assumed rotational Ω(t) and translational u(t) component; **b** calculated TAPS response regarding the detected rotation. One can see the existence a "false rotation signal" which amplitude depends on pendulums' differences (Jaroszewicz et al. 2003)

measurements (Nowożyński and Teisseyre 2002). Thus, the noise in the measurement is of one order of magnitude greater for TAPS than for SM-3. In practice, the attenuation difference for pendulums ($b_L \neq b_R$) of even a few per cent can be a source of a false rotational signal, especially in a region where the rotational component is small compared to the translational one, as is shown in the simulation presented in Fig. 4.

In order to overcome this problem, various techniques have been proposed for obtaining more reliable results for rotation measurements, however, with limited effectiveness. Chronologically these include a filtering procedure in the FFT domain (Teisseyre and Nagahama 1999) or the time domain (Nowożyński and Teisseyre 2002), a two-channel signal equalisation using a special calibration procedure before the measurement as shown in Fig. 3a bottom scheme (Teisseyre et al. 2003b), and a recorded data spline function approximation (Solarz et al. 2004). Currently, about ten TAPSs are used by the Institute of Geophysics PAS, Poland. TAPS is used for the registration of rotational events during earthquakes (Teisseyre 2002; Teisseyre et al. 2003b; Moriya and Teisseyre 2006), as well as seismic activity connected with an induced detonation in mine regions (Teisseyre and Boratyński 2002; Jaroszewicz et al. 2005; Wiszniowski 2006). However, TAPS does not meet defined in Sect. 2 requirements for seismological (S1–S4) and engineering (E1–E4) applications, as can be seen from the data presented in Table 2.

3.2 *Rotational Seismometers Using Pairs of Classical Geophones*

It is the second example of a practical realisation of Galitzin idea based on commercially available geophones which are called the "Rotaphone" (Brokešová et al. 2009; Brokešová and Málek 2010). This is a set of geophones mounted on a rigid

Table 2 Overview of rotational seismometers measuring rotation indirectly. Only the most important parameters are listed (Jaroszewicz et al. 2016)

Parameter	Unit	TAPS (Skarzyński and Krajewski 2016)	Rotaphone	
			6DOF (Brokešová 2014)	D (Brokešová et al. 2016)
Frequency range	Hz	7×10^{-1}–50[a, b]	2–60[b]	2–80[b]
Sampling frequency	Hz	100	250	250
Sensitivity[c]	rad/s	1×10^{-7}	2.16×10^{-9}	3.77×10^{-9}
Maximum rate	rad/s	1×10^{-1}	2.87×10^{-1}	3.17×10^{-2}
Dynamic range	dB	120	120	120
Paired sensor spacing	m	0.28	0.30	0.40
Operating temperature	°C	−10 to 45	−20 to 40	−40 to 100[d]
Weight	kg	15	9.5	15.3
Dimensions (L × W × H)	mm	450 × 180 × 350	350 × 350 × 430	445[e] × 112
Sensors: (pcs × type)	–	2 × SM-3	12 × SM-6	16 × SM-6
Natural frequency	Hz	4.5	4.5	4.5
A/D converter:	type	Sigma-Delta	4 × Tedia	1 × EE & S
• dynamic	Bit	26	28	24
• range	V	±10	±2.5	±1 or ±2.5

[a]Modified according to the recorder MK-6 by IG PAS
[b]the instrument generally operates in a high-frequency range (above the natural frequency of the sensors used)
[c]understood as an expression for the smallest signal that can be resolved (Havskov and Alguacil 2016 p. 79)
[d]data for geophone SM-6
[e]disc diameter

base horizontally or vertically either way. The system shown in Fig. 5a contains 12 geophones mounted at the edges of a rigid tube with a distance of 0.3 m is named the six-degree-of-freedom (6DOF). Taking into consideration a different configuration application, low-noise geophones (SM-6, Sensor Nederland B.V.) and better analogue/digital (A/D) converters, this version has increased sensitivity compared to the first prototypes (Brokešová et al. 2012a; Brokešová and Málek 2013). The final configuration of this system is named Rotaphone-D which uses 16 (eight horizontal and eight vertical) SM-6 geophones around a disc with a separation of the paired

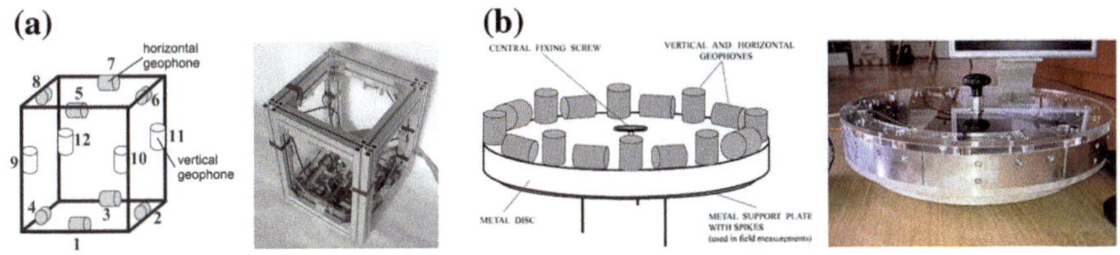

Fig. 5 Schematic diagrams of the Rotaphones and general views: **a** 6DOF prototype II (Brokešová and Málek 2015); **b** Rotaphone-D (Brokešová et al. 2016)

geophones of 0.4 m, as shown in Fig. 5b (Brokešová et al. 2016). The separating distances were chosen to correlate with a specific wavelength of interest. Due to the rigidity of the used frames, the components of rotation rate are calculated as (Brokešová et al. 2009):

$$\Omega_x = \frac{\partial v_z}{\partial y} = -\frac{\partial v_y}{\partial z},\ \Omega_y = \frac{\partial v_x}{\partial z} = -\frac{\partial v_z}{\partial x},\ \Omega_z = \frac{\partial v_x}{\partial y} = -\frac{\partial v_y}{\partial x} \tag{4}$$

where x, y, z are the Cartesian coordinates, as shown in Fig. 2, and v_i is the suitable time derivative of the displacement components measured by geophone.

Since both vertical and horizontal sensors are used, the rotation rate is determined by more than one geophone pair. Such approach allows for a very precise in situ calibration of the geophones which improves the signal/noise ratio of both translation and rotation, so that the rotation rate can be calculated more accurately. A summary of data for the Rotaphone devices described above is presented in Table 2. These devices have been used to record many tens of seismic events both induced by natural sources (weak earthquakes with a measured rotation of the order of 10^{-6} rad/s) (Brokešová and Málek 2010, 2015; Brokešová et al. 2012b; Brokešová 2014) and anthropogenic sources (blasts with measured rotation of the order of 10^{-3} rad/s) (Brokešová et al. 2012b; Brokešová and Málek 2013, 2015). With regard to the requirements defined in Sect. 2, the Rotaphones are close to fulfilling the requirements for seismological applications; however, their frequency ranges are still too narrow, and for this reason they should be treated as short-period systems.

4 Rotational Seismometers for a Direct Measurement of Rotation

According to the systems' review (Jaroszewicz et al. 2016) in general, there are three main technologies of constructing this type of rotational seismometer: mechanical, electrochemical and optical.

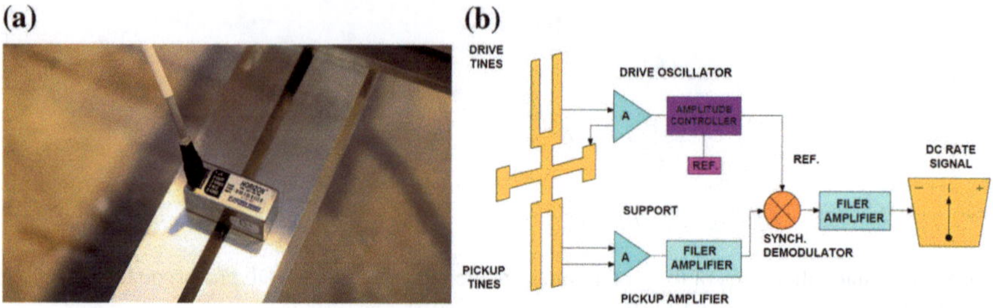

Fig. 6 The Horizon™ MEMS angular rate sensor: **a** general view of the HZ1-100-100; **b** scheme of operation (Systron 2016c)

4.1 Rotational Seismometer Using Mechanical Sensor Technology

Mechanical devices are based on the Micro-Electro-Mechanical System (MEMS) originally developed for manufacturing integrated accelerometers for airborne applications. Based on this idea, the Systron Donner Inertial (Concord, CA, USA) produces the Horizon™ (Systron 2016a), a small, high reliability, solid-state angular rotation sensor shown in Fig. 6a. According to the data available on the producer webpage (Systron 2016b), main elements of this device are piezoelectric quartz tines (see Fig. 6b). The tines are driven by an oscillator to vibrate at a precise amplitude which causes their moving forward and away from one another at a high frequency. This vibration causes a drive fork to become sensitive to the rate of an angular motion about an axis parallel to its tines which causes a sine wave of torque in the sensor's tines, resulting from the oscillating torque of drive tines at this frequency. Electrical output signals are produced by a pickup amplifier as pickup tines respond to the oscillating torque by moving in and out of plane regarding scheme shown in Fig. 6b. These signals are amplified and converted into a DC signal proportional to the rate using a synchronous demodulator. The manufacturer recommends using commercially available high-accuracy dynamic signal acquisition modules which provide an easy way to store data and carry out an analysis.

The analysis of the main parameters for the wide-range device, the HZ1-200-100, are summarised in Table 3. It can be seen that this can be used only as an additional device for a laboratory investigation of engineering applications in rotational seismology (Zembaty et al. 2013).

Table 3 Overview of no optical rotational seismometers measuring rotation directly. Only the most important parameters are listed (Jaroszewicz et al. 2016)

Parameter	Unit	HZ1-200-100 (Systron 2016c)	R-1 (Eentec 2016)	R-2 (AST LLC AST 2016)
Axial		Uniaxial	Triaxial	Triaxial
Sensitivity[a]	rad/s/$\sqrt{}$Hz	4.4×10^{-4}	1.2×10^{-7}	0.6×10^{-7}
Clip level[b]	rad/s	3.49	0.10	0.40
Dynamic range	dB	78	110	117
Frequency band	Hz	>60	0.05–20	0.03–50
Optional extended		n/a	0.03–50	0.01–100
Scale factor[c]/optional	V/rad/s	0.57(±2%)/n/a	$50/2 \times 10^2$	$50/5–2 \times 10^2$
Operating temperature	°C	−40 to +71	−15 to +55 (extended −45 to +55)	
Output signal	V	+0.5 to +4.5	±5, ±2.5	±20 differential
Calibration (S.F. deviation from 20/22 °C)	%/°C	<0.08	<0.03	Internal calibration electronics
Shock survival	g	200	200	200
Power supply	VDC	8–12	9–14	9–18
Supply current	mA	<20	20	30
Power consumption	W	0.24	0.28	0.54
Weight	kg	<0.06	1.0	1.5
Dimensions (L × W × H)	mm	58.3 × 25.3 × 25.3	120 × 120 × 90	120 × 120 × 100

[a]For an unambiguous comparison with data in Table 2, this is an output noise for SNR = 1, also defined as a resolution @ 1 Hz in (rad/s)
[b]identical to the maximum rate in Table 2
[c]understood as the gain of the instrument (Havskov and Alguacil 2016 p. 79)

4.2 *Rotational Seismometers Using Electrochemical Sensor Technology*

These devices use a fluid as an inertial mass which motion is detected using multilayer platinum electrodes with a spacing of a few tenths of a millimetre according to the scheme shown in Fig. 7a (Bernauer et al. 2012). The fluid is an ion-rich electrolyte and is free to move. At both ends of the channel, an elastic diaphragm allows for fluid motion. When a DC voltage is applied to the electrodes, it produces an ion concentration gradient between them. Due to the conductivity of the electrolyte, the bias voltage and its associated current produce an ion concentration gradient only

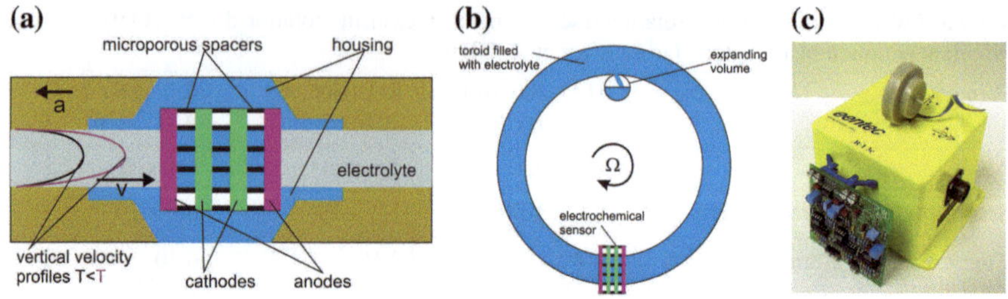

Fig. 7 The electrochemical rotational seismometer: **a** schematic diagram of the MET transducer (Bernauer et al. 2012); **b** a rotational sensor mechanical system design (MGD 2016); **c** the Eentec R-1 (MGD 2016)

between the electrodes. As the system is accelerated by a ground motion, the fluid flows relative to the electrodes and this yields a change in current, proportional to the fluid velocity and to the ion concentration. The above process is known as the Molecular Electronic Transfer (MET) technique (Abramovich and Kharlamov 2001). This technique is especially appropriate for pure rotational seismometers shown in Fig. 7b, c (MGD 2016). The sensor has a toroidal channel filled with electrolyte. When the sensor rotates, the liquid is forced through the sensor MET placed across the channel, converting the liquid motion into an electrical output. The expansion volume is used to compensate the expansion of the liquid due to temperature.

Several models of such seismometers can be found on the company's website for instance: PMD Scientific Inc., Weatogue, CT, USA (www.pmdsci.com), Eentec, Kirkwood, MO, USA (www.eentec.com), R-sensors LLC, Moscow, Russia (www.r-sensors.ru), AST LLC, Moscow, Russia (www.seismometers.ru). The manufacturers claim their seismometers are suitable for portable field use because they have no springs, hinges or moving mechanical parts (except the fluid), but information about their field testing is very limited (Wassermann et al. 2009; Nigbor et al. 2009; Bernauer et al. 2012; Lee et al. 2012). For this reason, we decided to present in Table 3 only two devices: R-1 manufactured by Eentec (2016) and the prototype R-2, probably developed by this company in cooperation with the small Russian company AST (AST LLC 2016).

As can be seen, the Eentec rotational seismometers have parameters which almost match those expected for seismological and engineering applications presented in Sect. 2. However, fulfilling the last requirements from Table 1 is still problematic. The R-1 testing for linear and cross-axis sensitivity showed that its linear sensitivity of 6×10^{-5} rad/s/(m/s^2) and 2% cross-axis sensitivity are conservative at the maximum value (Nigbor et al. 2009) and were twice as high as expected (Lee et al. 2012). The same doubt remains about the calibration quality, especially in the lower (<1 Hz) frequency range (Wassermann et al. 2009), since the frequency response does not have a flat shape, and at frequencies above 1 Hz the dynamic range is of 80 dB instead of the claimed value of above 110 dB (Nigbor et al. 2009). Finally, deviations from the nominal value of 27% and 18% in the scale factor values for R-1 and R-2 in a temperature range of 20–50 °C have been measured (Bernauer et al. 2012), giving

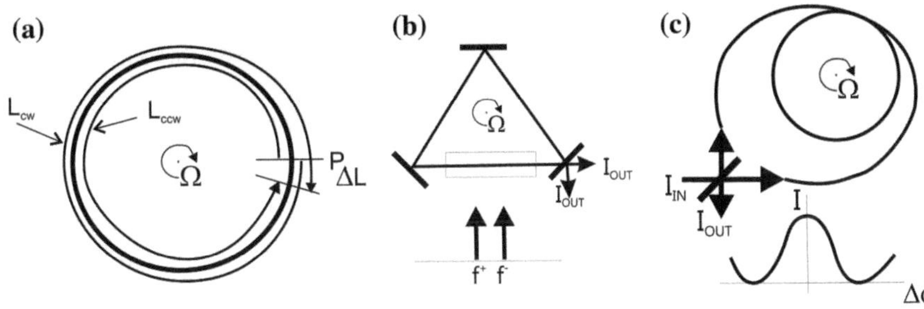

Fig. 8 The von Laue-Sagnac effect in a ring interferometer rotating with respect to an inertial frame of reference: **a** interferometric systems for its detection; **b** active method in the ring-laser approach; **c** passive method in a fibre-optic approach. Notation: L_{cw}, L_{ccw}—distances in clockwise and counterclockwise directions; I_{IN}, I_{OUT}—intensities of input and output beams (Jaroszewicz et al. 2006a)

rise to the suggestion that the liquid-based technology requires further improvement for reliable field measurements.

Despite the above reservations, the installed R-1 has, for instance, recorded several hundred local earthquakes and two explosions in Taiwan (Lee et al. 2009b). The largest peak rotational rate (0.63×10^{-3} rad/s for the vertical component with a dominant frequency band of about 2.5–5.5 Hz) recorded at the HGSD station was from an earthquake with a magnitude of 5.1 at 13:40 UTC July 23rd, 2007.

4.3 *Rotational Seismometers Using Optical Sensor Technology*

Since such device uses light for measurement which is a non-mass object, the problem with mass inertia (nonlinear frequency response) is solved from definition, directed to the best solution for a rotational seismometer. Physically all of them operate based on the von Laue-Sagnac effect (Sagnac 1913) and use an optical gyro configuration. This effect can be observed in any loop interferometer, as shown in Fig. 8a, since the optical path length difference ΔL experienced by light propagating in opposite directions along the interferometer which is rotating with rate $\mathbf{\Omega}$ is (Post 1967):

$$\Delta L = \frac{4}{c_0} \boldsymbol{A} \cdot \mathbf{\Omega}, \tag{5}$$

where $\mathbf{A}$ is the vector of the geometrical area enclosed by the wave path, c_0 is the velocity of light in a vacuum and $\mathbf{\Omega}$ is the rotation vector. It can be seen that the von Laue-Sagnac effect depends on the scalar product of two vectors ($\mathbf{A}$, $\mathbf{\Omega}$), therefore effect system detects only rotational components with an axis perpendicular to the geometrical area enclosed by the wave path; this axis can be positioned freely according to this area (Post 1967). In general, the result of the von Laue-Sagnac effect is extremely small; for instance, the Earth's rate of rotation (0.26 rad/h) gives a magnitude of ΔL equal to 9.7×10^{-15} m for an area of 10^{-2} m^2. Hence, ring laser and fibre-optic type systems (Fig. 8b, c) are technical implementations of the loop interferometer for an appropriate detection of the rotation rate magnitude expected for rotational seismology (see four requirement in Table 1).

The ring-laser approach for the measurement of the von Laue-Sagnac effect is the loop interferometer which includes an optical amplifier within the resonator (Rosenthal 1962). This type of an amplifier enables laser oscillation at f^q along the (q=+ for clockwise) and (q=− for counterclockwise) directions within the resonator (lower part of Fig. 8b). In the presence of the rotation $\mathbf{\Omega}$, the frequency difference Δf is given by:

$$\Delta f = f^+ - f^- = \frac{4A}{\lambda P}(\mathbf{n}, \mathbf{\Omega}), \tag{6}$$

where λ is the optical wavelength of the laser oscillator, $\mathbf{n}$ is the normal vector to the laser beam plane and P is the perimeter enclosed by the beam path. The ring-laser approach using a He–Ne amplifier (Aronowitz 1971) was the first successful Ring-Laser Gyroscope (RLG) and is now used in a number of navigation systems. Its implementation for seismological research has been proposed in various systems, but the main ones are three systems presented in Fig. 9: the C-II (Stedman 1997) and GEO ring-lasers (Rowe et al. 1999) in Christchurch, New Zealand, as well as the G-ring laser in Wettzell, Germany (Schreiber et al. 2001). It should be noticed that such systems for applications in seismic studies compared to another systems discussed above possess two fundamental advantages: they measure absolute rotation with regard to the local universe, and they do not depend on accelerated masses. In particular, this last property ensures an extremely wide dynamic range of operation, from a few 10^{-6} Hz for geophysical signals up to more than 10 Hz as obtained from regional earthquakes (Cochard et al. 2006). Since at present the G-ring laser has the best signal-to-noise performance, Table 4 contains the comparison with other optical rotational seismometers.

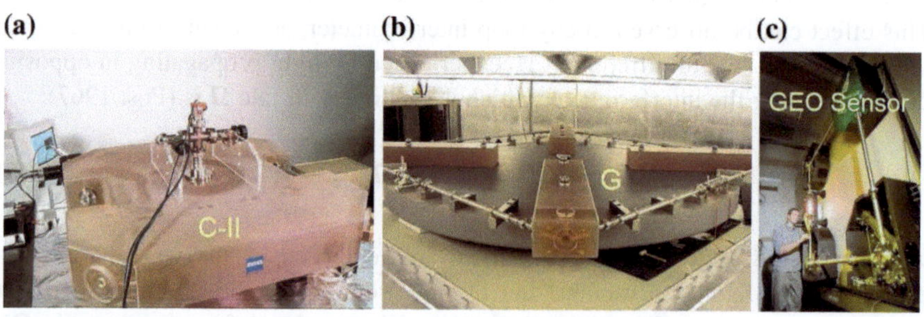

Fig. 9 The ring laser rotational seismometer (Wells and Schreiber 2016): **a** C-II; **b** G; **c** GEO. The first two are horizontally installed, whereas the last one is vertical

Table 4 Overview of optical rotational seismometers with RLG and FOG configurations. Only the most important parameters have been listed (Jaroszewicz et al. 2016)

Parameter	Unit	G-Ring (Schreiber et al. 2006)	μFORS-1 (Schreiber et al. 2009b; Litef 2016)	LCG[a] (Bernauer et al. 2012)	AFORS-1 (Jaroszewicz et al. 2011; Jaroszewicz et al. 2012a, b)	BlueSeis-3A (Bernauer et al. 2016; iXBlue 2016)
Axial		Uniaxial	Uniaxial	Triaxial	Uniaxial	Triaxial
Sensitivity[b]	rad/s/√Hz	9×10^{-11}	3×10^{-5}	6.3×10^{-7}	4×10^{-9}	2×10^{-8}
Maximum rate	rad/s	1	17.5	No data	6.4×10^{-3}	0.1
Dyn. range	dB	280	115	No data	124	135
Freq. band	Hz	0.003–10	No data	DC–100	0.83–106.15	DC–100
S. F. error[c]	%/°C	Not observed	$\leq 0.05(1\sigma)$	Not observed	No data	<0.01
Oper. temp.	°C	Constant	−40 to 77	No data	−10 to 50	−10 to 50
Shock survival	g	No data	250	10	No data	No data
Power supply	VDC	High	±5, 3.3	24	12	24
Power cons.	W	High	2.5	25	<24	<20
Weight	kg	No data	0.137	2.7	18	20
Dimensions (L × W × H)	mm	Area equal to 16 m^2	22 × 73 × 58	278 × 102 × 128	700 diameter × 160	300 × 300 × 280
Sampling rate	Hz	4	5–1000	200	212	up to 200

[a]LCG-Demonstrator based on the LCR-1000 gyrocompass AHRS
[b]for an unambiguous comparison with data in Table 2, this is an output noise for SNR = 1 defined also as a resolution @ 1 Hz in [rad/s]
[c]defined also as the temperature sensitivity of scale factor

Since ring-laser rotational seismometers are optimised for the detection of very weak rotational signals at extremely low frequencies, these are only stationary devices mounted in stabilised environmental conditions such as temperature, pressure and low vibration conditions. For the above reasons, they generally fulfill only the first and S4 requirements discussed in Sect. 2.

However, for the last two decades they have been sufficient for practical use in the detection of rotation events in both strong and weak earthquakes (Cochard et al. 2006; Schreiber et al. 2006, 2009a; Widmer-Schnidrig and Zurm 2009).

Another device, and probably the most promising, is an optical rotational seismometer based on the FOG (LeFevre 2014); its basic configuration is schematically illustrated in Fig. 9c. For L fibre length wound in a coil of D diameter, the phase shift is produced between counter-propagating light of the magnitude $\Delta\varphi$, given by (Vali and Shorthill 1976):

$$\Delta\varphi = \frac{2\pi LD}{\lambda_0 c_0}\Omega, \tag{7}$$

where Ω is the rotation component along the axis perpendicular to the fibre-optic loop, and λ_0 is the wavelength of the light in a vacuum. In other words, the sensitivity of the Sagnac interferometer in this approach is enhanced not only by increasing the physical sensor loop diameter, but also by increasing the total length of the used fibre.

The first successful application of this type of system for seismological research was an approach using a classical FOG (Takeo et al. 2009; Jaroszewicz et al. 2003). The next generation was FORS-II, installed in the Ojcow Observatory, Poland (Jaroszewicz et al. 2006b) for the investigation of rotational events which had a resolution of 4.3×10^{-8} rad/s @ 1 Hz. Limited information can be found in literature on other applications of the commercial FOG as a rotational seismometer. Bernauer et al. (2012) described a laboratory investigation of temperature stability of the LCG-Demonstrator based on LCR-100 AHRS (Northrop Grumman LITEF GmbH, Germany), shown in Fig. 10a, with parameters presented in Table 4. Within a temperature range between 20 and 50 °C, these authors observed no scale factor error, whereas the Allan deviation indicated an amplitude-modulated white noise in periods from 0.1 to 500 s. However, 25 W power consumption and rather low sensitivity of the LCG-Demonstrator limit this device mainly to rotational engineering applications in the authors' opinion. Similar conclusions can be drawn from a laboratory investigation and a field test of the μFORS-1 device at a wind generator (Schreiber et al. 2009b) shown in Fig. 10b with the parameters presented in Table 4. With regard to the discussion by Jaroszewicz et al. (2016) the main reason for this limitation is probably related to the fact that commercial FOGs have integrated electronics which are optimised to measure angle changes but not rotational rates. In order to avoid this problem, new systems with special electronics have been proposed. The first is our Autonomous Fibre-Optic Rotational Seismograph (AFORS-1) which is characterised in Table 4. This device, shown in Fig. 10c, has been used continuously in the Książ Observatory, Poland since July 21st, 2010. It records seismic events which

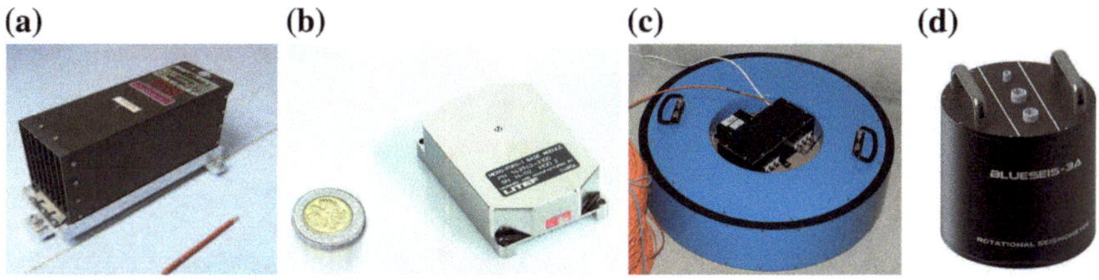

Fig. 10 Rotational seismometers based on FOG: **a** LCG-demonstrator (Bernauer et al. 2012); **b** μFORS-1; **c** AFORS-1; **d** BlueSies-3A (iXBlue 2016)

are stored on the spot together with data from two sets of TAPS for comparison of their recordings (Jaroszewicz et al. 2011, 2012a, b; Kurzych et al. 2014), as well as sending this to a FORS-Telemetric Server via GPS. Figure 11 presents an example of such recording connected with the Honshu earthquake (M = 9.0) recorded in Książ at 6:58 am. of March 11th, 2011. The main advantage of the AFORS-1 is the possibility of using a system fully remotely controlled via the internet (Jaroszewicz et al. 2012b). However, its main disadvantages are a frequency band which is too low, and a maximum detectable rate of a few mrad/s which limits AFORS application to the seismological area of interest. In view of this, the next rotational seismometer, known as the Fibre-Optic System for Rotational Events and phenomena Monitoring (FOSREM) described in Sect. 5, has been proposed (Jaroszewicz et al. 2015) as a device for seismological and engineering applications.

Finally, the iXBlue (Cedex, France) was presented at meetings in the first half of 2016 (Bernauer et al. 2016; iXBlue 2016). The prototype includes a broadband and high-grade triaxial device BlueSeis-3A; its parameters are listed in Table 4 and a general view is shown in Fig. 10d. An analysis of the parameters claimed for this device shows that it is a rotational seismometer for low self-noise and broadband measurement. The manufacturer has announced an extensive laboratory testing for later in 2016 so this device may be available by 2017.

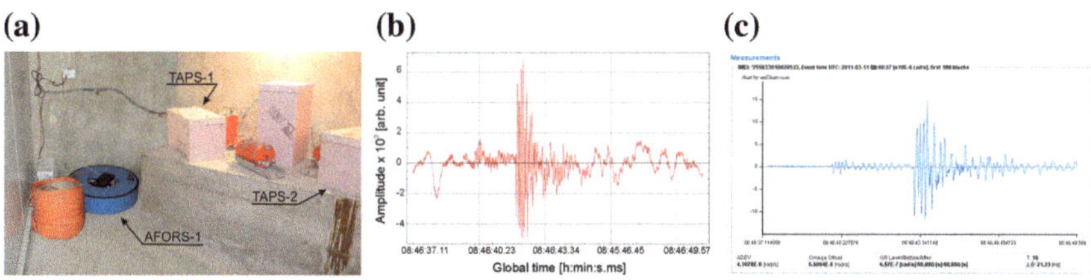

Fig. 11 Seismogram recorded in Książ from Honshu earthquake (M = 9.0) at 6:58 am, 03-11-2011; **a** general view of AFORS-1 and TAPSs installed in Książ **b** rotational event recorded indirectly by TAPS; **c** rotational event recorded directly by AFORS-1

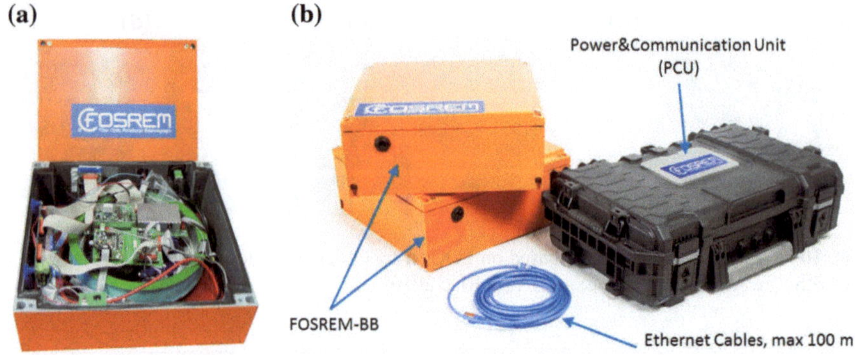

Fig. 12 FOSREM: **a** view of FOSREM-BB; **b** multi-sensor synchronous measuring system based on FOSREM-BB

5 FOSREM as a System for All Rotational Seismology Area of Interest

The FOSREM (see Fig. 12a) consists of two parts: optical and electronic. The optical part is constructed using a so-called minimum configuration of FOG (LeFevre 2014), and is designed to operate in depolarized light in the sensor loop (Jaroszewicz et al. 2015, 2016; Kurzych et al. 2016). Since the sensor loop contains a 5.000 m standard single-mode fibre (SMF-28) wound on a 0.215 m diameter spool with total losses in the optical part in the range of 16 dB, applied the 10 mW light source (SLED) reaches theoretical sensitivity equal to 2×10^{-8} rad/s/Hz$^{1/2}$. The electronic unit calculates and records rotational data through the use of an open-loop synchronous detection in a digital form using a 32-bit DSP. This involves specific electronic solutions (Kowalski et al. 2015) using signal processing to directly determine the component of rotation using the following formula (Kurzych et al. 2016):

$$\Omega = S_o \cdot \arctan[S_e \cdot (A_{1\omega}/A_{2\omega})], \quad (8)$$

where S_e, S_o are the electronic and optical constants of the FOSREM determined during the calibration procedure, and $A_{1\omega}$ and $A_{2\omega}$ are the first and second amplitudes of the harmonic output signal. The FOSREM version BB is designed to work as a multi-sensor synchronous measuring system including three FOSREM systems and a Power and Communication Unit (PCU) (see Fig. 12b). The connection provides data transmission and power supply over a single standard STP cable within a distance of 100 m. Additional four such systems can operate in a single network, with system remote control and data storage to a central cloud-based system via the internet or GSM/GPS.

The FOSREM enables the measurement of only the rotational component, over a wide range of signal amplitudes (from 2.06×10^{-8} rad/s to a few rad/s), as well as in a wide frequency band from DC to 328.12/n Hz (n = 1, ..., 128). It is a

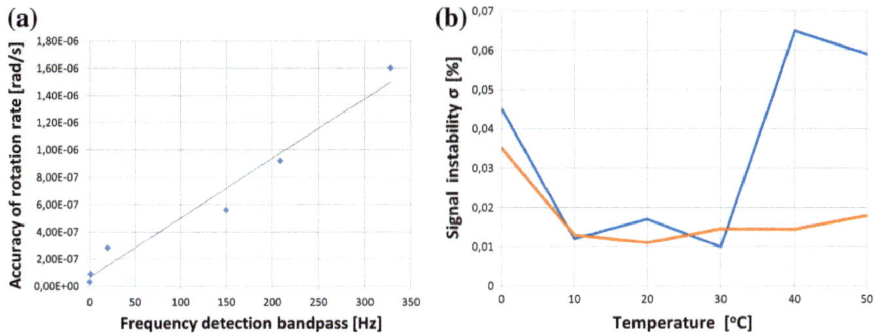

Fig. 13 The FOSREM laboratory investigation: **a** system accuracy depending on frequency band; **b** temperature instability of system work. Curve (blue color)—cooling process, curve (orange color)—heating process

360 mm × 360 mm × 160 mm dimension system with the weight below 10 kg and the power supply PoE 48 V from PCU. These aspects, combined with the remote control of the electronic module possible via the internet mean that the FOSREMs are portable and autonomous devices. The system accuracy has been evaluated based on the measurement of the defined constant angular velocity of the Earth in Warsaw, Poland ($\Omega_E = 4.45 \times 10^{-5}$ rad/s). According to the laboratory test (see Fig. 13) the FOSREM accuracies are in the range of 3×10^{-8} rad/s to 1.6×10^{-6} for the above mentioned frequency bandpass, as well as its work is stable during cooling and heating processes within a temperature range of 0–50 °C with temperature sensitivity of the scale factor <0.03%/°C.

The FOSREM has been additionally testing rotational events' recording on base of digitised data from the past real seismic earthquakes such as registered in California on May 18th, 1940 in El Centro and on October 17th, 1989 at Loma Prieta. Normally, a typical shaking table is used for the generation of horizontal shaking as a realistic simulation of movements that occur during earthquakes. For testing of FOSREM, the table was modified using an additional part, as shown in Fig. 14a. This configuration allows to change the linear shaking table movement v(t) from an applied record of accelerations in the angle φ(t) between the ground and the beam supported by the ground and to achieve the rotation Ω for the sensors installed on the moving beam (Fig. 14b) as (Jaroszewicz et al. 2016):

$$\Omega = \frac{d\varphi\,(t)}{dt}_{\,|\varphi(t)=\mathrm{ctan}^{-1}\left(\frac{X-dX}{H}\right)} = \cdots = \frac{H}{L^2} v\,(t) = 0.0365 v\,(t) \left[\frac{rad}{s}\right] \quad (9)$$

In order to carry out the test, based the on above mentioned digitised data, the FOSREM-BB was mounted together with two rotational sensors of the type of Horizon HZ1-100-100 (parameters as listed in Table 3 for HZ1-200-100, with the clip level limited to 1.74 rad/s and a scale factor deviation of twice as large (Systron 2016b)) as shown in Fig. 14b. The test used a sinusoidal excitation (Fig. 15a), a

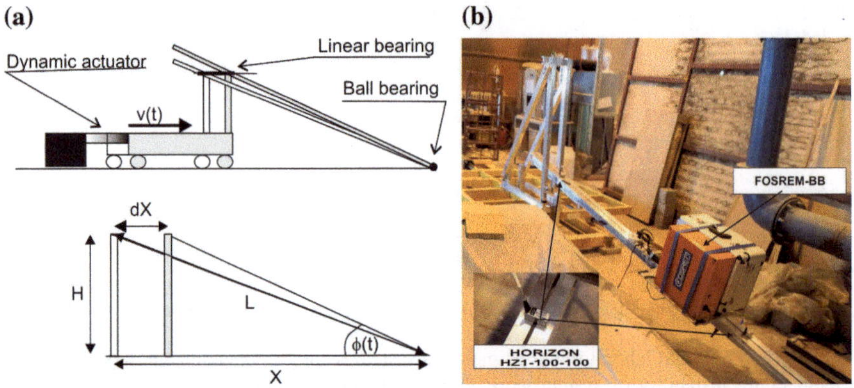

Fig. 14 Modified shaking table: **a** scheme and trigonometric dependence for Eq. (9) (H = 0.5 m, L = 3.7 m); **b** general view of shaking table with mounted FOSREM and HZ1-100-100

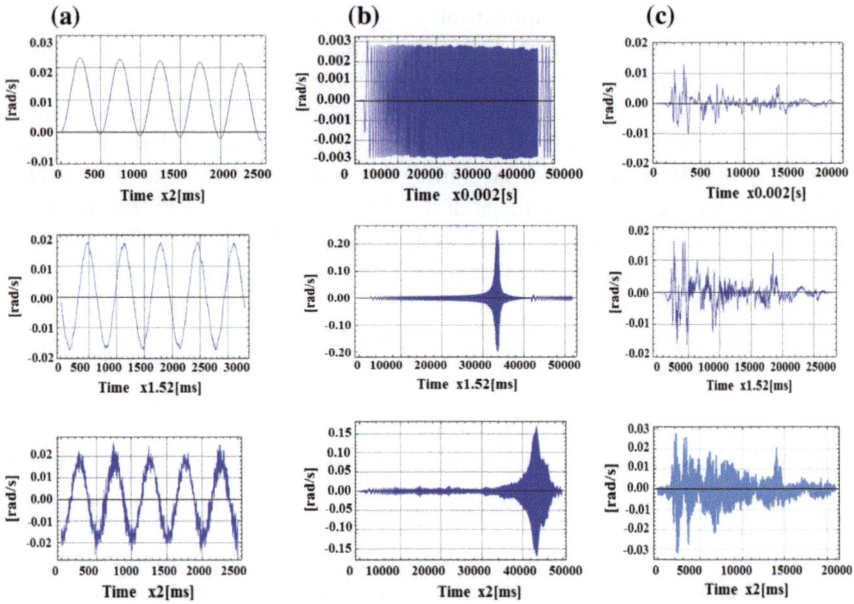

Fig. 15 Data obtained by various devices for **a** sine excitation; **b** sweep sine from 0.25 to 10 Hz; **c** El Centro earthquake (Jaroszewicz et al. 2016)

sweep sine excitation in the frequency band from 0.25 to 10 Hz (Fig. 15b), and a simulation of the real seismic earthquakes registered in California on May 18th, 1940 in El Centro (Fig. 15c). The first line in Fig. 15 presents the rotation rate forced by a shaking table evaluated on the basis of the installed accelerometer; the following two lines in Fig. 15 represent the data for FOSREM-BB and Horizon, respectively.

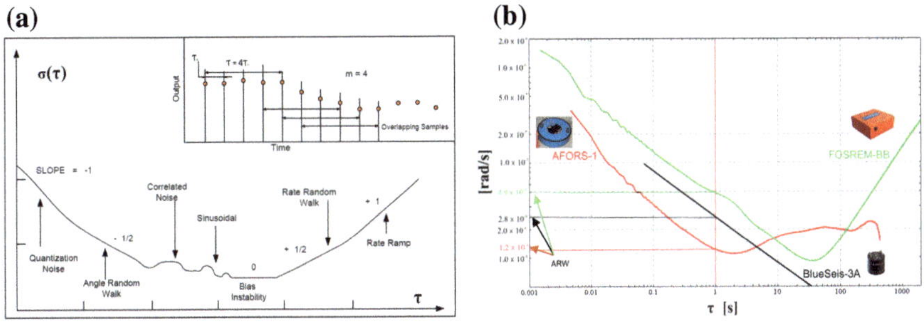

Fig. 16 The Allan variance analysing: **a** different noise processes on an Allan variance plot (Freescale Semiconductor Inc 2015); **b** Allan variance curve for fibre-optic rotational seismometers

The results indicate a good correlation between the real rotation used for the driven table and recorded by FOSREM-BB (line 1 and 2 in Fig. 15, respectively). The presented data also clearly show that the commercially available rotational sensor Horizon has lower sensitivity than the FOSREM which is reflected in the more illegible trace of the recorded signal, especially for lower values of a signal amplitude. A sweep sine excitation indicated nonlinear characteristics of the rotation rate induced in the set-up which was connected with resonance characteristics of the beam with a natural frequency of about 7.15 Hz. The advantage of this phenomenon is the possibility of reaching high values of the maximum rotation rate detected by FOSREM-BB in the range of 0.25 rad/s.

Finally, an Allan deviation plot has been carried out for noise identification in a rotational seismometer based on a fibre-optic technology with the Allan variance defined as (Freescale Semiconductor Inc 2015):

$$\sigma^2(\tau) = \frac{1}{2m^2(\mathrm{N}-2\mathrm{m})}\sum_{j=1}^{N-2m}\left\{\sum_{i=K}^{j+m-1}\left[\overline{\Omega}_{K+m}(\tau)-\overline{\Omega}_K(\tau)\right]^2\right\},\quad \overline{\Omega}_K = \frac{1}{\tau}\int\limits_{K\tau_0}^{K\tau_0+\tau}\Omega(t)\,dt, \tag{10}$$

where: N is the number of samples, τ_0 is the sample period, m is the averaging factor, $\overline{\Omega}$ is the average output rate samples across each cluster (see top window in Fig. 16a).

Specification of the Allan variance analysis of a time domain signal $\Omega(\mathrm{t})$ consists of computing its root of Allan variance or Allan deviation as a function of different averaging times t and then analysing characteristic regions and log-log scale slopes of the Allan deviation curves to identify different noise modes as is shown in Fig. 16a.

As one can see from Fig. 16b for three above described fibre-optic rotational seismometers the while noise/random walk appear (the Allan variance plot as a slope with gradient −0.5). From this, the angle random walk (the Allan variance for t = 1 s) equal to 4.9×10^{-8} rad/s$^{1/2}$, 2.8×10^{-8} rad/s$^{1/2}$, 1.2×10^{-8} rad/s$^{1/2}$ was obtained for FOSREM-BB, BlueSeis-3A and AFORS-1, respectively. It should be noticed that for FOSREM this is a good correlation with theoretical sensitivity, as well as system accuracy mentioned above. From the above analysis it can be concluded that the parameters presented for FOSREM meet all the requirements for a rotational motion sensor for both seismological and engineering applications; thus, this sensor is the most promising for rotational seismology applications.

It should be emphasised that the applied measurement method for FOSREM (also for AFORS) provides a completely different technical approach than the one used in FOG. In FOG system, the parameter of interest is the angle direction obtained by applying a suitable integration procedure to the rotation rate recorded using the von Laue-Sagnac effect. For this reason if such commercial FOG is used for the construction of a rotational seismometer, the rotation rate is a suitable time derivative of the output signal which introduces additional errors. The FOSREM electronic unit protects the rotation rate detection derived in Eq. (7), and is, thus, free from such errors. Moreover, FOSREM uses a specific averaging method for initial data recording which suitably eliminates drift phenomena (Jaroszewicz et al. 2006b; Kurzych et al. 2016).

6 Utility of the Rotational Seismometers for Seismotectonic Studies

In 1907, Albert Abraham Michelson had the honour to be the first American to receive the Nobel Prize in Physics "for his optical precision instruments and the spectroscopic and metrological investigations carried out with their aid". Michelson was one of the leading physicists which preferred experimental verifications much better than any theoretical investigations and frequently used the sentence: "experiment is tool of crime". We agree with this approach, but experiment needs special devices suitable for precise measurements, especially now—in this century where we observe a technological revolution opening new horizons for our knowledge as the gravity waves' detection, for example.

The traditional approach to the seismological investigation developed more than one hundred years ago seems to be not enough for the proper explanation of the observed rotational events and effects. A new theoretical approach using the advantageous mathematical structures is interesting, but needs experimental verification according to Michelson's idea. The rotational seismometers provide a unique approach to study a new idea in a wide range of seismotectonic disciplines, and we believe that the presented review in terms of principles of their operation and compared technical parameters will be useful for these studies.

7 Conclusions

The review presented here clearly shows that devices which use mass inertia as their main physical principle of operation are likely to be limited in their future utility due to their limited sensitivity, dynamic range and operational band, and this is mainly related to their nonlinear frequency characteristics. Arising from the above fundamental constraint are devices which use light for their operation. The von Laue-Sagnac effect is currently a useful basis as a physical principle for the construction of a rotational seismometer, as one can see from the parameters presented by existing devices. The main advantage of this type of sensor is its complete insensitivity to linear motion and its direct measurement of rotational speed. The development of the optical gyroscope nearly half a century ago offers an excellent technological and technical solution for the construction of an optical rotational seismometer. Despite the incredible sensitivity of ring-laser rotational seismometers, their dimensions, power consumption, and environment instability mean that such devices are best suited for a stationary research into fundamental geophysical phenomena.

The review presented here shows that fibre-optic rotational seismometers are the most attractive option, since their parameters can meet all the requirements of the various areas of interest within rotational seismology. Unfortunately, the direct application of the commercially available FOGs does not fulfil these requirements. In view of this, new types of devices are required, and BlueSeis-3A and FOSREM are the first of these. FOSREM, presented in this paper, fulfils all the technical requirements described in Sect. 2 (see Table 1) for a rotational motion detection, in both seismological observatories and engineering constructions. It is a remotely controlled sensor which is portable and works autonomously. Additionally, the use of a cloud system by FOSREM allows for the integration of dozen of sensors in a worldwide network, each transferring data to the central cloud-based system. The data can be viewed and analysed from anywhere in the world via the internet. The authors believe that the further application of such kind of a rotational seismometer in the investigation of rotational seismology effects will contribute to the provision of interesting and useful data.

Acknowledgements This work was carried out with the financial support of the Polish Ministry of Science and Higher Education project POIR.04.02.00-14-A003/16: "EPOS—System Obserwacji Płyty Europejskiej", The National Science Center project 2016/23/N/ST10/02508 as well as the Statutory Activity of the Military University of Technology PBS-654.

References

Abramovich JA, Kharlamov AV (2001) Electrochemical transducer and a method for fabricating the same. Electrochemical sensors transducer. US Patent US6,576,103 B2, 8 Aug 2001

Anderson JG (2003) Strong-motion seismology. In: Lee WHK, Kanamori H, Jennings PC, Kisslinger C (eds) The international handbook of earthquake and engineering seismology, part B, 1st edn. Academic Press, Amsterdam, pp 937–965

Aronowitz F (1971) The laser gyro. In: Ross M (ed) Laser applications, vol 1. Academic Press, New York, pp 133–200

AST LLC (2016) Rotational seismometers R-2, R3. In: Data sheet presented at the 4th international working group on rotational seismology meeting, Tutzing, 20–23 June 2016

Bernauer F, Wassermann J, Igel H (2012) Rotational sensors—a comparison of different sensor types. J Seismol 16:595–602

Bernauer F, Wassermann J, Guattari F, Igel H (2016) Testing a prototype broadband fiber-optic gyro. In: Proceeding of the 4th international working group on rotational seismology meeting, Tutzing, 20–23 June 2016. http://www.rotational-seismology.org/events/workshops/Presentations_2016/presentations/Bernauer_talk. Accessed 10 Aug 2016

Bodin P, Gomberg J, Singha SK, Santoyo M (1997) Dynamic deformations of shallow sediments in the Valley of Mexico, part 1: three dimensional strains and rotations recorded on a seismic array. Bull Seismol Soc Am 87:528–539

Brokešová J (2014) Short-period seismic rotations and translations recorded by Rotaphone. Habilitation thesis. Department of Geophysics Faculty of Mathematics and Physics Charles University

Brokešová J, Málek J (2015) Six-degree-of-freedom near-source seismic motions II: examples of real seismogram analysis and S-wave velocity retrieval. J Seismol 19:511–539

Brokešová J, Málek J, Štrunc J (2009) Rotational seismic sensor system, seismic measuring set containing that system, and seismic survey method. Czech Republic Patent CZ301217 B6, 8 Dec 2009

Brokešová J, Málek J (2010) New portable sensor system for rotational seismic motion measurements. Rev Sci Instrum 81:084501

Brokešová J, Málek J, Evans JR (2012a) Rotaphone, a new self-calibrated six-degree-of-freedom seismic sensor. Rev Sci Instrum 83:086108

Brokešová J, Málek J, Kolínský P (2012b) Rotaphone, a mechanical seismic sensor system for field rotation rate measurements and its in situ calibration. J Sesimol 16:603–621

Brokešová J, Málek J (2013) Rotaphone, a self-calibrated six-degree-of-freedom seismic sensor and its strong-motion records. Seismol Res Lett 84:737–744

Brokešová J, Málek J, Evans JR (2016) Rotaphone-D—a new six-degree-of freedom short-period seismic sensor: features, parameters, field records. In: Proceeding of the 4th international working group on rotational seismology meeting, Tutzing, 20–23 June 2016. http://www.rotational-seismology.org/events/workshops/Presentations_2016/presentations/Brokesova_talk. Accessed 10 Aug 2016

Carey SW (1983) Earth expansion and the null Universe. In: Carey SW (ed) Proceedings of the expanding earth symposium, Sydney, 1981, pp 365–372

Cochard A, Igel H, Schuberth B, Suryanto W, Velikoseltsev A, Schreiber KU, Wassermann J, Scherbaum F, Vollmer D (2006) Rotation motions in seismology, theory, observation, simulation. In: Teisseyre R, Takeo M, Majewski E (eds) Earthquake source asymmetry, structural media and rotation effects. Springer, Heidelberg, pp 391–412

Droste Z, Teisseyre R (1976) Rotational and displacemental components of ground motion as deduced from data of the azimuth system of seismographs. Pub Inst Geophys Polish Acad Sci 97:157–167

Eentec (2016) High resolution rotational seismometer model R-1. http://www.eentec.com/R-1_data_new.htm. Accessed 7 Jul 2016

Evans JR, Igel HI, Knopftt L, Teng TI, Trifunac MD (2007) Rotational seismology and engineering—online proceedings for the first international workshop. US Geol Surv Open-File Rep 1144:36–37

Farrell WE (1969) A gyroscope seismometer: measurements during the Borrego earthquake. Bull Seismol Soc Am 59:1239–1245

Freescale Semiconductor Inc (2015) Allan variance: noise analysis for gyroscopes. Application Note AN5087 Rev. 0.2/2015

Galitzin BB (1912) Lectures on seismometry. Russian Academy of Sciences, St. Petersburg, Russia (In Russian)

Graizer VM (1991) Inertial seismometry methods. Izv USSR Acad Sci Phys Solid Earth 27:51–61

Havskov J, Alguacil G (2016) Instrumentation in earthquake seismology, 2nd edn. Springer, Heidelberg

Huang BS (2003) Ground rotational motions of the 1991 Chi-Chi, Taiwan earthquake as inferred from dense array observations. Geophys Res Lett 30:1307–1310

Igel H, Brokesova J, Evans J, Zembaty Z (2012) Preface to the special issue on advances in rotational seismology: Instrumentation, theory, observations and engineering. J Seismol 16:571–572

iXBlue (2016) BlueSeis-3A rotational seismometer. In: Proceeding of the 4th international working group on rotational seismology meeting, Tutzing, 20–23 June 2016. http://www.rotational-seismology.org/events/workshops/Presentations_2016/Posters/Blueseis-3a. Accessed 10 Aug 2016

Jaroszewicz LR, Krajewski Z, Solarz L, Marć P, Kostrzyński T (2003) A new area of the fiber-optic Sagnac interferometer application. In: Proceedings of the international microwave and optoelectronics conference IMOC-2003, Iguazu Falls, 20–23 Sept 2003, pp 661–666

Jaroszewicz LR, Krajewski Z, Solarz L, Teisseyre R (2005) Application of the FORS-II for investigation of the seismic rotation waves. Proc SPIE 5776:385–393

Jaroszewicz LR, Krajewski Z, Solarz L (2006a) Absolute rotation measurement based on the Sagnac effect. In: Teisseyre R, Takeo M, Majewski E (eds) Earthquake source asymmetry, structural media and rotation effects. Springer, Heidelberg, pp 413–438

Jaroszewicz LR, Krajewski Z, Solarz L, Teisseyre R (2006b) Application of the fibre-optic Sagnac interferometer in the investigation of seismic rotational waves. Meas Sci Technol 17:1186–1193

Jaroszewicz LR, Krajewski Z, Kowalski H, Mazur G, Zinówko P, Kowalski J (2011) AFORS autonomous fibre-optic rotational seismograph: design and application. Acta Geophys 59:578–596

Jaroszewicz LR, Krajewski Z, Teisseyre KP (2012a) Usefulness of AFORS—autonomous fibre-optic rotational seismograph for investigation of rotational phenomena. J Seismol 16:573–586

Jaroszewicz LR, Krajewski Z, Teisseyre KP (2012b) Fibre-optic Sagnac interferometer as seismograph for direct monitoring of rotation events. In: D'Amico S (ed) Earthquake research and analysis: statistical studies, observations and planning. InTech Open Access Pub, Rijeka, pp 335–354

Jaroszewicz LR, Kurzych A, Krajewski Z, Kowalski JK, Teisseyre KP (2015) FOSREM—fibre-optic system for rotational events and phenomena monitoring: construction, investigation and area of application. In: Zembaty Z, De Stefano M (eds) Seismic behaviour and design of irregular and complex civil structures II. Springer, Berlin, pp 49–64

Jaroszewicz LR, Kurzych A, Krajewski Z, Marć P, Kowalski JK, Bobra P, Zembaty Z, Sakowicz B, Jankowski R (2016) Review of the usefulness of various rotational seismometers with laboratory results of fibre-optic ones tested for engineering applications. Sensors 16:2161

Ju L, Blair DG, Zhao C (2000) Detection of gravitational waves. Rep Prog Phys 63:1317–1427

Kalakan E, Graizer V (2007) Coupled tilt and translational ground motion response spectra. J Struct Eng 133:609–619

Kharin DA, Simonov LI (1969) VBPP seismometer for separate registration of translational motion and rotations. Seism Instrum 5:51–66 (In Russian)

Kowalski JK, Jaroszewicz LR, Krajewski Z, Kurzych A, Marć P (2015) Measurement method and system for measuring amplitude of first two harmonics of signal derived from SAGNAC system. Patent PCT/IB2015/059521, 10 Dec 2015

Kozák JT (2006) Development of earthquake rotational effect study. In: Teisseyre R, Takeo M, Majewski E (eds) Earthquake source asymmetry, structural media and rotation effects. Springer, Heidelberg, pp 3–10

Kozák JT (2009) Tutorial on earthquake rotational effects: historical examples. Bull Seismol Soc Am 99:998–1010

Kurzych A, Jaroszewicz LR, Krajewski Z, Teisseyre KP, Kowalski JK (2014) Fibre optic system for monitoring rotational seismic phenomena. Sensors 14:5459–5469

Kurzych A, Kowalski JK, Sakowicz B, Krajewski Z, Jaroszewicz LR (2016) The laboratory investigation of the innovative sensor for torsional effects in engineering structures' monitoring. Opto-Electron Rev 24:134–143

Lantz B, Schofield R, O'Reilly B, Clark DE, DeBra D (2009) Review: requirements for a ground rotation sensor to improve advanced LIGO. Bull Seismol Soc Am 99:980–989
Lee WHK (2009) A glossary of rotational seismology. Bull Seismol Soc Am 99:1082–1090
Lee WHK, Celebi M, Todorovska M, Igel H (2009a) Introduction to the special issue on rotational seismology and engineering applications. Bull Seismol Soc Am 99:945–957
Lee WHK, Huang BS, Langston CA, Lin CJ, Liu CC, Shin TC, Teng TL, Wu CF (2009b) Review: Progress in rotational ground-motion observations from explosions and local earthquakes in Taiwan. Bull Seismol Soc Am 99:958–967
Lee WHK, Evans JR, Huang BS, Hut CR, Lin CJ, Liu CC, Nigbor RL (2012) Measuring rotational ground motions in seismological practice. In: Bormann P (ed) New manual of seismological observatory practice 2 (NMSOP-2). Deutsches GeoForschungs Zentrum GFZ, Potsdam, pp 1–27
LeFevre HC (2014) The fiber optic gyroscope, 2nd edn. Artech House, Norwood
Litef (2016) μFORS-36 m/-1 Fiber optic rate sensors. http://www.northropgrumman.litef.com/fileadmin/downloads/Datenblaetter/Datenblatt_uFors-36_-1.pdf. Accessed 2 Oct 2016
Majewski E (2009) Spinors and twistors in the description of rotational seismic waves and spin and twist solitons. Bull Seismol Soc Am 99:1137–1146
Mallet R (1862) Great Neapolitan earthquake of 1857: The first principles of observational seismology as developed in the report to the royal society of London of the expedition made by command of the society into the interior of the Kingdom of Naples to investigate the circumstances of the great earthquake of December 1857, vol 2. Chapman and Hall, London
McGuire R (2008) Probabilistic seismic hazard analysis: early history. Earthq Eng Struct D 37:329–338
MGD (2016) Wraszczatelnyie Seismiczeskie Datcziki i MGD Obratnaia Swiaz. Accessed 1 Oct 2016
Merkel A, Tournat V, Gusev V (2008) Experimental evidence of rotational elastic waves in granular photonic crystals. Phys Rev Lett 107:225502
Moriya T, Teisseyre R (2006) Design of rotation seismometer and non-linear behaviour of rotation components of earthquakes. In: Teisseyre R, Takeo M, Majewski E (eds) Earthquake source asymmetry, structural media and rotation effects. Springer, Heidelberg, pp 439–450
Mustafa A (2015) Earthquake engineering—from engineering seismology to optimal seismic design of engineering structures. InTech, Rijeka
Newmark NM, Hall HJ (1969) Seismic design criteria for nuclear reactor facilities. In: B4 Proceedings of the 4th world conference on earthquake engineering, pp 37–50. Santiago
Newmark NM, Rosenblueth E (1971) Fundamentals of earthquake engineering. Prentice Hall, New York
Nigbor RL (1994) Six-degree-of-freedom ground motion measurement. Bull Seismol Soc Am 84:1665–1669
Nigbor RL, Evans JR, Hutt ChR (2009) Laboratory and field testing of commercial rotational seismometers. Bull Seismol Soc Am 99:1215–1227
Nowożyński K, Teisseyre KP (2002) Time-domain filtering of seismic rotation waves. Acta Geophys Pol 51:51–61
Olivera CS, Bolt BA (1989) Rotational components of surface strong ground motion. Earthq Eng Struct Dyn 18:517–526
Post EJ (1967) Sagnac effect. Rev Mod Phys 39:475–494
Rosenthal AH (1962) Regenerative circulatory multiple-beam interferometry for the study of light propagation effect. J Opt Soc Am 52:1143–1148
Rowe CH, Schreiber KU, Cooper SJ, King BT, Poulten M, Stedman GE (1999) Design and operation of a very large ring laser gyroscope. Appl Opt 38:2516–2523
Sagnac G (1913) L'ether lumineux demontre par l'effet du vent relatif d'Etherdanus un interferometre en rotation uniforme. C R Acad Sci 95:708–710 (In French)
Schreiber KU, Schneider M, Rowe CH, Stedman GE, Schlüter W (2001) Aspects of ring lasers as local earth rotation sensors. Surv Geophys 22:603–611

Schreiber KU, Stedman GE, Igel H, Flaws A (2006) Ring laser gyroscopes as rotation sensors for seismic wave studies. In: Teisseyre R, Takeo M, Majewski E (eds) Earthquake source asymmetry, structural media and rotation effects. Springer, Heidelberg, pp 377–390
Schreiber KU, Hautmann JN, Velikoseltsev A, Wassermann J, Igel H, Otero J, Vernon F, Wells JPR (2009a) Ring laser measurements of ground rotations for seismology. Bull Seismol Soc Am 99:1190–1198
Schreiber KU, Velikoseltsev A, Carr AJ, Franco-Anaya R (2009b) The application of fiber optic gyroscope for measurement of rotations in structural engineering. Bull Seismol Soc Am 99:1207–1214
Skrzyński A (Institute of Geophysics PAS, Warsaw, Poland), Krajewski Z (Military University of Technology, Warsaw, Poland) (2016) The main parameters of the two antiparallel pendulum seismometers. Personal communication
Solarz L, Krajewski Z, Jaroszewicz LR (2004) Analysis of seismic rotations detected by two antiparallel seismometers: spine function approximation of rotation and displacement velocities. Acta Geophys Pol 52:198–217
Spudich P, Steck LK, Hellweg M, Fletcher JB, Baker LM (1995) Transient stress at Parkfield, California produced by the M 7.4 Landers earthquake of June 28, 1992: observations from the UPSAR dense seismograph array. J Geophys Res 100:675–690
Stedman GE (1997) Ring laser tests of fundamental physics and geophysics. Rep Progr Phys 60:615–688
Suryanto W, Igel H, Wassermann J, Cochard A, Schuberth B, Vollmer D, Scherbaum F, Schreiber U, Velikoseltsev A (2006) First comparison of array-derived rotational ground motions with direct ring laser measurements. Bull Seismol Soc Am 96:2059–2071
Systron (2016a) Horizon™ MEMS quartz angular rate sensor. http://www.systron.com/sites/default/files/964120_n-hz1.pdf. Accessed 10 Aug 2016
Systron (2016b) Horizon quartz MEMS technology. http://www.systron.com/support/technology. Accessed 17 Sept 2016
Systron (2016c) Horizon user's guide. http://www.systron.com/sites/default/files/964013-rev_d_horizon_user_guide.pdf. Accessed 20 Aug 2016
Takeo M, Ueda H, Matzuzawa T (2009) Development of high-gain rotational-motion seismograph. Research grant 11354004, pp 5–29. Earthquake Research Institute, University of Tokyo, Tokyo
Teisseyre R, Boratyński W (2002) Continuum with self-rotation nuclei: evolution of defect fields and equation of motion. Acta Geophys Pol 50:223–230
Teisseyre R, Nagahama H (1999) Micro-inertia continuum: rotations and semi-waves. Acta Geophys Pol 47:259–272
Teisseyre R (2002) Continuum with defect and self-rotation fields. Acta Geophys Pol 50:51–68
Teisseyre R, Nagahama H, Majewski E (2008) Physics of asymmetric continuum: extreme and fracture process earthquake source asymmetry, structural media and rotation effects. Springer, Heidelberg
Teisseyre R, Suchcicki J, Teisseyre KP (2003a) Recording of seismic rotation waves: reliability analysis. Acta Geophys Pol 51:37–50
Teisseyre R, Suchcicki J, Teisseyre KP, Wiszniowski J, Palangio P (2003b) Seismic rotation waves: basic elements of the theory and recordings. Ann Geophys 46:671–685
Teisseyre R, Takeo M, Majewski E (2006) Earthquake source asymmetry, structural media and rotation effects. Springer, Heidelberg
Trifunac MD (1979) A note on surface strains associated with incident body waves. Bull Eur Assoc Earthq Eng 5:85–95
Trifunac MD (1982) A note on rotational components of earthquake motions on ground surface for incident body waves. Soil Dyn Earthq Eng 1:11–19
Trifunac MD (2006) Effects of torsional and rocking excitations on the response of structures. In: Teisseyre R, Takeo M, Majewski E (eds) Earthquake source asymmetry, structural media and rotation effects. Springer, Heidelberg, pp 569–582
Trifunac MD (2009) Review: rotations in structural response. Bull Seismol Soc Am 99:968–979

University of Munich (2016) Presentation on seismotectonics from the department of geophysics. http://www.geophysik.uni-muenchen.de/~igel/Lectures/Sedi/sedi_tectonics.ppt. Accessed 10 Aug 2016

Vali V, Shorthill RW (1976) Fiber ring interferometer. Appl Opt 15:1099–1100

Wang H, Igel H, Gallovič F, Cochard A (2009) Source and basin effects on rotational ground motions: comparison with translations. Bull Seismol Soc Am 99:1162–1173

Wassermann J, Lehndorfer S, Igel H, Schreiber U (2009) Performance test of a commercial rotational motions sensor. Bull Seismol Soc Am 99:1449–1456

Widmer-Schnidrig R, Zurm W (2009) Perspectives for ring laser gyroscopes in low-frequency seismology. Bull Seismol Soc Am 99:1199–1206

Wells JP, Schreiber KU (2016) Rotation sensing with lasers. In: Proceedings of the 4th international working group on rotational seismology meeting, Tutzing. http://www.rotational-seismology.org/events/workshops/Presentations_2016/presentations/Wells_talk. Accessed 10 Aug 2016

Wiszniowski J (2006) Rotation and twist motion recording—couple pendulum and rigid seismometer system. In: Teisseyre R, Takeo M, Majewski E (eds) Earthquake source asymmetry, structural media and rotation effects. Springer, Heidelberg, pp 451–470

Zembaty ZB (2009) Tutorial on surface rotations from the wave passage effects—stochastic approach. Bull Seismol Soc Am 99:1040–1049

Zembaty Z, Kokot S, Bobra P (2013) Application of rotation rate sensors in an experiment of stiffness 'reconstruction'. Smart Mater Struct 22:077001

Zembaty Z, Mutke G, Nawrocki D, Bobra P (2016) Rotational ground motion records from induced seismic events. Seismol Res Lett 88:13–22

MIX
Papier aus verantwortungsvollen Quellen
Paper from responsible sources
FSC® C105338

If you have any concerns about our products, you can contact us on
ProductSafety@springernature.com

In case Publisher is established outside the EU, the EU authorized representative is:
Springer Nature Customer Service Center GmbH
Europaplatz 3, 69115 Heidelberg, Germany

Printed by Libri Plureos GmbH
in Hamburg, Germany